Vertex Algebras and Algebraic Curves

Second Edition

Mathematical
Surveys
and
Monographs

Volume 88

Vertex Algebras and Algebraic Curves

Second Edition

Edward Frenkel
David Ben-Zvi

American Mathematical Society

Editorial Board

2000 *Mathematics Subject Classification.* Primary 17B69; Secondary 81R10, 81T40, 17B65, 17B67, 17B68, 14D20, 14D21, 14H10, 14H60, 14H81.

For additional information and updates on this book, visit
www.ams.org/bookpages/surv-88

Library of Congress Cataloging-in-Publication Data
Frenkel, Edward, 1968–
    Vertex algebras and algebraic curves / Edward Frenkel, David Ben-Zvi.—2nd ed.
        p. cm. — (Mathematical surveys and monographs, ISSN 0076-5376 ; v. 88)
    Includes bibliographical references and index.
    ISBN 0-8218-3674-9 (alk. paper)
    1. Vertex operator algebras. 2. Curves, Algebraic. I. Ben-Zvi, David, 1974– II. Title.
III. Mathematical surveys and monographs ; no. 88.

QA326 .F76   2004
512′.55—dc22   2004051904

Contents

Preface to the Second Edition

This is a new edition of the book, substantially rewritten and expanded. We would like to mention the most important changes that we have made.

Throughout the book we have dropped the requirement that a vertex algebra be $\mathbb{Z}$–graded with finite-dimensional graded components. The exposition of associativity and operator product expansion (Chapter 3 of the old edition) has been completely redone. A new chapter has been added (Chapter 4) in which we discuss in more detail the Lie algebra $U(V)$ attached to a vertex algebra V. In particular, we show that when V is the affine Kac-Moody vertex algebra $V_k(\mathfrak{g})$, the natural map from $U(V_k(\mathfrak{g}))$ to a completion $\widetilde{U}_k(\mathfrak{g})$ of the universal enveloping algebra of $\widehat{\mathfrak{g}}$ of level k is a Lie algebra homomorphism. We also define for an arbitrary vertex algebra a topological associative algebra $\widetilde{U}(V)$ (when $V = V_k(\mathfrak{g})$ this algebra is isomorphic to $\widetilde{U}_k(\mathfrak{g})$).

In Chapter 5 (Chapter 4 of the old edition) we show that there is an equivalence between the category of V–modules and the category of smooth $\widetilde{U}(V)$–modules. We have added a new section in Chapter 5 in which we introduce twisted modules associated to vertex algebras equipped with an automorphism of finite order. The following chapter, Chapter 6 (Chapter 5 of the old edition), has also been rewritten. We have added a new motivational section at the beginning of the chapter and have supplied a direct algebraic proof of coordinate-independence of the connection on the vertex algebra bundle. In Chapter 19 (old Chapter 18) on chiral algebras we have added a new motivational section and examples of chiral algebras that do not arise from vertex algebras. We have also explained how to attach to modules and twisted modules over vertex algebras certain modules over the corresponding chiral algebras.

Finally, we have added a new Chapter 20, on factorization algebras and factorization spaces. Factorization algebras, introduced by A. Beilinson and V. Drinfeld, provide a purely geometric reformulation of the definition of vertex algebras. Here we present an informal introduction to factorization algebras and give various examples. The most interesting examples come from factorization spaces, such as the Beilinson–Drinfeld Grassmannians which are moduli spaces of bundles on a curve equipped with trivializations away from finitely many points. We explain how the concept of factorization naturally leads us to Hecke correspondences on bundles and to the geometric Langlands correspondence. At the end of the chapter we discuss the chiral Hecke algebras introduced by Beilinson and Drinfeld which provide a tool for establishing the geometric Langlands conjecture. This chapter brings together and illuminates the material of several other chapters of the book.

We wish to thank those who kindly pointed out to us various typos in the first edition and pointed out additional references, especially, Michel Gros, Kenji

Iohara, Jim Lepowsky, David Nadler, Kiyokazu Nagatomo, Marcus Rosellen, Leon Takhtajan and A.J. Tolland.

In the course of preparation of the second edition E. Frenkel was partially supported by the NSF, DARPA and the Packard Foundation.

Preface to the First Edition

The present book is an introduction to the theory of vertex algebras with a particular emphasis on the relationship between vertex algebras and the geometry of algebraic curves. It is based on the lecture courses given by Edward Frenkel at Harvard University in the Fall of 1996 and at UC Berkeley in the Spring of 1999. The notes of these lectures were taken by David Ben-Zvi. They were subsequently improved and expanded by both authors. The main goal of this book is to introduce the concept of vertex algebra in a coordinate-independent way, and to define the spaces of conformal blocks attached to an arbitrary vertex algebra and a smooth algebraic curve, possibly equipped with some extra geometric data. From this point of view vertex algebras appear as the algebraic objects that encode the local geometric structure of various moduli spaces associated with algebraic curves.

In the fifteen years that have passed since they were introduced by R. Borcherds, vertex algebras have turned out to be extremely useful in many areas of mathematics. They are by now ubiquitous in the representation theory of infinite-dimensional Lie algebras. They have also found applications in such fields as algebraic geometry, the theory of finite groups, modular functions, topology, integrable systems, and combinatorics. The theory of vertex algebras also serves as the rigorous mathematical foundation for two-dimensional conformal field theory and string theory, extensively studied by physicists.

In the literature there exist two essentially different approaches to vertex algebras. The first is algebraic, following the original definition of Borcherds [**B1**]. It has been developed by I. Frenkel, J. Lepowsky, and A. Meurman [**FLM**] and more recently by V. Kac [**Kac3**]. Vertex operators appear here as formal power series acting on graded vector spaces. The second approach is geometric and more abstract: this is the theory of chiral algebras and factorization algebras developed by A. Beilinson and V. Drinfeld [**BD4**]. In this approach the main objects of study are $\mathcal{D}$–modules on powers of algebraic curves equipped with certain operations. Chiral algebras have non-linear versions called factorization spaces which encode various intricate structures of algebraic curves and bundles on them.

The present book aims to bridge the gap between the two approaches. It starts with the algebraic definition of vertex algebras, which is close to Borcherds', and essentially coincides with that of [**FKRW, Kac3**]. The key point is to make vertex operators coordinate-independent, thus effectively getting rid of the formal variable. This is achieved by attaching to each vertex algebra a vector bundle with a flat connection on the (formal) disc, equipped with an intrinsic operation. The formal variable is restored when we choose a coordinate on the disc; the fact that the operation is independent of this choice follows from the vertex algebra axioms. Once this is done and we obtain a coordinate-independent object, we can study the interplay between vertex algebras and various geometric structures related to algebraic curves, bundles and moduli spaces.

In particular, we attach to each vertex algebra and any pointed algebraic curve the spaces of coinvariants and conformal blocks. When we vary the curve X and other data on X (such as G–bundles), these spaces combine into a sheaf on the relevant moduli space. One can gain new insights into the structure of moduli spaces from the study of these sheaves.

The language of the book gradually changes from that of formal power series as in [**FLM, Kac3**] to that of bundles, sheaves, and connections on algebraic curves. Our goal however is to avoid using sophisticated techniques, such as the theory of $\mathcal{D}$–modules, as much as possible. In particular, we present most of the material without mentioning the "$\mathcal{D}$–word". Only at the end of the book do we use rudiments of $\mathcal{D}$–module theory when describing the relationship between vertex algebras and the Beilinson–Drinfeld chiral algebras, and the sheaves of coinvariants. Ultimately, the formalism developed in this book will enable us to relate the algebraic theory of vertex algebras to the geometric theory of factorization algebras and factorization spaces.

The first five chapters of this book contain a self-contained elementary introduction to the algebraic theory of vertex algebras and modules over them. We motivate all definitions and results, give detailed proofs and consider numerous examples. This part of the book is addressed mainly to beginners. No prerequisites beyond standard college algebra are needed to understand it.

In Chapters 6–10 we develop the geometric approach to vertex algebras. Here some familiarity with basic notions of algebraic geometry should be helpful. We have tried to make the exposition as self-contained as possible, so as to make it accessible for non-experts.

Next, we review in Chapters 11–16 various constructions and applications of vertex algebras, such as, the free field realization of affine Kac-Moody algebras, solutions of the Knizhnik–Zamolodchikov equations, and the Drinfeld–Sokolov reduction. We also study quasi-classical analogues of vertex algebras, called vertex Poisson algebras.

The last four chapters of the book are more algebro-geometrically oriented. Here we construct the sheaves of coinvariants on the moduli spaces of curves and bundles and introduce the chiral algebras and factorization algebras following Beilinson and Drinfeld. In particular, we show how to attach to any quasi-conformal vertex algebra a chiral algebra on an arbitrary smooth algebraic curve. We discuss various examples of factorization algebras and factorization spaces, including the Beilinson–Drinfeld Grassmannians. We also give a brief overview of the geometric Langlands correspondence.

This book may be used by the beginners as an entry point to the modern theory of vertex algebras and its geometric incarnations, and by more experienced readers as a guide to advanced studies in this beautiful and exciting field.

Acknowledgments

We are grateful to Ivan Mirković and Matthew Szczesny for their careful reading of drafts of this book and detailed comments which helped us improve the exposition and correct errors. We owe thanks to Matthew Emerton, George Glauberman, Namhoon Kim, Mark Kisin, Manfred Lehn, Evgeny Mukhin, Markus Rosellen,

Christoph Sorger, Joost van Hamel, and Weiqiang Wang, who made valuable suggestions and pointed out various typos. We also thank Ralph Sizer of the AMS for his sharp editorial work and Vladimir Frenkel for his careful drawing of the figures appearing in this book.

Edward Frenkel wishes to thank Boris Feigin for his longtime collaboration, and Alexander Beilinson and Vladimir Drinfeld for many illuminating discussions.

The authors gratefully acknowledge the support that they received in the course of writing this book from the Packard Foundation and the NSF.

Introduction

Some history and motivation

Vertex operators appeared in the early days of string theory as local operators describing propagation of string states. Mathematical analogues of these operators were discovered in the representation theory of affine Kac–Moody algebras in the works of J. Lepowsky–R. Wilson [**LW**] and I. Frenkel–V. Kac [**FK**]. In order to formalize the emerging structure, and motivated in particular by the I. Frenkel–J. Lepowsky–A. Meurman construction of the Moonshine Module of the Monster group, R. Borcherds gave the definition of vertex algebra in [**Bo2**]. The foundations of the theory were subsequently laid down in [**FLM, FHL**]; in particular, it was shown in [**FLM**] that the Moonshine Module indeed possessed a vertex algebra structure.

In the meantime, A. Belavin, A. Polyakov and A. Zamolodchikov [**BPZ**] initiated the study of two-dimensional conformal field theory (CFT). Vertex algebras can be seen in retrospect as the mathematical equivalent of the chiral symmetry algebras of CFT. Moreover, the key property of associativity of vertex algebras is equivalent to the property of operator product expansion in CFT, which goes back to the pioneering works of A. Polyakov and K. Wilson. Thus, vertex algebras may be thought of as the mathematical language of two-dimensional conformal field theory.

In recent years, in the course of their study of conformal field theories and string theories, physicists have come up with astonishing conjectures predicting and relating to each other various geometric invariants: mirror symmetry, Gromov-Witten invariants, Seiberg-Witten theory, etc. While many of these conjectures have been proved rigorously afterwards, the process of making these predictions remains for the most part a mystery for mathematicians. It is based on the usage of tools, such as the path integral, that have so far resisted rigorous mathematical formulation. The theory of vertex algebras, on the other hand, provides a rigorous mathematical foundation for two-dimensional conformal field theory and string theory from the Hamiltonian (i.e., operator) point of view. Namely, the space of states of the chiral sector of a CFT may be described as a representation of a vertex algebra, and chiral correlation functions may be considered as sections of various vector bundles on the moduli spaces of pointed curves. Thus, vertex algebras provide a natural point of entry for a mathematician into the world of conformal field theory and string theory.

The interaction between conformal field theory and algebraic geometry has already produced remarkable results, such as the computation of the dimensions of the spaces of "non-abelian theta-functions" by means of the Verlinde formula. These results came from examining particular examples of conformal field theories, such as the Wess–Zumino–Witten models. The physical theories that have led to new predictions in enumerative geometry (such as mirror symmetry) are also based on conformal field theories (such as the sigma models of Calabi-Yau manifolds). One may hope that further development of the theory of vertex algebras will allow us to build a suitable framework for understanding these theories as well. The first steps in this direction are already being made. For example, F. Malikov, V. Schechtman and A. Vaintrob [**MSV**] have recently constructed a sheaf of vertex superalgebras on an arbitrary smooth algebraic variety, called the chiral de Rham complex (see § 18.5), which may hopefully be used in understanding the new "stringy" invariants, such as the elliptic genera.

In another important development, A. Beilinson and V. Drinfeld have recently introduced a geometric version of vertex algebras which they call chiral algebras [**BD4**]. Chiral algebras give rise to some novel concepts and techniques which are likely to have a profound impact on algebraic geometry. The formalism of vertex and chiral algebras appears to be particularly suitable for the construction of the conjectural geometric Langlands correspondence between $\mathcal{D}$–modules on the moduli space of G–bundles on a smooth projective curve X over $\mathbb{C}$, and flat $^L G$–bundles on X, where G is a reductive algebraic group and $^L G$ is the Langlands dual group (see [**BD3**]). We will see two examples of such constructions: one involves the affine Kac-Moody vertex algebra of critical level (see § 18.4) and the other involves the chiral Hecke algebra (see § 20.5).

These applications present ample evidence for the relevance of vertex algebras not only in representation theory, where they originated, but also in other fields, such as algebraic geometry. In this book, we make the first steps towards reformulating the theory of vertex algebras in a way suitable for algebro-geometric applications.

What is a vertex algebra?

In a nutshell, a vertex algebra is a vector space V equipped with a vector $|0\rangle$ and an operation

$$Y : V \to \operatorname{End} V[[z^{\pm 1}]],$$

assigning to each $A \in V$ a formal power series, called a *vertex operator*,

(0.0.1)
$$Y(A, z) = \sum_{n \in \mathbb{Z}} A_{(n)} z^{-n-1},$$

where each $A_{(n)}$ is a linear operator on V, so that for any $v \in V$, we have $A_{(n)} v = 0$ for $n \gg 0$.

These data must satisfy a short list of axioms (see Chapter 1), the most important of which is the *locality* axiom. It states that for any $A, B \in V$, the formal power series in two variables, obtained by composing $Y(A, z)$ and $Y(B, w)$ in two possible ways, are equal to each other, possibly after multiplying them with a large enough power of $(z - w)$. In other words, the commutator $[Y(A, z), Y(B, w)]$ is a formal distribution supported on the diagonal $z = w$.

If we ask instead that the equality $Y(A, z)Y(B, w) = Y(B, w)Y(A, z)$ holds even before multiplying by a power of $(z - w)$, then we obtain the structure equivalent to that of a commutative associative algebra with a unit and a derivation. Thus, vertex algebras may be thought of as "meromorphic" generalizations of commutative algebras (for more on this point of view, see [**B3**]).

The first examples of (non-commutative) vertex algebras are the induced representations of infinite-dimensional Lie algebras related to the punctured disc. The simplest of them is the Heisenberg Lie algebra $\mathcal{H}$, which is a central extension of the commutative Lie algebra $\mathbb{C}((t))$ of functions on the punctured disc. It has a topological basis consisting of the elements $b_n = t^n, n \in \mathbb{Z}$, and the central element **1**. The Fock representation π of $\mathcal{H}$ is induced from the trivial one-dimensional representation of its Lie subalgebra $\mathbb{C}[[t]] \oplus \mathbf{1}$, and carries the structure of a vertex algebra. This vertex algebra is generated, in an appropriate sense, by the vertex operator

$$b(z) = \sum_{n \in \mathbb{Z}} b_n z^{-n-1},$$

which is nothing but the generating function of the basis elements of $\mathcal{H}$. All other vertex operators in π may be expressed in terms of $b(z)$ using the operations of normally ordered product and differentiation with respect to z.

It is straightforward to generalize this construction to the affine Kac–Moody algebras, which are central extensions of the Lie algebras $\mathfrak{g}((t))$ of functions on the punctured disc with values in a simple Lie algebra $\mathfrak{g}$, and the Virasoro algebra, which is a central extension of the Lie algebra $\operatorname{Der} \mathbb{C}((t))$ of vector fields on the punctured disc. In each case, the generating functions of basis elements of the Lie algebra play the role of the generating vertex operators in the corresponding vertex algebra. The formalism of vertex algebras provides a compact and uniform way for handling these generating functions. For example, the operator product expansion gives us a convenient tool for recording the commutation relations between the Fourier coefficients of various vertex operators.

But the formalism of vertex algebras goes far beyond computational convenience. In fact, the most interesting examples of vertex algebras, such as the lattice vertex algebras, the Monster Moonshine vertex algebra, and the $\mathcal{W}$–algebras, are not "finitely generated" by Lie algebras. There are still a finite number of vertex operators, which generate each of these vertex algebras, but their Fourier coefficients are no longer closed under the commutator. Therefore they are not easily accessible without the formalism of vertex algebras.

There is a special class of vertex algebras, which play a prominent role in conformal field theory, called rational vertex algebras. These vertex algebras are distinguished by the property that the category of their modules is semi-simple, with finitely many simple objects, up to isomorphism. Examples of rational vertex algebras are the lattice vertex algebras, the Moonshine Module vertex algebra, and the integrable vacuum modules over affine Kac–Moody algebras. We briefly review them in Chapter 5.

It is expected that the category of representations of a rational vertex algebra carries the rich structure of a modular tensor category. In particular, one can attach to it a collection of vector bundles with projectively flat connection on the moduli spaces of stable pointed curves. Though this picture is still conjectural for a general rational vertex algebra, in the genus one case it has already led to the following beautiful result of Y. Zhu [**Z1**]: the characters of all simple modules over a rational vertex algebra, considered as functions on the upper-half plane, span a representation of $SL_2(\mathbb{Z})$.

Coordinate-independence

In order to make the connection between vertex algebras and algebraic curves we first have to make vertex operators coordinate-independent. This is in fact one of the main goals of this book. To explain what this means, consider the affine Kac–Moody algebra $\widehat{\mathfrak{g}}$, which is a central extension of the Lie algebra $\mathfrak{g}((t))$. Let X be a smooth projective curve over $\mathbb{C}$, and x a point of X. Denote by $\mathcal{O}_x$ the completed local ring at x, and by $\mathcal{K}_x$ its field of fractions. Then we have isomorphisms $\mathcal{O}_x \simeq \mathbb{C}[[t]]$ and $\mathcal{K}_x \simeq \mathbb{C}((t))$, but they are not canonical. To specify such an isomorphism, we need to fix a formal coordinate t at x, for which we usually do not have a preferred choice. In applications it is often important to deal with the central extension of the Lie algebra $\mathfrak{g} \otimes \mathcal{K}_x$, rather than of $\mathfrak{g}((t))$. We should think of this extension as the affine Lie algebra $\widehat{\mathfrak{g}}_x$, "attached to the point x". Then we need to know how elements of $\mathfrak{g} \otimes \mathcal{K}_x$ are realized if we choose a different coordinate t' at x. This is of course obvious: if an element of $\mathfrak{g} \otimes \mathcal{K}_x$ appears as $J \otimes f(t)$ with respect to the coordinate t, then with respect to another coordinate t' it will appear as $J \otimes f(\rho(t'))$, where ρ is the change of variables from t' to t such that $t = \rho(t')$. Thus, ρ is an element of the group $\operatorname{Aut} \mathcal{O}$ of continuous automorphisms of $\mathcal{O} = \mathbb{C}[[t]]$, which acts simply transitively on the set of all formal coordinates at x. We conclude that we need to know how elements of the affine Lie algebra transform under the natural action of the group $\operatorname{Aut} \mathcal{O}$.

In the context of vertex algebras, we look at the generating function

$$(0.0.2) \qquad J^a(z) = \sum_{n \in \mathbb{Z}} J^a_{(n)} z^{-n-1},$$

where $J^a_{(n)} = J^a \otimes t^n \in \mathfrak{g} \otimes \mathcal{K}_x$, for some basis $\{J^a\}$ of $\mathfrak{g}$. These are the vertex operators generating the corresponding vertex algebra. It follows from the above discussion that the expression $J^a(t)dt$ transforms as a one-form on the punctured disc $D_x^\times = \operatorname{Spec} \mathcal{K}_x$.

But what is the transformation formula for a general vertex operator $Y(A, z)$ in a general vertex algebra? Can one describe the entire vertex operation Y in a coordinate-independent way, so that it makes sense in the neighborhood of any point on a complex curve, not equipped with a preferred coordinate? In Chapter 6 we answer these questions. The key to the intrinsic description of the vertex algebra structure is the incorporation of the action of the group of changes of coordinates $\operatorname{Aut} \mathcal{O}$ (and more generally, the action of the entire Virasoro algebra) into the vertex algebra structure. In other words, we want the Lie algebra $\operatorname{Der} \mathcal{O}$ of $\operatorname{Aut} \mathcal{O}$ to act on a vertex algebra V via Fourier coefficients of a vertex operator (i.e., natural basis elements of $\operatorname{Der} \mathcal{O}$ should act as the linear operators $A_{(n)}$ attached to some $A \in V$, in the notation of formula (0.0.1)). By exponentiating these Fourier coefficients we obtain what we call an action of $\operatorname{Aut} \mathcal{O}$ on V by "internal symmetries".

This naturally leads us to the notion of conformal vertex algebra – one that contains, among the vertex operators, the generating function of the basis elements of the Virasoro algebra. To a conformal vertex algebra V we assign a vector bundle $\mathcal{V}$ on an arbitrary smooth curve as follows. Consider the $\operatorname{Aut} \mathcal{O}$–bundle $\mathcal{A}ut_X$ over X whose fiber over $x \in X$ consists of all formal coordinates at x. Define the associated vector bundle $\mathcal{V}$ on X as

$$\mathcal{V} = \mathcal{A}ut_X \underset{\operatorname{Aut} \mathcal{O}}{\times} V.$$

Denote by $\mathcal{V}_x$ the fiber of $\mathcal{V}$ at a given point $x \in X$. The coordinate-free approach of this book is based on the interpretation of the vertex operation Y in a conformal vertex algebra V as an intrinsic section $\mathcal{Y}_x$ of the dual bundle $\mathcal{V}^*$ on the punctured disc $D_x^\times$ with values in $\operatorname{End}\mathcal{V}_x$.

It is easier to describe the corresponding linear map

$$\mathcal{Y}_x^\vee : \Gamma(D_x^\times, \mathcal{V} \otimes \Omega) \to \operatorname{End}\mathcal{V}_x$$

defined by the formula $s \mapsto \operatorname{Res}_x\langle \mathcal{Y}_x, s\rangle$. Namely, if we choose a coordinate z at x, then we can trivialize $\mathcal{V}|_{D_x}$ and identify $\mathcal{V}_x$ with V. Then

$$\mathcal{Y}_x^\vee(A \otimes z^n dz) = \operatorname{Res}_{z=0} Y(A, z)z^n dz,$$

for any $A \in V$ and $n \in \mathbb{Z}$.

Furthermore, we show that the vector bundle $\mathcal{V}$ carries a flat connection, with respect to which the section $\mathcal{Y}_x$ is horizontal.

Conformal blocks and coinvariants

Having reformulated the vertex operation in a coordinate-independent way, we can make contact with the global structure of algebraic curves and give a global geometric meaning to vertex operators. This naturally leads us to the concepts of conformal blocks and coinvariants.

Let us illustrate this concept on the example of affine Kac–Moody algebras. Denote by $\mathfrak{g}_{\mathrm{out}}(x)$ the Lie algebra of $\mathfrak{g}$–valued regular functions on $X\backslash x$. Then $\mathfrak{g}_{\mathrm{out}}(x)$ naturally embeds into $\mathfrak{g} \otimes \mathcal{K}_x$, and even into $\widehat{\mathfrak{g}}_x$. Given a $\widehat{\mathfrak{g}}_x$–module $\mathcal{M}_x$, we define its space of coinvariants as the quotient $\mathcal{M}_x/(\mathfrak{g}_{\mathrm{out}}(x) \cdot \mathcal{M})$ and its dual, the space of conformal blocks, as $\operatorname{Hom}_{\mathfrak{g}_{\mathrm{out}}(x)}(\mathcal{M}_x, \mathbb{C})$. Now recall that the basis elements of $\widehat{\mathfrak{g}}$ can be organized into vertex operators $J^a(z)$ given by formula (0.0.2), and that $J^a(z)dz$ is naturally a one-form on $D_x^\times$. Using the residue theorem, we can reformulate the notion of conformal blocks as follows: $\varphi \in \mathcal{M}_x^*$ is a conformal block if and only if for any $A \in \mathcal{M}_x$ the one-forms $\langle\varphi, J^a(z) \cdot A\rangle dz$, which are *a priori* defined only on $D_x^\times$, may be extended to regular one-forms on the entire $X\backslash x$.

The vertex operators in a general vertex algebra transform in a more complicated way than mere one-forms. Therefore it makes sense to consider all of them at once, i.e., the entire vertex operation $\mathcal{Y}_x$, which transforms as a section of $\mathcal{V}^*$. We will say that a linear functional φ on $\mathcal{V}_x$ is a conformal block if for any $A \in \mathcal{V}_x$, the section $\langle\varphi, \mathcal{Y}_x \cdot A\rangle \in \Gamma(D_x^\times, \mathcal{V}^*)$, which is *a priori* defined only on $D_x^\times$, can be extended to a regular section of $\mathcal{V}^*$ on $X\backslash x$.

This definition may be generalized to a broader setting when we have several distinct points on X, with a V–module attached to each of them (see Chapters 9 and 10).

The spaces of conformal blocks play an important role in the theory of vertex algebras. Even the vertex operation Y itself may be viewed as a special case of conformal blocks: in the situation when we have three points on $\mathbb{P}^1$ with two insertions of V and one insertion of the contragredient module $V^\vee$. From this point of view, various properties of vertex algebras may be interpreted geometrically in terms of degenerations of the pointed projective line (see § 10.4).

In general, the spaces of conformal blocks may be thought of as invariants of pointed algebraic curves, attached to a vertex algebra. For instance, in the case of the affine algebra $\widehat{\mathfrak{g}}$, the space of conformal blocks corresponding to a curve X with a single insertion of an integrable representation of $\widehat{\mathfrak{g}}$ may be identified with the space of "non-abelian theta-functions" on the moduli space of G–bundles on X. It is this type of result that we have in mind when we talk about the interplay between vertex algebras and algebraic curves.

This brings us to the connection between vertex algebras and moduli spaces.

Sheaves of coinvariants

An interesting question is to understand how the spaces of coinvariants and conformal blocks change as we vary the positions of the points and the complex structure on our curve. It turns out that the spaces of coinvariants can be organized into a sheaf on the moduli space $\mathfrak{M}_{g,n}$ of smooth n–pointed projective curves of genus g. Moreover, this sheaf carries the structure of a (twisted) $\mathcal{D}$–module on $\mathfrak{M}_{g,n}$. In other words, we can differentiate sections of this sheaf.

The key property of the moduli space $\mathfrak{M}_{g,n}$ used in defining the $\mathcal{D}$–module structure is its "Virasoro uniformization". Consider for simplicity the case $n = 1$. It turns out that the Lie algebra $\operatorname{Der}\mathbb{C}((t))$ acts transitively on the larger moduli space $\widehat{\mathfrak{M}}_{g,1}$ of triples (X, x, z), where X is a smooth projective curve of genus g, x is a point of X, and z is a formal coordinate at x. So $\widehat{\mathfrak{M}}_{g,1}$ looks like a homogeneous space for the Lie algebra $\operatorname{Der}\mathcal{K}$. More precisely, we can say that $\widehat{\mathfrak{M}}_{g,1}$ carries a transitive action of the Harish-Chandra pair $(\operatorname{Der}\mathcal{K}, \operatorname{Aut}\mathcal{O})$. By applying the formalism of Harish-Chandra localization to a conformal vertex algebra, we define a (twisted) $\mathcal{D}$–module structure on the sheaf of coinvariants on $\widehat{\mathfrak{M}}_{g,1}$ (see Chapter 17).

This construction may be generalized to other moduli spaces, such as the moduli space $\mathfrak{M}_G(X)$ of G–bundles on a smooth projective curve X, or the Picard variety (the moduli space of line bundles on X). In particular, applying this construction to affine Kac-Moody vertex algebra of critical level, we obtain $\mathcal{D}$–modules on $\mathfrak{M}_G(X)$ parameterized by ${}^L G$–opers on X, where ${}^L G$ is the Langlands dual group to G. These $\mathcal{D}$–modules are the Hecke eigensheaves whose existence is predicted by the geometric Langlands conjectures (see § 18.4).

The above localization construction may be further generalized. Suppose we are given a vertex algebra V and a Harish-Chandra pair $(\mathfrak{G}, \mathfrak{G}_+)$ of what we call internal symmetries of V. This means that the action of the Lie algebra $\mathfrak{G}$ is induced by the Fourier coefficients of vertex operators on V. Suppose also that $\mathfrak{G}_+$ (resp., $\mathfrak{G}$) consists of symmetries of certain geometric data Φ on the standard (resp., punctured) disc. Then to any such datum $\mathcal{T}$ on a smooth projective curve X we can attach a twisted space of coinvariants $H^{\mathcal{T}}(X, x, V)$. Moreover, these spaces with varying $\mathcal{T}$ may be combined into a (twisted) $\mathcal{D}$–module $\Delta(V)$ on the moduli space $\mathfrak{M}_\Phi$ of the data Φ on X.

For instance, we can take as Φ the complex structure on X, or a G–bundle on X, or a line bundle on X. Then by this construction, any vertex algebra equipped with the action of the corresponding Harish-Chandra pair by internal symmetries gives rise to a $\mathcal{D}$–module on the corresponding moduli space. The above general picture suggests that vertex algebras play the role of local algebraic objects governing deformations of curves and various geometric data on them. It raises the possibility that more exotic vertex algebras, such as the $\mathcal{W}$–algebras introduced in Chapter 15, may also correspond to some still unknown moduli spaces.

In special cases the space of conformal blocks may be identified with the ring of functions on the formal neighborhood of a point in the corresponding moduli space or the space of global sections of an appropriate line bundle on the moduli space. Conformal blocks may therefore be used to gain insights into the structure of moduli spaces.

Chiral algebras and factorization algebras

As we already mentioned above, A. Beilinson and V. Drinfeld have recently introduced the notion of chiral algebra. A chiral algebra on a smooth curve X is a $\mathcal{D}$–module on X equipped with a certain operation. This operation was designed as a geometric analogue of the operation of operator product expansion. It was expected that any vertex algebra V would give rise to a "universal chiral algebra", i.e., an assignment of a chiral algebra to any curve in such a way that their fibers are isomorphic to V. However, as far as we know, this has not been proved, except for the case when the curve is of genus zero, equipped with a global coordinate, when this was shown by Y.-Z. Huang and J. Lepowsky [**HL**].

For general curves, one first needs to develop the formalism of vertex algebras in a coordinate-independent way. This is done in this book, and using this formalism, we show in Chapter 19 that each (quasi-conformal) vertex algebra indeed gives rise to a universal chiral algebra.

Perhaps, the most exciting and far-reaching part of the theory of chiral algebras is the Beilinson–Drinfeld description of chiral algebras as factorization algebras, i.e., sheaves on the Ran space of finite subsets of a curve, satisfying certain compatibilities. The formalism of factorization algebras is closely related to that of functional realizations discussed in Chapter 10.

In Chapter 20 we give the definition of factorization algebras and explain their connection to chiral algebras and vertex algebras following Beilinson and Drinfeld [**BD4**]. The beauty of factorization algebras is that they have "non-linear" analogues, factorization spaces which arise in "nature". For instance, the Beilinson–Drinfeld Grassmannians which are the moduli spaces of G–bundles on a smooth curve X together with trivializations away from finitely many points of X form a typical factorization space. One can obtain factorization algebras (and hence chiral algebras) on X by "linearizing" factorization spaces, i.e., by considering the cohomology of various sheaves on them. This formalism provides a novel method of constructing chiral and vertex algebras that underscores their intimate relation to moduli spaces. As one of the applications of this method, Beilinson and Drinfeld have produced a fascinating object: the chiral Hecke algebra which we review at the end of Chapter 20. It is expected that the chiral Hecke algebra may be used for the construction of the Hecke eigensheaves whose existence is predicted by the geometric Langlands conjecture.

Overview of the book

This book may be roughly broken down into four large parts. Each of them may be read independently from the rest of the book (although there are many connections between them).

The first part, comprising Chapters 1–5, contains a self-contained elementary introduction to the algebraic theory of vertex algebras, intended mainly for beginners. We give the definition of vertex algebras, study their properties, and consider numerous examples.

In Chapter 1 we lay down the basics of the vertex algebra formalism, including several approaches to the crucial property of locality of fields. We then formulate the axioms of vertex algebras and discuss the example of commutative vertex algebras. In Chapter 2 we consider in great detail the vertex algebras associated to the Heisenberg, affine Kac–Moody, and Virasoro Lie algebras. Chapter 3 is devoted to the Goddard uniqueness theorem and the associativity property of vertex algebras. Using this property, we introduce the concept of the operator product expansion (OPE). Many examples of OPE are worked out in detail in this chapter. In Chapter 4 we attach to any vertex algebra a Lie algebra $U(V)$ and an associative algebra $\widetilde{U}(V)$. The functors sending V to $U(V)$ and $\widetilde{U}(V)$ allow us to relate vertex algebras to more familiar objects, Lie algebras and associative algebras (though they are much larger and more difficult to study than the vertex algebras themselves). At the end of the chapter we present two other applications of the associativity property: we prove a strong form of the Reconstruction Theorem and explain a construction (called functional realization) which assigns to any linear functional on a vertex algebra a collection of meromorphic functions on the symmetric powers of the disc ("n–point functions"). In Chapter 5 we introduce the notions of a module and a twisted module over a vertex algebra and study the general properties of modules. In particular, we explain that a module over a vertex algebra V is the same as a (smooth) module over the associative algebra $\widetilde{U}(V)$ introduced in Chapter 4. We then define the lattice vertex algebras and the free fermionic vertex superalgebra, and establish the boson–fermion correspondence. We introduce the concept of rational vertex algebra and discuss various examples. We also outline several constructions of vertex algebras, such as the coset construction and the theory of orbifolds.

The second, and the central, part of the book consists of Chapters 6–10. In these chapters we develop the geometric theory of vertex algebras and conformal blocks.

In Chapter 6 we give a coordinate-independent realization of the vertex operation. We introduce the group $\operatorname{Aut} \mathcal{O}$ and show how to make it act on a conformal vertex algebra by internal symmetries. Using this action, we attach to any conformal vertex algebra V a vector bundle $\mathcal{V}$ on any smooth algebraic curve. We prove that the vertex operation Y gives rise to a canonical section $\mathcal{Y}_x$ of $\mathcal{V}^*|_{D_x^\times}$ with values in $\operatorname{End} \mathcal{V}_x$, for any $x \in X$. Finally, we equip $\mathcal{V}$ with a flat connection and show that the section $\mathcal{Y}_x$ is horizontal.

A generalization of this construction, involving more general internal symmetries of vertex algebras (such as those generated by affine Kac–Moody algebras), is given in Chapter 7. Various examples of the geometric structures defined in Chapters 6 and 7 are discussed in Chapter 8. In particular, we explain how such familiar geometric objects as projective and affine connections appear naturally in the geometric theory of vertex algebras.

We give the definitions of the spaces of conformal blocks and coinvariants in Chapter 9. We start with the simplest case of the Heisenberg vertex algebra, for which conformal blocks may be defined in a rather elementary way. We then gradually proceed to define the spaces of one-point conformal blocks and coinvariants for general vertex algebras. The generalization to the case of several points, with arbitrary module insertions, is given in Chapter 10. Functorial properties and functional realizations of the spaces of conformal blocks are also discussed in this chapter. In particular, we explain that the vertex operation may itself be viewed as a special conformal block.

The third part of the book, Chapters 11–16, contains some important constructions and applications of vertex algebras. In Chapters 11 and 12 we motivate and then define the free field realization of affine Kac–Moody algebras following the works of M. Wakimoto [**Wak**] and B. Feigin–E. Frenkel [**FF1, FF2, F6**]. Then in Chapters 13 and 14 we use this realization to obtain an explicit description of the spaces of conformal blocks for affine Kac–Moody algebras in the genus zero case. This gives us a systematic way to produce the Schechtman–Varchenko integral solutions of the Knizhnik–Zamolodchikov equations [**FFR**]. Next, we introduce in Chapter 15 a family of vertex algebras, called $\mathcal{W}$–algebras, using the quantum Drinfeld–Sokolov reduction and the screening operators [**FF7, FF8, dBT**]. The $\mathcal{W}$–algebras became important recently because of their role in the geometric Langlands correspondence (see below). The $\mathcal{W}$–algebras generalize the Virasoro vertex algebra, but in contrast to the Virasoro vertex algebra, a general $\mathcal{W}$–algebra is not generated by a Lie algebra. Thus, the formalism of vertex algebras developed in this book is really essential in this case.

Poisson analogues of vertex algebras, in particular, of the $\mathcal{W}$–algebras, are discussed in detail in Chapter 16. At the beginning of the chapter we introduce the notions of vertex Lie algebras and vertex Poisson algebras. Important features of the formalism of vertex Poisson algebra can be traced back to the work by I.M. Gelfand, L.A. Dickey and others on the Hamiltonian structure of integrable hierarchies of soliton equations (see, e.g., [**Di**]). We explain how to obtain non-trivial examples of vertex Poisson algebras by taking suitable classical limits of vertex algebras, in particular, of the Virasoro and Kac–Moody vertex algebras.

Classical limits of vertex algebras give rise to hamiltonian structures of familiar integrable systems, such as the KdV hierarchy, which is closely related to the Virasoro algebra. In the mid-80s, V. Drinfeld and V. Sokolov [**DS**] attached an integrable hierarchy of soliton equations generalizing the KdV system to an arbitrary simple Lie algebra. The objects encoding the Hamiltonian structure of these hierarchies are classical limits of the $\mathcal{W}$–algebras, which may be viewed as the algebras of functions of the spaces of opers on the disc. The concept of an oper, which is implicit in [**DS**], has been generalized by A. Beilinson and V. Drinfeld to the case of an arbitrary curve [**BD2, BD3**]. We recall their definition of opers and explain the connection with the classical Drinfeld–Sokolov reduction and $\mathcal{W}$–algebras.

Finally, in the more advanced last four chapters of the book, we consider "localization" of vertex algebras on the moduli spaces and the relationship between vertex algebras and the Beilinson–Drinfeld formalism of chiral algebras and factorization algebras.

In Chapter 17 we review the formalism of Harish-Chandra localization following A. Beilinson and J. Bernstein. We apply it to construct sheaves of coinvariants of vertex algebras on the moduli spaces of pointed curves. The key to this construction is the "Virasoro uniformization" of the moduli spaces of pointed curves. This construction is generalized in Chapter 18 to the case of moduli spaces of bundles, for which we have an analogous "Kac–Moody uniformization". We discuss the sheaves of coinvariants corresponding to $\widehat{\mathfrak{g}}$–modules of critical level and their relationship to the geometric Langlands correspondence, following Beilinson and Drinfeld [**BD3**]. In establishing this correspondence, the classical $\mathcal{W}$–algebras and the center of a

Kac–Moody vertex algebra play an important role. At the end of Chapter 18 we briefly review the construction of the chiral de Rham complex [**MSV**].

In Chapter 19, we give a motivated introduction to the Beilinson–Drinfeld theory of chiral algebras and Lie* algebras. In particular, we show that any quasi-conformal vertex algebra gives rise to a chiral algebra on an arbitrary smooth curve. Finally, in Chapter 20 we introduce factorization algebras and explain their connection with the vertex and chiral algebras. We then discuss the motivating example of a factorization space, the Beilinson–Drinfeld Grassmannian, describing the modifications of a G–bundle on a curve. This and other naturally occurring factorization spaces serve as a geometric source for vertex algebras, through the process of linearization, which we illustrate with several examples. We conclude with a brief sketch of the role of factorization in the study of Hecke correspondences on the moduli spaces of bundles. We introduce the notion of a Hecke eigensheaf on the moduli space of G–bundles and formulate the geometric Langlands conjecture. We then sketch the Beilinson–Drinfeld construction of the chiral Hecke algebra.

The Appendix contains useful information about ind–schemes, $\mathcal{D}$–modules, and Lie algebra cohomology.

All objects in this book (vector spaces, algebras, curves, schemes, etc.) are considered over the field $\mathbb{C}$ of complex numbers, unless specified otherwise. It would be highly desirable to develop the arithmetic theory of vertex algebras, as indicated by the existing works on this subject [**KSU1, KSU2, Mo**].

Corrections and updates to the material of this book will be posted at

$$\texttt{http://www.math.berkeley.edu/}{\sim}\texttt{frenkel}$$

Bibliography

We end each chapter of the book with bibliographical notes. They are meant to indicate the sources of the results presented in the chapter as well as some references for further reading (sometimes the sources are mentioned inside the chapter, and sometimes this is deferred until the end). We apologize in advance for any omissions or inaccurate attributions.

Many of the results presented in this book are original and previously unpublished. These results are pointed out in the corresponding bibliographical notes.

For a brief summary of the material in this book, the reader may consult the Séminaire Bourbaki talk [**F5**] given by the first author in June of 2000.

Several other books on the theory of vertex algebras are available at present. The first book on the subject, by I. Frenkel, J. Lepowsky, and A. Meurman [**FLM**], laid the foundations of the theory with the applications to the Monster group as a motivation. Further results were presented in [**FHL**] and [**DL**]. A connection between the algebraic theory and the geometric approach to conformal field theory in genus zero through operads is discussed, among other things, in Y.-Z. Huang's book [**H1**]. The recent book by V. Kac [**Kac3**] is an excellent introduction to the subject, containing a number of interesting examples and new techniques, such as the theory of conformal algebras and their cohomology theory. The foundations of the theory of chiral algebras and factorization algebras are laid out in the book [**BD4**] by A. Beilinson and V. Drinfeld. Finally, a good source for the study of conformal field theory from the physics perspective is the book [**dFMS**].

CHAPTER 1

Definition of Vertex Algebras

In this chapter we give the definition of vertex algebras and establish their first properties. We first develop a formalism of formal power series needed to introduce the notion of locality of vertex operators. Then we give several definitions of locality, algebraic as well as analytic. After that we give the axioms of vertex algebra and discuss the simplest example, that of a commutative vertex algebra. More interesting examples are given in the next chapter and in Chapters 5 and 15.

1.1. Formal distributions

Let R be a $\mathbb{C}$–algebra.

1.1.1. Definition. An R–valued formal power series (or formal distribution) in variables $z_1, z_2, \ldots, z_n$ is an arbitrary (finite or infinite) series

$$(1.1.1) \qquad A(z_1, \ldots, z_n) = \sum_{i_1 \in \mathbb{Z}} \cdots \sum_{i_n \in \mathbb{Z}} A_{i_1, \ldots, i_n} z_1^{i_1} \cdots z_n^{i_n},$$

where each $A_{i_1, \ldots, i_n} \in R$. These series form a vector space, which is denoted by $R[[z_1^{\pm 1}, \ldots, z_n^{\pm 1}]]$.

If $P(z_1, \ldots, z_n) \in R[[z_1^{\pm 1}, \ldots, z_n^{\pm 1}]]$ and $Q(w_1, \ldots, w_m) \in R[[w_1^{\pm 1}, \ldots, w_m^{\pm 1}]]$, then their product is a well-defined element of $R[[z_1^{\pm 1}, \ldots, z_n^{\pm 1}, w_1^{\pm 1}, \ldots, w_m^{\pm 1}]]$.

In general, a product of two elements of $R[[z_1^{\pm 1}, \ldots, z_n^{\pm 1}]]$ does not make sense, since individual coefficients of the product are infinite sums of coefficients of the factors. However, the product of a formal power series by a Laurent *polynomial* (i.e., a series (1.1.1) such that $A_{i_1, \ldots, i_n} = 0$ for all but finitely many n–tuples $i_1, \ldots, i_n$) is always well-defined.

1.1.2. Power series as distributions. Given a formal power series in one variable, $f(z) = \sum_{i \in \mathbb{Z}} a_i z^i$, we define its residue (at 0) as

$$\mathrm{Res}\, f(z)dz = \mathrm{Res}_{z=0}\, f(z)dz = a_{-1}.$$

Note that if $R = \mathbb{C}$ and $f(z)$ is the Laurent series of a meromorphic function defined on a disc around 0, having poles only at 0, then

$$\mathrm{Res}_{z=0}\, f(z)dz = \frac{1}{2\pi i} \int f(z)dz,$$

where the integral is taken over a closed curve winding once around 0. To simplify notation, we will henceforth suppress the factor $2\pi i$ from all contour integrals.

Any formal power series $f(z) = \sum_{n \in \mathbb{Z}} f_n z^n$ in $\mathbb{C}[[z^{\pm 1}]]$ defines a linear functional on the space of Laurent polynomials $\mathbb{C}[z, z^{-1}]$ (in other words, a distribution on $\mathbb{C}^{\times}$) whose value on $g \in \mathbb{C}[z, z^{-1}]$ equals

$$\mathrm{Res}_{z=0} f(z)g(z)dz.$$

Similarly, formal power series in several variables define functionals on rational functions. For this reason we sometimes refer to them as "formal distributions".

1.1.3. Delta-function. An important example of a formal power series in two variables z, w is the *formal delta-function* $\delta(z - w)$, which is by definition

$$(1.1.2) \qquad\qquad \delta(z - w) = \sum_{m \in \mathbb{Z}} z^m w^{-m-1}.$$

Its coefficients $a_{mn} = \delta_{m,-n-1}$ are supported on the diagonal $m + n = -1$, and hence it can be multiplied by an arbitrary formal power series in *one* variable (i.e., depending only on z or only on w). Indeed, for such a series $A(w)$, we obtain

$$A(w)\delta(z - w) = \sum_{k \in \mathbb{Z}} A_k w^k \sum_{m \in \mathbb{Z}} z^m w^{-m-1} = \sum_{m,n \in \mathbb{Z}} A_{m+n+1} z^m w^n,$$

so each coefficient is well-defined. Furthermore, the formula above shows that as formal power series in z, w,

$$(1.1.3) \qquad\qquad A(z)\delta(z - w) = A(w)\delta(z - w),$$

which motivates the terminology "delta-function".

We obtain from formula (1.1.3) that

$$(1.1.4) \qquad\qquad (z - w)\delta(z - w) = 0$$

and, by induction,

$$(1.1.5) \qquad\qquad (z - w)^{n+1} \partial_w^n \delta(z - w) = 0.$$

In fact, one has the following abstract characterization of $\delta(z - w)$ and its derivatives.

1.1.4. Lemma ([Kac3]). *Let $f(z, w)$ be a formal power series in $R[[z^{\pm 1}, w^{\pm 1}]]$ satisfying $(z - w)^N f(z, w) = 0$ for a positive integer N. Then $f(z, w)$ can be written uniquely as a sum*

$$(1.1.6) \qquad\qquad \sum_{i=0}^{N-1} g_i(w) \partial_w^i \delta(z - w), \qquad g_i(w) \in R[[w^{\pm 1}]].$$

1.1.5. Proof. Formula (1.1.4) implies that $(z - w)^N \delta(z - w) = 0$ for any positive integer N. Differentiating this formula with respect to w, we obtain by induction that any element of the form (1.1.6) satisfies the equation $(z - w)^N f(z, w) = 0$.

Conversely, suppose that $f(z, w)$ satisfies this equation. Writing $f(z, w) = \sum_{n,m \in \mathbb{Z}} f_{n,m} z^n w^m$, we obtain the following relation on the coefficients $f_{n,m}$:

$$(1.1.7) \qquad\qquad (\Delta^N f)_{n,m} = 0, \qquad n, m \in \mathbb{Z},$$

where $(\Delta f)_{n,m} = f_{n-1,m} - f_{n,m-1}$. Note that each of the equations (1.1.7) involves only the coefficients on the same "diagonal", i.e., those of the form $f_{k,p-k}$ with fixed $p \in \mathbb{Z}$. Therefore a general solution is a sum $f(z, w) = \sum_{p \in \mathbb{Z}} f^{(p)}(z, w)$, where the series $f^{(p)}(z, w)$ is "supported" on the pth diagonal, i.e., $f_{n,m}^{(p)} = 0$ unless $n + m = p$. Restricted to the pth diagonal, the system (1.1.7) becomes a difference equation of order N. Hence the corresponding space of solutions is N–dimensional. But we already have N solutions supported on the pth diagonal: $w^{p+i+1}\partial_w^i\delta(z-w), i = 0, \ldots, N-1$. Their coefficients are polynomials in n of degree i, so these solutions are linearly independent, and hence span the space of solutions of (1.1.7) restricted to the pth diagonal. Therefore we obtain that the general solution of (1.1.7) is of the form (1.1.6).

1.1.6. Remark. Recall that a module M over the ring $\mathcal{O}(X)$ of regular functions on an affine variety X is said to be supported on a subvariety $Y \subset X$, if each $f \in M$ is annihilated by some power of the ideal of functions vanishing on Y. The above lemma means that the series of the form $f(w)\partial_w^i\delta(z-w)$, where $f(w) \in \mathbb{C}[[w^{-1}]]$ and $i \in \mathbb{Z}_+$, span the maximal $\mathbb{C}[z^{\pm 1}, w^{\pm 1}]$–submodule of $\mathbb{C}[[z^{\pm 1}, w^{\pm 1}]]$ supported on the diagonal $\{z = w\}$ in $(\mathbb{C}^\times)^2$.

1.1.7. Analytic interpretation. Formula (1.1.4) can also be interpreted as follows. Let us write

$$(1.1.8) \qquad \delta(z-w) = \underbrace{\frac{1}{z}\sum_{n\geq 0}\left(\frac{w}{z}\right)^n}_{\delta_-} + \underbrace{\frac{1}{z}\sum_{n>0}\left(\frac{z}{w}\right)^n}_{\delta_+}.$$

When z and w take complex values, such that $|z| > |w|$, the series $\delta(z-w)_-$ converges to the meromorphic function $\dfrac{1}{z-w}$. On the other hand, when $|z| < |w|$, the series $\delta(z-w)_+$ converges to $-\dfrac{1}{z-w}$. Formula (1.1.4) means that their sum is "supported at $z = w$".

To make this more precise analytically, let us set w equal to a non-zero complex number. Then $\delta(z-w)$ becomes a formal power series in one variable z, and hence defines a functional on $\mathbb{C}[z, z^{-1}]$. The value of this functional on $g(z) \in \mathbb{C}[z, z^{-1}]$ equals the value of g at w, so this functional can be considered as the "delta-function at the point w". On the other hand, $\delta(z-w)_\pm$ also give rise to functionals on $\mathbb{C}[z, z^{-1}]$, whose values of $g(z) \in \mathbb{C}[z, z^{-1}]$ are equal to

$$\mp \lim_{\epsilon\to+0}\int_{|z|=1}\frac{1}{z-wq^{\mp\epsilon}}g(z)dz,$$

where $|q| < 1$. Thus, we can write

$$\delta(z-w)_\pm = \mp\lim_{\epsilon\to+0}\frac{1}{z-wq^{\mp\epsilon}},$$

in the sense of distributions. Such distributions are called the boundary values of holomorphic functions. Then formula (1.1.8) means that $\delta(z-w)$ is the difference of boundary values of functions holomorphic in the interior and the exterior of the circle $|w| = $ const (see the picture).

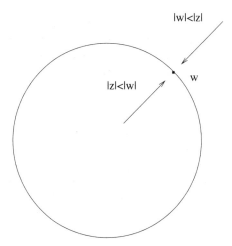

This is analogous to the Sokhotsky–Plemelj formula

$$2\pi i\,\delta(x) = \frac{1}{x + i0} - \frac{1}{x - i0},$$

well-known in complex analysis. Here in the right hand side we take the difference between the boundary values of the function $\frac{1}{x}$ on the real line from the upper and lower half-planes (denoted $\frac{1}{x + i0}$ and $\frac{1}{x - i0}$, respectively). The result is the delta–distribution on the real line at 0 (up to the factor $2\pi i$).

1.1.8. Algebraic reformulation. For any $\mathbb{C}$–algebra R, we denote by $R[[z]]$ the space of R–valued *formal Taylor series* in z. Its elements are series $\sum_{n\geq 0} a_n z^n$, where $a_n \in R$ for all $n \geq 0$. This space is naturally an algebra.

The space $R((z))$ of R–valued *formal Laurent series* in z is by definition the space of series $\sum_{n\in\mathbb{Z}} a_n z^n$, where $a_n \in R$ for all n, and there exists $N \in \mathbb{Z}$ such that $a_n = 0, \forall n \leq N$ (in other words, the series is finite in the "negative direction"). Note that $R((z))$ is an algebra, and if R is a field, then $R((z))$ is also a field.

Denote by $\mathbb{C}((z))((w))$ the space $R((w))$, where $R = \mathbb{C}((z))$. In other words, this is the space of Laurent series in w whose coefficients are Laurent series in z. Then the series $\delta(z - w)_-$ belongs to $\mathbb{C}((z))((w))$ (actually, even to $\mathbb{C}[z^{-1}][[w]]$). This is an algebraist's way of saying that $\delta(z - w)_-$ is the expansion of $\frac{1}{z - w}$ in the domain $|z| > |w|$ (i.e., in positive powers of w/z). Similarly, $\delta(z - w)_+$ belongs to $\mathbb{C}((w))((z))$, and it is the expansion of $-\frac{1}{z - w}$ in the domain $|z| < |w|$ (i.e., in positive powers of z/w).

Denote by $\mathbb{C}((z, w))$ the field of fractions of $\mathbb{C}[[z, w]]$; its elements may be viewed as meromorphic functions in two formal variables. This field has two natural topologies, in which the basis of open neighborhoods of 0 consists of all elements of the form $w^N f(z, w), N \in \mathbb{Z}$ (resp., $z^N f(z, w), N \in \mathbb{Z}$), where $f(z, w)$ does not contain w (resp., z) in the denominator. The completions of $\mathbb{C}((z, w))$ with respect to these topologies are $\mathbb{C}((z))((w))$ and $\mathbb{C}((w))((z))$, respectively. Thus, $\mathbb{C}((z))((w))$ contains expansions of meromorphic functions near the z axis, and $\mathbb{C}((w))((z))$ contains expansions of meromorphic functions near the w axis (see the picture).

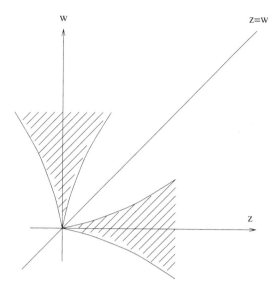

We obtain two embeddings

(1.1.9) $\mathbb{C}((z))((w)) \hookleftarrow \mathbb{C}((z,w)) \hookrightarrow \mathbb{C}((w))((z)).$

For example, $\delta(z-w)_-$ and $-\delta(z-w)_+$ are the images of $\dfrac{1}{z-w} \in \mathbb{C}((z,w))$ under these embeddings.

Now observe that both $\mathbb{C}((z))((w))$ and $\mathbb{C}((w))((z))$ are embedded into the space $\mathbb{C}[[z^{\pm 1}, w^{\pm 1}]]$ of formal power series in two variables. If we start with a Laurent polynomial $f(z,w) \in \mathbb{C}((z,w))$ (i.e., a finite sum of monomials in $z^{\pm 1}, w^{\pm 1}$), then its images in $\mathbb{C}[[z^{\pm 1}, w^{\pm 1}]]$ through the two embeddings will coincide. But this is not true for a general element of $\mathbb{C}((z,w))$. For instance, $\delta(z-w)_-$ and $-\delta(z-w)_+$ are clearly different elements of $\mathbb{C}[[z^{\pm 1}, w^{\pm 1}]]$, even though they come from the same element $\dfrac{1}{z-w}$ of $\mathbb{C}((z,w))$. The difference between them is our formal delta-function $\delta(z-w)$.

In fact,

$$\mathbb{C}((z))((w)) \cap \mathbb{C}((w))((z)) = \mathbb{C}[[z,w]][z^{-1}, w^{-1}]$$

(see the picture below), so any polar term other than z^{-1} and w^{-1} (such as $\dfrac{1}{z-w}$) will have different expansions in $\mathbb{C}((z))((w))$ and $\mathbb{C}((w))((z))$.

However, if we multiply $\dfrac{1}{z-w}$ by $(z-w)$, we obtain a (finite) polynomial, namely 1. Hence if we multiply both $\delta(z-w)_-$ and $-\delta(z-w)_+$ by $(z-w)$, we obtain the same element $1 \in \mathbb{C}[[z^{\pm 1}, w^{\pm 1}]]$. Therefore $(z-w)\delta(z-w) = 0$ (as we expect from a delta-function).

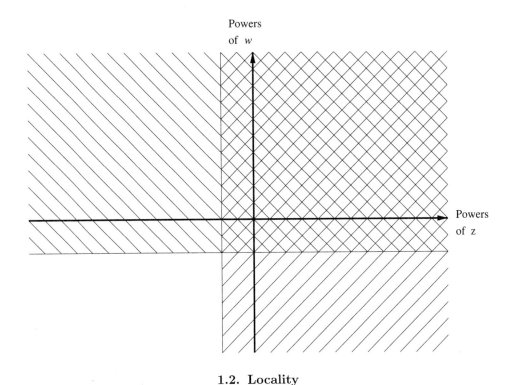

1.2. Locality

1.2.1. Fields. Let V be a vector space over $\mathbb{C}$. Denote by $\operatorname{End} V$ the algebra of linear operators on V. A formal power series

$$(1.2.1) \qquad A(z) = \sum_{j \in \mathbb{Z}} A_j z^{-j} \in \operatorname{End} V[[z^{\pm 1}]]$$

is called a *field* on V if for any $v \in V$ we have: $A_j \cdot v = 0$ for large enough j. In other words, $A(z) \cdot v$ is an element of $V((z))$, the space of formal Laurent series with coefficients in V (i.e., it has only finitely many terms with negative powers of z). Fields on V form a vector space denoted by $\mathcal{F}(V)$.

Now suppose that V is $\mathbb{Z}$–graded,

$$V = \bigoplus_{n \in \mathbb{Z}} V_n.$$

Recall that in this case a linear operator $\phi : V \to V$ is called homogeneous of degree m, if $\phi(V_n) \subset V_{n+m}$ for all n.

A (homogeneous) field of conformal dimension $\Delta \in \mathbb{Z}$ is by definition a field (1.2.1) on V, where each A_j is a homogeneous linear operator on V of degree $-j+\Delta$. Suppose that $V_k = 0$ for all k less than a fixed integer K. Then if $v \in V_n$ is a vector of degree n, then $\deg(A_j \cdot v) = n - j + \Delta$, and so $A_j \cdot v = 0$ for all $j > n + K + \Delta$. It follows that for an arbitrary $v \in V$ there is an integer N such that $A_j \cdot v = 0$ for $j > N$. Thus

$$(1.2.2) \qquad A(z) \cdot v = \sum_{j \in \mathbb{Z}} (A_j \cdot v) z^{-j} \in V((z)),$$

so $A(z)$ automatically satisfies the condition of being a field.

If $A(z)$ is a field of conformal dimension Δ, then

$$\partial_z A(z) = \sum_{j \in \mathbb{Z}} (-j) A_j z^{-j-1}$$

is a field of conformal dimension $\Delta + 1$.

If we substitute for the formal variable z a non-zero complex number (also denoted by z), then (1.2.2) yields an infinite sum of different homogeneous components lying in the subspaces $V_n, n \in \mathbb{Z}$. In other words, we obtain a vector in the direct product

$$\overline{V} = \prod_{n \in \mathbb{Z}} V_n.$$

Suppose that $V_n = 0$ for $n < K$, i.e.,

$$V = \bigoplus_{n=K}^{\infty} V_n.$$

Then the space $\overline{V}$ is the completion of V in the topology, which has a basis of open neighborhoods of 0 of the form

$$V_{\geq k} = \bigoplus_{n=k}^{\infty} V_n, k \geq K.$$

Thus, we see that for any non-zero complex number $z \in \mathbb{C}^\times$, $A(z)$ can be considered as a linear operator from V to $\overline{V}$.

1.2.2. Composing fields. We would like to formulate a reasonable condition of commutativity (which we will refer to as locality) of two fields, $A(z)$ and $B(w)$, acting on V.

Observe that for any vector $v \in V$ and any linear functional $\varphi : V \to \mathbb{C}$, the matrix element $\langle \varphi, A(z)v \rangle$ of a field $A(z)$ is a Laurent power series (so the degrees in z are bounded from below).

Given another field, $B(w)$, we consider the composition $A(z)B(w)$ as an $\operatorname{End} V$-valued formal power series in z, w. Given $v \in V$ and $\varphi \in V^*$ (where V^* denotes the vector space of all linear functionals on V), consider the matrix element

$$\langle \varphi, A(z)B(w)v \rangle \in \mathbb{C}[[z^{\pm 1}, w^{\pm 1}]].$$

From the definition of a field given in § 1.2.1, we see that this formal power series actually belongs to $\mathbb{C}((z))((w))$. Indeed, the degrees of w in this series are bounded from below, and if we write $B(w) = \sum_{j \in \mathbb{Z}} B_j w^{-j}$, then the coefficient in front of w^{-j} equals $\langle \varphi, A(z)B_j v \rangle$ and hence belongs to $\mathbb{C}((z))$.

Similarly, the matrix coefficient of the product of these fields taken in the opposite order, $\langle \varphi, B(w)A(z)v \rangle$ belongs to $\mathbb{C}((w))((z))$.

If we impose the naive condition that the fields $A(z)$ and $B(w)$ commute, i.e., the above two matrix elements are equal for all v and φ, then all of these matrix coefficients will have to belong to the intersection of $\mathbb{C}((z))((w))$ and $\mathbb{C}((w))((z))$ in $\mathbb{C}[[z^{\pm 1}, w^{\pm 1}]]$, i.e., to the space $\mathbb{C}[[z, w]][z^{-1}, w^{-1}]$.

This is a very strong condition (see § 1.4 below), so we try to relax it in some way. Our experience with formal power series (see (1.1.9)) suggests that it is reasonable to require that the two expressions be expansions of one and the same rational function of z and w, or a more general element of $\mathbb{C}((z, w))$, the field of fractions of $\mathbb{C}[[z, w]]$. We then impose the additional condition that this function has poles only at $z = 0$, $w = 0$, and on the diagonal $z = w$. This leads us to the following definition.

1.2.3. Definition. Two fields $A(z)$ and $B(w)$ acting on a vector space V are said to be *local* with respect to each other if for every $v \in V$ and $\varphi \in V^*$, the matrix elements

$$\langle \varphi, A(z)B(w)v \rangle \quad \text{and} \quad \langle \varphi, B(w)A(z)v \rangle$$

are expansions of one and the same element

$$f_{v,\varphi} \in \mathbb{C}[[z,w]][z^{-1}, w^{-1}, (z-w)^{-1}]$$

in $\mathbb{C}((z))((w))$ and $\mathbb{C}((w))((z))$, respectively, and the order of pole of $f_{v,\varphi}$ in $(z-w)$ is uniformly bounded for all v, φ.

1.2.4. A reformulation. The last condition may be reformulated as saying that there exists $N \in \mathbb{Z}_+$ such that

$$(z-w)^N f_{v,\varphi} \in \mathbb{C}[[z,w]][z^{-1}, w^{-1}]$$

for all v, φ. But then the expansions of $(z-w)^N f_{v,\varphi}$ in $\mathbb{C}((z))((w))$ and $\mathbb{C}((w))((z))$ are equal to each other. Therefore if $A(z)$ and $B(w)$ are local with respect to each other, then

$$(z-w)^N A(z)B(w) = (z-w)^N B(w)A(z),$$

or equivalently, $(z-w)^N [A(z), B(w)] = 0$, where $[A, B] \overset{\text{def}}{=} AB - BA$. The following proposition shows that the converse is also true.

1.2.5. Proposition. *Two fields $A(z), B(w)$ are local if and only if there exists $N \in \mathbb{Z}_+$ such that*

(1.2.3) $$(z-w)^N [A(z), B(w)] = 0$$

as a formal power series in $\operatorname{End} V[[z^{\pm 1}, w^{\pm 1}]]$.

1.2.6. Proof. Suppose that the fields $A(z)$ and $B(w)$ satisfy the condition of the proposition. Then we obtain that

$$(z-w)^N \langle \varphi, A(z)B(w)v \rangle = (z-w)^N \langle \varphi, B(w)A(z)v \rangle$$

for all $v \in V, \varphi \in V^*$. Following the argument of § 1.2.2, we find that both are equal to some $P(z, w) \in \mathbb{C}[[z,w]][z^{-1}, w^{-1}]$. Therefore the formal power series $f(z, w) = \langle \varphi, A(z)B(w)v \rangle$ is a solution of the equation

(1.2.4) $$(z-w)^N f(z, w) = P(z, w).$$

But we know from § 1.2.2 that $f(z, w)$ belongs to $\mathbb{C}((z))((w))$, which is a field, so equation (1.2.4) has a unique solution in $\mathbb{C}((z))((w))$, namely,

$$P(z, w)\delta(z-w)_-^N$$

(i.e., the expansion of $P(z, w)/(z-w)^N$ in positive powers of w/z). Thus, the series $\langle \varphi, A(z)B(w)v \rangle$ is the expansion of

$$f_{v,\varphi} \overset{\text{def}}{=} P(z, w)/(z-w)^N$$

in $\mathbb{C}((z))((w))$. Likewise, we obtain that $\langle \varphi, B(w)A(z)v \rangle$ must be the expansion of $P(z, w)/(z-w)^N$ in $\mathbb{C}((w))((z))$. This completes the proof.

1.2.7. Remark. Suppose that V is $\mathbb{Z}$–graded, with the grading bounded from below, i.e.,

$$\bigoplus_{n=K}^{\infty} V_n$$

for some $K \in \mathbb{Z}$. Introduce the restricted dual space,

$$V^{\vee} = \bigoplus_{n=K}^{\infty} V_n^*.$$

If $\varphi \in V^*$ belongs to $V^{\vee}$, then

$$\langle \varphi, A(z)v \rangle \in \mathbb{C}[z^{\pm 1}] \quad \text{and} \quad \langle \varphi, A(z)B(w)v \rangle \in \mathbb{C}[z^{\pm 1}]((w)).$$

The locality condition may be reformulated as saying that for any $v \in V$ and $\varphi \in V^{\vee}$, the matrix elements

$$\langle \varphi, A(z)B(w)v \rangle \quad \text{and} \quad \langle \varphi, B(w)A(z)v \rangle$$

are expansions of the same rational function

$$f_{v,\varphi} \in \mathbb{C}[z^{+1}, w^{\pm 1}, (z-w)^{-1}],$$

in $\mathbb{C}[z^{\pm 1}]((w))$ and $\mathbb{C}[w^{\pm 1}]((z))$, respectively, and the order of pole of $f_{v,\varphi}$ in $(z-w)$ is uniformly bounded for all v, φ. In other words, for large enough N and any given v and φ,

$$(z-w)^N \langle \varphi, A(z)B(w)v \rangle \quad \text{and} \quad (z-w)^N \langle \varphi, B(w)A(z)v \rangle$$

are equal to one and the same Laurent polynomial in $\mathbb{C}[z^{\pm 1}, w^{\pm 1}]$.

The reader may be wondering why we allow poles only on the diagonal $z = w$, and not on more general curves. This is indeed possible. For instance, in the theory of quantum groups it is natural to allow poles on "shifted" diagonals $z = wq^n$, where q is a parameter (see [**FR, EK**]). However, the diagonal is the only "coordinate-independent" divisor (other than the lines $z = 0, w = 0$, which are included anyway).

1.2.8. Analytic reformulation. In this section we continue under the assumption that V is $\mathbb{Z}$–graded with the grading bounded from below. Let $A(z)$ be a field acting on V. Recall from § 1.2.1 that for each $z \in \mathbb{C}^{\times}$, $A(z)$ is a linear operator from V to its completion $\overline{V}$ defined in § 1.2.1. Hence the operators $A(z)$ and $B(w)$ (for $w \in \mathbb{C}^{\times}$) act between different spaces, namely, from V to $\overline{V}$. How can we compose such operators? To answer this question, it is convenient to represent a linear operator $P : V \to \overline{V}$ by an infinite collection of linear operators

$$P_i^j : V_i \hookrightarrow V \to \overline{V} \twoheadrightarrow V_j.$$

If Q is another such operator, with components Q_i^j, we may attempt to define the composition PQ componentwise, by setting

$$(P \circ Q)_i^j = \sum_{k \in \mathbb{Z}_+} P_k^j Q_i^k.$$

The problem here is that this is an infinite sum of operators, so that the image of a vector in V_i is an infinite sum of vectors in V_j, which does not *a priori* make sense.

We will say that the composition $P \circ Q$ *exists* if for any $\varphi \in V_j^*$ and $v \in V_i$, the series

$$\sum_{k \in \mathbb{Z}_+} \langle \varphi, P_k^j Q_i^k v \rangle$$

converges absolutely. In this case we indeed obtain well-defined operators $(P \circ Q)_i^j : V_i \to V_j$ and hence a linear operator

$$P \circ Q : V \to \overline{V}.$$

Now suppose that $A(z)$ and $B(w)$ are local with respect to each other. Then according to Remark 1.2.7, for any $v \in V$, $\varphi \in V^\vee$, and sufficiently large N, the matrix element $(z - w)^N \langle \varphi, A(z)B(w)v \rangle$ is a polynomial in $z^{\pm 1}$ and $w^{\pm 1}$, and so it may be evaluated for arbitrary non-zero values of z and w. Thus $(z - w)^N A(z)B(w)$ is a well-defined linear operator from V to $\overline{V}$, depending holomorphically on $z, w \in \mathbb{C}^\times$.

This property of mutually local fields allows us to give a reformulation of locality from the analytic point of view. Observe that the image of a meromorphic function $f(z, w) \in \mathbb{C}[z^{\pm 1}, w^{\pm 1}, (z - w)^{-1}]$ in $\mathbb{C}[z^{\pm 1}]((w))$ is a series which converges in the domain $|z| > |w|$, while the image of $f(z, w)$ in $\mathbb{C}[w^{\pm 1}]((z))$ converges in the domain $|w| > |z|$. Therefore we obtain the following proposition. Let us call a collection $\{P_i^j\}_{i,j \in \mathbb{Z}_+}$ of meromorphic matrix-valued functions on $\mathbb{C}^2$, which are holomorphic away from $M \subset \mathbb{C}^2$, an operator–valued meromorphic function on $\mathbb{C}^2$ with singularities along M.

1.2.9. Proposition. *Two fields $A(z)$, $B(z)$ are local with respect to each other if and only if*

(1) *For any $z, w \in \mathbb{C}^\times$ with $|z| > |w|$, the composition $A(z)B(w)$ exists, and can be analytically continued to an operator–valued meromorphic function $R(A(z)B(w))$ on $\mathbb{C}^2$, with singularities at $\{z = 0\}, \{w = 0\}$ and $\{z = w\}$, such that the order of the pole at $z = w$ is uniformly bounded.*

(2) *For $|w| > |z|$ the composition $B(w)A(z)$ exists and can be analytically continued to an operator–valued meromorphic function $R(B(w)A(z))$ on $\mathbb{C}^2$, with singularities at $\{z = 0\}, \{w = 0\}$ and $\{z = w\}$, such that the order of the pole at $z = w$ is uniformly bounded.*

(3) $R(A(z)B(w)) = R(B(w)A(z))$.

1.3. Definition of a vertex algebra

Now we are ready to give the definition of a vertex algebra.

1.3.1. Definition. A *vertex algebra* is a collection of data:

- (*space of states*) a vector space V;
- (*vacuum vector*) a vector $|0\rangle \in V$;
- (*translation operator*) a linear operator $T : V \to V$;
- (*vertex operators*) a linear operation

$$Y(\cdot, z) : V \to \text{End}\, V[[z^{\pm 1}]]$$

taking each $A \in V$ to a field acting on V,

$$Y(A, z) = \sum_{n \in \mathbb{Z}} A_{(n)} z^{-n-1}.$$

These data are subject to the following axioms:

- (*vacuum axiom*) $Y(|0\rangle, z) = \mathrm{Id}_V$. Furthermore, for any $A \in V$ we have

$$Y(A, z)|0\rangle \in V[[z]],$$

so that $Y(A, z)|0\rangle$ has a well–defined value at $z = 0$, and

$$Y(A, z)|0\rangle \mid_{z=0} = A.$$

In other words, $A_{(n)}|0\rangle = 0, n \geq 0$, and $A_{(-1)}|0\rangle = A$.
- (*translation axiom*) For any $A \in V$,

$$[T, Y(A, z)] = \partial_z Y(A, z)$$

and $T|0\rangle = 0$.
- (*locality axiom*) All fields $Y(A, z)$ are local with respect to each other.

A vertex algebra is called $\mathbb{Z}$–graded if V is a $\mathbb{Z}$–graded vector space, $|0\rangle$ is a vector of degree 0, T is a linear operator of degree 1, and for $A \in V_m$ the field $Y(A, z)$ has conformal dimension m (i.e., $\deg A_{(n)} = -n + m - 1$).

1.3.2. Remarks. 1. It is easy to adopt the above definition of vertex algebra to the supercase. Then V is a superspace with a decomposition $V = V_{\bar{0}} \oplus V_{\bar{1}}$, and the above structures and axioms should be modified as follows: if $A \in V_{\bar{i}}$, then all Fourier coefficients of $Y(A, z)$ should be endomorphisms of V of parity $\bar{i}$; $|0\rangle$ should be an element of $V_{\bar{0}}$; T should have even parity; and we should insert a sign in the definition of locality:

$$(1.3.1) \qquad (z - w)^N Y(A, z) Y(B, w) = (-1)^{p(A)p(B)} (z - w)^N Y(B, w) Y(A, z),$$

for some $N \in \mathbb{Z}_+$, where $p(A)$ denotes the parity of $A \in V$. The gradation may take half-integral values on $V_{\bar{1}}$ (see [**Kac3**] for more details). In this book, we mostly consider purely even vertex superalgebras, but general vertex superalgebras are also very important; for instance, $N = 2$ superconformal vertex superalgebras are relevant to physical models appearing in mirror symmetry.

2. Essentially all vertex algebras that we consider in this book are $\mathbb{Z}$–graded (and most of them are even $\mathbb{Z}_+$–graded with finite-dimensional graded components). An example of a vertex algebra which does not admit a $\mathbb{Z}$–grading is given in § 18.4.6.

3. Note that according to the translation axiom, the action of T on V is completely determined by the operation Y, namely, $T \cdot A = A_{(-2)}|0\rangle$. However, the above system of axioms with T as an independent structure appears to us to be more natural than the one without T.

4. The vacuum axiom implies that the map $V \to \mathrm{End}\, V$ defined by the formula $A \mapsto A_{(-1)}$ is injective. Therefore Y is also injective.

1.3.3. Analogy between algebras and vertex algebras. The axioms of a vertex algebra may look rather complicated, but in fact they are quite natural generalizations of the axioms of a associative commutative algebra with a unit. (Moreover, we will see in § 1.4 that a vertex algebra, in which all vertex operators are power series in positive powers of z only, is the same as a commutative algebra with a derivation.) Indeed, such an algebra may be defined as a vector space V along with a linear operator $Y : V \to \mathrm{End}(V)$ of degree 0 and an element $1 \in V$ such that $Y(B)1 = B$ and $Y(A)Y(B) = Y(B)Y(A)$. The linear operator Y defines a product structure by the formula $Y(A)B = A \cdot B$. The identity $Y(A)Y(B) =$

$Y(B)Y(A)$ implies both the commutativity and the associativity of the product. The former is obvious, and to see the latter, observe that $Y(A)Y(B) = Y(B)Y(A)$ gives $A \cdot (B \cdot C) = B \cdot (A \cdot C)$ for all $C \in V$. Hence

$$A \cdot (B \cdot C) = A \cdot (C \cdot B) = C \cdot (A \cdot B) = (A \cdot B) \cdot C.$$

We can think of the vertex operation $Y(\cdot, z)$ in two ways: as defining an infinite collection of "products" labeled by integers (the Fourier coefficients of the fields $Y(A, z)$), or by "points of the punctured disc". The locality axiom can be viewed as an analogue of the axiom of commutativity (with a proper regularization of the compositions of operators), and the first half of the vacuum axiom is an analogue of the axiom of unit. Just as in the case of commutative algebras, there is an analogue of the associativity property for vertex algebras, which follows from the locality and other axioms of a vertex algebra (see § 3.2).

It is straightforward to define homomorphisms between vertex algebras, vertex subalgebras, ideals and quotients.

1.3.4. Definition. A *vertex algebra homomorphism* ρ between vertex algebras

$$(V, |0\rangle, T, Y) \to (V', |0\rangle', T', Y')$$

is a linear map $V \to V'$ mapping $|0\rangle$ to $|0\rangle'$, intertwining the translation operators, and satisfying

$$\rho(Y(A, z)B) = Y(\rho(A), z)\rho(B).$$

A *vertex subalgebra* $V' \subset V$ is a T–invariant subspace containing the vacuum vector, and satisfying $Y(A, z)B \in V'((z))$ for all $A, B \in V'$ (with the induced vertex algebra structure).

A *vertex algebra ideal* $I \subset V$ is a T–invariant subspace satisfying $Y(A, z)B \in I((z))$ for all $A \in I$ and $B \in V$. The skew-symmetry property proved in Proposition 3.2.5 below implies that then $Y(B, z)A \in I((z))$ as well (with A, B as before) so that I is automatically a "two–sided" ideal. It follows that for any proper ideal I, V/I inherits a natural quotient vertex algebra structure.

1.3.5. Exercise. Show that if $\rho : V \to V'$ is a vertex algebra homomorphism, then the image of ρ is a vertex subalgebra of V' and its kernel is a vertex algebra ideal in V.

1.3.6. Lemma. *For two vertex algebras* $(V_1, |0\rangle_1, T_1, Y_1)$ *and* $(V_2, |0\rangle_2, T_2, Y_2)$, *the data* $(V_1 \otimes_{\mathbb{C}} V_2, |0\rangle_1 \otimes |0\rangle_2, T_1 \otimes 1 + 1 \otimes T_2, Y)$, *where*

$$Y(A_1 \otimes A_2, z) = Y_1(A_1, z) \otimes Y_2(A_2, z)$$

defines a vertex algebra called the tensor product of V_1 *and* V_2.

1.4. First example: commutative vertex algebras

A vertex algebra is called *commutative* if all vertex operators $Y(A, z), A \in V$, commute with each other (i.e., we have $N = 0$ in formula (1.2.3)).

Suppose we are given a commutative vertex algebra V. Then for any $A, B \in V$ we have

$$Y(A, z) \cdot B = Y(A, z)Y(B, w)|0\rangle|_{w=0} = Y(B, w)Y(A, z)|0\rangle|_{w=0}.$$

But by the vacuum axiom, the last expression has no negative powers of z. Therefore $Y(A, z)B \in V[[z]]$ for all $A, B \in V$, so $Y(A, z) \in \text{End } V[[z]]$ for all $A \in V$. Conversely, suppose that we are given a vertex algebra V in which $Y(A, z) \in \text{End } V[[z]]$ for all $A \in V$. Observe that if the equality $(z-w)^N f_1(z, w) = (z-w)^N f_2(z, w)$ holds for $f_1, f_2 \in \mathbb{C}[[z, w]]$ and $N \in \mathbb{Z}_+$, then necessarily $f_1(z, w) = f_2(z, w)$. Therefore we obtain that $[Y(A, z), Y(B, w)] = 0$ for all $A, B \in V$, so V is commutative.

Thus, a commutative vertex algebra may alternatively be defined as one in which all vertex operators $Y(A, z)$ are regular at $z = 0$.

Denote by Y_A the endomorphism of a commutative vertex algebra V which is the constant term of $Y(A, z)$, $A \in V$, and define a bilinear operation $\circ$ on V by setting $A \circ B = Y_A \cdot B$. By construction, $Y_A Y_B = Y_B Y_A$. As explained in § 1.3.3, this implies both commutativity and associativity of $\circ$. Hence we obtain a commutative and associative product on V. Furthermore, the vacuum vector $|0\rangle$ is a unit, and the operator T is a derivation with respect to this product. Thus, we have the structure of a commutative algebra with a derivation on V.

Conversely, let V be a commutative algebra with a unit and finite-dimensional homogeneous components. Suppose that V is equipped with a derivation T of degree 1. Then V carries a canonical vertex algebra structure. Namely, we take the unit of V as the vacuum vector $|0\rangle$, and define $Y(A, z)$ to be the operator of multiplication by

$$\sum_{n \geq 0} \frac{z^n}{n!}(T^n A) = e^{zT} A.$$

It is straightforward to check that all axioms of a commutative vertex algebra are satisfied.

Thus, we see that the notion of commutative vertex algebra is essentially equivalent to that of commutative algebra with a unit and a derivation.

1.4.1. Remark. The operator T may be viewed as the generator of infinitesimal translations of the parameter z on the formal additive group. Therefore commutative vertex algebras are commutative algebras with an action of the formal additive group. Thus, we may think of general vertex algebras as "meromorphic" generalizations of commutative algebras with an action of the formal additive group. This point of view has been developed by Borcherds [**B3**], who showed that vertex algebras are "singular commutative rings" in a certain category. He has also considered generalizations of vertex algebras, replacing the formal additive group by other (formal) groups or Hopf algebras.

In this book, rather than considering z as the coordinate on a one–dimensional group, we consider it as a formal coordinate at a point of a smooth algebraic curve. As we will see, the axioms of vertex algebra will eventually allow us to eliminate this coordinate and to give a coordinate–independent description of the vertex algebra structure.

1.5. Bibliographical notes

The first mathematical definition of vertex algebra was given by R. Borcherds [**B1**]. The theory was further developed by I. Frenkel, J. Lepowsky and A. Meurman [**FLM**] (see also [**FHL, DL**]). The works [**B1**] and [**FLM**] were motivated by the study of the Monster group. For an introduction to the subject, see the book [**Kac3**] by V. Kac.

We adopt the system of axioms from [**FKRW**] and [**Kac3**]. The equivalence between these axioms and the original axioms of [**B1, FLM**] is explained in [**Kac3**] (see also [**DL, Li1**]).

The definition of locality given by formula (1.2.3) first appeared in [**DL**], where it was shown that it can be used as a replacement for the original axioms of [**B1, FLM**] (see also [**Li1**]). The calculation in § 1.3.3 is taken from [**Li1**].

CHAPTER 2

Vertex Algebras Associated to Lie Algebras

In this chapter we start the exploration of vertex algebras with the examples of (non-commutative) vertex algebras associated with infinite-dimensional Lie algebras, such as the Heisenberg, affine Kac–Moody, and Virasoro algebras. These vertex algebras appear as chiral symmetry algebras of two-dimensional conformal field theories: the free bosonic theory, the Wess–Zumino–Witten model and the Belavin–Polyakov–Zamolodchikov minimal models, respectively. From the point of view of the geometric formalism of vertex algebras developed in this book, these vertex algebras appear as the algebraic objects responsible for the local symmetries of various geometric structures on algebraic curves: line bundles, G bundles and complex structures, respectively.

2.1. Heisenberg Lie algebra

2.1.1. Central extensions. Let $\mathfrak{g}$ be a *Lie algebra*, i.e., a vector space equipped with a bilinear operation $[\cdot,\cdot] : \mathfrak{g} \otimes \mathfrak{g} \to \mathfrak{g}$ satisfying the following:

- (skew–symmetry) $[x,y] = -[y,x]$.
- (Jacobi identity) $[x,[y,z]] + [y,[z,x]] + [z,[x,y]] = 0$.

A *central extension* of $\mathfrak{g}$ is an exact sequence

$$(2.1.1) \qquad\qquad 0 \to \mathfrak{a} \to \widehat{\mathfrak{g}} \to \mathfrak{g} \to 0$$

of Lie algebras, where $\mathfrak{a} \subset \widehat{\mathfrak{g}}$ is central (that is, $[a,x] = 0$ for all $a \in \mathfrak{a}, x \in \widehat{\mathfrak{g}}$).

We will primarily be interested in one–dimensional central extensions, where $\mathfrak{a} \simeq \mathbb{C}\mathbf{1}$ with a generator $\mathbf{1}$. Choose a splitting of $\widehat{\mathfrak{g}}$ as a vector space, $\widehat{\mathfrak{g}} \simeq \mathbb{C}\mathbf{1} \oplus \mathfrak{g}$. Then the Lie bracket on $\widehat{\mathfrak{g}}$ gives rise to a Lie bracket on $\mathbb{C}\mathbf{1} \oplus \mathfrak{g}$ such that $\mathbb{C}\mathbf{1}$ is central and

$$[x,y]_{\mathrm{new}} = [x,y]_{\mathrm{old}} + c(x,y)\mathbf{1},$$

where $x,y \in \mathfrak{g}$ and $c : \mathfrak{g} \otimes \mathfrak{g} \to \mathbb{C}$ is a linear map. The above formula defines a Lie bracket on $\widehat{\mathfrak{g}}$ if and only if c satisfies

- $c(x,y) = -c(y,x)$.
- $c(x,[y,z]) + c(y,[z,x]) + c(z,[x,y]) = 0$.

Such expressions are called *two–cocycles* on $\mathfrak{g}$.

Now suppose that we are given two extensions $\widehat{\mathfrak{g}}_1$ and $\widehat{\mathfrak{g}}_2$ of the form (2.1.1) with $\mathfrak{a} = \mathbb{C}\mathbf{1}$ and a Lie algebra isomorphism ϕ between them which preserves their subspaces $\mathbb{C}\mathbf{1}$. Choosing splittings $\imath_1 : \mathbb{C}\mathbf{1} \oplus \mathfrak{g} \to \widehat{\mathfrak{g}}_1$ and $\imath_2 : \mathbb{C}\mathbf{1} \oplus \mathfrak{g} \to \widehat{\mathfrak{g}}_2$ for these extensions, we obtain the two-cocycles c_1 and c_2. Define a linear map $f : \mathfrak{g} \to \mathbb{C}$ as the composition of $\imath_1$, ϕ and the projection $\widehat{\mathfrak{g}}_2 \to \mathbb{C}\mathbf{1}$ induced by $\imath_2$. A straightforward calculation gives

$$c_2(x,y) = c_1(x,y) + f([x,y]).$$

Thus, we obtain that the set of isomorphism classes of one–dimensional central extensions of $\mathfrak{g}$ is the quotient of the vector space of two-cocycles on $\mathfrak{g}$ by the subspace of those two-cocycles c for which

$$c(x, y) = f([x, y]), \qquad \forall x, y \in \mathfrak{g},$$

for some $f : \mathfrak{g} \to \mathbb{C}$ (such cocycles are called coboundaries). This quotient space is the second Lie algebra cohomology group $H^2(\mathfrak{g}, \mathbb{C})$, see § A.4.

The central extensions that we consider below (corresponding to the Heisenberg, Kac-Moody and Virasoro algebras) will later be interpreted geometrically using connections and projective structures (see Chapter 8 and §§ 16.4, 16.5) and using moduli spaces of curves and bundles (see §§ 17.3.17 and 18.1.12).

2.1.2. Definition of the Heisenberg Lie algebra. Consider the vector space $\mathbb{C}((t))$ of formal Laurent series in one variable as a commutative Lie algebra. We define the *Heisenberg Lie algebra* $\mathcal{H}$ as the central extension

$$(2.1.2) \qquad\qquad 0 \to \mathbb{C}\mathbf{1} \to \mathcal{H} \to \mathbb{C}((t)) \to 0$$

with the cocycle

$$(2.1.3) \qquad\qquad c(f, g) = -\operatorname{Res}_{t=0} f dg.$$

One checks easily that (2.1.3) defines a two–cocycle and hence a central extension of $\mathbb{C}((t))$. The definition (2.1.3) of the cocycle using residues of one–forms rather than coefficients of power series is clearly independent of the choice of local coordinate t. Thus we may define a Heisenberg Lie algebra canonically as a central extension of the space of functions on a punctured disc $D^\times$ which is not endowed with a specific choice of formal coordinate t. This fact will eventually allow us to give a global meaning to the Heisenberg Lie algebra on arbitrary algebraic curves (see § 9.1.8).

There is another version of the Heisenberg Lie algebra: the one-dimensional central extension $\mathcal{H}'$ of the commutative Lie algebra of Laurent *polynomials* $\mathbb{C}[t, t^{-1}]$ defined by the two-cocycle (2.1.3). This Lie algebra has an obvious basis: $b_n = t^n, n \in \mathbb{Z}$, and the central element $\mathbf{1}$. The Lie algebra $\mathcal{H}$ does not possess such a simple basis. However, $\mathcal{H}$ is the completion of $\mathcal{H}'$ with respect to the topology in which the basis of open neighborhoods of 0 is formed by the subspaces $t^N \mathbb{C}[t], N \in \mathbb{Z}$. Hence it makes sense to say that that $\mathcal{H}$ is *topologically* generated by $b_n = t^n, n \in \mathbb{Z}$, and $\mathbf{1}$. Note that the Lie bracket is continuous with respect to the above topology, and so $\mathcal{H}$ is a complete topological Lie algebra.

By a representation of a topological Lie algebra like $\mathcal{H}$ we will understand (unless noted otherwise) a continuous representation in a vector space V equipped with the *discrete* topology. In the case of $\mathcal{H}$, this is equivalent to the requirement that for any $v \in V$, we have $t^N \mathbb{C}[[t]] \cdot v = 0$ for some $N \in \mathbb{Z}$. Giving such a representation of $\mathcal{H}$ is the same as giving a representation of $\mathcal{H}'$ such that for any $v \in V$, we have $t^N \mathbb{C}[t] \cdot v = 0$ for some $N \in \mathbb{Z}$, because then the action of $\mathcal{H}'$ on V may be extended to $\mathcal{H}$ by continuity.

The advantage of $\mathcal{H}$ over $\mathcal{H}'$ is that $\mathcal{H}$ is preserved under changes of the coordinate t, while $\mathcal{H}'$ is not. This property will become important later on when we develop a coordinate-independent approach to vertex algebras (see, e.g., Chapter 9). However, when studying concrete representations, it is more convenient to view them as representations of the Lie algebra $\mathcal{H}'$.

We find from formula (2.1.3) that

$$\begin{aligned}
c(t^n, t^m) &= -\operatorname{Res}_{t=0} t^n dt^m \\
&= -m \operatorname{Res}_{t=0} t^{n+m-1} dt \\
&= -m\delta_{n,-m} = n\delta_{n,-m}.
\end{aligned}$$

Hence the generators satisfy the following relations:

$$(2.1.4) \qquad [b_n, b_m] = n\delta_{n,-m}\mathbf{1}, \qquad [\mathbf{1}, b_n] = 0.$$

Recall that the *universal enveloping algebra* of a Lie algebra $\mathfrak{g}$, denoted by $U(\mathfrak{g})$, is by definition the quotient of the tensor algebra $T(\mathfrak{g}) = \bigoplus_{n \geq 0} \mathfrak{g}^{\otimes n}$ by the two-sided ideal generated by the elements of the form $x \otimes y - y \otimes x - [x, y]$, for all $x, y \in \mathfrak{g}$ (in other words, we identify the commutator of two elements in $\mathfrak{g} \subset T(\mathfrak{g})$ with their Lie bracket). Note that $U(\mathfrak{g})$ is an associative algebra with a unit, and that any representation of $\mathfrak{g}$ is automatically a $U(\mathfrak{g})$–module. Let $\{x_i\}_{i \in I}$ be an ordered basis of $\mathfrak{g}$. By the Poincaré–Birkhoff–Witt theorem, the lexicographically ordered monomials $x_{i_1} \ldots x_{i_n}$, where $i_1 \preceq \ldots \preceq i_n$, form a basis of $U(\mathfrak{g})$. This basis is called the *Poincaré–Birkhoff–Witt basis* (or PBW basis for short).

The universal enveloping algebra $U(\mathcal{H}')$ of $\mathcal{H}'$ is an associative algebra with generators $b_n, n \in \mathbb{Z}$, and $\mathbf{1}$, and relations (for $n, m \in \mathbb{Z}$)

$$b_n b_m - b_m b_n = n\delta_{n,-m}\mathbf{1}, \qquad b_n \mathbf{1} - \mathbf{1} b_n = 0.$$

Introduce a topology on $U(\mathcal{H}')$, in which the basis of open neighborhoods of 0 is formed by the left ideals of the subspaces $t^N \mathbb{C}[t] \subset \mathcal{H}' \subset U(\mathcal{H}'), N \in \mathbb{Z}$. The completion of $U(\mathcal{H}')$ with respect to this topology will be denoted by $\widetilde{U}(\mathcal{H})$. In more concrete terms, elements of $\widetilde{U}(\mathcal{H})$ may be described as (possibly infinite) series of the form $R_0 + \sum_{n \geq 0} P_n b_n$, where R_0 and the P_n's are arbitrary (finite) elements of $U(\mathcal{H}')$. It is easy to check that $\widetilde{U}(\mathcal{H})$ is an associative algebra with a unit.

Note that $\widetilde{U}(\mathcal{H})$ contains as a subalgebra (but is not equal to) the universal enveloping algebra $U(\mathcal{H})$. The latter consists of finite linear combinations of products of elements of $\mathcal{H}$. Moreover, $\widetilde{U}(\mathcal{H})$ is the completion of $U(\mathcal{H})$ with respect to the topology in which the basis of open neighborhoods of 0 is formed by the left ideals of the subspaces $t^N \mathbb{C}[[t]] \subset \mathcal{H} \subset U(\mathcal{H}), N \in \mathbb{Z}$.

It is clear that any representation of $\mathcal{H}$ of the type discussed above is automatically a $\widetilde{U}(\mathcal{H})$–module. In this chapter we consider representations of $\mathcal{H}$ on which $\mathbf{1}$ acts as the identity operator. These are the same as the representations of the *Weyl algebra* $\widetilde{\mathcal{H}}$, which is defined as the quotient of $\widetilde{U}(\mathcal{H})$ by the two-sided ideal generated by $(\mathbf{1} - 1)$. Here $\mathbf{1}$ stands for the central element of $\mathcal{H} \subset \widetilde{U}(\mathcal{H})$, and 1 is the unit element of $\widetilde{U}(\mathcal{H})$.

2.1.3. The Fock representation. We wish to construct a representation of the Weyl algebra $\widetilde{\mathcal{H}}$ which is "as small as possible". Unlike commutative algebras, $\widetilde{\mathcal{H}}$ does not possess one–dimensional representations. Indeed, in $\widetilde{\mathcal{H}}$ the commutator of $b_n, n \neq 0$, and b_{-n} gives a non-zero multiple of the unit element,

$$(2.1.5) \qquad b_n b_{-n} - b_{-n} b_n = n,$$

so it is impossible to have a representation on which both $b_n, n \neq 0$, and b_{-n} act by zero.

Consider the subalgebra $\widetilde{\mathcal{H}}_+ \subset \widetilde{\mathcal{H}}$ generated by $b_0, b_1, b_2, \ldots$. It is a commutative subalgebra, and hence has a trivial one–dimensional representation (in fact, $\widetilde{\mathcal{H}}_+$ is a maximal commutative subalgebra of $\widetilde{\mathcal{H}}$). We may then define a representation of $\widetilde{\mathcal{H}}$ as the induced representation from this representation of $\widetilde{\mathcal{H}}_+$:

$$\pi \overset{\text{def}}{=} \operatorname{Ind}_{\widetilde{\mathcal{H}}_+}^{\widetilde{\mathcal{H}}} \mathbb{C} = \widetilde{\mathcal{H}} \underset{\widetilde{\mathcal{H}}_+}{\otimes} \mathbb{C}$$

(note that $\widetilde{\mathcal{H}}_+$ acts on $\widetilde{\mathcal{H}}$ from the right). It is called the *Fock representation* of $\widetilde{\mathcal{H}}$ (or of $\mathcal{H}$).

Let $\widetilde{\mathcal{H}}_-$ be the commutative subalgebra of $\mathcal{H}$ generated by $b_n, n < 0$. It follows from the Poincaré–Birkhoff-Witt theorem that $\widetilde{\mathcal{H}} \simeq \widetilde{\mathcal{H}}_+ \otimes \widetilde{\mathcal{H}}_-$. Therefore

$$\pi \simeq \widetilde{\mathcal{H}}_- = \mathbb{C}[b_{-1}, b_{-2}, \ldots].$$

Under this isomorphism, the generators b_n, $n < 0$, simply act on π by multiplication. To find the action of the $b_n, n \geq 0$, on an element of $\mathbb{C}[b_m]_{m<0}$, we use the relation (2.1.5) to "move" b_n through the $b_m, m < 0$, and the rule $b_n \cdot 1 = 0, n \geq 0$ (where $1 \in \mathbb{C}[b_m]_{m<0}$), which follows from the definition. We then obtain by induction that b_n, $n > 0$, acts as the derivation $n \dfrac{\partial}{\partial b_{-n}}$ and b_0 acts by 0 on π.

The operators b_n with $n < 0$ are known in this context as creation operators, since they "create the state b_n from the vacuum 1", while the operators b_n with $n \geq 0$ are the annihilation operators, repeated application of which will "kill" any vector in π.

2.2. The vertex algebra structure on π

We wish to endow the Fock space π with the structure of a $\mathbb{Z}_+$–graded vertex algebra. This involves the following data (see Definition 1.3.1):

- Vacuum vector: $|0\rangle = 1$, considered as a monomial of degree zero.
- The translation operator T is defined by the rules $T \cdot 1 = 0$ and $[T, b_i] = -ib_{i-1}$.
- $\mathbb{Z}_+$–gradation: π has a basis of monomials of the form $b_{j_1} \ldots b_{j_k}$ where $j_1 \leq j_2 \leq \cdots \leq j_k < 0$. We assign to this monomial degree $-\sum j_i$. In other words, we set $\deg 1 = 0, \deg b_j = -j$.

Note that these formulas uniquely determine T by induction on the degree of monomials:

$$T \cdot b_k m = b_k \cdot T \cdot m + [T, b_k] \cdot m$$

for any monomial m, and so

$$T \cdot b_{j_1} \ldots b_{j_k} = -\sum_{i=1}^{k} j_i b_{j_1} \ldots b_{j_i - 1} \ldots b_{j_k}.$$

This formula for T is motivated by the representation $b_n \mapsto t^n$, $T \mapsto -\partial_t$.

2.2.1. Defining the vertex operators. We now need to define the main ingredient of the vertex algebra structure, namely, the vertex operation $Y(\cdot, z)$. Let us look at the picture of the first few homogeneous components of π.

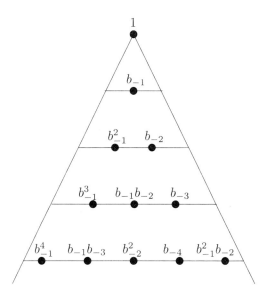

To the vacuum vector $|0\rangle = 1$, we are required to assign $Y(|0\rangle, z) = \mathrm{Id}$, so there is no ambiguity here. The key definition is that of the field $Y(b_{-1}, z)$, which as we will see generates π in an appropriate sense. Let us denote $Y(b_{-1}, z)$ by $b(z)$ for brevity. Since $\deg b_{-1} = 1$, the field $b(z)$ needs to have conformal dimension one. We now set

$$b(z) = \sum_{n \in \mathbb{Z}} b_n z^{-n-1},$$

where b_n is considered as an endomorphism of π. Since $\deg b_n = -n$, $b(z)$ is indeed a field of conformal dimension one. Note that $b(z)$ is nothing but a generating function for the generators b_n of the Heisenberg Lie algebra $\mathcal{H}$.

Next, we proceed to define $Y(b_{-2}, z)$ as the field

$$\partial_z b(z) = \sum_{n \in \mathbb{Z}} (-n - 1) b_n z^{-n-2}.$$

It has conformal dimension 2, as required, since $\deg b_{-2} = 2$. This formula may be motivated by the fact that $b_{-2} \sim t^{-2}$ equals $-\partial_t \cdot t^{-1}$. Following the same route, we obtain by induction

$$Y(b_{-k}, z) = \frac{1}{(k-1)!} \partial_z^{k-1} b(z).$$

Besides b_{-2}, the other generator in degree 2 is b_{-1}^2, to which one is tempted to assign

$$b(z)^2 = \sum_{n \in \mathbb{Z}} \left(\sum_{k+l=n} b_k b_l \right) z^{-n-2}.$$

Let us check the behavior of this expression. Consider, for instance, the sum

(2.2.1)
$$\sum_{k+l=10} b_k b_l$$

appearing as the z^{-12} coefficient in $b(z)^2$. Given any vector $x \in \pi$, we have $b_n x = 0$ for all but finitely many annihilation operators b_n ($n \geq 0$) (since these are differentiation operators and x is a polynomial). Thus in (2.2.1), except for finitely many terms, either $b_k \cdot x = 0$ or $b_l \cdot x = 0$ (see the picture).

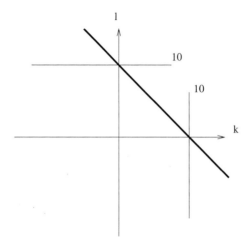

The Heisenberg relations (2.1.4) imply that b_k and b_l commute whenever $k+l \neq 0$; hence we may rearrange the terms in our sum so that an annihilation operator always appears to the right of a creation operator. It follows that the operator (2.2.1), when applied to any $x \in \pi$, yields a finite sum. Thus, the z^{-12} coefficient in $b(z)^2$ is a well-defined operator on π.

The same argument applies to the z^i coefficient of $b(z)^2$ for any $i \neq -2$. But what about the coefficient

$$(2.2.2) \qquad\qquad \sum_{k+l=0} b_k b_l$$

in front of z^{-2}? In this case b_k and b_l do not commute, so the above argument does not apply. We rewrite (2.2.2) as

$$\sum_{k \in \mathbb{Z}} b_k b_{-k} = \sum_{k<0} -k b_k \frac{\partial}{\partial b_k} + \sum_{k>0} k \frac{\partial}{\partial b_{-k}} b_{-k}$$

(recall that b_0 acts by 0 on π). The first sum again becomes a finite sum when applied to any vector in π. But each term in the second sum first multiplies by b_{-k} and then differentiates with respect to the same b_{-k}, producing an infinite sum $\sum\limits_{k>0} k$ when applied to the vector $1 \in \pi$ (or a similar infinite sum for any other vector of π). This sum obviously does not converge. Thus, the z^{-2} coefficient of $b(z)^2$ is not well-defined, and hence $b(z)^2$ does not make sense as a field.

In this case only one Fourier coefficient of the square posed problems, because most of the generators of the Heisenberg Lie algebra commute with each other. In more complicated algebras, many more such problems will arise.

There is a standard way to remove the infinities in products of fields. Namely, we rearrange the factors in each "problematic" term in such a way that the annihilation operators always act before the creation operators. Then the infinite sums arising from repeatedly creating and annihilating the same state will not occur. This procedure is known as *normal ordering*.

In the case at hand, we define the normally ordered product of $b(z)$ with itself, denoted by $:b(z)b(z):$, as the following formal power series in z:

$$:b(z)b(z): \ = \sum_{n \in \mathbb{Z}} :b_k b_l: z^{-n-2},$$

where we set

$$:b_k b_l: \ \overset{\text{def}}{=} \begin{cases} b_l b_k, & \text{if} \quad l = -k, k \geq 0, \\ b_k b_l, & \text{otherwise.} \end{cases}$$

Thus, taking the normally ordered product simply amounts to the instruction to replace everywhere the terms $b_k b_{-k}$ for $k > 0$ by $b_{-k} b_k$.

This procedure does not change the coefficients of $b(z)^2$ in front of $z^n, n \neq -2$, but the coefficient in front of z^{-2} now becomes

$$\sum_{k \in \mathbb{Z}} :b_k b_{-k}: = 2 \sum_{k < 0} -k b_k \frac{\partial}{\partial b_k},$$

which is a well-defined linear operator on π.

Now we give the general definition of normally ordered product of two fields. Although equivalent to the above definition in the case when both fields are equal to $b(z)$, it is more complicated. The reason is that the commutation relations between the Fourier coefficients of the field $b(z)$ are very simple.

2.2.2. Definition. The *normally ordered product* of the fields

$$A(z) = \sum_{n \in \mathbb{Z}} A_{(n)} z^{-n-1}, \qquad B(w) = \sum_{m \in \mathbb{Z}} B_{(m)} w^{-m-1}$$

is defined as the formal power series

$$:A(z)B(w): = \sum_{n \in \mathbb{Z}} \left(\sum_{m < 0} A_{(m)} B_{(n)} z^{-m-1} + \sum_{m \geq 0} B_{(n)} A_{(m)} z^{-m-1} \right) w^{-n-1}$$

$$= A(z)_+ B(w) + B(w) A(z)_-,$$

where for a formal power series $f(z) = \sum_{n \in \mathbb{Z}} f_n z^n$, we write

$$f(z)_+ = \sum_{n \geq 0} f_n z^n, \qquad f(z)_- = \sum_{n < 0} f_n z^n.$$

The proof of the following lemma is left to the reader.

2.2.3. Lemma. (1) *For any $v \in V$ and $\varphi \in V^*$, the matrix element*

$$\langle \varphi, :A(z)B(w):v \rangle$$

belongs to $\mathbb{C}[[z,w]][z^{-1}, w^{-1}]$.

For any $z, w \in \mathbb{C}^{\times}$, $:A(z)B(w):$ is a well-defined linear operator $V \to \overline{V}$.

(2) *The specialization of $:A(z)B(w):$ at $w = z$ is a well-defined field which is called the* normally ordered product *of $A(z)$ and $B(z)$. If $A(z)$ and $B(z)$ are homogeneous of conformal dimensions Δ_B and Δ_B, respectively, then $:A(z)B(z):$ is also homogeneous of conformal dimension $\Delta_A + \Delta_B$.*

(3)
$$:A(w)B(w): = \text{Res}_{z=0}(\delta(z-w)_- A(z)B(w) + \delta(z-w)_+ B(w)A(z)).$$

2.2.4. Remark. In general, the operation of normally ordered product is neither commutative nor associative. We follow the convention that the normal ordering is read from right to left, so that by definition

$$:A(z)B(z)C(z): = :A(z)(:B(z)C(z):):.$$

2.2.5. General fields. With normal ordering in hand we may now proceed to define our field $Y(b_{-1}^2, z)$ as

$$Y(b_{-1}^2, z) = :b(z)^2: .$$

It is clear from the above discussion that this is a well–defined field.

The general assignment of vertex operators combines the preceding two cases of $b_j, j < 0$, and b_{-1}^2. Namely, we define

$$(2.2.3) \quad Y(b_{j_1} b_{j_2} \ldots b_{j_k}, z) = \frac{1}{(-j_1 - 1)! \ldots (-j_k - 1)!} :\partial_z^{-j_1 - 1} b(z) \ldots \partial_z^{-j_k - 1} b(z): .$$

At first glance, the operation of normal ordering introduced above and formula (2.2.3) appear rather *ad hoc*. It is not clear why we should define the vertex operation $Y(\cdot, z)$ this way. However, according to Corollary 3.3.5 below, the vertex operation $Y(\cdot, z)$ is completely determined by the assignment $Y(b_{-1}, z) = b(z)$. More precisely, by Theorem 4.4.1, there exists (at most) one vertex algebra structure on π such that $Y(b_{-1}, z) = b(z)$, namely, the one given by formula (2.2.3). The reason is that the Fourier coefficients of the field $b(z)$ generate π from the vacuum vector 1. Thus, the operation of normal ordering is not an arbitrary convention, but a natural consequence of the structure of a vertex algebra.

2.2.6. Remark. In contrast to the general case, the operation of normally ordered product of the fields $\partial_z^n b(z)$ and its derivatives is commutative and associative. Moreover, there is a simpler definition of the normally ordered product in the Heisenberg case, which we now give.

First define $:b_{n_1} \ldots b_{n_k}:$ as the monomial obtained from $b_{n_1} \ldots b_{n_k}$ by moving all b_{n_i} with $n_i < 0$ to the left of all b_{n_j} with $n_j \geq 0$ (in other words, moving all "creation operators" $b_n, n < 0$, to the left and all "annihilation operators" $b_n, n \geq 0$, to the right). The important fact that makes this definition correct is that the operators b_n with $n < 0$ (resp., $n \geq 0$) commute among themselves. Hence it does not matter how we order the creation (resp., annihilation) operators among themselves. Note that this property does not hold for other vertex algebras that we consider below (affine, Virasoro, etc.), and so this definition does not work in the general case.

Now define $:\partial_z^{m_1} b(z) \ldots \partial_z^{m_k} b(z):$ as the power series in z obtained from the ordinary product $\partial_z^{m_1} b(z) \ldots \partial_z^{m_k} b(z)$ by replacing each term $b_{n_1} \ldots b_{n_k}$ by $:b_{n_1} \ldots b_{n_k}:$. We leave it to the reader to check that this definition is equivalent to the general definition of the normally ordered product of fields given above.

2.3. Checking vertex algebra axioms

We wish to establish that with the definitions of the previous section, π indeed satisfies the axioms of a vertex algebra.

2.3.1. The locality of $b(z)$. The only really problematic axiom is the requirement that all fields be mutually local. Tackling this problem simultaneously for all fields is a daunting computational task, so we would rather begin by checking locality for our "generating" field $b(z)$ and then start developing tools that will enable us to derive locality in an inductive fashion from some simple properties of generating fields. The verification of locality of $b(z)$ with itself is similar in flavor to the previous calculation with normal ordering. We start out by writing

$$
\begin{aligned}
b(z)b(w) &= \sum_{n,m} b_n b_m z^{-n-1} w^{-m-1} \\
&= \underbrace{\sum_{N\in\mathbb{Z}\setminus\{0\}}\sum_{n+m=N} b_n b_m z^{-n-1} w^{-m-1}}_{\Sigma_{\neq 0}} + \underbrace{\sum_{n\in\mathbb{Z}} b_{-n} b_n z^{n-1} w^{-n-1}}_{\Sigma_0} .
\end{aligned}
$$

As we observed in calculating the ill–fated expression $b(z)^2$, the terms of $\Sigma_{\neq 0}$ are products $b_n b_m$ in which b_n, b_m commute with each other. Thus we may switch the order of the factors with impunity to a normally ordered form, and hence $\Sigma_{\neq 0} = :\Sigma_{\neq 0}:$. Again the difficulty lies with Σ_0.

Let us write $\Sigma_0 = :\Sigma_0: + (\Sigma_0 - :\Sigma_0:)$. Normal ordering does not change the positive part of the sum, which already has annihilation operators on the right:

$$
:\sum_{n\geq 0} b_{-n} b_n z^{n-1} w^{-n-1}: = \sum_{n\geq 0} b_{-n} b_n z^{n-1} w^{-n-1} .
$$

In the negative part however, applying normal ordering means switching the order of the factors:

$$
:\sum_{n<0} b_{-n} b_n z^{n-1} w^{-n-1}: = \sum_{n<0} b_n b_{-n} z^{n-1} w^{-n-1} .
$$

Hence

$$
\Sigma_0 = :\Sigma_0: + \sum_{n<0} [b_{-n}, b_n] z^{n-1} w^{-n-1} = :\Sigma_0: + \sum_{m>0} m z^{-m-1} w^{m-1}.
$$

Summarizing, we obtain

$$
b(z)b(w) = :b(z)b(w): + \sum_{m>0} m z^{-m-1} w^{m-1} .
$$

Now observe that the series

$$
\tag{2.3.1} \sum_{m>0} m z^{-m-1} w^{m-1}
$$

is the expansion in $\mathbb{C}((z))((w))$ of the rational function $1/(z-w)^2$. Hence we write

$$
\tag{2.3.2} b(z)b(w) = \frac{1}{(z-w)^2} + :b(z)b(w): ,
$$

where $1/(z-w)^2$ stands for its expansion in $\mathbb{C}((z))((w))$, i.e., in positive powers of w/z.

To compute the product in the opposite order, $b(w)b(z)$, we simply switch z and w in the above formulas. We then obtain the series

$$
\tag{2.3.3} \sum_{m>0} m w^{-m-1} z^{m-1},
$$

which is the expansion of the same rational function $1/(z-w)^2$, but now in $\mathbb{C}((w))((z))$, i.e., in positive powers of z/w. Hence we write

$$(2.3.4) \qquad b(w)b(z) = \frac{1}{(z-w)^2} + :b(w)b(z): \, ,$$

where $1/(z-w)^2$ now stands for its expansion in $\mathbb{C}((w))((z))$.

In addition, we have

$$:b(z)b(w): \, = \, :b(w)b(z):$$

(as elements of $\operatorname{End} V[[z^{\pm 1}, w^{\pm 1}]]$). Indeed, almost all generators b_k, b_l commute, and in the only remaining case $k = -l$, the factor with positive number appears to the right after ordering, regardless of how they started out.

By Lemma 2.2.3,(1), for any $v \in V, \varphi \in V^*$, the matrix elements

$$\langle \varphi, :b(z)b(w):v \rangle = \langle \varphi, :b(w)b(z):v \rangle$$

are well-defined elements of $\mathbb{C}[[z, w]][z^{-1}, w^{-1}]$. Comparing formulas (2.3.3) and (2.3.4), we obtain that the fields $b(z)$ and $b(w)$ satisfy the conditions of Definition 1.2.3 (the order of pole of any matrix element at $z = w$ is clearly bounded by 2). Therefore $b(z)$ and $b(w)$ are mutually local.

2.3.2. Alternative derivation. We may also establish locality using the formulation of Proposition 1.2.5. Namely, we find that

$$[b(z), b(w)] = \sum_{n,m \in \mathbb{Z}} [b_n, b_m] z^{-n-1} w^{-m-1} = \sum_{n \in \mathbb{Z}} [b_n, b_{-n}] z^{-n-1} w^{n-1}$$

$$= \sum_{n \in \mathbb{Z}} n z^{-n-1} w^{n-1} = \partial_w \delta(z-w).$$

We have already established in formula (1.1.5) that $(z-w)^2 \partial_w \delta(z-w) = 0$. This implies that $(z-w)^2 [b(z), b(w)] = 0$, and hence the field $b(z)$ is local with itself.

Finally, let us discuss the locality in analytic terms. Observe that the series (2.3.1) converges in the domain $|z| > |w|$ to the rational function $1/(z-w)^2$, while the series (2.3.3) converges to the same function in the domain $|w| > |z|$. According to Lemma 2.2.3,(1), $:b(z)b(w):$ is a well-defined operator–valued meromorphic function in z and w with poles at $\{z = 0\}$ and $\{w = 0\}$. Thus, we have established that $b(z)b(w)$ converges in the domain $|z| > |w|$ to an operator valued function, which can be analytically continued to

$$(2.3.5) \qquad R(b(z)b(w)) = :b(z)b(w): + \frac{1}{(z-w)^2} \; .$$

Repeating the same calculation for the product in the opposite order (this amounts to simply switching z and w everywhere), we obtain also that

$$R(b(w)b(z)) = :b(w)b(z): + \frac{1}{(w-z)^2} \; .$$

Thus, we again see that the fields $b(z)$ and $b(w)$ are mutually local from Proposition 1.2.9.

2.3.3. General case. In order to establish that other fields in the Heisenberg vertex algebra are local, we avoid the brute–force calculation of other operator products, and instead prove a general result that shows that if $A(z), B(z), C(z)$ are fields which are pairwise mutually local, then the normally ordered product $:A(z)B(z):$ is local with respect to $C(z)$.

2.3.4. Dong's Lemma. *If $A(z), B(z), C(z)$ are three mutually local fields, then the fields $:A(z)B(z):$ and $C(z)$ are also mutually local.*

2.3.5. Proof. By assumption we may find r so that for all $s \geq r$,

$$
\begin{aligned}
(w-z)^s A(z)B(w) &= (w-z)^s B(w)A(z),\\
(u-z)^s A(z)C(u) &= (u-z)^s C(u)A(z),\\
(u-w)^s B(w)C(u) &= (u-w)^s C(u)B(w).
\end{aligned}
$$

We wish to find an integer N such that

$$(w-u)^N :A(w)B(w):C(u) = (w-u)^N C(u):A(w)B(w): .$$

By part (3) of Lemma 2.2.3, this will follow from the statement

$$(2.3.6) \quad (w-u)^N(\delta(z-w)_- A(z)B(w) + \delta(z-w)_+ B(w)A(z))C(u)$$
$$= (w-u)^N C(u)(\delta(z-w)_- A(z)B(w) + \delta(z-w)_+ B(w)A(z)).$$

Let us take $N = 3r$. Writing

$$(w-u)^{3r} = (w-u)^r \sum_{s=0}^{2r} \binom{2r}{s}(w-z)^s (z-u)^{2r-s},$$

we see that the terms in the left hand side of (2.3.6) with $r < s \leq 2r$ vanish: one factor of $(z-w)$ kills the sum $\delta(z-w)_- + \delta(z-w)_+ = \delta(z-w)$, while we will still have at least r such factors, allowing us to switch the order of $A(z), B(w)$ by their locality. The terms with $0 \leq s \leq r$, on the other hand, have $(z-u)$ appearing to a power of at least r, allowing us to move $C(u)$ through $A(z)$, while also still having $(w-u)$ to the rth power, so that we can move $C(u)$ through $B(w)$. The same phenomena occur on the right hand side of (2.3.6): the terms with $r < s \leq 2r$ will vanish, and the other terms give us the same expression as what we now have on the left hand side. Thus we have established (2.3.6), and hence the lemma.

2.3.6. Back to π. Now we are ready to prove that π is a $\mathbb{Z}_+$–graded vertex algebra. Let us summarize all the structures on π that we have defined up to now:

- Vacuum vector: $|0\rangle = 1$.
- The translation operator T is defined by the rules $T \cdot 1 = 0$ and $[T, b_i] = -ib_{i-1}$.
- The vertex operators are defined by the formula

$$Y(b_{j_1} b_{j_2} \ldots b_{j_k}, z) = \frac{1}{(-j_1-1)! \ldots (-j_k-1)!} :\partial_z^{-j_1-1} b(z) \ldots \partial_z^{-j_k-1} b(z): .$$

- $\mathbb{Z}_+$–gradation: $\deg b_{j_1} \ldots b_{j_k} = -\sum j_i$.

2.3.7. Theorem. *The Fock representation π with the above structures satisfies the axioms of a $\mathbb{Z}_+$–graded vertex algebra.*

2.3.8. Proof. First we verify that the field $Y(b_{j_1} b_{j_2} \ldots b_{j_k}, z)$ defined above is homogeneous of conformal dimension $-(j_1 + \ldots + j_k)$, using Lemma 2.2.3,(2). Now we check the axioms of a vertex algebra.

The statement $Y(|0\rangle, z) = \mathrm{Id}$ follows from our definition. The remainder of the vacuum axiom,

$$(2.3.7) \qquad\qquad \lim_{z \to 0} Y(A, z)|0\rangle = A,$$

follows by induction. We start with the case $A = b_{-1}$, where

$$Y(b_{-1}, z)|0\rangle = \sum_{n \in \mathbb{Z}} b_n z^{-n-1}|0\rangle.$$

But all non–negative b_n annihilate the vacuum, so the limit is well defined, while the constant coefficient is indeed b_{-1}. Next, according to the above definition, the vertex operator associated to each polynomial in the b_i's is a normally ordered product of derivatives of our basic field $b(z)$. Thus we just need to check that if (2.3.7) holds for the field

$$Y(A, z) = \sum_{n \in \mathbb{Z}} A_{(n)} z^{-n-1},$$

then it holds for the field

$$Y(b_{-k}A, z) = \frac{1}{(k-1)!} :\partial_z^{k-1} b(z) Y(A, z): , \qquad k > 0.$$

But by definition of the normally ordered product,

$$\frac{1}{(k-1)!} :\partial_z^{k-1} b(z) Y(A, z):$$

$$= \frac{1}{(k-1)!} \sum_{m \in \mathbb{Z}} \left(\sum_{n \leq -k} (-n-1)(-n-2)\ldots(-n-k+1) b_n A_{(m-n)} \right.$$

$$\left. + \sum_{n \geq 0} (-n-1)(-n-2)\ldots(-n-k+1) A_{(m-n)} b_n \right) z^{-k-m-1}.$$

The second sum kills $|0\rangle$, and by our inductive assumption, the first sum gives a power series with positive powers of z only, with the constant term

$$b_{-k} A_{(-1)}|0\rangle = b_{-k} A.$$

To check the translation axiom, we first note that $T|0\rangle = 0$ by our definition. Next, since $[T, b_j] = -j b_{j-1}$, we have $[T, b(z)] = \partial_z b(z)$. Continuing in the same way, we derive $[T, \partial_z^n b(z)] = \partial_z^{n+1} b(z)$. Then one checks explicitly that the Leibniz rule holds for the normally ordered product:

$$\partial_z :A(z)B(z): = :\partial_z A(z)B(z): + :A(z)\partial_z B(z):$$

using the residue definition of the normal ordering given in Lemma 2.2.3,(3). This implies that if $[T, \cdot]$ acts as ∂_z on two fields, it will act like this on their normally ordered product. By induction, this implies the full translation axiom.

Finally, locality of any two fields follows by a similar induction from locality of the fields $\partial_z^n b(z)$ and $\partial_w^m b(w)$ with arbitrary $n, m \geq 0$ using Dong's lemma. To see that $\partial_z^n b(z)$ and $\partial_w^m b(w)$ are mutually local, recall from § 2.3.1 that $b(z)$ and $b(w)$ are local. But if $A(z), B(w)$ are local, then $(z-w)^N [A(z), B(w)] = 0$ for some N.

Differentiating this formula with respect to z, and multiplying the result by $(z - w)$, we obtain

$$(z - w)^{N+1}[\partial_z A(z), B(w)] = 0,$$

so $\partial_z A(z)$ and $B(w)$ are mutually local. Hence by induction $\partial_z^n b(z)$ and $\partial_w^m b(w)$ are indeed local for any $n, m \geq 0$, and this completes the proof.

2.3.9. Remark. For $\kappa \in \mathbb{C}$, let π^κ be the $\mathcal{H}$–module induced from the one-dimensional representation of $\mathcal{H}_+ = \mathbb{C}[[t]] \oplus \mathbb{C}\mathbf{1}$ on which $\mathbb{C}[[t]]$ acts by 0 and $\mathbf{1}$ acts by multiplication by κ. Then the above formulas still define a vertex algebra structure on π^κ. When $\kappa = 0$, this vertex algebra becomes commutative, and corresponds to the commutative algebra $\mathbb{C}[b_{-1}, b_{-2}, \ldots]$ with the derivation T (in the sense of § 1.4).

2.3.10. Generalization. The above proof is easily generalized to yield the following result, which provides a "generators–and–relations" approach to the construction of vertex algebras.

Let V be a vector space, $|0\rangle$ a non-zero vector, and T an endomorphism of V. Let S be a countable ordered set and $\{a^\alpha\}_{\alpha \in S}$ a collection of vectors in V. Suppose we are also given fields

$$a^\alpha(z) = \sum_{n \in \mathbb{Z}} a^\alpha_{(n)} z^{-n-1}$$

such that the following conditions hold:

(1) For all α, $a^\alpha(z)|0\rangle = a^\alpha + z(\ldots)$.
(2) $T|0\rangle = 0$ and $[T, a^\alpha(z)] = \partial_z a^\alpha(z)$ for all α.
(3) All fields $a^\alpha(z)$ are mutually local.
(4) V has a basis of vectors

$$a^{\alpha_1}_{(j_1)} \ldots a^{\alpha_m}_{(j_m)}|0\rangle,$$

where $j_1 \leq j_2 \leq \ldots \leq j_m < 0$, and if $j_i = j_{i+1}$, then $\alpha_i \preceq \alpha_{i+1}$ with respect to the given order on the set S.

(This basis should be compared with the PBW basis.)

2.3.11. Reconstruction Theorem (preliminary version). *Under the above assumptions, the assignment*

$$(2.3.8) \quad Y(a^{\alpha_1}_{(j_1)} \ldots a^{\alpha_m}_{(j_m)}|0\rangle, z)$$

$$= \frac{1}{(-j_1 - 1)! \ldots (-j_m - 1)!} : \partial_z^{-j_1 - 1} a^{\alpha_1}(z) \ldots \partial_z^{-j_m - 1} a^{\alpha_m}(z):$$

defines a vertex algebra structure on V. Moreover, if V is a $\mathbb{Z}$–graded vector space, $|0\rangle$ has degree 0, the vectors a^α are homogeneous, T has degree 1, and the fields $a^\alpha(z)$ have conformal deminsion $\deg a^\alpha$, then V is a $\mathbb{Z}$–graded vertex algebra.

2.3.12. Proof. It is practically identical to the proof in the Heisenberg algebra case, where we had $V = \pi_0 \simeq \mathbb{C}[b_{-1}, b_{-2}, \ldots]|0\rangle$, $a^\alpha = b_{-1}$, and $a^\alpha(z) = b(z) = \sum_{n \in \mathbb{Z}} b_n z^{-n-1}$.

2.3.13. Remark. An essential feature of the Heisenberg Fock representation π is that it is generated by a vector killed by one "half" of the Heisenberg Lie algebra, and freely generated (as a polynomial algebra) by the other half. Moreover, these generators are all expressible as derivatives of a single generator of degree one, b_{-1}. As a result, we obtain a vertex algebra where all information is essentially contained in the vertex operator associated to the vector $b_{-1} \in \pi$. More general vertex algebras that we study below have a similar structure of induced representation generated by a vector killed by the "positive half" of a (Lie) algebra. Furthermore, the entire structure is generated by a finite number of basic fields, as in Theorem 2.3.11.

The Reconstruction Theorem will be strengthened below in Theorem 4.4.1 in two ways: first, we will replace condition (4) of § 2.3.10 that the ordered monomials form a basis of V by the weaker property that they span V, and second, we will show that under these conditions the above vertex algebra structure on V is unique.

2.4. Affine Kac–Moody algebras and their vertex algebras

2.4.1. Central extension of a loop algebra. The Heisenberg Lie algebra was defined as a central extension of the formal loop algebra of a one–dimensional commutative Lie algebra $\mathbb{C}$. The affine Kac–Moody algebras are defined in a similar fashion. Let $\mathfrak{g}$ be a finite–dimensional simple Lie algebra $\mathfrak{g}$ over $\mathbb{C}$. We define the formal *loop algebra* of $\mathfrak{g}$,

$$L\mathfrak{g} = \mathfrak{g}((t)) = \mathfrak{g} \otimes \mathbb{C}((t)),$$

as the Lie algebra with the commutator

$$[A \otimes f(t), B \otimes g(t)] = [A, B] \otimes f(t)g(t).$$

Note that the same definition would make sense if we replace $\mathbb{C}((t))$ by any commutative $\mathbb{C}$–algebra R. The Lie algebra axioms for $\mathfrak{g} \otimes R$ follow from the fact that $\mathfrak{g}$ is a Lie algebra and R is a commutative algebra.

To define the affine Kac–Moody algebra $\widehat{\mathfrak{g}}$ we need to digress for a moment and discuss invariant bilinear forms. Recall that a bilinear form on a Lie algebra $\mathfrak{g}$ is called *invariant* if

$$([x, y], z) + (y, [x, z]) = 0, \qquad \forall x, y, z \in \mathfrak{g}.$$

This formula is the infinitesimal version of the formula $(\operatorname{Ad} g \cdot y, \operatorname{Ad} g \cdot z) = (y, z)$ expressing the invariance of the form with respect to the Lie group action (with g being an element of the corresponding Lie group). It is well-known that the vector space of non-degenerate invariant bilinear forms on a finite–dimensional simple Lie algebra $\mathfrak{g}$ is one-dimensional. One can produce such a form starting with any faithful finite-dimensional representation $\rho_V : \mathfrak{g} \to \operatorname{End} V$ of $\mathfrak{g}$ by the formula

$$(x, y) = \operatorname{Tr}_V(\rho_V(x)\rho_V(y)).$$

The standard choice is the adjoint representation, which gives rise to the *Killing form* $(\cdot, \cdot)_K$. However, we prefer a different normalization. For example, for $\mathfrak{g} = \mathfrak{sl}_n$, we would like to set $(x, y) = \operatorname{Tr}_{\mathbb{C}^n}(xy)$, the trace in the defining representation, which is "smaller" than the adjoint representation.

Let $\mathfrak{h}$ be the Cartan subalgebra of $\mathfrak{g}$, and $\mathfrak{h}^*$ its dual. Following [**Kac2**], we define the bilinear form $(\cdot, \cdot)$ on $\mathfrak{g}$ as the multiple of $(\cdot, \cdot)_K$ such that with respect to the induced bilinear form on $\mathfrak{h}^*$ the square of length of the maximal root of $\mathfrak{g}$ is equal to 2. We will call this form *canonically normalized*. It is known (see [**Kac2**]) that

$$(\cdot, \cdot) = \frac{1}{2h^\vee} (\cdot, \cdot)_K,$$

where $h^\vee$ is the *dual Coxeter number* of $\mathfrak{g}$ (see [**Kac2**] for the definition; e.g., for $\mathfrak{g} = \mathfrak{sl}_n$, $h^\vee = n$).

We now define the affine Kac–Moody algebra $\widehat{\mathfrak{g}}$ as a central extension

(2.4.1) $$0 \to \mathbb{C}K \to \widehat{\mathfrak{g}} \to L\mathfrak{g} \to 0.$$

As a vector space, $\widehat{\mathfrak{g}} \simeq L\mathfrak{g} \oplus \mathbb{C}K$, with the commutation relations $[K, \cdot] = 0$ (so K is central) and

(2.4.2) $$[A \otimes f(t), B \otimes g(t)] = [A, B] \otimes f(t)g(t) - (\mathrm{Res}_{t=0}\, f dg)(A, B)K.$$

Note the similarity of this definition with (2.1.3) defining the Heisenberg central extension. As in that case, the above formula is independent of the choice of coordinate t. We will exploit this fact below (see § 6.4.9).

The Kac–Moody cocycle is non-trivial, i.e., $\widehat{\mathfrak{g}}$ cannot be split as a Lie algebra into a direct sum $L\mathfrak{g} \oplus \mathbb{C}K$. This cocycle represents a non-zero class in the cohomology group $H^2(L\mathfrak{g}, \mathbb{C})$, which as we know parameterizes central extensions (see § 2.1.1). In fact, it is known that the vector space $H^2(L\mathfrak{g}, \mathbb{C})$ is canonically isomorphic to the space of non–degenerate invariant bilinear forms on $\mathfrak{g}$, and hence is one–dimensional. Thus, the Kac–Moody extension is a *universal* central extension of $L\mathfrak{g}$, i.e., any other central extension $\widetilde{\mathfrak{g}}$ of $\mathfrak{g}$ admits a Lie algebra homomorphism $\widehat{\mathfrak{g}} \to \widetilde{\mathfrak{g}}$.

A geometric characterization of the Kac–Moody central extension (in terms of flat connections on the punctured disc) is presented in Corollary 16.4.7.

2.4.2. The vacuum representation. Inside the loop algebra $L\mathfrak{g} = \mathfrak{g}((t))$ there is a "positive" Lie subalgebra $\mathfrak{g}[[t]] = \mathfrak{g} \otimes \mathbb{C}[[t]]$. If $f, g \in \mathbb{C}[[t]]$, then $\mathrm{Res}_{t=0}\, f dg = 0$. Hence the central extension becomes trivial when restricted to this subspace, and so $\mathfrak{g}[[t]]$ and $\mathfrak{g}[[t]] \oplus \mathbb{C}K$ are *Lie subalgebras* of $\widehat{\mathfrak{g}}$.

Now consider the one–dimensional representation $\mathbb{C}_k$ of $\mathfrak{g}[[t]] \oplus \mathbb{C}K$ on which $\mathfrak{g}[[t]]$ acts by 0 and K acts as multiplication by a scalar $k \in \mathbb{C}$. We define the *vacuum representation of level k* of $\widehat{\mathfrak{g}}$ as the representation induced from $\mathbb{C}_k$:

(2.4.3) $$V_k(\mathfrak{g}) = \mathrm{Ind}_{\mathfrak{g}[[t]] \oplus \mathbb{C}K}^{\widehat{\mathfrak{g}}} \mathbb{C}_k = U(\widehat{\mathfrak{g}}) \underset{U(\mathfrak{g}[[t]] \oplus \mathbb{C}K)}{\otimes} \mathbb{C}_k,$$

where $U(\widehat{\mathfrak{g}})$ denotes the universal enveloping algebra of $\widehat{\mathfrak{g}}$. More generally, we will say that a module M over $\widehat{\mathfrak{g}}$ has *level $k \in \mathbb{C}$*, if K acts on M as multiplication by k.

By the Poincaré–Birkhoff–Witt theorem, the direct sum decomposition

$$\widehat{\mathfrak{g}} = (\mathfrak{g} \otimes t^{-1}\mathbb{C}[t^{-1}]) \oplus (\mathfrak{g}[[t]] \oplus \mathbb{C}K)$$

(as a vector space only!) gives us the isomorphisms of vector spaces

$$U(\widehat{\mathfrak{g}}) \simeq U(\mathfrak{g} \otimes t^{-1}\mathbb{C}[t^{-1}]) \otimes U(\mathfrak{g}[[t]] \oplus \mathbb{C}K),$$

$$V_k(\mathfrak{g}) \simeq U(\mathfrak{g} \otimes t^{-1}\mathbb{C}[t^{-1}]).$$

2.4.3. Remark. While the construction of the vacuum modules $V_k(\mathfrak{g})$ looks parallel to the construction of the Fock representation π^κ of the Heisenberg Lie algebra given in Remark 2.3.9, there are important differences. First of all, the representations π^κ are practically indistinguishable from each other when $\kappa \neq 0$. Indeed, if we simply rescale our generators $b_n \mapsto \kappa^{-1/2} b_n$, then with respect to the new generators π^κ becomes identified with $\pi^1 = \pi$. In contrast, the structure of $\widehat{\mathfrak{g}}$–modules $V_k(\mathfrak{g})$ for different values of k may be very different. Second, π^0 is a commutative vertex algebra (see § 1.4), while $V_0(\mathfrak{g})$ is not.

2.4.4. Vertex algebra structure. We now wish to define a vertex algebra structure on the vacuum representation. Let $\{J^a\}_{a=1,\dots,\dim\mathfrak{g}}$ be an ordered basis of $\mathfrak{g}$. Split the extension $\widehat{\mathfrak{g}}$ as a vector space. For any $A \in \mathfrak{g}$ and $n \in \mathbb{Z}$, we denote

$$A_n \overset{\text{def}}{=} A \otimes t^n \in L\mathfrak{g}.$$

Then the elements $J^a_n, n \in \mathbb{Z}$, and $\mathbf{K}$ form a (topological) basis for $\widehat{\mathfrak{g}}$, while the elements $J^a_n, n \geq 0$, and $\mathbf{K}$ form a basis for the "positive" subalgebra from which we induced $V_k(\mathfrak{g})$. The commutation relations read

$$(2.4.4) \qquad [J^a_n, J^b_m] = [J^a, J^b]_{n+m} + n(J^a, J^b)\delta_{n,-m}K.$$

Denote by v_k the image of $1 \otimes 1 \in U\widehat{\mathfrak{g}} \otimes \mathbb{C}_k$ in V_k. By the Poincaré–Birkhoff–Witt theorem, $V_k(\mathfrak{g})$ has a PBW basis of monomials of the form

$$J^{a_1}_{n_1} \dots J^{a_m}_{n_m} v_k,$$

where $n_1 \leq n_2 \leq \dots \leq n_m < 0$, and if $n_i = n_{i+1}$, then $a_i \leq a_{i+1}$. We will refer to such monomials as lexicographically ordered. Here is the picture of the first few "layers" (i.e., homogeneous components) of $V_k(\mathfrak{g})$:

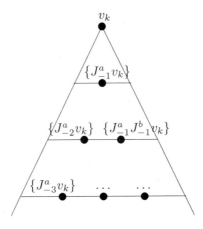

To define a $\mathbb{Z}_+$–graded vertex algebra structure on $V_k(\mathfrak{g})$ we proceed as in the Heisenberg case:

- Vacuum vector: $|0\rangle = v_k$.
- Translation operator: $Tv_k = 0$, $[T, J^a_n] = -nJ^a_{n-1}$.

- Vertex operators: $Y(v_k, z) = \mathrm{Id}$,

$$Y(J^a_{-1} v_k, z) = J^a(z) = \sum_{n \in \mathbb{Z}} J^a_n z^{-n-1},$$

and in general,

$$(2.4.5) \quad Y(J^{a_1}_{n_1} \dots J^{a_m}_{n_m} v_k, z)$$

$$= \frac{1}{(-n_1 - 1)! \dots (-n_m - 1)!} :\partial_z^{-n_1 - 1} J^{a_1}(z) \dots \partial_z^{-n_m - 1} J^{a_m}(z): \; .$$

- $\mathbb{Z}_+$–gradation:

$$\deg J^{a_1}_{n_1} \dots J^{a_m}_{n_m} v_k = -\sum_{i=1}^m n_i.$$

2.4.5. Theorem. $V_k(\mathfrak{g})$ *is a $\mathbb{Z}_+$–graded vertex algebra.*

2.4.6. Proof. We use Theorem 2.3.11. As our generating set a^α we take the vectors $J^a_{-1} v_k$ with associated fields $J^a(z)$. The Fourier components $J^a_n, n \geq 0$, kill the vacuum vector $|0\rangle = v_k$; hence $J^a(z)|0\rangle = J^a_{-1}|0\rangle + z(\dots)$. This implies that condition (1) of Theorem 2.3.11 is satisfied. It is clear from the definition that T satisfies condition (2). Finally, we calculate

$$[J^a(z), J^b(w)] = \sum_{n \in \mathbb{Z}} \sum_{m \in \mathbb{Z}} [J^a_n, J^b_m] z^{-n-1} w^{-m-1}$$

$$= \sum_{n,m \in \mathbb{Z}} [J^a, J^b]_{n+m} z^{-n-1} w^{-m-1} + \sum_{n \in \mathbb{Z}} n (J^a, J^b) k z^{-n-1} w^{n-1}$$

$$= \sum_{l \in \mathbb{Z}} [J^a, J^b]_l \left(\sum_{n \in \mathbb{Z}} z^{-n-1} w^n \right) w^{-l-1} + (J^a, J^b) k \sum_{n \in \mathbb{Z}} n z^{-n-1} w^{n-1}$$

$$(2.4.6) \qquad = [J^a, J^b](w)\delta(z - w) + (J^a, J^b)k \partial_w \delta(z - w).$$

It then follows that for any a and b,

$$(z - w)^2 [J^a(z), J^b(w)] = 0,$$

so locality holds for these fields. Thus, condition (3) of Theorem 2.3.11 is also satisfied. Condition (4) is satisfied by the Poincaré-Birkhoff-Witt theorem. Applying Theorem 2.3.11, we obtain that $V_k(\mathfrak{g})$ is a vertex algebra.

2.4.7. Remark. If we write J^a_n as $J^a \otimes t^n$, the field $J^a(z)$ takes the form $\sum_n (J^a \otimes t^n) z^{-n-1}$, which we can write by abuse of notation as $J^a \otimes \delta(t - z)$. Thus we may think of $J^a(z)$ as representing an operator–valued delta-function acting on $V_k(\mathfrak{g})$.

2.5. The Virasoro vertex algebra

2.5.1. The Virasoro algebra. From now on we will use the notation $\mathcal{K}$ for $\mathbb{C}((t))$ and $\mathcal{O}$ for $\mathbb{C}[[t]]$.

Consider the Lie algebra $\operatorname{Der} \mathcal{K} = \mathbb{C}((t))\partial_t$ of (continuous) derivations of $\mathcal{K}$. The *Virasoro algebra* is by definition the central extension of $\operatorname{Der} \mathcal{K}$:

$$0 \to \mathbb{C}C \to Vir \to \operatorname{Der} \mathcal{K} \to 0,$$

defined by the commutation relations

$$(2.5.1) \qquad [f(t)\partial_t, g(t)\partial_t] = (fg' - f'g)\partial_t - \frac{1}{12}\left(\operatorname{Res}_{t=0} fg'''dt\right) C.$$

It is known that this extension is universal. It has (topological) generators C, and

$$L_n = -t^{n+1}\partial_t, \qquad n \in \mathbb{Z},$$

satisfying the relations that C is central and

$$(2.5.2) \qquad [L_n, L_m] = (n - m)L_{n+m} + \frac{n^3 - n}{12}\delta_{n,-m}C.$$

The last term comes from

$$\operatorname{Res}_{t=0} t^{n+1}(m + 1)m(m - 1)t^{m-2}dt$$

(the factor $1/12$ is there for historic reasons: in this normalization, the central element C acts as the identity operator in the simplest representation of Vir on π, which is described in § 2.5.9 below).

Unlike the Heisenberg or Kac–Moody cocycles, the cocycle used to define the Virasoro algebra is *not* coordinate-independent. Thus, it is not clear that a coordinate-independent Virasoro algebra exists. Fortunately, it does exist, as explained in § 19.6.3 below. However, there is no canonical splitting of the corresponding extension, even as a vector space. This explains why the above formula depends on the choice of the coordinate t.

We will say that a module M over the Virasoro algebra has *central charge* $c \in \mathbb{C}$, if C acts on M by multiplication by c.

2.5.2. Remark. The Virasoro algebra plays a prominent role in the theory of vertex algebras and in conformal field theory. Its Lie subalgebra $\operatorname{Der} \mathcal{O} = \mathbb{C}[[t]]\partial_t$ may be viewed as the Lie algebra of infinitesimal changes of variables on the disc $D = \operatorname{Spec} \mathbb{C}[[t]]$. The action of $\operatorname{Der} \mathcal{O}$ on a vertex algebra V by "internal symmetries" will be used in Chapter 6 to attach to V a vector bundle with connection on any smooth curve. Moreover, the full action of the Virasoro algebra on V will allow us to obtain the $\mathcal{D}$–module structure on sheaves of coinvariants on the moduli spaces of pointed curves (see Chapter 17).

2.5.3. We may write the Virasoro relations succinctly using a generating function

$$T(z) = \sum_{n \in \mathbb{Z}} L_n z^{-n-2}$$

(the convention on the power of z will become natural when we realize $T(z)$ as a field of conformal dimension 2).

2.5.4. Lemma.
$$[T(z), T(w)] = \frac{C}{12}\partial_w^3 \delta(z-w) + 2T(w)\partial_w \delta_w(z-w) + \partial_w T(w) \cdot \delta(z-w)$$

as formal power series in $z^{\pm 1}, w^{\pm 1}$.

2.5.5. Proof. We have
$$
\begin{aligned}
[T(z), T(w)] &= \sum_{n,m}(n-m)L_{n+m}z^{-n-2}w^{-m-2} + C\sum_n \frac{n^3-n}{12}z^{-n-2}w^{n-2} \\
&= \sum_{j,l} 2lL_j w^{-j-2}z^{-l-1}w^{l-1} + \sum_{j,l}(-j-2)L_j w^{-j-3}z^{-l-1}w^l \\
&\quad + \frac{C}{12}\sum_l l(l-1)(l-2)z^{-l-1}w^{l-3} \\
&= 2T(w)\partial_w \delta(z-w) + \partial_w T(w) \cdot \delta(z-w) + \frac{C}{12}\partial_w^3 \delta(z-w)
\end{aligned}
$$

(here we have made the substitution $j = n+m, l = n+1$).

2.5.6. Definition of the vertex algebra. We wish to associate to the Virasoro algebra a vertex algebra. Our experience with the Heisenberg and affine Kac–Moody algebras suggests that the vertex algebra should be defined on an induced representation of the Virasoro algebra. Recall the identification $L_n \sim -t^{n+1}\partial_t$, and note that $\operatorname{Der}\mathcal{O} = \mathbb{C}[[t]]\partial_t$, topologically spanned by $L_n, n \geq -1$, is a Lie subalgebra of the Virasoro algebra (in other words, the restriction of the two–cocycle to this Lie subalgebra is zero). It is natural to pick the induced representation in which the generating vector $|0\rangle$ satisfies $L_n|0\rangle = 0$ for all $n \geq -1$. More precisely, let $U(Vir)$ be the universal enveloping algebra of Vir. For each $c \in \mathbb{C}$ we define the induced representation
$$\mathrm{Vir}_c = \mathrm{Ind}_{\operatorname{Der}\mathcal{O}\oplus\mathbb{C}C}^{Vir}\mathbb{C}_c = U(Vir) \underset{U(\operatorname{Der}\mathcal{O}\oplus\mathbb{C}C)}{\otimes} \mathbb{C}_c,$$

where C acts as multiplication by c and $\operatorname{Der}\mathcal{O}$ acts by zero on the one-dimensional module $\mathbb{C}_c$. Note that Vir_c has central charge c as a module over the Virasoro algebra.

By the Poincaré–Birkhoff–Witt theorem, Vir_c has a PBW basis consisting of monomials of the form

(2.5.3) $$L_{j_1}\dots L_{j_m}v_c, \qquad j_1 \leq j_2 \leq \dots \leq j_m \leq -2.$$

Here v_c is the image of $1 \otimes 1 \in U(Vir) \otimes \mathbb{C}_c$ in the induced representation, and we take it to be the vacuum vector of the vertex algebra. We define a $\mathbb{Z}_+$–gradation on Vir_c by the formulas $\deg L_n = -n, \deg v_c = 0$. The picture of the first few "layers" of Vir_c is given below.

As the translation operator we take $T = L_{-1}$. Finally, we use the Reconstruction Theorem 2.3.11 to define the vertex operators. Namely, we set
$$Y(L_{-2}v_c, z) \overset{\text{def}}{=} T(z) = \sum_{n\in\mathbb{Z}} L_n z^{-n-2}.$$

This is the generating field of Vir_c. It has conformal dimension 2, because $\deg L_{-2}v_c = 2$ by the above definition of gradation on Vir_c. Next, we define the vertex operators $Y(A, z)$ for the PBW monomials of the form (2.5.3) according to the prescription of Theorem 2.3.11:

$$Y(L_{j_1} \ldots L_{j_m} v_c, z) = \frac{1}{(-j_1 - 2)!} \cdots \frac{1}{(-j_m - 2)!} :\partial_z^{-j_1 - 2} T(z) \ldots \partial_z^{-j_m - 2} T(z): \, .$$

Lemma 2.5.4 implies that

$$(z - w)^4 [T(z), T(w)] = 0.$$

Hence the generating field $T(z)$ is local with itself. The Reconstruction Theorem 2.3.11 then implies that Vir_c with the above structures is a vertex algebra for any $c \in \mathbb{C}$. Note that the gradation on Vir_c coincides with the gradation by the eigenvalues of L_0.

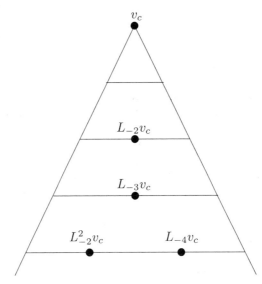

2.5.7. Conformal vertex algebras. As we mentioned above, it is important to have the Lie algebra $\mathrm{Der}\,\mathcal{O}$ of infinitesimal changes of coordinates (and even better, the whole Virasoro algebra) act on a given vertex algebra by "internal symmetries". This property is formalized by the following definition.

2.5.8. Definition. A $\mathbb{Z}$–graded vertex algebra V is called *conformal*, of central charge $c \in \mathbb{C}$, if we are given a non-zero *conformal vector* $\omega \in V_2$ such that the Fourier coefficients L_n^V of the corresponding vertex operator

$$(2.5.4) \qquad\qquad Y(\omega, z) = \sum_{n \in \mathbb{Z}} L_n^V z^{-n-2}$$

satisfy the defining relations of the Virasoro algebra with central charge c, and in addition we have $L_{-1}^V = T$, $L_0^V |_{V_n} = n\,\mathrm{Id}$.

Note that according to our notation scheme, $\omega_{(n)} = L_{n-1}^V$.

2.5.9. Examples. The Virasoro vertex algebra Vir_c is clearly conformal, with central charge c and conformal vector $\omega = L_{-2}v_c$.

The Heisenberg vertex algebra π comes with a natural one–parameter family of conformal structures. Given $\lambda \in \mathbb{C}$, set

$$\omega_\lambda = \frac{1}{2}b_{-1}^2 + \lambda b_{-2}.$$

Then ω_λ is a conformal vector with central charge $c_\lambda = 1 - 12\lambda^2$. In particular, we obtain a family of representations of the Virasoro algebra on π (thus, if $\lambda = 0$ we obtain a representation with central charge 1, which explains our normalization of the two-cocycle defining the Virasoro algebra). To see this we need to check that the Fourier coefficients of the field

$$Y(\omega_\lambda, z) = \frac{1}{2}{:}b(z)^2{:} + \lambda\partial_z b(z) = \sum_{n \in \mathbb{Z}} \omega_{\lambda,n} z^{-n-2}$$

satisfy the Virasoro relations, that $\omega_{\lambda,-1} = T$, and that $\omega_{\lambda,0}$ is the degree operator.

The interested reader should try to check the Virasoro relations by a brute–force calculation and then look in § 3.4.8 for a "shortcut" using the operator product expansion (OPE).

The other two conditions are easy to verify. Indeed, we have

$$\omega_{\lambda,-1} = \int \omega_\lambda(z)\,dz = \frac{1}{2}\sum_{i+j=-1} b_i b_j$$

(note that the total derivative term in the expression for ω_λ does not contribute to this Fourier coefficient). An easy check shows that this operator kills the vacuum, and thus

$$[\omega_{\lambda,-1}, b_k] = -kb_{k-1}.$$

Therefore $\omega_{\lambda,-1} = T$. Finally,

$$\omega_{\lambda,0} = \int \omega_\lambda(z)z\,dz = \frac{1}{2}\sum_{i+j=0} {:}b_i b_j{:} = \sum_{n>0} b_{-n}b_n = \sum_{n>0} nb_{-n}\frac{\partial}{\partial b_{-n}}.$$

This is the "Euler vector field", whose action of $\pi = \mathbb{C}[b_n]_{n<0}$ coincides with the gradation operator.

A rather unexpected fact, proved in Lemma 3.4.5, is that the above two conditions are actually sufficient for proving that ω_λ is a conformal vector (albeit not sufficient to determine the central charge c) – so we don't have to check the Virasoro commutation relations. More precisely, the only extra information one needs is that $L_2 \cdot \omega_\lambda$ is proportional to the vacuum vector (with coefficient of proportionality being $c/2$). But since $|0\rangle$ spans the homogeneous component of π of degree 0, we obtain that $L_2 \cdot \omega_\lambda$ must be proportional to the vacuum vector. In order to find the value of the central charge, we still need to compute $L_2 \cdot \omega_\lambda$, which is relatively easy to do (see § 3.4.8).

2.5.10. The Segal-Sugawara construction. The vertex algebra $V_k(\mathfrak{g})$ has a natural conformal vector, called the Segal-Sugawara vector, when $k \neq -h^\vee$, where $h^\vee$ denotes the dual Coxeter number of $\mathfrak{g}$. Recall from § 2.4.2 that we have an isomorphism of vector spaces

$$V_k(\mathfrak{g}) \simeq U(\mathfrak{g} \otimes t^{-1}\mathbb{C}[t^{-1}])v_k.$$

Pick a basis $\{J^a\}_{a=1,\dots,d}$ of $\mathfrak{g}$ (where $d = \dim \mathfrak{g}$), and let $\{J_a\}_{a=1,\dots,d}$ be its dual basis with respect to the invariant bilinear form $(\cdot, \cdot)$ (see § 2.4.1).

We write

$$J^a(z) = \sum_{n \in \mathbb{Z}} J^a_n z^{-n-1}, \qquad J_a(z) = \sum_{n \in \mathbb{Z}} J_{a,n} z^{-n-1}.$$

Set

$$(2.5.5) \qquad S = \frac{1}{2} \sum_{a=1}^{d} J_{a,-1} J^a_{-1} v_k$$

(clearly, it is independent of the choice of basis $\{J^a\}$). Then for $k \neq -h^\vee$, $S/(k+h^\vee)$ is a conformal vector in $V_k(\mathfrak{g})$, of central charge

$$c(k) = \frac{k \cdot \dim \mathfrak{g}}{k + h^\vee}.$$

We call it the *Segal-Sugawara conformal vector*. Thus, the Fourier coefficients of the field

$$(2.5.6) \qquad \frac{1}{k+h^\vee} Y(S, z) = \frac{1}{2(k+h^\vee)} \sum_{a=1}^{d} :J_a(z) J^a(z):$$

generate an action of the Virasoro algebra with central charge $c(k)$ on $V_k(\mathfrak{g})$.

Checking the Virasoro relations directly is rather difficult, so we will do it later in § 3.4.8 using the operator product expansion.

Further, the resulting Virasoro operators satisfy the following commutation relations with the elements of $\widehat{\mathfrak{g}}$:

$$(2.5.7) \qquad [L_n, J^a_m] = -m J^a_{n+m}.$$

The above commutation relations and Proposition 4.2.2 proved below imply that the $\widehat{\mathfrak{g}}$–action on any $\mathbb{Z}$–graded $\widehat{\mathfrak{g}}$–module with gradation bounded from below (on all such modules the Fourier coefficients of the field (2.5.6) are well-defined) may be extended to the action of the semi-direct product of $\widehat{\mathfrak{g}}$ and the Virasoro algebra (naturally acting on $\widehat{\mathfrak{g}}$ by exterior infinitesimal automorphisms).

Formula (2.5.7) will be proved in § 3.4.8 using the formalism of operator product expansion.

2.6. Bibliographical notes

The definition of Lie algebra cohomology and a discussion of its relationship to central extensions are contained in D.B. Fuchs' book [**Fu**] (see also § A.4). For more information on Heisenberg, affine Kac-Moody and Virasoro algebras, the reader may consult [**Kac2, KR, FZe, EFK**].

The first published proof of Dong's Lemma 2.3.4 appeared in [**Li1**] (Proposition 3.2.7). The Reconstruction Theorem 2.3.11 is due to [**FKRW**].

CHAPTER 3

Associativity and Operator Product Expansion

In this chapter we establish two important results: the Goddard uniqueness theorem and the associativity property of vertex algebras. The associativity property allows us to introduce an important tool in the theory of vertex algebras: the operator product expansion (OPE). We use the OPE to compute the commutation relations between the Fourier coefficients of vertex operators from a vertex algebra V. The commutator of two such Fourier coefficients is a linear combination of other Fourier coefficients. At the end of the chapter we illustrate the general formula for the OPE and the commutation relations by concrete examples.

3.1. Goddard's uniqueness theorem

The Goddard uniqueness theorem that we present in this section is important in that it explains how rigid the condition of locality really is. Essentially, the property of locality allows one to reconstruct a vertex operator from the mere knowledge of how it acts on the vacuum vector.

3.1.1. Goddard's Uniqueness Theorem. *Let V be a vertex algebra, $A(z)$ a field on V. Suppose there exists a vector $a \in V$ such that*

$$A(z)|0\rangle = Y(a, z)|0\rangle$$

and $A(z)$ is local with respect to $Y(b, z)$ for all $b \in V$. Then $A(z) = Y(a, z)$.

3.1.2. Proof. Take $b \in V$. Using the formulation of locality from Proposition 1.2.5, we obtain the following equalities in $V[[z^{\pm 1}, w^{\pm 1}]]$ for large enough N:

$$(z - w)^N A(z) Y(b, w)|0\rangle = (z - w)^N Y(b, w) A(z)|0\rangle$$
$$= (z - w)^N Y(b, w) Y(a, z)|0\rangle = (z - w)^N Y(a, z) Y(b, w)|0\rangle.$$

By the vacuum axiom, both $(z - w)^N Y(a, z) Y(b, w)|0\rangle$ and $(z - w)^N A(z) Y(b, w)|0\rangle$ are well-defined at $w = 0$. Setting $w = 0$, and using $Y(b, w)|0\rangle|_{w=0} = b$, we obtain the equality $Y(a, z)b = A(z)b$ for all $b \in V$, which is what we wanted.

3.1.3. Lemma.

$$Y(a, z)|0\rangle = e^{zT} a.$$

3.1.4. Proof. By the vacuum axiom, $a_{(m)}|0\rangle = 0$ for $m \geq 0$, and $a_{(-1)}|0\rangle = a$. We need to prove that

(3.1.1) $$a_{(-n-1)}|0\rangle = \frac{1}{n!} T^n \cdot a, \qquad n > 0.$$

By the translation axiom,

$$\partial_z Y(a, z)|0\rangle = [T, Y(a, z)]|0\rangle = TY(a, z)|0\rangle.$$

47

Equating the coefficients in front of $z^{n-1}, n > 0$, we find that $na_{(-n-1)}|0\rangle = Ta_{(-n)}|0\rangle$. Since $a_{(-1)}|0\rangle = a$, by the vacuum axiom, formula (3.1.1) follows by induction.

3.1.5. Remark. In view of Lemma 3.1.3, the hypothesis of Theorem 3.1.1 can be rephrased as the statement that $A(z)|0\rangle = e^{zT}a$. This in turn is equivalent to the formulas

$$\partial_z A(z)|0\rangle = TA(z)|0\rangle, \qquad A(z)|0\rangle|_{z=0} = a.$$

In other words, $A(z)$ satisfies a differential equation in z determined by the operator T, and its "initial condition" at $z = 0$ equals a. Thus, Goddard's theorem means that the vertex operator $Y(a, z)$ is uniquely characterized as the local field satisfying these properties. This relation between vectors in V and fields is known as the state–field correspondence in conformal field theory.

3.1.6. Corollary. *For all $a \in V$, $Y(Ta, z) = \partial_z Y(a, z)$.*

3.1.7. Proof. We have

$$\partial_z^2 Y(a, z)|0\rangle = \partial_z([T, Y(a, z)]|0\rangle) = \partial_z(TY(a, z)|0\rangle) = T\partial_z Y(a, z)|0\rangle,$$

and

$$\partial_z Y(a, z)|0\rangle|_{z=0} = [T, Y(a, z)]|0\rangle|_{z=0} = TY(a, z)|0\rangle|_{z=0} = Ta.$$

In view of Remark 3.1.5, the corollary now follows from Theorem 3.1.1 applied to the field $A(z) = \partial_z Y(a, z)$.

3.1.8. Remark. Note the difference between the above formula $Y(Ta, z) = \partial_z Y(a, z)$ and the formula $[T, Y(a, z)] = \partial_z Y(a, z)$, which is part of the axioms of a vertex algebra.

3.2. Associativity

In this section we derive an "associativity" property of vertex algebras. We already know from the locality axiom of vertex algebras that for any vertex algebra V and vectors $A, B, C \in V$, the expressions

(3.2.1) $$\begin{aligned} Y(A, z)Y(B, w)C &\in V((z))((w)) \quad \text{and} \\ Y(B, w)Y(A, z)C &\in V((w))((z)) \end{aligned}$$

are the expansions, in their respective domains, of the same element of

$$V[[z, w]][z^{-1}, w^{-1}, (z - w)^{-1}].$$

As we explained in § 1.3.3, the notion of locality may be viewed as an analogue of the notion of commutativity for ordinary algebras, which may be written as the identity

$$Y(A)Y(B) = Y(B)Y(A).$$

Here A and B are arbitrary elements of an algebra V and $Y : V \to \operatorname{End} V$ is the operation defined by the formula $Y(A)B = A \cdot B$. We explained in § 1.3.3 that this identity implies both commutativity and associativity of the product on V. But now let us observe that the associativity may be written in the form

$$Y(A)Y(B)C = Y(Y(A)B)C$$

for all $A, B, C \in V$.

We would like to obtain an analogue of the last identity for vertex algebras. So we add a third expression to the pair (3.2.1) which satisfies a similar property. It is natural to call this property the *associativity property* for vertex algebras.

3.2.1. Theorem. *Any vertex algebra V satisfies the following associativity property: for all $A, B, C \in V$ the three expressions*

$$Y(A, z)Y(B, w)C \in V((z))((w)),$$
$$Y(B, w)Y(A, z)C \in V((w))((z)), \text{ and}$$
$$Y(Y(A, z - w)B, w)C \in V((w))((z - w))$$

are the expansions, in their respective domains, of the same element of

$$V[[z, w]][z^{-1}, w^{-1}, (z - w)^{-1}].$$

3.2.2. Remark. A geometric interpretation of the associativity property in terms of conformal blocks on the projective line is given in § 10.4.11. This property will enable us to derive the Jacobi identity for chiral algebras associated with vertex algebras (see the proof of Theorem 19.3.3).

In analytic terms, the associativity property can be stated as follows: for any $A, B \in V$, the formal power series

$$Y(Y(A, z - w)B, w) = \sum_{n \in \mathbb{Z}} Y(A_{(n)} \cdot B, w)(z - w)^{-n-1}$$

in $(z - w)^{\pm 1}$ and $w^{\pm 1}$ converges in the domain $|w| > |z - w|$ and can be analytically continued to a meromorphic operator-valued function on $\mathbb{C}^2$ with poles at $z = 0, w = 0$, and $z = w$, denoted by $R(Y(Y(A, z - w)B, w))$. Moreover,

(3.2.2) $$R(Y(A, z)Y(B, w)) = R(Y(Y(A, z - w)B, w)).$$

Now we will prove Theorem 3.2.1. The proof is based on the following two assertions.

3.2.3. Lemma.

$$e^{wT}Y(A, z)e^{-Tw} = Y(A, z + w)$$

in $\operatorname{End} V[[z^{\pm 1}, w^{\pm 1}]]$, *where by negative powers of* $(z + w)^{-1}$ *we understand their expansions in* $\mathbb{C}((z))((w))$, *i.e., in positive powers of* w/z.

3.2.4. Proof. By the translation axiom, $[T, Y(A, z)] = \partial_z Y(A, z)$. Hence

$$e^{wT}Y(A, z)e^{-Tw} = \sum_{n=0}^{\infty} \frac{1}{n!}(\operatorname{ad} wT)^n \cdot Y(A, z) = \sum_{n=0}^{\infty} \frac{w^n}{n!}\partial_z^n Y(A, z),$$

where we use the notation $\operatorname{ad} B \cdot C = [B, C]$. To complete the proof, we use the formal variable version of the Taylor formula: for any algebra R and a formal power series $f(z) \in R[[z^{\pm 1}]]$, we have the following identity in $R[[z^{\pm 1}, w^{\pm 1}]]$:

$$\sum_{n=0}^{\infty} \frac{w^n}{n!}\partial_z^n f(z) = f(z + w).$$

3.2.5. Proposition (skew-symmetry). *The identity*

(3.2.3) $$Y(A, z)B = e^{zT}Y(B, -z)A$$

holds in $V((z))$.

3.2.6. Proof. First, note that for any field $\phi(z)$ on V, the product $e^{zT}\phi(z)$ is a well-defined element of $\text{End}\,V[[z^{\pm 1}]]$, so the statement of the lemma makes sense.

By locality,

$$(z-w)^N Y(A,z)Y(B,w)|0\rangle = (z-w)^N Y(B,w)Y(A,z)|0\rangle,$$

for large enough N. Using Lemma 3.1.3, we obtain

$$(z-w)^N Y(A,z)e^{wT}B = (z-w)^N Y(B,w)e^{zT}A.$$

Applying Lemma 3.2.3, we find that

$$(z-w)^N Y(A,z)e^{wT}B = (z-w)^N e^{zT}Y(B,w-z)A.$$

where by $(w-z)^{-1}$ we understand its expansion in $\mathbb{C}((w))((z))$. We may always choose N large enough so that the right hand side of this formula does not contain any negative powers of $w-z$. Then it becomes an identity in $V((z))[[w]]$, in which we can set $w=0$, and then divide by z^N. This gives us formula (3.2.3).

3.2.7. Proof of Theorem 3.2.1. By Proposition 3.2.5, we have the following equality in $\text{End}\,V[[z^{\pm 1}, w^{\pm 1}]]$:

$$Y(A,z)Y(B,w)C = Y(A,z)e^{wT}Y(C,-w)B = e^{Tw}(e^{-Tw}Y(A,z)e^{wT})Y(C,-w)B.$$

Note that since $e^{\pm wT}$ has only positive powers of w, and $Y(A,z)e^{wT}Y(C,-w)B \in V((z))((w))$, the last equality makes sense.

Using Lemma 3.2.3 in the right hand side of this formula, we obtain

$$(3.2.4) \qquad Y(A,z)Y(B,w)C = e^{wT}Y(A,z-w)Y(C,-w)B,$$

where by $(z-w)^{-1}$ we understand its expansion in positive powers of w/z. Thus, $Y(A,z)Y(B,w)C$ and $e^{wT}Y(A,z-w)Y(C,-w)B$ are equal to the expansions of the same element of $V[[z,w]][z^{-1}, w^{-1}, (z-w)^{-1}]$ in $V((z))((w))$ and $V((z-w))((w))$, respectively.

On the other hand, consider

$$Y(Y(A,z-w)B,w)C = \sum_{n\in\mathbb{Z}}(z-w)^{-n-1}Y(A_{(n)}\cdot B,w)C$$

as a formal power series in $(z-w)$ and w. By Proposition 3.2.5,

$$Y(A_{(n)}\cdot B,w)C = e^{wT}Y(C,-w)A_{(n)}B.$$

Hence

$$(3.2.5) \qquad Y(Y(A,z-w)B,w)C = e^{wT}Y(C,-w)Y(A,z-w)B,$$

as formal power series in $(z-w)$ and w. By locality, the right hand side of (3.2.5) is the expansion in $V((w))((z-w))$ of an element of $V[[z,w]][z^{-1}, w^{-1}, (z-w)^{-1}]$. Therefore the left hand side of (3.2.5) has the same properties.

By applying the locality axiom to the right hand sides of (3.2.4) and (3.2.5), we obtain that $Y(A,z)Y(B,w)C$ and $Y(Y(A,z-w)B,w))C$ are expansions of the same element of $V[[z,w]][z^{-1}, w^{-1}, (z-w)^{-1}]$ in $V((z))((w))$ and $V((w))((z-w))$, respectively. Theorem 3.2.1 follows.

3.3. Operator product expansion

Consider the expression $Y(A, z)Y(B, w)C$. By locality, it is the expansion in $V((z))((w))$ of an element of the space

$$V[[z, w]][z^{-1}, w^{-1}, (z-w)^{-1}].$$

Let us denote this element by $f_{A,B,C}$. By Theorem 3.2.1, the expansion of $f_{A,B,C}$ in $V((w))((z-w))$ is equal to $Y(Y(A, z-w)B, w)C$. The corresponding embedding

$$V[[z, w]][z^{-1}, w^{-1}, (z-w)^{-1}] \to V((w))((z-w))$$

is obtained by sending z to $w + (z - w)$ and z^{-1} to $(w + (z - w))^{-1}$ considered as a power series in positive powers of $(z - w)/w$. Thus, we find that applying this procedure to $f_{A,B,C}$ we obtain $Y(Y(A, z-w)B, w)C$. By abusing notation, we may write this as

$$Y(A, z)Y(B, w)C = Y(Y(A, z - w)B, w)C, \qquad A, B, C \in V,$$

or as

$$(3.3.1) \qquad Y(A, z)Y(B, w)C = \sum_{n \in \mathbb{Z}} \frac{Y(A_{(n)} \cdot B, w)}{(z - w)^{n+1}} \, C, \qquad A, B, C \in V,$$

where the left hand side should be understood in the above sense.

Formula (3.3.1) is called the *operator product expansion* (or OPE for short). From the physics point of view, it expresses the important property in quantum field theory that the product of two fields at nearby points can be expanded in terms of other fields and the small parameter $z - w$.

In this section we explain how to derive from formula (3.3.1) the commutation relations between the Fourier coefficients of the fields $Y(A, z)$ and $Y(B, w)$ and explain formula (2.3.8) for the vertex operators in the Reconstruction Theorem 2.3.11. The first step is the following

3.3.1. Proposition ([Kac3]). *Let $\phi(z)$ and $\psi(w)$ be arbitrary fields on a vector space V. Then the following are equivalent:*

(1) *There is an identity in* $\mathrm{End}\, V[[z^{\pm 1}, w^{\pm 1}]]$:

$$(3.3.2) \qquad [\phi(z), \psi(w)] = \sum_{j=0}^{N-1} \frac{1}{j!} \gamma_j(w) \partial_w^j \delta(z - w),$$

where $\gamma_j(w), j = 0, \ldots, N - 1$, are some fields.

(2) *$\phi(z)\psi(w)$ (resp., $\psi(w)\phi(z)$) equals*

$$(3.3.3) \qquad \sum_{j=0}^{N-1} \frac{\gamma_j(w)}{(z - w)^{j+1}} + {:}\phi(z)\psi(w){:},$$

where $1/(z - w)$ is expanded in positive powers of w/z (resp., z/w).

(3) *$\phi(z)\psi(w)$ converges for $|z| > |w|$ to the expression in (3.3.3), and $\psi(w)\phi(z)$ converges for $|w| > |z|$ to the same expression.*

3.3.2. Proof. First we derive (2) from (1). Given a formal power series $f(z) = \sum_{n \in \mathbb{Z}} f_n z^n$, we will write, as before,

$$f(z)_+ = \sum_{n \geq 0} f_n z^n, \qquad f(z)_- = \sum_{n < 0} f_n z^n.$$

On the left hand side of (3.3.2), z appears only in the term $\phi(z)$. Thus, taking the negative Fourier coefficients with respect to z, we obtain

$$[\phi(z)_-, \psi(w)] - (\text{R. H. S. of } (3.3.2))_-$$

and similarly for $(\cdot)_+$. But since $:\phi(z)\psi(w): = \psi(w)\phi(z)_- + \phi(z)_+\psi(w)$ while $\phi(z)\psi(w) = \phi(z)_-\psi(w) + \phi(z)_+\psi(w)$, we see that

$$[\phi(z)_-, \psi(w)] = \phi(z)\psi(w) - :\phi(z)\psi(w): .$$

Thus,

$$(\text{RHS of } (3.3.2))_- = \sum_{j=0}^{N-1} \frac{1}{j!} \gamma_j(w) \partial_w^j \delta(z - w)_-.$$

Since $\frac{1}{j!}\partial_w^j \delta(z - w)_-$ is the expansion of $1/(z - w)^{j+1}$ in $\mathbb{C}((z))((w))$, we obtain formula (3.3.3). The formula for $\psi(w)\phi(z)$ is obtained in the same way by taking the positive part of the commutator $[\psi(w), \phi(z)]$ (again with respect to z), which equals

$$[\psi(w), \phi(z)_+] = \psi(w)\phi(z) - :\phi(z)\psi(w): .$$

To derive part (1) from part (2), observe that when we take the commutator between $\phi(z)$ and $\psi(w)$, the normally ordered terms will cancel each other, and we will be left with the combination of terms of the form

$$\sum_{j=0}^{N-1} \left(\frac{1}{j!} \gamma_j(w)(\partial_w^j \delta(z - w)_- + \partial_w^j \delta(z - w)_+) \right) = \text{RHS of } (3.3.2).$$

The equivalence of (2) and (3) is clear.

3.3.3. Remark. We may interchange the roles of $\phi(z)$ and $\psi(w)$, and write $[\psi(w), \phi(z)]$ as a sum

$$\sum_{j=0}^{N-1} \frac{1}{j!} \widetilde{\gamma}_j(z) \partial_z^j \delta(z - w),$$

where now the fields appearing in the right hand side are taken at z, not w. Then according to Proposition 3.3.1, we obtain

$$(3.3.4) \qquad \psi(w)\phi(z) = \sum_{j=0}^{N-1} \frac{\widetilde{\gamma}_j(z)}{(w - z)^{j+1}} + :\psi(w)\phi(z): .$$

Informally speaking, in (3.3.3) we did expansion around w, whereas now we expand around z. While the right hand side of formula (3.3.4) equals the expression in formula (3.3.3), the individual terms appearing in them (and hence in the OPEs $\phi(z)\psi(w)$ and $\psi(w)\phi(z)$) are different in general: $\gamma_j(z) \neq \widetilde{\gamma}_j(z)$ and $:\phi(z)\psi(w): \neq :\psi(w)\phi(z):$.

This fact makes verification of locality of $\phi(z)$ and $\psi(w)$ using direct computation of the power series expansions of $\phi(z)\psi(w)$ and $\psi(w)\phi(z)$ often very difficult. It is usually a lot easier to check locality of $\phi(z)$ and $\psi(z)$ by computing the commutator $[\phi(z), \psi(w)]$. By Proposition 3.3.1 and Lemma 1.1.4, locality will follow if the result is an expression involving derivatives of the delta-function. Moreover, we then also obtain formulas for the OPE of the two fields using Proposition 3.3.1.

Comparing formula (3.3.2) with the OPE formula (3.3.1), we can derive some useful identities. First we use the part of the OPE which is regular at $z = w$ to obtain formulas for the normally ordered product of two vertex operators. After that we will use the singular part of the OPE to obtain commutation relations between the Fourier coefficients of vertex operators.

3.3.4. Normally ordered product and OPE. Using the OPE (3.3.1) and Corollary 3.1.6, we can now explain formula (2.3.8) for the vertex operators in the Reconstruction Theorem 2.3.11. Let $Y(A, z)$ and $Y(B, w)$ be two vertex operators. Using the locality formula (1.2.3) and Lemma 1.1.4, we obtain

$$(3.3.5) \qquad [Y(A, z), Y(B, w)] = \sum_{j=0}^{N-1} \frac{1}{j!} \gamma_j(w) \partial_w^j \delta(z - w),$$

where $\gamma_j(w), j = 0, \ldots, N - 1$, are some fields. Using Proposition 3.3.1, we find that

$$(3.3.6) \qquad Y(A, z)Y(B, w) = \sum_{j=0}^{N-1} \frac{\gamma_j(w)}{(z - w)^{j+1}} + :Y(A, z)Y(B, w): \, ,$$

where by $(z - w)^{-1}$ we understand its expansion in positive powers of w/z. Hence for any $C \in V$ the series $Y(A, z)Y(B, w)C \in V((z))((w))$ is an expansion of

$$\left(\sum_{j=0}^{N-1} \frac{\gamma_j(w)}{(z - w)^{j+1}} + :Y(A, z)Y(B, w): \right) C \in V[[z, w]][z^{-1}, w^{-1}, (z - w)^{-1}].$$

Using the Taylor formula, we obtain that the expansion of this element of

$$V[[z, w]][z^{-1}, w^{-1}, (z - w)^{-1}]$$

in $V((w))((z - w))$ is equal to

$$(3.3.7) \qquad \left(\sum_{j=0}^{N-1} \frac{\gamma_j(w)}{(z - w)^{j+1}} + \sum_{m \geq 0} \frac{(z - w)^m}{m!} :\partial_w^m Y(A, w) \cdot Y(B, w): \right) C.$$

The coefficient in front of $(z - w)^k, k \in \mathbb{Z}$, in the right hand side of (3.3.7) must be equal to the corresponding term in the right hand side of formula (3.3.1). Let us first look at the terms with $k \geq 0$. Comparison of the two formulas gives

$$(3.3.8) \qquad Y(A_{(n)} \cdot B, z) = \frac{1}{(-n - 1)!} :\partial_z^{-n-1} Y(A, z) \cdot Y(B, z): \, , \qquad n < 0.$$

Since $Y(B, z)|0\rangle = e^{zT}B$, by Lemma 3.1.3, we find that $B_{(-n-1)}|0\rangle = \frac{1}{n!}T^n B$. Applying Corollary 3.1.6 several times, we find that

$$(3.3.9) \qquad Y(B_{(-n-1)}|0\rangle, z) = \frac{1}{n!} \partial_z^n Y(B, z).$$

Using formulas (3.3.8) and (3.3.9), we obtain the following corollary by induction on k (recall that our convention is that the normal ordering is nested from right to left):

3.3.5. Corollary. *For any* $A^1, \ldots, A^k \in V$, *and* $n_1, \ldots, n_k < 0$, *we have*

$$Y(A^1_{(n_1)} \ldots A^k_{(n_k)} |0\rangle, z)$$

$$= \frac{1}{(-n_1 - 1)!} \cdots \frac{1}{(-n_k - 1)!} :\partial_z^{-n_1 - 1} Y(A^1, z) \cdot \ldots \cdot \partial_z^{-n_k - 1} Y(A^k, z): .$$

This gives us formula (2.3.8) under the assumptions of Theorem 2.3.11.

3.3.6. Commutation relations. Now we compare the coefficients in front of $(z - w)^k, k < 0$, in formulas (3.3.1) and (3.3.7). We find that

$$\gamma_j(w) = Y(A_{(j)} \cdot B, w), \qquad j \geq 0,$$

and so formula (3.3.6) can be rewritten as

$$(3.3.10) \qquad Y(A, z)Y(B, w) = \sum_{n \geq 0} \frac{Y(A_{(n)} \cdot B, w)}{(z - w)^{n+1}} + :Y(A, z)Y(B, w): .$$

Note that unlike formula (3.3.1), this identity makes sense in $\mathrm{End}\, V[[z^{-1}, w^{-1}]]$ if we expand $(z - w)^{-1}$ in positive powers of w/z.

Now Proposition 3.3.1 implies the following commutation relations:

$$(3.3.11) \qquad [Y(A, z), Y(B, w)] = \sum_{n \geq 0} \frac{1}{n!} Y(A_{(n)} \cdot B, w) \partial_w^n \delta(z - w).$$

Expanding both sides of (3.3.11) as formal power series and using the equality

$$\frac{1}{n!} \partial_w^n \delta(z - w) = \sum_{m \in \mathbb{Z}} \binom{m}{n} z^{-m-1} w^{m-n},$$

we obtain the following identity for the commutators of Fourier coefficients of arbitrary vertex operators:

$$(3.3.12) \qquad [A_{(m)}, B_{(k)}] = \sum_{n \geq 0} \binom{m}{n} (A_{(n)} \cdot B)_{(m+k-n)}.$$

Here, by definition, for any $m \in \mathbb{Z}$,

$$\binom{m}{n} = \frac{m(m-1) \ldots (m-n+1)}{n!}, \quad n \in \mathbb{Z}_{>0}; \qquad \binom{m}{0} = 1.$$

In particular, we see that only the terms in the OPE (3.3.1) which are singular on the diagonal $z = w$ contribute to the commutator. Thus, we will often write formula (3.3.10) in the form

$$Y(A, z)Y(B, w) = \sum_{n \geq 0} \frac{Y(A_{(n)} \cdot B, w)}{(z - w)^{-n-1}} + \mathrm{reg.},$$

emphasizing the singular terms in the OPE.

Formula (3.3.12) also shows that the commutator of Fourier coefficients of two fields is a linear combination of Fourier coefficients of other fields (namely, the ones corresponding to the vectors $A_{(n)} \cdot B$). Therefore we obtain the following result:

3.3.7. Proposition. *Let* V *be a vertex algebra. Then the span in* $\mathrm{End}\, V$ *of all Fourier coefficients of all vertex operators in* V *is a Lie algebra.*

An easy corollary of formula (3.3.12) is the construction of vertex algebra automorphisms using residues of fields.

3.3.8. Corollary. *For any $A \in V$, the residue $A_{(0)} = \int Y(A, z)dz$ of $Y(A, z)$ is an infinitesimal automorphism of the vertex algebra V, that is,*

$$[A_{(0)}, Y(B, w)] = Y(A_{(0)} \cdot B, w).$$

Examples of residues of fields are the operator $T = L_{-1} = \int T(z)dz$ in a conformal vertex algebra, and the operators $J_0^a = \int J^a(z)dz$ in the vertex algebra $V_k(\mathfrak{g})$ giving the action of $\mathfrak{g}$ on $V_k(\mathfrak{g})$. An important class of residue operators consists of the differentials in the BRST construction, see § 5.7.3.

3.3.9. Remark. The commutation relations (3.3.12) are equivalent to the relations (3.3.11). The latter imply that the fields $Y(A, z)$ and $Y(B, z)$ are local with respect to each other, by Lemma 1.1.4 and Proposition 1.2.5. Therefore we obtain that in the axioms of a vertex algebra given in Definition 1.3.1 we can replace the locality axiom by the relations (3.3.12) for all $A, B \in V$.

3.3.10. Contour integrals and the Borcherds identity. One can also derive the commutation relations (3.3.12) analytically from the Cauchy formula for contour integrals.

According to the analytic version of the locality property given in Proposition 1.2.9, the compositions $Y(A, z)Y(B, w)$ and $Y(B, w)Y(A, z)$ are well-defined in the domain $|z| > |w|$ (resp., $|w| > |z|$) and can be analytically continued to the same operator–valued meromorphic function in z, w with poles at $z = 0, w = 0$, and $z = w$. Let $R > \rho > r > 0$ be real numbers. Denote by C_z^a the circle on the z–plane of radius a centered at 0.

Let $f(z, w)$ be a rational function which has poles only at $z = 0, w = 0$ and $z = w$. Then the integrals

$$\int_{C_w^\rho} \int_{C_z^R} Y(A, z)Y(B, w)f(z, w)\, dzdw, \qquad \int_{C_w^\rho} \int_{C_z^r} Y(B, w)Y(A, z)f(z, w)\, dzdw$$

are well–defined. Moreover, by the locality property,

$$\int_{C_w^\rho} \int_{C_z^R} Y(A, z)Y(B, w)f(z, w)\, dzdw - \int_{C_w^\rho} \int_{C_z^r} Y(B, w)Y(A, z)f(z, w)\, dzdw$$

$$= \int_{C_w^\rho} \int_{C_z^R - C_z^r} R(Y(A, z)Y(B, w))f(z, w)\, dzdw.$$

Using the analytic version of the associativity property given in Remark 3.2.2 (see formula (3.2.2)), we obtain that the last integral equals

$$\int_{C_w^\rho} \int_{C_z^R - C_z^r} R(Y(Y(A, z - w)B, w))f(z, w)\, dzdw$$

$$= \int_{C_w^\rho} \int_{C_z^R - C_z^r} \sum_{n \in \mathbb{Z}} Y(A_{(n)} \cdot B, w)(z - w)^{-n-1}f(z, w)\, dzdw.$$

But inside the annulus with boundary $C_z^R - C_z^r$ the integrand, considered as a function in z with fixed w, may only have poles at the point $z = w$. Therefore we can replace the contour $C_z^R - C_z^r$ by a small circle $C^\delta(w)$ of radius δ around w (see the picture).

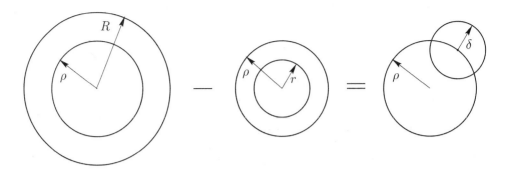

Thus, we obtain the identity

$$\int_{C_w^\rho}\int_{C_z^R} Y(A,z)Y(B,w)f(z,w)\,dzdw - \int_{C_w^\rho}\int_{C_z^r} Y(B,w)Y(A,z)f(z,w)\,dzdw$$

$$= \int_{C_w^\rho}\int_{C_z^\delta(w)}\sum_{n\in\mathbb{Z}}(z-w)^{-n-1}Y(A_{(n)}\cdot B,w)f(z,w)\,dzdw,$$

for any $f \in \mathbb{C}[z^{\pm 1}, w^{\pm 1}, (z-w)^{-1}]$.

If we choose

$$f(z,w) = z^m w^k,$$

then we obtain, using the Cauchy formula,

$$[A_{(m)}, B_{(k)}] = \sum_{n\geq 0}\int_{C_w^\rho} w^k dw \int_{C_z^\delta(w)} dz\, \frac{z^m}{(z-w)^{n+1}} Y(A_{(n)}\cdot B, w)$$

$$= \sum_{n\geq 0}\binom{m}{n}(A_{(n)}\cdot B)_{(m+k-n)}.$$

This is the same commutator identity as formula (3.3.12).

More generally, we can take $f(z,w) = z^m w^k (z-w)^l$. Then we obtain the *Borcherds identity* :

$$\sum_{j\geq 0}\binom{l}{j}\left((-1)^j A_{(m+l-j)}B_{(k+j)} - (-1)^{j+l}B_{(k+l-j)}A_{(m+j)}\right)$$

$$= \sum_{n\geq 0}\binom{m}{n}(A_{(n+l)}\cdot B)_{(m+k-n)}.$$

Borcherds used a special case of this identity as one of the axioms in his original definition of vertex algebras [**B1**].

3.4. Examples of OPE

3.4.1. Heisenberg vertex algebra. Let us compare the two sides of formula (3.3.1) in the case when $A = B = b_{-1}$ in the Heisenberg vertex algebra π. We have already computed the left hand side in formula (2.3.2). Using the Taylor formula for $b(z)$, we obtain from formula (2.3.2) that the expansion of $b(z)b(w)$ in $\mathrm{End}\,\pi((w))((z-w))$ is equal to

$$b(z)b(w) = \frac{1}{(z-w)^2} + \sum_{m\geq 0}\frac{1}{m!}{:}\partial_w^m b(w)b(w){:}(z-w)^m.$$

For the right hand side of formula (3.3.1) we find that

$$\sum_{n\in\mathbb{Z}} \frac{Y(b_nb_{-1},w)}{(z-w)^{n+1}} = \frac{Y(b_1b_{-1},w)}{(z-w)^2} + \sum_{n\leq -1} Y(b_nb_{-1},w)(z-w)^{-n-1}$$

$$= \frac{1}{(z-w)^2} + \sum_{m\geq 0} \frac{1}{m!} :\partial_w^m b(w)\cdot b(w): (z-w)^m.$$

The two sides, considered as power series in the space $\mathrm{End}\,\pi((w))((z-w))$, are indeed equal to each other. Now formula (3.3.12) gives

$$[b_m,b_k] = \sum_{n\geq 0} \binom{m}{n} (b_nb_{-1})_{(m+k-n)} = m\delta_{m,-k},$$

as expected.

3.4.2. Kac–Moody vertex algebra. Next consider the case of the vertex algebra $V_k(\mathfrak{g})$, with $A = J_{-1}^a v_k$ and $B = J_{-1}^b v_k$. The left hand side of formula (3.3.1) is then computed using (2.4.6) and Proposition 3.3.1:

$$(3.4.1) \qquad J^a(z)J^b(w) = \frac{k(J^a,J^b)}{(z-w)^2} + \frac{[J^a,J^b](w)}{z-w} + :J^a(z)J^b(w): .$$

The right hand side equals

$$\sum_{n\in\mathbb{Z}} \frac{Y(J_n^a J_{-1}^b v_k,w)}{(z-w)^{n+1}}$$

$$= \frac{Y(J_1^a J_{-1}^b v_k,w)}{(z-w)^2} + \frac{Y(J_0^a J_{-1}^b v_k,w)}{z-w} + \sum_{n\leq -1} Y(J_n^a J_{-1}^b,w)(z-w)^{-n-1}$$

$$= \frac{k(J^a,J^b)}{(z-w)^2} + \frac{[J^a,J^b](w)}{z-w} + \sum_{m\geq 0} \frac{1}{m!} :\partial_z^m J^a(w)\cdot J^b(w): (z-w)^m.$$

The last sum is simply the Taylor expansion of the last term in (3.4.1). Thus, the identity (3.3.1) indeed holds in this case.

Applying formula (3.3.12), one obtains precisely the Kac–Moody commutation relations (2.4.4):

$$[J_m^a,J_l^b] = \sum_{n\geq 0} \binom{m}{n} (J_n^a J_{-1}^b v_k)_{(m+l-n)}$$

$$= [J^a,J^b]_{m+l} + m(J^a,J^b)k\delta_{m,-l}.$$

3.4.3. Conformal vertex algebras. Consider the case of a conformal vertex algebra V with a conformal vector ω. The left hand side of formula (3.3.1) for $A = B = \omega$ is obtained by combining Lemma 2.5.4 and Proposition 3.3.1:

$$(3.4.2) \qquad T(z)T(w) = \frac{c/2}{(z-w)^4} + \frac{2T(w)}{(z-w)^2} + \frac{\partial_w T(w)}{z-w} + :T(z)T(w): .$$

On the other hand, the right hand side of (3.3.1) equals

$$\sum_{n\in\mathbb{Z}} \frac{Y(L_n\omega,w)}{(z-w)^{n+2}} = \frac{Y(L_2\omega,w)}{(z-w)^4} + \frac{Y(L_1\omega,w)}{(z-w)^3}$$

$$+ \frac{Y(L_0\omega,w)}{(z-w)^2} + \frac{Y(L_{-1}\omega,w)}{z-w} + \sum_{n\leq -2} Y(L_n\omega,w)(z-w)^{-n-2}$$

$$= \frac{c/2}{(z-w)^4} + \frac{2T(w)}{(z-w)^2} + \frac{\partial_w T(w)}{z-w} + :T(z)T(w): \,,$$

where we used the commutation relations of the Virasoro algebra and the Taylor formula. In particular, the factor $c/2$ is due to the fact that $L_2 L_{-2} v_c = \frac{1}{2} c v_c$. Indeed,

$$L_2 L_{-2} v_c = [L_2, L_{-2}] v_c = \left(4L_0 + \frac{1}{12}(8-2)C\right) v_c = \frac{1}{2} c v_c.$$

The resulting formula coincides with formula (3.4.2), as expected.

The reader should check that by applying formula (3.3.12) to the above OPE one obtains precisely the commutation relations (2.5.2) of the Virasoro algebra.

3.4.4. Characterization of conformal vertex algebras. Corollary 3.3.5 allows us to obtain the following characterization of conformal vertex algebras (see Definition 2.5.8).

3.4.5. Lemma. *A $\mathbb{Z}$-graded vertex algebra V is conformal, of central charge c, if and only if it contains a non-zero vector $\omega \in V_2$ such that the Fourier coefficients L_n^V of the corresponding vertex operator*

$$Y(\omega, z) = \sum_{n \in \mathbb{Z}} L_n^V z^{-n-2}$$

satisfy $L_{-1}^V = T$ and $L_2^V \cdot \omega = \frac{c}{2}|0\rangle$.

Moreover, in that case there is a unique vertex algebra homomorphism $\mathrm{Vir}_c \to V$ such that $v_c \mapsto |0\rangle$ and $L_{-2} v_c \mapsto \omega$.

3.4.6. Proof. If V is a conformal vertex algebra, then the above conditions are clearly satisfied. So it remains to prove the converse statement.

By Definition 2.5.8, in a conformal vertex algebra we have $L_{-1}^V = T$, and L_0^V is the gradation operator. Using these facts and the associativity property, from Theorem 3.2.1, we obtain

$$Y(\omega, z) Y(\omega, w) = \sum_{n \in \mathbb{Z}} \frac{Y(L_n^V \omega, w)}{(z-w)^{n+2}}$$

(3.4.3)
$$= \frac{\frac{c}{2} Y(|0\rangle, w)}{(z-w)^4} + \frac{Y(L_1^V \omega, w)}{(z-w)^3} + \frac{2Y(\omega, w)}{(z-w)^2} + \frac{Y(T\omega, w)}{z-w} + \mathrm{reg}\,.$$

By Corollary 3.1.6, we have $Y(T\omega, z) = \partial_z Y(\omega, z)$. Thus, the above expression is the same as the operator product expansion for the Virasoro field $T(z)$ that was found in § 3.4.3, except for the term $Y(L_1^V \omega, w)/(z-w)^3$.

Suppose that $Y(L_1^V \omega, w) \neq 0$. Then in $Y(\omega, z) Y(\omega, w)$ we will have the term $Y(L_1^V \omega, w)/(z-w)^3$. Now let us look at the product in the opposite order, $Y(\omega, w) Y(\omega, z)$. On the one hand, we can compute $Y(\omega, w) Y(\omega, z)$ by simply exchanging z and w in the above formula. On the other hand, by locality, $Y(\omega, w) Y(\omega, z)$ and $Y(\omega, z) Y(\omega, w)$ should both be the expansions of the same rational function. But switching z and w we obtain the term $-Y(L_1^V \omega, z)/(z-w)^3$, and it is clear that because of the minus sign the locality property cannot be satisfied. Therefore we must have $Y(L_1^V \omega, w) = 0$, and so $L_1^V \omega = 0$ as desired.

Now we recover the commutation relations between the operators L_n^V from the singular part of the right hand side of formula (3.4.3) using formula (3.3.12). Then we find that the operators $L_n^V, n \in \mathbb{Z}$, indeed generate an action of the Virasoro algebra with central charge c on V.

Finally, define a linear map $\mathrm{Vir}_c \to V$ by sending each PBW monomial basis vector (2.5.3) to the vector $L^V_{j_1} \ldots L^V_{j_m} |0\rangle \in V$ (in particular, $L_{-2}v_c \mapsto \omega$). In order to show that this map is a homomorphism of vertex algebras, we need to check that

$$Y(L^V_{j_1} \ldots L^V_{j_m} |0\rangle, z) = \frac{1}{(-j_1 - 2)!} \cdots \frac{1}{(-j_m - 2)!} :\partial_z^{-j_1-2} Y(\omega, z) \ldots \partial_z^{-j_m-2} Y(\omega, z):.$$

But this follows immediately from Corollary 3.3.5. Now observe that if μ is any homomorphism of vertex algebras $\mathrm{Vir}_c \to V$ such that $\mu(v_c) = |0\rangle, \mu(L_{-2}v_c) = \omega$, then it is necessarily a homomorphism of modules over the Virasoro algebra. But then $\mu(L_{j_1} \ldots L_{j_m} v_c) = L^V_{j_1} \ldots L^V_{j_m} |0\rangle$, so μ is uniquely determined by these properties. This completes the proof.

3.4.7. Remark. Let V be a conformal vertex algebra. According to the lemma, there exists a non-trivial vertex algebra homomorphism $\mathrm{Vir}_c \to V$ (in the sense that the image is not spanned by $|0\rangle \in V$). The image of this homomorphism is a vertex subalgebra of V which is isomorphic to a quotient of Vir_c. It is known that Vir_c is irreducible for c not equal to one of the values $c(p, q)$ given by formula (4.4.3) below. Hence if $c \neq c(p, q)$, then V contains Vir_c as a vertex subalgebra. Otherwise, V contains either $\mathrm{Vir}_{c(p,q)}$ or its quotient.

3.4.8. Back to Segal-Sugawara. Recall the Segal-Sugawara vector S given by formula (2.5.5). We want to show that for $k \neq -h^\vee$ the vector $\omega = \frac{1}{k+h^\vee} S$ is a conformal vector in $V_k(\mathfrak{g})$. In particular, this would imply that the Fourier coefficients of the field $Y(\omega, z) = \frac{1}{k+h^\vee} Y(S, z)$ satisfy the commutation relations of the Virasoro algebra. According to Lemma 3.4.5, it suffices to show that the Fourier coefficients of the field $\frac{1}{k+h^\vee} Y(S, z)$ (which we simply denote by L_n) satisfy $L_{-1} = T$, L_0 is the gradation operator, and $L_2 \cdot \omega = \frac{c}{2}|0\rangle$.

By definition of the gradation operator and the operator T on $V_k(\mathfrak{g})$ (see § 2.4.4), the first two conditions are equivalent to the conditions $S_{-1} \cdot v_k = 0$ and $S_0 \cdot v_k = 0$, which obviously hold because $Y(S, z)$ has conformal dimension two, and the commutation relations (2.5.7):

$$(3.4.4) \qquad \left[\frac{1}{k+h^\vee} S_n, J^b_m \right] = -m J^b_{n+m}$$

for $n = 0, 1$. To verify these commutation relations (for all n), we compute the OPE $S(z)J^b(w)$, where $S(z) \overset{\text{def}}{=} Y(S, z) = \sum_{n \in \mathbb{Z}} S_n z^{-n-2}$. For degree reasons, $S_n \cdot J^b_{-1} v_k = 0$ for all $n > 1$. Further,

$$S_1 \cdot J^b_{-1} v_k = \frac{1}{2} \sum_a J^a_0 J_{a,1} \cdot J^b_{-1} v_k + \frac{1}{2} \sum_a J^a_1 J_{a,0} \cdot J^b_{-1} v_k.$$

Each summand in the first term gives 0, and the second term gives

$$\frac{k}{2} \sum_a (J^a, [J_a, J^b]) v_k = 0$$

due to the invariance of the bilinear form. It remains to calculate $S_0 \cdot J^b_{-1} v_k$ and $S_{-1} \cdot J^b_{-1} v_k$. We find that

$$(3.4.5) \qquad S_0 \cdot J^b_{-1} v_k = \sum_a \left(\frac{1}{2} J^a_0 J_{a,0} + J^a_{-1} J_{a,1} \right) \cdot J^b_{-1} v_k.$$

The first term of (3.4.5) equals $\frac{1}{2}cas(\mathfrak{g})J^b_{-1}v_k$, where $cas(\mathfrak{g})$ is the eigenvalue of the Casimir operator $\sum_a J^a J_a$ on $\mathfrak{g}$. Recall that $\{J^a\}$ and $\{J_a\}$ are dual bases with respect to the bilinear form which equals $\frac{1}{2h^\vee}$ of the Killing form (see § 2.4.1). If $\{\widetilde{J}^a\}$ and $\{\widetilde{J}_a\}$ are dual bases with respect to the Killing form, then $\sum_a J^a J_a = 2h^\vee \sum_a \widetilde{J}^a \widetilde{J}_a$. It follows from the definition of the Killing form that $\sum_a \widetilde{J}^a \widetilde{J}_a$ acts on $\mathfrak{g}$ as the identity operator. Therefore $\frac{1}{2}cas(\mathfrak{g}) = \frac{1}{2}2h^\vee = h^\vee$.

The second term of (3.4.5) equals $\sum_a kJ^a_{-1}(J_a, J^b)v_k = kJ^b_{-1}v_k$. Thus, we obtain

$$S_0 \cdot J^b_{-1}v_k = (k + h^\vee)J^b_{-1}v_k.$$

A similar calculation also gives

$$S_{-1} \cdot J^b_{-1}v_k = (k + h^\vee)J^b_{-2}v_k.$$

Summarizing, we obtain the OPE

$$S(z)J^a(w) = (k + h^\vee)\left(\frac{J^a(w)}{(z-w)^2} + \frac{\partial_w J^a(w)}{z-w}\right) + \text{reg}.$$

Thus, we find that if $k = -h^\vee$, the OPE between $S(z)$ and $J^a(w)$ is regular, and so

$$[S_n, J^a_m] = 0, \qquad n, m \in \mathbb{Z}.$$

This implies that the S_n's are central elements of the completed universal enveloping algebra $\widetilde{U}_{-h^\vee}(\mathfrak{g})$ of $\widehat{\mathfrak{g}}$ at $k = -h^\vee$ (see § 4.2 for the precise definition). When $k \neq -h^\vee$, we have

$$(3.4.6) \qquad \frac{1}{k+h^\vee}S(z)J^a(w) = \frac{J^a(w)}{(z-w)^2} + \frac{\partial_w J^a(w)}{z-w} + \text{reg}.$$

Using formula (3.3.12), we derive from this OPE the commutation relations (3.4.4).

To complete the proof that $\frac{1}{k+h^\vee}S$ is a conformal vector with central charge $c(k) = k\dim\mathfrak{g}/(k+h^\vee)$, we need to compute $S_2 \cdot S$. Using the commutation relations (3.4.4), we find that

$$S_2 \cdot S = S_2 \cdot \frac{1}{2}\sum_a J_{a,-1}J^a_{-1}v_k = \frac{k+h^\vee}{2}\sum_a J_{a,1}J^a_{-1}v_k = \frac{k\dim\mathfrak{g}(k+h^\vee)}{2}v_k.$$

Therefore $L_2 \cdot \omega = \frac{c(k)}{2}v_k$, as desired.

Thus, we have proved that if $k \neq -h^\vee$, then $\omega = \frac{1}{k+h^\vee}S$ is a conformal vector. In particular, this implies that the Fourier coefficients of the field $\frac{1}{k+h^\vee}S(z)$ acting on $V_k(\mathfrak{g})$ satisfy the commutation relations of the Virasoro algebra. But actually the action of these operators is well-defined on any $\widehat{\mathfrak{g}}$–module M of level $k \neq -h^\vee$ satisfying the following condition: for any $x \in M$ there exists $N \in \mathbb{Z}$ such that

$$J^a_n \cdot x = 0, \qquad n \geq N.$$

Such modules will be called *smooth* in § 5.1.5. It follows from Proposition 4.2.2 proved below that the coefficients of the field $\frac{1}{k+h^\vee}S(z)$ acting on any smooth $\widehat{\mathfrak{g}}$–module satisfy the Virasoro algebra relations. Thus, any smooth $\widehat{\mathfrak{g}}$–module of level $k \neq -h^\vee$ automatically carries an action of the Virasoro algebra.

Moreover, the commutation relations (3.4.4) show that the actions of $\widehat{\mathfrak{g}}$ and Vir on $V_k(\mathfrak{g})$ combine into the action of the semi-direct product $Vir \ltimes \widehat{\mathfrak{g}}$ (here Vir acts on $\widehat{\mathfrak{g}}$ through $\mathrm{Der}\,\mathbb{C}((t))$) by infinitesimal automorphisms). Proposition 4.2.2 shows that the same is true for any smooth $\widehat{\mathfrak{g}}$–module of level $k \neq -h^\vee$.

3.4.9. In the same way as above we prove that $\omega_\lambda = \frac{1}{2}b_{-1}^2 + \lambda b_{-2}$ is a conformal vector with central charge $1 - 12\lambda^2$ in the Heisenberg vertex algebra π. Indeed, according to the discussion in § 2.5.9, we already know that the Fourier coefficients L_{-1} and L_0 of the field $Y(\omega_\lambda, z)$ are equal to the operator T and the gradation operator, respectively. Thus, it remains to compute $L_2 \cdot \omega_\lambda$. We find that

$$L_2 \cdot \omega_\lambda = \left(\frac{1}{2}b_1^2 + \sum_{n>1} b_{-n+2}b_n - 3\lambda b_2 \right) \omega_\lambda$$

$$= \frac{1}{2}b_1^2 \cdot \frac{1}{2}b_{-1}^2 - 3\lambda b_2 \cdot \lambda b_{-2} = \frac{1}{2}(1 - 12\lambda^2),$$

as desired.

3.5. Bibliographical notes

The original reference to P. Goddard's uniqueness Theorem is [**Go**]. The proofs in § 3.1 are borrowed from [**Kac3**].

Associativity was originally derived from locality in [**FHL, DL, Li1**]. Proposition 3.3.1 is due to V. Kac [**Kac3**]. The derivation of the commutation relations (3.3.12) using contour integrals in § 3.3.10 is taken from [**FLM**], Appendix A (see also the earlier work [**Fr**] of I. Frenkel).

CHAPTER 4

Applications of the Operator Product Expansion

In this chapter we consider applications of the associativity property and the operator product expansion. We use the formula for the commutation relations of the Fourier coefficients of vertex operators derived in the previous chapter to attach to any vertex algebra V a Lie algebra denoted by $U(V)$ and an associative algebra denoted by $\widetilde{U}(V)$. In the case when $V = V_k(\mathfrak{g})$, the Kac-Moody vertex algebra, the algebra $\widetilde{U}(V_k(\mathfrak{g}))$ is isomorphic to a completion of the universal enveloping algebra of $\widehat{\mathfrak{g}}$ with the central element identified with $k \in \mathbb{C}$. The functors sending V to $U(V)$ and $\widetilde{U}(V)$ allow us to relate vertex algebras to more familiar objects, Lie algebras and associative algebras (though they are much larger and more difficult to study than the vertex algebras themselves). We will show in the next chapter that modules over a vertex algebra V are the same as (smooth) modules over the associative algebra $\widetilde{U}(V)$.

At the end of the chapter we present two other applications of the associativity property. Namely, we strengthen the Reconstruction Theorem 2.3.11 by dropping the assumption that the PBW type monomials form a basis in the vertex algebra. We also introduce the "n–point functions" which enable us to give a "functional realization" of the dual space to any vertex algebra. This realization is a first step towards a rigorous mathematical formulation of chiral correlation functions of conformal field theory (see § 10.3).

4.1. A Lie algebra attached to a vertex algebra

According to Proposition 3.3.7, the span in End V of all Fourier coefficients of all vertex operators of V is a Lie algebra. In fact, more is true: formula (3.3.12) defines a Lie algebra structure on the "abstract" vector space $U'(V)$ spanned by the Fourier coefficients $A_{(n)}, A \in V, n \in \mathbb{Z}$, of vertex operators, subject to the relations $(TA)_{(n)} = -nA_{(n-1)}$. Thus, one can attach a Lie algebra to each vertex algebra, in a canonical way. This Lie algebra will play an important role in our study of vertex algebras. For example, we will show in Proposition 5.1.6 that V–modules are the same as $U'(V)$–modules satisfying some natural conditions. In Chapters 6 and 7 we will use the action of various Lie subalgebras of a completion $U(V)$ of $U'(V)$ in order to attach to V some geometric objects on algebraic curves. In Chapters 9 and 10 we will employ these structures to define the spaces of coinvariants and conformal blocks. In Chapter 19 we will explain a coordinate-independent meaning of the Lie algebra $U(V)$.

4.1.1. The Lie algebras $U'(V)$ and $U(V)$. Consider the linear operator

$$\partial = T \otimes \mathrm{Id} + \mathrm{Id} \otimes \partial_t$$

on the vector space $V \otimes \mathbb{C}[t, t^{-1}]$. Set

$$U'(V) = (V \otimes \mathbb{C}[t, t^{-1}]) / \operatorname{Im} \partial.$$

Denote the projection of the element $A \otimes t^n \in V \otimes \mathbb{C}[t, t^{-1}]$ onto $U'(V)$ by $A_{[n]}$. Then $U'(V)$ is spanned by $A_{[n]}, A \in V, n \in \mathbb{Z}$, subject to the relations

(4.1.1) $(TA)_{[n]} = -nA_{[n-1]}.$

Since the relation $(TA)_{(n)} = -nA_{(n-1)}$ holds in $\operatorname{End} V$ by Corollary 3.1.6, we have a well-defined linear map $U'(V) \to \operatorname{End} V$, sending $A_{[n]}$ to $A_{(n)}$.

We define a bilinear operation

$$
\begin{aligned}
(V \otimes \mathbb{C}[t, t^{-1}])^{\otimes 2} &\rightarrow \quad U'(V), \\
(A \otimes t^m) \otimes (B \otimes t^k) &\mapsto \quad \sum_{n \geq 0} \binom{m}{n} (A_{(n)} \cdot B)_{[m+k-n]}.
\end{aligned}
$$

Using the identity

$$[T, A_{(n)}] = -nA_{(n-1)}$$

in $\operatorname{End} V$, which follows from the translation axiom, one checks easily that this map factors through $U'(V)^{\otimes 2}$. Hence we obtain a well-defined linear map $[,] : U'(V)^{\otimes 2} \to U'(V)$ given by

(4.1.2) $[A_{[m]}, B_{[k]}] = \sum_{n \geq 0} \binom{m}{n} (A_{(n)} \cdot B)_{[m+k-n]}.$

If V is $\mathbb{Z}$–graded, then we introduce a $\mathbb{Z}$–gradation on $U'(V)$, by setting, for each homogeneous element A of V, $\deg A_{[n]} = -n + \deg A - 1$. The map (4.1.2) preserves this gradation.

We also consider a completion $U(V)$ of $U'(V)$ with respect to the natural topology on $\mathbb{C}[t, t^{-1}]$:

$$U(V) = (V \otimes \mathbb{C}((t))) / \operatorname{Im} \partial.$$

It is spanned by linear combinations $\sum_{n \geq N} f_n A_{[n]}$, $f_n \in \mathbb{C}, A \in V, N \in \mathbb{Z}$, modulo the relations which follow from the identity (4.1.1).

We have a linear map $U(V) \to \operatorname{End} V$,

$$\sum_{n \geq N} f_n A_{[n]} \mapsto \int Y(A, z) f(z) dz,$$

where $f(z) = \sum_{n \geq N} f_n z^n \in \mathbb{C}((z))$. The map $U'(V)^{\otimes 2} \to U'(V)$ given by formula (4.1.2) is clearly continuous. Hence it gives rise to a map $U(V)^{\otimes 2} \to U(V)$.

4.1.2. Theorem. *The bracket (4.1.2) defines Lie algebra structures on $U'(V)$ and on $U(V)$. Furthermore, the natural maps $U'(V) \to \operatorname{End} V$ and $U(V) \to \operatorname{End} V$ are Lie algebra homomorphisms.*

4.1.3. Proof. Let $U(V)_0$ be the quotient $V / \operatorname{Im} T$. Denote the image of $A \in V$ in $U(V)_0$ by $A_{[0]}$. Define a linear map $U(V)_0^{\otimes 2} \to U(V)_0$ by the formula

$$A_{[0]} \otimes B_{[0]} \mapsto (A_{(0)} \cdot B)_{[0]}.$$

We claim that this map defines a Lie algebra structure on $U(V)_0$. Indeed, the skew-symmetry property of this map follows from the identity

$$A_{(0)} \cdot B = -B_{(0)} \cdot A \quad \operatorname{mod} T,$$

which is obtained from the z^{-1} coefficient of Proposition 3.2.5. The Jacobi identity follows from the identity

$$[A_{(0)}, B_{(0)}] = (A_{(0)} \cdot B)_{(0)},$$

which is a special case of formula (3.3.12).

Now observe that the commutative algebra $\mathbb{C}[t, t^{-1}]$ carries a natural $\mathbb{Z}$–gradation such that $\deg t = -1$, and a derivation ∂_t of degree 1. Therefore, we may define an operation Y on $\mathbb{C}[t, t^{-1}]$ following the procedure of § 1.4. This makes $\mathbb{C}[t, t^{-1}]$ into a $\mathbb{Z}$–graded commutative vertex algebra. Hence the tensor product $V \otimes \mathbb{C}[t, t^{-1}]$ also carries a vertex algebra structure that is $\mathbb{Z}$–graded if V is $\mathbb{Z}$–graded.

By construction, we have $U'(V) = U(V \otimes \mathbb{C}[t, t^{-1}])_0$. Therefore $U'(V)$ is a Lie algebra.

Define a topology on $U'(V)$ by taking as the basis of open neighborhoods of 0 the subspaces $U(V)_M, M > 0$, spanned by $\sum_{n \geq M} f_n A_{[n]}, f_n \in \mathbb{C}, A \in V$. It is clear from formula (4.1.2) that the Lie bracket on $\bar{U}(V)$ is continuous in this topology. Clearly, $U(V)$ is the completion of $U'(V)$ with respect to this topology and the bracket on $U(V)$ is induced from that on $U'(V)$ by continuity. Hence it defines the structure of a complete topological Lie algebra on $U(V)$.

This completes the proof.

It follows from the construction that the assignments $V \mapsto U'(V)$ and $V \mapsto U(V)$ define functors from the category of vertex algebras to the categories of Lie algebras and complete topological Lie algebras, respectively.

4.1.4. Remark. The homomorphisms $U'(V) \to \operatorname{End} V$ and $U(V) \to \operatorname{End} V$ are usually injective, but not always so. For example, suppose that V contains a vector B different from $|0\rangle$ such that $A_{(m)} B = 0$ for all $A \in V, m \geq 0$ (such elements are called central; see § 5.7.2 below for the definition of the center of a vertex algebra). For instance, the Segal-Sugawara vector S defined by formula (2.5.5) is such an element when $V = V_{-h^\vee}(\mathfrak{g})$ (see § 3.4.8). Then it follows from formula (4.1.2) that $B_{(n)}, n \in \mathbb{Z}$, are central elements of the Lie algebra $U'(V)$ for all $n \in \mathbb{Z}$.

If $TA \neq 0$, then $A_{[n]}, n \geq 0$, is a non-zero element of $U'(V)$. But $A_{(n)}, n \geq 0$, acts by 0 on $|0\rangle$ and hence on the entire V, and so $A_{[n]} \in U'(V), n \geq 0$, maps to 0 in $\operatorname{End} V$. Hence Theorem 4.1.2 is stronger than Proposition 3.3.7.

4.2. $U(V)$ and a completion of the universal enveloping algebra

In the case when V is generated by a Lie algebra $\mathcal{G}$, such as the Heisenberg, Kac–Moody, or Virasoro algebra, the Lie algebra $U(V)$ embeds into a completion of the universal enveloping algebra of $\mathcal{G}$. For the sake of definiteness, in this section we consider the case of the vertex algebra $V_k(\mathfrak{g})$ associated to the affine Kac–Moody algebra $\widehat{\mathfrak{g}}$ (other Lie algebras may be treated in exactly the same manner). We define the completed enveloping algebra $\widetilde{U}_k(\widehat{\mathfrak{g}})$ of $\widehat{\mathfrak{g}}$ of level k and show that the natural map $U(V_k(\mathfrak{g})) \to \widetilde{U}_k(\widehat{\mathfrak{g}})$ is a Lie algebra homomorphism.

We then define an analogue of $\widetilde{U}_k(\widehat{\mathfrak{g}})$ for an arbitrary vertex algebra V. This is a complete topological associative algebra denoted by $\widetilde{U}(V)$ (so that $\widetilde{U}(V_k(\mathfrak{g})) \simeq \widetilde{U}_k(\widehat{\mathfrak{g}})$). This algebra is important: we will see in Chapter 5 that a module over a vertex algebra V is the same as a (smooth) module over $\widetilde{U}(V)$.

4.2.1. Definition of the completion. Let $U_k(\widehat{\mathfrak{g}})$ be the quotient of the universal enveloping algebra $U(\widehat{\mathfrak{g}})$ by the two-sided ideal generated by $(K - k)$, where $k \in \mathbb{C}$. Define a topology on $U_k(\widehat{\mathfrak{g}})$ in which a basis of open neighborhoods of 0 is formed by the left ideals generated by $\mathfrak{g} \otimes t^N \mathbb{C}[[t]], N \geq 0$. It is straightforward to check that the operation of multiplication is continuous in this topology, and so the completion

$$\widetilde{U}_k(\mathfrak{g}) = \varprojlim U_k(\mathfrak{g})/(U_k(\mathfrak{g}) \cdot \mathfrak{g} \otimes t^N \mathbb{C}[[t]])$$

is a complete topological algebra.

By the definition of the vertex operation Y on $V_k(\mathfrak{g})$ given in § 2.4.4, all Fourier coefficients of vertex operators from $V_k(\mathfrak{g})$ may be expressed as (possibly infinite) series in terms of the basis elements A_n of $\widehat{\mathfrak{g}}$. Moreover, one checks by induction that each of these series becomes finite modulo the left ideal $U_k(\mathfrak{g}) \cdot \mathfrak{g} \otimes t^N \mathbb{C}[[t]]$, for all $N \geq 0$. Therefore we obtain a natural linear map $U(V_k(\mathfrak{g})) \to \widetilde{U}_k(\widehat{\mathfrak{g}})$. Consider $\widetilde{U}_k(\widehat{\mathfrak{g}})$ as a Lie algebra with the Lie bracket given by the formula $[A, B] = AB - BA$. We wish to prove that the map $U(V_k(\mathfrak{g})) \to \widetilde{U}_k(\widehat{\mathfrak{g}})$ is a Lie algebra homomorphism.

Note that we have a natural homomorphism of Lie algebras $\widetilde{U}_k(\widehat{\mathfrak{g}}) \to \mathrm{End}\, V_k(\mathfrak{g})$ obtained from the action of $\widetilde{U}_k(\widehat{\mathfrak{g}})$ on $V_k(\mathfrak{g})$. The composition

$$U(V_k(\mathfrak{g})) \to \widetilde{U}_k(\widehat{\mathfrak{g}}) \to \mathrm{End}\, V_k(\mathfrak{g})$$

defined by the formula

$$\sum_n f_n A_{[n]} \mapsto \sum_n f_n A_{(n)}$$

is also a Lie algebra homomorphism as we saw in Theorem 4.1.2. Hence the statement of the proposition would follow if we showed that this composition $U(V_k(\mathfrak{g})) \to \mathrm{End}\, V_k(\mathfrak{g})$ is injective. But this is false for $k = -h^\vee$, because $U(V_k(\mathfrak{g}))$ contains central elements, such as the Segal-Sugawara operators S_n (see Remark 4.1.4); those of them that annihilate the vacuum vector act on $V_k(\mathfrak{g})$ identically by zero. But since the Lie bracket in $U(V_k(\mathfrak{g}))$ is polynomial in k, we would still obtain the desired statement if we could prove that the map $U(V_k(\mathfrak{g})) \to \mathrm{End}\, V_k(\mathfrak{g})$ is injective for generic k.

This suggests a possible way to prove that the above map is indeed a Lie algebra homomorphism. Here we give a different proof.

4.2.2. Proposition. *The map* $U(V_k(\mathfrak{g})) \to \widetilde{U}_k(\widehat{\mathfrak{g}})$ *is a Lie algebra homomorphism.*

4.2.3. Proof. Recall from Lemma 2.2.3,(3) that for any two fields $A(z)$ and $B(z)$ we have

(4.2.1) $:A(w)B(w): = \mathrm{Res}_{z=0}(\delta(z - w)_- A(z)B(w) + \delta(z - w)_+ B(w)A(z)),$

where $\delta(z-w)_-$ (resp., $-\delta(z-w)_+$) is the expansion of $(z-w)^{-1}$ in positive (resp., negative) powers of w/z. Let us define now for $m \geq 0$

$$A(w)_{(m)}B(w) \overset{\text{def}}{=} \mathrm{Res}_{z=0}\left((z - w)^m A(z)B(w) - (z - w)^m B(w)A(z)\right).$$

In what follows we will also consider the above formula for $m = -1$, where the right hand side will be understood to be the right hand side of formula (4.2.1), so that by definition $A(w)_{(-1)}B(w) = :A(w)B(w):$.

Note that we have the following identity:

$$(4.2.2) \qquad \mathrm{Res}_{z=0}(z-w)^n \frac{1}{m!} \partial_w^m \delta(z-w) = \delta_{n,m},$$

for all $n, m \geq 0$.

Therefore if $A(z)$ and $B(z)$ are two mutually local fields, so that we have

$$[A(z), B(w)] = \sum_{m \geq 0} \frac{1}{m!} C_m(w) \partial_w^m \delta(z-w)$$

by Lemma 1.1.4, then

$$C_m(w) = A(w)_{(m)} B(w), \qquad m \geq 0.$$

The following lemma is the key step in the proof of Proposition 4.2.2.

4.2.4. Lemma. *For any mutually local fields $A(z), B(z)$ and $C(z)$ and $m \geq 0$ we have*

$$(4.2.3) \quad A(w)_{(m)} {:} B(w) C(w) {:}$$

$$= \sum_{j=0}^{m} \binom{m}{j} (A(w)_{(j)} B(w))_{(m-1-j)} C(w) + {:} B(w)(A(w)_{(m)} C(w)) {:} \,.$$

4.2.5. Completion of the proof of Proposition 4.2.2. We need to show that for any $A, B \in V$ the commutator of $A_{[m]}$ and $B_{[k]}$, considered as elements of the completion $\widetilde{U}_k(\widehat{\mathfrak{g}})$, is given by formula (4.1.2). Let $\widetilde{Y}(A, z)$ and $\widetilde{Y}(B, z)$ be the corresponding vertex operators with coefficients viewed as elements of $\widetilde{U}_k(\widehat{\mathfrak{g}})$. Applying the Dong Lemma 2.3.4, we find that all the $\widetilde{U}_k(\widehat{\mathfrak{g}})$–valued formal power series of the form $\widetilde{Y}(A, z)$ are mutually local. Therefore, by Lemma 1.1.4, the commutator between them has the form

$$[\widetilde{Y}(A, z), \widetilde{Y}(B, w)] = \sum_{m \geq 0} \frac{1}{m!} C_m(w) \partial_w^m \delta(z-w),$$

and

$$C_j(w) = \widetilde{Y}(A, w)_{(m)} \widetilde{Y}(B, w), \qquad m \geq 0.$$

Thus, we need to show that for any $m \geq 0$ we have

$$\widetilde{Y}(A, w)_{(m)} \widetilde{Y}(B, w) = \widetilde{Y}(A_{(m)} B, w).$$

Without loss of generality we may assume that A and B are monomials in the generators $J_n^a, n < 0$, applied to the vacuum vector $v_k \in V_k(\mathfrak{g})$. We will prove this formula by induction on the degrees of these monomials, considered as polynomials in J_n^a. For the monomials of degree 1 this follows from the comparison of the commutation relations in $\widehat{\mathfrak{g}}$ and the OPE between the generating fields $J^a(z)$ and $J^b(w)$ (see § 3.4.2). Now suppose that we know this formula for all monomials A and C of degrees less than or equal to some fixed positive degrees, and let us prove it for the monomials A and $J_n^a C, n < 0$. Then we have

$$\widetilde{Y}(J_n^a C, z) = \frac{1}{(-n-1)!} {:} \partial_z^n J^a(z) \cdot \widetilde{Y}(C, z) {:} \,, \qquad n < 0,$$

and so we may apply formula (4.2.3).

We apply Lemma 4.2.4 in the case when $A(z)$ is $\widetilde{Y}(A, z)$, $B(z)$ is $J^a(z)$ and $C(z)$ is $\widetilde{Y}(B, z)$. Then using formula (4.2.3) and our inductive assumption, we find that

$$
\begin{aligned}
\widetilde{Y}(A, w)_{(m)}\widetilde{Y}(J_n^a C, w) &= \sum_{j=0}^{m} \binom{m}{j} (\widetilde{Y}(A, w)_{(j)}\widetilde{Y}(J_n^a v_k, w))_{(m-1-j)}\widetilde{Y}(B, w) \\
&+ \ :\widetilde{Y}(J_n^a v_k, w)(\widetilde{Y}(A, w)_{(m)}\widetilde{Y}(B, w)): \\
&= \sum_{j=0}^{m} \binom{m}{j} \widetilde{Y}((A_{(j)} \cdot J_n^a v_k)_{(m-1-j)} \cdot B, w) \\
&+ \ :\widetilde{Y}(J_n^a v_k, w)\widetilde{Y}(A_{(m)} \cdot B, w): \ .
\end{aligned}
$$

Note that $A_{(j)} \cdot J_n^a v_k$ may be written as a linear combination of monomials of degrees less than or equal to that of A, so our inductive assumption indeed applies. Now we use formula (3.3.12) to rewrite the last sum as

$$
\widetilde{Y}([A_{(m)}, J_n^a]B, w) + \widetilde{Y}(J_n^a A_{(m)} B, w) = \widetilde{Y}(A_{(m)} \cdot J_n^a B, w),
$$

as desired. This completes the proof of Proposition 4.2.2.

4.2.6. Proof of Lemma 4.2.4. We need to compute

$$
(4.2.4) \qquad A(w)_{(m)}:B(w)C(w): = \operatorname{Res}_{z=0}(z - w)^m[A(z), :B(w)C(w):].
$$

Using formula (4.2.1) for the normally ordered product and the commutator formula

$$
[A(z), B(w)] = \sum_{m \geq 0} \frac{1}{m!} A(w)_{(m)} B(w)\partial_w^m \delta(z - w),
$$

we rewrite the right hand side of (4.2.4) as the sum of two terms:

$$
\operatorname{Res}_{z=0}\operatorname{Res}_{x=0}(z - w)^m \left(\sum_{j \geq 0} B(x)A(w)_{(j)}C(w)\partial_w^j\delta(z - w)\delta(x - w)_- + \right.
$$

$$
\left. \sum_{j \geq 0} A(w)_{(j)}C(w)B(x)\partial_w^j\delta(z - w)\delta(x - w)_+ \right)
$$

and

$$
\operatorname{Res}_{z=0}\operatorname{Res}_{x=0}(z - w)^m \left(\sum_{j \geq 0} A(x)_{(j)}B(x)C(w)\partial_x^j\delta(z - x)\delta(x - w)_- + \right.
$$

$$
\left. \sum_{j \geq 0} C(w)A(x)_{(j)}B(x)\partial_x^j\delta(z - x)\delta(x - w)_+ \right).
$$

Using the identity (4.2.2), we find that the first term is equal to

$$
:B(w)(A(w)_{(m)}C(w)): \ .
$$

For the second term we use the formula (4.2.2), the binomial expansion

$$
(z - w)^m = \sum_{j=0}^{m} \binom{m}{j} (z - x)^j(x - w)^{m-j}
$$

and the identity $(x - w)^k \delta(x - w)_{\pm} = \mp (x - w)^{k-1}$ for all $k > 0$. This gives us the summation term in formula (4.2.3). This completes the proof.

4.2.7. Remark. The Lie algebra $U(V_k(\mathfrak{g}))$ was called the "local completion" of $U_k(\widehat{\mathfrak{g}})$ in [**FF7**]. To explain this terminology, we recall the notion of local functional. Consider the vector space $\mathbb{C}((t))$. A local functional on it is a functional $\mathbb{C}((t)) \to \mathbb{C}$ whose value on any $f \in \mathbb{C}((t))$ is a linear combination of expressions

$$\int P(f(t), f'(t), \ldots) \phi(t) dt,$$

where P is a polynomial, and $\phi(t) \in \mathbb{C}((t))$. The word "local" here expresses the fact that the value on $f(t)$ depends only on the coefficients of the expansion of $f(t)$ at $t = 0$. Now observe that elements of $\widetilde{U}_k(\widehat{\mathfrak{g}})$ which belong to $U(V_k(\mathfrak{g}))$ may be written as

$$\int :P(J^a(z), \partial_z J^a(z), \ldots): \phi(z) dz,$$

where P is a polynomial. Hence we refer to them as "local" elements of $\widetilde{U}_k(\widehat{\mathfrak{g}})$.

It can be shown that the homomorphism $U(V_k(\mathfrak{g})) \to \widetilde{U}_k(\widehat{\mathfrak{g}})$ is injective for all $k \in \mathbb{C}$, but we will not prove it here (see [**FF7**], Lemma 7).

4.3. An associative algebra attached to a vertex algebra

Note that the image of $U(V_k(\mathfrak{g}))$ in $\widetilde{U}_k(\widehat{\mathfrak{g}})$ is not closed under multiplication. For instance, $U(V_k(\mathfrak{g}))$ contains $\widehat{\mathfrak{g}}$ as a Lie subalgebra. It is spanned by $A_n = (A_{-1} v_k)_{[n]}, A \in \mathfrak{g}, n \in \mathbb{Z}$, and $(v_k)_{[-1]}$. But $U(V_k(\mathfrak{g}))$ does not contain products of elements of $\widehat{\mathfrak{g}}$, such as $A_n B_k$. In fact, all quadratic elements in $U(V_k(\mathfrak{g}))$ are infinite series.

However, we can obtain $\widetilde{U}_k(\widehat{\mathfrak{g}})$ as a completion of the universal enveloping algebra of $U(V_k(\mathfrak{g}))$ modulo certain natural relations. This construction may in fact be performed for any vertex algebra V.

4.3.1. Definition of $\widetilde{U}(V)$. Denote by $U(U(V))$ the universal enveloping algebra of the Lie algebra $U(V)$. Define its completion

$$\widetilde{U}(U(V)) = \varprojlim U(U(V))/I_N,$$

where I_N is the left ideal generated by $A_{[n]}, A \in V, n > N$. Let $\widetilde{U}(V)$ be the quotient of $\widetilde{U}(U(V))$ by the two-sided ideal generated by the Fourier coefficients of the series

$$Y[A_{(-1)}B, z] - :Y[A, z]Y[B, z]:, \qquad A, B \in V,$$

where we set

$$Y[A, z] = \sum_{n \in \mathbb{Z}} A_{[n]} z^{-n-1},$$

and the normal ordering is defined in the same way as for the vertex operators $Y(A, z)$.

Clearly, $\widetilde{U}(V)$ is a complete topological associative algebra (with the topology for which a basis of open neighborhoods of 0 is given by the completions of the left ideals $I_N, N > 0$). The assignment $V \mapsto \widetilde{U}(V)$ gives rise to a functor from the category of vertex algebras to the category of complete topological associative algebras.

The algebra $\widetilde{U}(V)$ may be viewed as a generalization of the completed enveloping algebra $\widetilde{U}_k(\widehat{\mathfrak{g}})$ because, as the following lemma shows, for $V = V_k(\mathfrak{g})$ we have $\widetilde{U}(V_k(\mathfrak{g})) \simeq \widetilde{U}_k(\widehat{\mathfrak{g}})$.

4.3.2. Lemma. *There is a natural isomorphism $\widetilde{U}(V_k(\mathfrak{g})) \simeq \widetilde{U}_k(\widehat{\mathfrak{g}})$ for all $k \in \mathbb{C}$.*

4.3.3. Proof. We need to define mutually inverse algebra homomorphisms between $\widetilde{U}(V_k(\mathfrak{g}))$ and $\widetilde{U}_k(\widehat{\mathfrak{g}})$. The Lie algebra homomorphism $U(V_k(\mathfrak{g})) \to \widetilde{U}_k(\widehat{\mathfrak{g}})$ of Proposition 4.2.2 gives rise to a homomorphism of the universal enveloping algebra of the Lie algebra $U(V_k(\mathfrak{g}))$ to $\widetilde{U}_k(\widehat{\mathfrak{g}})$. It is clear that under this homomorphism any element of the ideal I_N is mapped to the left ideal of $U(V_k(\mathfrak{g}))$ generated by $A_n, A \in \mathfrak{g}, n > M$, for sufficiently large M. Therefore this homomorphism extends to a homomorphism from the completion of the universal enveloping algebra of $U(V_k(\mathfrak{g}))$ to $\widetilde{U}_k(\widehat{\mathfrak{g}})$. But according to the definitions, this map sends the series $Y[A_{(-1)}B, z]$ precisely to the series $:Y[A, z]Y[B, z]:$ for all $A, B \in V_k(\mathfrak{g})$. Hence we obtain a homomorphism $\widetilde{U}(V_k(\mathfrak{g})) \to \widetilde{U}_k(\widehat{\mathfrak{g}})$.

To construct the inverse homomorphism, recall that $\widehat{\mathfrak{g}}$ is naturally a Lie subalgebra of $U(V_k(\mathfrak{g}))$. Hence we have a homomorphism from $U_k(\widehat{\mathfrak{g}})$ to the universal enveloping algebra of the Lie algebra $U(V_k(\mathfrak{g}))$. This homomorphism extends by continuity to a homomorphism from $\widetilde{U}_k(\widehat{\mathfrak{g}})$ to the completion of the universal enveloping algebra of $U(V_k(\mathfrak{g}))$, and hence to $\widetilde{U}(V_k(\mathfrak{g}))$. Now observe that this map is surjective, because each series $Y[A, z]$ is a linear combination of normally ordered products of the generating series $J^a(z) = \sum_{n \in \mathbb{Z}} J^n_a z^{-n-1}$ of the elements J^a_n of $\widehat{\mathfrak{g}}$. Furthermore, it is clear from the construction that the composition $\widetilde{U}_k(\widehat{\mathfrak{g}}) \to \widetilde{U}(V_k(\mathfrak{g})) \to \widetilde{U}_k(\widehat{\mathfrak{g}})$ is the identity. Therefore the two maps indeed define mutually inverse isomorphisms.

4.4. Strong reconstruction theorem

Recall the Reconstruction Theorem 2.3.11, which enabled us to construct a vertex algebra from a vector space V, a collection a^α of vectors in V, and the associated fields

$$a^\alpha(z) = \sum_{n \in \mathbb{Z}} a^\alpha_{(n)} z^{-n-1}.$$

We required in the statement of Theorem 2.3.11 that the lexicographically ordered monomials in the Fourier coefficients of these fields form a basis of V. Now we will relax this condition and prove a stronger version of the reconstruction theorem. The argument that allows us to do that is parallel to the argument used above in the proof of Goddard's uniqueness theorem.

4.4.1. Reconstruction Theorem. *Let V be a vector space equipped with the structures of § 2.3.10 satisfying the conditions (1)–(3) of § 2.3.10 and the condition*

(4′) V is spanned by the vectors

(4.4.1) $a^{\alpha_1}_{(j_1)} \ldots a^{\alpha_m}_{(j_m)} |0\rangle, \qquad j_i < 0.$

Then these structures together with the vertex operation given by formula (2.3.8) give rise to a well-defined vertex algebra structure on V. Moreover, this is the unique vertex algebra structure on V satisfying conditions (1)–(3), (4′) and such that $Y(a^\alpha, z) = a^\alpha(z)$.

4.4.2. Proof. From the proof of Theorem 2.3.11 we see that all fields

$$(4.4.2) \qquad \frac{1}{(-j_1 - 1)! \ldots (-j_m - 1)!} :\partial_z^{-j_1 - 1} a^{\alpha_1}(z) \ldots \partial_z^{-j_m - 1} a^{\alpha_m}(z):$$

are mutually local. By induction, it is easy to see from conditions (1) and (2) that if we apply such a field to $|0\rangle$ and set $z = 0$, we obtain the vector

$$a^{\alpha_1}_{(j_1)} \ldots a^{\alpha_m}_{(j_m)} |0\rangle.$$

Further, by our assumptions, the fields $Y(a^\alpha, z)$ satisfy condition (2) from § 2.3.10. Since $[T, \partial_z] = 0$, this implies that each field $Y(a^\alpha, z)$ satisfies

$$[T, \partial_z^n Y(a^\alpha, z)] = \partial_z^{n+1} Y(a^\alpha, z).$$

Hence we obtain by induction that each field $B(z)$ of the form (4.4.2) satisfies

$$[T, B(z)] = \partial_z B(z).$$

Since $T|0\rangle = 0$, we have

$$\partial_z B(z)|0\rangle = T B(z)|0\rangle.$$

Now let us choose a basis among the monomials (4.4.1) and define the vertex algebra structure on V by formula (2.3.8) extended to arbitrary vectors in V by linearity. Then we show, in exactly the same way as in the proof of Theorem 2.3.11, that the resulting structure satisfies the axioms of vertex algebra.

If we choose another basis among the monomials (4.4.1), we may in principle obtain a different vertex algebra structure. But the vertex operators in both vertex algebras will be mutually local by the Dong Lemma 2.3.4. Furthermore, by the above calculations, if $Y_1(A, z)$ and $Y_2(A, z)$ are the vertex operators assigned to $A \in V$ by the two vertex algebra structures, then

$$Y_1(A, z)|0\rangle = Y_2(A, z)|0\rangle = A,$$

and

$$\partial_z Y_1(A, z)|0\rangle = T Y_1(A, z)|0\rangle, \qquad \partial_z Y_2(A, z)|0\rangle = T Y_2(A, z)|0\rangle.$$

Therefore we obtain from the Goddard Uniqueness theorem (see the reformulation in Remark 3.1.5) that $Y_1(a, z) = Y_2(A, z)$. Hence the two vertex algebra structures coincide.

Finally, we have to show that this vertex algebra structure is unique. This amounts to showing that formula (2.3.8) is uniquely determined by the assignment $Y(a^\alpha, z) = a^\alpha(z)$. But this follows from Corollary 3.3.5. This completes the proof.

4.4.3. Application. Consider the Kac-Moody vertex algebra $V_k(\mathfrak{g})$ as a $\widehat{\mathfrak{g}}$-module. It is reducible when $k \in \mathbb{Z}_+$; namely, it contains a proper submodule I_k generated by the vector $(e^{\alpha_m}_{-1})^{k+1} v_k$, where α_{m} is the maximal root of $\mathfrak{g}$ (see § 11.2.3 for the definition of e^α). The quotient by this submodule is known to be irreducible. We denote it by $L_k(\mathfrak{g})$. If $k = 0$, then $L_k(\mathfrak{g})$ is the one-dimensional trivial representation of $\widehat{\mathfrak{g}}$.

So we consider the case when $k \in \mathbb{Z}_{>0}$. Then the lexicographically ordered monomials $J^{a_1}_{n_1} \ldots J^{a_m}_{n_m} v_k$ (see § 2.4.4) span $L_k(\mathfrak{g})$ from the vacuum, though they are no longer linearly independent. Nevertheless, the "generating" subspace $(\mathfrak{g} \otimes t^{-1}) v_k$ of $V_k(\mathfrak{g})$ maps isomorphically into $L_k(\mathfrak{g})$. According to Theorem 4.4.1, the definition of the vertex algebra structure on $V_k(\mathfrak{g})$ given in § 2.4.4 still makes sense for $L_k(\mathfrak{g})$, and thus we obtain a vertex algebra structure on $L_k(\mathfrak{g})$. It follows that I_k is an ideal of the vertex algebra $V_k(\mathfrak{g})$ (see Definition 1.3.4) and $L_k(\mathfrak{g})$ is the quotient of $V_k(\mathfrak{g})$ by this ideal. Moreover, since $L_k(\mathfrak{g})$ is irreducible as a $\widehat{\mathfrak{g}}$–module, $L_k(\mathfrak{g})$ is a simple vertex algebra, i.e., it does not contain proper non-zero vertex algebra ideals.

The fact that $L_k(\mathfrak{g})$ is a vertex algebra (equivalently, I_k is an ideal of $V_k(\mathfrak{g})$) has an interesting corollary: for any $A \in I_k$, all Fourier coefficients of the corresponding field $Y(A, z)$, defined in formula (2.4.5), act by zero on $L_k(\mathfrak{g})$. For example, since $(e^{\alpha_m}_{-1})^{k+1} v_k \in I_k$, we find that all Fourier coefficients of the field $(e^{\alpha_m}(z))^{k+1}$ act by zero on $L_k(\mathfrak{g})$ (try to prove this without using the vertex algebra formalism).

As another example, consider the Virasoro vertex algebra Vir_c. It follows from the results of [**Kac1**, **FFu**] that Vir_c is reducible as a module over the Virasoro algebra if and only if

$$(4.4.3) \qquad c = c(p, q) \stackrel{\mathrm{def}}{=} 1 - 6(p - q)^2/pq, \qquad p, q > 1, (p, q) = 1.$$

Let $L_{c(p,q)}$ be the irreducible quotient of $\mathrm{Vir}_{c(p,q)}$. In the same way as above, we show using Theorem 4.4.1 that $L_{c(p,q)}$ is a vertex algebra which is a simple quotient of $\mathrm{Vir}_{c(p,q)}$.

4.5. Correlation functions

In this section we discuss a "multi–point" generalization of the OPE formula (3.3.1) and the locality property.

Recall that V^* denotes the space of all linear functionals on V. Given a set $A_1, \ldots, A_n$ of elements of V, and $\varphi \in V^*, v \in V$, we consider the formal power series in $\mathbb{C}[[z_1^{\pm 1}, \ldots, z_n^{\pm 1}]]$,

$$(4.5.1) \qquad \langle \varphi, Y(A_1, z_1) \ldots Y(A_n, z_n) v \rangle,$$

which we call the *n–point functions*.

Since $v = Y(v, z)|0\rangle|_{z=0}$ by the vacuum axiom, we obtain

$$\langle \varphi, Y(A_1, z_1) \ldots Y(A_n, z_n) v \rangle = \langle \varphi, Y(A_1, z_1) \ldots Y(A_n, z_n) Y(v, z_{n+1})|0\rangle|_{z_{n+1}=0}.$$

Therefore it is sufficient to consider the n–point functions (4.5.1) with $v = |0\rangle$. The properties of these functions are summarized by the following:

4.5.1. Theorem. *Let $A_1, \ldots, A_n \in V$. For any $\varphi \in V^*$, and any permutation σ on n elements, the formal power series*

$$(4.5.2) \qquad \langle \varphi, Y(A_{\sigma(1)}, z_{\sigma(1)}) \ldots Y(A_{\sigma(n)}, z_{\sigma(n)})|0\rangle$$

is the expansion in $\mathbb{C}((z_{\sigma(1)})) \ldots ((z_{\sigma(n)}))$ of an element $f^\sigma_{A_1, \ldots, A_n}(z_1, \ldots, z_n)$ of

$$\mathbb{C}[[z_1, \ldots, z_n]][(z_i - z_j)^{-1}]_{i \neq j}.$$

Moreover, this element satisfies the following properties:

(1) $f^\sigma_{A_1, \ldots, A_n}(z_1, \ldots, z_n)$ *does not depend on σ (hence from now on we suppress the index σ).*

(2) *For all $i \neq j$ the expression*

$$f_{Y(A_i, z_i - z_j) A_j, A_1, \ldots, \widehat{A_i}, \ldots, \widehat{A_j}, \ldots, A_n}(z_j, z_1, \ldots, \widehat{z_i}, \ldots, \widehat{z_j}, \ldots, z_n),$$

considered as a formal Laurent power series in $(z_i - z_j)$ with coefficients in

$$\mathbb{C}[[z_1, \widehat{z_i}, \ldots, \widehat{z_j}, \ldots, z_n]][(z_k - z_l)^{-1}]_{k \neq l; k, l \neq i, j},$$

is the expansion of

$$f_{A_1,\ldots,A_n}(z_1,\ldots,z_n) \in \mathbb{C}[[z_1,\ldots,z_n]][(z_i - z_j)^{-1}]_{i \neq j}.$$

(3)

$$\partial_{z_i} f_{A_1,\ldots,A_n}(z_1,\ldots,z_n) = f_{A_1,\ldots,TA_i,\ldots,A_n}(z_1,\ldots,z_n).$$

4.5.2. Proof. Recall that $\langle \varphi, Y(A,z)v \rangle \in \mathbb{C}((z))$ for any $A, v \in V$ and $\varphi \in V^*$. Therefore by induction the series (4.5.2) belongs to $\mathbb{C}((z_{\sigma(1)}))\ldots((z_{\sigma(n)}))$. On the other hand, by the locality axiom (in the form of equation (1.2.3)), there exists an even integer $N_{ij} \in 2\mathbb{Z}_+$ such that

$$(z - w)^{N_{ij}}[Y(A_i, z), Y(A_j, w)] = 0.$$

Hence after multiplying the expression (4.5.2) by $\prod_{i<j}(z_{\sigma(i)} - z_{\sigma(j)})^{N_{ij}}$ we obtain a series which is independent of σ. Moreover, it contains only positive powers of all variables by the vacuum axiom: $Y(A, z)|0\rangle \in V[[z]]$. Thus, we obtain that

$$f^\sigma_{A_1,\ldots,A_n}(z_1,\ldots,z_n) \prod_{i<j}(z_{\sigma(i)} - z_{\sigma(j)})^{N_{ij}} \in \mathbb{C}[[z_1,\ldots,z_n]].$$

This equation has a unique solution in the field $\mathbb{C}((z_{\sigma(1)}))\ldots((z_{\sigma(n)}))$. This proves part (1). Part (2) follows from (1) and the associativity property from Theorem 3.2.1. Part (3) follows from the identity $Y(TA, z) = \partial_z Y(A, z)$ proved in Corollary 3.1.6.

4.5.3. n–point functions. To summarize, we attach to each $\varphi \in V^*$ a collection of (meromorphic) n–point functions

$$(4.5.3) \qquad \langle \varphi, Y(A_1, z_1)\ldots Y(A_n, z_n)|0\rangle$$

on $D^n = \operatorname{Spec} \mathbb{C}[[z_1,\ldots,z_n]]$ with poles only on the diagonals. By Theorem 4.5.1, these functions satisfy a symmetry condition, a "bootstrap" condition (which describes the behavior of the n–point function near the diagonals in terms of the $(n-1)$–point functions), and a "horizontality" condition (a system of first order differential equations).

Let $\mathcal{F}_n$ be the vector space of all collections

$$(4.5.4) \quad \{f_{A_1,\ldots,A_m}(z_1,\ldots,z_m) \in \mathbb{C}[[z_1,\ldots,z_m]][(z_i - z_j)^{-1}]_{i \neq j} |$$
$$1 \leq m \leq n, A_i \in V\}$$

satisfying conditions (1)–(3) of Theorem 4.5.1. Define linear maps $\kappa_n : V^* \to \mathcal{F}_n$ and $\mu_n : \mathcal{F}_n \to V^*$ as follows. The map κ_n sends $\varphi \in V^*$ to the collection of n–point functions (4.5.3). The map μ_n sends the collection (4.5.4) to the functional φ on V defined by the formula $\langle \varphi, A \rangle = f_A(0)$. This formula is motivated by the definition of the series $f(A) \in \mathbb{C}[[z]]$: $f(A) = \varphi, Y(A, z)|0\rangle$ and the vacuum axiom of vertex algebras: $Y(A, z)|0\rangle|_{z=0} = A$.

4.5.4. Theorem. *The maps κ_n and μ_n are mutually inverse isomorphisms of V^* and the space of n–point functions $\mathcal{F}_n$ for each $n \geq 1$.*

4.5.5. Proof. It follows immediately from the definitions that $\mu_n \circ \kappa_n = \mathrm{Id}$. Hence the theorem will follow if we show that μ_n is injective. Suppose we have a collection of functions (4.5.4) satisfying conditions (1)–(3) of Theorem 4.5.1 and such that $f_A(0) = 0, \forall A \in V$. Condition (3) then implies that $(\partial_z f_A(z))|_{z=0} = 0, \forall A \in V$. Hence we obtain that $f_A(z) \equiv 0, \forall A \in V$. According to condition (2), for any $A_1, A_2 \in V$, the function $f_{A_1,A_2}(z_1, z_2)$ can be written as a power series in $(z_1 - z_2)^{\pm 1}$ whose coefficients are of the form $f_A(z_2)$. Since the latter are all 0, so is $f_{A_1,A_2}(z_1, z_2)$. Continuing by induction, we obtain, using condition (2), that all functions in the collection (4.5.4) are equal to 0. This completes the proof.

4.5.6. Functional realization. The theorem gives us a "functional realization" of the dual space to any vertex algebra V in terms of n–point functions.

Suppose that V is generated by fields such that the singular terms in their OPEs are linear combinations of the same fields and their derivatives (in this case these fields span a vertex Lie algebra, and V is the enveloping vertex algebra of this vertex Lie algebra, as explained in § 16.1 below). For example, vertex algebras obtained from Lie algebras, such as Heisenberg, affine Kac-Moody and Virasoro, satisfy this property. Then we can simplify the above functional realization by considering only the n–point functions of the generating fields.

Consider for example the Heisenberg vertex algebra. Then the generating field $b(z)$ satisfies the above conditions. Hence for each $\varphi \in \pi^*$ we may restrict ourselves to the n–point functions

$$(4.5.5) \qquad \omega_n(z_1, \ldots, z_n) = \langle \varphi, b(z_1) \ldots b(z_n) |0\rangle.$$

By Theorem 4.5.1 and the OPE (2.3.2), these functions are symmetric in $z_1, \ldots, z_n$ and satisfy the bootstrap condition

$$(4.5.6) \qquad \omega_n(z_1, \ldots, z_n) = \frac{\omega_{n-2}(z_1, \ldots, \widehat{z_i}, \ldots, \widehat{z_j}, \ldots, z_n)}{(z_i - z_j)^2} + \text{regular}.$$

Let Ω_∞ be the vector space of infinite collections $(\omega_n)_{n \geq 0}$, where

$$\omega_n \in \mathbb{C}[[z_1, \ldots, z_n]][(z_i - z_j)^{-1}]_{i \neq j}$$

satisfy the above conditions. Using the functions (4.5.5), we obtain a map $\pi^* \to \Omega_\infty$. We will show in § 10.5 that this map is an isomorphism (see Proposition 10.5.3).

4.6. Bibliographical notes

The definition of the Lie algebras $U(V)_0$ and $U'(V)$ is due to R. Borcherds, and our proof of Theorem 4.1.2 follows [**B1**]. Lemma 4.2.4 is due to V. Kac [**Kac3**]. The definition of the associative algebra $\widetilde{U}(V)$ and Proposition 4.2.2 are due to E. Frenkel.

The Reconstruction Theorem 4.4.1 appeared first in [**FKRW**], Prop. 3.1 (see also [**Kac3**]). Theorem 4.5.1 is due to I. Frenkel, Y.-Z. Huang and J. Lepowsky [**FHL**].

CHAPTER 5

Modules over Vertex Algebras and More Examples

In this chapter we give the definition of a module over a vertex algebra and study the general properties of modules. In particular, we explain that a module over a vertex algebra V is the same as a (smooth) module over the associative algebra $\widetilde{U}(V)$ introduced in § 4.3.1.

We use the modules to construct new vertex algebras. The idea is to try and define a vertex algebra structure on the direct sum of a vertex algebra V and its module (so that the old vertex algebra V would be a vertex subalgebra of the new one). We demonstrate how to do this explicitly in the case when $V = \pi$, the Heisenberg vertex algebra. In this case, the vertex algebra structure on π may be extended to that on the direct sum of its Fock modules π_λ, where λ runs over the one-dimensional integral lattice $\sqrt{N}\mathbb{Z}$, N being a positive integer. This construction may be generalized by replacing $\sqrt{N}\mathbb{Z}$ by an arbitrary integral lattice. Along the way, we establish the boson-fermion correspondence, which is an isomorphism between the vertex superalgebra associated to the lattice $\mathbb{Z}$ and the free fermionic vertex superalgebra.

The example of a lattice vertex algebra is quite different from those that we discussed in Chapter 2 which may have created the false impression that all vertex algebras can be constructed from Lie algebras. Here we survey examples of vertex algebras which do not arise as induced representations of Lie algebras (another series of examples, the $\mathcal{W}$–algebras, will be considered in Chapter 15). In particular, we discuss the so-called rational vertex algebras. These vertex algebras have finitely many simple modules (up to isomorphism). They arise in physically meaningful models of conformal field theory, such as the Belavin–Polyakov–Zamolodchikov minimal models or the Wess–Zumino–Witten models. The lattice vertex algebras are also examples of rational vertex algebras. A beautiful example of a rational vertex algebra is the Moonshine Module vertex algebra constructed by I. Frenkel, J. Lepowsky and A. Meurman, which motivated Borcherds' definition of vertex algebras.

Next, we define twisted modules over vertex algebras equipped with an automorphism of finite order. The twisted modules may be used to construct ordinary modules over the invariant subalgebra of a vertex algebra with respect to a finite group of automorphisms. This is referred to as the orbifold construction. We also present two other constructions which enable one to construct new vertex algebras from the existing ones: the coset and the BRST constructions. An example of the former is the Goddard–Kent–Olive construction which enables one to obtain the unitary minimal Virasoro vertex algebra from the vertex algebras associated to integrable representations of $\widehat{\mathfrak{sl}}_2$.

5.1. Modules over vertex algebras

5.1.1. Definition. Let $(V, |0\rangle, T, Y)$ be a vertex algebra. A vector space M is called a V–*module* if it is equipped with an operation $Y_M : V \to \operatorname{End} M[[z^{\pm 1}]]$ which assigns to each $A \in V$ a field

$$Y_M(A, z) = \sum_{n \in \mathbb{Z}} A^M_{(n)} z^{-n-1}$$

on M subject to the following axioms:

- $Y_M(|0\rangle, z) = \operatorname{Id}_M$;
- for all $A, B \in V$, $C \in M$ the three expressions

$$
\begin{aligned}
Y_M(A, z) Y_M(B, w) C &\in M((z))((w)), \\
Y_M(B, w) Y_M(A, z) C &\in M((w))((z)), \text{ and} \\
Y_M(Y(A, z - w) B, w) C &\in M((w))((z - w))
\end{aligned}
$$

are the expansions, in their respective domains, of the same element of

$$M[[z, w]][z^{-1}, w^{-1}, (z - w)^{-1}].$$

If V is $\mathbb{Z}$–graded, then a V–module M is called *graded* if M is a $\mathbb{C}$–graded vector space and for $A \in V_m$ the field $Y_M(A, z)$ has conformal dimension m, i.e., the operator $A^M_{(n)}$ is homogeneous of degree $-n + m - 1$.

Note that by shifting a given gradation on M by any complex number, one obtains a new gradation on M.

For an example of a module which does not admit a gradation, see § 18.4.6.

These axioms imply that V is a module over itself. Furthermore, if $W \to V$ is a homomorphism of vertex algebras and M is a V–module, then M also acquires the structure of a W–module. Thus, we obtain a functor from the category of V–modules to the category of W–modules.

We have obvious notions of a submodule and quotient module. A module M whose only submodules are 0 and itself is called *simple* or *irreducible*.

5.1.2. Proposition.
(1) $Y_M(TA, z) = \partial_z Y_M(A, z)$.
(2) *All fields* $Y_M(A, z)$ *are mutually local.*

5.1.3. Proof. We apply the second axiom in the case when $B = |0\rangle$. Then for any $C \in M$ the expansion of the corresponding element of

$$M[[z, w]][z^{-1}, w^{-1}, (z - w)^{-1}]$$

in $M((w))((z - w))$ is on the one hand equal to

$$\sum_{n \geq 0} \frac{1}{n!} \partial_w Y_M(A, w)(z - w)^n,$$

and on the other hand to

$$\sum_{n \geq 0} Y_M(A_{(-n-1)}|0\rangle, w)(z - w)^n.$$

Part (1) of the proposition now follows by comparing the $(z - w)$ coefficients on both sides and recalling that $A_{(-2)}|0\rangle = TA$ (see Remark 1.3.2.3).

To prove part (2), let N be a positive integer such that
$$(z-w)^N Y(A, z-w)B \in V[[z-w]].$$
Then the second axiom implies that for any C the formal power series
$$(z-w)^N Y_M(A,z)Y_M(B,w)C \qquad \text{and} \qquad (z-w)^N Y_M(B,w)Y_M(A,z)C$$
are expansions of the same element of $M[[z,w]][z^{-1}, w^{-1}]$ in $M((z))((w))$ and $M((w))((z))$, respectively. Hence they must be equal, and so the fields $Y_M(A,z)$ and $Y_M(B,z)$ are mutually local.

5.1.4. Remark. In the second axiom of Definition 5.1.1 it is sufficient to require only that the two series
$$\begin{aligned} Y_M(A,z)Y_M(B,w)C &\in M((z))((w)), \text{ and} \\ Y_M(Y(A,z-w)B,w)C &\in M((w))((z-w)) \end{aligned}$$
are the expansions of the same element, denote it by $f_{A,B,C}$, of
$$M[[z,w]][z^{-1}, w^{-1}, (z-w)^{-1}].$$

Indeed, suppose this is so, and let us show that $Y_M(B,w)Y_M(A,z)C$ is equal to the expansion of $f_{A,B,C}$ in $M((w))((z))$. By the above assertion, we find that $Y_M(B,w)Y_M(A,z)C$ and $Y_M(Y(B,w-z)A,z)C$ are the expansion of an element
$$g_{A,B,C} \in M[[z,w]][z^{-1}, w^{-1}, (z-w)^{-1}],$$
in $M((w))((z))$ and $M((z))((z-w))$, respectively. Note that the embeddings
$$M((z))((z-w)) \hookleftarrow M[[z,w]][z^{-1}, w^{-1}, (z-w)^{-1}] \hookrightarrow M((w))((z-w))$$
are intertwined by a map (actually, an isomorphism)
$$M((z))((z-w)) \to M((w))((z-w))$$
sending $f(z, z-w)$ to
$$\sum_{n \geq 0} \frac{(z-w)^n}{n!} \partial_w^n f(w, z-w)$$
(here and below we consider ∂_w as a partial derivative). Hence the expansion of $g_{A,B,C}$ in $M((w))((z-w))$ is equal to
$$\sum_{n \geq 0} \frac{(z-w)^n}{n!} \partial_w^n Y_M(Y(B, w-z)A, w)C$$
$$= \sum_{n \geq 0} \frac{(z-w)^n}{n!} Y_M(T^n Y(B, w-z)A, w)C = Y_M(e^{(z-w)T} Y(B, w-z)A, w)C,$$
by Proposition 5.1.2,(1) (note that in the proof of Proposition 5.1.2,(1) we only used that $Y_M(A,z)Y_M(B,w)C$ and $Y_M(Y(A, z-w)B, w)C$ are the expansions of the same element). But by the skew-symmetry property, Proposition 3.2.5, we have
$$e^{(z-w)T} Y(B, w-z)A = Y(A, z-w)B.$$
Hence we find that the expansion of $g_{A,B,C}$ in $M((w))((z-w))$ is equal to
$$Y_M(Y(A, z-w)B, w)C,$$
and so $g_{A,B,C} = f_{A,B,C}$, as desired.

5.1.5. $\widetilde{U}(V)$**–modules and** V**–modules.** In § 4.1.1 we attached to each vertex algebra V a Lie algebra $U'(V)$. Recall that the Lie algebra $U'(V)$ is spanned by $A_{[n]}, A \in V, n \in \mathbb{Z}$, subject to the relations $(TA)_{[n]} = -nA_{[n-1]}$.

Recall the notation

$$Y[A, z] = \sum_{n \in \mathbb{Z}} A_{[n]} z^{-n-1} \in U'(V)[[z^{\pm 1}]]$$

introduced in § 4.3.1. These series satisfy the relations

(5.1.1) $$\partial_z Y[A, z] = Y[TA, z],$$

like the vertex operators $Y(A, z)$.

We call a $U'(V)$–module *smooth* if for any vector $v \in M$ and any $A \in V$ we have $A_{[n]}v = 0$ for large enough n. Then we define, in the same way as in Definition 2.2.2, the normally ordered product $:Y[A, z]Y[B, z]:$, which is a well-defined formal powers series in $z^{\pm 1}$ with coefficients in $\operatorname{End} M$.

Note that the action of $U'(V)$ on a smooth module extends uniquely to an action of the Lie algebra $U(V)$ which is the completion of $U'(V)$ introduced in § 4.1.1. It then also extends to an action of the universal enveloping algebra of $U(V)$ and its completion $\widetilde{U}(U(V))$ introduced in § 4.3.1. Recall that

$$\widetilde{U}(U(V)) = \varprojlim U(U(V))/I_N,$$

where I_N is the left ideal generated by $A_{[n]}, A \in V, n > N$.

We will say that a $U'(V)$–module M is *coherent*, if it is smooth and the following relations hold in $\operatorname{End} M[[z^{\pm 1}]]$: $Y[|0\rangle, z] = \operatorname{Id}_M$, and

(5.1.2) $$Y[A_{(-1)} \cdot B, z] = \,:Y[A, z]Y[B, z]: \,, \qquad \forall A, B \in V.$$

Such a module carries an action of the quotient of $\widetilde{U}(U(V))$ by the left ideal generated by the relations (5.1.2), which is precisely the associative algebra $\widetilde{U}(V)$ introduced in § 4.3.1. Let us call a $\widetilde{U}(V)$–module M smooth if for any vector $v \in M$ and any $A \in V$ we have $A_{[n]}v = 0$ for large enough n. It is clear from the definitions that the category of coherent $U'(V)$–modules is equivalent to the category of smooth $\widetilde{U}(V)$–modules.

5.1.6. Theorem. *There is an equivalence between the category of V–modules, the category of coherent $U'(V)$–modules and the category of smooth $\widetilde{U}(V)$–modules.*

5.1.7. Proof. Since the last two categories are obviously equivalent, it suffices to prove that the category of V–modules is equivalent to the category of coherent $U'(V)$–modules.

Suppose that M is a V–module. Define a linear map $U'(V) \to \operatorname{End} M$ by sending $A_{[n]}$ to $A^M_{(n)}$ (and hence $Y[A, z]$ to $Y_M(A, z)$). By Proposition 5.1.2,(1), $(TA)^M_{(n)} = -nA^M_{(n)}$, so this map is well-defined. By Proposition 5.1.2,(2), the fields $Y_M(A, z)$ are mutually local. Hence by Lemma 1.1.4,

$$[Y_M(A, z), Y_M(B, w)] = \sum_{j=0}^{N-1} \frac{1}{j!} \gamma_j(w) \partial_w^j \delta(z - w)$$

for some fields $\gamma_j(w)$ on M. Applying Proposition 3.3.1 and using the second axiom of Definition 5.1.1 in the same way as in § 3.3.6, we obtain that

$$(5.1.3) \qquad [Y_M(A,z), Y_M(B,w)] = \sum_{n\geq 0} \frac{1}{n!} Y_M(A_{(n)} \cdot B, w) \partial_w^n \delta(z-w),$$

and so

$$(5.1.4) \qquad Y_M(A,z)Y_M(B,w) = \sum_{n\geq 0} \frac{Y_M(A_{(n)} \cdot B, w)}{(z-w)^{n+1}} + :Y_M(A,z)Y_M(B,w): \, .$$

By definition of the Lie bracket on $U'(V)$ (see formula (4.1.2)), we have

$$(5.1.5) \qquad [Y[A,z], Y[B,w]] = \sum_{n\geq 0} \frac{1}{n!} Y[A_{(n)} \cdot B, w] \partial_w^n \delta(z-w).$$

Comparing formulas (5.1.3) and (5.1.5), we obtain that the map $U'(V) \to \operatorname{End} M$ is a Lie algebra homomorphism. Hence we obtain a $U'(V)$–module structure on M. By Definition 5.1.1, the series $Y_M(A,z)$ is a field on M (see § 1.2.1). Hence M is a smooth $U'(V)$–module. In addition, using the second axiom of Definition 5.1.1 and formula (5.1.4) in the same way as in § 3.3.4, we find that

$$Y_M(A_{(-1)} \cdot B, z) = :Y_M(A,z)Y_M(B,z): \, , \qquad \forall A, B \in V.$$

Therefore M is a coherent $U'(V)$–module.

Conversely, let M be a $U'(V)$–module. Define the structure of a V–module on M by setting $Y_M(A,z), A \in V$, to be equal to $Y[A,z]$, viewed as an element of $\operatorname{End} M[[z^{\pm 1}]]$. The first axiom of V–module follows from the above definition of coherent $U'(V)$–module. Applying Proposition 3.3.1 to the commutator (5.1.5), we obtain for any $C \in M$ the following identity in $M((z))((w))$:

$$Y[A,z]Y[B,w]C = \sum_{n\geq 0} \frac{Y[A_{(n)} \cdot B, w]}{(z-w)^{n+1}} C + :Y[A,z]Y[B,w]: C,$$

where $1/(z-w)$ is expanded in positive powers of w/z. Therefore $Y[A,z]Y[B,w]C$ is the expansion in $M((z))((w))$ of an element of $M[[z,w]][z^{-1}, w^{-1}, (z-w)^{-1}]$. Expanding this element in $M((w))((z-w))$ and using formula (5.1.1), we have

$$(5.1.6) \qquad \sum_{n\geq 0} \frac{Y[A_{(n)} \cdot B, w]}{(z-w)^{n+1}} C + \sum_{m\geq 0} \frac{1}{m!} :Y[T^m A, w]Y[B, w](z-w)^m: C.$$

But since M is a coherent $U'(V)$–module, we have the identity (5.1.2). This identity, together with formula (5.1.1), implies that

$$Y[A_{(-m-1)} \cdot B, z] = \frac{1}{m!} :Y[T^m A, w]Y[B, w]: \, , \qquad \forall A, B \in V, m \geq 0.$$

Therefore the series (5.1.6) is equal to $Y[Y(A, z-w)B, w]C$. Hence $Y[A,z]Y[B,w]C$ and $Y[Y(A, z-w)B, w]C$ are expansions of the same element of

$$M[[z,w]][z^{-1}, w^{-1}, (z-w)^{-1}].$$

By Remark 5.1.4, $Y[B,w]Y[A,z]C$ is then also automatically the expansion of that element. This shows that the second axiom of Definition 5.1.1 holds and completes the proof.

5.1.8. Special cases. In the case when the vertex algebra V is the induced representation of a Lie algebra, such as the Heisenberg, Kac–Moody or Virasoro algebra, the associative algebra $\widetilde{U}(V)$ is nothing but the completion of the universal enveloping algebra of the underlying Lie algebra. For example, in the case of the affine Kac-Moody algebra $\widehat{\mathfrak{g}}$ this is the algebra $\widetilde{U}_k(\widehat{\mathfrak{g}})$ defined in § 4.2.1 (see Lemma 4.3.2). It is clear that a smooth module over this algebra is the same as a smooth module over the underlying Lie algebra. For example, a smooth $\widehat{\mathfrak{g}}$–module is a $\widehat{\mathfrak{g}}$–module M such that for any $v \in M$, there exists $N \in \mathbb{Z}$ such that $J_n^a \cdot v = 0$ for all $n \geq N$. Thus, we obtain from Proposition 5.1.6 that a smooth $\widehat{\mathfrak{g}}$–module of level k is automatically a $V_k(\mathfrak{g})$–module.

Conversely, if M is a $V_k(\mathfrak{g})$–module, then the Fourier coefficients of the fields $Y_M(A_{-1}v_k), A \in \mathfrak{g}$, give rise to the structure of a smooth $\widehat{\mathfrak{g}}$–module of level k on M.

Lemma 3.4.5 and Proposition 5.1.6 imply that any module over a conformal vertex algebra V with central charge c is automatically a smooth module over the Virasoro vertex algebra Vir_c and over the Virasoro algebra with the same central charge. This leads us to the following definition.

5.1.9. Definition. Let V be a conformal vertex algebra with conformal vector ω, and M a V–module. Then M is called a *conformal V–module* if the Fourier coefficient L_0^M of the field

$$Y_M(\omega, z) = \sum_{n \in \mathbb{Z}} L_n^M z^{-n-2}$$

acts semi-simply on M. In that case it defines a gradation on M.

Note that L_0^M automatically acts semi-simply on any irreducible V–module, and so any irreducible module over a conformal vertex algebra is automatically conformal.

We remark also that for any module M over a conformal vertex algebra the operator L_{-1}^M gives us a translation operator on M which satisfies $[L_{-1}^M, Y_M(A, z)] = Y_M(TA, z) = \partial_z Y_M(A, z)$.

5.2. Vertex algebras associated to one-dimensional integral lattices

Let W be a vertex algebra, and V is a vertex subalgebra of W. Then W is a module over V. We may think of W as a V–module with the additional property that the vertex algebra structure on V extends to that on W. It is natural to try to construct such an extension W by adding to V a direct sum of its modules.

In the next three sections we present such a construction in the case when V is the vertex algebra associated to a Heisenberg Lie algebra. This leads us to the definition of the lattice vertex algebras. We start out with the case of a one-dimensional integral lattice. Then we introduce the fermionic vertex superalgebra and prove the boson-fermion correspondence. After that we define the vertex algebras associated to arbitrary positive integral lattices. As we will see later, the lattice vertex algebras are the simplest examples of the so-called rational vertex algebras.

5.2.1. Modules over the Weyl algebra. Recall the Weyl algebra $\widetilde{\mathcal{H}}$ from § 2.1.2. For $\lambda \in \mathbb{C}$, let π_λ be the $\widetilde{\mathcal{H}}$–module generated by a vector $|\lambda\rangle$ such that

$$b_n|\lambda\rangle = 0, \quad n \geq 0, \qquad b_0|\lambda\rangle = \lambda|\lambda\rangle.$$

Then π_λ is isomorphic to $\mathbb{C}[b_n]_{n<0}|\lambda\rangle$ as a vector space, on which the generators $b_n, n < 0$, act as multiplication by b_n, the generators $b_n, n > 0$, act as derivations $n\partial/\partial b_{-n}$, and b_0 acts as scalar multiplication by λ.

The algebra $\widetilde{\mathcal{H}}$ is $\mathbb{Z}$–graded with the following degree assignment: $\deg b_n = -n$. According to the discussion of § 5.1.8, $\mathbb{Z}$–graded $\widetilde{\mathcal{H}}$–modules with gradation bounded from below are the same as $\mathbb{Z}$–graded modules over the vertex algebra π with gradation bounded from below. For instance, π_λ is a $\mathbb{Z}$–graded module over both of them, with the gradation defined by the formulas $\deg|\lambda\rangle = 0, \deg b_n = -n$, so that it takes only non-negative values.

5.2.2. Lemma. *Any irreducible $\mathbb{Z}$–graded $\widetilde{\mathcal{H}}$–module with gradation bounded from below is isomorphic to π_λ, for some $\lambda \in \mathbb{C}$.*

5.2.3. Proof. Let us show first that π_λ is irreducible. Any vector $x \in \pi_\lambda$ can be written as $P|\lambda\rangle$, where P is a polynomial in $b_n, n < 0$. For any non-constant polynomial there exists $n < 0$ such that $\partial P/\partial b_n \neq 0$. Hence there exists $m > 0$ such that $b_m \cdot x \neq 0$. The polynomial P is a sum of homogeneous terms of various degrees, and so is $b_m P$. But since the degree of $b_m, m < 0$, is negative, the maximal degree of a non-zero homogeneous term of $b_m P$ is smaller than that of P. Hence there exist $m_1, m_2, \ldots, m_k > 0$ such that $b_{m_1} b_{m_2} \ldots b_{m_k} x \neq 0$ and has degree zero. Hence $b_{m_1} b_{m_2} \ldots b_{m_k} x = \alpha|\lambda\rangle$ for a non-zero scalar α. But $|\lambda\rangle$ generates π_λ as an $\widetilde{\mathcal{H}}$–module. Therefore for any non-zero $P, Q \in \pi_\lambda$, there exists an element A of $\widetilde{\mathcal{H}}$ such that $A \cdot P = Q$, i.e., π_λ is irreducible.

Now let M be an irreducible $\mathbb{Z}$–graded $\widetilde{\mathcal{H}}$–module with gradation bounded from below. Let $\widetilde{M}$ be the non-zero homogeneous component of M of lowest possible degree. Since $\deg b_n = -n$, it follows that b_0 preserves $\widetilde{M}$ and $b_n \cdot \widetilde{M} = 0, \forall n > 0$. If $\widetilde{M}'$ is a proper subspace of $\widetilde{M}$ stable under b_0, then $\widetilde{\mathcal{H}} \cdot \widetilde{M}' = \mathbb{C}[b_n]_{n<0} \cdot \widetilde{M}'$ is a proper $\widetilde{\mathcal{H}}$–submodule of M. Since M is irreducible, we obtain that $\dim \widetilde{M} = 1$, and b_0 acts on $\widetilde{M}$ as multiplication by a scalar $\lambda \in \mathbb{C}$. But then $M \simeq \pi_\lambda$.

5.2.4. Virasoro action. Since π is a conformal vertex algebra with the conformal vector $\frac{1}{2}b_{-1}^2$, we obtain an action of the corresponding Virasoro algebra on each π_λ:

$$(5.2.1) \qquad L_n = \frac{1}{2}\sum_{m\in\mathbb{Z}} :b_m b_{n-m}: , \qquad n \in \mathbb{Z},$$

with central charge 1.

Now let N be a positive integer. Set

$$V_{\sqrt{N}\mathbb{Z}} = \bigoplus_{m\in\mathbb{Z}} \pi_{m\sqrt{N}}.$$

Then $V_{\sqrt{N}\mathbb{Z}}$ is a conformal π–module, with the gradation operator equal to the operator L_0 defined by formula (5.2.1). We refer to this module structure as *canonical*.

In the proof of the following proposition, we will use the fact (left to the reader) that the Reconstruction Theorem 4.4.1 is valid in the case of vertex superalgebras, with the following modification of the operation of normal ordering. Namely, if

$A(z)$ and $B(z)$ are two fields of parity α and β, we define

$$:A(z)B(w): = \sum_{n \in \mathbb{Z}} \left(\sum_{m < 0} A_{(m)} B_{(n)} z^{-m-1} + (-1)^{\alpha \beta} \sum_{m \geq 0} B_{(n)} A_{(m)} z^{-m-1} \right) w^{-n-1}$$

(compare with Definition 2.2.2).

5.2.5. Proposition. *For any even N (resp., odd N), $V_{\sqrt{N}\mathbb{Z}}$ carries a structure of conformal vertex algebra (resp., superalgebra) such that $\pi_0 = \pi$ is a conformal vertex subalgebra of $V_{\sqrt{N}\mathbb{Z}}$, and the structure of $V_{\sqrt{N}\mathbb{Z}}$ as a conformal π–module induced by the vertex algebra structure on $V_{\sqrt{N}\mathbb{Z}}$ coincides with the canonical structure.*

5.2.6. Proof. By definition, $V_{\sqrt{N}\mathbb{Z}}$ is generated from the vectors $|\lambda\rangle, \lambda \in \sqrt{N}\mathbb{Z}$, under the action of the generators $b_n, n < 0$, of $\widetilde{\mathcal{H}}$. By our assumptions, we already know that $Y(b_{-1}|0\rangle, z) = b(z)$. According to the Reconstruction Theorem 4.4.1, in order to define the vertex algebra structure on $V_{\sqrt{N}\mathbb{Z}}$, it suffices to define the fields

$$V_\lambda(z) \overset{\text{def}}{=} Y(|\lambda\rangle, z), \qquad \lambda \in \sqrt{N}\mathbb{Z},$$

satisfying the conditions of Theorem 4.4.1.

By our assumptions, the conformal vector in $V_{\sqrt{N}\mathbb{Z}}$ should be $\frac{1}{2}b_{-1}^2 \in \pi \subset V_{\sqrt{N}\mathbb{Z}}$. Hence we obtain that the gradation on $V_{\sqrt{N}\mathbb{Z}}$ should be given by the formulas $\deg b_n = -n, \deg |\lambda\rangle = \lambda^2/2$, and the translation operator on $V_{\sqrt{N}\mathbb{Z}}$ should be $T = \frac{1}{2}\sum_{n \in \mathbb{Z}} b_n b_{-1-n}$. In particular, $T \cdot |\lambda\rangle = \lambda b_{-1}|\lambda\rangle$. By Corollary 3.1.6,

$$\partial_z Y(|\lambda\rangle, z) = Y(T \cdot |\lambda\rangle, z) = \lambda Y(b_{-1}|\lambda\rangle).$$

Using Corollary 3.3.5, we find that

$$(5.2.2) \qquad\qquad \partial_z V_\lambda(z) = \lambda :b(z)V_\lambda(z): .$$

By formula (3.3.10),

$$b(z)V_\lambda(w) = \frac{Y(b_0|\lambda\rangle, w)}{z - w} + :b(z)V_\lambda(w): = \frac{\lambda V_\lambda(w)}{z - w} + :b(z)V_\lambda(w): ,$$

from which we find the commutation relations:

$$(5.2.3) \qquad\qquad [b_n, V_\lambda(z)] = \lambda z^n V_\lambda(z).$$

In particular, $[b_0, V_\lambda(z)] = \lambda V_\lambda(z)$, and so for any $A \in \pi_\mu$,

$$V_\lambda(z) \cdot A \in \pi_{\lambda + \mu}((z)).$$

Let us write

$$(5.2.4) \qquad\qquad V_\lambda(z) = \sum_{n \in \mathbb{Z}} V_\lambda[n] z^{-n - \lambda^2/2}.$$

Since $V_\lambda(z)$ should be a field of conformal dimension $\lambda^2/2$, the linear operator $V_\lambda[n]$ should have degree $-n$. By definition, $\deg |\nu\rangle = \nu^2/2, \deg |\lambda + \nu\rangle = (\lambda + \nu)^2/2$. Comparing degrees, we obtain that for any $\lambda, \nu \in \sqrt{N}\mathbb{Z}$ we must have

$$(5.2.5) \qquad\qquad V_\lambda[s] \cdot |\nu\rangle = 0, \qquad s > -\lambda^2/2 - \lambda\nu,$$

$$(5.2.6) \qquad V_\lambda[-\lambda^2/2 - \lambda\nu] \cdot |\nu\rangle = c_{\lambda,\nu}|\lambda + \nu\rangle,$$

where $c_{\lambda,\nu}$ are some complex numbers (as yet undetermined). The vacuum axiom requires that

$$(5.2.7) \qquad\qquad c_{\lambda,0} = 1, \qquad \forall \lambda \in \sqrt{N}\mathbb{Z}.$$

Formulas (5.2.2)–(5.2.6) completely determine the fields $V_\lambda(z), \lambda \in \sqrt{N}\mathbb{Z}$, on $V_{\sqrt{N}\mathbb{Z}}$. Indeed, once we know (5.2.5) and (5.2.6), we find $V_\lambda[s] \cdot |\nu\rangle$ with $s < -\lambda^2/2 - \lambda\nu$ by induction using formulas (5.2.4) and (5.2.2). We can then compute

$$V_\lambda[s] \cdot b_{j_1} \dots b_{j_m} |\nu\rangle$$

for arbitrary $s, j_1, \dots, j_m$ using the commutation relations (5.2.3).

Actually, it is easy to write down an explicit formula for the fields $V_\lambda(z)$. Let S_λ be the shift operator $\pi_\mu \to \pi_{\lambda+\mu}$ defined by the conditions

$$S_\lambda|\nu\rangle = c_{\lambda,\nu}|\nu + \lambda\rangle; \qquad [S_\lambda, b_n] = 0, \quad n \neq 0.$$

Then

$$(5.2.8) \qquad V_\lambda(z) = S_\lambda z^{\lambda b_0} \exp\left(-\lambda \sum_{n<0} \frac{b_n}{n} z^{-n}\right) \exp\left(-\lambda \sum_{n>0} \frac{b_n}{n} z^{-n}\right).$$

We leave it to the reader to check that this field satisfies the equations (5.2.2)–(5.2.6).

Note that for $\lambda = m\sqrt{N}$, when we restrict $V_\lambda(z)$ to $\pi_{k\sqrt{N}}$, the factor $z^{\lambda b_0}$ becomes z^{mkN}, so the above formula is well-defined on $V_{\sqrt{N}\mathbb{Z}}$. In fact, formula (5.2.8) defines a power series in integral powers of z only on the Fock modules with highest weights from an integral lattice. That is the reason why we consider an integral lattice $\sqrt{N}\mathbb{Z}$.

By construction, the Fourier coefficients of the fields $V_{m\sqrt{N}}(z), m \in \mathbb{Z}$, and $b(z)$ generate $V_{\sqrt{N}\mathbb{Z}}$ from the vacuum vector $|0\rangle$. Moreover, it is clear from our construction of these fields that they satisfy the vacuum axiom. Hence it remains to check that these fields satisfy the conditions (2) and (3) of the Reconstruction Theorem 4.4.1 (see § 2.3.10). We already know that $b(z)$ satisfies condition (2). Using formula (5.2.8) and the commutation relations between T and the b_n's, we obtain

$$[T, V_\lambda(z)] = \lambda{:}b(z)V_\lambda(z){:} = \partial_z V_\lambda(z),$$

by formula (5.2.2). Hence condition (2) is satisfied for $V_\lambda(z)$ as well.

The locality of $b(z)$ with itself has been proved before. The locality of $b(z)$ and $V_\lambda(z)$ follows from the commutation relations (5.2.3) written in the form

$$[b(z), V_\lambda(w)] = \lambda V_\lambda(w)\delta(z - w).$$

Finally, we check the locality between $V_\lambda(z)$ and $V_\mu(z)$, where $\lambda, \mu \in \sqrt{N}\mathbb{Z}$. Note that

$$\left[\sum_{n>0} \frac{b_n}{n} z^{-n}, \sum_{m<0} \frac{b_m}{m} w^{-m}\right] = -\sum_{n>0} \frac{1}{n}\left(\frac{w}{z}\right)^n = \log\left(1 - \frac{w}{z}\right).$$

From this we derive that

$$(5.2.9) \qquad V_\lambda(z)V_\mu(w) = (z - w)^{\lambda\mu}{:}V_\lambda(z)V_\mu(w){:},$$

where

$${:}V_\lambda(z)V_\mu(w){:} = S_\lambda S_\mu z^{(\lambda+\mu)b_0}.$$

$$\cdot \exp\left(-\sum_{n<0}\left(\lambda\frac{b_n}{n}z^{-n}+\mu\frac{b_n}{n}w^{-n}\right)\right)\exp\left(-\sum_{n>0}\left(\lambda\frac{b_n}{n}z^{-n}+\mu\frac{b_n}{n}w^{-n}\right)\right).$$

Here by $(z-w)^{\lambda\mu}$ we understand the product of $z^{\lambda\mu}$ and the expansion of $(1-w/z)^{\lambda\mu}$ in positive powers of w/z (if $\lambda\mu < 0$).

The locality of the fields $V_\lambda(z)$ and $V_\mu(w)$ (in the "super" sense, see formula (1.3.1)) means that there exists an integer M such that

$$(z-w)^M V_\lambda(z)V_\mu(w) = (-1)^{p(\lambda)p(\mu)}(z-w)^M V_\mu(w)V_\lambda(z),$$

where $p(\lambda)$ denotes the parity of the vector $|\lambda\rangle$ (and of the field $V_\lambda(z)$). From formula (5.2.9) we find that this condition is satisfied if and only if

$$S_\lambda S_\mu = (-1)^{p(\lambda)p(\mu)+\lambda\mu}S_\mu S_\lambda.$$

This is equivalent to

$$(5.2.10)\qquad c_{\lambda,\nu+\mu}c_{\mu,\nu} = (-1)^{p(\lambda)p(\mu)+\lambda\mu}c_{\mu,\nu+\lambda}c_{\lambda,\nu}, \qquad \forall\lambda,\mu,\nu\in\sqrt{N}\mathbb{Z}$$

(here we extend the definition of $c_{\lambda,\mu}$ by setting $c_{0,\nu} = 1$ for all ν). If $\lambda = m\sqrt{N}$ and $\mu = k\sqrt{N}$, we obtain from (5.2.10) that $p(m\sqrt{N}) = m^2 N \bmod 2$. Thus, in order to satisfy locality we have to define the superspace structure on $V_{\sqrt{N}\mathbb{Z}}$ in such a way that all vectors in $\pi_{m\sqrt{N}}$ have parity $m^2 N \bmod 2$. Hence if N is even, $V_{\sqrt{N}\mathbb{Z}}$ is purely even, and if N is odd, then the parity of $\pi_{m\sqrt{N}}$ equals $m \bmod 2$.

Thus, for any collection of complex numbers $\{c_{\lambda,\nu}\}_{\lambda,\nu\in\sqrt{N}\mathbb{Z}}$ satisfying equations (5.2.10) and (5.2.7) (required by the locality axiom and the vacuum axiom, respectively) the fields $V_\lambda(z)$ given by formula (5.2.8) equip $V_{\sqrt{N}\mathbb{Z}}$ with the structure of a vertex superalgebra. The most interesting solution of these equations corresponds to the case when the numbers $c_{\sqrt{N},m\sqrt{N}}$, are non-zero for all $m \neq 0$. In this case we can normalize the vectors $|m\sqrt{N}\rangle, m \neq 0,1$, in such a way that $c_{\sqrt{N},m\sqrt{N}} = 1$ for all $k \in \mathbb{Z}$. This condition and equations (5.2.10) and (5.2.7) uniquely determine the remaining numbers; namely, $c_{\lambda,\nu} = 1$ for all $\lambda,\nu\in\sqrt{N}\mathbb{Z}$. Thus, we obtain a solution of equations (5.2.10) and (5.2.7), and therefore a vertex superalgebra structure on $V_{\sqrt{N}\mathbb{Z}}$ satisfying the conditions of Proposition 5.2.5. This proves the proposition.

We remark that there exist other solutions of equations (5.2.10) and (5.2.7), but they are degenerate in the sense that some of the numbers $c_{\lambda,\nu}$ are equal to 0. Therefore from now on by the vertex superalgebra structure on $V_{\sqrt{N}\mathbb{Z}}$ we will understand this particular structure, where all $c_{\lambda,\nu} = 1$.

5.2.7. Remark. The fields $V_\lambda(z)$ are called *bosonic vertex operators*. Historically, they were the first vertex operators originally introduced in string theory.

5.2.8. Other conformal vectors. In the above description of $V_{\sqrt{N}\mathbb{Z}}$ we used the conformal vector $\omega_0 = \frac{1}{2}b_{-1}^2$. There are other choices of conformal vectors in $V_{\sqrt{N}\mathbb{Z}}$. It is clear that a conformal vector of $V_{\sqrt{N}\mathbb{Z}}$ must belong to π_0, for otherwise the Fourier coefficients of the corresponding vertex operator would not preserve any given Fock subspace $\pi_{\sqrt{N}m}$ of $V_{\sqrt{Z}}$. We know that conformal vectors in $\pi_0 = \pi$ are of the form $\omega_\lambda = \frac{1}{2}b_{-1}^2 + \lambda b_{-2}$, where $\lambda \in \mathbb{C}$ (with central charge $1 - 12\lambda^2$). The corresponding operator L_{-1} does not depend on λ and hence coincides with the translation operator defined above. But we also need the eigenvalues of the operator L_0 on $V_{\sqrt{N}\mathbb{Z}}$ to be integers (or half-integers if N is odd). We check that L_0 acts on

$|\mu\rangle$ by multiplication by $\frac{1}{2}\mu(\mu - 2\lambda)$, while the commutator $[L_0, b_n] = -nb_n$ does not change. Hence λ should be in $\frac{\sqrt{N}}{2}\mathbb{Z}$.

Thus, ω_λ is a well-defined conformal vector in the vertex algebra $V_{\sqrt{N}\mathbb{Z}}$ for any $\lambda \in \frac{\sqrt{N}}{2}\mathbb{Z}$ (with the appropriately defined gradation).

5.3. Boson–fermion correspondence

In this section we establish the boson-fermion correspondence, which is an isomorphism between the lattice vertex superalgebra $V_\mathbb{Z}$ defined in the previous section and the free fermionic vertex superalgebra.

5.3.1. Free fermionic vertex superalgebra. Consider the vector space $\mathbb{C}((t)) \oplus \mathbb{C}((t))dt$ with the inner product induced by the residue pairing. Let $\mathcal{C}l$ be the Clifford algebra associated to it. It has topological generators $\psi_n = t^n, \psi_n^* = t^{n-1}dt, n \in \mathbb{Z}$, satisfying the anti-commutation relations

$$(5.3.1) \qquad [\psi_n, \psi_m]_+ = [\psi_n^*, \psi_m^*]_+ = 0, \qquad [\psi_n, \psi_m^*]_+ = \delta_{n,-m}.$$

Here we use the notation $[A, B]_+ = AB + BA$.

Denote by $\bigwedge$ the *fermionic Fock representation* of $\mathcal{C}l$, generated by a vector $|0\rangle$, such that

$$(5.3.2) \qquad \psi_n|0\rangle = 0, n \geq 0, \qquad \psi_n^*|0\rangle = 0, n > 0.$$

Thus,

$$(5.3.3) \qquad \bigwedge \simeq \bigwedge(\psi_n)_{n<0} \otimes \bigwedge(\psi_n^*)_{n\leq 0}|0\rangle,$$

where $\bigwedge(a_i)_{i \in I}$ denotes the exterior algebra with generators $a_i, i \in I$. The monomials

$$(5.3.4) \qquad \psi_{n_1} \ldots \psi_{n_k} \psi_{m_1}^* \ldots \psi_{m_l}^*|0\rangle,$$

where $n_1 < n_2 < \ldots < n_k < 0$ and $m_1 < m_2 < \ldots m_l \leq 0$, form a basis of $\bigwedge$. We define a superspace structure on $\bigwedge$ by setting the parity of the monomial (5.3.4) to be $(k + l)$ mod 2.

We now introduce the structure of a $\mathbb{Z}_+$–graded vertex superalgebra (see Remark 1.3.2.1) on $\bigwedge$ as follows:

- Vacuum vector: $|0\rangle$.
- Translation operator: $T|0\rangle = 0$, and
$$[T, \psi_n] = -n\psi_{n-1}, \qquad [T, \psi_n^*] = -(n-1)\psi_{n-1}^*.$$

- Vertex operators: $Y(|0\rangle, z) = \text{Id}$,

$$Y(\psi_{-1}|0\rangle, z) = \psi(z) \stackrel{\text{def}}{=} \sum_{n \in \mathbb{Z}} \psi_n z^{-n-1}, \qquad Y(\psi_0^*|0\rangle, z) = \psi^*(z) \stackrel{\text{def}}{=} \sum_{n \in \mathbb{Z}} \psi_n^* z^{-n},$$

$$Y(\psi_{n_1} \ldots \psi_{n_k} \psi_{m_1}^* \ldots \psi_{m_l}^*|0\rangle, z) = \prod_{i=1}^{k} \frac{1}{(-n_i - 1)!} \prod_{j=1}^{l} \frac{1}{(-m_j)!} \cdot$$

$$:\partial_z^{-n_1-1}\psi(z) \ldots \partial_z^{-n_k-1}\psi(z)\partial_z^{-m_1}\psi^*(z) \ldots \partial_z^{-m_l}\psi^*(z):.$$

- $\mathbb{Z}_+$–gradation: $\deg \psi_{n_1} \ldots \psi_{n_k} \psi_{m_1}^* \ldots \psi_{m_l}^*|0\rangle = -\sum_{i=1}^{k} n_i - \sum_{j=1}^{l} m_j$.

Reconstruction Theorem 2.3.11 (adapted to the case of vertex superalgebras) implies that the above formulas indeed provide $\bigwedge$ with the structure of a vertex superalgebra. The proof, as before, essentially boils down to checking locality between the generating fields. In the supercase, the locality property is expressed by formula (1.3.1). By construction, the parity of the fields $\psi(z)$ and $\psi^*(z)$ is equal to $\bar{1}$. From the anti–commutation relations (5.3.1) in the Clifford algebra we find that

$$[\psi(z), \psi(w)]_+ = [\psi^*(z), \psi^*(w)]_+ = 0, \qquad [\psi(z), \psi^*(w)]_+ = \delta(z - w),$$

and the last formula implies $(z - w)[\psi(z), \psi^*(w)]_+ = 0$. This proves locality of the fields $\psi(z), \psi^*(z)$ and completes the proof that $\bigwedge$ is a vertex superalgebra.

In physics literature the fields $\psi(z)$ and $\psi^*(z)$ are traditionally denoted by $b(z)$ and $c(z)$, and the corresponding conformal field theory is called the bc–system.

Now recall the vertex superalgebra $V_{\mathbb{Z}}$ defined in Proposition 5.2.5. In the following theorem we will use the $\mathbb{Z}_+$–gradation on $V_{\mathbb{Z}}$ defined by the Fourier coefficient L_0 of the Virasoro vertex operator $Y(\omega_{-1/2}, z)$ (rather than $Y(\omega_0, z)$ which was used in Proposition 5.2.5). This gradation is determined by the formulas $\deg |m\rangle = \frac{1}{2}m(m - 1)$, $\deg b_n = -n$.

5.3.2. Theorem (Boson-fermion Correspondence). *There is an isomorphism of vertex superalgebras $\bigwedge \simeq V_{\mathbb{Z}}$.*

5.3.3. Proof. We find from formula (5.2.9) the following OPEs in $V_{\mathbb{Z}}$:

$$(5.3.5) \qquad V_1(z)V_{-1}(w) = \frac{1}{z - w}{:}V_1(z)V_{-1}(w){:} = \frac{1}{z - w} + \text{reg.},$$

$$(5.3.6) \qquad V_{\pm 1}(z)V_{\pm 1}(w) = (z - w){:}V_1(z)V_{-1}(w){:} = \text{reg.}$$

Introduce linear operators ϕ_n and ϕ_n^*, $n \in \mathbb{Z}$, on $V_{\mathbb{Z}}$ by the formulas

$$V_1(z) = \sum_{n \in \mathbb{Z}} \phi_n^* z^{-n}, \qquad V_{-1}(z) = \sum_{n \in \mathbb{Z}} \phi_n z^{-n-1}.$$

In the same way as in § 3.3.6 we derive from formulas (5.3.5), (5.3.6) the anti-commutation relations between the operators ϕ_n, ϕ_m^*:

$$[\phi_n, \phi_m]_+ = [\phi_n^*, \phi_m^*]_+ = 0, \qquad [\phi_n, \phi_m^*]_+ = \delta_{n,-m}.$$

Thus, these operators satisfy the anti-commutation relations (5.3.1) of the Clifford algebra $\mathcal{C}l$, and hence define a representation of $\mathcal{C}l$ on $V_{\mathbb{Z}}$. Moreover, we find from the defining formula (5.2.8) for the fields $V_{\pm 1}(z)$ that

$$\phi_n|0\rangle = 0, n \geq 0, \qquad \phi_n^*|0\rangle = 0, n > 0.$$

These are the same conditions as those satisfied by the vacuum vector $|0\rangle$ of the $\mathcal{C}l$–module $\bigwedge$. Hence we obtain a map of $\mathcal{C}l$–modules $\rho : \bigwedge \to V_{\mathbb{Z}}$ sending $|0\rangle$ to $|0\rangle$. But $\bigwedge$ is an irreducible $\mathcal{C}l$–module. This is proved in the same way as the irreducibility of π_λ in § 5.2.1: for each $x \in \bigwedge$, there exists a monomial m in $\psi_n, n \geq 0$, and $\psi_n^*, n > 0$, such that $m \cdot x = |0\rangle$. Since $|0\rangle$ generates $\bigwedge$, we see that any vector in $\bigwedge$ can be obtained from any other vector by the action of $\mathcal{C}l$.

Therefore the map ρ is injective. Let us show that it is also surjective. Using formula (5.2.8), we obtain

$$\rho(\psi_n \psi_{n+1} \dots \psi_{-1}|0\rangle) = |n\rangle, \qquad n < 0,$$
$$\rho(\psi_{-n+1}^* \psi_{-n+2}^* \dots \psi_0^*|0\rangle) = |n\rangle, \qquad n > 0.$$

Furthermore, we find from formula (5.3.5) (using Taylor expansion) that

(5.3.7) $:V_1(z)V_{-1}(z): = b(z).$

It follows that if a vector $A \in V_{\mathbb{Z}}$ is in the image of ρ, then any vector of the form $P \cdot A$, where $P \in \widetilde{\mathcal{H}}$, is also in the image of ρ. But we have seen that each $|n\rangle, n \in \mathbb{Z}$, is in the image of ρ. Therefore we obtain that each $\pi_n, n \in \mathbb{Z}$, is in the image of ρ, and therefore ρ is surjective.

Thus, ρ is an isomorphism of vector spaces (moreover, this isomorphism preserves $\mathbb{Z}$–gradations). Therefore we obtain that both $\bigwedge$ and $V_{\mathbb{Z}}$ are irreducible modules over the Clifford algebra $\mathcal{C}l$. By the Reconstruction Theorem 4.4.1 (adapted to the supercase), the vertex superalgebra structure on $\bigwedge$ (resp., $V_{\mathbb{Z}}$) is determined by the fields $\psi(z)$ and $\psi^*(z)$ (resp., $V_{\pm 1}(z)$). By construction, ρ intertwines the actions of these fields. Therefore ρ is an isomorphism of vertex superalgebras.

5.3.4. Conformal vectors. Recall from § 5.2.8 that the possible conformal vectors in $V_{\mathbb{Z}}$ are $\omega_\lambda = \frac{1}{2}b_{-1}^2 + \lambda b_{-2}, \lambda \in \frac{1}{2}\mathbb{Z}$. Let us find the corresponding conformal vectors in $\bigwedge$. According to formula (5.3.7),

$$\rho^{-1}(b(z)) = h(z) \overset{\text{def}}{=} :\psi^*(z)\psi(z): .$$

Hence we need to compute $(\frac{1}{2}h_{-1}^2 + \lambda h_{-2})|0\rangle$ in $\bigwedge$. We have

$$h_{-1}^2|0\rangle = \sum_{n\in\mathbb{Z}} :\psi_n^*\psi_{-n-1}: \cdot \; \psi_0^*\psi_{-1}|0\rangle = (\psi_{-1}^*\psi_{-1} + \psi_{-2}\psi_0^*)|0\rangle,$$

$$h_{-2}|0\rangle = (\psi_{-1}^*\psi_{-1} - \psi_{-2}\psi_0^*)|0\rangle.$$

Hence

(5.3.8) $\rho^{-1}(\omega_\lambda) = (\mu\psi_{-1}^*\psi_{-1} + (1-\mu)\psi_{-2}\psi_0^*)|0\rangle,$

where $\mu = \lambda + \frac{1}{2}$. The corresponding vertex operator equals

(5.3.9) $T_\mu(z) = \mu:\partial_z\psi^*(z)\,\psi(z): + (1-\mu):\partial_z\psi(z)\,\psi(z)^*: .$

The Fourier coefficients of this vertex operator generate a representation of the Virasoro algebra on $\bigwedge$ with central charge

(5.3.10) $1 - 12\lambda^2 = -2(6\mu^2 - 6\mu + 1).$

The above isomorphism $\bigwedge \simeq V_{\mathbb{Z}}$ is an isomorphism of conformal vertex superalgebras if we choose $\omega_{-1/2}$ as the conformal vector in $V_{\mathbb{Z}}$. Therefore we find that $\bigwedge$ as defined in § 5.3.1 is a conformal vertex superalgebra with conformal vector $\psi_{-2}\psi_0^*|0\rangle$.

We can also choose as the conformal vector of $\bigwedge$ the vector (5.3.8) with an arbitrary integral or half-integral μ. In that case we must also redefine the gradation on $\bigwedge$ as follows: $\deg \psi_n = -\mu - n, \deg \psi_n^* = \mu - n$.

5.3.5. Application: Jacobi triple product identity. Let us compute the characters of $\bigwedge$ and $V_{\mathbb{Z}}$. Introduce an additional $\mathbb{Z}$–gradation, called *charge*, on $\bigwedge$, setting charge $\psi_n^* = -$ charge $\psi_n = 1$, charge $|0\rangle = 0$. Let $\bigwedge_{n,m}$ be the subspace of $\bigwedge$ on which the gradation takes value n and the charge takes value m (note that $\bigwedge = \bigoplus_{n,m} \bigwedge_{n,m}$). The character of $\bigwedge$ is by definition the formal power series in two variables

$$\mathrm{ch} \bigwedge = \sum_{n,m} \dim \bigwedge_{n,m} q^n u^m.$$

Recall that $\deg \psi_n = \deg \psi_n^* = -n, \deg |0\rangle = 0$. Using the isomorphism (5.3.3), we obtain that

$$\operatorname{ch} \bigwedge = \prod_{n>0} (1 + uq^{n-1})(1 + u^{-1}q^n).$$

Now we evaluate the character of $V_{\mathbb{Z}}$. Recall that we have $\deg b_n = -n, \deg |m\rangle = m(m-1)/2$. We also find that under the identification $V_{\mathbb{Z}} \simeq \bigwedge$ we have charge $b_n = 0$, charge $|m\rangle = m$. Therefore

$$\operatorname{ch} V_{\mathbb{Z}} = \sum_{m \in \mathbb{Z}} u^m q^{m(m-1)/2} \prod_{n>0} (1 - q^n)^{-1}.$$

Equating the two formulas, multiplying both sides by $\prod_{n>0}(1 - q^n)$, and replacing u by $-u$, we arrive at the *Jacobi triple product identity*:

$$\prod_{n>0} (1 - q^n)(1 - uq^{n-1})(1 - u^{-1}q^n) = \sum_{m \in \mathbb{Z}} (-1)^m u^m q^{m(m-1)/2}.$$

5.3.6. Other interesting cases. According to Theorem 5.3.2, the vertex superalgebra $V_{\mathbb{Z}}$ is isomorphic to the fermionic vertex superalgebra. Consider next the vertex algebra $V_{\sqrt{2}\mathbb{Z}}$. One can show that it is isomorphic to the vertex algebra $L_1(\mathfrak{sl}_2)$, i.e., the irreducible quotient of $V_1(\mathfrak{sl}_2)$, see § 4.4.3. Let $\{e, h, f\}$ be the standard basis of $\mathfrak{sl}_2$. Recall that the vectors $e_{-1}v_1, h_{-1}v_1, f_{-1}v_1 \in L_1(\mathfrak{sl}_2)$ generate $L_1(\mathfrak{sl}_2)$ in the sense of Theorem 4.4.1. Under the isomorphism $L_1(\mathfrak{sl}_2) \simeq V_{\sqrt{2}\mathbb{Z}}$, they are identified with the vectors $|\sqrt{2}\rangle, \sqrt{2}b_{-1}|0\rangle$, and $|-\sqrt{2}\rangle \in V_{\sqrt{2}\mathbb{Z}}$, respectively. The action of the corresponding fields makes $V_{\sqrt{2}\mathbb{Z}}$ into an $\widehat{\mathfrak{sl}}_2$–module which is isomorphic to $L_1(\mathfrak{sl}_2)$. This module is called the basic representation of $\widehat{\mathfrak{sl}}_2$.

Finally, we remark that the vertex superalgebra $V_{\sqrt{3}\mathbb{Z}}$ is related to the $N = 2$ superconformal algebra in a similar way.

5.4. Lattice vertex algebras

5.4.1. The Heisenberg Lie algebra associated to a lattice. The construction of the vertex superalgebra $V_{\sqrt{N}\mathbb{Z}}$ can be generalized by replacing the rank one lattice $\sqrt{N}\mathbb{Z}$ with an arbitrary positive definite integral lattice.

Thus, let L be a lattice of finite rank equipped with a symmetric bilinear form $(\cdot, \cdot) : L \times L \to \mathbb{Z}$ such that $(\lambda, \lambda) > 0$ for all $\lambda \in L \backslash \{0\}$.

Set $\mathfrak{h} = L \otimes_{\mathbb{Z}} \mathbb{C}$. The bilinear form on L induces a bilinear form on $\mathfrak{h}$, for which we use the same notation. Let $\widehat{\mathfrak{h}}$ be the central extension of $\mathfrak{h}((t))$,

$$0 \to \mathbb{C}\mathbf{1} \to \widehat{\mathfrak{h}} \to \mathfrak{h}((t)) \to 0,$$

with the commutation relations

$$[A \otimes f(t), B \otimes g(t)] = -(A, B)(\operatorname{Res} f(t)g'(t)dt)\mathbf{1}.$$

This is the Heisenberg Lie algebra associated to $\mathfrak{h}$ equipped with the bilinear form $(\cdot, \cdot)$.

Define the Weyl algebra $\widetilde{\mathcal{H}}_L$ as the (completed) enveloping algebra of $\widehat{\mathfrak{h}}$ modulo the relation $\mathbf{1} = 1$. It has (topological) generators $h_n, h \in \mathfrak{h}, n \in \mathbb{Z}$, and relations

$$[h_n, g_m] = n(h, g)\delta_{n, -m}.$$

For $\lambda \in \mathfrak{h}$, define the Fock representation π_λ of $\widetilde{\mathcal{H}}_L$, generated by the vector $|\lambda\rangle$, such that
$$h_n|\lambda\rangle = 0, \quad n > 0; \qquad h_0|\lambda\rangle = (\lambda, h)|\lambda\rangle.$$
The Fock representation π_0 carries a vertex algebra structure, defined in the same way as in the case when $\dim \mathfrak{h} = 1$ (see Theorem 2.3.7).

5.4.2. Vertex superalgebra V_L. Set
$$V_L = \bigoplus_{\lambda \in L} \pi_\lambda.$$

Define a $\mathbb{Z}$–gradation on V_L by the formulas $\deg h_n = -n, \deg |\lambda\rangle = (\lambda, \lambda)/2$. Note that since L is positive definite, the gradation takes only non-negative values.

The vertex superalgebra structure on V_L is defined in the same way as in the case $L = \sqrt{N}\mathbb{Z}$. We have obvious analogues of formulas (5.2.2)–(5.2.6). These formulas determine $Y(|\lambda\rangle, z)$ uniquely. Namely, $Y(|\lambda\rangle, z)$ is the field $V_\lambda(z)$ given by formula (5.2.8), where the numbers $c_{\lambda,\mu}$ satisfy the equations (5.2.10) and (5.2.7). One checks easily that these equations are equivalent to the following system:

(5.4.1) $c_{\lambda,0} = c_{0,\lambda} = 1,$

(5.4.2) $c_{\lambda,\mu} = (-1)^{p(\lambda)p(\mu)+(\lambda,\mu)}c_{\mu,\lambda},$

(5.4.3) $c_{\mu,\nu}c_{\mu+\nu,\lambda} = c_{\mu,\nu+\lambda}c_{\nu,\lambda}$

for all $\lambda, \mu, \nu \in L$, where $p(\lambda)$ stands for the parity of $|\lambda\rangle$ (and hence of π_λ). In particular, substituting $\lambda = \mu$ in the second formula, we obtain that $p(\lambda) = (\lambda, \lambda) \bmod 2$. This should then be the parity of the subspace π_λ of V_L. In particular, if L is an even lattice, i.e., such that $(\lambda, \lambda) \in 2\mathbb{Z}, \forall \lambda \in L$, then V_L is purely even.

5.4.3. The non-degenerate case. Let us consider the non-degenerate case, when all $c_{\lambda,\mu} \neq 0$. Then the equations (5.4.1) and (5.4.3) mean that $\{c_{\lambda,\mu}\}$, considered as a function $L \times L \to \mathbb{C}^\times$, is a two-cocycle φ of the additive group L with coefficients in $\mathbb{C}^\times$. Since we are interested in the vertex superalgebra structures on V_L up to isomorphism, we may multiply the vectors $|\lambda\rangle$ by arbitrary non-zero constants a_λ. Then $c_{\lambda,\mu}$ will transform to $a_\lambda a_\mu a_{\lambda+\mu}^{-1} c_{\lambda,\mu}$. As a result, the two-cocycle φ will change by a coboundary. Thus, a cohomology class in $H^2(L, \mathbb{C}^\times)$, satisfying the extra condition (5.4.2), determines an isomorphism class of vertex superalgebra structure on V_L. One can show that such a cohomology class is unique (see, e.g., [**Kac3**], § 5.5). Moreover, it can be chosen so that it takes values in $\mathbb{Z}_2 = \{\pm 1\}$, and therefore corresponds to an extension
$$0 \to \mathbb{Z}_2 \to \widetilde{L} \to L \to 0.$$

In summary, under the condition that all $c_{\lambda,\mu} \neq 0$ there exists a unique (up to isomorphism) vertex superalgebra structure on V_L. The resulting vertex superalgebra is called the *lattice vertex superalgebra* associated to L (if L is an even lattice, then V_L is a vertex algebra). As a vector space, this vertex superalgebra is canonically isomorphic to $\pi_0 \otimes \mathbb{C}_\varphi[L]$, where $\mathbb{C}_\varphi[L]$ denotes the twisted group algebra $\mathbb{C}[\widetilde{L}]/(1 - (-1))$, where (-1) is the element of $\mathbb{Z}_2 = \{\pm 1\} \subset \widetilde{L}$.

The vector $|\lambda\rangle$ is identified with the element e^λ of $\mathbb{C}_\varphi[L]$. The vertex operators $Y(|\lambda\rangle, z) = V_\lambda(z)$ are given by formula (5.2.8), where S_λ is just the operator of left multiplication by e^λ acting on $\mathbb{C}_\varphi[L]$. The vertex operators $Y(h_{-1}|0\rangle, z)$ are given by the usual formula $h(z) = \sum_{n \in \mathbb{Z}} h_n z^{-n-1}$, and the general vertex operators are obtained from these as in the Reconstruction Theorem 4.4.1. Finally, V_L has a conformal vector

$$\omega = \frac{1}{2} \sum_{i=1}^{\dim \mathfrak{h}} (h_i)_{-1}(h^i)_{-1},$$

where $\{h_i\}$ and $\{h^i\}$ are dual bases in $\mathfrak{h}$ with respect to our bilinear form. The corresponding central charge equals $\dim \mathfrak{h} = \operatorname{rank} L$.

5.4.4. Remark. If L is the root lattice of a simply-laced simple Lie algebra $\mathfrak{g}$, then according to the I. Frenkel–Kac–Segal construction [**FK, Seg1**], V_L is isomorphic to the vertex algebra $L_1(\mathfrak{g})$ (the irreducible quotient of $V_k(\mathfrak{g})$ at $k = 1$).

5.4.5. Modules. Let $L' = \{\lambda \in L \otimes_{\mathbb{Z}} \mathbb{Q} | (\lambda, \mu) \in \mathbb{Z}, \forall \mu \in L\}$ be the dual lattice to L. For each $\gamma \in L'/L$, consider the vector space

$$V_L^\gamma = \bigoplus_{\lambda \in \gamma + L} \pi_\lambda.$$

Note that the vertex operators $V_\lambda(z), \lambda \in L$, give rise to well-defined fields on each V_L^γ because $(\lambda, \mu) \in \mathbb{Z}$ for all $\lambda \in \gamma + L$ and $\mu \in L$. Hence for any $A \in V_L$, the vertex operator $Y(A, z)$ makes sense as a field on V_L^γ. It is straightforward to check that these fields define the structure of a conformal V_L–module on V_L^γ. Moreover, it is shown in [**D1**] that $V_L^\gamma, \gamma \in L'/L$, are all simple V_L–modules up to an isomorphism (and they are not pairwise isomorphic). In particular, we see that the vertex algebra V_L has finitely many non-isomorphic simple modules. Vertex algebras of this type are called rational.

5.5. Rational vertex algebras

5.5.1. Definition. A conformal vertex algebra V (see Definition 2.5.8) is called *rational* if every $\mathbb{Z}_+$–graded V–module is completely reducible (i.e., isomorphic to a direct sum of simple V–modules).

5.5.2. Properties. It is shown in [**DLM2**] that this condition implies that

(1) V has *finitely many* inequivalent simple $\mathbb{Z}_+$–graded modules;
(2) the graded components of each simple $\mathbb{Z}_+$–graded V–module are finite-dimensional.

If M is a simple $\mathbb{Z}_+$–graded V–module, then the Virasoro operator L_0^M on M is automatically semi-simple and hence defines a gradation on M. Any other $\mathbb{Z}_+$–gradation on M will necessarily coincide with it up to a shift by a complex number. The above properties allow us to attach to a $\mathbb{Z}_+$–graded simple V–module M its *character*

$$\operatorname{ch} M = \operatorname{Tr}_M q^{L_0^M - c/24} = \sum_\alpha \dim M_\alpha q^{\alpha - c/24},$$

where M_α is the subspace of M on which L_0^M acts by multiplication by α, and c is the central charge of V. Set $q = e^{2\pi i \tau}$.

Now let $C_2(V)$ be the subspace of V spanned by all elements of the form $A_{(-2)} \cdot B$ for all $A, B \in V$. Then a rational vertex algebra V is said to satisfy the C_2 *cofiniteness condition* if

(1) $\dim V/C_2(V) < \infty$;
(2) every vector in V can be written as a linear combination of vectors of the form $L_{n_1} \dots L_{n_k} A, n_i < 0$, where A satisfies $L_n A = 0$ for all $n > 0$.

Y. Zhu [**Z1**] has shown that if V satisfies the C_2 cofiniteness condition, then the characters give rise to holomorphic functions in τ on the upper-half plane $\mathbb{H}_+ = \{\tau \in \mathbb{C} \mid \mathrm{Im}\,\tau > 0\}$. The group $SL_2(\mathbb{Z})$ acts on $\mathbb{H}_+$ in the standard way. Zhu proved the following remarkable fact.

5.5.3. Theorem. *Let V be a rational vertex algebra satisfying the C_2 cofiniteness condition, and let $\{M_1, \ldots, M_n\}$ be the set of all inequivalent simple $\mathbb{Z}$–graded V–modules (up to an isomorphism). Then the vector space spanned by $\mathrm{ch}\,M_i, i = 1, \ldots, n$, is invariant under the action of $SL_2(\mathbb{Z})$.*

5.5.4. Remark. This result has the following heuristic explanation. To each vertex algebra we can attach certain $\mathcal{D}$–modules on the moduli spaces of curves (see Chapter 17), whose fibers are the spaces of coinvariants (see Definition 9.2.7). It is expected that the spaces of coinvariants of a rational vertex algebra satisfying Zhu's condition are finite-dimensional. Therefore the corresponding $\mathcal{D}$–module on the moduli space of curves (see Chapter 17) is a vector bundle of finite rank with a (projectively) flat connection. In the special case of elliptic curves, the characters of simple V–modules should form the basis of the space of horizontal sections of the corresponding bundle on the moduli space of elliptic curves $\mathbb{H}_+/SL_2(\mathbb{Z})$ near $\tau = \infty$. The $SL_2(\mathbb{Z})$ action of Theorem 5.5.3 is then just the monodromy action on these sections.

5.5.5. Examples. Here are some examples of rational vertex algebras.

(1) Let L be an even positive definite lattice in a real vector space W. In § 5.4 we attached to it a vertex algebra V_L. It is proved in [**D1**] that this vertex algebra is rational, and its inequivalent simple modules are parameterized by L'/L, where L' is the dual lattice. The characters of these modules are the theta-functions corresponding to L. The vertex algebra V_L is the chiral symmetry algebra of the free bosonic conformal field theory compactified on the torus W/L.

(2) Let $\mathfrak{g}$ be a simple Lie algebra and $k \in \mathbb{Z}_+$. Let $L_k(\mathfrak{g})$ be the irreducible quotient of the $\widehat{\mathfrak{g}}$–module $V_k(\mathfrak{g})$. We have shown in § 4.4.3 that $L_k(\mathfrak{g})$ is a vertex algebra. It is proved in [**FZ**] that it is a rational vertex algebra whose modules are the integrable representations of $\widehat{\mathfrak{g}}$ of level k. The transformation formula for characters of these modules is given in [**KP**]. The corresponding conformal field theory is the Wess-Zumino-Witten model (see [**dFMS**]).

(3) Let $L_{c(p,q)}$ be the irreducible quotient of $\mathrm{Vir}_{c(p,q)}$ (as a module over the Virasoro algebra), where $c(p,q)$ is given by formula (4.4.3). Then $L_{c(p,q)}$ is a vertex algebra (see § 4.4.3). It is proved in [**Wan**] that this is a rational vertex algebra whose simple modules form the "minimal model" of conformal field theory defined by Belavin, Polyakov, and Zamolodchikov [**BPZ**]. For the transformation formula of characters, see, e.g., [**dFMS**].

(4) The Moonshine Module vertex algebra $V^{\natural}$ (see § 5.7.1 below).

5.6. Twisted modules

An automorphism of a vertex algebra V is by definition a linear operator $X : V \to V$ such that $X \cdot |0\rangle = |0\rangle$, $[X, T] = 0$, and $Y(X \cdot A, z) = XY(A, z)X^{-1}$, for all $A \in V$.

For a finite order automorphism σ of V, one defines a σ–twisted V–module by modifying the above axioms of V–module in such a way that $Y_M(A, z)$ has monodromy λ^{-1} around the origin, if $\sigma \cdot A = \lambda v$ (see [**DLM2**]).

5.6.1. Definition. Let σ be an automorphism of order N of a vertex algebra V. A vector space M^σ is called a σ–*twisted V–module* (or simply a twisted module) if it is equipped with an operation

$$Y_{M^\sigma} : V \to \operatorname{End} M^\sigma[[z^{\pm \frac{1}{N}}]],$$

$$A \mapsto Y_{M^\sigma}(A, z^{\frac{1}{N}}) = \sum_{n \in \frac{1}{N}\mathbb{Z}} A^{M^\sigma}_{(n)} z^{-n-1}$$

such that for any $v \in M^\sigma$ we have $A^{M^\sigma}_{(n)} v = 0$ for large enough n. This operation must satisfy the following axioms:

- $Y_{M^\sigma}(|0\rangle, z^{\frac{1}{N}}) = \operatorname{Id}_{M^\sigma}$.
- For any $A, B \in V$ and $v \in M^\sigma$, there exists an element

$$f_v \in M^\sigma[[z^{\frac{1}{N}}, w^{\frac{1}{N}}]][z^{-\frac{1}{N}}, w^{-\frac{1}{N}}, (z-w)^{-1}]$$

  such that the formal power series

$$Y_{M^\sigma}(A, z^{\frac{1}{N}}) Y_{M^\sigma}(B, w^{\frac{1}{N}}) v,$$

$$Y_{M^\sigma}(B, w^{\frac{1}{N}}) Y_{M^\sigma}(A, z^{\frac{1}{N}}) v, \quad \text{and}$$

$$Y_{M^\sigma}(Y(A, z-w)B, w^{\frac{1}{N}}) v$$

  are expansions of f_v in

$$M^\sigma((z^{\frac{1}{N}}))((w^{\frac{1}{N}})), \quad M^\sigma((w^{\frac{1}{N}}))((z^{\frac{1}{N}})), \quad \text{and} \quad M^\sigma((w^{\frac{1}{N}}))((z-w)),$$

  respectively.
- If $A \in V$ is such that $\sigma(A) = e^{\frac{2\pi i m}{N}} A$, then $A^{M^\sigma}_{(n)} = 0$ unless $n \in \frac{m}{N} + \mathbb{Z}$.

5.6.2. Example. Recall the Heisenberg vertex algebra π. As a vector space, it is an irreducible Fock module over the Heisenberg Lie algebra $\mathcal{H}$ with the generators $b_n, n \in \mathbb{Z}$, and $\mathbf{1}$ and the commutation relations (2.1.4). Consider an automorphism σ of $\mathcal{H}$ of order two sending b_n to $-b_n$. It induces an automorphism of the vertex algebra π also denoted by σ. We will construct a σ–twisted π–module.

For that we need to introduce a twisted version of $\mathcal{H}$. This is a Heisenberg Lie algebra $\mathcal{H}^\sigma$ with generators $\widetilde{b}_n, n \in \frac{1}{2} + \mathbb{Z}$, and $\mathbf{1}$ and commutation relations

$$[\widetilde{b}_n, \widetilde{b}_m] = n\delta_{n,-m}\mathbf{1}.$$

Let $\mathcal{H}^\sigma_+$ be its Lie subalgebra generated by $\widetilde{b}_n, n > 0$.

Define the irreducible $\mathcal{H}^\sigma$–module

$$\pi^\sigma = \operatorname{Ind}^{\mathcal{H}^\sigma}_{\mathcal{H}^\sigma_+ \oplus \mathbb{C} \cdot \mathbf{1}} \mathbb{C},$$

where $\mathbb{C}$ is the one–dimensional representation of $\mathcal{H}^\sigma_+ \oplus \mathbb{C} \cdot \mathbf{1}$ on which $\mathcal{H}^\sigma_+$ acts by zero and $\mathbf{1}$ acts as the identity. Then π^σ has the structure of a σ–twisted π–module. The operation Y_{π^σ} is given by the following formulas:

$$(5.6.1) \qquad Y_{\pi^\sigma}(b_{-1}, z) = \widetilde{b}(z) = \sum_{n \in \frac{1}{2}+\mathbb{Z}} \widetilde{b}_n z^{-n-1};$$

and the twisted vertex operator assigned to an arbitrary vector $v \in \pi$ is (see [**FLM, D2**])

$$(5.6.2) \qquad Y_{\pi^\sigma}(v, z) = W_{\pi^\sigma}(\exp \Delta_z \cdot v, z),$$

where

$$W_{\pi^\sigma}(b_{n_1}\ldots b_{n_k}, z) = \frac{1}{(-n_1-1)!}\cdots\frac{1}{(-n_k-1)!} :\partial_z^{-n_1-1}\widetilde{b}(z)\ldots\partial_z^{-n_k-1}\widetilde{b}(z):$$

and

$$\Delta_z = \sum_{m,n\geq 0} c_{mn} b_m b_n z^{-m-n}$$

where the constants c_{mn} are determined by the formula

$$\sum_{m,n\geq 0} c_{mn} x^m y^n = -\log\left(\frac{(1+x)^{1/2}+(1+y)^{1/2}}{2}\right)$$

Thus, the twisted module structure is rather complicated, but it is uniquely determined by the assignment (5.6.1), which is a twisted bosonic field.

5.6.3. A Lie algebra associated to an automorphism. The axioms of twisted modules imply, in the same way as in the proof of Proposition 5.1.2, that

$$Y_{M^\sigma}(TA, z^{\frac{1}{N}}) = \partial_z Y_{M^\sigma}(A, z^{\frac{1}{N}}).$$

The second axiom of twisted modules also implies that for any $A, B \in V$ the corresponding twisted vertex operators are mutually local in the sense that there exists a non-negative integer N such that

$$(z-w)^N Y_{M^\sigma}(A, z^{\frac{1}{N}}) Y_{M^\sigma}(B, w^{\frac{1}{N}}) = (z-w)^N Y_{M^\sigma}(B, w^{\frac{1}{N}}) Y_{M^\sigma}(A, z^{\frac{1}{N}}).$$

In the same way as in the case of ordinary vertex operators (see § 3.3.6) this implies the following commutation relations between the coefficients of the twisted vertex operators:

$$(5.6.3) \qquad [A^{M^\sigma}_{(m)}, B^{M^\sigma}_{(k)}] = \sum_{n\geq 0}\binom{m}{n}(A_{(n)}\cdot B)^{M^\sigma}_{(m+k-n)}.$$

In the case of ordinary vertex operators we used the formula for the commutation relations between their Fourier coefficients to define in § 4.1.1 a Lie algebra structure on

$$U'(V) = V \otimes \mathbb{C}[t^{\pm 1}]/\operatorname{Im}(T\otimes 1 + 1\otimes\partial_t).$$

Now we use formula (5.6.3) to construct a twisted version of the Lie algebra $U'(V)$ corresponding to an automorphism σ. Namely, define an operator $\widetilde{\sigma}$ on $V \otimes \mathbb{C}[t^{\pm\frac{1}{N}}]$ by the formula

$$\widetilde{\sigma}(A\otimes t^{\frac{n}{N}}) = e^{-2\pi i\frac{n}{N}}\sigma(A)\otimes t^{\frac{n}{N}}.$$

Let $(V\otimes\mathbb{C}[t^{\pm\frac{1}{N}}])^{\widetilde{\sigma}}$ be the invariant subspace with respect to the action of $\widetilde{\sigma}$ and set

$$U'_\sigma(V) = (V\otimes\mathbb{C}[t^{\pm\frac{1}{N}}])^{\widetilde{\sigma}}/\operatorname{Im}(T\otimes 1 + 1\otimes\partial_t).$$

Then $U'_\sigma(V)$ is spanned by elements of the form $A^{M^\sigma}_{[n]}$ which are the projections of $A\otimes t^{\frac{n}{N}}$ with $A\in V$ such that $\sigma(A) = e^{\frac{2\pi i m}{N}}A$. We define a Lie bracket on $U'_\sigma(V)$ by formula (5.6.3).

It is easy to see that any σ–twisted V–module is a smooth module over the Lie algebra $U'_\sigma(V)$, and hence over its completion $U_\sigma(V)$ defined in the same way as $U(V)$. However, the notion of a coherent $U'_\sigma(V)$–module (and, likewise, the definition of the twisted analogue of the associative algebra $\widetilde{U}(V)$) is more subtle. Indeed, as we can see even in the simplest example given above, the formula for the twisted vertex operator associated to the element of V of the form $A_{(-1)}B$ is rather complicated (though in principle it can be obtained from the second axiom of twisted modules).

Recall that in the case when $V = V_k(\mathfrak{g})$ the $V_k(\mathfrak{g})$–modules are the same as smooth modules over the affine Kac-Moody algebra $\widehat{\mathfrak{g}}$. If σ is a finite order automorphism of $\mathfrak{g}$, then it gives rise to an automorphism of the vertex algebra $\widehat{\mathfrak{g}}$. One can show that the σ–twisted $V_k(\mathfrak{g})$–modules are the same as smooth modules over the twisted affine Kac-Moody algebra associated to σ.

5.7. Constructing new vertex algebras

In this section we discuss three ways to construct new vertex algebras from the existing ones.

5.7.1. Orbifolds and the Monster.
Let G be a group acting by automorphisms on a vertex algebra V. Then the space V^G of G–invariants in V is a vertex subalgebra of V. When G is a finite group, *orbifold theory* allows one to construct V^G–modules from *twisted V–modules*.

If G is a finite group of automorphisms of a vertex algebra V, and M^g is a g–twisted V–module for $g \in G$, then M^g is an ordinary module over the vertex subalgebra V^G of G–invariants in V. Conjecturally, all V^G–modules can be obtained as V^G–submodules of the g–twisted V–modules for $g \in G$.

Let us explain in more detail the situation when the vertex algebra V is *holomorphic*, i.e., it is a rational vertex algebra with a unique simple module, namely itself. Suppose that the cyclic group G generated by an element g of order n acts by automorphisms of V. Then the following pattern is expected to hold (see, e.g., [**D3**]):

(1) for each $h \in G$ there is a unique simple h–twisted V–module V_h;
(2) V_h has a natural G–action, so we can write $V_h = \bigoplus_{i=0}^{n-1} V_h(i)$, where $V_h(i) = \{v \in V_h | g \cdot v = e^{2\pi i k/n} v\}$; then each $V_h(i)$ is a simple V^G–module, and for varying h and i these are all simple V^G–modules;
(3) the vector space $\widetilde{V} = \bigoplus_{h \in G} V_h(0)$ carries a canonical vertex algebra structure, which is holomorphic.

A beautiful example of this pattern is the construction of the *Moonshine Module* vertex algebra $V^\natural$ by I. Frenkel, Lepowsky and Meurman [**FLM**]. In this case V is the vertex algebra V_Λ associated to the Leech lattice Λ (it is self-dual, hence V_Λ is holomorphic, see Example 5.5.5,(1)), and g is constructed from the involution -1 on Λ. Then $V^\natural = V_\Lambda(0) \oplus V_{\Lambda,g}(0)$ is a vertex algebra.

A remarkable fact is that the group of automorphisms of $V^\natural$ is the Monster group [**FLM**]. Moreover, $V^\natural$ is holomorphic [**D3**]. Conjecturally, $V^\natural$ is the unique holomorphic vertex algebra V with central charge 24 and such that $V_1 = 0$. Its character is the modular function $j(\tau) - 744$. More generally, for each element x of the Monster group, consider the Thompson series $\mathrm{Tr}_{V^\natural} x q^{L_0-1}$. Conway and Norton had conjectured that there exists a natural graded representation of the Monster group with the character equal to $j(\tau) - 744$ such that the Thompson series are Hauptmoduls for genus zero subgroups of $SL_2(\mathbb{R})$. After $V^\natural$ was constructed in [**FLM**] it was believed that this graded representation is $V^\natural$. In this formulation, the Conway-Norton conjecture was first proved in [**FLM**] in the case when x commutes with a certain involution, and later in the general case (by a different method) by R. Borcherds [**B2**].

5.7.2. Coset construction. Let V be a vertex algebra, and W a vector subspace of V. Denote by $C(V, W)$ the vector subspace of V spanned by vectors v such that $Y(A, z) \cdot v \in V[[z]]$ for all $A \in W$.

Then $C(V, W)$ is a vertex subalgebra of V. Indeed, this condition implies that all Fourier coefficients of the fields $Y(B, z)$ for $B \in C(V, W)$ commute with all Fourier coefficients of the fields $Y(A, z)$ for $A \in W$. Therefore, the fields $Y(B, z), B \in C(V, W)$, preserve the subspace $C(V, W)$. Thus, one obtains a well-defined vertex operation on $C(V, W)$. Using the formula $[T, Y(A, z)] = \partial_z Y(A, z)$, we obtain that $C(V, W)$ is preserved by T. Moreover, $C(V, W)$ is obviously $\mathbb{Z}_+$-graded and contains $|0\rangle$. The vertex algebra axioms for V imply that these structures on $C(V, W)$ satisfy the vertex algebra axioms.

The resulting vertex algebra $C(V, W)$ is called the *coset vertex algebra* of the pair (V, W). In the case $W = V$, we obtain that the fields $Y(B, z), B \in C(V, V)$, commute with each other and with all other fields in the strong sense (i.e., with $N = 0$ in formula (1.2.3)). Hence the vertex algebra $\mathcal{Z}(V) = C(V, V)$ is commutative (see § 1.4). It is natural to call it the *center* of V. Note that the Fourier coefficients of the fields $Y(B, z), B \in \mathcal{Z}(V)$, lie in the center of the Lie algebra $U(V)$.

An example of a coset vertex algebra is provided by the *Goddard–Kent–Olive construction* [**GKO**]. To explain this construction, we start with the following general facts.

Let $\mathfrak{g}$ be a reductive Lie algebra, decomposed into a direct sum $(\bigoplus_{i=1}^{m} \mathfrak{g}_i) \oplus \mathfrak{r}$, where each $\mathfrak{g}_i$ is a simple Lie algebra, and $\mathfrak{r}$ is an abelian Lie algebra (the center of $\mathfrak{g}$). For any invariant bilinear form ψ on $\mathfrak{g}$ we define the one-dimensional central extension $\widehat{\mathfrak{g}}$ of $\mathfrak{g}((t))$ with the commutation relations given by formula (2.4.2). Then $\widehat{\mathfrak{g}}$ is the direct sum of the central extensions of the Lie algebras $\mathfrak{g}_i((t)), i = 1, \ldots, m$, and $\mathfrak{r}((t))$, with their central elements identified. Here the central extension of $\mathfrak{g}_i((t))$ is isomorphic to the affine Kac–Moody algebra $\widehat{\mathfrak{g}}_i$, and the central extension of $\mathfrak{r}((t))$ is the Heisenberg Lie algebra corresponding to the bilinear form $\psi|_{\mathfrak{r}}$ defined as in § 5.4.1.

Define a vertex algebra $V_\psi(\mathfrak{g})$ in the same way as in the case when $\mathfrak{g}$ is an affine Kac–Moody algebra (see § 2.4.2). Then $V_\psi(\mathfrak{g})$ is the tensor product of vertex algebras corresponding to $\widehat{\mathfrak{g}}_i, i = 1, \ldots, m$ (of levels k_i equal to the ratio between $\psi|_{\mathfrak{g}_i}$ and the canonical bilinear form on $\mathfrak{g}_i$), and the Heisenberg vertex algebra associated to the Heisenberg Lie algebra $\widehat{\mathfrak{r}}$ defined as in § 5.4.2.

Let $\omega_{\mathfrak{g}}$ be the sum of the normalized Segal-Sugawara vectors corresponding to each $\mathfrak{g}_i$ and the conformal vector $\frac{1}{2} \sum_i (b_{-1}^i)^2$ of $\mathfrak{r}$, where $\{b^i\}$ is an orthonormal basis of $\mathfrak{r}$ with respect to $\psi|_{\mathfrak{r}}$. Then $\omega_{\mathfrak{g}}$ is a conformal vector in $V_\psi(\mathfrak{g})$ with central charge

$$c(\mathfrak{g}, \psi) = \sum_{i=1}^{m} \frac{k_i \dim \mathfrak{g}_i}{k_i + h_i^\vee} + \dim \mathfrak{r}.$$

Furthermore, by construction, for any $A \in \mathfrak{g}$ we have

(5.7.1) $$[L_n^{\mathfrak{g}}, A_m] = -m A_{n+m},$$

where the $L_n^{\mathfrak{g}}$'s are the corresponding Virasoro operators.

Now let $\mathfrak{h}$ be a reductive Lie subalgebra of $\mathfrak{g}$, and let ψ' be the invariant bilinear form on $\mathfrak{h}$ obtained by restriction of ψ to $\mathfrak{h}$. Denote by $\widehat{\mathfrak{h}}$ the central extension of

$\mathfrak{h}((t))$ with the cocycle determined by ψ'. Then $\widehat{\mathfrak{h}}$ is a Lie subalgebra of $\widehat{\mathfrak{g}}$, and $V_{\psi'}(\mathfrak{h})$ is naturally a vertex subalgebra of $V_\psi(\mathfrak{g})$. Denote by $\omega_{\mathfrak{h}}$ the corresponding conformal vector in $V_{\psi'}(\mathfrak{h})$.

Let $\omega_{\mathfrak{g}/\mathfrak{h}} = \omega_{\mathfrak{g}} - \omega_{\mathfrak{h}}, L_n^{\mathfrak{g}/\mathfrak{h}} = L_n^{\mathfrak{g}} - L_n^{\mathfrak{h}}$. Then formula (5.7.1) gives

$$(5.7.2) \qquad\qquad [L_n^{\mathfrak{g}/\mathfrak{h}}, A_m] = 0, \qquad \forall A \in \mathfrak{h}.$$

It is straightforward to check using formula (5.7.2) that the operators $L_n^{\mathfrak{g}/\mathfrak{h}}, n \in \mathbb{Z}$, satisfy the commutation relations of the Virasoro algebra with central charge

$$c(\mathfrak{g}, \mathfrak{h}, \psi) = c(\mathfrak{g}, \psi) - c(\mathfrak{h}, \psi').$$

Hence $\omega_{\mathfrak{g}/h}$ is a conformal vector, and by Lemma 3.4.5 it gives rise to a non-trivial homomorphism of vertex algebras $\mathrm{Vir}_{c(\mathfrak{g},\mathfrak{h},\psi)} \to V_\psi(\mathfrak{g})$. By formula (5.7.2), the image of this homomorphism belongs to $C(V_\psi(\mathfrak{g}), V_{\psi'}(\mathfrak{h}))$. Hence $C(V_\psi(\mathfrak{g}), V_{\psi'}(\mathfrak{h}))$ is a conformal vertex algebra with conformal vector $\omega_{\mathfrak{g}/h}$, and the image of the homomorphism $\mathrm{Vir}_{c(\mathfrak{g},\mathfrak{h},\psi)} \to V_\psi(\mathfrak{g})$ is a vertex subalgebra of $V_\psi(\mathfrak{g})$.

There is another version of this construction, in which we replace the vacuum modules $V_\psi(\mathfrak{g})$ and $V_{\psi'}(\mathfrak{h})$ by their irreducible quotients $L_\psi(\mathfrak{g})$ and $L_{\psi'}(\mathfrak{h})$. Consider the special case when $\mathfrak{g} = \mathfrak{sl}_2 \oplus \mathfrak{sl}_2$ and ψ is the direct sum of the canonical bilinear form on the first factor and k times the canonical form on the second factor, where k is a positive integer. Then $L_\psi(\mathfrak{g}) = L_1(\mathfrak{sl}_2) \otimes L_k(\mathfrak{sl}_2)$. Now let $\mathfrak{h}$ be the diagonal $\mathfrak{sl}_2$ subalgebra of $\mathfrak{sl}_2 \oplus \mathfrak{sl}_2$. Then ψ' equals $(k+1)$ times the canonical form, and so $V_{\psi'}(\mathfrak{h}) = L_{k+1}(\mathfrak{sl}_2)$. It follows from the results of Goddard, Kent and Olive [**GKO**] that in this case the coset vertex algebra is isomorphic to the rational Virasoro vertex algebra:

$$C(L_1(\mathfrak{sl}_2) \otimes L_k(\mathfrak{sl}_2), L_{k+1}(\mathfrak{sl}_2)) \simeq L_{c(k+2,k+3)}$$

(see § 4.4.3 for the definition of $L_{c(p,q)}$). For other examples of cosets, see [**dFMS**].

5.7.3. BRST construction. According to Corollary 3.3.8, the residues $A_{(0)}$ of vertex operators $Y(A, z), A \in V$, are infinitesimal automorphisms of the vertex (super)algebra V. Now suppose that $V^\bullet = \bigoplus_{i\in\mathbb{Z}} V^i$ is a vertex superalgebra with an additional $\mathbb{Z}$–gradation (shown by the upper index). Let $A \in V^1$ be a vector of degree 1 with respect to this gradation and such that $A_{(0)}^2 = 0$. Then $(V^\bullet, A_{(0)})$ is a complex, and its cohomology is a vertex superalgebra.

Indeed, the condition that A is homogeneous implies that $A_{(0)}$ is a homogeneous linear operator with respect to the vertex algebra gradation on $V^\bullet$. Hence $\mathrm{Ker}\, A_{(0)}$ and $\mathrm{Im}\, A_{(0)}$ are graded subspaces of $V^\bullet$. In addition, $[T, A_{(0)}] = (TA)_{(0)} = 0$, so T also preserves $\mathrm{Ker}\, A_{(0)}$ and $\mathrm{Im}\, A_{(0)}$. Finally, by Corollary 3.3.8,

$$[A_{(0)}, Y(B, w)] = Y(A_{(0)} \cdot B, w).$$

This implies that all vertex operators $Y(B, z), B \in \mathrm{Ker}\, A_{(0)}$, preserve $\mathrm{Ker}\, A_{(0)}$, so that $\mathrm{Ker}\, A_{(0)}$ is a vertex subalgebra of $V^\bullet$. Moreover, $\mathrm{Im}\, A_{(0)}$ is an ideal in $\mathrm{Ker}\, A_{(0)}$, and so the quotient is a vertex superalgebra.

Important examples of such complexes are provided by *topological vertex superalgebras* introduced by B. Lian and G. Zuckerman [**LZ1**]. In this case the cohomology is a graded commutative vertex superalgebra, but it has the additional structure of a Batalin–Vilkovisky algebra. For example, Lian and Zuckerman have shown in [**LZ3**] that Borcherds' monster Lie algebra (which Borcherds used in his proof [**B2**] of the Conway–Norton conjecture mentioned in § 5.7.1) may be obtained as the cohomology of a topological vertex superalgebra.

Another example is the BRST complex of quantum hamiltonian reduction. In Chapter 15 we define it in the case of quantum Drinfeld–Sokolov reduction, following [**FF7**]. This construction leads to the definition of an important class of vertex algebras, called W–algebras.

5.8. Bibliographical notes

The notion of a module over a vertex algebra was introduced by R. Borcherds [**B1**] (see also [**FLM, FHL**]). Theorem 5.1.6 is due to E. Frenkel.

Lattice vertex algebras were first defined by Borcherds [**B1**] (see also [**Fr**]). They are studied in detail in [**FLM, Kac3**]. Boson-fermion correspondence has remarkable applications to the theory of solitons, due to E. Date, M. Jimbo, M. Kashiwara and T. Miwa (see [**DJKM**]).

The notion of rational vertex algebra was introduced by Y. Zhu [**Z1**]. C. Dong, H. Li and G. Mason [**DLM2**] obtained the characterization of rational vertex algebras that we use in Definition 5.5.1. Zhu's results on the modular invariance of characters were further elucidated in [**Miy**].

Zhu [**Z1**] has attached to each vertex algebra V an associative algebra $A(V)$ such that simple V–modules are in one-to-one correspondence with simple $A(V)$–modules. This was generalized to the case of superalgebras in [**KW**]. Zhu's algebra gives one an efficient tool for classifying simple modules over vertex algebras. For more on Zhu's algebras, the reader is referred to [**FZ, Wan, DLM1, DLM2, DLM3, BN, NT**].

The systematic study of twisted modules was initiated in [**FLM**], where twisted vertex operators were used in the construction of the Moonshine Module vertex algebra (see Chapter 9 of [**FLM**] and the works [**Le1, Le2**]). The notion of the twisted module was formulated in [**FFR, D2**] following [**FLM**]. The notion of twisted module over a vertex algebra was introduced in [**FeFR, D2**]. For more on twisted modules and orbifolds, see [**D2, D3, DLM2, DM, Li2, Tu, FrSz1**]. Topological vertex algebras and related structures are studied in [**LZ1, LZ2, LZ3, Ge, KVZ, HZ**].

CHAPTER 6

Vertex Algebra Bundles

Up to this point, we have discussed fields and vertex algebras in the language of formal power series. Thus we have considered the vertex operation $Y : V \rightarrow \operatorname{End} V[[z^{\pm 1}]]$, or, equivalently, a collection of matrix elements

$$\langle \varphi, Y(A, z)v \rangle \in \mathbb{C}((z)).$$

For fixed $v \in V, \varphi \in V^*$ and varying $A \in V$ we may view this collection as a section of a vector bundle over the punctured disc $D^\times = \operatorname{Spec} \mathbb{C}((z))$ with fiber V^*. Therefore we may consider Y as an $\operatorname{End} V$–valued section of this vector bundle. The question is whether one can define this bundle in such a way that this section is canonical, i.e., independent of the choice of z, considered as a coordinate on the disc. In order to do this, we need a precise description of the transformation properties of Y under changes of coordinates.

In this chapter we explain how to do this in the case when the vertex algebra V carries an action of the group $\operatorname{Aut} \mathcal{O}$ of changes of coordinates by "internal symmetries". This means that the action of $\operatorname{Aut} \mathcal{O}$ comes about by exponentiation of the action of the corresponding Lie algebra $\operatorname{Der}_0 \mathcal{O}$ via Fourier coefficients of a vertex operator acting on V (as in conformal vertex algebras). Application of the OPE formula then allows us to find the precise transformation formula (due originally to Y.-Z. Huang) for all vertex operators under the action of $\operatorname{Aut} \mathcal{O}$. We then use this formula to give an intrinsic geometric meaning to Y. Namely, we attach to each conformal vertex algebra V (and even to more general vertex algebras that we call quasi-conformal) a vector bundle $\mathcal{V}$ on an arbitrary smooth complex curve X. Let x be a point of X, and let $D_x^\times$ be the punctured disc around x (see § 6.5). Then Y gives rise to a section $\mathcal{Y}_x$ of the dual bundle $\mathcal{V}^*|_{D_x^\times}$ with values in $\operatorname{End} \mathcal{V}_x$ (where $\mathcal{V}_x$ is the fiber of $\mathcal{V}$ at x). This geometric realization of Y will eventually enable us to give a global geometric meaning to vertex operators on arbitrary algebraic curves.

At the end of this chapter we define a flat connection on $\mathcal{V}$ and show that $\mathcal{Y}_x$ is a horizontal section with respect to this connection.

6.1. Motivation

Consider the affine Kac-Moody algebra introduced in § 2.4. This is the central extension of the formal loop algebra $L\mathfrak{g} = \mathfrak{g} \otimes \mathbb{C}((z))$. The commutation relations (2.4.2) involve multiplication in $\mathbb{C}((z))$ and taking the residue of a one-form in $\mathbb{C}((z))dz$. Both operations are coordinate-independent, and hence so are the commutation relations. In other words, if we make a change of coordinates $z \mapsto w(z)$, the commutation relations will not change. This enables us to define an affine Kac-Moody algebra in the situation where the field of Laurent series $\mathbb{C}((z))$ is replaced by any complete topological algebra that is isomorphic (but not canonically!) to $\mathbb{C}((z))$.

6.1.1. Abstract discs. To put this in a more geometric context, let us consider the "space" or, more precisely, the scheme underlying the $\mathbb{C}$–algebra $\mathbb{C}[[z]]$. We call this scheme the *standard disc* and denote it by D. In other words, we view $\mathbb{C}[[z]]$ as the ring of (complex valued) functions on the affine scheme

$$D \overset{\text{def}}{=} \operatorname{Spec} \mathbb{C}[[z]].$$

As a topological space, D is very simple: it has two points, the "origin" and the "generic point" (the punctured disc), corresponding to the maximal ideal $z\mathbb{C}[[z]]$ and the zero ideal, respectively. However, there are many non-trivial morphisms between D and other schemes. Indeed, a morphism from D to an affine scheme $Z = \operatorname{Spec} R$, where R is a $\mathbb{C}$–algebra, is nothing but a homomorphism of algebras $R \to \mathbb{C}[[z]]$. If Z is a curve, such a homomorphism (actually, an embedding) may be constructed by realizing $\mathbb{C}[[z]]$ as a completion of R. Geometrically, this should be viewed as identifying the disc D with the formal neighborhood of a point on the curve Z (see § A.1 for more details).

In what follows it is often instructive to think of D as a small complex-analytic disc (keeping in mind that its "radius" is infinitesimally small).

Now, an *abstract disc* is by definition the affine scheme $\operatorname{Spec} \widehat{\mathcal{R}}$, where $\widehat{\mathcal{R}}$ is a $\mathbb{C}$–algebra isomorphic to $\mathbb{C}[[z]]$. The difference between an abstract disc and the standard disc $D = \operatorname{Spec} \mathbb{C}[[z]]$ is that the standard disc carries a coordinate z; in other words, the maximal ideal $z\mathbb{C}[[z]]$ has a preferred generator, z. In contrast, there is usually no preferred generator in the maximal ideal of $\widehat{\mathcal{R}}$, and hence no preferred coordinate on an abstract disc.

For example, let X be a smooth complex curve, and x a point of X. Denote by $\mathcal{O}_x$ the completion of the local ring of x. Then $\mathcal{O}_x$ is non-canonically isomorphic to $\mathcal{O} = \mathbb{C}[[z]]$. To specify such an isomorphism, or equivalently, an isomorphism between

$$D_x \overset{\text{def}}{=} \operatorname{Spec} \mathcal{O}_x$$

and $D = \operatorname{Spec} \mathbb{C}[[z]]$, we need to choose a formal coordinate t_x at x, i.e., a topological generator of the maximal ideal $\mathfrak{m}_x$ of $\mathcal{O}_x$. But in general there is no preferred formal coordinate at x. So D_x is an abstract disc. Actually, these are all the abstract discs that we will need, so from now on we will only consider such discs, even though all of our constructions in this chapter work equally well for an arbitrary abstract disc.

6.1.2. Lie algebras and modules attached to discs. We would like to think that the "standard" affine Kac-Moody algebra $\widehat{\mathfrak{g}}$ is attached to the standard disc $D = \operatorname{Spec} \mathbb{C}[[z]]$, and we wish to attach such a Lie algebra to any abstract disc such as D_x, where x is a point of a smooth curve. Let $\mathcal{O}_x$ be the corresponding complete local ring and $\mathcal{K}_x$ its field of fractions. If we choose a coordinate t_x on D_x, then we obtain isomorphisms $\mathcal{O}_x \simeq \mathbb{C}[[t_x]]$ and $\mathcal{K}_x \simeq \mathbb{C}((t_x))$. Now we define the affine Kac-Moody algebra $\widehat{\mathfrak{g}}_x$ attached to D_x as the one-dimensional central extension of the Lie algebra $\mathfrak{g} \otimes \mathcal{K}_x$ with the commutation relations given by formula (2.4.2). To use these commutation relations, we need to choose a coordinate t_x on D_x and identify $\mathcal{K}_x$ with $\mathbb{C}((t_x))$. But as we explained above, the commutation relations do not depend on this choice: if we choose another coordinate t'_x, the result will be the same.

Next, we define the vacuum module $\mathcal{V}_k(\mathfrak{g})_x$ attached to D_x. Observe that $\mathfrak{g} \otimes \mathcal{O}_x$ is a Lie subalgebra of $\mathfrak{g} \otimes \mathcal{K}_x$ and $\mathfrak{g} \otimes \mathcal{O}_x \oplus \mathbb{C}K$ is a Lie subalgebra of $\widehat{\mathfrak{g}}_x$. Let $\mathbb{C}_k$ be the one-dimensional module over $\mathfrak{g} \otimes \mathcal{O}_x \oplus \mathbb{C}K$ on which $\mathfrak{g} \otimes \mathcal{O}_x$ acts by 0 and K by multiplication by $k \in \mathbb{C}$. Then we set

$$(6.1.1) \qquad \mathcal{V}_k(\mathfrak{g})_x = \mathrm{Ind}_{\mathfrak{g} \otimes \mathcal{O}_x \oplus \mathbb{C}K}^{\widehat{\mathfrak{g}}_x} \mathbb{C}_k.$$

Again, $\mathcal{V}_k(\mathfrak{g})_x$ is isomorphic to $V_k(\mathfrak{g})$ (which is the vacuum module attached to the standard disc), but not canonically. However, $\mathcal{V}_k(\mathfrak{g})_x$ carries a canonical action of the Lie algebra $\widehat{\mathfrak{g}}_x$.

In application it is important to have a unified theory of all vacuum modules $\mathcal{V}_k(\mathfrak{g})_x$ attached to arbitrary points of arbitrary smooth curves. Therefore it is natural to ask what kind of structure the vertex algebra structure on $V_k(\widehat{\mathfrak{g}})$ induces on all of the $\mathcal{V}_k(\mathfrak{g})_x$'s. In other words, what is the coordinate-independent meaning of vertex operators?

Recall that the vertex algebra $V_k(\mathfrak{g})$ is generated by the vertex operators

$$(6.1.2) \qquad J^a(z) = \sum_{n \in \mathbb{Z}} J_n^a z^{-n-1},$$

where the J^a's are elements of a basis of $\mathfrak{g}$, and $J_n^a = J^a \otimes z^n \in \widehat{\mathfrak{g}}$. The operators J_n^a naturally act on $V_k(\mathfrak{g})$, and $J^a(z)$ is the generating function of these operators.

In the case of an arbitrary vacuum module $\mathcal{V}_k(\mathfrak{g})_x$ we have an action of the Lie algebra $\widehat{\mathfrak{g}}_x$. In particular, we have an action of the subspace $J^a \otimes \mathcal{K}_x \subset \widehat{\mathfrak{g}}_x$ that is isomorphic to $\mathcal{K}_x$. We claim that the vertex operator (6.1.2) gives rise to a canonical differential on the punctured disc

$$D_x^\times \overset{\mathrm{def}}{=} \mathrm{Spec}\,\mathcal{K}_x$$

with values in $\mathrm{End}\,\mathcal{V}_k(\mathfrak{g})_x$. Before explaining this, we recall the notion of differential.

6.1.3. Differentials. Let Δ be an integer. A Δ–differential on a smooth curve is by definition a section of the Δth tensor power of the canonical line bundle Ω (see § 8.3.1). The 1–differentials are called simply differentials, or one-forms.

If we choose a local coordinate z we may trivialize $\Omega^{\otimes\Delta}$ by the non-vanishing section $(dz)^{\otimes\Delta}$. For brevity, from now on we will write Ω^Δ and dz^Δ. Any section s of Ω^Δ may then be written as $f(z)(dz)^\Delta$. If we choose another coordinate $w = \rho(z)$, then the same section will be written as $g(w)(dw)^\Delta$, where

$$(6.1.3) \qquad f(z) = g(\rho(z))(\rho'(z))^\Delta.$$

Now let us suppose that we have a section s of Ω^Δ whose representation by a function does not depend on the choice of local coordinate, i.e., $g(w) = f(w)$ and so $f(z) = f(\rho(z))(\rho'(z))^\Delta$ for any change of variable $\rho(z)$. In that case we can say that $f(z)dz^\Delta$ is a canonical Δ–differential. Of course, we know that canonical Δ–differentials cannot exist when $\Delta \neq 0$ (for instance, if $\Delta = 1$ that would mean that our curve is equipped with a canonical metric). If $\Delta = 0$, then the constant functions $f(z) = \mathrm{const}$ are the only canonical sections of $\Omega^0 = \mathcal{O}$. However, if we consider sections of Ω^Δ with values in a vector space that itself transforms non-trivially under changes of coordinates, canonical sections may exist.

Denote the space of differentials on $D_x^\times$ by Ω_x.

6.1.4. Lemma. *Suppose we are given a linear map* $\rho : \mathcal{K}_x \to \operatorname{End} W$, *where* W *is a vector space, such that for any* $v \in W$ *we have* $\rho(\mathfrak{m}_x)^N \cdot v = 0$ *for large enough* N, *where* $\mathfrak{m}_x$ *is the maximal ideal of* $\mathcal{O}_x$. *Choose a formal coordinate* t_x *on* D_x, *i.e., a generator of* $\mathfrak{m}_x$. *Then*

$$\varpi_x = \sum_{n \in \mathbb{Z}} \rho(t_x^n) t_x^{-n-1} dt_x$$

is a canonical $\operatorname{End} W$*-valued differential on* $D_x^\times$, *i.e., it is independent of the choice of coordinate* t_x.

6.1.5. Proof. If we choose a coordinate t_x on D_x, then we obtain an isomorphism $\Omega_x \simeq \mathbb{C}((t_x))dt_x$. We have a pairing

$$\mathcal{K}_x \times \Omega_x \to \mathbb{C},$$

$$f, \omega \mapsto \operatorname{Res}_x f\omega,$$

where Res_x is the residue map (see § 9.1.2). This pairing identifies Ω_x with the topological dual vector space to $\mathcal{K}_x$.

If we choose a coordinate t_x and write $f\omega$ as $\sum_n f_n t_x^n dt_x$, then $\operatorname{Res}_x f\omega = f_{-1}$. But this pairing is independent of the choice of t_x.

Using the above pairing, we may reformulate the assertion of the lemma as follows. For any $f \in \mathcal{K}_x$ the linear operator on W defined by the formula $\operatorname{Res}_x f\varphi_x$ is coordinate-independent. But it follows from the definition of ϖ_x that this operator is nothing but the operator of the action of $\rho(f)$ on W. Hence it is coordinate-independent by our assumption. This completes the proof.

The lemma has the following immediate consequence:

6.1.6. Corollary. *Given a coordinate* t_x *on* D_x, *define an* $\operatorname{End} V_k(\mathfrak{g})_x$*-valued differential on* $D_x^\times$ *by the formula*

$$J^a(t_x)dt_x = \sum_{n \in \mathbb{Z}} J_n^a t_x^{-n-1} dt_x,$$

where J_n^a *is the operator on* $V_k(\mathfrak{g})_x$ *corresponding to the action of*

$$J^a \otimes t_x^n \in J^a \otimes \mathbb{C}((t_x)) \simeq J^a \otimes \mathcal{K}_x \subset \widehat{\mathfrak{g}}_x,$$

where we identify $V_k(\mathfrak{g})_x$ *with* $V_k(\mathfrak{g})$ *using the coordinate* t_x. *Then* $J^a(t_x)dt_x$ *is a canonical* $\operatorname{End} V_k(\mathfrak{g})_x$*-valued differential on* $D_x^\times$.

6.1.7. General vertex algebras. The corollary explains the meaning of the generating vertex operators $J^a(z)$ of the vertex algebra $V_k(\mathfrak{g})$ for vacuum modules attached to arbitrary discs: $J^a(z)dz$ is a canonically defined $\operatorname{End} V_k(\mathfrak{g})_x$-valued one-form on any formal punctured disc $D_x^\times$ (for any choice of formal coordinate z at x). We would like to have a similar statement for arbitrary vertex operators from $V_k(\mathfrak{g})$. More generally, we would like to develop a similar construction for a general vertex algebra V. Namely, we wish to attach to any disc a certain twist $\mathcal{V}_x$ of V, so that V is attached to the standard disc, and for any coordinate t_x on D_x we have an isomorphism

$$\imath_{t_x, x} : V \xrightarrow{\sim} \mathcal{V}_x.$$

We then wish to associate to vertex operators on V sections of some bundles on $D_x^\times$.

The system of isomorphisms $\imath_{t_x,x}$ should satisfy a natural compatibility condition: if t_x and t'_x are two coordinates on D_x such that $t'_x = \rho(t_x)$, then we obtain an automorphism $\imath_{t'_x,x}^{-1} \circ \imath_{t_x,x}$ of V. The condition is that the assignment

$$\rho(z) \mapsto \imath_{t'_x,x}^{-1} \circ \imath_{t_x,x}$$

defines a representation on V of the group Aut $\mathcal{O}$ of changes of coordinates. If this condition is satisfied, then $\mathcal{V}_x$ gets canonically identified with the twist of V by the Aut $\mathcal{O}$–torsor of formal coordinates at x.

In the next section we will explain these notions in more detail and give the precise definitions. This will enable us to construct the twists $\mathcal{V}_x$ for an arbitrary quasi-conformal vertex algebra V and ultimately to give a coordinate-independent interpretation of the vertex algebra structure on V.

6.2. The group Aut $\mathcal{O}$

Let $\mathcal{O}$ denote the complete topological $\mathbb{C}$–algebra $\mathbb{C}[[z]]$, and let Aut $\mathcal{O}$ be the group of continuous automorphisms of $\mathcal{O}$. Such an automorphism is completely determined by its action on the topological generator z of $\mathbb{C}[[z]]$. Thus, any element ρ of Aut $\mathcal{O}$ may be represented by $\rho(z) \in \mathbb{C}[[z]]$. Clearly, ρ is a continuous automorphism of $\mathbb{C}[[z]]$ if and only if $\rho(z)$ is a formal power series in z of the form $a_1 z + a_2 z^2 + \ldots$, with $a_1 \in \mathbb{C}^\times$. In what follows we identify Aut $\mathcal{O}$ with the set of such series. The composition law is then given by the formula

$$(\rho_1 * \rho_2)(z) = \rho_2(\rho_1(z)).$$

For example,

$$(z + z^2) * (z + z^3) = (z + z^2) + (z + z^2)^3.$$

These automorphisms take z to all possible topological generators of $\mathcal{O}$.

Let us describe the structure of Aut $\mathcal{O}$ in more detail. Consider the following Lie groups and Lie algebras:

$$
\begin{array}{ccc}
\mathrm{Aut}_+ \mathcal{O} = \{z + a_2 z^2 + \ldots\} & \quad & \mathrm{Der}_+ \mathcal{O} = z^2 \mathbb{C}[[z]]\partial_z \\
\cap & & \cap \\
\mathrm{Aut}\, \mathcal{O} = \{a_1 z + a_2 z^2 + \ldots, a_1 \neq 0\} & & \mathrm{Der}_0\, \mathcal{O} = z\mathbb{C}[[z]]\partial_z \\
& & \cap \\
& & \mathrm{Der}\, \mathcal{O} = \mathbb{C}[[z]]\partial_z
\end{array}
$$

Denote by $\mathbb{G}_m$ the multiplicative group (this is the algebraic group whose group of $\mathbb{C}$–points is $\mathbb{C}^\times$).

6.2.1. Lemma.

(1) Aut $\mathcal{O}$ *is a semi-direct product of* $\mathbb{G}_m$ *and* $\mathrm{Aut}_+ \mathcal{O}$.
(2) $\mathrm{Aut}_+ \mathcal{O}$ *has the structure of a prounipotent proalgebraic group.*
(3) $\mathrm{Lie}(\mathrm{Aut}\, \mathcal{O}) = \mathrm{Der}_0\, \mathcal{O}$, $\mathrm{Lie}(\mathrm{Aut}_+ \mathcal{O}) = \mathrm{Der}_+ \mathcal{O}$, *and the exponential map* $\exp : \mathrm{Der}_+ \mathcal{O} \to \mathrm{Aut}_+ \mathcal{O}$ *is an isomorphism.*

6.2.2. Proof.

It is clear from the definition that $\mathrm{Aut}_+ \mathcal{O}$ is a normal subgroup of Aut $\mathcal{O}$, and that the quotient is isomorphic to $\mathbb{G}_m$. Furthermore, we obtain that the adjoint action of $\mathbb{G}_m$ on $\mathrm{Aut}_+ \mathcal{O}$ is given by $h \cdot \rho(z) = h\rho\left(\frac{z}{h}\right)$.

Next, there is a family of projections

$$\mathrm{Aut}_+ \mathcal{O} \to \mathrm{Aut}_+ (\mathcal{O}/z^n \mathcal{O})$$

expressing $\text{Aut}_+ \mathcal{O}$ as a projective limit

$$\text{Aut}_+ \mathcal{O} = \varprojlim \text{Aut}_+(\mathcal{O}/z^n \mathcal{O})$$

of unipotent groups. Hence $\text{Aut}_+ \mathcal{O}$ is a prounipotent algebraic group. Similarly, $\text{Der}_+ \mathcal{O} = \varprojlim \text{Der}_+(\mathcal{O}/z^n \mathcal{O})$. The exponential map is an isomorphism at each step, giving an isomorphism on the projective limits as desired. Note that we have

$$\exp\left(\rho \partial_z\right)(z) = z + \rho + \frac{1}{2}\rho \rho' + \dots, \qquad \rho \partial_z \in \text{Der}_+ \mathcal{O}.$$

6.2.3. Digression: Group functors. At this stage we notice that something is wrong with the group $\text{Aut}\, \mathcal{O}$: one expects the derivations of a ring to form the Lie algebra of its group of automorphisms, but $\text{Der}\, \mathcal{O}$ is *bigger* than $\text{Der}_0 \mathcal{O} = \text{Lie}(\text{Aut}\, \mathcal{O})$. It seems that we are unable to exponentiate the infinitesimal shift ∂_z.

This anomaly is resolved by considering the group of automorphisms of $\mathcal{O}$ not as a set, but as a *group functor* over $\mathbb{C}$ (see § A.1.2). In other words, rather than studying the group of continuous automorphisms of $\mathbb{C}[[z]]$, we should consider the corresponding group functor on $\mathbb{C}$–algebras and see if it is represented by a group scheme or a group ind-scheme. This group functor is defined as follows: it assigns to a $\mathbb{C}$–algebra R the group of continuous automorphisms of the complete topological R–algebra $R[[z]]$. One easily verifies that these automorphisms are of the form

$$z \mapsto r_0 + r_1 z + r_2 z^2 + \dots,$$

where r_0 is *nilpotent* and r_1 is a unit in R (if $R = \mathbb{C}$, then there are no nilpotents, so we obtain the transformations we considered above). This group functor is represented by the *ind–scheme*

$$\underline{\text{Aut}}\mathcal{O} = \varinjlim \text{Spec}\, \mathbb{C}[a_0, a_1, a_1^{-1}, a_2, a_3, \dots]/(a_0^N).$$

Thus we can exponentiate ∂_z, but only in "nilpotent directions". Now we indeed recover the full space of derivations $\text{Der}\, \mathcal{O}$ as the Lie algebra of the ind–group $\underline{\text{Aut}}\mathcal{O}$.

Note that $\underline{\text{Aut}}\mathcal{O}$ is not representable by a scheme. Rather, it is an extension of the formal additive group $\widehat{\mathbb{G}}_a = \text{Spf}\, \mathcal{O}$ (the group of translations on the formal disc) by the group scheme $\text{Aut}\, \mathcal{O}$. In particular, the formal additive group $\widehat{\mathbb{G}}_a$ is identified with the quotient $\underline{\text{Aut}}\mathcal{O}/\text{Aut}\, \mathcal{O}$.

Thus, if we allow ourselves to consider rings with nilpotents, we can see automorphisms of $\mathcal{O}$ which are "invisible" over $\mathbb{C}$. The existence of these automorphisms is important: we will use them to construct flat connections on vertex algebra bundles in § 17.1.3.

From the point of view of this discussion, $\text{Aut}\, \mathcal{O}$ is really the group of automorphisms of the *pointed* disc, i.e., it represents the functor which assigns to a $\mathbb{C}$ algebra R the group of continuous automorphisms of $R[[t]]$ that preserve its maximal ideal $tR[[t]]$ (defining the "zero point" of the disc).

6.3. Exponentiating vector fields

Given a vertex algebra and a field $Y(A, z)$, we may perform the substitution $z \mapsto \rho(z)$, giving rise to a new field $Y(A, \rho(z))$. We now seek an action

$$\text{Aut}\, \mathcal{O} \times V \to V, \qquad (\rho, A) \mapsto R(\rho)A,$$

such that $Y(A, \rho(z))$ is related to $Y(R(\rho)A, z)$ in some reasonable way. In other words, we seek to incorporate the action of changes of variables into the structure of the vertex algebra. It is natural to try to construct an action of $\operatorname{Aut}\mathcal{O}$ on V by exponentiating the action of $\operatorname{Der}_0 \mathcal{O}$. This raises the question as to which $\operatorname{Der}_0 \mathcal{O}$–modules can be exponentiated to $\operatorname{Aut}\mathcal{O}$.

Recall that $\operatorname{Aut}\mathcal{O}$ is a semi-direct product of $\operatorname{Aut}_+ \mathcal{O}$ with $\mathbb{G}_m$. Suppose we are given a representation of the Lie algebra

$$\operatorname{Lie}\mathbb{G}_m \simeq \mathbb{C} \cdot z\partial_z$$

on a vector space V. Then this representation can be exponentiated to a representation of the multiplicative group $\mathbb{G}_m$ (i.e., an action of $\mathbb{C}^\times$ on V) if and only if the action of $z\partial_z$ on V is diagonalizable and its eigenvalues are integers. In that case we define the $\mathbb{C}^\times$–action on V by having $a \in \mathbb{C}^\times$ act by a^n on the eigenvectors of $z\partial_z$ with eigenvalue n. We can then write $a \mapsto a^{z\partial_z}$. Thus we obtain the usual equivalence between $\mathbb{C}^\times$–actions and $\mathbb{Z}$–gradations on a vector space.

Next, we have the exponential map $\operatorname{Der}_+ \mathcal{O} = z^2\mathbb{C}[[z]]\partial_z \to \operatorname{Aut}_+ \mathcal{O}$, which is an isomorphism. We may exponentiate a $\operatorname{Der}_+ \mathcal{O}$–action on a vector space M to $\operatorname{Aut}_+ \mathcal{O}$, if it is locally nilpotent, i.e., for any $v \in M$ and $x \in \operatorname{Der}_+ \mathcal{O}$, $x^N \cdot v = 0$ for sufficiently large N. This guarantees that the formula for the action of $\exp x$ on v is a finite sum.

Thus, we obtain that an action of $\operatorname{Der}_0 \mathcal{O}$ on a module M can be exponentiated to an action of $\operatorname{Aut}\mathcal{O}$ if the following conditions are satisfied:

(1) The action of the Euler vector field $z\partial_z$ is diagonalizable with integral eigenvalues.
(2) The action of $\operatorname{Der}_+ \mathcal{O}$ is locally nilpotent.

6.3.1. Explicit formulas. A conformal vertex algebra V (see Definition 2.5.8) is equipped with an action of the Virasoro algebra, and hence its Lie subalgebra $\operatorname{Der}_0 \mathcal{O}$, which is the Lie algebra of $\operatorname{Aut}\mathcal{O}$.

Let us suppose that the $\mathbb{Z}$–gradation on V by the operator L_0 is bounded from below, i.e., $V = \bigoplus_{n=K}^\infty V_n$ for some $K \in \mathbb{Z}$. Then the action of the Lie subalgebra $\operatorname{Der}_+ \mathcal{O}$ is locally nilpotent, since the vector field $z^{j+1}\partial_z, j > 0$, acts as the operator $-L_j$, which has degree $-j$. Furthermore, $z\partial_z$ acts as the operator $-L_0$, which is the gradation operator, and therefore is diagonalizable with integral eigenvalues. Thus, the action of $\operatorname{Der}_0 \mathcal{O}$ on a conformal vertex algebra V can be exponentiated to an action of $\operatorname{Aut}\mathcal{O}$.

To write this action more explicitly, recall that any element of $\operatorname{Aut}\mathcal{O}$ can be considered as a power series

$$(6.3.1) \qquad z \mapsto f(z) = a_1 z + a_2 z^2 + \dots, \qquad a_1 \neq 0.$$

Given f, we can find $v_i \in \mathbb{C}, i \geq 0$, such that

$$f(z) = \exp\left(\sum_{i>0} v_i z^{i+1}\partial_z\right) v_0^{z\partial_z} \cdot z$$

(here $v_0^{z\partial_z} z = v_0 z$, since $z\partial_z \cdot z = z$). Thus

$$\exp\left(\sum_{i>0} v_i z^{i+1}\partial_z\right) v_0 z = v_0 z + \sum_{i>0} v_i v_0 z^{i+1} + \frac{1}{2}\left(\sum_{i>0} v_i z^{i+1}\partial_z\right) \cdot \sum_{i>0} v_i v_0 z^{i+1} + \dots$$

Comparing this series term by term with (6.3.1) gives rise to the system of equations

$$
\begin{aligned}
v_0 &= a_1, \\
v_1 v_0 &= a_2, \\
v_2 v_0 + v_1^2 v_0 &= a_3, \quad \text{etc.,}
\end{aligned}
$$

which may be solved recursively for all v_i. Identifying a_i with the Taylor coefficient $\frac{1}{n!} f^{(n)}(0)$ of $f(z)$, we obtain

$$
\begin{aligned}
v_0 &= f'(0), \\
(6.3.2) \qquad v_1 &= \frac{1}{2} \frac{f''(0)}{f'(0)}, \\
v_2 &= \frac{1}{6} \frac{f'''(0)}{f'(0)} - \frac{1}{4} \left(\frac{f''(0)}{f'(0)} \right)^2, \quad \text{etc.}
\end{aligned}
$$

Under the representation $z^{j+1} \partial_z \mapsto -L_j$ we define a linear operator $R(f) : V \to V$ by

$$
(6.3.3) \qquad R(f) = \exp\left(-\sum_{j>0} v_j L_j \right) v_0^{-L_0}.
$$

Note that for each $B \in V_n$ we have $v_0^{-L_0} B = v_0^{-n} B$, and $R(f)B$ is a sum of finitely many terms. Moreover, by construction, the subspaces $V_{\leq m} \overset{\text{def}}{=} \bigoplus_{n=K}^m V_n$ are stable under all operators $R(f)$, $f \in \text{Aut}\, \mathcal{O}$. Thus we obtain:

6.3.2. Lemma. *The assignment $f \mapsto R(f)$ defines a representation of $\text{Aut}\, \mathcal{O}$ on V (in other words, $R(f * g) = R(f)R(g)$) which is the inductive limit of the representations $V_{\leq m}, m \geq K$.*

6.3.3. Commutation relations. We wish to understand the relation between the fields $Y(A, z)$ and $Y(A, f(z))$ in a general conformal vertex algebra. First we will compute the infinitesimal action, using the Virasoro field

$$
T(z) = \sum_{n \in \mathbb{Z}} L_n z^{-n-2}.
$$

For the "exponentiated" version see Lemma 6.5.6 below.

By the associativity property (see formula (3.3.1)), for any $A \in V$,

$$
T(z)Y(A, w) = \sum_{m \in \mathbb{Z}} Y(L_m A, w)(z - w)^{-m-2}.
$$

Hence we obtain from formula (3.3.12) that

$$
\begin{aligned}
(6.3.4) \qquad [L_n, Y(A, w)] &= \sum_{m \geq -1} \binom{n+1}{m+1} Y(L_m A, w) w^{n-m} \\
&= \sum_{m \geq -1} \frac{1}{(m+1)!} (\partial_w^{m+1} w^{n+1}) Y(L_m A, w).
\end{aligned}
$$

More generally, given a vector field

$$
v(z) \partial_z = \sum_{n \geq -1} v_n z^{n+1} \partial_z \in \text{Der}\, \mathcal{O},
$$

we assign to it the operator $\mathbf{v} = -\sum_{n \geq -1} v_n L_n$, and obtain from the above formula

$$(6.3.5) \qquad [\mathbf{v}, Y(A, w)] = -\sum_{m \geq -1} \frac{1}{(m+1)!} (\partial_w^{m+1} v(w)) Y(L_m A, w).$$

Suppose that we are given a vertex algebra V equipped with an action of the Lie algebra Der $\mathcal{O}$ satisfying the commutation relations (6.3.5) and such that it can be exponentiated to an action of Aut $\mathcal{O}$. We will show that in this case we can attach to V a vector bundle $\mathcal{V}$ on any smooth curve X, and that the matrix elements of vertex operators on V give rise to a collection of intrinsic (i.e., coordinate-independent) sections of the dual bundle $\mathcal{V}^*$ in the neighborhood of any point. We introduce special terminology for such vertex algebras.

6.3.4. Definition. A vertex algebra is called *quasi-conformal* if it carries an action of Der $\mathcal{O}$ such that formula (6.3.5) holds for any $A \in V$, the element $L_{-1} = -\partial_z$ acts as the translation operator T, $L_0 = -z\partial_z$ acts semi-simply with integral eigenvalues, and the Lie subalgebra Der$_+$ $\mathcal{O}$ acts locally nilpotently.

The above computation shows that any conformal vertex algebra on which Der$_+$ $\mathcal{O}$ acts locally nilpotently (in particular, one that is $\mathbb{Z}$–graded with gradation bounded from below) is automatically quasi-conformal. But there are quasi-conformal vertex algebras that are not conformal. For example, the vertex algebra $V_{-h^\vee}(\mathfrak{g})$, which will be discussed later in § 18.4, is quasi-conformal, but not conformal. Another class of examples comes from commutative vertex algebras (see § 1.4), which certainly cannot be conformal.

Let V is a commutative vertex algebra. Recall from § 1.4 that a commutative vertex algebra carries an algebra structure, so that $Y(A, z)$ is the operator of multiplication by $e^{zT} A$ for any $A \in V$.

6.3.5. Lemma. *A commutative vertex algebra V is quasi-conformal if and only if it is equipped with an action of the Lie algebra Der $\mathcal{O}$ by derivations of the algebra structure such that L_{-1} is the translation operator T on V, L_0 acts semi-simply with integral eigenvalues, and the Lie subalgebra Der$_+$ $\mathcal{O}$ acts locally nilpotently.*

6.3.6. Proof. Let V be a commutative vertex algebra satisfying the above conditions. Then $[L_n, T] = (n+1)L_{n-1}$. Hence we find that for any $A \in V$

$$[L_n, Y(A, w)] = [L_n, e^{wT} A] = \sum_{k \geq 0} \frac{w^k}{k!} [L_n, T^k A]$$

$$= \sum_{m \geq -1} \binom{n+1}{m+1} w^{n-m} e^{wT} (L_m A) = \sum_{m \geq -1} \binom{n+1}{m+1} w^{n-m} Y(L_m A, w)$$

(here by an element of V we understand the corresponding multiplication operator on V, and we use the fact that the L_n's are derivations of the algebra V). Thus we obtain formula (6.3.4), which implies (6.3.5). Hence V is quasi-conformal. Verification of the converse statement is straightforward and is left to the reader.

6.4. Primary fields

Before giving a coordinate-independent description of all vertex operators at once, we start with the simplest case of vertex operators corresponding to one-dimensional Aut $\mathcal{O}$–submodules of V. These are called primary fields.

6.4.1. Definition. A vector A in a quasi-conformal vertex algebra V is called *primary* of conformal dimension $\Delta \in \mathbb{Z}_+$ if

$$L_n \cdot A = 0, \quad n > 0, \qquad L_0 \cdot A = \Delta A.$$

Such vectors are also known as singular, or null, vectors for the Virasoro action.

A vector is primary if and only if it generates a one–dimensional $\mathrm{Der}_0\, \mathcal{O}$ (and hence Aut $\mathcal{O}$) sub-representation of V.

The vertex operator $Y(A, z)$ corresponding to a primary vector A is called a *primary field*. According to formula (3.3.1), it has the following OPE with $T(z)$:

$$T(z)A(w) = \frac{\Delta A(w)}{(z-w)^2} + \frac{\partial_w A(w)}{z-w} + \mathrm{reg}.$$

6.4.2. Examples. The vector $b_{-1} \in \pi$ (and hence the field $b(z)$) is primary if we choose the conformal vector $\omega_0 = \frac{1}{2}b_{-1}^2$. Likewise, the vector $J_{-1}^a v_k \in V_k(\mathfrak{g})$ is primary with respect to the Segal-Sugawara conformal vector. All of these vectors have conformal dimension 1.

The fields $\psi(z)$ and $\psi^*(z)$ in the fermionic vertex superalgebra $\bigwedge$ introduced in § 5.3.1 are primary with respect to the conformal vector (5.3.8), of conformal dimensions $1 - \mu$ and μ, respectively.

Next, observe that the Virasoro vertex algebra Vir_c contains primary vectors of positive conformal dimension (that is, different from v_c) if and only if Vir_c is reducible as a module over the Virasoro algebra. Indeed, suppose that Vir_c contains a proper submodule W (so W cannot contain v_c). Since Vir_c is $\mathbb{Z}_+$–graded with respect to the action of L_0, so is W. Hence the degrees of elements of W are bounded from below by some $N > 0$. Any homogeneous vector in W of minimal degree N must be killed by all annihilation operators L_n, $n > 0$, and thus it is a primary vector of conformal dimension N. Conversely, the submodule of Vir_c generated by a primary vector of conformal dimension $N > 0$ will not contain elements of lower degrees, and hence will be proper.

The values of c for which Vir_c is reducible (and hence contains a primary vector) are exactly the values $c(p, q)$ listed in formula (4.4.3). The submodule generated by this vector is irreducible, and so is the quotient $L_{c(p,q)}$ of $\mathrm{Vir}_{c(p,q)}$ by this submodule, mentioned in §§ 4.4.3 and 5.5.5.

6.4.3. Changes of coordinates. Consider the primary field $Y(A, z)$ associated to a primary vector A of conformal dimension Δ. Equation (6.3.5) now gives

$$(6.4.1) \qquad \begin{aligned} [\mathbf{v}, Y(A, w)] &= -v(w)Y(L_{-1}A, w) - \Delta \partial_w v(w) \cdot Y(A, w) \\ &= -v(w)\partial_w Y(A, w) - \Delta \partial_w v(w) \cdot Y(A, w). \end{aligned}$$

Our next goal is to show that $Y(A, z)$ behaves under changes of the coordinate z as a Δ–differential on the punctured disc $D^\times = \mathrm{Spec}\,\mathbb{C}((z))$ with values in the endomorphisms of a twist of V.

6.4.4. Proposition. *If* $Y(A, z)$ *is a primary field, then*

$$Y(A, z) = R(\rho)Y(A, \rho(z))R(\rho)^{-1}(\rho'(z))^{\Delta}.$$

6.4.5. Proof. Let $\mathcal{F}(V)$ be the space of fields on V. Recall from § 1.2.1 that $\mathcal{F}(V)$ consists of all formal power series $B(z) \in \text{End } V[[z^{\pm 1}]]$ such that $B(z)v \in V((z))$ for all $v \in V$. In particular, all vertex operators belong to $\mathcal{F}(V)$. Given $\rho(z) = a_1 z + a_2 z^2 + \ldots \in \text{Aut } \mathcal{O}$, we write

$$\rho(z)^{-1} = a_1^{-1}z^{-1}(1 + a_2 a_1^{-1}z + \ldots)^{-1} \in z^{-1}\mathbb{C}[[z]].$$

Then for any $B(z) \in \mathcal{F}(V)$, the series $B(\rho(z))$ is also a well-defined element of $\mathcal{F}(V)$. Define a linear operator $\widetilde{R}(\rho)$ on $\mathcal{F}(V)$ by the formula

$$\widetilde{R}(\rho) \cdot B(z) = R(\rho)B(\rho(z))R(\rho)^{-1}(\rho'(z))^{\Delta}.$$

One checks easily that $\widetilde{R}(\rho * \mu) = \widetilde{R}(\rho)\widetilde{R}(\mu)$, and so $f \to \widetilde{R}(f)$ is a representation of the Lie group $\text{Aut } \mathcal{O}$ on $\mathcal{F}(V)$. We need to prove that the primary field $Y(A, z)$ is invariant under this representation. Since the exponential map $\text{Der}_0 \, \mathcal{O} \to \text{Aut } \mathcal{O}$ is surjective, it suffices to prove that the corresponding infinitesimal action of the Lie algebra $\text{Der}_0 \, \mathcal{O}$ on $Y(A, z)$ is trivial.

Let us compute this infinitesimal action. Write

$$\rho(z) = z + \epsilon v(z) + o(\epsilon), \qquad R(\rho) = \text{Id} + \epsilon \mathbf{v} + o(\epsilon),$$

where $v(z)\partial_z$ is a vector field, and $\mathbf{v}$ is the corresponding element of the Virasoro algebra. Ignoring powers of ϵ greater than 1, we obtain

$$
\begin{aligned}
\widetilde{R}(\rho) \cdot Y(A, z) &= (\text{Id} + \epsilon \mathbf{v})Y(A, z + \epsilon v(z))(\text{Id} - \epsilon \mathbf{v})((1 + \epsilon v(z)'))^{\Delta} \\
&= Y(A, z) + \epsilon([\mathbf{v}, Y(A, z)] + v(z)\partial_z Y(A, z) + \Delta v'(z)Y(A, z)) \\
&= Y(A, z),
\end{aligned}
$$

thanks to (6.4.1). This proves the proposition.

Now we want to formulate precisely what kind of object $Y(A, z)dz^{\Delta}$ is. For that we need to introduce the notions of torsors and twists.

6.4.6. Torsors and twists. Let G be a group, and S a non–empty set. Then S is called a G–*torsor*, if it is equipped with a simply transitive *right* action of G, i.e., given $x, y \in S$, there exists a unique $g \in G$ such that $x \cdot g = y$ (we recall that for a right action we have $x \cdot (gh) = (x \cdot g) \cdot h$). G–torsors are also known as principal homogeneous spaces over G. Equivalently, a G–torsor is a principal G–bundle over one point. The choice of any $x \in S$ allows us to identify S with G by sending $x \cdot g$ to g.

Given a G–module V and a G–torsor S, we define the S–*twist* of V as the set

$$V_S = S \underset{G}{\times} V = S \times V/\{(s \cdot g, v) \sim (s, gv)\}.$$

Given $x \in S$, we may identify V with V_S, by $v \mapsto (x, v)$. This identification depends on the choice of x. However, since G acts on V by linear operators, the vector space structure induced by the above identification does not depend on the choice of x. Thus V_S is canonically a vector space. If one thinks of S as a principal G–bundle over a point, then V_S is simply the associated vector bundle corresponding to V. In fact, any structure on V (such as a bilinear form or multiplicative structure) that is preserved by G will be inherited by V_S. Note however that V_S does *not* carry a G–action unless G is commutative.

The main examples of torsors that we consider in this book are constructed using abstract discs (see § 6.1.1).

Consider the disc $D_x = \operatorname{Spec} \mathcal{O}_x$, where $\mathcal{O}_x$ is the completed local ring of a smooth curve X at a point $x \in X$ (see § 6.1.1). Recall that by definition a coordinate on D_x, or equivalently a formal coordinate at x, is a topological generator of the unique maximal ideal $\mathfrak{m}_x \subset \mathcal{O}_x$. Let $\mathcal{A}ut_x$ be the set of all coordinates on D_x. It comes equipped with a natural right action of the group $\operatorname{Aut} \mathcal{O}$: if $t_x \in \mathcal{A}ut_x$, and $\rho \in \operatorname{Aut} \mathcal{O}$, then $\rho(t_x) \in \mathcal{A}ut_x$. Furthermore, $(\rho * \mu)(t) = \mu(\rho(t))$, so this indeed defines a right action of $\operatorname{Aut} \mathcal{O}$ on $\mathcal{A}ut_x$. It is clear that this action is simply transitive. Therefore $\mathcal{A}ut_x$ is an $\operatorname{Aut} \mathcal{O}$–torsor.

Let V be a quasi-conformal vertex algebra (see Definition 6.3.4). Then the action of the Lie algebra $\operatorname{Der} \mathcal{O}$ on V can be exponentiated to an action of the group $\operatorname{Aut} \mathcal{O}$, and we can form the twist

$$\mathcal{V}_x \overset{\text{def}}{=} \mathcal{A}ut_x \underset{\operatorname{Aut} \mathcal{O}}{\times} V.$$

For each formal coordinate z at x, any element of $\mathcal{V}_x$ may be written uniquely as a pair (z, v), where $v \in V$.

Recall that we denote by $\mathcal{K}_x$ the field of fractions of $\mathcal{O}_x$, and by $D_x^\times = \operatorname{Spec} \mathcal{K}_x$ the punctured disc at x. A Δ–differential ϖ on $D_x^\times$ is just a Δ–differential on D_x, possibly having a pole of finite order at x.

6.4.7. Proposition. *Let A be a primary vector of degree Δ in a quasi-conformal vertex algebra V. Let z be a coordinate on D_x. Define a Δ–differential ϖ_x on $D_x^\times$ with values in $\operatorname{End} \mathcal{V}_x$ as follows: identify $\operatorname{End} \mathcal{V}_x$ with $\operatorname{End} V$ using the coordinate z, and set*

$$\varpi_x = Y(A, z) dz^\Delta.$$

Then ϖ_x does not depend on the choice of coordinate z, i.e., we obtain a canonical $\operatorname{End} \mathcal{V}_x$–valued Δ–differential on D_x.

Equivalently, for any $\widetilde{v} \in \mathcal{V}_x$ and $\widetilde{\varphi} \in \mathcal{V}_x^*$, define a Δ–differential on $D_x^\times$ by the formula $\langle \varphi, Y(A, z) \cdot v \rangle dz^\Delta$, where $\widetilde{v} = (z, v)$, $\widetilde{\varphi} = (z, \varphi)$. Then this Δ–differential is independent of z.

6.4.8. Proof. Let $w = \rho(z)$ be another coordinate on D_x. Then we obtain another Δ–differential on $D_x^\times$ defined by the formula $Y(A, w) dw^\Delta$, where we identify $\operatorname{End} \mathcal{V}_x$ with $\operatorname{End} V$ using the coordinate w. Now we want to rewrite this Δ–differential in terms of the coordinate z and to show that it coincides with $Y(A, z) dz^\Delta$.

If a vector in $\mathcal{V}_x$ is equal to (z, v), i.e., it is identified with $v \in V$ using the coordinate z, then the same vector equals $(w, R(\rho)^{-1} \cdot v)$, i.e., it is identified with $R(\rho)^{-1} \cdot v \in V$ using the coordinate w. Therefore if we have an operator on $\mathcal{V}_x$ which is equal to S under the identification $\operatorname{End} \mathcal{V}_x \simeq \operatorname{End} V$ using the coordinate w, then this operator equals $R(\rho) S R(\rho)^{-1}$ under the identification $\operatorname{End} \mathcal{V}_x \simeq \operatorname{End} V$ using the coordinate z. Thus, in terms of the coordinate z, the differential $Y(A, w) dw^\Delta$ becomes

$$R(\rho) Y(A, \rho(z)) R(\rho)^{-1} (\rho'(z))^\Delta dz^\Delta.$$

This expression coincides with $Y(A, z) dz^\Delta$ by Proposition 6.4.4, and this proves the proposition.

6.4.9. Example. Consider the action of Aut $\mathcal{O}$ on

$$V_k(\mathfrak{g}) = \mathrm{Ind}_{\mathfrak{g} \otimes \mathbb{C}[[z]] \oplus \mathbb{C}K}^{\widehat{\mathfrak{g}}} \mathbb{C}_k$$

induced by the natural action of Aut $\mathcal{O}$ on $\widehat{\mathfrak{g}}$ stabilizing its Lie subalgebra $\mathfrak{g} \otimes \mathbb{C}[[z]] \oplus \mathbb{C}K$. This action is obtained by exponentiating the action of Der $\mathcal{O}$ on $V_k(\mathfrak{g})$, which for $k \neq -h^\vee$ comes from the conformal vertex algebra structure on $V_k(\mathfrak{g})$ given by the Segal-Sugawara vector (see § 3.4.8). Therefore this action satisfies the relations (6.3.5) for $k \neq -h^\vee$. But since these commutation relations have no pole at $k \neq -h^\vee$, we find that the action of Der $\mathcal{O}$ on $V_k(\mathfrak{g})$ satisfies the axioms of quasi-conformal vertex algebra even for $k = -h^\vee$. Thus, for any $k \in \mathbb{C}$ the natural Aut $\mathcal{O}$–action on $V_k(\mathfrak{g})$ comes from a Der $\mathcal{O}$–action that defines a quasi-conformal structure on $V_k(\mathfrak{g})$. In particular, Proposition 6.4.7 is applicable in the situation where $V = V_k(\mathfrak{g})$ and $A = J_{-1}^a v_k$. The corresponding primary field of conformal dimension one is the vertex operator $J^a(z)$.

Now recall the affine Kac–Moody algebra $\widehat{\mathfrak{g}}_x$ attached to the disc D_x and its vacuum module $\mathcal{V}_k(\mathfrak{g})_x$ introduced in § 6.1.2. It is clear from the definitions that we have canonical identifications

$$\mathcal{A}ut_x \underset{\mathrm{Aut}\,\mathcal{O}}{\times} \mathbb{C}[[z]] = \mathcal{O}_x, \qquad \mathcal{A}ut_x \underset{\mathrm{Aut}\,\mathcal{O}}{\times} \mathbb{C}((z)) = \mathcal{K}_x.$$

Hence we obtain a canonical identification

$$\mathcal{V}_k(\mathfrak{g})_x = \mathcal{A}ut_x \underset{\mathrm{Aut}\,\mathcal{O}}{\times} V_k(\mathfrak{g}),$$

so that the vacuum module $\mathcal{V}_k(\mathfrak{g})_x$ over $\widehat{\mathfrak{g}}_x$ introduced in § 6.1.2 is a special case of the general construction of Aut $\mathcal{O}$–twists introduced in § 6.4.6. Moreover, the statement of Proposition 6.4.7 for $V = V_k(\mathfrak{g})$ and $A = J_{-1}^a v_k$ becomes the statement of Corollary 6.1.6. Thus, the above construction of $\mathcal{V}_x$ solves the first problem proposed in § 6.1.7: constructing a twist of a vertex algebra V attached to any point of an arbitrary smooth curve. In the next section we will solve the second problem: give a coordinate-independent interpretation of the entire vertex algebra structure.

6.5. The main construction

We now explain how to convert the whole vertex operation Y into an intrinsic object on any abstract disc. The construction is based on a remarkable principal bundle for the group Aut $\mathcal{O}$, which naturally exists on an arbitrary smooth curve and on any disc.

Let X be a smooth complex curve, and x a point of X. Recall that $\mathcal{O}_x$ denotes the completion of the local ring at x, $\mathfrak{m}_x$ denotes its maximal ideal, and $\mathcal{K}_x$ the field of fractions of $\mathcal{O}_x$. We denote by D_x (resp., $D_x^\times$) the disc (resp., the punctured disc) at x, defined as $\mathrm{Spec}\,\mathcal{O}_x$ (resp., $\mathrm{Spec}\,\mathcal{K}_x$).

A formal coordinate t_x at x is by definition a coordinate on $D_x = \mathrm{Spec}\,\mathcal{O}_x$ (see § 6.1.1), i.e., a topological generator of $\mathfrak{m}_x$. A choice of formal coordinate is the same as a choice of isomorphism between $\mathcal{O} = \mathbb{C}[[z]]$ and $\mathcal{O}_x$ (or between $D = \mathrm{Spec}\,\mathcal{O}$ and D_x). Recall that in § 6.4.6 we denoted by $\mathcal{A}ut_x$ the set of all formal coordinates at x.

The group $\text{Aut}\,\mathcal{O}$ acts naturally on $\mathcal{A}ut_x$, making it into an $\text{Aut}\,\mathcal{O}$–torsor. Given an $\text{Aut}\,\mathcal{O}$–module V, in § 6.4.6 we denoted by $\mathcal{V}_x$ the $\mathcal{A}ut_x$–twist of V,

$$\mathcal{V}_x = \mathcal{A}ut_x \underset{\text{Aut}\,\mathcal{O}}{\times} V.$$

We would like to organize the torsors $\mathcal{A}ut_x$ into a principal $\text{Aut}\,\mathcal{O}$–bundle over X, and the vector spaces $\mathcal{V}_x$ into a vector bundle over X. Consider the set of pairs (x, t_x), where $x \in X$ and t_x is a formal coordinate at x. One can show that this is the set of points of a scheme $\mathcal{A}ut_X$ over X (of infinite type), which may be defined as an open subscheme of the jet scheme JX of X (see §§ 9.4.4 and 11.3.3). It represents the functor which associates to a $\mathbb{C}$–algebra R the set of morphisms $\text{Spec}\,R[[z]] \to X$ with non-vanishing differential at $z = 0$.

Moreover, the projection $\mathcal{A}ut_X \twoheadrightarrow X$ is a principal $\text{Aut}\,\mathcal{O}$–bundle, locally trivial in the Zariski topology (see § 6.5.2 below). The fiber of this bundle at x is the $\text{Aut}\,\mathcal{O}$–torsor $\mathcal{A}ut_x$.

6.5.1. Definition. Given a finite-dimensional $\text{Aut}\,\mathcal{O}$–module V, let

$$\mathcal{V}_X = \mathcal{A}ut_X \underset{\text{Aut}\,\mathcal{O}}{\times} V$$

be the vector bundle associated to V and $\mathcal{A}ut_X$.

Thus, $\mathcal{V}_X$ is a vector bundle of finite rank over X whose fiber at $x \in X$ is $\mathcal{V}_x$. We will usually write $\mathcal{V}$ for $\mathcal{V}_X$ when there is no ambiguity.

6.5.2. Coordinates and the Zariski topology. In the analytic topology, we can cover the curve X by open subsets which are isomorphic to open subsets of $\mathbb{C}$. On each of these coordinate charts we then obtain a global coordinate, i.e., a map to $\mathbb{C}$ with non-vanishing differential. This global coordinate t induces a formal coordinate $t - t(x)$ at each point x of the chart. Hence we obtain a trivialization of the bundle $\mathcal{A}ut_X \to X$. Thus we obtain that $\mathcal{A}ut_X \to X$ is a locally trivial bundle over X in the analytic topology.

Now observe that in order to trivialize the restriction of $\mathcal{A}ut_X$ to an open subset U, it is sufficient to be given an *étale* map $f : U \to \mathbb{C}$ (that is, a representation of U as an unramified cover of an open subset of $\mathbb{C}$). Given such a map f and a point $x \in U$, the function $f - f(x)$ has non-vanishing differential at x, and thus identifies the disc D_x with the standard disc D around 0 in $\mathbb{C}$ and the torsor $\mathcal{A}ut_x$ with the group $\text{Aut}\,\mathcal{O}$.

For any smooth point x on a curve X we can find a Zariski open set $U \subset X$ containing x and an étale map $f : U \to \mathbb{C}$. Namely, take a meromorphic function on X with non-vanishing derivative at x, and let $U \subset X$ be the complement of the divisor of df. If X is a smooth curve, it can be covered by Zariski open subsets U_α, each equipped with an étale map $f_\alpha : U_\alpha \to \mathbb{C}$. Then on each U_α we obtain a trivialization of $\mathcal{A}ut_X$. Thus, $\mathcal{A}ut_X$ is a locally trivial principal $\text{Aut}\,\mathcal{O}$–bundle in the Zariski topology. Therefore the vector bundle $\mathcal{V}$ associated to any $\text{Aut}\,\mathcal{O}$–module V is Zariski locally trivial.

Henceforth by a local coordinate on X we will understand an étale map from a Zariski open subset U to $\mathbb{C}$, inducing formal coordinates on the discs D_x for all $x \in U$. We may then study bundles like $\mathcal{V}$ using charts and transition functions, as one is accustomed to doing in the analytic setting.

6.5.3. Vertex algebra bundle. For an $\operatorname{Aut}\mathcal{O}$–module V which has a filtration by finite-dimensional submodules $V^i, i \geq 0$, we consider the directed system $(\mathcal{V}_X^i)$ of vector bundles of finite rank on X and embeddings $\mathcal{V}_X^i \hookrightarrow \mathcal{V}_X^j, i \leq j$. We will denote this system simply by $\mathcal{V}_X$, thinking of it as a vector bundle of infinite rank over X (an inductive limit of bundles of finite rank). The sheaf of sections of $\mathcal{V}_X$ is a quasi-coherent $\mathcal{O}_X$–module. Likewise, by $\mathcal{V}_X^*$ we will understand the inverse system of bundles $(\mathcal{V}_X^i)^*$ and surjections $(\mathcal{V}_X^j)^* \twoheadrightarrow (\mathcal{V}_X^i)^*, i \leq j$, thinking of it as a projective limit of bundles of finite rank (note that the sheaf of sections of $\mathcal{V}_X^*$ is not quasi-coherent).

Let V be a quasi-conformal vertex algebra. Then the operator $L_0 = -z\partial_z \in \operatorname{Der}\mathcal{O}$ gives us a $\mathbb{Z}$–gradation on V. Let us assume for simplicity that

$$V = \bigoplus_{n=K}^{\infty} V_n, \qquad \dim V_n < \infty.$$

Then we have a filtration $V_{\leq m} = \bigoplus_{n=K}^{m} V_n$ of V by finite-dimensional $\operatorname{Aut}\mathcal{O}$–submodules. Hence we can define $\mathcal{V}_X$ and $\mathcal{V}_X^*$ as inductive and projective limits of vector bundles of finite rank, respectively, over any smooth curve X. We call them the *vertex algebra bundle* and its dual.

The above definitions of $\mathcal{A}ut_X$ and $\mathcal{V}_X$ may be adapted to the case when X is replaced by the disc $D = \operatorname{Spec}\mathbb{C}[[z]]$, or $D_x = \operatorname{Spec}\mathcal{O}_x$. In this case $\mathcal{A}ut_{D_x} = \mathcal{A}ut_X|_{D_x}$ and $\mathcal{V}_{D_x} = \mathcal{V}_X|_{D_x}$.

Now we come to the main point of this chapter.

Informally, we can view the vertex operation $Y(\cdot, z)$ as an operation which takes an element of V and assigns to it an endomorphism of V, for each z in the punctured disc $D^\times$. In order to be able to define Y for the disc $D_x^\times$, not equipped with a coordinate, we need to give a coordinate–free description of this operation.

Let $\mathcal{A}ut_x$ be the $\operatorname{Aut}\mathcal{O}$–torsor of coordinates at $x \in X$. Recall that

$$\mathcal{V}_x = \mathcal{A}ut_x \underset{\operatorname{Aut}\mathcal{O}}{\times} V$$

is the fiber of $\mathcal{V}|_{D_x}$ at $x \in X$. We will now define an $\operatorname{End}\mathcal{V}_x$–valued meromorphic section $\mathcal{Y}_x$ of the bundle $\mathcal{V}^*$ on the punctured disc $D_x^\times$. To be precise, by such a section we understand an operation

$$v, \varphi, s \mapsto \langle \varphi, \mathcal{Y}_x(s) \cdot v \rangle$$

assigning an element of $\mathcal{K}_x$ (i.e., a function on $D_x^\times$), denoted by $\langle \varphi, \mathcal{Y}_x(s) \cdot v \rangle$, to each triple: $v \in \mathcal{V}_x, \varphi \in \mathcal{V}_x^*$, and a regular section s of $\mathcal{V}|_{D_x}$. This operation must be linear in v and φ and $\mathcal{O}_x$–linear in s. This operation is nothing but the operation of taking the matrix coefficients of our section. From now on, when we say "$\operatorname{End}\mathcal{V}_x$–valued section", we will understand the collection of its matrix coefficients defined as above.

Here is the construction of $\mathcal{Y}_x$: pick a coordinate z on the disc D_x. Then we obtain a trivialization of our bundle,

$$\imath_z : V[[z]] \xrightarrow{\sim} \Gamma(D_x, \mathcal{V}),$$

which we call the z–trivialization. We also obtain trivializations of the fiber: $V \xrightarrow{\sim} \mathcal{V}_x$ and its dual: $V^* \xrightarrow{\sim} \mathcal{V}_x^*$. We will denote the image of $v \in V$ in $\mathcal{V}_x$ (resp., the image of of $\varphi \in V^*$ in $\mathcal{V}_x^*$) under this trivialization by (z, v) (resp., (z, φ)).

To define our section with respect to these trivializations through its matrix coefficients we need to attach an element of $\mathbb{C}((z))$ to each triple: $(z, v) \in \mathcal{V}_x$, $(z, \varphi) \in \mathcal{V}_x^*$, and a section $\imath_z(s)$ of $\mathcal{V}|_{D_x}$ (where $s \in V[[z]]$). Furthermore, this operation must be $\mathbb{C}$–linear in v and φ and $\mathbb{C}[[z]]$–linear in s. Hence it is sufficient to assign a function to the triples v, φ, s, where s is a constant section equal to $A \in V \subset V[[z]]$ in the z–trivialization.

6.5.4. Theorem–Definition. *Define an* $\operatorname{End} \mathcal{V}_x$*–valued section* $\mathcal{Y}_x$ *of* $\mathcal{V}^*$ *on* $D_x^\times$ *by the formula*

$$\langle (z, \varphi), \mathcal{Y}_x(\imath_z(A)) \cdot (z, v) \rangle = \langle \varphi, Y(A, z)v \rangle,$$

where z *is a coordinate on* D_x*. Then the section* $\mathcal{Y}_x$ *is canonical, i.e., independent of the choice of coordinate* z *on* D_x*.*

6.5.5. Proof. Let $w = \rho(z)$ be another coordinate. Then we construct analogously a section $\widetilde{\mathcal{Y}}_x$ by the formula

$$\langle (w, \widetilde{\varphi}), \widetilde{\mathcal{Y}}_x(\imath_w(\widetilde{A})) \cdot (w, \widetilde{v}) \rangle = \langle \widetilde{\varphi}, Y(\widetilde{A}, w)\widetilde{v} \rangle.$$

We need to show that, as an $\operatorname{End} \mathcal{V}_x$–valued section of $\mathcal{V}^*$, this section coincides with the section $\mathcal{Y}_x$.

We already explained in the proof of Proposition 6.4.7 that

$$(z, v) = (w, R(\rho)^{-1} \cdot v), \qquad (z, \varphi) = (w, \varphi \cdot R(\rho)).$$

Now we need to find out what the section $\imath_z(A)$ of $\mathcal{V}|_{D_x}$ looks like in the w–trivialization of $\mathcal{V}|_{D_x}$.

It is instructive to do this first in the analytic setting, when D_x is a small analytic disc around x. Then the coordinate z on D_x induces a local coordinate $z - y$ at each point $y \in D_x$. The coordinate w on D_x induces a local coordinate $w - \rho(y)$ at the same point. These two coordinates are related by an automorphism given by $\rho_y(t) \in \mathbb{C}[[t]]$ defined by the formula $w - \rho(y) = \rho_y(z - y)$. This formula implies that

$$\rho_y(t) = \rho(y + t) - \rho(y).$$

When we trivialize $\mathcal{V}|_{D_x}$ using the coordinate z, we identify the fiber $\mathcal{V}_y$ of $\mathcal{V}$ at y with V via $(z - y, A) \mapsto A$. We then have

$$(z - y, A) \sim (w - \rho(y), R(\rho_y)^{-1}A).$$

In other words, in the w–trivialization of $\mathcal{V}|_{D_x}$, the value of the section $\imath_z(A)$ at $\rho(y)$ is $R(\rho_y)^{-1}A$. Thus, we obtain that the section $\imath_z(A)$ of $\mathcal{V}|_{D_x}$ becomes in the w–trivialization the section $\imath_w(R(\rho_z)^{-1}A)$.

To see that the same formula for $\imath_z(A)$ is true in our formal setting (when $D_x = \operatorname{Spec} \mathcal{O}_x$), it suffices to show that it is true at the level of n–jets, for all $n > 0$. In that case, we consider z and w as n–jets of coordinates at x, and consider n–jets of our sections. But any n–jet of a coordinate or a section at x may be extended to a coordinate or a section on a small analytic disc around x (or a small enough Zariski neighborhood of x), in which case the formula is true as we have shown above.

Collecting these formulas together, we obtain

$$\langle (z, \varphi), \widetilde{\mathcal{Y}}_x(\imath_z(A)) \cdot (z, v) \rangle = \langle \varphi, R(\rho)Y(R(\rho_z)^{-1}A, \rho(z))R(\rho)^{-1}v \rangle.$$

Hence coordinate-independence of our section $\mathcal{Y}_x$ boils down to the following formula:

$$\langle \varphi, Y(A, z)v \rangle = \langle \varphi, R(\rho)Y(R(\rho_z)^{-1}A, \rho(z))R(\rho)^{-1}v \rangle,$$

or, equivalently,

$$Y(A, z) = R(\rho)Y(R(\rho_z)^{-1}A, \rho(z))R(\rho)^{-1}$$

for all $A \in V$.

This is the content of the following lemma, originally due to Y.-Z. Huang [**H1**]:

6.5.6. Lemma.

(6.5.1) $Y(A, z) = R(\rho)Y(R(\rho_z)^{-1}A, \rho(z))R(\rho)^{-1}, \qquad A \in V.$

6.5.7. Proof. Consider the vector space $\mathrm{Hom}(V, \mathcal{F}(V))$, where $\mathcal{F}(V)$ is the space of fields on V introduced in § 1.2.1. In particular, the vertex operation $Y(\cdot, z)$ belongs to $\mathrm{Hom}(V, \mathcal{F}(V))$. For each $\rho \in \mathrm{Aut}\,\mathcal{O}$, consider the linear operator T_ρ on $\mathrm{Hom}(V, \mathcal{F}(V))$ defined by the right hand side of formula (6.5.1):

$$(T_\rho \cdot X)(A, z) = R(\rho)X(R(\rho_z)^{-1}A, \rho(z))R(\rho)^{-1}.$$

It is easy to see that this action preserves $\mathrm{Hom}(V, \mathcal{F}(V))$. Moreover, the operators T_ρ define a representation of the Lie group $\mathrm{Aut}\,\mathcal{O}$ on $\mathrm{Hom}(V, \mathcal{F}(V))$. This follows from the fact that R is a representation of $\mathrm{Aut}\,\mathcal{O}$ on V and from the formula

$$\mu_z * \rho_{\mu(z)} = (\mu * \rho)_z$$

(recall that by definition, $\rho_z(t) = \rho(z+t) - \rho(z)$). The lemma will follow if we show that $Y(\cdot, z)$ is stable under this action. Since the exponential map $\mathrm{Der}_0\,\mathcal{O} \to \mathrm{Aut}\,\mathcal{O}$ is surjective, it suffices to show that the infinitesimal action of the Lie algebra $\mathrm{Der}_0\,\mathcal{O}$ is trivial.

So we write $\rho(z) = \exp(v(z)\partial_z) \cdot z$, where

$$v(z)\partial_z = -\sum_{n \geq 0} v_n z^{n+1}\partial_z.$$

We denote the operator $\sum_{n \geq 0} v_n L_n$ corresponding to $v(z)\partial_z$ by $\mathbf{v}$, and set $Q_\epsilon = T_{\exp(\epsilon v \partial_z)}$. We then claim that

(6.5.2) $Q_\epsilon Y(A, z) = Y(A, z) + \epsilon([\mathbf{v}, Y(A, z)] + v(z)\partial_z Y(A, z)$

$$+ \sum_{m \geq 0} \frac{1}{(m+1)!}(\partial_z^{m+1}v(z))Y(L_m A, z)) + o(\epsilon).$$

The commutator term comes from the conjugation by $R(\rho)$. The derivative term is the first order contribution of the change of variables $Y(A, \rho(z))$, and can be rewritten as $v(z)Y(L_{-1}A, z)$. The third summand comes from the ϵ–linear term in $R(\rho_z)^{-1}$, or equivalently, the negative of the ϵ–linear term in $R(\rho_z)$. To see that, we expand $\rho_z(t)$ in powers of ϵ:

$$\rho_z(t) = \rho(z+t) - \rho(z) = z + t + \epsilon v(z+t) + \ldots - \rho(z)$$

$$= t + \epsilon \sum_{m > 0} \frac{1}{m!}\partial_z^m v(z)t^m + \ldots.$$

Since L_m corresponds to $-t^{m+1}\partial_t$, we obtain that the ϵ–linear term in $R(\rho_z)^{-1}$ is indeed equal to the third summand in formula (6.5.2).

Comparing (6.5.2) with (6.3.5), it follows that the ϵ–linear term in the right hand side of formula (6.5.2) vanishes, and the lemma follows. This also completes the proof of Theorem 6.5.4.

6.5.8. A dual description of $\mathcal{Y}_x$. Let Ω_x be the space of differentials on the punctured disc $D_x^\times$ (see § 6.1.3). We have a canonical residue map $\mathrm{Res}_x : \Omega_x \to \mathbb{C}$ (see § 9.1.2), which gives rise to a perfect pairing

$$\Gamma(D_x^\times, \mathcal{V}^*) \times \Gamma(D_x^\times, \mathcal{V} \otimes \Omega) \to \mathbb{C},$$

sending (ϕ, μ) to $\mathrm{Res}_x \langle \phi, \mu \rangle$.

For each section μ of $\mathcal{V} \otimes \Omega$ over $D_x^\times$, we obtain using this pairing a linear operator O_μ on $\mathcal{V}_x$ given by the formula

$$O_\mu = \mathrm{Res}_x \langle \mathcal{Y}_x, \mu \rangle.$$

Thus, we obtain a well-defined linear map

(6.5.3) $\mathcal{Y}_x^\vee : \Gamma(D_x^\times, \mathcal{V} \otimes \Omega) \to \mathrm{End}\, \mathcal{V}_x$

sending μ to O_μ. If we choose a formal coordinate z at x, and a section $\mu = A \otimes z^n dz$ of $\mathcal{V} \otimes \Omega$ with respect to the z–trivialization, then $O_\mu = A_{(n)}$.

The data of the map $\mathcal{Y}_x^\vee$ are equivalent to the data of the section $\mathcal{Y}_x$. The advantage of $\mathcal{Y}_x^\vee$ is that to define it we do not need to assume that the $\mathbb{Z}$–gradation on V is bounded from below or that the graded components are finite-dimensional, which we needed to make sense of the dual bundle $\mathcal{V}^*$. We only need to assume that V is a quasi-conformal vertex algebra.

6.5.9. The structure of the bundle $\mathcal{V}$. Theorem 6.5.4 tells us that, given a quasi-conformal vertex algebra V and a smooth curve X, we can construct a vector bundle $\mathcal{V}$ over X and a canonical section $\mathcal{Y}_x$ of $\mathcal{V}^*|_{D_x^\times}$ with values in $\mathrm{End}\, \mathcal{V}_x$ for any $x \in X$.

Note that the vacuum vector $|0\rangle \in V$ is $\mathrm{Aut}\, \mathcal{O}$–invariant, and hence defines a canonical vector in the fiber $\mathcal{V}_x$. By the vacuum axiom, $Y(A, z)|0\rangle \in V[[z]]$. Therefore $\mathcal{Y}_x|0\rangle$ is a *regular* section of $\mathcal{V}^*|_{D_x}$ (over the unpunctured disc) with values in $\mathcal{V}_x$.

Suppose $A \in V_\Delta$ is a primary vector. Then $\mathbb{C}A \subset V$ is an $\mathrm{Aut}\, \mathcal{O}$–submodule of V, and conversely, one–dimensional $\mathrm{Aut}\, \mathcal{O}$ submodules of V are spanned by primary vectors. The twisting construction of Definition 6.5.1 defines a functor from $\mathrm{Aut}\, \mathcal{O}$–modules to vector bundles over a given smooth curve X. Thus $\mathbb{C} \cdot A$ gives rise to a line subbundle

(6.5.4) $\mathcal{A}ut_X \underset{\mathrm{Aut}\, \mathcal{O}}{\times} \mathbb{C}A \subset \mathcal{V}.$

What is this line bundle?

We have

$$\rho_y(t) = \rho'(y)t + \frac{1}{2}\rho''(y)t^2 + \dots.$$

Now we obtain from the formula $\rho_y(t) = \exp\left(\sum_{j>0} \nu_j(y)t^{j+1}\partial_t\right)\nu_0(y)^{t\partial_t}t$ that $\nu_0(y) = \rho'(y)$. Hence $R(\rho_y)^{-1}A = \rho'(y)^\Delta A$. This implies that the line bundle (6.5.4) is isomorphic to $\Omega^{-\Delta}$, because $(dy)^{-\Delta} = \rho'(y)^\Delta(d\rho(y))^{-\Delta}$. Thus we obtain an inclusion $\Omega^{-\Delta} \hookrightarrow \mathcal{V}$.

Dually, we obtain a surjection $\mathcal{V}^* \twoheadrightarrow \Omega^\Delta$. The section $\mathcal{Y}_x$ projects onto an $\operatorname{End}\mathcal{V}_x$–valued Δ–differential, which is nothing but the Δ–differential constructed in Proposition 6.4.7.

Suppose that $V = \bigoplus_{n=K}^{\infty} V_n$, where $\dim V_n < \infty$ for all n. Then $V_{\leq m} = \bigoplus_{n=K}^{m} V_n$ is preserved by $\operatorname{Aut}\mathcal{O}$. The exact sequences of $\operatorname{Aut}\mathcal{O}$–modules

$$0 \to V_{\leq(m-1)} \to V_{\leq m} \to V_m \to 0$$

gives rise to an exact sequence of vector bundles

$$0 \to \mathcal{V}_{\leq(m-1)} \to \mathcal{V}_{\leq m} \to \mathcal{V}_m \to 0.$$

The space V_m, considered as a quotient of $V_{\leq m}$ by $V_{\leq(m-1)}$, is a direct sum of one–dimensional representations of $\operatorname{Aut}\mathcal{O}$. Hence

$$\mathcal{V}_m \simeq (\Omega^{-m})^{\oplus \dim V_m}.$$

It follows that $\mathcal{V}$ is filtered by an increasing union of subbundles of finite rank, with successive quotients isomorphic to finite direct sums of negative powers of the canonical bundle. The dual bundle $\mathcal{V}^*$ is the projective limit of its quotient subbundles, which are successive extensions of finite direct sums of positive powers of the canonical bundle. The extensions that appear in assembling $\mathcal{V}$ from these graded pieces contain a lot of interesting geometric information. In Chapter 8 we study the simplest examples of these extensions.

6.6. A flat connection on the vertex algebra bundle

6.6.1. Definition of the flat connection. Let S be a smooth complex variety and $\mathcal{E} \to S$ a holomorphic vector bundle over S. We will use the same notation $\mathcal{E}$ for the sheaf of holomorphic sections of $\mathcal{E}$. Denote by Ω the sheaf of differentials on S. A holomorphic *connection* ∇ on $\mathcal{E}$ is a $\mathbb{C}$–linear map $\nabla : \mathcal{E} \to \mathcal{E} \otimes \Omega$ satisfying the Leibniz rule

$$\nabla(f\phi) = f\nabla(\phi) + \phi \otimes df$$

for any holomorphic function f. Suppose that $U \subset S$ is open, and ξ is a holomorphic vector field on U. We attach to ξ the operator

$$\nabla_\xi : \mathcal{E} \to \mathcal{E}, \qquad \nabla_\xi(\phi) = \langle \nabla\phi, \xi \rangle,$$

so we can differentiate sections of $\mathcal{E}$ along holomorphic vector fields on S. (Moreover, a holomorphic connection may be extended in a unique way to an ordinary connection by using as the operators ∇_ξ, where ξ is an anti-holomorphic vector field, the $\bar{\partial}$–operators defining the holomorphic structure on $\mathcal{E}$.) One defines a connection in the algebraic setting in a similar way.

The connection ∇ is called *flat* if $\xi \mapsto \nabla_\xi$ is a Lie algebra homomorphism, i.e.,

$$\nabla_{[\xi,\eta]} = [\nabla_\xi, \nabla_\eta]$$

for any two holomorphic vector fields ξ, η on an arbitrary open subset U (note that if the holomorphic connection ∇ is flat, then the corresponding ordinary connection is also flat). If the complex dimension of the base S is 1, as it is in our setting, then any holomorphic connection is flat. For more information on connections, the reader is referred to § A.2.

A section ϕ of $\mathcal{E}$ is called *horizontal* if $\nabla_\xi \phi = 0$ for all ξ, i.e., $\nabla\phi = 0$.

An example of a flat connection is the trivial line bundle $\mathcal{E}$, with the connection given by the de Rham differential: $\nabla = d$, $\nabla_\xi = \xi$. Choosing a local coordinate z, we may write $d = \partial_z \otimes dz$. The horizontal sections are the constant functions.

6.6.2. The case of the vertex algebra bundle. Now let V be a quasi-conformal vertex algebra and $\mathcal{V}$ the corresponding vector bundle on a smooth curve X. We wish to define an (algebraic) connection on $\mathcal{V}$. This connection will automatically be flat, since X is one-dimensional as an algebraic variety. Recall that the definition of the bundle $\mathcal{V}$ utilized the action of the Lie algebra $\mathrm{Der}_0\, \mathcal{O} = z\mathbb{C}[[z]]\partial_z$ on V through the operators $L_n, n \geq 0$. Now we will use the operator L_{-1}, corresponding to the vector field $-\partial_z$, to define a connection on $\mathcal{V}$. The following result is proved by direct computation. In Chapter 17 we will give another, more conceptual, proof. Note that according to the discussion of § 6.5.2, the definition given below is valid in the Zariski topology.

6.6.3. Theorem. *Let $U \subset X$ be open, z a coordinate on U. Define $\nabla : \mathcal{V} \to \mathcal{V} \otimes \Omega$ on U by the formula*
$$\nabla_{\partial_z} = \partial_z + L_{-1},$$
where we use the trivialization of the bundle $\mathcal{V}$ induced by the coordinate z. Then ∇ is a canonically defined connection on $\mathcal{V}$ (independent of the choice of coordinate z).

6.6.4. Proof. Let $w = \rho(z)$ be another coordinate. Then we define another connection operator $\widetilde{\nabla}$ with
$$\widetilde{\nabla}_{\partial_w} = \partial_w + L_{-1},$$
where we use the trivialization $\imath_w : V \times U \xrightarrow{\sim} \mathcal{V}|_U$ induced by w. Let us rewrite $\widetilde{\nabla}$ in terms of z and the trivialization $\imath_z : V \times U \xrightarrow{\sim} \mathcal{V}|_U$ induced by z. As we showed in the proof of Theorem 6.5.4, the section $\imath_z(s)$ of $\mathcal{V}|_U$ becomes in the w–trivialization the section $\imath_w(R(\rho_z)^{-1}s)$, where as before
$$\rho_z(t) = \rho(z + t) - \rho(z).$$
Therefore with respect to the trivialization induced by z we have
$$\widetilde{\nabla}_{\partial_z} = R(\rho_z)\partial_z R(\rho_z)^{-1} + \rho'(z)R(\rho_z)L_{-1}R(\rho_z)^{-1}.$$
To prove the theorem we need to prove the identity
$$(6.6.1) \qquad R(\rho_z)\partial_z R(\rho_z)^{-1} + \rho'(z)R(\rho_z)L_{-1}R(\rho_z)^{-1} = \partial_z + L_{-1}.$$

It is easy to see from the definition of the homomorphism R that
$$R(\rho_z)\partial_z R(\rho_z)^{-1} = \partial_z + \sum_{n \geq -1} F_n(z)L_n$$
and
$$R(\rho_z)L_{-1}R(\rho_z)^{-1} = \sum_{n \geq -1} G_n(z)L_n$$
for some power series $F_n(z)$ and $G_n(z)$. Furthermore, we find that
$$-\sum_{n \geq -1} F_n(z)t^{n+1}\partial_t \cdot t = (\partial_z \cdot \rho_z^{-1}(u))|_{u=\rho_z(t)} = (\partial_z\rho(z+t))^{-1}\rho'(z) - 1$$
and
$$-\sum_{n \geq -1} G_n(z)t^{n+1}\partial_t \cdot t = -(\partial_u \cdot \rho_z^{-1}(u))|_{u=\rho_z(t)} = (\partial_z\rho(z+t))^{-1}.$$

This gives us the identity (6.6.1).

6.6.5. Remark. L. Takhtajan has remarked (private communication) that one may view (6.6.1) as a differential equation on $R(\rho_z)$, namely,

$$\frac{\partial R(\rho_z)}{\partial z} = \rho'(z)R(\rho_z)L_{-1} - L_{-1}R(\rho_z).$$

Its unique solution with the initial condition $R(\rho_z)|_{z=0} = R(\rho)$ is

(6.6.2) $$R(\rho_z) = \exp(-zL_{-1})R(\rho)\exp(\rho(z)L_{-1}).$$

Then the above proof implies that the right hand side is well-defined, and hence gives us an "explicit" formula for $R(\rho_z)$.

Conversely, one can obtain another proof of Theorem 6.6.3 by giving an independent proof of formula (6.6.2). Such a proof was suggested by L. Takhtajan: let us write

$$\rho(t) = \exp\left(\sum_{i>0} v_i t^{i+1} \partial_t\right) v_0^{t\partial_t} \cdot t.$$

Then using the formula $\exp(a\partial_t) \cdot f(t) = f(t+a)$ we find that

$$\rho_z(t) = \rho(z+t) - \rho(z) = \exp(z\partial_t)\exp\left(\sum_{i>0} v_i t^{i+1}\partial_t\right) v_0^{t\partial_t}\exp(-\rho(z)\partial_t)\cdot t.$$

But we know that $\rho_z(t)$ may be written in the form

$$\rho_z(t) = \exp\left(\sum_{i>0} v_i(z)t^{i+1}\partial_t\right) v_0(z)^{t\partial_t}\cdot t.$$

Therefore in any representation of the Lie algebra Der $\mathcal{O}$ the corresponding operator $R(\rho_z)$ given by formula (6.6.2) may be expressed as

$$R(\rho_z) = \exp\left(-\sum_{i>0} v_i(z)L_i\right) v_0(z)^{-L_0}$$

(i.e., without the operator L_{-1}). Hence the right hand side of formula (6.6.2) is indeed well-defined.

6.6.6. Dual connection. A connection ∇ on a vector bundle $\mathcal{E}$ gives rise to a unique connection ∇^* on the dual bundle $\mathcal{E}^*$ with the property

$$\langle \nabla^*\eta, \xi\rangle + \langle \eta, \nabla\xi\rangle = d\langle \eta, \xi\rangle.$$

Consider the vector bundle $\mathcal{V}$ associated to a quasi-conformal vertex algebra V, with a flat connection as defined above. Then the bundle $\mathcal{V}^*$ acquires a connection which is locally of the form

$$\nabla^* = d - L_{-1} \otimes dz.$$

Recall that we have also defined a canonical section $\mathcal{Y}_x$ of $\mathcal{V}^*$ with values in End $\mathcal{V}_x$ over the punctured disc $D_x^\times$.

6.6.7. Proposition. *We have $\nabla^*\mathcal{Y}_x = 0$, i.e., $\mathcal{Y}_x$ is a horizontal* End $\mathcal{V}_x$- *valued section of $\mathcal{V}^*|_{D_x^\times}$. In other words, for any $v \in \mathcal{V}_x$ and $\varphi \in \mathcal{V}_x^*$, $\langle\varphi, \mathcal{Y}_x \cdot v\rangle$ is a horizontal section of $\mathcal{V}^*$.*

6.6.8. Proof. Using the trivialization induced by a coordinate z, we can write the value of $\langle \varphi, \mathcal{Y}_x \cdot v \rangle$ on a constant section $\imath_z(A)$ of $\mathcal{V}$ as $\langle \varphi, Y(A, z)v \rangle$. Differentiating, we obtain that the value of $\nabla^* \langle \varphi, \mathcal{Y}_x \cdot v \rangle$ on A equals

$$\langle \varphi, (\partial_z Y(A, z) - Y(L_{-1}A, z)) \cdot v \rangle dz = 0,$$

by Corollary 3.1.6, as desired.

Recall the map

$$\mathcal{Y}_x^\vee : \Gamma(D_x^\times, \mathcal{V} \otimes \Omega) \to \operatorname{End} \mathcal{V}_x$$

introduced in § 6.5.8. Let us call a section of $\mathcal{V} \otimes \Omega$ a *total derivative* if it belongs to the image of $\nabla : \mathcal{V} \to \mathcal{V} \otimes \Omega$.

6.6.9. Corollary. *The map $\mathcal{Y}_x^\vee$ vanishes on the total derivatives and hence factors through a map $U(\mathcal{V}_x) \to \operatorname{End} \mathcal{V}_x$, where*

$$U(\mathcal{V}_x) = \Gamma(D_x^\times, \mathcal{V} \otimes \Omega)/\operatorname{Im} \nabla.$$

Observe that if we choose a coordinate z at x and trivialize $\mathcal{V}|_{D_x}$, then we can identify $U(\mathcal{V}_x)$ with the Lie algebra $U(V)$ introduced in § 4.1.1. This gives $U(\mathcal{V}_x)$ a Lie algebra structure, which *a priori* depends on the choice of z. The following result follows from Corollary 19.4.14 (to prove it we need to develop the theory of chiral algebras which we do in Chapter 19).

6.6.10. Theorem. *The Lie algebra structure on $U(\mathcal{V}_x)$ is coordinate-independent, and the map $\mathcal{Y}_x^\vee : U(\mathcal{V}_x) \to \operatorname{End} \mathcal{V}_x$ is a Lie algebra homomorphism.*

6.6.11. Remark. To be precise, $\mathcal{V}$ is not an inductive limit of flat bundles, since $\nabla(\mathcal{V}_{\leq m}) \subset \mathcal{V}_{\leq(m+1)} \otimes \Omega$, and likewise, $\mathcal{V}^*$ is not a projective limit of flat bundles. A proper way to say this is that $\mathcal{V}$ and $\mathcal{V}^*$ are $\mathcal{D}$–modules on X (note that $\mathcal{V}^*$ is not quasi-coherent as an $\mathcal{O}$–module).

6.7. Bibliographical notes

The definition of the ind–group $\underline{\operatorname{Aut}}\mathcal{O}$ given in § 6.2.3 is due to A. Beilinson and V. Drinfeld [**BD3**].

Further results on the representation theory of the Virasoro algebra can be found in [**Kac2, KR, FFu**]. The minimal models were defined in [**BPZ**]. They are discussed in [**dFMS**].

The construction of $\mathcal{A}ut_X$ is part of the "formal geometry" developed by I.M. Gelfand, D. Kazhdan and D. Fuchs (see [**GK, GKF, BR**]) as a method of applying representation theory of infinite–dimensional Lie algebras to finite–dimensional geometry. Definition 6.5.1 follows A. Beilinson and V. Drinfeld [**BD3**]. A similar definition was proposed by E. Witten in his study of conformal field theory [**Wi**].

The action of $\operatorname{Aut} \mathcal{O}$ on a vertex algebra was considered by Y.-Z. Huang in [**H1**]. Lemma 6.5.6 was originally proved by Y.-Z. Huang in [**H1**] by a different method (see [**H2, Le, Gab**] for related work).

The notion of quasi-conformal vertex algebra, Theorems 6.5.4 and 6.6.3 and the above proof of Lemma 6.5.6 are due to E. Frenkel.

CHAPTER 7

Action of Internal Symmetries

We have seen in the previous chapter that the vertex operation Y is covariant under the $\operatorname{Aut}\mathcal{O}$–action (see formula (6.5.1)). Therefore Y gives rise to a well-defined operation on the twist of V by any $\operatorname{Aut}\mathcal{O}$–torsor (see Theorem 6.5.4). This "twisting property" of Y follows from the fact that $\operatorname{Aut}\mathcal{O}$ acts on V by "internal symmetries", i.e., by exponentiation of Fourier coefficients of the Virasoro vertex operator $T(z)$. In this chapter we show that vertex algebras exhibit a similar twisting property with respect to any group of internal symmetries, i.e., a group (or more generally, an ind–group) $\mathcal{G}$ obtained by exponentiation of Fourier coefficients of vertex operators.

We start this chapter with a discussion of the special case when $\mathcal{G}$ is the semi-direct product $\operatorname{Aut}\mathcal{O} \ltimes G(\mathcal{O})$. This group acts by internal symmetries on vertex algebras with $\widehat{\mathfrak{g}}$–structures. After that we discuss the twisting property in general. Using the commutation relations between Fourier coefficients of vertex operators obtained in § 3.3.6, we derive an analogue of formula (6.5.1) for the transformation of Y under the action of any group of internal symmetries. This formula implies that Y gives rise to a well-defined operation on the twist of V by any $\mathcal{G}$–torsor.

In the last section we consider a coordinate-independent description of the n–point functions and an analogue of the section $\mathcal{Y}_x$ for modules over vertex algebras.

7.1. Affine algebras, revisited

Let $\mathfrak{g}$ be a simple Lie algebra, and G the corresponding simply-connected Lie group. Recall the affine Kac–Moody algebra $\widehat{\mathfrak{g}}$, which is a central extension of the formal loop algebra $\mathfrak{g}(\mathcal{K}) = \mathfrak{g}((t))$, see § 2.4.1. In this section we outline the geometric theory of vertex algebras with $\widehat{\mathfrak{g}}$–symmetries, which is analogous to the theory of conformal vertex algebras (i.e., those with Virasoro symmetry). Our protagonists are the following Lie groups and Lie algebras, which are analogous to those described at the beginning of Chapter 6:

$$
\begin{array}{cc}
G^{+}(\mathcal{O}) & \mathfrak{g}^{+}(\mathcal{O}) \\
\cap & \cap \\
G(\mathcal{O}) & \mathfrak{g}(\mathcal{O})
\end{array}
$$

Here $\mathfrak{g}^{+}(\mathcal{O}) = \mathfrak{g} \otimes t\mathbb{C}[[t]]$ and $G^{+}(\mathcal{O}) = \{g \in G(\mathcal{O}) | g \equiv 1 \bmod t\mathbb{C}[[t]]\}$.

The group $G^{+}(\mathcal{O})$ is a prounipotent proalgebraic group, which is the inverse limit of the algebraic groups associated to the nilpotent Lie algebras $\mathfrak{g} \otimes t\mathbb{C}[t]/t^{n}$. It follows that any locally nilpotent representation of the Lie algebra $\mathfrak{g}^{+}(\mathcal{O})$ may be exponentiated to $G^{+}(\mathcal{O})$. Suppose $\mathfrak{g}(\mathcal{O})$ acts on a vector space V in such a way that $\mathfrak{g}^{+}(\mathcal{O})$ acts locally nilpotently and V decomposes into a direct sum of finite-dimensional representations under the action of the constant Lie subalgebra

$\mathfrak{g} \subset \mathfrak{g}(\mathcal{O})$, so that $\mathfrak{g}$–action is integrable to G. Then the action of $\mathfrak{g}(\mathcal{O})$ on V may be exponentiated to $G(\mathcal{O})$.

For instance, consider the vacuum representation $V_k(\mathfrak{g})$ of $\widehat{\mathfrak{g}}$. The $\mathfrak{g}^+(\mathcal{O})$–action on $V_k(\mathfrak{g})$ decreases degrees, and hence is locally nilpotent. Since $\mathfrak{g}$ commutes with the gradation operator L_0, it preserves the homogeneous components of V, all of which are finite-dimensional. Therefore we obtain that $V_k(\mathfrak{g})$ decomposes under $\mathfrak{g}$ into a direct sum of finite–dimensional representations. Thus, $V_k(\mathfrak{g})$ is naturally a representation of $G(\mathcal{O})$.

The analogue of the notion of conformal vertex algebra, with $\widehat{\mathfrak{g}}$ playing the role of the Virasoro algebra, is described as follows.

7.1.1. Definition. We say that a vertex algebra V has a $\widehat{\mathfrak{g}}$–*structure* of level k if there is an injection $\alpha : \mathfrak{g} \hookrightarrow V$ such that the Fourier coefficients of the vertex operators $Y(\alpha(A), z)$, $A \in \mathfrak{g}$, generate an action of $\widehat{\mathfrak{g}}$ on V of level k.

We have the following analogue of Lemma 3.4.5:

7.1.2. Lemma. *A vertex algebra V has a $\widehat{\mathfrak{g}}$–structure of level k if and only if there is a non-zero (and then automatically injective) map $\alpha : \mathfrak{g} \to V$ such that the Fourier coefficients $J_n^{a,V}$ of the corresponding vertex operators*

$$Y(\alpha(J^a), z) = \sum_{n \in \mathbb{Z}} J_n^{a,V} z^{-n-1}$$

satisfy

$$J_n^{a,V} \cdot \alpha(J^b) = 0, \quad n > 1; \quad J_1^{a,V} \cdot \alpha(J^b) = k(J^a, J^b)|0\rangle; \quad J_0^{a,V} \cdot \alpha(J^b) = \alpha([J^a, J^b]).$$

Moreover, in that case there is a unique vertex algebra homomorphism $V_k(\mathfrak{g}) \to V$ such that $v_k \mapsto |0\rangle$ and $J_{-1}^a v_k \mapsto \alpha(J^a)$.

7.1.3. Semi-direct product. Note that the Lie algebra Der $\mathbb{C}((t))$ (and hence the Virasoro algebra Vir) acts naturally on $\widehat{\mathfrak{g}}$ by outer derivations. We will say that a conformal vertex algebra V has a *compatible* $\widehat{\mathfrak{g}}$–structure if the corresponding actions of Vir and $\widehat{\mathfrak{g}}$ on V combine into the action of the semi-direct product $Vir \ltimes \widehat{\mathfrak{g}}$ (note that in this case we also have an action of $Vir \ltimes \widehat{\mathfrak{g}}$ on any V–module).

For example, recall that if $k \neq -h^\vee$, then $V_k(\mathfrak{g})$ is a conformal vertex algebra via the Segal-Sugawara construction (see § 2.5.10). According to formula (2.5.7) this conformal structure is compatible with the tautological $\widehat{\mathfrak{g}}$–structure on V. If V is any vertex algebra with a $\widehat{\mathfrak{g}}$–structure of level $k \neq -h^\vee$, then by Lemma 7.1.2 we have a homomorphism $V_k(\mathfrak{g}) \to V$. The image of the Segal-Sugawara vector in V is then a conformal vector in V. Thus, we obtain a conformal structure on V, which is obviously compatible with the $\widehat{\mathfrak{g}}$–structure. In this case the central charge of the Virasoro algebra is determined by k: it is equal to $c(k) = k \dim \mathfrak{g}/(k + h^\vee)$. In general, however, we may have compatible conformal and $\widehat{\mathfrak{g}}$–structure on V such that the central charge of the Virasoro algebra is different from $c(k)$.

Similarly, we say that a quasi-conformal vertex algebra V has a compatible $\widehat{\mathfrak{g}}$–structure if the corresponding actions of Der $\mathcal{O}$ and $\widehat{\mathfrak{g}}$ on V combine into an action of the semi-direct product Der $\mathcal{O} \ltimes \widehat{\mathfrak{g}}$. An example is the vertex algebra $V_{-h^\vee}(\mathfrak{g})$.

7.1.4. Kac–Moody twisting. The action of $\operatorname{Aut}\mathcal{O}$ on a quasi-conformal vertex algebra V gives rise to a vector bundle $\mathcal{V}$ which is the twist of V by the $\operatorname{Aut}\mathcal{O}$–torsor $\mathcal{A}ut_X$ of formal coordinates on a smooth curve X. The bundle $\mathcal{V}$ provides the setting for a coordinate-independent description of the vertex operators in V. Namely, by Theorem 6.5.4, we obtain an intrinsic $\operatorname{End}\mathcal{V}_x$–valued section $\mathcal{Y}_x$ of $\mathcal{V}^*|_{D_x^\times}$ for any $x \in X$.

Now we would like to construct a similar bundle in the case of a quasi-conformal vertex algebra V with a compatible $\widehat{\mathfrak{g}}$–structure. Suppose that the action of $\mathfrak{g}(\mathcal{O})$ on V can be exponentiated to a $G(\mathcal{O})$–action. Then this $G(\mathcal{O})$–action may be extended to an action of the semi-direct product $\operatorname{Aut}\mathcal{O} \ltimes G(\mathcal{O})$, and we want to twist V by an $\operatorname{Aut}\mathcal{O} \ltimes G(\mathcal{O})$–torsor on X. Further, we want the vertex operators to give rise to an intrinsic section of such a bundle in the neighborhood of any point of X.

Suitable $\operatorname{Aut}\mathcal{O} \ltimes G(\mathcal{O})$–torsors can be constructed from G–bundles. Let $\mathcal{P}$ be a principal G–bundle on X. For $x \in X$, consider the set $\widetilde{\mathcal{P}}_x$ of trivializations of the restriction of $\mathcal{P}$ to D_x. The group $G(\mathcal{O}_x)$ naturally acts on these trivializations, and this action makes $\widetilde{\mathcal{P}}_x$ into a $G(\mathcal{O})_x$–torsor.

Now let $\widehat{\mathcal{P}}_x$ be the set of pairs (z,s) where z is a formal coordinate at x and s is a trivialization of $\mathcal{P}|_{D_x}$. Then the group $\operatorname{Aut}\mathcal{O} \ltimes G(\mathcal{O})$ acts on $\widehat{\mathcal{P}}_x$. Indeed, given a point (z,s) of $\widehat{\mathcal{P}}_x$, we trivialize $\mathcal{P}|_{D_x}$ and identify $G(\mathcal{O}_x)$ with $G(\mathcal{O})$. We can then act on (z,s) by an element of $\operatorname{Aut}\mathcal{O} \ltimes G(\mathcal{O})$ by changing the trivialization of $\mathcal{P}|_{D_x}$ and changing the coordinate z. This action makes $\widehat{\mathcal{P}}_x$ into an $\operatorname{Aut}\mathcal{O} \ltimes G(\mathcal{O})$–torsor. Furthermore, it is easy to see that there exists a principal $\operatorname{Aut}\mathcal{O} \ltimes G(\mathcal{O})$–bundle $\widehat{\mathcal{P}}$ over X (locally trivial in the Zariski topology) whose fiber at $x \in X$ is $\widehat{\mathcal{P}}_x$.

7.1.5. Definition. Let M be an $\operatorname{Aut}\mathcal{O} \ltimes G(\mathcal{O})$–module. For a principal G–bundle $\mathcal{P}$ on X, the $\mathcal{P}$–twist of M is by definition the vector bundle

$$\mathcal{M}^{\mathcal{P}} = \widehat{\mathcal{P}} \underset{\operatorname{Aut}\mathcal{O} \ltimes G(\mathcal{O})}{\times} M.$$

The fiber of $\mathcal{M}^{\mathcal{P}}$ at x will be denoted by $\mathcal{M}_x^{\mathcal{P}}$.

Now let us take as M our vertex algebra V satisfying the above conditions. We then define an $\operatorname{End}\mathcal{V}_x^{\mathcal{P}}$–valued section $\mathcal{Y}_x^{\mathcal{P}}$ of $(\mathcal{V}^{\mathcal{P}})^*|_{D_x^\times}$ as follows. Choose a formal coordinate z at x and a trivialization s of $\mathcal{P}|_{D_x}$. Then we obtain trivializations of the bundles $\widehat{\mathcal{P}}$ and $\mathcal{V}^{\mathcal{P}}$ over D_x. As in § 6.6.3 we construct a flat connection on $\mathcal{V}^{\mathcal{P}}$ such that in this trivialization the connection operator is given by by the formula $\nabla_{\partial_z} = \partial_z + L_{-1}$. Following the argument of § 17.1.2, one shows that this connection does not depend on the choice of z.

Now given $v \in \mathcal{V}_x^{\mathcal{P}}$ and $\varphi \in (\mathcal{V}_x^{\mathcal{P}})^*$, we consider them as elements of V and V^*, respectively. We then set the corresponding matrix element of $\mathcal{Y}_x^{\mathcal{P}}$ evaluated on the constant section $A \in V$ of $\mathcal{V}^{\mathcal{P}}|_{D_x}$ with respect to this trivialization to be equal to $\langle \varphi, Y(A,z)v \rangle$.

7.1.6. Theorem. *Let V be a quasi-conformal vertex algebra with a compatible $\widehat{\mathfrak{g}}$–structure such that the action of $\mathfrak{g}(\mathcal{O})$ can be exponentiated to $G(\mathcal{O})$. Then the vertex operation Y on V gives rise to a canonical $\operatorname{End}\mathcal{V}_x^{\mathcal{P}}$–valued horizontal section $\mathcal{Y}_x^{\mathcal{P}}$ of $(\mathcal{V}^{\mathcal{P}})^*|_{D_x^\times}$. We also have a canonical map $\Gamma(D_x^\times, \mathcal{V}^{\mathcal{P}})^{\cdot} \to \operatorname{End}\mathcal{V}_x^{\mathcal{P}}$, which vanishes on the total derivatives.*

This statement follows from a more general result that will be established below in Theorem 7.2.9.

7.2. The general twisting property

In Chapter 6, we studied the action of changes of coordinates on a conformal vertex algebra using the OPE of the field $T(z)$ with other fields. In particular, formula (6.5.1) and its infinitesimal form (6.3.5) describe the transformation rules for vertex operators $Y(A, z)$ under the action of Aut $\mathcal{O}$. In this section, we generalize these arguments to describe how vertex operators transform under an arbitrary group of internal symmetries of the vertex algebra. In fact, these general transformation rules given in formulas (7.2.2) and (7.2.3) are equivalent to the commutation relations (3.3.12) (and hence to the locality axiom, by Remark 3.3.9).

In Theorem 6.5.4 we used the formula for the transformation of Y under the group Aut $\mathcal{O}$ to show that Y gives rise to an intrinsic section $\mathcal{Y}_x$ of $\mathcal{V}^*|_{D_x^\times}$, where $\mathcal{V}$ is a twist of V by an Aut $\mathcal{O}$–torsor. Likewise, we will use the transformation property of Y under a general group of internal symmetries to derive Theorem 7.2.6, which asserts that Y gives rise to a section of the twist of V^* by a $\mathcal{G}$–torsor. Thus, given any group of internal symmetries of a vertex algebra, such as the group Aut $\mathcal{O}$ or Aut $\mathcal{O} \ltimes G(\mathcal{O})$, there is a well-defined vertex operation on the twist of V by a torsor over this group. An important example of this construction is the twisting by principal G–bundles, which was discussed in § 7.1.4. Another example will be considered in § 18.5.

7.2.1. Internal symmetries. Recall the Lie algebra $U(V)$ defined in § 4.1.1, which can be thought of as the completion of the span of all Fourier coefficients of vertex operators from V. We have a natural homomorphism of Lie algebras $U(V) \to \operatorname{End} V$. Namely, to each $A \in V$ and any Laurent series

$$(7.2.1) \qquad\qquad a(z) = \sum_{k \in \mathbb{Z}} a_k z^k,$$

we assign the linear operator on V,

$$\mathbf{a} = \sum_{k \in \mathbb{Z}} a_k A_{(k)}$$

$$= \int a(z) Y(A, z) dz = \int a(z) \sum_{n \in \mathbb{Z}} A_{(n)} z^{-n-1} dz.$$

The operators of this form span the image of the Lie algebra $U(V)$ in $\operatorname{End} V$. For the sake of brevity, in this chapter we will use the same notation $U(V)$ for this image.

By definition, a *Lie algebra of internal symmetries* of V is any Lie subalgebra of $U(V)$. Examples are the Lie algebras Vir, $\operatorname{Der} \mathcal{O}$, $\operatorname{Der}_0 \mathcal{O}$ in a conformal vertex algebra, and the Lie algebras $\widehat{\mathfrak{g}}, \mathfrak{g}(\mathcal{O}), \mathfrak{g}^+(\mathcal{O})$ in a vertex algebra with a $\widehat{\mathfrak{g}}$–structure (or the corresponding semi-direct products).

Let us discuss whether we can exponentiate the action of such Lie algebras on V. We will assume for the remainder of this section that V is a $\mathbb{Z}$–graded vertex algebra with gradation bounded from below. If A is a homogeneous element of V, of degree Δ_A, then the Fourier coefficients $A_{(n)}$ of the vertex operator $Y(A, z)$ with $n \geq \Delta_A$ have negative degrees. Hence they act locally nilpotently on V, and the formal exponential $\exp A_{(n)}$ is a well-defined linear operator on V. Denote by

$U(V)_+$ the Lie subspace of $U(V)$ spanned by elements $\mathbf{a}$ given by formula (7.2.1), where the summation is over $k \geq \Delta_A$. By inspection of the Lie bracket formula (3.3.12) we obtain that $U(V)_+$ is a Lie subalgebra of $U(V)$. The action of $U(V)_+$ can be exponentiated to an action of a prounipotent proalgebraic group, which we denote by $\mathcal{G}(V)_+$.

The Fourier coefficients $A_{(\Delta_A-1)}$, which have degree 0, may sometimes be exponentiated as well (e.g., the operator L_0). But $A_{(n)}, n < \Delta_A - 1$, are creation operators and cannot be exponentiated. Hence they cannot generate an action of a group. However, note that the operator $\exp \mathbf{a}$ would be well-defined if all coefficients $a_k, k < \Delta_A$, were nilpotent. Hence the Lie algebra $U(V)$ generates an action of an ind–group, which we denote by $\mathcal{G}(V)$. By definition, the group of R–points of $\mathcal{G}(V)$, where R is a commutative $\mathbb{C}$–algebra, is the group of automorphisms of $V \otimes R$ generated by the operators $\exp \mathbf{a}$, where $\mathbf{a}$ is given by formula (7.2.1) and the coefficients $a_k, k < \Delta_A$, are nilpotent elements of R (if $A_{(\Delta_A-1)}$ can be exponentiated, we relax this condition to $k < \Delta_A - 1$).

By definition, an (ind)–group of internal symmetries of V is a subgroup of $\mathcal{G}(V)_+$ (resp., $\mathcal{G}(V)$). Examples of groups of internal symmetries are $\mathrm{Aut}\,\mathcal{O}$ and $G(\mathcal{O})$. The ind–group $\underline{\mathrm{Aut}}\mathcal{O}$ from § 6.2.3 is an example of an ind–group of internal symmetries.

7.2.2. Commutation relations.

The simplest internal symmetries of a vertex algebra V are the residues $A_{(0)} : V \to V$ of vertex operators $Y(A, z), A \in V$. According to Corollary 3.3.8, they satisfy the commutation relations

$$[A_{(0)}, Y(B, w)] = Y(A_{(0)} \cdot B, w),$$

and hence they are infinitesimal automorphisms of V. We will now derive an analogous formula for an arbitrary element $\mathbf{a}$ of $U(V)$ instead of $A_{(0)}$.

Given a Laurent power series $a(z)$, define a formal power series $\mathbf{a}_w$ in $w^{\pm 1}$ with coefficients in $\mathrm{End}\,V$ by the formula

$$\mathbf{a}_w = \mathrm{Res}_{z=w}\, a(z) Y(A, z-w) dz = \sum_{n \in \mathbb{Z}} \mathrm{Res}_{z=w}\, a(z) \frac{A_{(n)}}{(z-w)^{n+1}} dz$$

$$= \sum_{n \geq 0} \frac{1}{n!} \partial_w^n a(w)\, A_{(n)}.$$

Recall from § 3.3 that the associativity property of V enables us to calculate the commutators of Fourier coefficients of vertex operators. Specifically, we have

$$[A_{(k)}, Y(B, w)] = \sum_{n \geq 0} \mathrm{Res}_{z=w} \frac{z^k}{(z-w)^{n+1}} Y(A_{(n)} \cdot B, w) dz$$

for any $B \in V$. It follows that

$$\begin{aligned}
[\mathbf{a}, Y(B, w)] &= \left[\sum_{k \in \mathbb{Z}} a_k A_{(k)}, Y(B, w)\right] \\
&= \mathrm{Res}_{z=w} \sum_{k,n} \frac{a_k z^k}{(z-w)^{n+1}} Y(A_{(n)} \cdot B, w) dz \\
&= \sum_{n \geq 0} \frac{1}{n!} \partial_w^n a(w) \cdot Y\left(A_{(n)} \cdot B, w\right).
\end{aligned}$$

Thus we have the following generalization of formula (6.3.5):

$$(7.2.2) \qquad\qquad [\mathbf{a}, Y(B, w)] = Y(\mathbf{a}_w \cdot B, w).$$

Note that this formula is equivalent to formulas (3.3.11) and (3.3.12) (which, according to Remark 3.3.9, may be taken as a replacement of the locality axiom in the definition of vertex algebras).

7.2.3. Exponential form. Let R be a $\mathbb{C}$–algebra, and assume that $a(z) \in R((z))$ is such that $\exp \mathbf{a}$ makes sense as an automorphism of $V \otimes R$. In that case, following the argument in the proof of Lemma 6.5.6, we obtain the following identity, which is analogous to formula (6.5.1):

$$(7.2.3) \qquad\qquad Y(B, w) = e^{\mathbf{a}} Y(e^{-\mathbf{a}_w} \cdot B, w) e^{-\mathbf{a}}.$$

Now we wish to interpret the identity (7.2.3), in the case when $a(z) \in \mathbb{C}[[z]]$, in a way similar to that of the identity (6.5.1). We will assume from now on that V is a quasi-conformal vertex algebra.

Consider the trivial vector bundle with the fiber $V((z))$ on $D = \operatorname{Spec} \mathbb{C}[[w]]$, with the connection operator $\partial_w + L_{-1}$. This connection commutes with the operator $\partial_z + L_{-1}$ acting along the fibers. Hence it preserves $\operatorname{Im}(\partial_z + L_{-1})$ and descends to an operator on

$$U(V)[[w]] = V((z))[[w]] / \operatorname{Im}(\partial_z + L_{-1}).$$

Since $[L_{-1}, Y(A, z)] = Y(L_{-1}A, z)$, this connection operator is equal to $\partial_w + \operatorname{ad} L_{-1}$ (where $\operatorname{ad} L_{-1}$ is a linear operator on $U(V)$). We now find from the formula $\partial_z Y(A, z) = [L_{-1}, Y(A, z)]$ that

$$(\partial_w + \operatorname{ad} L_{-1})(\mathbf{a}_w) = (\partial_w + \operatorname{ad} L_{-1}) \left(\operatorname{Res}_{z=w} a(z) Y(A, z - w) dz \right) = 0.$$

Therefore we obtain:

7.2.4. Lemma. $\mathbf{a}_w$ *can be characterized as the unique horizontal section of the trivial vector bundle on D with the fiber $U(V)$ and the connection $\partial_w + \operatorname{ad} L_{-1}$, whose value at $0 \in D$ equals $\mathbf{a}$.*

7.2.5. Twisting by a $\mathcal{G}$–torsor. Let $\mathcal{G}$ be a group of internal symmetries of V such that the corresponding Lie algebra is stable under the operator $\operatorname{ad} L_{-1}$ acting on $U(V)$. Let $\mathcal{T}_0$ be a $\mathcal{G}$–torsor. We associate to it a canonical $\mathcal{G}$–bundle $\mathcal{T}$ with a flat connection on the disc $D = \operatorname{Spec} \mathbb{C}[[z]]$, whose fiber at 0 is $\mathcal{T}_0$ (note that a flat bundle on the disc is uniquely determined by its zero fiber). Let $V^{\mathcal{T}} = \mathcal{T} \underset{\mathcal{G}}{\times} V$ be the twist of V by $\mathcal{T}$.

Now define a section $\mathcal{Y}^{\mathcal{T}}$ of $(V^{\mathcal{T}})^*$ with values in $\operatorname{End} V^{\mathcal{T}_0}$ as follows. Pick a trivialization $\imath_0 : \mathcal{G} \xrightarrow{\sim} \mathcal{T}_0$. It extends canonically to an isomorphism of flat bundles $\imath : D \times \mathcal{G} \xrightarrow{\sim} \mathcal{T}$, where the trivial bundle $D \times \mathcal{G}$ carries the connection $\nabla_{\partial_w} = \partial_w + L_{-1}$. To define $\mathcal{Y}^{\mathcal{T}}$ using this trivialization, we need to specify the matrix elements $\langle \varphi, \mathcal{Y}^{\mathcal{T}}(\imath(A)) v \rangle$ for all $v \in V, \varphi \in V^*$ and $A \in V$ (considered as the constant section of $D \times \mathcal{G}$).

7.2.6. Theorem. *Define an* $\operatorname{End} V^{\mathcal{T}_0}$*-valued section* $\mathcal{Y}^{\mathcal{T}}$ *of* $(V^{\mathcal{T}})^*$ *by the formula*

$$\langle \varphi, \mathcal{Y}^{\mathcal{T}}(\imath(A)) \cdot v \rangle = \langle \varphi, Y(A,w)v \rangle.$$

Then the section $\mathcal{Y}^{\mathcal{T}}$ *is independent of the choice of the trivialization* $\imath_0$*, and is horizontal with respect to the connection on* $(V^{\mathcal{T}})^*$ *induced by the flat connection on* $\mathcal{T}$.

7.2.7. Proof. We use the same argument as in the proof of Theorem 6.5.4. More precisely, if we choose another trivialization $\imath'_0$ of $\mathcal{T}_0$ which differs from $\imath_0$ by $e^{-\mathbf{a}} \in \mathcal{G}$, then we should replace $v \mapsto e^{-\mathbf{a}} \cdot v$ and $\varphi \mapsto \varphi \cdot e^{\mathbf{a}}$. By construction, the isomorphism $\imath$ preserves flat connections. Hence by Lemma 7.2.4, the section $\imath'(A)$ becomes $\imath(e^{-\mathbf{a}_w} \cdot A)$. Formula (7.2.3) then shows that the section $\mathcal{Y}^{\mathcal{T}}$ is independent of the trivialization $\imath_0$.

7.2.8. Incorporating changes of coordinate. The drawback of the previous construction is that we had to fix a coordinate w on the disc. In order make it coordinate-independent, we need to impose an additional condition on the group $\mathcal{G}$. Recall that by our assumption V is a quasi-conformal vertex algebra. Then $\operatorname{Der} \mathcal{O}$ acts on $U(V)$ preserving the Lie algebra structure. Suppose that the Lie algebra of $\mathcal{G}$ is normalized by $\operatorname{Der} \mathcal{O}$ (for example, $\mathcal{G} = G(\mathcal{O})$ satisfies this condition). Then the group $\widetilde{\mathcal{G}} = \operatorname{Aut} \mathcal{O} \ltimes \mathcal{G}$ acts on V.

Recall the $\operatorname{Aut} \mathcal{O}$–bundle $\mathcal{A}ut_X$ on a smooth curve X. Define the group scheme $\mathcal{G}_X$ over X as the $\mathcal{A}ut_X$–twist of $\mathcal{G}$:

$$\mathcal{G}_X = \mathcal{A}ut_X \underset{\operatorname{Aut} \mathcal{O}}{\times} \mathcal{G}.$$

In the same way as in § 6.6.3 we construct a flat connection on $\mathcal{G}_X$. If we choose a coordinate z on D_x, then the restriction $\mathcal{G}_{D_x}$ of the bundle $\mathcal{G}_X$ to D_x gets trivialized, and we define the connection operator by the formula $\nabla_{\partial_z} = \partial_z + L_{-1}$. Following the argument of § 17.1.2, one shows that this connection does not depend on the choice of z.

Let $\mathcal{V} = \mathcal{A}ut_X \underset{\operatorname{Aut} \mathcal{O}}{\times} V$, as before. Then $\mathcal{G}_X$ acts on $\mathcal{V}$, and so $\mathcal{G}_{D_x}$ acts on $\mathcal{V}|_{D_x}$. Suppose we are given a $\mathcal{G}_{D_x}$–torsor $\mathcal{T}$. Define a vector bundle $\mathcal{V}^{\mathcal{T}}$ on D_x by the formula

$$\mathcal{V}^{\mathcal{T}} = \mathcal{T} \underset{\mathcal{G}_{D_x}}{\times} (\mathcal{V}|_{D_x}).$$

Then $\mathcal{T}$ acquires a canonical flat connection such that for any trivialization of $\mathcal{T}_x$, the flat bundle $\mathcal{T}$ gets identified with $\mathcal{G}_{D_x}$. In addition, the vector bundle $\mathcal{V}^{\mathcal{T}}$ inherits a connection from $\mathcal{T}$.

Now let us fix a coordinate z on D_x. Using Theorem 7.2.6, we then obtain an $\operatorname{End} V_x^{\mathcal{T}}$–valued section of $(V^{\mathcal{T}})^*$ on $D_x^\times$. Following the argument in the proof of Theorem 6.5.4, we show that this section is independent of the choice of the coordinate z. Thus, we obtain the following general result:

7.2.9. Theorem. *If* $\mathcal{G}$ *be a group of internal symmetries of a quasi-conformal vertex algebra* V *satisfying the above assumption, then for any* $\mathcal{G}_{D_x}$*-torsor* $\mathcal{T}$ *there is a canonical* $\operatorname{End} V_x^{\mathcal{T}}$*-valued horizontal section* $\mathcal{Y}_x^{\mathcal{T}}$ *of* $(\mathcal{V}^{\mathcal{T}})^*$ *on* $D_x^\times$*. Equivalently, there is a canonical map* $\Gamma(D_x^\times, \mathcal{V}^{\mathcal{T}} \otimes \Omega) \to \operatorname{End} V_x^{\mathcal{T}}$*, which vanishes on the total derivatives.*

7.2.10. The case of $G(\mathcal{O})$. Let us apply Theorem 7.2.9 in the case when V is a quasi-conformal vertex algebra with a compatible $\hat{\mathfrak{g}}$–structure, and such that the action of $\mathfrak{g}(\mathcal{O})$ can be exponentiated to $\mathcal{G} = G(\mathcal{O})$. Then $\mathcal{G}_X$ is the group scheme over X whose fiber at $x \in X$ is $G(\mathcal{O}_x)$. To a G–bundle $\mathcal{P}$ on X, we attach a $\mathcal{G}_X$–torsor $\widetilde{\mathcal{P}}$ whose fiber at $x \in X$ consists of all trivializations of $\mathcal{P}|_{D_x}$. Hence we can use $\widetilde{\mathcal{P}}|_{D_x}$ as $\mathcal{T}$ in the discussion of § 7.2.8. Now observe that

$$\widetilde{\mathcal{P}} \underset{\mathcal{G}_X}{\times} \mathcal{V} = \widehat{\mathcal{P}} \underset{\mathrm{Aut}\,\mathcal{O} \ltimes G(\mathcal{O})}{\times} V,$$

where $\widehat{\mathcal{P}}$ is the torsor defined in § 7.1.4. Therefore Theorem 7.2.9 becomes Theorem 7.1.6 in this case (to be precise, the notation for the section $\mathcal{Y}_x^{\mathcal{P}}$ should be $\mathcal{Y}_x^{\widetilde{\mathcal{P}}}$ to make it compatible with the notation of Theorem 7.2.9).

7.3. Description of the n–point functions and modules

In this section we generalize the constructions of Chapter 6 to the case of several points and to the case of modules over vertex algebras.

7.3.1. n–point functions, revisited. Let X be a smooth curve, and $\mathcal{V}$ the vector bundle on X corresponding to a quasi-conformal vertex algebra V. Denote by $\mathcal{V}^{\boxtimes n} = \mathcal{V} \boxtimes \cdots \boxtimes \mathcal{V}$ the n–fold external tensor power of $\mathcal{V}$. The bundle $\mathcal{V}^{\boxtimes n}$ is a union of vector bundles of finite rank. By its dual bundle, denoted by $(\mathcal{V}^{\boxtimes n})^*$, we will understand the projective limit of the corresponding dual bundles of finite rank.

Let x be a point of X and D_x the disc around x. If we choose a formal coordinate z at x, then $D_x = \operatorname{Spec}\mathbb{C}[[z]]$. Set

$$D_x^n \;=\; \operatorname{Spec}\mathbb{C}[[z_1,\ldots,z_n]].$$

Note that $D_x^n \subset X^n$ does not depend on the choice of coordinate z.

A section of the dual vector bundle $(\mathcal{V}^{\boxtimes n})^*$ over D_x^n with poles on the pairwise diagonals is by definition a $\mathbb{C}[[z_1,\ldots,z_n]]$–linear map from the space of sections of $\mathcal{V}^{\boxtimes n}$ on D_x^n to $\mathbb{C}[[z_1,\ldots,z_n]][(z_i - z_j)^{-1}]_{i \neq j}$. The space of such sections will be denoted by $\Gamma(D_x^n, (\mathcal{V}^{\boxtimes n})^*(\infty\Delta))$, where Δ is the union of all pairwise diagonals $z_i = z_j, i \neq j$.

We will now attach such a section to any element of $\mathcal{V}_x^*$. Using the formal coordinate z on D_x, we trivialize $\mathcal{V}$, so that

$$\Gamma(D_x^n, \mathcal{V}^{\boxtimes n}) \simeq V^{\otimes n}[[z_1,\ldots,z_n]].$$

In order to define a section of $(\mathcal{V}^{\boxtimes n})^*(\infty\Delta)$ over D_x^n it then suffices to attach an element of $\mathbb{C}[[z_1,\ldots,z_n]][(z_i - z_j)^{-1}]_{i\neq j}$ to each constant section $A_1 \otimes \ldots \otimes A_n$ of $\mathcal{V}^{\boxtimes n}$, where $A_i \in V$. Given $\varphi \in \mathcal{V}_x^* \simeq V^*$, we attach to it a section of $(\mathcal{V}^{\boxtimes n})^*(\infty\Delta)$ by the rule

$$(7.3.1) \qquad A_1 \otimes \ldots \otimes A_n \mapsto \langle \varphi, Y(A_1, z_1)\ldots Y(A_n, z_n)|0\rangle,$$

where the right hand side is considered as an element of $\mathbb{C}[[z_1,\ldots,z_n]][(z_i - z_j)^{-1}]_{i\neq j}$ (see Theorem 4.5.1).

Using Theorem 6.5.4, we obtain

7.3.2. Theorem. *The above map $\mathcal{V}_x^* \to \Gamma(D_x^n, (\mathcal{V}^{\boxtimes n})^*(\infty\Delta))$ is independent of the choice of coordinate z on D_x.*

7.3.3. Flat connection. The connection ∇ on $\mathcal{V}$ can also be generalized in a straightforward manner to the case of multiple points.

For $i = 1, \ldots, n$, let z_i be a formal coordinate on X near x_i, and $\partial/\partial z_i$ the corresponding coordinate vector field. Since these generate the tangent space at $(x_1, \ldots, x_n) \in X^n$, in order to define a connection on $\mathcal{V}^{\boxtimes n}$, it suffices to specify how the vector fields ∂_{z_i}, $i = 1, \ldots, n$, act on the sections of $\mathcal{V}^{\boxtimes n}$. Introduce the operators ∇_i, $i = 1, \ldots, n$, on $\mathcal{V}^{\boxtimes n}$ by the formula

$$\nabla_i = \partial_{z_i} + L_{-1}^{(i)}.$$

Here we adopt the notational convention that if P is an operator on $\mathcal{V}$, then $P^{(i)}$ denotes this operator acting on the ith factor.

7.3.4. Proposition. *The assignment $\nabla : \partial_{z_i} \mapsto \nabla_i$ defines a flat connection on the vector bundle $\mathcal{V}^{\boxtimes n}$ which is independent of the choices of coordinates z_i.*

Moreover, the sections of $(\mathcal{V}^{\boxtimes n})^*(\infty\Delta)$ defined in Theorem 7.3.2 are horizontal with respect to the dual connection.

7.3.5. Proof. The independence of the connection on the coordinates z_i follows from § 17.1.6 below. The horizontality of the sections defined in Theorem 7.3.2 follows from Theorem 4.5.1,(3).

7.3.6. Coordinate-independent description of modules. Let V be a conformal vertex algebra and M a conformal V–module (see Definitions 5.1.1 and 5.1.9). Then M is a module over the Virasoro algebra, and hence over the Lie algebra Der $\mathcal{O}$. If the eigenvalues of L_0 on M are integers, then the $\mathrm{Der}_0\, \mathcal{O}$–action on M can be exponentiated to an Aut $\mathcal{O}$–action. Hence we obtain a vector bundle $\mathcal{M} = \mathcal{A}ut_X \underset{\mathrm{Aut}\,\mathcal{O}}{\times} M$ on any smooth curve X. In the same way as in the case of V, one constructs a coordinate-independent version of the operation $Y_M : V \to \mathrm{End}\, M[[z^{\pm 1}]]$. Moreover, Proposition 5.1.2,(1) implies that the section of $\mathcal{V}^*|_{D_x^\times}$ that we obtain is horizontal with respect to the connection ∇ defined in § 6.6.

7.3.7. Theorem. *The operation Y_M gives rise to a canonical horizontal section $\mathcal{Y}_{M,x}$ of $\mathcal{V}^*|_{D_x^\times}$ with values in $\mathrm{End}\, \mathcal{M}_x$.*

Equivalently, we have a canonical map

$$\mathcal{Y}_{M,x}^\vee : \Gamma(D_x^\times, \mathcal{V} \otimes \Omega) \to \mathrm{End}\, \mathcal{M}_x$$

sending μ to $\mathrm{Res}_x\langle \mathcal{Y}_{M,x}, \mu \rangle$ (cf. § 6.5.8). This map vanishes on the total derivatives (cf. Corollary 6.6.9).

7.3.8. Flat connection. The construction of the flat connections on $\mathcal{V}$ and $\mathcal{V}^{\boxtimes n}$ also carries over to the case of a conformal V–module M with integral eigenvalues of L_0. Namely, we define a flat connection ∇_M on $\mathcal{M}$ by the formula $d + L_{-1}^M$. In the same way as in § 17.1.7 one shows that this definition is coordinate-independent. Furthermore, the section $\mathcal{Y}_{M,x}$ is automatically horizontal. If $M_1, \ldots, M_n$ are V–modules with integral eigenvalues L_0, then we construct a flat connection on the vector bundle $\mathcal{M}_1 \boxtimes \cdots \boxtimes \mathcal{M}_n$ over X^n by setting

$$\nabla_i = \frac{\partial}{\partial z_i} + (L_{-1}^M)^{(i)}.$$

The above connection will be used to define the Knizhnik–Zamolodchikov connection in Chapter 13.

7.3.9. The case of non-integral gradation.

Let M be a representation of $\mathrm{Der}\,\mathcal{O}$ on which $\mathrm{Der}_+\,\mathcal{O}$ acts locally nilpotently but the eigenvalues of L_0 are not integral. In that case the representation of $\mathrm{Der}_+\,\mathcal{O}$ on M can be exponentiated to a representation of $\mathrm{Aut}_+\,\mathcal{O}$, but the action of $\mathrm{Der}_0\,\mathcal{O}$ cannot be exponentiated to an action of $\mathrm{Aut}\,\mathcal{O}$. Therefore we cannot attach to M a vector bundle $\mathcal{M}$ on a smooth curve as before. However, we can attach to it a bundle on a $\mathbb{C}^\times$–bundle over X.

This bundle, denoted $\widetilde{X}$, is the space of pairs (x,τ) where $x \in X$ and τ is a non-zero tangent vector at x. Thus, $\widetilde{X}$ is the total space of the principal $\mathbb{C}^\times$–bundle underlying the tangent bundle to X. It may also be described as the bundle associated to the principal $\mathrm{Aut}\,\mathcal{O}$–bundle $\mathcal{A}ut_X$ under the natural action of $\mathrm{Aut}\,\mathcal{O}$ on $\mathrm{Aut}\,\mathcal{O}/\mathrm{Aut}_+\,\mathcal{O} \simeq \mathbb{C}^\times$. Note that the one-jet of every formal coordinate z at x defines a non-zero tangent vector at x. Hence we have a map $\mathcal{A}ut_X \to \widetilde{X}$ which is a principal $\mathrm{Aut}_+\,\mathcal{O}$–bundle over $\widetilde{X}$. Its fiber $\mathcal{A}ut_{x,\tau}$ at (x,τ) consists of the formal coordinates z at x such that the one-jet of z equals τ.

Now we attach, to any $\mathrm{Aut}_+\,\mathcal{O}$–module M, a vector bundle $\mathcal{M}$ over $\widetilde{X}$, by setting

$$\mathcal{M} = \mathcal{A}ut_X \underset{\mathrm{Aut}_+\,\mathcal{O}}{\times} M.$$

The fibers of this bundle are $\mathcal{A}ut_{x,\tau}$–twists of M.

We also have a coordinate-independent description of the operation $Y_M : V \to \mathrm{End}\,M[[z,z^{-1}]]$. Fix a point (x,τ) of $\widetilde{X}$. Denote by $\mathcal{M}_{(x,\tau)}$ the fiber of $\mathcal{M}$ at (x,τ). The following theorem is proved in the same way as Theorem 6.5.4.

7.3.10. Theorem. *The operation Y_M gives rise to a canonical horizontal section $\mathcal{Y}_{M,(x,\tau)}$ of $\mathcal{V}^*$ over $D_x^\times$ with values in $\mathrm{End}\,\mathcal{M}_{(x,\tau)}$.*

7.3.11. Modules over vertex algebras with a $\widehat{\mathfrak{g}}$–structure.

Suppose that V also carries a compatible $\widehat{\mathfrak{g}}$–structure. Let M be a V–module. Then it carries an action of the Lie algebra $\mathrm{Der}\,\mathcal{O} \ltimes \mathfrak{g}(\mathcal{O})$. Suppose that the eigenvalues of L_0 on M are integers and that the action of $\mathfrak{g}$ on M is integrable to G. Then the action of the Lie algebra $\mathrm{Der}\,\mathcal{O} \ltimes \mathfrak{g}(\mathcal{O})$ on M can be exponentiated to an action of $\mathrm{Aut}\,\mathcal{O} \ltimes G(\mathcal{O})$. For a curve X and a G–bundle $\mathcal{P}$ on X, we obtain a vector bundle $\mathcal{M}^{\mathcal{P}}$ on X (see Definition 7.1.5). Then for each $x \in X$ the operation Y_M gives rise to a canonical section $\mathcal{Y}^{\mathcal{P}}_{M,x}$ of $(\mathcal{V}^{\mathcal{P}})^*|_{D_x^\times}$ with values in $\mathrm{End}\,\mathcal{M}^{\mathcal{P}}_x$ as in Theorem 7.1.6.

In the general case, M only carries an action of the group $\mathrm{Aut}_+\,\mathcal{O} \ltimes G^+(\mathcal{O})$. Choose a non-zero tangent vector τ at x and an isomorphism $\phi : G \simeq \mathcal{P}_x$. Let $\widehat{\mathcal{P}}^\phi_{(x,\tau)}$ be the set of pairs (z,s), where z is a formal coordinate at x compatible with τ, and s is a trivialization of $\mathcal{P}|_{D_x}$ compatible with ϕ. Then $\widehat{\mathcal{P}}_{\tau,\phi}$ is a torsor for the group $\mathrm{Aut}_+\,\mathcal{O} \ltimes G^+(\mathcal{O})$, and we define $\mathcal{M}^{\mathcal{P},\phi}_{(x,\tau)}$ to be the corresponding twist of M. Then we obtain a section $\mathcal{Y}^{\mathcal{P},\phi}_{M,(x,\tau)}$ of $(\mathcal{V}^{\mathcal{P}})^*$ over $D_x^\times$ with values in $\mathrm{End}\,\mathcal{M}^{(\mathcal{P},\phi)}_{(x,\tau)}$.

7.4. Bibliographical notes

Theorem 7.1.6, Theorem 7.2.6, and Theorem 7.2.9 are due to the authors. Theorem 7.3.7 and Theorem 7.3.10 are due to E. Frenkel. A similar coordinate-independent interpretation of twisted modules was obtained by E. Frenkel and M. Szczesny [**FrSz1**].

CHAPTER 8

Vertex Algebra Bundles: Examples

In this chapter we consider explicit examples of various subbundles of the vertex algebra bundles constructed in the previous two chapters. Our goal is to demonstrate that already these simple examples contain interesting structures, such as affine and projective connections on curves and flat connections on principal G–bundles, which correspond to the Heisenberg, Virasoro, and Kac–Moody algebras, respectively.

8.1. The Heisenberg algebra and affine connections

As our first example we take the Heisenberg vertex algebra π, with the conformal vector

$$\omega_\lambda = \frac{1}{2}b^2_{-1} + \lambda b_{-2}.$$

This structure induces an action of the group Aut $\mathcal{O}$ on π. Following Definition 6.5.1, we attach to π a vector bundle on a smooth curve X, which we denote by $^\lambda\Pi_X$, or simply by $^\lambda\Pi$.

8.1.1. A rank two subbundle. The first piece $\pi_{\leq 0}$ of our filtration on $\pi = \mathbb{C}[b_n]_{n<0}$ is the one-dimensional subspace spanned by the vacuum vector 1. This is a trivial representation of the group Aut $\mathcal{O}$. Hence it gives rise to a trivial line subbundle $\mathcal{O}_X \subset {}^\lambda\Pi_X$. The next piece in the filtration, $\pi_{\leq 1}$, consists of elements of degree less than or equal to 1. We have the exact sequence

$$0 \to \mathbb{C}1 \to \pi_{\leq 1} \to \mathbb{C}b_{-1} \to 0.$$

The action of $\mathrm{Der}_0\,\mathcal{O}$ on $\pi_{\leq 1}$ is given by

$$L_n \cdot 1 = 0, \qquad n \geq 0,$$

(8.1.1) $\qquad L_n b_{-1} = 0, \quad n \geq 2, \qquad L_1 b_{-1} = -2\lambda \cdot 1, \qquad L_0 b_{-1} = b_{-1}.$

These formulas give rise to the operator product expansion

(8.1.2) $\qquad T(z)b(w) = \dfrac{-2\lambda}{(z-w)^3} + \dfrac{b(w)}{(z-w)^2} + \dfrac{\partial_w b(w)}{z-w} + \mathrm{reg}.$

The subspace $\pi_{\leq 1}$ gives rise to a rank two subbundle of $^\lambda\Pi$, which we denote by $\mathcal{B}_\lambda$. According to the discussion in § 6.5.9, we have the following exact sequence:

$$0 \to \mathcal{O}_X \to \mathcal{B}_\lambda \to \Theta_X \to 0,$$

where $\Theta_X = \Omega^{-1}$ is the tangent sheaf of X.

133

8.1.2. Transition functions. Let us describe the bundle $\mathcal{B}_\lambda$ explicitly. For that we cover X by open subsets U_α with coordinate functions $z_\alpha : U_\alpha \to \mathbb{C}$ and coordinate transition functions $f_{\alpha\beta}$ such that $z_\alpha = f_{\alpha\beta}(z_\beta)$. Note that this can be done in the Zariski topology as well as in the analytic topology, if we interpret coordinate functions as étale maps to $\mathbb{C}$ (these suffice to trivialize $\mathcal{A}ut_X$ by § 6.5.2). In this case, however, one should be careful, since the coordinate transition functions $f_{\alpha\beta}$ may be multivalued.

On each U_α we trivialize $\mathcal{B}_\lambda$, $s_\alpha : \mathcal{B}_\lambda \xrightarrow{\sim} \mathcal{O}_{U_\alpha} \oplus \mathcal{O}_{U_\alpha}$. Defining the bundle $\mathcal{B}_\lambda$ is the same as defining the transition functions $\phi_{\alpha\beta} = s_\alpha \circ s_\beta^{-1}$ for each non-empty intersection $U_\alpha \cap U_\beta$, satisfying the cocycle condition $\phi_{\alpha\gamma} = \phi_{\alpha\beta}\phi_{\beta\gamma}$ on non-empty triple intersections $U_\alpha \cap U_\beta \cap U_\gamma$. Given a point $x \in U_\alpha$, we obtain a local coordinate $Z_\alpha = z_\alpha - z_\alpha(x)$ at x, hence a trivialization of $\mathcal{B}_\lambda$ near x. If this point lies in U_β as well, then we obtain another local coordinate $Z_\beta = z_\beta - z_\beta(x)$ at x, and hence another trivialization of $\mathcal{B}_\lambda$. The transition function $f_x = Z_\alpha \circ Z_\beta^{-1}$ on coordinates gives rise to the transition function $\phi_{\alpha\beta}$ on the bundle $\mathcal{B}_\lambda$, which acts on the fiber at x as $R(f_x)^{-1}$.

Expanding in Taylor series at x as in § 6.3, we obtain

$$
\begin{aligned}
f_x(z) &= f(x+z) - f(x) = f'(x)z + \frac{f''(x)}{2}z^2 + \dots \\
&= \exp\left(\sum_{j\geq 1} v_j(x)z^{j+1}\partial_z\right) v_0(x)^{z\partial_z} \cdot z \\
&= (1 + v_1(x)z^2\partial_z + \dots)v_0(x) \cdot z,
\end{aligned}
$$

where $f = f_{\alpha\beta}$. Hence we find that $v_1(x) = \frac{1}{2}\frac{f''(x)}{f'(x)}$ (cf. (6.3.2)). Using formula (8.1.1) for the action of L_n's on b_{-1}, we find that

$$
\begin{aligned}
R(f_x) \cdot b_{-1} &= \left(1 - \frac{1}{2}\frac{f''}{f'}L_1 + \dots\right)(f')^{-1}b_{-1} \\
&= (f')^{-1}b_{-1} + \lambda\frac{f''}{(f')^2}1
\end{aligned}
$$

Therefore in the basis $\{1, b_{-1}\}$ we may write

$$
R(f_x) = \begin{pmatrix} 1 & \lambda\frac{f''}{(f')^2} \\ 0 & (f')^{-1} \end{pmatrix},
$$

$$
R(f_x)^{-1} = \begin{pmatrix} 1 & -\lambda\frac{f''}{f'} \\ 0 & f' \end{pmatrix}.
$$

It follows that our transition functions can be written as

$$
\phi_{\alpha\beta} = \begin{pmatrix} 1 & -\lambda\Theta_1(f_{\alpha\beta}) \\ 0 & \frac{\partial f_{\alpha\beta}}{\partial z_\beta} \end{pmatrix},
$$

where $\Theta_1 \cdot f \overset{\text{def}}{=} \frac{f''}{f'}$ (the derivative is taken with respect to z_β).

8.1.3. The transformation property of $b(z)$. The dual bundle to $\mathcal{B}_\lambda$ is an extension

(8.1.3) $$0 \to \Omega \to \mathcal{B}_\lambda^* \to \mathcal{O}_X \to 0.$$

A section of $\mathcal{B}_\lambda^*$ is a pair of functions $\begin{pmatrix} g_\alpha^1 \\ g_\alpha^2 \end{pmatrix}$ over each open set U_α satisfying

(8.1.4) $$\begin{pmatrix} g_\beta^1 \\ g_\beta^2 \end{pmatrix} = \begin{pmatrix} 1 & 0 \\ -\lambda \Theta_1(f_{\alpha\beta}) & \frac{\partial f_{\alpha\beta}}{\partial z_\beta} \end{pmatrix} \begin{pmatrix} g_\alpha^1 \\ g_\alpha^2 \end{pmatrix}.$$

According to Theorem 6.5.4, we have an $\mathrm{End}\,^\lambda \Pi_x^*$–valued section $\mathcal{Y}_x$ of $^\lambda \Pi^*|_{D_x^\times}$. It projects onto a section of any quotient bundle of $^\lambda \Pi^*$, in particular, of $\mathcal{B}_\lambda^*$. Recalling the definition of $\mathcal{Y}_x$, we obtain that the corresponding section of $\mathcal{B}_\lambda^*|_{D_x^\times}$ reads

$$\begin{pmatrix} Y(1,z) \\ Y(b_{-1},z) \end{pmatrix} = \begin{pmatrix} \mathrm{Id} \\ b(z) \end{pmatrix}.$$

Using formula (8.1.4), we obtain the transformation formula for $b(z)$ under coordinate changes:

(8.1.5) $$b(z) = R(\rho)b(\rho(z))R(\rho)^{-1}\rho'(z) - \lambda \frac{\rho''(z)}{\rho'(z)}\,\mathrm{Id}\,.$$

In particular, if $\lambda = 0$, then $b(z)dz$ transforms as an $\mathrm{End}\,^\lambda \Pi_x$–valued one-form on $D_x^\times$ for any $x \in X$ (where z is a formal coordinate at x). For $\lambda \neq 1$, $b(z)$ gives rise to an $\mathrm{End}\,^\lambda \Pi_x$–valued section of $\mathcal{B}_\lambda^*$ on $D_x^\times$, which projects onto the section $1 \otimes \mathrm{Id}$ of the quotient $\mathcal{O}_X \otimes \mathrm{End}\,^\lambda \Pi_x$, and transforms according to formula (8.1.5) under changes of coordinates.

We want to understand the geometric meaning of such sections. To simplify the situation, let us first ignore the fact that our section is $\mathrm{End}\,^\lambda \Pi_x$–valued. Observe that a section of $\mathcal{B}_\lambda^*$ which projects onto the constant section 1 of the quotient $\mathcal{O}_X$ may be interpreted as a splitting of the bundle $\mathcal{B}_\lambda^*$. Indeed, we can use such a section to lift 1 from $\mathcal{O}_X$, and then, by $\mathcal{O}_X$–linearity, we lift any section of $\mathcal{O}_X$. Conversely, a splitting of $\mathcal{B}_\lambda^*$ gives us the desired section, as the image of $1 \in \mathcal{O}_X$ under the splitting. The set of splittings, if non-empty, is a torsor over the space of one-forms, since we can always add a one-form to a splitting, and the difference between two splittings naturally gives us a one-form.

In order to describe the splittings of $\mathcal{B}_\lambda^*$, we introduce the notion of λ–connections (see § 6.6.1 for the definition of ordinary connections). For $\lambda \in \mathbb{C}$, a λ–*connection* on a vector bundle $\mathcal{E}$ is an operator $\nabla : \mathcal{E} \to \mathcal{E} \otimes \Omega$ on sections satisfying the λ–Leibniz rule, $\nabla(f\xi) = f\nabla(\xi) + \lambda \xi \otimes df$. Locally, a λ–connection can be written as $\nabla = \lambda d + \eta$, where η is a one–form.

Note that for $\lambda \neq 0$, there is a one–to–one correspondence between λ–connections and ordinary connections, which maps a λ–connection ∇ to the connection $\lambda^{-1}\nabla$. When $\lambda = 0$, a λ–connection is simply a one–form, considered as an $\mathcal{O}_X$–linear operator $\mathcal{E} \to \mathcal{E} \otimes \Omega$.

8.1.4. Definition. An *affine λ–connection* on X is a λ–connection on the canonical line bundle Ω. For $\lambda = 1$ they are called simply affine connections.

We remark that sometimes by an affine connection one understands a connection on the tangent bundle Θ_X; by duality, the two notions are equivalent.

8.1.5. Lemma. *There is a natural bijection between the set of splittings of the bundle $\mathcal{B}_\lambda^*$ on X and the set of affine λ–connections on X.*

8.1.6. Proof. Writing down a λ–connection on Ω amounts to assigning to each U_α an expression $\nabla_\alpha = \lambda \partial_{z_\alpha} + \eta_\alpha$, where η_α is a function. Since the connection operator maps $\Omega \to \Omega^{\otimes 2}$, on overlaps $U_\alpha \cap U_\beta$ we have

$$\nabla_\alpha(g_\alpha) = \left(\frac{\partial z_\alpha}{\partial z_\beta}\right)^{-2} \nabla_\beta \left(\frac{\partial z_\alpha}{\partial z_\beta} g_\alpha\right).$$

Hence

$$\lambda \partial_{z_\alpha} g_\alpha + \eta_\alpha g_\alpha = \lambda \left(\frac{\partial z_\alpha}{\partial z_\beta}\right)^{-1} \partial_{z_\beta} g_\alpha + \lambda \Theta_1(z_\alpha) g_\alpha + \eta_\beta \left(\frac{\partial z_\alpha}{\partial z_\beta}\right)^{-1} g_\alpha,$$

$$\eta_\beta = \eta_\alpha \frac{\partial z_\alpha}{\partial z_\beta} - \lambda \Theta_1(z_\alpha),$$

and so

$$\begin{pmatrix} 1 \\ \eta_\beta \end{pmatrix} = \begin{pmatrix} 1 & 0 \\ -\lambda \Theta_1(z_\alpha) & \frac{\partial z_\alpha}{\partial z_\beta} \end{pmatrix} \begin{pmatrix} 1 \\ \eta_\alpha \end{pmatrix}.$$

8.1.7. Remark. If $\lambda \in \mathbb{Z}$, then λ–connections on Ω transform in the same way as ordinary connections on Ω^λ. To see this, we repeat the calculation of Lemma 8.1.5 for a connection $\partial_{z_\alpha} + \eta_\alpha$ on Ω^λ, observing that

$$\left(\frac{\partial z_\alpha}{\partial z_\beta}\right)^{-(\lambda+1)} \frac{\partial}{\partial z_\beta} \left(\frac{\partial z_\alpha}{\partial z_\beta}\right)^\lambda = \lambda \Theta_1(z_\alpha).$$

We will see in § 8.4 that this interpretation of Lemma 8.1.5 is in fact more natural. Affine connections can be further interpreted as affine structures.

8.1.8. Definition. An *affine chart* on a curve X is an open (analytic) cover $\{U_\alpha\}$ of X with coordinates $z_\alpha : U_\alpha \to \mathbb{C}$ such that $f_{\alpha\beta}$ is an affine transformation, i.e., $z_\alpha = a z_\beta + b$. Two affine charts on X are called equivalent if their union is also an affine chart. The equivalence classes of affine charts are called *affine structures*.

8.1.9. Lemma. *There is a canonical bijection between the set of affine connections on X and the set of affine structures on X.*

8.1.10. Proof. Let us choose a coordinate atlas on X. An affine connection gives rise to a collection of first order operators

$$\nabla_\alpha = \partial_{z_\alpha} + \eta_\alpha : \Omega \to \Omega^2$$

on each open subset U_α. Denote by ψ_α a horizontal section of ∇_α, i.e., a solution of the equation $\nabla_\alpha \cdot \psi_\alpha = 0$. This is a one–form; hence, choosing a reference point $x \in U_\alpha$, we may integrate it to obtain a function $y_\alpha(z) = \int_x^z \psi_\alpha$ on U_α. By passing to a more refined coordinate atlas if needed, we can assume without loss of generality that the solution ψ_α does not vanish anywhere on U_α, and hence y_α has a non-vanishing derivative everywhere on U_α. Thus, after refining the atlas further, if necessary, we obtain a new coordinate atlas $\{(U_\alpha, y_\alpha)\}$ on X. We claim that it gives us an affine structure on X.

Indeed, the only ambiguities that we have introduced in the construction are the choice of the solution ψ_α, which is unique up to a constant non-zero multiple, and the choice of reference point x, which shifts y_α by a constant. Therefore the transition functions $y_\alpha \circ y_\beta^{-1}$ on non-empty intersections $U_\alpha \cap U_\beta$ are necessarily affine transformations.

Conversely, given an affine structure, represented by the affine chart $\{(U_\alpha, y_\alpha)\}$, we construct an affine connection by setting $\nabla_\alpha = \partial_{y_\alpha}$. Since the transition functions are affine transformations, hence annihilated by the operator Θ_1, we see that this gives a well-defined affine connection.

8.1.11. The field $b(z)$ as an affine connection. Returning to the original question of describing the field $b(z)$ as a geometric object for $\lambda \neq 0$, we can now say that $\partial_z + \frac{1}{\lambda}\langle \varphi, b(z)v \rangle$ is an affine connection on $D_x^\times$ for any $v \in {}^\lambda\Pi_x, \varphi \in {}^\lambda\Pi_x^*$. In other words, $\partial_z + \frac{1}{\lambda}b(z)$ is naturally an $\mathrm{End}\,{}^\lambda\Pi_x$–valued affine connection on the punctured disc $D_x^\times$.

This example was based on the simplest two–dimensional $\mathrm{Aut}\,\mathcal{O}$–invariant subbundle of the simplest conformal vertex algebra. In fact the only specific information we needed about b_{-1} is contained in the formulas (8.1.1), so the same results hold for other vectors of degree one in a vertex algebra. For our next example, we study the geometric interpretation of vectors of degree two, for which we have a natural candidate – namely, the conformal vector itself.

8.2. The Virasoro algebra and projective connections

Let V be a conformal vertex algebra, with conformal vector $\omega \in V_2$ and central charge c. The corresponding Virasoro field $T(z) = Y(\omega, z)$ is interpreted in conformal field theory as the stress-energy tensor. We will find below its transformation properties under changes of coordinate z.

8.2.1. The bundle $\mathcal{T}_c$. By Lemma 3.4.5, the vector ω transforms under the action of $\mathrm{Der}_0\,\mathcal{O}$ as follows:

$$(8.2.1) \qquad L_n\omega = 0, \quad n > 2, n = 1; \qquad L_2\omega = \frac{c}{2}|0\rangle, \qquad L_0\omega = 2\omega.$$

Hence the vector space $\mathbb{C}|0\rangle \oplus \mathbb{C}\omega$ is $\mathrm{Aut}\,\mathcal{O}$–stable, and gives rise to a rank two subbundle $\mathcal{T}_c$ of $\mathcal{V}$, which is an extension

$$(8.2.2) \qquad 0 \to \mathcal{O}_X \to \mathcal{T}_c \to \Omega^{-2} \to 0.$$

In order to compute the transition functions of the bundle $\mathcal{T}_c$ we need to find the restriction of the operator $R(f_x)^{-1}$ to $\mathbb{C}|0\rangle \oplus \mathbb{C}\omega$. We proceed in the same way as in the case of affine connections (see § 8.1.2), using formula (6.3.2):

$$v_2 = \frac{1}{6}\frac{f'''(0)}{f'(0)} - \frac{1}{4}\left(\frac{f''(0)}{f'(0)}\right)^2.$$

Then we find that

$$\begin{aligned}
R(f_x)\cdot\omega &= \left(1 - \frac{1}{2}\frac{f''}{f'}L_1 - \left(\frac{1}{6}\frac{f'''}{f'} - \frac{1}{4}\left(\frac{f''}{f'}\right)^2\right)L_2 - \dots\right)(f')^{-2}\cdot\omega \\
&= (f')^{-2}\omega - \frac{c}{12}\left(\frac{f'''}{f'} - \frac{3}{2}\left(\frac{f''}{f'}\right)^2\right)(f')^{-2}|0\rangle.
\end{aligned}$$

It follows that the restriction of the operator $R(f_x)^{-1}$ to $\mathbb{C}|0\rangle \oplus \mathbb{C}\omega$ equals

$$(8.2.3) \qquad R(f_x)^{-1} = \begin{pmatrix} 1 & \frac{c}{12}\Theta_2(f) \\ 0 & (f')^2 \end{pmatrix},$$

where

(8.2.4)
$$\Theta_2(f) \overset{\text{def}}{=} \frac{f'''}{f'} - \frac{3}{2}\left(\frac{f''}{f'}\right)^2$$

is the *Schwarzian derivative*.

Thus, the transition function for the bundle $\mathcal{T}_c$ reads

$$\phi_{\alpha\beta} = \begin{pmatrix} 1 & \frac{c}{12}\Theta_2(f_{\alpha\beta}) \\ 0 & \left(\frac{\partial f_{\alpha\beta}}{\partial z_\beta}\right)^2 \end{pmatrix},$$

These transition functions automatically satisfy the cocycle condition by construction.

8.2.2. Transformation formula for $T(z)$**.** Consider the dual bundle $\mathcal{T}_c^*$, which is an extension

(8.2.5)
$$0 \to \Omega^2 \to \mathcal{T}_c^* \to \mathcal{O}_X \to 0.$$

The section $\mathcal{Y}_x$ gives rise to an End $\mathcal{V}_x$–valued section of $\mathcal{T}_c^*$ over the punctured disc $D_x^\times$. In terms of the basis $\{|0\rangle, \omega\}$ of $\mathbb{C}|0\rangle \oplus \mathbb{C}\omega$ this section is

$$\begin{pmatrix} Y(|0\rangle, z) \\ Y(\omega, z) \end{pmatrix} = \begin{pmatrix} \mathrm{Id} \\ T(z) \end{pmatrix}.$$

Thus, we find that the Virasoro field $T(z)$ transforms as follows:

$$T(z) = R(f)T(f(z))R(f)^{-1}(f'(z))^2 + \frac{c}{12}\Theta_2(f(z)).$$

For $c = 0$ our bundle $\mathcal{T}_c^*$ splits and $T(z)(dz)^2$ behaves as a quadratic differential. In general, for any point $x \in X$ and a formal coordinate z at x, the field $T(z)$ gives rise to an End $\mathcal{V}_x$–valued section of the rank two bundle $\mathcal{T}_c^*|_{D_x^\times}$, which projects onto the constant section 1 of $\mathcal{O}_X$.

A section of $\mathcal{T}_c^*|_{D_x^\times}$ which projects onto $1 \in \Gamma(\mathcal{O}_X)$ is the same as a splitting of $\mathcal{T}_c|_{D_x^\times}$. If non-empty, the set of such splittings is a torsor over the space of quadratic differentials. We want to explain the geometric meaning of these splittings.

Let us pick a square root of the canonical bundle, i.e., a bundle of $1/2$–forms $\Omega^{\frac{1}{2}}$ on X. Such line bundles are known as *theta characteristics* on X, and they exist for any smooth curve X. For instance, if X is projective, then the degree $2g - 2$ of Ω is always even, and it is easy to see that any line bundle of even degree has a square root. The ratio of any two square roots of Ω is a square root of the trivial line bundle on X. Thus the set of theta characteristics is a torsor over the group of square roots of the trivial line bundle. This group is identified with $H^1(X, \mathbb{Z}/2\mathbb{Z})$, and so it has order 2^{2g}.

Given a differential operator D acting from sections of a vector bundle $\mathcal{E}$ to sections of a vector bundle $\mathcal{F}$ on a curve X, we will define in § 8.3.4 the adjoint differential operator $D^t : \mathcal{F}^* \otimes \Omega \to \mathcal{E}^* \otimes \Omega$. In particular, if $\mathcal{E} = \Omega^{-\frac{1}{2}}, \mathcal{F} = \Omega^{\frac{3}{2}}$, we find that the adjoint operator acts between the same line bundles, so it makes sense to ask for such an operator to be self-adjoint.

In a local coordinate z, a self–adjoint second–order differential operator $\Omega^{-\frac{1}{2}} \to \Omega^{\frac{3}{2}}$ with symbol α can be written as

$$\alpha\partial_z^2 + q(z),$$

since taking the adjoint changes the sign of ∂_z.

When the symbol is 1, an operator of this form is known as a *projective connection*. The following statement is proved in Lemma 8.3.10 below.

8.2.3. Lemma. *There is a bijection between the set of splittings of $\mathcal{T}_c^*$ and the set of self–adjoint second–order differential operators $\rho : \Omega^{-\frac{1}{2}} \to \Omega^{\frac{3}{2}}$ with symbol $\frac{c}{6}$. In a local coordinate z, the differential operator corresponding to a section $(1, q(z))$ of $\mathcal{T}_c^*$ equals $\frac{c}{6}\partial_z^2 + q(z)$.*

Thus, for $c \neq 0$, $\partial_z^2 + \frac{6}{c}T(z)$ transforms as an $\mathrm{End}\,\mathcal{V}_x$–valued projective connection on the punctured disc $D_x^\times$, while for $c = 0$, $T(z)dz^2$ transforms as an $\mathrm{End}\,\mathcal{V}_x$–valued quadratic differential on $D_x^\times$.

As was the case with affine connections, projective connections have a natural geometric interpretation.

8.2.4. Definition. A *projective chart* on X is by definition a coordinate atlas on X whose transition functions are Möbius transformations $z \mapsto \dfrac{az + b}{cz + d}$. Two projective charts are called equivalent if their union is also a projective chart. The equivalence classes of projective charts are called *projective structures*.

8.2.5. Lemma. *There is a bijection between the set of projective connections on X and the set of projective structures on X.*

8.2.6. Proof. Given a projective connection, we obtain, for a given coordinate atlas $\{(U_\alpha, z_\alpha)\}$ on X, a collection of second order operators $\rho_\alpha = \partial_{z_\alpha}^2 + q_\alpha$ on each U_α. The differential equation

$$(8.2.6) \qquad (\partial_{z_\alpha}^2 + q_\alpha)\eta_\alpha = 0$$

locally possesses two linearly independent solutions η_α^1 and η_α^2. Passing to a more refined atlas, we can assume without loss of generality that η_α^2 and the Wronskian of $\eta_\alpha^1, \eta_\alpha^2$ do not vanish on U_α. Then the ratio $\mu_\alpha = \frac{\eta_\alpha^1}{\eta_\alpha^2}$ is a well-defined function on U_α with non-vanishing differential. Hence we obtain a new coordinate atlas $\{(U_\alpha, \mu_\alpha)\}$. We claim that this defines a projective structure on X. Indeed, the only ambiguity that we have introduced into the construction is the choice of a basis in the space of solutions of the equation (8.2.6). Changing the basis results in a Möbius transformation of μ_α. Therefore the coordinate transition functions in our atlas are Möbius transformations.

Conversely, given a projective structure, with chart $\{(U_\alpha, z_\alpha)\}$, we define a projective connection by setting $\rho_\alpha = \partial_{z_\alpha}^2$ for each α. Since Möbius transformations are precisely the holomorphic functions annihilated by the Schwarzian derivative Θ_2, we see that this gives a well-defined projective connection.

8.2.7. Projective connections and opers. Projective connections and projective structures have an important interpretation as flat connections on distinguished rank two bundles on curves. To see that, observe that a projective structure on X gives rise to a principal $PGL_2(\mathbb{C})$–bundle $\mathcal{P}$ on X. Namely, on the charts of our projective atlas we consider the trivial bundle $U_\alpha \times PGL_2(\mathbb{C})$, and take as transition function on the overlaps $U_\alpha \cap U_\beta$ the coordinate transformation from z_α to z_β considered as an element of $PGL_2(\mathbb{C})$. Since these transition functions are *constant* elements of $PGL_2(\mathbb{C})$, we find that $\mathcal{P}$ comes equipped with a flat connection.

The group $PGL_2(\mathbb{C})$ acts on $\mathbb{P}^1$. Hence we attach to $\mathcal{P}$ a $\mathbb{P}^1$–bundle $\mathbb{P}^1_{\mathcal{P}} \overset{\text{def}}{=} \mathcal{P} \underset{PGL_2(\mathbb{C})}{\times} \mathbb{P}^1$. This bundle is endowed with a distinguished global section μ, given locally by the chart maps $\mu_\alpha : U_\alpha \to \mathbb{P}^1$. By construction of $\mathbb{P}^1_{\mathcal{P}}$, these local sections are glued into a global section. Therefore the principal bundle $\mathcal{P}$ has a reduction $\mathcal{P}_B$ to the subgroup $B \subset PGL_2$ of upper–triangular matrices,

$$B = \left\{ \begin{pmatrix} a & b \\ 0 & a^{-1} \end{pmatrix} \right\} \subset PGL_2(\mathbb{C}).$$

This is the group of transformations of $\mathbb{P}^1$ preserving the point ∞. The reduction to B comes by considering the stabilizer of μ.

Finally, in any of the charts of our atlas, the maps μ_α have nowhere vanishing derivatives. This means that our distinguished section μ is nowhere horizontal, i.e., it is transversal to the flat connection. Let $(\mathfrak{sl}_2)_{\mathcal{P}}$ be the rank three vector bundle on X associated to $\mathcal{P}$ under the adjoint representation of PGL_2. This is simply the bundle whose fibers are the vector spaces of global vector fields on the fibers of our $\mathbb{P}^1$–bundle $\mathbb{P}^1_{\mathcal{P}}$. The reduction $\mathcal{P}_B$ gives rise to a rank two subbundle $(\mathfrak{b})_{\mathcal{P}_B}$, whose fibers are the spaces of vector fields along the fibers of $\mathbb{P}^1_{\mathcal{P}}$, which vanish at μ. The transversality of our section then becomes the statement that, using μ and the connection, the tangent space to X at any point is identified with the fiber of $(\mathfrak{sl}_2/\mathfrak{b})_{\mathcal{P}_B}$.

8.2.8. Definition. An $\mathfrak{sl}_2$–*oper* (or *indigenous bundle*) on X is a principal $PGL_2(\mathbb{C})$–bundle $\mathcal{P}$, equipped with a reduction $\mathcal{P}_B$ to B and a connection which gives rise to an isomorphism between the tangent bundle of X and $(\mathfrak{sl}_2/\mathfrak{b})_{\mathcal{P}_B}$.

8.2.9. Lemma. *There is a natural equivalence between projective structures and $\mathfrak{sl}_2$–opers.*

8.2.10. Proof. We have described the construction of an $\mathfrak{sl}_2$–oper from a projective structure. The inverse construction is given by a "period map". Let $(\mathcal{P}, \mathcal{P}_B, \nabla)$ be an $\mathfrak{sl}_2$–oper. Given a point $x \in X$, pick a simply–connected neighborhood U of x. Consider the $\mathbb{P}^1$–bundle $\mathbb{P}^1_{\mathcal{P}}$ associated to $\mathcal{P}$. The connection ∇ allows us to identify $\mathbb{P}^1_{\mathcal{P}}$ over U with the fiber of $\mathbb{P}^1_{\mathcal{P}}$ at x, which we further identify with $\mathbb{P}^1$. The reduction $\mathcal{P}_B$ defines a section of $\mathbb{P}^1_{\mathcal{P}}$, which under the above trivialization gives a map $\mu_x : U \to \mathbb{P}^1$. Without loss of generality we can assume that the image is contained in $\mathbb{C} = \mathbb{P}^1 \backslash \infty$. Thus we obtain a coordinate function on U. The oper condition on ∇ now implies that this map has nowhere vanishing derivative, and hence is injective. Altering the choice of identification $\mathbb{P}^1_{\mathcal{P}}|_x \simeq \mathbb{P}^1$ or of the base point $x \in U$ will only change the map μ_x by a Möbius transformation. Thus, we can cover X by open sets equipped with coordinates such that the transition functions are Möbius transformations. This defines a projective structure on X, as desired.

8.2.11. Remark. In § 16.6 we will define $\mathfrak{g}$–*opers* generalizing the above notion of $\mathfrak{sl}_2$–opers to an arbitrary simple Lie algebra $\mathfrak{g}$. While $\mathfrak{sl}_2$–opers are related to the Virasoro algebra as we have seen above, $\mathfrak{g}$–opers are related to generalizations of the Virasoro algebra called $\mathcal{W}$–algebras.

8.2.12. Existence of affine and projective connections. Now we discuss the natural question as to when affine and projective connections exist on a curve X, and if they exist, how many of them there are.

We have seen above that an affine (respectively, projective) connection defines a splitting of $\mathcal{B}_\lambda^*$ (respectively, $\mathcal{T}_c^*$). The extensions (8.1.3) and (8.2.5) define classes in the cohomology groups $H^1(X, \Omega)$ and $H^1(X, \Omega^2)$, respectively. The functions $\Theta_1(f_{\alpha\beta})$ and $\Theta_2(f_{\alpha\beta})$ on $U_\alpha \cap U_\beta$ may be viewed as Čech cocycles representing these classes.

If X is open (i.e., an affine curve), then all higher cohomology groups vanish. Therefore our extensions can be split, and thus affine and projective connections always exist in this case. The space of affine connections on X is then a torsor over the space of one-forms on X. In the case of projective connections, the situation is similar: the space of projective connections on X is a torsor over the space of quadratic differentials on X.

Suppose now that X is complete (i.e., a projective curve). Then we have $H^1(X, \Omega) = \mathbb{C}$, so our extension may be non-trivial. To check if it is, recall that affine connections are holomorphic connections on the sheaf Ω. Since all holomorphic connections on a curve are necessarily flat, a line bundle can have a connection only if its degree is zero. Since the degree of the canonical bundle Ω is $2g - 2$, where g is the genus of X, there are no affine connections on X for $g \neq 1$. In fact, the Čech one-cocycle $\{\Theta_1(f_{\alpha\beta})\}$ that is used in the definition of $\mathcal{B}_\lambda^*$ is a generator of $H^1(X, \Omega)$ in this case.

When the genus of X is 1, X is a complex torus and the canonical line bundle Ω may be trivialized. Hence it admits connections. It is easy to construct them using affine structures, which can be obtained using its covering by the plane: we simply choose, in an arbitrary way, sections of such a covering over small enough open subsets of the torus. The space of affine connections on a torus X is thus a torsor over $H^0(X, \Omega) \simeq \mathbb{C}$.

The class controlling projective connections lives in $H^1(X, \Omega^2)$. For $g > 1$ the sheaf Ω is ample, and so $H^1(X, \Omega^2) = 0$. Thus projective connections exist in this case. Since affine transformations are special cases of Möbius transformations, any affine structure can also be viewed as a projective structure. Hence we see that genus 1 surfaces also carry projective structures. Finally, in the case of genus 0, our curve is $\mathbb{P}^1$. Therefore it carries a canonical projective structure. So we see that projective connections exist in all genera.

The space of projective connections is a torsor over $H^0(X, \Omega^2)$, the space of quadratic differentials on X. This space is $3g - 3$–dimensional if $g > 1$. In fact the uniformization theorem for such curves implies that there exists a canonical (and in most cases unique) *real* projective structure on any such X (one in which the transition functions can be chosen to be real Möbius transformations, with real coefficients). If $g = 1$, the space of projective connections is again a torsor over $\mathbb{C}$, and in the case of genus 0, there is a unique projective connection.

8.3. Kernel functions

In this section we present, following Beilinson and Schechtman [**BS**], a description of the vector bundles $\mathcal{B}_\lambda^*$ and $\mathcal{T}_c^*$ discussed in this chapter in terms of *kernel functions* on X^2 supported on the diagonal.

8.3.1. Definition of the module of differentials. First we recall Grothendieck's definition of the sheaf of differentials.

Let R be a commutative algebra over a field k. The R–module of differentials $\Omega_{R/k}$ is by definition the R–module generated by symbols df, $f \in R$, subject to the relations

$$d(f + g) = df + dg$$

and

$$d(fg) = f\,dg + g\,df.$$

Alternatively, $\Omega_{R/k}$ is defined as I/I^2, where I is the kernel of the multiplication map $I \to R \otimes_k R \to R$. The R–action on $\Omega_{R/k}$ is given by

$$a(r_1 \otimes r_2) = ar_1 \otimes r_2 = r_1 \otimes ar_2 \ (\mathrm{mod}\ I^2).$$

The de Rham differential

$$d : R \to \Omega_{R/k}$$

is given by

$$d(a) = a \otimes 1 - 1 \otimes a.$$

Geometrically, we may think of R as the coordinate ring $k[X]$ of a k–scheme $X = \mathrm{Spec}\,R$. Then $R \otimes_k R$ is the space of functions on $X \times X$, I consists of functions vanishing on the diagonal

$$\Delta : X \hookrightarrow X \times X,$$

and I^2 consists of those functions vanishing to order two on the diagonal.

If X is a curve, then Δ is a divisor, so we can identify the sheaf of differentials on a smooth curve X with

(8.3.1)
$$\Omega = \frac{\mathcal{O}_X \boxtimes \mathcal{O}_X(-\Delta)}{\mathcal{O}_X \boxtimes \mathcal{O}_X(-2\Delta)}|_\Delta.$$

8.3.2. The kernels. Now let X be a smooth complex curve. Given integers $a > b$, we define a sheaf on $X \times X$,

$$\mathcal{P}_{a,b} = \frac{\mathcal{O}_X \boxtimes \Omega(a\Delta)}{\mathcal{O}_X \boxtimes \Omega(b\Delta)}.$$

In other words, we consider "integral kernels" of the form $\psi(z, w)dw$ on $X \times X$, with poles of order at most a allowed on the diagonal, and quotient out by those with poles of order at most b. The sheaf $\mathcal{P}_{a,b}$ is set–theoretically supported on the diagonal in X^2. Scheme–theoretically, its support is $(a - b)\Delta$, the $(a - b - 1)$st infinitesimal neighborhood of the diagonal. Therefore the push-forwards of $\mathcal{P}_{a,b}$ onto the first and the second factors in $X \times X$ are canonically isomorphic to each other and to the *sheaf–theoretic* restriction of $\mathcal{P}_{a,b}$ to the diagonal Δ. We will use the same notation $\mathcal{P}_{a,b}$ for this restriction. By functoriality, it carries an $\mathcal{O}_X$–bimodule structure (with the left $\mathcal{O}_X$–module structure corresponding to the first factor) coming from the $\mathcal{O}_{X \times X}$–module structure on $\mathcal{P}_{a,b}$, via two embeddings $\mathcal{O}_X \hookrightarrow \mathcal{O}_{X \times X}$.

Let $\mathcal{D}$ be the sheaf of algebraic differential operators on X, and $\mathcal{D}_{\leq n}$ its subsheaf of differential operators of order $\leq n$. The inclusion $i : \mathcal{O}_X \to \mathcal{D}$ makes $\mathcal{D}$ into a bimodule over $\mathcal{O}_X$, via left and right multiplication. Hence $\mathcal{D}$ may be viewed as a quasi-coherent sheaf on $X \times X$. The following proposition is a coordinate-free reformulation of the Cauchy integral formula:

$$(8.3.2) \qquad \frac{1}{n!}\frac{\partial^n f}{\partial z^n}(z) = \mathrm{Res}_{w=z}\,\frac{f(w)dw}{(w-z)^{n+1}}.$$

8.3.3. Proposition. *There is a canonical isomorphism of $\mathcal{O}_X$–bimodules*

$$\mathcal{P}_{n+1,0} \xrightarrow{\sim} \mathcal{D}_{\leq n},$$

given by sending the "integral kernel" $\psi(z_1, z_2)dz_2$ to the differential operator δ_ψ defined by the formula

$$\delta_\psi(f(z_1)) = \mathrm{Res}_{z_2=z_1}\, f(z_2)\psi(z_1, z_2)dz_2.$$

Thus

$$\mathcal{D} \simeq \mathcal{P}_{\infty,0} = \varinjlim \mathcal{P}_{n,0}.$$

8.3.4. Adjoint operators. More generally, for vector bundles $\mathcal{E}$ and $\mathcal{F}$ on X, the sheaf of differential operators $\mathcal{D}_{\leq n}(\mathcal{E}, \mathcal{F})$ of order less than or equal to n from sections of $\mathcal{E}$ to $\mathcal{F}$ is isomorphic (as an $\mathcal{O}_X$–bimodule) to

$$(8.3.3) \qquad \frac{\mathcal{F}\boxtimes\mathcal{E}^\vee((n+1)\Delta)}{\mathcal{F}\boxtimes\mathcal{E}^\vee},$$

where $\mathcal{E}^\vee = \mathcal{E}^* \otimes \Omega$. Let $D : \mathcal{E} \to \mathcal{F}$ be an nth order differential operator, and ψ_D be the corresponding section of (8.3.3) on X^2. Applying the transposition to $-\psi_D$, we obtain a section of

$$\frac{\mathcal{E}^\vee\boxtimes\mathcal{F}((n+1)\Delta)}{\mathcal{E}^\vee\boxtimes\mathcal{F}},$$

and hence a differential operator $D^* : \mathcal{F}^\vee \to \mathcal{E}^\vee$, which is called the *adjoint operator* to D.

8.3.5. Connection with $\mathrm{Aut}\,\mathcal{O}$**–twists.** There is a close connection between the sheaves of kernels defined above and the sheaves of sections of bundles obtained from $\mathrm{Aut}\,\mathcal{O}$–modules via the construction of § 6.5. Namely, let $\Omega_\mathcal{O}(a)$ denote the $\mathrm{Aut}\,\mathcal{O}$–module of differentials on the punctured disc with poles at the origin of order at most a. For $b < a$, there is an embedding of $\mathrm{Aut}\,\mathcal{O}$–modules $\Omega_\mathcal{O}(b) \hookrightarrow \Omega_\mathcal{O}(a)$. Denote the quotient by $\Omega_\mathcal{O}(a, b)$. Consider the corresponding vector bundle

$$\mathcal{F}(a, b) = \mathcal{A}ut_X \underset{\mathrm{Aut}\,\mathcal{O}}{\times} \Omega_\mathcal{O}(a, b)$$

on a smooth curve X. The next lemma follows directly from the definitions.

8.3.6. Lemma. *The sheaf of sections of $\mathcal{F}(a, b)$ is canonically isomorphic to the sheaf $\mathcal{P}_{a,b}$ (with its left $\mathcal{O}_X$–module structure).*

For instance, the space $\Omega_\mathcal{O}(0, -r - 1)$ is the space of r–jets of differentials at a point, and $\mathcal{P}_{0,-r-1}$ is the sheaf of r–jets of differentials on X.

8.3.7. More generally, consider the space $\Omega_\mathcal{O}^n(a)$ of meromorphic n–differentials on the punctured disc with poles at the origin of order at most a. Let $\Omega_\mathcal{O}^n(a, b) = \Omega_\mathcal{O}^n(a)/\Omega_\mathcal{O}^n(b)$ for $b < a$. For $m \in \mathbb{Z}$, denote by $Char_m$ the one-dimensional $\mathrm{Aut}\,\mathcal{O}$–module on which $\mathrm{Aut}\,\mathcal{O}$ acts through a character of its quotient $\mathbb{C}^\times$, $x \mapsto x^m$. Recall from § 6.5.9 that the $\mathcal{A}ut_X$–twist of $Char_m$ is the line bundle Ω^m on X. Now denote the $\mathcal{A}ut_X$–twist of $\Omega_\mathcal{O}^n(a, b) \otimes Char_m$ by $\mathcal{F}_{m,n}(a, b)$.

8.3.8. Lemma. *The sheaf of sections of $\mathcal{F}_{m,n}(a,b)$ is canonically isomorphic to*

$$\frac{\Omega^m \boxtimes \Omega^n(a\Delta)}{\Omega^m \boxtimes \Omega^n(b\Delta)}$$

(with its left $\mathcal{O}_X$–module structure).

8.3.9. Another description of $\mathcal{B}_\lambda^*$ and $\mathcal{T}_c^*$. Using the above lemmas, we can give another description of the sheaves of sections of the bundles $\mathcal{B}_\lambda^*$ and $\mathcal{T}_c^*$ constructed above via the Aut $\mathcal{O}$–twisting construction from vertex algebras.

Consider the sheaf

$$(8.3.4) \qquad \frac{\Omega^2 \boxtimes \mathcal{O}_X(2\Delta)}{\Omega^2 \boxtimes \mathcal{O}_X}.$$

By Proposition 8.3.3, this is the sheaf of differential operators $\Omega \to \Omega^2$ of order less than or equal to 1. In particular, its sections of the form

$$(8.3.5) \qquad \left(\frac{\lambda}{(z_2 - z_1)^2} + \frac{\eta(z_1)}{z_2 - z_1} + \text{reg.} \right)(dz_1)^2, \qquad \lambda \in \mathbb{C},$$

are just the affine λ–connections introduced in Definition 8.1.4. By Lemma 8.3.8, the sheaf (8.3.4) is attached to the Aut $\mathcal{O}$–module $\mathcal{O}(2,0) \otimes Char_2$ with the basis $\{1/z^2, 1/z\}$. Recalling that $L_n = -z^{n+1}\partial_z$, we obtain that the action of the Lie algebra $\mathrm{Der}_0\, \mathcal{O}$ is given in this basis by the formulas

$$L_0 \cdot \frac{1}{z^2} = 0, \qquad L_0 \cdot \frac{1}{z} = -\frac{1}{z}, \qquad L_1 \cdot \frac{1}{z^2} = 2\frac{1}{z}.$$

Comparing with formulas (8.1.1), we obtain that for $\lambda \neq 0$ the Aut $\mathcal{O}$–module $\pi_{\leq 1}^* = \mathrm{span}\{1^*, b_{-1}^*\}$ is isomorphic to $\mathcal{O}(2,0) \otimes Char_2$ under the map $1^* \mapsto \lambda/z^2, b_{-1}^* \mapsto 1/z$. Therefore the sheaf of sections of the vector bundle $\mathcal{B}_\lambda^*$ (see (8.1.3)) associated to $\pi_{\leq 1}^*$ is isomorphic to the sheaf (8.3.4), and under this isomorphism the sections of $\mathcal{B}_\lambda^*$ projecting onto $1 \in \mathcal{O}_X$ are mapped to affine λ–connections (8.3.5). Thus, we obtain another proof of Lemma 8.1.5.

There is a similar description of the Virasoro bundle $\mathcal{T}_c^*$. Choose a square root $\Omega^{\frac{1}{2}}$ of the canonical bundle Ω. We want to use the above isomorphism to prove Lemma 8.2.3, which identifies splittings of $\mathcal{T}_c^*$ with second–order differential operators from $\Omega^{-\frac{1}{2}}$ to $\Omega^{\frac{3}{2}}$. By Proposition 8.3.3, these operators may be viewed as sections of the sheaf

$$(8.3.6) \qquad \frac{\Omega^{\frac{3}{2}} \boxtimes \Omega^{\frac{3}{2}}(3\Delta)}{\Omega^{\frac{3}{2}} \boxtimes \Omega^{\frac{3}{2}}}.$$

Under this identification, the self–adjoint second order operator with symbol γ,

$$\gamma \partial_z^2 + q(z) : \Omega^{-\frac{1}{2}} \to \Omega^{\frac{3}{2}},$$

becomes the section

$$(8.3.7) \qquad \left(\frac{2\gamma}{(z_2 - z_1)^3} + \frac{q(z_1)}{z_2 - z_1} + \text{reg.} \right) dz_1^{\frac{3}{2}} dz_2^{\frac{3}{2}}$$

of the sheaf (8.3.6) (see formula (8.3.2)). Lemma 8.2.3 is a consequence of the following:

8.3.10. Lemma. *There is a canonical bijection between the set of splittings of $\mathcal{T}_c^*$ and the set of anti-symmetric sections of the sheaf (8.3.6) with $\gamma = c/6$.*

8.3.11. Proof. Consider the Aut $\mathcal{O}$–module span$\{|0\rangle, \omega\}$ with the action of Der$_0$ $\mathcal{O}$ described by formula (8.2.1). By definition, $\mathcal{T}_c$ is the $\mathcal{A}ut_X$–twist of L_c.

By Lemma 8.3.8, the sheaf (8.3.6) is attached to the Aut $\mathcal{O}$–module $\Omega_{\mathcal{O}}^{\frac{3}{2}}(3, 0) \otimes Char_{\frac{3}{2}}$. This module contains a submodule spanned by $dz^{\frac{3}{2}}/z^3$ and $dz^{\frac{3}{2}}/z$, and the action of Der$_0$ $\mathcal{O}$ on it is given by the formulas

$$L_0 \cdot \frac{dz^{\frac{3}{2}}}{z^3} = 0, \qquad L_0 \cdot \frac{dz^{\frac{3}{2}}}{z} = -2\frac{dz^{\frac{3}{2}}}{z}, \qquad L_2 \cdot \frac{dz^{\frac{3}{2}}}{z^3} = -\frac{3}{2}\frac{dz^{\frac{3}{2}}}{z}.$$

Comparing with (8.2.1), we find that for $c \neq 0$ the Aut $\mathcal{O}$–module span$\{|0\rangle^*, \omega^*\}$ is isomorphic to the module span$\{dz^{\frac{3}{2}}/z^3, dz^{\frac{3}{2}}/z\}$ under the map $|0\rangle^* \mapsto \frac{c}{3}dz^{\frac{3}{2}}/z^3$, $\omega^* \mapsto dz^{\frac{3}{2}}/z$.

Therefore the sheaf of sections of the vector bundle $\mathcal{T}_c^*$ (see (8.2.2)) associated to the Aut $\mathcal{O}$–module span$\{|0\rangle^*, \omega^*\}$ is isomorphic to the sheaf of anti-symmetric sections of (8.3.6), and under this isomorphism the sections of $\mathcal{T}_c^*$ projecting onto $1 \in \mathcal{O}_X$ are mapped to the sections (8.3.7) with $\gamma = c/6$. This completes the proof.

One can also realize the sheaves $\mathcal{B}_\lambda$ and $\mathcal{T}_c$ in terms of differential kernel functions.

8.3.12. Definition. The sheaf of differential kernels on X is the projective limit

$$\mathcal{P}_{\infty, -\infty} = \varprojlim \mathcal{P}_{\infty, -n}.$$

8.3.13. Extension by ∂_z^{-1}. Note that we have $\mathcal{P}_{0,-1} \simeq \Omega$, and more generally, $\mathcal{P}_{n+1,n} \simeq \Omega^{-n}$. In the case $n > 0$ this follows from Proposition 8.3.3, and for other n can be derived from the isomorphism (8.3.1).

The extension $\mathcal{B}_\lambda \otimes \Omega$ (resp., $\mathcal{T}_c \otimes \Omega$) with non-zero λ and c may be interpreted as extension of functions (resp., vector fields) by $\mathcal{P}_{0,-1}$. Namely, consider the following diagram of short exact sequences:

(8.3.8)
$$\begin{array}{ccccccccc} & & & & & & 0 & & \\ & & & & & & \downarrow & & \\ & & & & & & \mathcal{O}_X & & \\ & & & & & & \downarrow & & \\ 0 & \to & \Omega & \to & \mathcal{P}_{2,-1} & \to & \mathcal{P}_{2,0} & \to & 0 \\ & & & & & & \downarrow & & \\ & & & & & & \Theta_X & & \\ & & & & & & \downarrow & & \\ & & & & & & 0 & & \end{array}$$

Observe that the vertical sequence has a canonical splitting. Indeed, we have identified $\mathcal{P}_{2,0}$ with the sheaf of differential operators of order less than or equal to one. But any vector field can be lifted to a unique first order differential operator which annihilates constant functions. Thus $\mathcal{P}_{2,0} \simeq \mathcal{O}_X \oplus \Theta_X$.

In the same way as above, one can show that the pull-back of the extension (8.3.8) to $\mathcal{O}_X$ is isomorphic to $\mathcal{B}_\lambda \otimes \Omega$ with $\lambda \neq 0$ (note that the bundles $\mathcal{B}_\lambda$ with non-zero λ are isomorphic to each other). Furthermore, the pull-back of the extension (8.3.8) to Θ_X is isomorphic to $\mathcal{T}_c \otimes \Omega$ with $c \neq 0$.

Finally, consider the three-dimensional subspace of the vertex algebra π spanned by the vectors $1, b_{-1}$, and ω_λ. This space is Aut $\mathcal{O}$–invariant. The reader can check that the sheaf of sections of the corresponding rank three vector bundle on X, tensored with Ω, is isomorphic to the sheaf $\mathcal{P}_{2,-1}$, if $\lambda \neq 0$ and $c = 1 - 12\lambda^2 \neq 0$.

8.4. The gauge action on the Heisenberg bundle

We can obtain interesting structures by twisting a vertex algebra V by torsors for groups of internal symmetries, other than the group $\operatorname{Aut} \mathcal{O}$ (see § 7.2.1). The simplest example is provided by the Heisenberg vertex algebra.

8.4.1. The case of $\lambda = 0$. Recall from Remark 2.3.9 that there is a natural one–parameter family of Heisenberg vertex algebras π^κ of level $\kappa \in \mathbb{C}^\times$. Consider the simplest conformal vector

$$(8.4.1) \qquad \omega_{0,\kappa} = \frac{1}{2\kappa} b_{-1}^2 \in \pi^\kappa,$$

and the corresponding vertex algebra bundle ${}^0\Pi^\kappa$ on a curve X. It has a rank two subbundle $\mathcal{B}_{0,\kappa}$ corresponding to the subspace $\pi^\kappa_{\leq 1} \subset \pi^\kappa$. According to the results of § 8.1, for $\lambda = 0$ the bundle $\mathcal{B}_{0,\kappa}$ splits as the direct sum $\Theta_X \oplus \mathcal{O}_X$. Therefore the field $b(z)$ transforms as a one-form.

The picture becomes more interesting if we enlarge the group $\operatorname{Aut} \mathcal{O}$ to the group $\operatorname{Aut} \mathcal{O} \ltimes \mathcal{O}^\times$ of internal symmetries. The action of the abelian group $\mathcal{O}^\times$ is obtained by exponentiating the action of its Lie algebra $\mathcal{O} = \mathcal{H}_+ \subset \mathcal{H}$ spanned by the Fourier coefficients $b_n, n \geq 0$, of $b(z)$ on π^κ. Recall that the generators $b_n, n \geq 0$, act on π^κ as annihilation operators, lowering the degree of vectors, and hence are locally nilpotent. Moreover, b_0 acts by zero on all of π^κ. Therefore the action of the Lie algebra $\mathcal{H}_+$ may be exponentiated to an action of the proalgebraic group $\mathcal{O}^\times = GL_1(\mathcal{O})$. To calculate this action explicitly, we proceed in the same way as we did when analyzing the $\operatorname{Aut} \mathcal{O}$–action in § 6.3. Namely, for $F(z) \in \mathcal{O}^\times$, we write

$$\begin{aligned}
F(z) &= F(0) + F'(0)z + \frac{1}{2}F''(0)z^2 + \dots \\
&= \exp(\beta_1 z + \beta_2 z^2 + \dots)\beta_0,
\end{aligned}$$

from which we can solve

$$\begin{aligned}
\beta_0 &= F(0), \\
\beta_1 &= \frac{F'}{F}(0), \quad \text{etc.}
\end{aligned}$$

Replacing the function $t^n \in \mathcal{O}$ by the operator $b_n \in \mathcal{H}_+$ (and recalling that b_0 acts by zero on π^κ), we find the action of $F(z) \in \mathcal{O}^\times$ on $b_{-1}|0\rangle \in \pi^\kappa$:

$$\begin{aligned}
F \cdot b_{-1}|0\rangle &= \left(1 + \frac{F'}{F}(0)b_1 + \dots\right) b_{-1}|0\rangle \\
&= b_{-1}|0\rangle + \kappa \frac{F'}{F}(0)|0\rangle.
\end{aligned}$$

Thus in the basis $\{|0\rangle, b_{-1}|0\rangle\}$ of the two–dimensional subspace $\pi^\kappa_{\leq 1} \subset \pi^\kappa$, the action of $F(z) \in \mathcal{O}^\times$ is given by the matrix

$$\begin{pmatrix} 1 & \kappa \frac{F'}{F}(0) \\ 0 & 1 \end{pmatrix}.$$

The actions of $\text{Aut}\,\mathcal{O}$ and $\mathcal{O}^\times$ on the vertex algebra π^κ combine into an action of the semi-direct product $\text{Aut}\,\mathcal{O} \ltimes \mathcal{O}^\times$. Given a line bundle $\mathcal{L}$ on X, let $\widehat{\mathcal{L}}$ be the $\text{Aut}\,\mathcal{O} \ltimes \mathcal{O}^\times$–torsor on X, whose fiber at $x \in X$ consists of pairs (z, s), where z is a formal coordinate at x, and s is a trivialization of $\mathcal{L}|_{D_x}$. Let $\Pi_{\mathcal{L}}^\kappa$ be the twist of π^κ by this torsor:

$$\Pi_{\mathcal{L}}^\kappa = \widehat{\mathcal{L}} \underset{\text{Aut}\,\mathcal{O} \ltimes \mathcal{O}^\times}{\times} \pi^\kappa.$$

Clearly, $\pi_{\leq 1}^\kappa$ is stable under the $\text{Aut}\,\mathcal{O} \ltimes \mathcal{O}^\times$–action. Hence it gives rise to a rank two subbundle $\mathcal{B}_{0,\kappa}(\mathcal{L})$ of $\Pi_{\mathcal{L}}^\kappa$.

The dual bundle $\mathcal{B}_{0,\kappa}(\mathcal{L})^*$ is again an extension of $\mathcal{O}_X$ by Ω, but the above calculation shows that it is no longer trivial. Indeed, let us choose a coordinate z_α and a trivialization s_α of $\mathcal{L}_\alpha$ on an open set $U_\alpha \subset X$. Then a section of $\mathcal{B}_{0,\kappa}(\mathcal{L})^*$ can be written as $\begin{pmatrix} g_\alpha^1(z_\alpha) \\ g_\alpha^2(z_\alpha) \end{pmatrix}$. If we change the trivialization s_α by multiplying it by a nowhere vanishing function $F(z_\alpha)$, then this section transforms to

$$\begin{pmatrix} 1 & 0 \\ \kappa \frac{F'}{F} & 1 \end{pmatrix} \begin{pmatrix} g_\alpha^1 \\ g_\alpha^2 \end{pmatrix}.$$

In particular, a section of the form $\begin{pmatrix} 1 \\ b(z_\alpha) \end{pmatrix}$ of $\mathcal{B}_{0,\kappa}(\mathcal{L})^*$ (i.e., a splitting of $\mathcal{B}_{0,\kappa}(\mathcal{L})^*$) transforms to $\begin{pmatrix} 1 \\ F \cdot b(z_\alpha) \end{pmatrix}$, where

$$(8.4.2) \qquad\qquad F \cdot b(z) = b(z) + \kappa \frac{F'(z)}{F(z)}.$$

On the other hand, let ∇ be a κ–connection on $\mathcal{L}$. Choosing a coordinate z and a trivialization of $\mathcal{L}$, we obtain a connection operator $\nabla_{\partial_z} = \kappa \partial_z + \phi(z)$, where $\phi(z)$ is a function. If we choose a new trivialization which differs from the old one by multiplication by a nowhere vanishing function $F(z)$, then the connection operator becomes

$$F(z)^{-1}(\kappa \partial_z + \phi(z))F(z) = \kappa \partial_z + \phi(z) + \kappa \frac{F'(z)}{F(z)}.$$

This formula expresses the *gauge transformations* of the connection operator under changes of trivialization of the underlying bundle (a similar formula may be written for any flat vector bundle). Comparing with formula (8.4.2), we obtain that for any point $x \in X$ and a formal coordinate z at x, the operator $\kappa \partial_z + b(z)$ transforms as an $\text{End}\,\Pi_{\mathcal{L},x}^\kappa$–valued κ–*connection* on $\mathcal{L}|_{D_x^\times}$. We also obtain

8.4.2. Lemma. *For any line bundle $\mathcal{L}$ on a smooth curve X, there is a natural bijection between the set of splittings of $\mathcal{B}_{0,\kappa}(\mathcal{L})^*$ and the set of κ-connections on $\mathcal{L}$.*

8.4.3. General λ. Now turn on the λ–parameter in the Heisenberg conformal structure, setting

$$(8.4.3) \qquad\qquad \omega_{\lambda,\kappa} = \frac{1}{2\kappa} b_{-1}^2 + \lambda b_{-2}.$$

This changes the Virasoro action (we now have $L_1 \cdot b_{-1} = -2\lambda\kappa$) and leaves intact the $\mathcal{O}^\times$–action. Given a line bundle $\mathcal{L}$, we define a vector bundle $^\lambda\Pi_{\mathcal{L}}^\kappa$ as above. Let $\mathcal{B}_{\lambda,\kappa}(\mathcal{L})$ be its rank two subbundle corresponding to $\pi_{\leq 1}^\kappa \subset \pi^\kappa$. Combining the above calculation with that of Remark 8.1.7, we find that $b(z)$ transforms as a κ–connection on $\Omega^\lambda \otimes \mathcal{L}$.

8.5. The affine Kac–Moody vertex algebras and connections

Let $V_k(\mathfrak{g})$ be the affine vertex algebra. The subspace $V_k(\mathfrak{g})_{\leq 1} = V_k(\mathfrak{g})_0 \oplus V_k(\mathfrak{g})_1$ is an extension

$$(8.5.1) \qquad 0 \to \mathbb{C}v_k \to V_k(\mathfrak{g})_{\leq 1} \to \mathfrak{g} \to 0,$$

where we have identified $\mathfrak{g}$ with $(\mathfrak{g} \otimes t^{-1})v_k = V_k(\mathfrak{g})_1$. The subspace $V_k(\mathfrak{g})_{\leq 1}$ is preserved by the $\operatorname{Aut}\mathcal{O}$–action (derived from the Segal-Sugawara construction, see § 2.5.10), and so gives rise to a vector bundle which fits in the exact sequence

$$(8.5.2) \qquad 0 \to \mathcal{O}_X \to \mathcal{V}_k(\mathfrak{g})_{\leq 1} \to \mathfrak{g} \otimes \Theta_X \to 0.$$

The commutation relations (2.5.7) show that the space $V_k(\mathfrak{g})_1$ is annihilated by the action of $\operatorname{Der}_+ \mathcal{O}$ (i.e., it consists of primary vectors). Therefore the extension (8.5.2) as well as the dual extension

$$(8.5.3) \qquad 0 \to \mathfrak{g}^* \otimes \Omega \to \mathcal{V}_k(\mathfrak{g})^*_{\leq 1} \to \mathcal{O}_X \to 0$$

are canonically split. The section $\mathcal{Y}_x$ gives rise to an $\operatorname{End} \mathcal{V}_k(\mathfrak{g})_x$–valued section of $\mathcal{V}_k(\mathfrak{g})^*_{\leq 1}|_{D_x^\times}$ projecting onto $1 \in \Gamma(\mathcal{O})$. Splitting the sequence (8.5.3), we identify this section with a $\mathfrak{g}^* \otimes \operatorname{End} \mathcal{V}_k(\mathfrak{g})_x$–valued one-form.

However, on $V_k(\mathfrak{g})$ we have an action of a larger group of internal symmetries, namely, $\operatorname{Aut}\mathcal{O} \ltimes G(\mathcal{O})$. This group preserves the subspace $\mathcal{V}_k(\mathfrak{g})_{\leq 1}$. Under the action of the group $G(\mathcal{O})$, the extension (8.5.1) does *not* split for $k \neq 0$, as one can readily see from the following formula for the action of the corresponding Lie algebra $\mathfrak{g}[[t]]$:

$$J_1^b \cdot (J_{-1}^a v_k) = [J_1^b, J_{-1}^a] \cdot v_k = (J^b, J^a)k v_k.$$

Thus, we find ourselves in a situation similar to the one discussed in the previous section, where the subspace $\pi_{\leq 1}^\kappa$ did not split as an $\mathcal{O}^\times$–module.

Let $\mathcal{P}$ be a G–bundle on X, and $\widehat{\mathcal{P}}$ the $\operatorname{Aut}\mathcal{O} \ltimes G(\mathcal{O})$–bundle over X introduced in Definition 7.1.5. Then we define $\mathcal{V}_k(\mathfrak{g})^{\mathcal{P}}$ and its rank two subbundle $\mathcal{V}_k(\mathfrak{g})^{\mathcal{P}}_{\leq 1}$ as in Definition 7.1.5. We have the exact sequence

$$(8.5.4) \qquad 0 \to \mathfrak{g}_{\mathcal{P}} \otimes \Omega \to (\mathcal{V}_k(\mathfrak{g})^{\mathcal{P}}_{\leq 1})^* \to \mathcal{O}_X \to 0,$$

where we use the notation $\mathfrak{g}_{\mathcal{P}} = \mathcal{P} \underset{G}{\times} \mathfrak{g}$, and identify $\mathfrak{g}$ with $\mathfrak{g}^*$ using the canonical invariant bilinear form. The following statement is proved in the same way as Lemma 8.4.2.

8.5.1. Lemma. *The space of splittings of $(\mathcal{V}_k(\mathfrak{g})^{\mathcal{P}}_{\leq 1})^*$ is naturally isomorphic to the space of k–connections on the G–bundle $\mathcal{P}$ over X.*

Thus, we obtain that the fields $J^a(z)$ combine into an $\operatorname{End} \mathcal{V}_k(\mathfrak{g})^{\mathcal{P}}_x$–valued k–connection on $\mathcal{P}|_{D_x^\times}$.

8.5.2. Remark. In § A.3.2 below we define the Atiyah algebroid $\mathcal{A}_{\mathcal{P}}$ of infinitesimal symmetries of a principal G–bundle $\mathcal{P}$ on X. This is an extension of Θ by $\mathfrak{g}_{\mathcal{P}} \otimes \mathcal{O}_X$. Lemma 8.5.1 implies that the sheaf $\mathcal{A}_{\mathcal{P}}$ is naturally isomorphic to $\mathcal{V}_k(\mathfrak{g})^{\mathcal{P}}_{\leq 1} \otimes \Theta_X$, and the Atiyah sequence is the extension (8.5.4) tensored with Θ_X.

8.5.3. Deformed Segal-Sugawara construction. The vertex algebra $V_k(\mathfrak{g})$ admits a family of conformal vectors, generalizing the Segal-Sugawara vector and analogous to the family ω_λ for the Heisenberg vertex algebra.

For $u \in \mathfrak{g}$, consider the vector $\omega_u = \omega + u_{-2}$, where $\omega = \frac{1}{k+h^\vee} S$ is the Segal-Sugawara conformal vector in $V_k(\mathfrak{g})$ introduced in § 2.5.10. One easily finds that the Segal-Sugawara OPE (3.4.6) is modified as follows:

$$Y(\omega_u, z)J^a(w) = -\frac{2k(u, J^a)}{(z-w)^3} + \frac{J^a(w) - [u, J^a](w)}{(z-w)^2} + \frac{\partial_w J^a(w)}{w-z} + \text{reg.}$$

Assume for simplicity that u is semi-simple, so it lies in a fixed Cartan subalgebra $\mathfrak{h} \subset \mathfrak{g}$, and pick the basis J^a to be a basis of eigenvectors for $\operatorname{ad} u$, so that $[u, J^a] = a(u)J^a$ for some $a(u) \in \mathbb{C}$. Then if J^a is not in $\mathfrak{h}$, we have

$$Y(\omega_u, z)J^a(w) = \frac{(1 - a(u))J^a(z)}{(z-w)^2} + \frac{\partial_w J^a(w)}{z-w} + \text{reg.},$$

while for $J^a = h \in \mathfrak{h}$,

$$Y(\omega_u, z)h(w) = \frac{-2k(u, h)}{(z-w)^3} + \frac{h(w)}{(z-w)^2} + \frac{\partial_w h(w)}{z-w} + \text{reg.}$$

Comparing these formulas with formula (8.1.2), we see that the field $h(z)$ transforms like the Heisenberg field $b(z)$, with respect to the conformal vector $\omega_\lambda \in \pi$, where $\lambda = (h, u)$. The other fields $J^a(z)$, on the other hand, remain primary but transform as operator–valued sections of $\Omega^{1-a(u)}$.

Altogether, the fields $J^a(z)$ combine into an operator-valued section of the vector bundle corresponding to $V_k(\mathfrak{g})^*_{\leq 1}$ under the new action of $\operatorname{Aut} \mathcal{O}$. In the case when $a(u) \in \mathbb{Z}$ for all a, this bundle is an extension of $\mathcal{O}_X$ by the vector bundle $\mathfrak{g}_{\mathcal{P}(u)} \overset{\text{def}}{=} \mathfrak{g} \underset{\mathbb{C}^\times}{\times} \Omega$, where the $\mathbb{C}^\times$–action on $\mathfrak{g}$ corresponds to the $\mathbb{Z}$–gradation defined by the operator $\operatorname{ad} u$ on $\mathfrak{g}$. Our section projects onto $1 \in \Gamma(\mathcal{O})$. Such sections may be identified with connections on the G_{ad}–bundle corresponding to $\mathfrak{g}_{\mathcal{P}(u)}$, where G_{ad} is the adjoint group of $\mathfrak{g}$.

8.6. Bibliographical notes

An excellent introduction to affine and projective connections on Riemann surfaces is chapter 9 of R. Gunning's book [**Gu**]. For a modern treatment of projective connections, see [**BiRa**].

The fact that the stress tensor transforms as a projective connection has been known since the advent of conformal field theory (see, e.g., [**BPZ, Seg2, T, Ra**]). The fact that the Heisenberg field transforms as an affine connection is less known, but the corresponding transformation formula may be found in the physics literature.

The discussion of kernel functions and extensions in § 8.3 follows the work of A. Beilinson and V. Schechtman [**BS**] (see also [**BMS, BiRa, BZB**]).

CHAPTER 9

Conformal Blocks I

In this chapter we use the coordinate-independent description of the vertex operation to give the definition of the spaces of coinvariants and conformal blocks associated to a quasi-conformal vertex algebra V and a smooth projective curve X with a point x. The idea is very simple. Let $\mathcal{V}$ be the vector bundle on X attached to V. We have shown in Chapter 6 that for each $A \in \mathcal{V}_x$ and $\varphi \in \mathcal{V}_x^*$, the vertex operation defines a canonical section $\langle \varphi, \mathcal{Y}_x \cdot A \rangle$ of $\mathcal{V}^*$ on $D_x^\times$. We will say that φ is a conformal block, if this section can be extended to a regular section of $\mathcal{V}^*$ on $X \backslash x$ for all $A \in \mathcal{V}_x$. Conformal blocks form a vector space, which should be thought of as an invariant attached to V, X and x. These spaces carry interesting information about the geometry of algebraic curves and various structures on them (such as vector bundles). The most interesting question, which will be discussed in the subsequent chapters, is how the spaces of conformal blocks behave as we vary the complex structure on X and other geometric data.

In the case when the vertex algebra V is generated by a Lie algebra (e.g., the Heisenberg Lie algebra), the spaces of conformal blocks can be defined in a simpler way. Because of this, we start with the case of the Heisenberg vertex algebra. We then gradually move from the definition of conformal blocks which works in the case of vertex algebras generated by a Lie algebra, to a general definition which can be used for an arbitrary vertex algebra. The simplest examples of coinvariants (for commutative vertex algebras) are discussed in § 9.4. The reader who wants to see examples of actual computations of coinvariants and conformal blocks before studying the general formalism may skip ahead to Chapter 13 after reading the first definition of conformal blocks for the Heisenberg algebra in § 9.1.

In this chapter and in the rest of the book, unless noted otherwise, by an algebraic curve we will always understand a smooth connected curve.

9.1. Defining conformal blocks for the Heisenberg algebra

9.1.1. Differentials. Recall the definition of the module of differentials $\Omega_{R/k}$ from § 8.3.1. In the case when R is the ring $\mathbb{C}(X)$ of rational functions on a curve X and $k = \mathbb{C}$, we denote the corresponding module $\Omega_{\mathbb{C}(X)/\mathbb{C}}$ of meromorphic differentials (or one–forms) on X by $\Omega(X)$.

Now let x be a point of X. We want to define the module of differentials on the disc D_x at x, so we take as our ring R the complete local ring $\mathcal{O}_x$. In this case the naive application of the definition from § 8.3.1 gives an $\mathcal{O}_x$–module with uncountably many generators. For instance, it is not true that $df(t) = f'(t)dt$, where t is a formal coordinate at x and f is an infinite power series in t. To eliminate this problem, we should either impose these relations by hand, or first define $\Omega_{\mathcal{O}_{x,n}/\mathbb{C}}$, where $\mathcal{O}_{x,n} = \mathcal{O}_x/\mathfrak{m}_x^{n+1}$ (here $\mathfrak{m}_x$ denotes the maximal ideal in $\mathcal{O}_x$), and then take the limit $\varprojlim \Omega_{\mathcal{O}_{x,n}/\mathbb{C}}$. We denote this limit by $\Omega_{\mathcal{O}_x}$.

The localization of $\Omega_{\mathcal{O}_x}$ with respect to $\mathfrak{m}_x$ is the topological module of differentials on the punctured disc $D_x^\times$. It corresponds to $R = \mathcal{K}_x$, the field of fractions of $\mathcal{O}_x$. We denote this module by $\Omega_{\mathcal{K}_x}$ or simply by Ω_x. A choice of a formal coordinate t at x identifies $\mathcal{K}_x$ with $\mathbb{C}((t))$ and $\mathcal{O}_x$ with $\mathbb{C}[[t]]$. Then Ω_x and $\Omega_{\mathcal{O}_x}$ get identified with $\mathbb{C}((t))dt$ and $\mathbb{C}[[t]]dt$, respectively.

9.1.2. The residue. Given a one–form $\omega_x \in \Omega_x$, written in terms of a formal coordinate t at x as

$$\omega_x = \sum_{i\in\mathbb{Z}} f_i t^i dt,$$

its *residue* is defined as

$$\mathrm{Res}_x \, \omega_x = f_{-1}.$$

It is easy to check that this definition is independent of the choice of the formal coordinate t.

Observe that the expansion of a one–form at a point x in X embeds $\Omega(X)$ into Ω_x for any $x \in X$. Residues satisfy the following fundamental property:

9.1.3. Residue Theorem. *Let X be a smooth complete curve. Then*

$$\sum_{x\in X} \mathrm{Res}_x \, \omega_x = 0$$

for any $\omega \in \Omega(X)$, where ω_x denotes the image of ω in Ω_x.

9.1.4. Heisenberg Lie algebra. Again, let X be a smooth complete curve over $\mathbb{C}$ and $x \in X$. In addition to these data, we temporarily fix a formal coordinate t at x.

Recall that the Heisenberg Lie algebra is an extension

$$0 \to \mathbb{C}\mathbf{1} \to \mathcal{H} \to \mathbb{C}((t)) \to 0$$

with the commutator defined by the formula

(9.1.1) $[f,g] = (-\mathrm{Res}_{t=0} \, f dg)\mathbf{1}.$

Let $\mathcal{H}_{\mathrm{out}}(x)$ be the space of regular functions on $X\backslash x$. Such a function gives rise to an element of $\mathcal{K}_x$. Choosing a formal coordinate t at x identifies $\mathcal{K}_x$ with $\mathbb{C}((t))$. Thus we obtain an embedding of $\mathcal{H}_{\mathrm{out}}(x)$ into $\mathbb{C}((t))$, sending functions to their Laurent series expansions at x in the coordinate t.

9.1.5. Lemma. *The embedding $\mathcal{H}_{\mathrm{out}}(x) \to \mathbb{C}((t))$ of (commutative) Lie algebras can be lifted to an embedding $\mathcal{H}_{\mathrm{out}}(x) \to \mathcal{H}$.*

9.1.6. Proof. We need to show that the commutator (9.1.1) vanishes for $f, g \in \mathcal{H}_{\mathrm{out}}(x)$. But for such f, g, the differential $f dg$ is meromorphic on X, with possible poles only at x. Thus

$$\mathrm{Res}_x \, f dg = \sum_{y\in X} \mathrm{Res}_y \, f dg.$$

The latter sum vanishes by Theorem 9.1.3; hence the commutator vanishes, as desired.

9.1.7. Preliminary definition of conformal blocks. Recall the Fock representation π of $\mathcal{H}$ defined in § 2.1.3. Since $\mathcal{H}_{out}(x)$ is a Lie subalgebra of $\mathcal{H}$, it acts on π.

The space of *coinvariants* associated to (X, x, t, π) is by definition the vector space

$$\widetilde{H}(X, x, t, \pi) = \pi/\mathcal{H}_{out}(x) \cdot \pi,$$

where $\mathcal{H}_{out}(x) \cdot \pi = \mathrm{span}\{f \cdot A | f \in \mathcal{H}_{out}(x), A \in \pi\}$.

The space of *conformal blocks* associated to (X, x, t, π) is by definition the vector space

$$\widetilde{C}(X, x, t, \pi) = \mathrm{Hom}_{\mathcal{H}_{out}(x)}(\pi, \mathbb{C})$$

of all $\mathcal{H}_{out}(x)$–invariant functionals on the Fock space π. Thus a functional $\varphi \in \pi^*$ is a conformal block if and only if $\langle \varphi, f \cdot A \rangle = 0$ for all $f \in \mathcal{H}_{out}(x)$ and $A \in \pi$. Therefore $\widetilde{C}(X, x, t, \pi)$ is the dual vector space to $\widetilde{H}(X, x, t, \pi)$.

9.1.8. Coordinate-independent version. Next, we wish to get rid of the choice of formal coordinate t in the above definitions. Observe that the commutation relations in $\mathcal{H}$ given by formula (9.1.1) are coordinate-independent. Therefore we may use them to define a central extension $\mathcal{H}_x$ of the commutative Lie algebra $\mathcal{K}_x$,

$$0 \to \mathbb{C}\mathbf{1} \to \mathcal{H}_x \to \mathcal{K}_x \to 0.$$

The restrictions of this central extension to the Lie subalgebras $\mathcal{O}_x$ and $\mathcal{H}_{out}(x)$ are trivial, and therefore both are (commutative) Lie subalgebras of $\mathcal{H}_x$. Consider the induced $\mathcal{H}_x$–module

$$(9.1.2) \qquad \Pi_x \overset{\text{def}}{=} \mathrm{Ind}_{\mathcal{O}_x \oplus \mathbb{C}\mathbf{1}}^{\mathcal{H}_x} \mathbb{C},$$

where $\mathbf{1}$ acts on $\mathbb{C}$ as the identity, and $\mathcal{O}_x$ acts by zero. If we choose a formal coordinate t at x, then we may identify $\mathcal{H}_x$ with $\mathcal{H}$ and Π_x with π.

Now the Lie algebra $\mathcal{H}_{out}(x)$ naturally acts on Π_x, and so we may give a coordinate-independent reformulation of the above definitions.

9.1.9. Definition. The space of *coinvariants* associated to (X, x, π) is the vector space $\widetilde{H}(X, x, \pi) = \Pi_x/\mathcal{H}_{out}(x) \cdot \Pi_x$.

The space of *conformal blocks* associated to (X, x, π) is the dual vector space $\widetilde{C}(X, x, \pi) = \mathrm{Hom}_{\mathcal{H}_{out}(x)}(\Pi_x, \mathbb{C})$ of all $\mathcal{H}_{out}(x)$–invariant functionals on Π_x.

9.1.10. Vertex operator version. Let us rephrase the definition of the action of $\mathcal{H}_x$ and $\mathcal{H}_{out}(x)$ on Π_x using the vertex algebra language. Suppose we have an element of $\mathcal{K} = \mathbb{C}((t))$,

$$f = \sum_{n \in \mathbb{Z}} f_n t^n.$$

The corresponding operator on π equals

$$\widetilde{f} = \sum_{n \in \mathbb{Z}} f_n b_n,$$

where b_n is the Heisenberg generator corresponding to t^n. In terms of the vertex operator

$$b(t) = \sum_{n \in \mathbb{Z}} b_n t^{-n-1}$$

this expression can be written as

$$\widetilde{f} = \mathrm{Res}_x\, f(t)b(t)dt.$$

Consider π as a conformal vertex algebra with conformal vector $\omega_0 = \frac{1}{2}b_{-1}^2$ (so the parameter $\lambda = 0$). Then π acquires an Aut $\mathcal{O}$–module structure. Denote by Π the vector bundle on X associated to π,

$$\Pi = \mathcal{A}ut_X \underset{\mathrm{Aut}\,\mathcal{O}}{\times} \pi,$$

and by Π_x its fiber at x; in other words, Π_x is the $\mathcal{A}ut_x$–twist of π, where $\mathcal{A}ut_x$ is the Aut $\mathcal{O}$–torsor of formal coordinates at x (see § 6.5). This Π_x coincides with the induced representation (9.1.2) (compare with § 6.4.9).

According to Proposition 6.4.7 (see also § 8.1.3), $b(t)dt$ is naturally a one–form on $D_x^\times$ with values in $\mathrm{End}\,\Pi_x$. As such, it does not depend on the choice of the coordinate t. In fact, $b(t)dt$ is the projection of the canonical $\mathrm{End}\,\Pi_x$–valued section $\mathcal{Y}_x$ of Π^* onto the quotient line bundle Ω of $\mathcal{B}_0^*$ (recall from § 8.1.3 that $\mathcal{B}_0^*$ splits canonically as $\mathcal{O}_X \oplus \Omega$, and so we have a surjection $\Pi^* \twoheadrightarrow \mathcal{B}_0^* \twoheadrightarrow \Omega$).

It follows then that for any $f \in \mathcal{H}_{\mathrm{out}}(x) = \mathbb{C}[X\backslash x]$,

$$\mathrm{Res}_x\, f(t)b(t)dt$$

is a well–defined linear operator on Π_x, independent of the choice of formal coordinate t. Thus we have a canonical map of Lie algebras

$$\mathcal{H}_{\mathrm{out}}(x) \to \mathrm{End}\,\Pi_x.$$

This gives us an action of $\mathcal{H}_{\mathrm{out}}(x)$ on Π_x, which we used above to define the spaces of coinvariants and conformal blocks.

9.2. Definition of conformal blocks for general vertex algebras

We want to generalize the above definition of Heisenberg algebra conformal blocks to the case of an arbitrary vertex algebra.

9.2.1. The case of Kac–Moody and Virasoro vertex algebras.
Consider first the vertex algebra $V_k(\mathfrak{g})$ associated to an affine Kac–Moody algebra $\widehat{\mathfrak{g}}$. To define conformal blocks for $V_k(\mathfrak{g})$, one proceeds in direct analogy with the Heisenberg case, replacing the field $b(t)$ by the vertex operators $J^a(t)$. By Proposition 6.4.7, $J^a(t)dt$ transforms as a one–forms on $D_x^\times$ with values in $\mathrm{End}\,V_k(\mathfrak{g})_x$, where $V_k(\mathfrak{g})_x$ is the $\mathcal{A}ut_x$–twist of $V_k(\mathfrak{g})$. We then obtain in the same way as above a canonical homomorphism of Lie algebras

$$\mathfrak{g}_{\mathrm{out}}(x) \overset{\mathrm{def}}{=} \mathfrak{g} \otimes \mathbb{C}[X\backslash x] \to \mathrm{End}\,V_k(\mathfrak{g})_x, \qquad J^a \otimes f \mapsto \mathrm{Res}_x\, f(t)J^a(t)dt.$$

This homomorphism may also be obtained using the definition of $V_k(\mathfrak{g})_x$ as the induced representation of the central extension of $\mathfrak{g} \otimes \mathcal{K}_x$ given in § 6.4.9 (note that in § 6.4.9 it was denoted by $V_k(\mathfrak{g})(D_x)$). Hence we may define the space of conformal blocks associated to $(X, x, V_k(\mathfrak{g}))$ as the space of $\mathfrak{g}_{\mathrm{out}}(x)$–invariant functionals $\mathrm{Hom}_{\mathfrak{g}_{\mathrm{out}}(x)}(V_k(\mathfrak{g})_x, \mathbb{C})$.

The reason this fairly simple construction works is that the Heisenberg and Kac–Moody vertex algebras are generated by small subspaces, which transform in the simplest possible way under the action of Aut $\mathcal{O}$. These subspaces give rise to easily describable quotients of the vector bundle $\mathcal{V}^*$ on X corresponding to the entire vertex algebra V. In the Heisenberg case, we take the Aut $\mathcal{O}$–submodule

$\mathbb{C}b_{-1} \subset \pi$, which leads to the quotient $\Pi^* \twoheadrightarrow \Omega$. As we will see below, we do not lose any information by passing to this quotient, since the Fourier coefficients b_n of $b(t)$ form a Lie algebra whose action generates π. Similarly, in the Kac–Moody case the vertex algebra

$$V_k(\mathfrak{g}) \simeq U(\mathfrak{g} \otimes t^{-1}\mathbb{C}[t^{-1}]) \cdot v_k$$

is generated by the Aut $\mathcal{O}$–invariant subspace $(\mathfrak{g} \otimes t^{-1})v_k$ of degree one. Thus we obtain the quotient

$$\mathcal{V}_k^* \twoheadrightarrow \mathfrak{g}^* \otimes \Omega,$$

which suffices in order to define conformal blocks.

In the case of a general vertex algebra, there is no analogous simple Aut $\mathcal{O}$–stable subspace with respect to which we can take invariant functionals. Already in the case of the Virasoro vertex algebra Vir_c the situation becomes more subtle. In this case, the obvious analogue of $\mathcal{H}_{\mathrm{out}}(x)$ and $\mathfrak{g}_{\mathrm{out}}(x)$ is the Lie algebra $\mathrm{Vect}(X\backslash x)$ of regular vector fields on $X\backslash x$. By analogy with the Heisenberg and Kac-Moody cases, we would like to define a homomorphism

$$(9.2.1) \qquad \mathrm{Vect}(X\backslash x) \to \mathrm{End}\,\mathrm{Vir}_{c,x}, \qquad \xi(t)\partial_t \mapsto \mathrm{Res}_x \xi(t)T(t)dt$$

(where t is a formal coordinate at x).

If the central charge $c = 0$, then the field $T(t)(dt)^2$ transforms as a quadratic differential, and so the residue pairing with vector fields makes sense. However, when $c \neq 0$, we *cannot* pair $T(t)$ with vector fields in an invariant fashion. As we saw in § 8.2, if $c \neq 0$ then $T(t)$ does not transform as a tensor. Rather, $\partial_t^2 + \frac{6}{c}T(t)$ is an operator–valued projective connection: it is naturally a section of a rank two bundle $\mathcal{T}_c^*$. Unlike the Heisenberg and Kac–Moody cases, the bundle generated by our basic field is a non–split extension in this case, and its sections do not transform as simple tensors. Thus we are forced to make use of the full rank two bundle $\mathcal{T}_c^*$, which can be paired invariantly only with sections of the rank two bundle $\mathcal{T}_c \otimes \Omega$.

In fact, we can split the bundle $\mathcal{T}_c \otimes \Omega$ by choosing a projective structure on X, i.e. an atlas, in which the transition functions are projective transformations, which are "invisible" to the Schwarzian derivative (see § 8.2). Then we can make $T(t)(dt)^2$ behave as a quadratic differential. In other words, if we fix a "background" projective connection $\rho(t)$, then its difference with any other projective connection (such as $T(t) - \rho(t)$) is a quadratic differential. However, there is no canonical way to choose a projective connection algebraically, so we prefer not to do that.

Actually, one can get around this problem using a result which will be discussed in § 19.6. Namely, it turns out that the sheaf $\widehat{\mathcal{T}}_c = (\mathcal{T}_c \otimes \Omega)/d\mathcal{O}_X$, which is an extension of Θ_X by $\Omega/d\mathcal{O}_X$, has a canonical Lie algebra structure, and its sections on any Zariski open subset $\Sigma \subset X$ act on $\mathcal{V}ir_{c,x}$. The residue theorem implies that the resulting homomorphism of Lie algebras $\Gamma(X\backslash x, \widehat{\mathcal{T}}_c) \to \mathrm{End}\,\mathcal{V}ir_{c,x}$ factors through $\mathrm{Vect}(X\backslash x)$, and hence the above map (9.2.1) is well-defined.

Thus, already in the Virasoro case we have to deal with Lie algebras coming from sections of more complicated bundles on our curve than we may have anticipated. In general we should expect a much more complicated structure. This suggests that the right thing to do for a general vertex algebra is to impose the invariance condition for *all* vertex operators simultaneously. In specific examples (as in the Heisenberg or Kac–Moody cases above), one can reduce this condition to the invariance with respect to a select few "generating" vertex operators without changing the space of conformal blocks, as we will see below.

9.2.2. From sections to operators. Let X be a smooth complete curve and x a point of X. Let V be a quasi-conformal vertex algebra, and $\mathcal{Y}_x$ the $\operatorname{End} V_x$–valued section of $\mathcal{V}^*$ on $D_x^\times$ defined in Theorem 6.5.4. Recall the dual map

$$\mathcal{Y}_x^\vee : \Gamma(D_x^\times, \mathcal{V} \otimes \Omega) \to \operatorname{End} V_x$$

defined using the residue pairing in § 6.5.8. Namely, the image of a section μ of $\mathcal{V} \otimes \Omega$ over $D_x^\times$ under the map $\mathcal{Y}_x^\vee$ is the linear operator O_μ on V_x given by the formula $O_\mu = \operatorname{Res}_x \langle \mathcal{Y}_x, \mu \rangle$.

Consider the complex

$$0 \to \mathcal{V} \xrightarrow{\nabla} \mathcal{V} \otimes \Omega \to 0,$$

where $\mathcal{V} \otimes \Omega$ is placed in cohomological degree 0 and $\mathcal{V}$ is placed in cohomological dimension -1. The differential ∇ is the connection operator defined in § 6.6 (recall that, with respect to a local coordinate z, $\nabla = d + L_{-1} \otimes dz$). This is the de Rham complex of the flat vector bundle $\mathcal{V}$, considered as a complex of sheaves on the curve X, shifted by -1 in cohomological dimensions (this convention will be clarified later in § 19.4.7).

We will denote by $h(\mathcal{V} \otimes \Omega)$ the sheaf of the 0th cohomology of the above shifted de Rham complex (equivalently, this is the sheaf of the first de Rham cohomology of the flat vector bundle $\mathcal{V}$). Thus for $\Sigma \subset X$, $\Gamma(\Sigma, h(\mathcal{V} \otimes \Omega))$ denotes the degree 0 cohomology of the restriction of this complex to Σ. If Σ is affine or $D_x^\times$, then $\Gamma(\Sigma, h(\mathcal{V} \otimes \Omega))$ is simply the quotient of $\Gamma(\Sigma, \mathcal{V} \otimes \Omega)$ by the image of ∇ (the total derivatives). We set

$$U(\mathcal{V}_x) \overset{\text{def}}{=} \Gamma(D_x^\times, h(\mathcal{V} \otimes \Omega)).$$

Note that in the case of Heisenberg and Kac-Moody vertex algebras, $U(\mathcal{V}_x)$ contains $\mathcal{H}_x$ and $\widehat{\mathfrak{g}}_x$, respectively, as Lie subalgebras.

Since the residue of a total derivative vanishes, the map $\mathcal{Y}_x^\vee$ factors through $U(\mathcal{V}_x)$, so we obtain a map

$$\mathcal{Y}_x^\vee : U(\mathcal{V}_x) \to \operatorname{End} V_x.$$

The image of this map is the span of all operators O_μ for $\mu \in \Gamma(D_x^\times, \mathcal{V} \otimes \Omega)$.

9.2.3. Lemma. $U(\mathcal{V}_x)$ *is a Lie algebra, and the map* $\mathcal{Y}_x^\vee : U(\mathcal{V}_x) \to \operatorname{End} V_x$ *is a homomorphism of Lie algebras.*

9.2.4. Proof. If we choose a formal coordinate t at x, then we can identify $U(\mathcal{V}_x)$ with the quotient

$$V \otimes \mathbb{C}((t)) / \operatorname{Im}(T \otimes \operatorname{Id} + \operatorname{Id} \otimes \partial_t),$$

which is the Lie algebra $U(V)$ introduced in § 4.1.1. The corresponding map $U(V) \to \operatorname{End} V$ is a Lie algebra homomorphism by Theorem 4.1.2. The fact that the Lie algebra structure and the homomorphism do not depend on the choice of the formal coordinate t follows from the results of § 19.4.13.

9.2.5. The Lie algebra $U_\Sigma(V)$. Given a Zariski open subset $\Sigma \subset X$, set

$$U_\Sigma(V) \overset{\text{def}}{=} \Gamma(\Sigma, h(\mathcal{V} \otimes \Omega)).$$

The restriction of sections from Σ to $D_x^\times$ gives us a map $U_\Sigma(\mathcal{V}) \to U(\mathcal{V}_x)$. We will prove in Theorem 19.4.9 below that the sheaf $h(\mathcal{V} \otimes \Omega)$ is a sheaf of Lie algebras on X. This implies (see Corollary 19.4.14):

9.2.6. Theorem. *For any Zariski open subset* $\Sigma \subset X$, $U_\Sigma(V)$ *is a Lie algebra and the map* $U_\Sigma(V) \to U(\mathcal{V}_x)$ *is a Lie algebra homomorphism, so that the image is a Lie subalgebra of* $U(\mathcal{V}_x)$.

Denote the image of the homomorphism $U_\Sigma(\mathcal{V}) \to U(\mathcal{V}_x)$ by $U_\Sigma(\mathcal{V}_x)$. Note that, in the case of Heisenberg and Kac-Moody vertex algebras, $U_\Sigma(\mathcal{V}_x)$ contains $\mathcal{H}_{\mathrm{out}}(x)$ and $\mathfrak{g}_{\mathrm{out}}(x)$, respectively.

Now we can define the spaces of coinvariants and conformal blocks for a general quasi-conformal vertex algebra V (note that this definition makes sense even without Theorem 9.2.6).

9.2.7. Definition. The *space of coinvariants associated to* (X, x, V) is the vector space $H(X, x, V) = \mathcal{V}_x/(U_{X\backslash x}(\mathcal{V}_x) \cdot \mathcal{V}_x)$.

The *space of conformal blocks associated to* (X, x, V) is the space of $U_{X\backslash x}(\mathcal{V}_x)$-invariant functionals on $\mathcal{V}_x$, $C(X, x, V) = \mathrm{Hom}_{U_{X\backslash x}(\mathcal{V}_x)}(\mathcal{V}_x, \mathbb{C})$.

9.2.8. It follows from the definition that $C(X, x, V)$ is the dual space to $H(X, x, V)$. We would like to reformulate the definition of the space of conformal blocks using the converse of the Residue Theorem 9.1.3.

Let $\mathcal{E}$ be a vector bundle on X. Then we have a pairing

$$\Gamma(D_x^\times, \mathcal{E} \otimes \Omega) \times \Gamma(D_x^\times, \mathcal{E}^*) \to \mathbb{C}$$

given by

$$(\mu, \nu) \mapsto \mathrm{Res}_x \langle \mu, \nu \rangle.$$

It is clear that this pairing is non-degenerate.

9.2.9. Strong Residue Theorem. *A section* $\nu \in \Gamma(D_x^\times, \mathcal{E}^*)$ *has the property that*

$$\mathrm{Res}_x \langle \mu, \nu \rangle = 0, \qquad \forall \mu \in \Gamma(X\backslash x, \mathcal{E} \otimes \Omega)$$

if and only if ν *can be extended to a regular section of* $\mathcal{E}^*$ *over* $X\backslash x$ *(i.e.* $\nu \in \Gamma(X\backslash x, \mathcal{E}^*)$).

9.2.10. Remark. The "if" part of the theorem is a consequence of the Residue Theorem 9.1.3, since the pairing of two sections regular outside of x must have all its residues at x, and all residues must add up to 0.

The "only if" part can be reformulated as the statement that the residue pairing (which is well-defined by the "if" part)

$$\Gamma(X\backslash x, \mathcal{E} \otimes \Omega) \times \big(\Gamma(D_x^\times, \mathcal{E}^*)/\Gamma(X\backslash x, \mathcal{E}^*)\big) \to \mathbb{C}$$

is *non-degenerate*.

Note that the above theorem can be generalized as follows: let $x_1, \ldots, x_n \in X$ be a set of distinct points. Then a section

$$\nu \in \bigoplus_{i=1}^n \Gamma(D_{x_i}^\times, \mathcal{E}^*)$$

has the property that

$$\sum_{i=1}^n \mathrm{Res}_{x_i} \langle \mu, \nu \rangle = 0, \qquad \forall \mu \in \Gamma(X\backslash\{x_1, \ldots, x_n\}, \mathcal{E} \otimes \Omega),$$

if and only if ν can be extended to a regular section of $\mathcal{E}$ over $X\backslash\{x_1, \ldots, x_n\}$ (i.e. $\nu \in \Gamma(X\backslash\{x_1, \ldots, x_n\}, \mathcal{E}^*)$).

9.2.11. Dual description. Returning to Definition 9.1.9 of conformal blocks for the Heisenberg algebra, recall that $\varphi \in \widetilde{C}(X, x, \pi)$ if and only if

$$\langle \varphi, \mathrm{Res}_x \, f(t) b(t) dt \cdot A \rangle = \mathrm{Res}_x \, f(t) \langle \varphi, b(t) \cdot A \rangle dt = 0 \qquad \forall A \in \Pi_x.$$

Hence by Theorem 9.2.9 the one–form $\langle \varphi, b(t) \cdot A \rangle dt$ (which as we know is independent of the choice of the formal coordinate t at x) on $D_x^\times$ can be extended to a regular one–form on $X \backslash x$. This allows us to reformulate the definition of $\widetilde{C}(X, x, \pi)$: $\varphi \in \widetilde{C}(X, x, \pi)$ if and and only if $\langle \varphi, b(t) \cdot A \rangle dt$ can be extended to a regular one–form on $X \backslash x$ for all $A \in \Pi_x$.

In the case of a general vertex algebra, we can also reformulate the definition of the space of conformal blocks using Theorem 9.2.9 as follows:

9.2.12. Definition. A linear functional $\varphi \in \mathcal{V}_x^*$ is a conformal block if and only if $\langle \varphi, \mathcal{Y}_x \cdot A \rangle$ can be extended to a regular section of $\mathcal{V}^*$ on $X \backslash x$ for all $A \in \mathcal{V}_x$.

9.2.13. Remark. According to Proposition 6.6.7, $\langle \varphi, \mathcal{Y}_x \cdot A \rangle$ is a *horizontal* section of $\mathcal{V}^*$ on the punctured disc for any $\varphi \in \mathcal{V}_x^*$ and $A \in \mathcal{V}_x$. Therefore if φ is a conformal block, then $\langle \varphi, \mathcal{Y}_x \cdot A \rangle$ is a horizontal section of $\mathcal{V}^*$ on $X \backslash x$. Denote this section by φ_A.

If W is an Aut $\mathcal{O}$–stable subspace of V, then φ_A can be projected onto a section of $\mathcal{W}^*$ over $X \backslash x$. Note that according to the vacuum axiom, $\varphi_{|0\rangle}$ is actually a regular section of $\mathcal{V}^*$ over all of X (and so is its projection onto $\mathcal{W}^*$). In particular, taking as W the two-dimensional subspace of V spanned by $|0\rangle$ and ω, we assign to each conformal block a projective connection on X (if $c \neq 0$). If we choose a formal coordinate z at x, then we can write this projective connection as $\partial_z^2 + \frac{6}{c} \langle \varphi, T(z) |0\rangle \rangle$. If $c = 0$, then we obtain a quadratic differential $\langle \varphi, T(z) |0\rangle \rangle (dz)^2$.

9.3. Comparison of the two definitions of conformal blocks

In the case of the Heisenberg algebra (and similarly, in the case of an affine algebra), we now have two definitions of conformal blocks:

$$\widetilde{C}(X, x, \pi) = \{\varphi \in \Pi_x^* | \langle \varphi, b(t_x) \cdot A \rangle dt_x \text{ extends to a section of } \Omega \text{ on } X \backslash x \text{ for all } A\}$$

(where t_x is a formal coordinate at x) and

$$C(X, x, \pi) = \{\varphi \in \Pi_x^* | \langle \varphi, \mathcal{Y}_x \cdot A \rangle \text{ extends to a section of } \Pi^* \text{ on } X \backslash x \text{ for all } A\}.$$

We will now show that these two definitions are equivalent.

Since Π^* projects onto Ω, any section of Π^* gives rise to a section of Ω. We know that the section $\mathcal{Y}_x$ maps to $b(t_x) dt_x$ under this projection. Therefore we obtain an embedding $C(X, x, \pi) \hookrightarrow \widetilde{C}(X, x, \pi)$. Now we construct a map in the opposite direction.

9.3.1. Two-point conformal blocks. First we prove a preliminary result that shows that the space of one-point conformal blocks is isomorphic to the space of two-point conformal blocks.

Let y be a point of X other than x. Set

$$\mathcal{H}_{\mathrm{out}}(x, y) = \mathbb{C}[X \backslash \{x, y\}].$$

Define a map $\mathcal{H}_{\mathrm{out}}(x, y) \to \mathrm{End}(\Pi_x \otimes \Pi_y)$ by the formula

$$f \to \widetilde{f} = (\mathrm{Res}_x \, f(t_x) b(t_x) dt_x) \otimes \mathrm{Id} + \mathrm{Id} \otimes (\mathrm{Res}_y \, f(t_y) b(t_y) dt_y).$$

It is clear that this map does not depend on the choice of formal coordinates t_x, t_y at x, y.

Define the space of coinvariants

$$\widetilde{H}(X, x, y, \pi \otimes \pi) = (\Pi_x \otimes \Pi_y)/\mathcal{H}_{\mathrm{out}}(x, y) \cdot (\Pi_x \otimes \Pi_y).$$

Its dual space is the space of *two-point conformal blocks*

$$\widetilde{C}(X, x, y, \pi \otimes \pi) = \mathrm{Hom}_{\mathcal{H}_{\mathrm{out}}(x,y)}(\Pi_x \otimes \Pi_y, \mathbb{C}).$$

Given $\widetilde{\varphi} \in \widetilde{C}(X, x, y, \pi \otimes \pi)$, let $\varphi = \widetilde{\varphi}|_{\Pi_x \otimes 1}$, where 1 is the vacuum vector in Π_y (note that since the vacuum vector is invariant under the action of $\mathrm{Aut}\,\mathcal{O}$, it is uniquely determined in Π_y). It is easy to see that $\varphi \in \widetilde{C}(X, x, \pi)$. Thus we obtain a linear map

(9.3.1)
$$\widetilde{C}(X, x, y, \pi \otimes \pi) \to \widetilde{C}(X, x, \pi).$$

9.3.2. Proposition. *The map* (9.3.1) *is an isomorphism.*

The proof is given at the end of this Chapter, in § 9.6.

Let us denote by $\widetilde{\varphi}_y$ the image of $\varphi \in \widetilde{C}(X, x, \pi)$ in $\widetilde{C}(X, x, y, \pi \otimes \pi)$ under the above isomorphism. We are now ready to show that the two definitions of (one–point) conformal blocks for the Heisenberg algebra are equivalent.

9.3.3. Theorem. *Let φ be a linear functional on Π_x such that $\langle \varphi, b(t_x) \cdot A \rangle dt_x$ can be extended to a regular one-form on $X \backslash x$ for all $A \in \Pi_x$. Then the section $\langle \varphi, \mathcal{Y}_x \cdot A \rangle$ of Π^* over $D_x^\times$ can be extended to a regular section $\widetilde{\varphi}_A$ of Π^* on $X \backslash x$, for all $A \in \Pi_x$.*

Thus, we have a natural isomorphism $\widetilde{C}(X, x, \pi) \simeq C(X, x, \pi)$.

9.3.4. Proof. We will construct the section $\widetilde{\varphi}_A$ starting from $\varphi \in \widetilde{C}(X, x, \pi)$, $A \in \Pi_x$, using the isomorphism $\widetilde{C}(X, x, \pi) \simeq \widetilde{C}(X, x, y, \pi \otimes \pi)$ of Proposition 9.3.2. However, in order to do that we first need to organize the spaces $\widetilde{C}(X, x, y, \pi \otimes \pi)$ into a vector bundle over $X \backslash x$ (i.e., we want to be able to "move" the point y along $X \backslash x$). This can be done as follows.

Let $\overset{\circ}{X}{}^n$ denote $X^n \backslash \Delta$, where Δ is the union of all pairwise diagonals in X^n. We have a natural projection $p_n : \overset{\circ}{X}{}^{n+1} \to \overset{\circ}{X}{}^n$, whose fiber over $(x_1, \ldots, x_n)$ is $X \backslash \{x_1, \ldots, x_n\}$. Let $\mathcal{O}^{(n)} = (p_n)_* \mathcal{O}_{\overset{\circ}{X}{}^{n+1}}$; this is a quasi–coherent sheaf on $\overset{\circ}{X}{}^n$, whose fiber at $(x_1, \ldots, x_n)$ is $\mathbb{C}[X \backslash \{x_1, \ldots, x_n\}]$. Let $\Pi_{(n)} = j_n^* \Pi^{\boxtimes n}$, where $j_n : \overset{\circ}{X}{}^n \hookrightarrow X^n$. Then $\mathcal{O}^{(n)}$ acts naturally on $\Pi_{(n)}$ in such a way that fiberwise we obtain the obvious action of $\mathbb{C}[X \backslash \{x_1, \ldots, x_n\}] = \mathcal{H}_{\mathrm{out}}(x_1, \ldots, x_n)$ on $\bigotimes_{i=1}^n \Pi_{x_i}$. Now introduce the sheaf of homomorphisms

$$\widetilde{C}^{(n)} = \mathcal{H}om_{\mathcal{O}^{(n)}}(\Pi_{(n)}, \mathcal{O}_{\overset{\circ}{X}{}^n}).$$

This is a sheaf of $\mathcal{O}$–modules on $\overset{\circ}{X}{}^n$, whose fibers are the spaces of n–point conformal blocks defined in the same way as in the case $n = 2$.

When $n = 2$ we can further restrict the sheaf $\widetilde{C}^{(2)}$ over $\overset{\circ}{X}{}^2$ to $x \times (X \backslash x)$. Then we obtain a subsheaf $\widetilde{C}_x^{(2)}$ of $(\Pi_x \otimes \Pi)^*$ over $X \backslash x$, whose fiber at $y \in X \backslash x$ is the space of conformal blocks $\widetilde{C}(X; x, y; \pi^{\otimes 2})$, as desired. The isomorphism of

Proposition 9.3.2 identifies the fibers of $\widetilde{C}_x^{(2)}$ with the vector space $\widetilde{C}(X, x, \pi)$, thus providing a trivialization of $\widetilde{C}_x^{(2)}$.

Now, given $\varphi \in \widetilde{C}(X, x, \pi)$, we obtain a section of $\widetilde{C}_x^{(2)}$ and hence of $(\Pi_x \otimes \Pi)^*$ on $X \backslash x$. Contracting it with $A \in \Pi_x$, we obtain a section of Π^* on $X \backslash x$. This is the desired section $\widetilde{\varphi}_A$. By construction, the value of $\widetilde{\varphi}_A$ on $B \in \Pi_y$ at $y \in X \backslash x$ equals $\widetilde{\varphi}_y(A \otimes B)$ (here we use the notation $\widetilde{\varphi}_y$ introduced after Proposition 9.3.2).

Now we want to show that the section $\widetilde{\varphi}_A$ is the analytic continuation of the section $\langle \varphi, \mathcal{Y}_x \cdot A \rangle$ on $D_x^\times$, i.e., that when y is close to x, $\widetilde{\varphi}_A = \langle \varphi, \mathcal{Y}_x \cdot A \rangle$. In the argument below, we will use analytic topology, but the same argument also works if x and y are infinitesimally nearby points in the sense of § A.2.3.

Choose a small analytic open neighborhood U of x with coordinate z such that $y \in U$. Thus we may take $t_x = z - x$ as a local coordinate at x, and $t_y = z - y$ as a local coordinate at $y \in U$. Let us identify Π_x and Π_y with π using these coordinates. It remains to prove the following:

9.3.5. Lemma. *For all $A, B \in \pi$, and $y \in U$,*

$$(9.3.2) \qquad \langle \widetilde{\varphi}_y, A \otimes B \rangle = \langle \widetilde{\varphi}_y, Y(B, y - x) \cdot A \otimes 1 \rangle$$
$$= \langle \varphi, Y(B, y - x) \cdot A \rangle$$

9.3.6. Proof. The second equality follows from Proposition 9.3.2.

The proof of the first equality proceeds by induction. First assume that $B = 1$ is the vacuum vector in π. In this case formula (9.3.2) obviously holds because $Y(1, z) = \mathrm{Id}$.

Denote by $\pi^{(m)}$ the subspace of π spanned by all monomials of the form $b_{i_1} \ldots b_{i_k}, k \leq m$. Suppose that we have proved formula (9.3.2) for all $B \in \pi^{(m)}$. The inductive step is to prove the equality for elements of the form $B = b_n \cdot B'$, for $B' \in \pi^{(m)}$. To do this we need the following lemma.

9.3.7. Lemma. *For any $A \in \Pi_x$ and $B \in \Pi_y$ there exists a regular one–form on $X \backslash \{x, y\}$ such that its restriction to $D_x^\times$ is $\langle \widetilde{\varphi}_y, (b(t_x) \cdot A) \otimes B \rangle dt_x$ and to $D_y^\times$ is $\langle \widetilde{\varphi}_y, A \otimes (b(t_y) \cdot B) \rangle dt_y$.*

9.3.8. Proof. Since $\widetilde{\varphi}_y$ is a conformal block, we obtain, using the definition of the $\mathcal{H}_{\mathrm{out}}(x, y)$–action on $\pi \otimes \pi$,

$$\langle \widetilde{\varphi}_y, f \cdot (A \otimes B) \rangle = \langle \widetilde{\varphi}_y, f_x \cdot A \otimes B \rangle + \langle \widetilde{\varphi}_y, A \otimes f_y \cdot B \rangle,$$

for all $f \in \mathbb{C}[X \backslash \{x, y\}]$. On the other hand,

$$\langle \widetilde{\varphi}_y, f_x \cdot A \otimes B \rangle + \langle \widetilde{\varphi}_y, A \otimes f_y \cdot B \rangle = \mathrm{Res}_x \langle \widetilde{\varphi}_y, f(t_x) b(t_x) \cdot A \otimes B \rangle dt_x$$
$$+ \mathrm{Res}_y \langle \widetilde{\varphi}_y, A \otimes f(t_y) b(t_y) \cdot B \rangle dt_y.$$

Hence

$$((\langle \widetilde{\varphi}_y, b(t_x) \cdot A \otimes B \rangle dt_x, \langle \widetilde{\varphi}_y, A \otimes b(t_y) \cdot B \rangle dt_y) \in \Omega_x \oplus \Omega_y$$

has the property that its residue pairing with any $f \in \mathbb{C}[X \backslash \{x, y\}]$ is zero. The Strong Residue Theorem 9.2.9 then implies that there exists a regular one-form on $X \backslash \{x, y\}$ whose expansions at x and y are

$$\langle \widetilde{\varphi}_y, b(t_x) \cdot A \otimes B \rangle dt_x \qquad \text{and} \qquad \langle \widetilde{\varphi}_y, A \otimes b(t_y) \cdot B \rangle dt_y,$$

respectively. This proves the statement of Lemma 9.3.7.

9.3.9. End of proof of Lemma 9.3.5. By our inductive hypothesis, we know that if $B \in \pi^{(m)}$, then

$$\langle \widetilde{\varphi}_y, (b(t_x) \cdot A) \otimes B \rangle = \langle \varphi, Y(B, y - x) b(t_x) \cdot A \rangle.$$

According to Lemma 9.3.7 above, we also have

$$\langle \widetilde{\varphi}_y, (b(t_x) \cdot A) \otimes B \rangle = \langle \widetilde{\varphi}_y, A \otimes (b(t_y) \cdot B) \rangle.$$

Using locality and associativity in the vertex algebra π, we obtain

$$
\begin{aligned}
\langle \varphi, Y(B, y - x) b(t_x) \cdot A \rangle &= \langle \varphi, b(t_x) Y(B, y - x) \cdot A \rangle \\
&= \langle \varphi, Y(b(t_x - (y - x)) \cdot B, y - x) \cdot A \rangle \\
&= \langle \varphi, Y(b(t_y) \cdot B, y - x) \cdot A \rangle.
\end{aligned}
$$

Combining these calculations, we obtain

$$\langle \widetilde{\varphi}_y, A \otimes b(t_y) \cdot B \rangle = \langle \varphi, Y(b(t_y) \cdot B, y - x) \cdot A \rangle.$$

Multiplying both sides by t_y^n and taking residues at y, we find that

$$\langle \widetilde{\varphi}_y, A \otimes b_n \cdot B \rangle = \langle \varphi, Y(b_n \cdot B, y - x) \cdot A \rangle.$$

This proves the inductive step and hence Lemma 9.3.5.

This also completes the proof of Theorem 9.3.3.

9.3.10. Remark. The same argument applies to show that the two definitions of conformal blocks agree for affine Kac–Moody algebras and the Virasoro algebra as well (see § 9.2.1). For more general vertex algebras, one may attempt a similar argument once provided with a set of generators spanning an Aut $\mathcal{O}$–stable subspace of the vertex algebra. However, this will be more difficult when the vertex algebra is not generated by a vertex Lie algebra (see § 16.1).

9.4. Coinvariants for commutative vertex algebras

Let V be a quasi-conformal commutative vertex algebra. Recall from § 1.4 and Lemma 6.3.5 that this means that V is a $\mathbb{Z}_+$–graded commutative algebra with a unit, equipped with the action of the Lie algebra Der $\mathcal{O}$ by derivations, such that L_0 is the gradation operator. The operator L_{-1} is then the translation operator, and $Y(A, z)$ is the operator of multiplication by $e^{zL_{-1}} A$.

By the general constructions of §§ 6.5 and 6.6, we attach to V a vector bundle $\mathcal{V}$ with a flat connection ∇ on any smooth curve. Moreover, it follows from the construction that the sheaf of sections of $\mathcal{V}$ is a sheaf of algebras (in particular, its fibers $\mathcal{V}_x$ are algebras), preserved by the connection ∇ (in the sense that the Leibniz rule holds). Therefore we can attach to $\mathcal{V}$ a scheme Spec $\mathcal{V}$ over X, whose fiber over $x \in X$ is Spec $\mathcal{V}_x$. This scheme is equipped with a flat connection, which we also denote by ∇.

Let $\Gamma_\nabla(X, \operatorname{Spec} \mathcal{V})$ be the scheme of global horizontal sections of Spec $\mathcal{V}$ over X. For any $\mathbb{C}$–algebra R, the set of R–points of $\Gamma_\nabla(X, \operatorname{Spec} \mathcal{V})$ is by definition the set of horizontal homomorphisms of sheaves of algebras $\mathcal{V} \to \underline{R}$, where $\underline{R}$ is the constant sheaf on X with the fiber R and the trivial connection.

9.4.1. Proposition. *The space of coinvariants $H(X, x, V)$ is a commutative algebra that is canonically isomorphic to the algebra of functions on $\Gamma_\nabla(X, \operatorname{Spec} \mathcal{V})$.*

9.4.2. Proof. In a commutative vertex algebra we have $Y(A, z)B \in V[[z]]$ for any $A, B \in V$. Therefore it follows from the definition that $H(X, x, V)$ is the quotient of $\mathcal{V}_x$ by the ideal generated by the subspace $U_{X \setminus x}(\mathcal{V}_x)|0\rangle \subset \mathcal{V}_x$. Thus, $H(X, x, V)$ is an algebra.

Let $\overline{\varphi}$ be an algebra homomorphism $H(X, x, V) \to \mathbb{C}$, and let φ be the corresponding homomorphism $\mathcal{V}_x \to \mathbb{C}$. Then φ is a conformal block. Note that for any functional $\varphi \in \mathcal{V}_x^*$, the section $\langle \varphi, \mathcal{Y}_x \cdot |0\rangle \rangle$ is just the horizontal section of $\mathcal{V}^*|_{D_x}$ which is equal to $\varphi \in \mathcal{V}_x^*$ at x. Since φ is a conformal block, we obtain from Definition 9.2.12 that this section extends to a global (and necessarily horizontal) section of $\mathcal{V}^*$ on the entire curve X. Denote this section by φ_X. Since φ is an algebra homomorphism, φ_X is a horizontal homomorphism $\mathcal{V} \to \underline{\mathbb{C}}$.

Conversely, given a horizontal homomorphism $\mathcal{V} \to \underline{\mathbb{C}}$, its restriction to the fiber $\mathcal{V}_x$ is a homomorphism $\mathcal{V}_x \to \mathbb{C}$, which is a conformal block, and therefore factors through $H(X, x, V)$. Thus, we obtain a canonical identification between homomorphisms $H(X, x, V) \to \mathbb{C}$ and horizontal homomorphisms $\mathcal{V} \to \underline{\mathbb{C}}$.

In the same way one proves that $\mathrm{Hom}(H(X, x, V), R) = \mathrm{Hom}_\nabla(\mathcal{V}, \underline{R})$ for any $\mathbb{C}$–algebra R. Therefore $\mathrm{Spec}\, H(X, x, V) = \Gamma_\nabla(X, \mathcal{V})$.

9.4.3. Example. Consider the commutative Heisenberg vertex algebra $\pi^0 = \mathbb{C}[b_n]_{n<0}$ defined in Remark 2.4.3. Define the action of $\mathrm{Der}\, \mathcal{O}$ on π^0 by the formulas

$$L_m \mapsto \sum_{n<0} (m - n) b_n \frac{\partial}{\partial b_{n-m}}.$$

Note that these operators appear in the limit of the action of $\mathrm{Der}\, \mathcal{O}$ on the conformal vertex algebra π^κ with the conformal vector $\frac{1}{2\kappa} b_{-1}^2$ when $\kappa \to 0$. This $\mathrm{Der}\, \mathcal{O}$–action makes π^0 a quasi-conformal vertex algebra. In fact, under this action π^0 gets identified with $\mathrm{Sym}(\mathcal{K}/\mathcal{O})$ and Π_x^0 is identified with $\mathrm{Sym}(\mathcal{K}_x/\mathcal{O}_x)$. Hence, using the residue pairing, we obtain $\mathrm{Spec}\, \pi^0 = \Omega_\mathcal{O}, \mathrm{Spec}\, \Pi_x^0 = \Omega_{\mathcal{O}_x}$. Therefore the scheme $\mathrm{Spec}\, \Pi^0$ on a curve X is nothing but the bundle of jets of sections of the canonical line bundle Ω on X.

The bundle of jets of sections of any vector bundle $\mathcal{E}$ on X carries a canonical connection. Moreover, the space of its global horizontal sections is just the space of ordinary sections of $\mathcal{E}$ over X. In our case, we obtain that $\Gamma_\nabla(X, \mathrm{Spec}\, \Pi^0) = H^0(X, \Omega)$. Therefore, according to Proposition 9.4.1, $H(X, x, \pi^0) = \mathrm{Fun}\, H^0(X, \Omega)$, the ring of polynomial functions on $H^0(X, \Omega)$.

This result may also be obtained in a simpler way using the first definition of coinvariants for the Heisenberg algebra. Since $\Pi_x^0 = \mathrm{Sym}(\mathcal{K}_x/\mathcal{O}_x)$, we find that

$$H(X, x, \pi^0) = \Pi_x/\mathbb{C}[X \setminus x] \cdot \Pi_x = \mathrm{Sym}\,(\mathbb{C}[X \setminus x] \setminus \mathcal{K}_x/\mathcal{O}_x) = \mathrm{Sym}\, H^1(X, \mathcal{O}_X),$$

because $H^1(X, \mathcal{O}_X) \simeq \mathbb{C}[X \setminus x] \setminus \mathcal{K}_x/\mathcal{O}_x$ (cf. Lemma 17.3.6). By the Serre duality, $H^0(X, \Omega) \simeq H^1(X, \mathcal{O}_X)^*$, so $\mathrm{Sym}\, H^1(X, \mathcal{O}_X) \simeq \mathrm{Fun}\, H^0(X, \Omega)$.

9.4.4. Jet schemes. As our next example, we consider the commutative vertex algebras associated to jet schemes. Let Z be an affine algebraic variety, $Z = \mathrm{Spec}\, A$, where $A = \mathbb{C}[x_1, \ldots, x_N]/(P_1, \ldots, P_k)$. Consider the ring of polynomials in infinitely many variables $\mathbb{C}[x_{1,n}, \ldots, x_{N,n}]_{n \leq 0}$. It carries a derivation T, defined by the formula $T \cdot x_{i,n} = -(n-1)x_{i,n-1}$. We identify the old variables x_i with $x_{i,0}$. Introduce a new ring

$$A_\infty \overset{\mathrm{def}}{=} \mathbb{C}[x_{1,n}, \ldots, x_{N,n}]_{n \leq 0}/(P_{1,n}, \ldots, P_{k,n})_{n \leq 0},$$

where $P_{i,n}$ is the coefficient of t^n in the series $P_i(x_1(t), \ldots, x_m(t))$, where $x_i(t) = \sum_{n \leq 0} x_{i,n} t^{-n}$ (equivalently, $P_{i,n} = \frac{1}{n!} T^n P_i$). The scheme of infinite type $JZ = \operatorname{Spec} A_\infty$ is called the *jet scheme* of Z. It is easy to check that for any $\mathbb{C}$–algebra R, we have $\operatorname{Hom}(\operatorname{Spec} R[[t]], Z) = \operatorname{Hom}(\operatorname{Spec} R, JZ)$. Therefore JZ represents the space of maps $D \to Z$.

Note that A_∞ is a commutative algebra with a derivation $T = -\partial_t$, so it is a commutative vertex algebra. Moreover, it can be given the structure of a quasi-conformal vertex algebra with the action of $\operatorname{Der} \mathcal{O}$ defined by the formulas

$$L_m \mapsto \sum_{i=1}^N \sum_{n \leq 0} (m - n + 1) x_{i,n} \frac{\partial}{\partial x_{i,n-m}}.$$

This action preserves the ideal generated by the $P_{i,n}$'s and hence descends to A_∞. The action of $\operatorname{Der}_0 \mathcal{O}$ can be exponentiated to an action of $\operatorname{Aut} \mathcal{O}$, and the Aut_x–twist $A_{\infty,x}$ of A_∞ is nothing but the algebra of functions on the scheme of maps $D_x \to Z$.

Let $\mathcal{A}_\infty$ be the vector bundle on a curve X which is the Aut_X–twist of A_∞. Then the scheme $\operatorname{Spec} \mathcal{A}_\infty$ gets identified with the scheme of jets of maps $X \to Z$, whose fiber at $x \in X$ consists of maps $D_x \to Z$. This scheme carries a natural connection. The scheme of horizontal sections $\Gamma_\nabla(X, \operatorname{Spec} \mathcal{A}_\infty)$ gets identified with the scheme of sections of the trivial bundle on X with the fiber $Z = \operatorname{Spec} A$. Since Z is an affine variety, this scheme is isomorphic to Z for any complete curve X. Hence by Proposition 9.4.1, the algebra of coinvariants $H(X, x, A_\infty)$ is nothing but the algebra A we started with.

Another example of the space of coinvariants of a commutative vertex algebra will be considered in § 18.4.7.

9.5. Twisted version of conformal blocks

In § 7.2.8 we discussed the twists of a vertex algebra V by torsors of groups of internal symmetries other than $\operatorname{Aut} \mathcal{O}$. We have shown that such twists possess an analogue of the section $\mathcal{Y}_x$. We can use this section to define the corresponding twisted versions of the spaces of conformal blocks. We demonstrate this explicitly in the case when the group of internal symmetries is $\operatorname{Aut} \mathcal{O} \ltimes G(\mathcal{O})$ (arising from an affine Kac–Moody algebra) or $\operatorname{Aut} \mathcal{O} \ltimes \mathcal{O}^\times$ (arising from the Heisenberg algebra). The general case follows along the same lines.

9.5.1. Twisted conformal blocks for affine algebras. Let V be a quasi-conformal vertex algebra with a compatible $\hat{\mathfrak{g}}$–structure (see Definition 7.1.1). Then V carries an action of the Lie algebra $\operatorname{Der} \mathcal{O} \ltimes \hat{\mathfrak{g}}$. Suppose that the action of its Lie subalgebra $\operatorname{Der}_0 \mathcal{O} \ltimes \mathfrak{g}(\mathcal{O})$ can be exponentiated to an action of the group $\operatorname{Aut} \mathcal{O} \ltimes G(\mathcal{O})$ on V. Then we say that this group acts on V by internal symmetries. To a principal G–bundle $\mathcal{P}$ on a smooth projective curve X we attach a principal $\operatorname{Aut} \mathcal{O} \ltimes G(\mathcal{O})$–bundle $\hat{\mathcal{P}}$ over X introduced in § 7.1.4. Denote by $V^{\mathcal{P}}$ the $\hat{\mathcal{P}}$–twist of V. Then according to Theorem 7.1.6 the vertex operation Y on V gives rise to a canonical $\operatorname{End} V_x^{\mathcal{P}}$–valued section $\mathcal{Y}_x^{\mathcal{P}}$ of $(V^{\mathcal{P}})^* |_{D_x^\times}$.

9.5.2. Definition. A linear functional on $V_x^{\mathcal{P}}$ is called a *twisted conformal block* if the section of $(V^{\mathcal{P}})^*$ over $D_x^\times$ defined by the formula $\langle \varphi, \mathcal{Y}_x^{\mathcal{P}} \cdot A \rangle$ can be extended to a regular section of $(V^{\hat{\mathcal{P}}})^*$ on $X \setminus x$ for all $A \in V_x^{\mathcal{P}}$. We denote the vector space spanned by the twisted conformal blocks by $C^{\mathcal{P}}(X, x, V)$.

9.5.3. The dual definition. According to § 17.1.10, the vector bundle $\mathcal{V}^{\mathcal{P}}$ carries a canonical flat connection. Define the sheaf $h(\mathcal{V}^{\mathcal{P}} \otimes \Omega)$ of its zeroth de Rham cohomology as in § 9.2.2. By Theorem 19.4.9, this is a sheaf of Lie algebras on X. In particular, the Lie algebra $U^{\mathcal{P}}(\mathcal{V}_x) \overset{\text{def}}{=} \Gamma(D_x^\times, h(\mathcal{V}^{\mathcal{P}} \otimes \Omega))$ is isomorphic to the Lie algebra $U(V)$ (see § 4.1.1).

The section $\mathcal{Y}_x^{\mathcal{P}}$ defines a linear map

$$\mathcal{Y}_x^{\mathcal{P}\vee} : \Gamma(D_x^\times, \mathcal{V}^{\mathcal{P}} \otimes \Omega) \to \operatorname{End} \mathcal{V}_x^{\mathcal{P}}$$

(see Theorem 7.1.6), which factors through $U^{\mathcal{P}}(\mathcal{V}_x)$. Given an open subset $\Sigma \subset X$, let $U_\Sigma^{\mathcal{P}}(V) = \Gamma(\Sigma, h(\mathcal{V}^{\mathcal{P}} \otimes \Omega))$. The restriction of sections gives us a map $U_\Sigma^{\mathcal{P}}(V) \to U(\mathcal{V}_x)$. Its image is a Lie subalgebra of $U^{\mathcal{P}}(\mathcal{V}_x)$, denoted by $U_\Sigma^{\mathcal{P}}(\mathcal{V}_x)$, which acts on $\mathcal{V}_x^{\mathcal{P}}$.

By the Strong Residue Theorem 9.2.9, $C^{\mathcal{P}}(X, x, V)$ is the space of $U_{X \backslash x}^{\mathcal{P}}(\mathcal{V}_x)$–invariant functionals on $\mathcal{V}_x^{\mathcal{P}}$. Its dual space is the space of coinvariants of $\mathcal{V}_x^{\mathcal{P}}$ under the action of $U_{X \backslash x}^{\mathcal{P}}(\mathcal{V}_x)$, denoted by $H^{\mathcal{P}}(X, x, V)$.

9.5.4. Example. Let $\mathfrak{g}_{\mathcal{P}}$ be the vector bundle $\mathcal{P} \underset{G}{\times} \mathfrak{g}$. The space $\mathfrak{g}_{\text{out}}^{\mathcal{P}}(x)$ of its sections over $X \backslash x$ is a Lie algebra which acts naturally on $\mathcal{V}_x^{\mathcal{P}}$ (note that $\mathfrak{g}_{\text{out}}^{\mathcal{P}}(x)$ is a Lie subalgebra of $U_{X \backslash x}^{\mathcal{P}}(\mathcal{V}_x)$).

In the case when $V = V_k(\mathfrak{g})$, the space of conformal blocks $C^{\mathcal{P}}(X, x, V_k(\mathfrak{g}))$ is isomorphic to the space of $\mathfrak{g}_{\text{out}}^{\mathcal{P}}(x)$–invariant functionals on $V_k(\mathfrak{g})_x^{\mathcal{P}}$, and

$$H^{\mathcal{P}}(X, x, V_k(\mathfrak{g})) = V_k(\mathfrak{g})_x^{\mathcal{P}} / \mathfrak{g}_{\text{out}}^{\mathcal{P}}(x) \cdot V_k(\mathfrak{g})_x^{\mathcal{P}}.$$

9.5.5. Twisted conformal blocks for the Heisenberg algebra. Let V be a vertex algebra equipped with a non-zero homomorphism (hence necessarily an embedding) $\pi \to V$. Then V carries an action of the Lie algebra $Vir \ltimes \mathcal{H}$. Suppose that the action of its Lie subalgebra $\operatorname{Der}_0 \mathcal{O} \ltimes \mathcal{O}$ can be exponentiated to an action of the group $\operatorname{Aut} \mathcal{O} \ltimes \mathcal{O}^\times$. Given a line bundle $\mathcal{L}$ on X, let $\widehat{\mathcal{L}}$ be the principal $\operatorname{Aut} \mathcal{O} \ltimes \mathcal{O}^\times$–bundle over X, whose fiber at x consists of pairs (z, s), where z is a formal coordinate at x and s is a trivialization of $\mathcal{L}|_{D_x}$. Denote by $\mathcal{V}^{\mathcal{L}}$ the $\widehat{\mathcal{L}}$–twist of V. Then the vertex operation Y on V gives rise to a canonical $\operatorname{End} \mathcal{V}_x^{\mathcal{L}}$–valued section $\mathcal{Y}_x^{\mathcal{L}}$ of $(\mathcal{V}^{\mathcal{L}})^*|_{D_x^\times}$.

Now for any line bundle $\mathcal{L}$ on X we define the space of twisted conformal blocks $C^{\mathcal{L}}(X, x, V)$ as the space of linear functionals $\varphi : \mathcal{V}_x^{\mathcal{L}} \to \mathbb{C}$ such that $\langle \varphi, \mathcal{Y}_x^{\mathcal{L}} \cdot A \rangle$ can be extended to a regular section of $(\mathcal{V}^{\mathcal{L}})^*$ on $X \backslash x$ for all $A \in \mathcal{V}_x^{\mathcal{L}}$. Its dual space is the space of coinvariants $H^{\mathcal{L}}(X, x, V)$.

9.6. Appendix. Proof of Proposition 9.3.2

We will prove the equivalent statement that the corresponding spaces of coinvariants $\widetilde{H}(X, x, \pi)$ and $\widetilde{H}(X, x, y, \pi \otimes \pi)$ are isomorphic.

To simplify matters, we choose coordinates t_x and t_y at x and y and identify Π_x and Π_y with π using these coordinates. Then $\widetilde{H}(X, x, \pi)$ (resp., $\widetilde{H}(X, x, y, \pi \otimes \pi)$) is identified with the 0th homology of the Lie algebra $\mathcal{H}_{\text{out}}(x)$ (resp., $\mathcal{H}_{\text{out}}(x, y)$) with coefficients in π (resp., $\pi \otimes \pi$, with the action of $\mathcal{H}_{\text{out}}(x, y)$ defined in the same way as in Lemma 9.1.5).

Consider the principal part at y map:

$$\mu : \mathbb{C}((t_x)) \oplus \mathbb{C}((t_y)) \twoheadrightarrow \mathbb{C}((t_y)) / \mathbb{C}[[t_y]] \simeq t_y^{-1} \mathbb{C}[t_y^{-1}].$$

The restriction of μ to $\mathcal{H}_{\text{out}}(x, y)$ is surjective, because by Riemann-Roch theorem there exists a meromorphic function on X with poles only at x and y with any given principal part at y. On the other hand, the kernel of μ equals $\mathcal{H}_{\text{out}}(x)$. Thus we have an exact sequence

$$(9.6.1) \qquad 0 \to \mathcal{H}_{\text{out}}(x) \to \mathcal{H}_{\text{out}}(x, y) \xrightarrow{\mu} t_y^{-1}\mathbb{C}[t_y^{-1}] \to 0.$$

The homology $H_i(\mathfrak{g}, M)$ of a Lie algebra $\mathfrak{g}$ with coefficients in a $\mathfrak{g}$–module M is computed via the Chevalley complex $M \otimes \bigwedge^\bullet(\mathfrak{g})$ (see § A.4). The 0th homology $H_0(\mathfrak{g}, M)$ is the space of coinvariants $M/\mathfrak{g} \cdot M$. In our case the Lie algebra is $\mathfrak{g} = \mathcal{H}_{\text{out}}(x, y)$, and the Chevalley complex is

$$C^\bullet = \pi_x \otimes \pi_y \otimes \bigwedge{}^\bullet(\mathcal{H}_{\text{out}}(x, y)),$$

with the differential $d : C^i \to C^{i-1}$ given by the formula

$$d = \sum f_i \otimes \psi_i^*.$$

Here $\{f_i\}$ is a basis in $\mathcal{H}_{\text{out}}(x, y)$ and $\{\psi_i^*\}$ is the dual basis of $\mathcal{H}_{\text{out}}(x, y)^*$ acting on $\bigwedge^\bullet \mathcal{H}_{\text{out}}(x, y)$ by contraction.

Choose pull–backs $z_n, n < 0$, of $t_y^n, n < 0$, in $\mathcal{H}_{\text{out}}(x, y)$ under μ. Because of the exactness of the sequence (9.6.1), we can choose a basis $\{f_i\}$ in $\mathcal{H}_{\text{out}}(x, y)$, which is a union of $\{z_n\}_{n<0}$ and a basis of $\mathcal{H}_{\text{out}}(x)$. In this basis we may decompose

$$d = d_x + \sum z_n \otimes \phi_n^*,$$

where d_x is the differential for $\mathcal{H}_{\text{out}}(x)$ and ϕ_n^* denotes the element of the dual basis to $\{f_i\}$ corresponding to z_n.

We need to show that the homologies of this complex are isomorphic to the homologies of the complex $\pi \otimes \bigwedge^\bullet(\mathcal{H}_{\text{out}}(x))$. The following is a standard "Shapiro's lemma" type argument.

Introduce an increasing filtration on π, letting $\pi^{(m)}$ be the span of all monomials of order less than or equal to m in $b_n, n < 0$. Now introduce a filtration $\{F_i\}$ on the Chevalley complex $C^\bullet$ by setting

$$F_i = \text{Span}\{A \otimes B \otimes C | B \in \pi^{(m)}, C \in \bigwedge{}^{i-m}(\mathcal{H}_{\text{out}}(x, y))\}.$$

Our differential preserves this filtration, because each element of $\mathcal{H}_{\text{out}}(x, y)$ acts either on the factor π corresponding to the point y by mapping $\pi^{(m)}$ to $\pi^{(m+1)}$, or on the factor π corresponding to the point x. It is clear that this filtration is preserved by the differential.

Consider now the spectral sequence associated to the filtered complex $C^\bullet$. The zeroth term E^0 is the associated graded space of the Chevalley complex, isomorphic to

$$(\pi_y \otimes \bigwedge{}^\bullet(\phi_n^*)_{n<0}) \otimes (\pi_x \otimes \bigwedge{}^\bullet(\mathcal{H}_{\text{out}}(x)).$$

The zeroth differential acts along the first factor of the above decomposition and is given by the formula

$$d^0 = \sum_{n<0} b_n \otimes \phi_n^*,$$

because on the graded module the operator z_n acts as $b_n, n < 0$. But π is isomorphic to the symmetric algebra with generators $b_n, n < 0$, and so our differential d^0 is simply the Koszul differential for this symmetric algebra.

Here by a Koszul complex we understand the tensor product of the symmetric algebra $\mathrm{Sym}(x_i)_{i \in I}$ and the exterior algebra $\bigwedge^\bullet(\phi_i)_{i \in I}$, with the differential given by the formula $\sum_{i \in I} x_i \otimes \phi_i^*$, where ϕ_i^* denotes the operator of contraction of ϕ_i. It is well-known that the zeroth homology of this complex is isomorphic to $\mathbb{C}$ (spanned by $1 \otimes 1$), and all other homologies vanish. Therefore all positive homologies of our differential d^0 vanish, while the zeroth homology is represented by the space $\pi_x \otimes \bigwedge^\bullet(\mathcal{H}_{\mathrm{out}}(x))$. Hence the first term E^1 of our spectral sequence is isomorphic, as a vector space, to the Chevalley complex of the homology of $\mathcal{H}_{\mathrm{out}}(x)$ with coefficients in π.

Moreover, $\mathcal{H}_{\mathrm{out}}(x)$ consists of functions that are regular at y, and hence act on π_y as linear combinations of $b_n, n \geq 0$, killing the vacuum vector 1 of π_y. Therefore the differential of E^1 coincides with the standard differential of the Chevalley complex for $(\mathcal{H}_{\mathrm{out}}(x), \pi)$. Since E^1 consists of a single row, we obtain that the homologies of E^1 are isomorphic to the homologies of $\mathcal{H}_{\mathrm{out}}(x, y)$ with coefficients in $\pi \otimes \pi$. Thus, we obtain the desired isomorphism

$$H_i(\mathcal{H}_{\mathrm{out}}(x, y), \pi \otimes \pi) \simeq H_i(\mathcal{H}_{\mathrm{out}}(x), \pi).$$

By the above construction, the dual map on conformal blocks

$$\widetilde{C}(X, x, y, \pi \otimes \pi) \to \widetilde{C}(X, x, \pi)$$

can be described as follows. Given a conformal block $\widetilde{\varphi} \in \widetilde{C}(X, x, y, \pi \otimes \pi)$, i.e., a linear map

$$\widetilde{\varphi} : \Pi_x \otimes \Pi_y \to \mathbb{C}$$

which is $\mathcal{H}_{\mathrm{out}}(x, y)$–invariant, we produce the conformal block for Π_x by evaluating $\widetilde{\varphi}$ at the vacuum vector in Π_y:

$$\varphi = \widetilde{\varphi}|_{\Pi_x \otimes \mathbf{1}}.$$

Proposition 9.3.2 is now proved.

9.7. Bibliographical notes

In the case of vertex algebras associated to Lie algebras (such as the Heisenberg, affine Kac-Moody, Virasoro) the definition of the spaces of coinvariants and conformal blocks is well-known (see, e.g., [**BS, BFM, Fe2, FFu2, TUY, Fel**]).

The definition of conformal blocks for general vertex algebras given in this chapter is new; it is due to E. Frenkel. In the framework of the theory of chiral algebras, an alternative definition has been given by A. Beilinson and V. Drinfeld [**BD4**] (see also [**Gai**]). The proof of Theorem 9.3.3 on the comparison of two definitions of conformal blocks is due to E. Frenkel.

Proposition 9.4.1 is implicit in [**BD3**]. The idea of the example of commutative vertex algebra from § 9.4.4 is borrowed from [**BD3**]. More information on jet schemes may be found in [**Bu**].

For the definition of the module of differentials and the residue theorem, the reader may consult [**Ha, Ser, Ta**]. An alternative definition of the residue has been given by J. Tate [**Ta**] (see also [**ADK**]). The proof of the strong residue theorem can be found in [**Ta**].

CHAPTER 10

Conformal Blocks II

In this chapter we continue the study of conformal blocks. We generalize the definition of the space of conformal blocks to the situation when we have several points on a curve, with insertions of V–modules. We then discuss various properties of the functors of conformal blocks and coinvariants. Finally, we introduce a "functional realization" point of view on the spaces of conformal blocks. This point of view allows us to make contact with the chiral correlation functions studied in conformal field theory. We also show that the vertex operation Y on a vertex algebra V may be interpreted in terms of the space of conformal blocks associated to $\mathbb{P}^1$ with three points.

10.1. Multiple points

The definition of the space of conformal blocks $C(X, x, V)$ for a general conformal vertex algebra V can be generalized to the case of multiple points at which we insert arbitrary conformal V–modules. Assume first that the eigenvalues of the operator L_0 on all of these modules are integers. To each such module M we attach a vector bundle $\mathcal{M}$ on any smooth curve X, which is the Aut_X–twist of M. By Theorem 7.3.7, we have an $\operatorname{End} \mathcal{M}_x$–valued section $\mathcal{Y}_{M,x}$ of $\mathcal{V}^*|_{D_x^\times}$.

10.1.1. Definition. Let V be a conformal vertex algebra, and $M_1, \ldots, M_n$ conformal V–modules attached to distinct points $x_1, \ldots x_n$ of a smooth projective curve X. The corresponding space of conformal blocks $C_V(X, (x_i), (M_i))_{i=1}^n$ is by definition the space of linear functionals φ on $\bigotimes_{i=1}^n \mathcal{M}_{i,x_i}$ such that for any $A_i \in \mathcal{M}_{i,x_i}, i = 1, \ldots, n$, the sections of $\mathcal{V}^*$ over $D_{x_i}^\times$ defined by the formula

$$\langle \varphi, A_1 \otimes \cdots \otimes (\mathcal{Y}_{M_i,x_i} \cdot A_i) \otimes \cdots \otimes A_n \rangle, \qquad i = 1, \ldots, n,$$

can be extended to the *same* regular section of $\mathcal{V}^*$ over $X \backslash \{x_1, \ldots, x_n\}$.

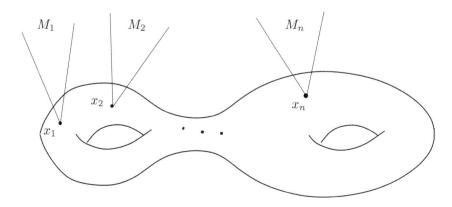

Informally, one can say that the local sections of $\mathcal{V}^*$ over the discs around the points x_i obtained by acting with vertex operators at those points can be "glued together" into a single meromorphic section of $\mathcal{V}^*$.

10.1.2. The dual definition. Recall the Lie algebra $U_\Sigma(V) = \Gamma(\Sigma, h(\mathcal{V}^r))$ defined in § 9.2.5 for an open subset Σ of X. Let $\Sigma = X \backslash \{x_1, \ldots, x_n\}$. The canonical map $\Gamma(\Sigma, h(\mathcal{V}^r)) \to \bigoplus_{i=1}^n \Gamma(D^\times_{x_i}, h(\mathcal{V}^r))$ gives us a Lie algebra homomorphism

$$U_\Sigma(V) \to \bigoplus_{i=1}^n U(\mathcal{V}_{x_i}).$$

In Theorem 7.3.7 we constructed an $\operatorname{End} \mathcal{M}_{i,x_i}$–valued section $\mathcal{Y}_{M,x_i}$ of $\mathcal{V}^*|_{D^\times_{x_i}}$ and a homomorphism of Lie algebras

$$\mathcal{Y}^\vee_{M,x_i} : U(\mathcal{V}_{x_i}) \to \operatorname{End} \mathcal{M}_{i,x_i},$$

which sends $\mu \in U(\mathcal{V}_{x_i})$ to $\operatorname{Res}_{x_i} \langle \mathcal{Y}_{M,x_i}, \mu \rangle$. Hence we obtain that the Lie algebra $\bigoplus_{i=1}^n U(\mathcal{V}_{x_i})$ acts on $\bigotimes_{i=1}^n \mathcal{M}_{i,x_i}$. Therefore $U_\Sigma(V)$ also acts on $\bigotimes_{i=1}^n \mathcal{M}_{i,x_i}$.

Using the Strong Residue Theorem (see Remark 9.2.10), we obtain that

$$C_V(X, (x_i), (M_i))_{i=1}^n = \operatorname{Hom}_{U_\Sigma(V)} \left(\bigotimes_{i=1}^n \mathcal{M}_{i,x_i}, \mathbb{C} \right).$$

We also define the space of coinvariants $H_V(X, (x_i), (M_i))_{i=1}^n$ as the quotient of $\bigotimes_{i=1}^n \mathcal{M}_{i,x_i}$ by the action of $U_\Sigma(V)$.

10.1.3. The case of non-integral gradation. Now suppose that the eigenvalues of L_0 on the modules M_i are not integers. Then each M_i carries an action of the subgroup $\operatorname{Aut}_+ \mathcal{O}$ of $\operatorname{Aut} \mathcal{O}$. In order to define the space of conformal blocks we need to choose non-zero tangent vectors τ_i at x_i. Let $\mathcal{M}_{(x_i, \tau_i)}$ be the $\operatorname{Aut}_{x_i, \tau_i}$–twist of M_i defined in § 7.3.9. According to Theorem 7.3.10, we have a canonical $\operatorname{End} \mathcal{M}_{(x_i, \tau_i)}$–valued section $\mathcal{Y}_{M_i,(x_i,\tau_i)}$ of $\mathcal{V}^*|_{D^\times_x}$. We then define the space of conformal blocks $C_V(X, (x_i), (\tau_i), (M_i))_{i=1}^n$ in the same way as in Definition 10.1.1 using the sections $\mathcal{Y}_{M_i,(x_i,\tau_i)}$. The dual space is the space $H_V(X, (x_i), (\tau_i), (M_i))_{i=1}^n$ of coinvariants, which is defined as in § 10.1.2.

10.1.4. Modular functor. For any conformal vertex algebra V and a pointed algebraic curve equipped with a non-zero tangent vector at each point, we obtain a functor from the Cartesian product of n copies of the category of conformal V–modules to the category of vector spaces,

$$(M_i)_{i=1}^n \mapsto C_V(X, (x_i), (M_i))_{i=1}^n$$

(or, alternatively, $H_V(X, (x_i), (M_i))_{i=1}^n$). This is a version of the *modular functor* corresponding to V (see, e.g., [**Seg2, Fe2, Gaw, Z2, BaKi**]). It is known that in the case of rational vertex algebras $L_k(\mathfrak{g})$ or $L_{c(p,q)}$ the spaces of conformal blocks are finite-dimensional (see [**FFu2, BFM, TUY**]). Moreover, the functor can be extended to the category of pointed stable curves (i.e., curves with nodal singularities and finite groups of automorphisms) with module insertions. It then satisfies a factorization property, which expresses the space of conformal blocks associated to a singular curve with a double point in terms of conformal blocks associated to its normalization [**BFM, TUY**]. The same is expected to be true for general rational vertex algebras.

10.1.5. Twisted conformal blocks. The definition of the twisted space of conformal blocks given in § 9.5 also has a multiple point generalization.

Let V be a conformal vertex algebra with a compatible $\widehat{\mathfrak{g}}$–structure (see Definition 7.1.1). Then V and any V–module M are modules over the Lie algebra $Vir \ltimes \widehat{\mathfrak{g}}$, and in particular over its Lie subalgebra $\mathrm{Der}_0 \, \mathcal{O} \ltimes \mathfrak{g}(\mathcal{O})$.

Let M be a V–module such that as a module over $\mathfrak{g} \subset \mathfrak{g}(\mathcal{O})$ it is a direct sum of finite-dimensional representations and the eigenvalues of L_0 are integers. Then the action of the Lie algebra $\mathrm{Der}_0 \, \mathcal{O} \ltimes \mathfrak{g}(\mathcal{O})$ on M can be exponentiated to an action of the group $\mathrm{Aut} \, \mathcal{O} \ltimes G(\mathcal{O})$. Recall that for any principal G–bundle $\mathcal{P}$ on X we have defined the $\mathrm{Aut} \, \mathcal{O} \ltimes G(\mathcal{O})$–bundle $\widehat{\mathcal{P}}$ and the $\widehat{\mathcal{P}}$–twist $M^{\mathcal{P}}$ of M on X (see Definition 7.1.5). For each point $x \in X$, we have a canonical $\mathrm{End}\, M_x^{\mathcal{P}}$–valued section $\mathcal{Y}_{M,x}^{\mathcal{P}}$ of $(\mathcal{V}^{\mathcal{P}})^*|_{D_x^\times}$ (see Theorem 7.1.6).

Let $M_1, \ldots, M_n$ be V–modules satisfying the above conditions, and $x_1, \ldots, x_n$ a collection of distinct points of X.

10.1.6. Definition. The space $C_V^{\mathcal{P}}(X, (x_i), (M_i))_{i=1}^n$ of $\mathcal{P}$–twisted conformal blocks is the space of linear functionals φ on $\bigotimes_{i=1}^n M_{i,x_i}^{\mathcal{P}}$ such that for any $A_i \in M_{i,x_i}^{\mathcal{P}}, i = 1, \ldots, n$, the sections of $(\mathcal{V}^{\mathcal{P}})^*$ over $D_{x_i}^\times$ defined by the formula

$$\langle \varphi, A_1 \otimes \cdots \otimes (\mathcal{Y}_{M_i,x_i}^{\mathcal{P}} \cdot A_i) \otimes \cdots \otimes A_n \rangle, \qquad i = 1, \ldots, n,$$

can be extended to the same regular section of $(\mathcal{V}^{\mathcal{P}})^*$ over $X \backslash \{x_1, \ldots, x_n\}$.

The space of coinvariants $H_V^{\mathcal{P}}(X, (x_i), (M_i))_{i=1}^n$ is the quotient of $\bigotimes_{i=1}^n M_{i,x_i}^{\mathcal{P}}$ by the action of the Lie algebra $U_\Sigma^{\mathcal{P}}(V)$ that was introduced in § 9.5.3, where $\Sigma = X \backslash \{x_1, \ldots, x_n\}$.

10.1.7. General case. In general, a V–module M carries an action of the group $\mathrm{Aut}_+ \, \mathcal{O} \ltimes G^+(\mathcal{O})$. Let $\mathcal{P}$ be a G–bundle on X. Given a point x of X, a non-zero tangent vector to X at x and an isomorphism $\phi : G \simeq \mathcal{P}_x$, we have the twist $M_{(x,\tau)}^{\mathcal{P},\phi}$ of M by the $\mathrm{Aut}_+ \, \mathcal{O} \ltimes G^+(\mathcal{O})$–torsor $\widehat{\mathcal{P}}_{(x,\tau)}^\phi$ (see § 7.3.11). Then we have an $\mathrm{End}\, M_{(x,\tau)}^{\mathcal{P},\phi}$–valued section $\mathcal{Y}_{M,(x,\tau)}^{\mathcal{P},\phi}$ of $(\mathcal{V}^{\mathcal{P}})^*|_{D_x}$. Using these sections, we assign to an n–tuple of V–modules $M_1, \ldots, M_n$ the twisted spaces of conformal blocks and coinvariants in the same way as above.

Similar definitions may also be given if we replace the affine algebra $\widehat{\mathfrak{g}}$ by the Heisenberg Lie algebra $\mathcal{H}$.

10.2. Functoriality of conformal blocks

In this section we analyze functorial properties of conformal blocks.

First of all, observe that if $M_i' \to M_i, i = 1, \ldots, n$, are homomorphisms of conformal V–submodules, then we have a linear map

$$C_V(X, (x_i), (M_i))_{i=1}^n \to C_V(X, (x_i), (M_i'))_{i=1}^n.$$

Now let $V \to W$ be a homomorphism of conformal vertex algebras. Denote by $\mathcal{V}$ and $\mathcal{W}$ the vector bundles on a smooth curve X attached to V and W, respectively. We have natural maps of bundles $\mathcal{V} \to \mathcal{W}$ and $\mathcal{W}^* \to \mathcal{V}^*$.

Let M be a conformal W–module. Then it is automatically a conformal V–module. By Theorem 7.3.7, the action of W on M gives rise to the $\mathrm{End}\, M_x$–valued section $\mathcal{Y}_{M,x}^W$ of $\mathcal{W}^*|_{D_x^\times}$. The projection of this section under the map $\mathcal{W}^* \to \mathcal{V}^*$ is an $\operatorname{End}\mathcal{M}_x$–valued section of $\mathcal{V}^*|_{D_x^\times}$. This section is nothing but the section $\mathcal{Y}_{M,x}^V$ associated to M, considered as a V–module.

If $M_1, \ldots, M_n$ are conformal W–modules and $\varphi \in C_W(X, (x_i), (M_i))_{i=1}^n$, then according to the definition, for any $A_i \in \mathcal{M}_{i,x_i}, i = 1, \ldots, n$, the sections

$$(10.2.1) \qquad \langle \varphi, A_1 \otimes \cdots \otimes (\mathcal{Y}_{M_i,x_i}^W \cdot A_i) \otimes \cdots \otimes A_n \rangle, \qquad i = 1, \ldots, n,$$

of $\mathcal{W}^*|_{D_{x_i}^\times}$ can be extended to a regular section $\varphi_{A_1,\ldots,A_n}$ of $\mathcal{W}^*$ over $X \backslash \{x_1, \ldots, x_n\}$. The projection of $\varphi_{A_1,\ldots,A_n}$ onto $\mathcal{V}^*$ gives us a section of $\mathcal{V}^*$ over $X \backslash \{x_1, \ldots, x_n\}$. By construction, the restriction of this section to D_{x_i} is equal to

$$\langle \varphi, A_1 \otimes \cdots \otimes (\mathcal{Y}_{M_i,x_i}^V \cdot A_i) \otimes \cdots \otimes A_n \rangle,$$

Hence the functional $\varphi : \bigotimes_{i=1}^n \mathcal{M}_{x_i} \to \mathbb{C}$ is automatically a conformal block for V. Thus, we obtain

10.2.1. Lemma. *Let $V \to W$ be a homomorphism of conformal vertex algebras. Then for any n–tuple of conformal W–modules there is an embedding*

$$C_W(X, (x_i), (M_i))_{i=1}^n \hookrightarrow C_V(X, (x_i), (M_i))_{i=1}^n.$$

10.2.2. Example. This result is important in applications, as illustrated by the following example (we also use it in Chapter 14, where we obtain conformal blocks of an affine Lie algebra from those of a Heisenberg Lie algebra).

By Lemma 3.4.5, for a conformal vertex algebra V we have a homomorphism of vertex algebras $\mathrm{Vir}_c \to V$. Lemma 10.2.1 then implies that there is an embedding

$$C_V(X, (x_i), (M_i))_{i=1}^n \hookrightarrow C_{\mathrm{Vir}_c}(X, (x_i), (M_i))_{i=1}^n.$$

The space on the right hand side is by definition the space of $U_\Sigma(\mathrm{Vir}_c)$–invariant functionals on $\bigotimes_{i=1}^n \mathcal{M}_{x_i}$, where $\Sigma = X \backslash \{x_1, \ldots, x_n\}$ (see § 10.1.2). According to the discussion of § 19.6.5, the Lie algebra $\mathrm{Vect}(X \backslash \{x_1, \ldots, x_n\})$ of meromorphic vector fields on X with poles only at $x_1, \ldots, x_n$ acts on $\bigotimes_{i=1}^n \mathcal{M}_{x_i}$. Therefore we obtain that any V–conformal block is automatically invariant with respect to the Lie algebra $\mathrm{Vect}(X \backslash \{x_1, \ldots, x_n\})$.

For instance, let us take as V the Kac–Moody vertex algebra $V_k(\mathfrak{g})$ with $k \neq -h^\vee$. Then the space $C_{V_k(\mathfrak{g})}(X, (x_i), (M_i))_{i=1}^n$ is equal to the space of $\mathfrak{g} \otimes \mathbb{C}[X \backslash \{x_1, \ldots, x_n\}]$–invariant functionals on $\bigotimes_{i=1}^n \mathcal{M}_{x_i}$ (see Remark 9.3.10). Thus we obtain that any $\mathfrak{g} \otimes \mathbb{C}[X \backslash \{x_1, \ldots, x_n\}]$–invariant functional on $\bigotimes_{i=1}^n \mathcal{M}_{x_i}$ is automatically $\mathrm{Vect}(X \backslash \{x_1, \ldots, x_n\})$–invariant.

This statement is non-trivial, because there is no embedding of the Lie algebra $\mathrm{Vect}(X \backslash \{x_1, \ldots, x_n\})$ into the Lie algebra $\mathfrak{g} \otimes \mathbb{C}[X \backslash \{x_1, \ldots, x_n\}]$ or its universal enveloping algebra. Indeed, the elements of the Virasoro algebra are expressed quadratically in terms of the generators of the Kac–Moody algebra by the Sugawara formula. But this only gives us an embedding of the Virasoro algebra into the big Lie algebra $U(V_k(\mathfrak{g}))$ of all Fourier coefficients of vertex operators from $V_k(\mathfrak{g})$. Hence $\mathrm{Vect}(X \backslash \{x_1, \ldots, x_n\})$ (and $U_\Sigma(\mathrm{Vir}_c)$) embeds into the Lie algebra $U_\Sigma(V_k(\mathfrak{g}))$, where $\Sigma = X \backslash \{x_1, \ldots, x_n\}$. However, the elements of $\mathrm{Vect}(X \backslash \{x_1, \ldots, x_n\})$ inside $U_\Sigma(V_k(\mathfrak{g}))$ cannot be obtained from the elements of $\mathfrak{g} \otimes \mathbb{C}[X \backslash \{x_1, \ldots, x_n\}]$.

10.3. Chiral correlation functions

Assume for simplicity that the eigenvalues of the Virasoro operator $L_0^{M_i}$ on the modules $M_i, i = 1, \ldots, n$, are integers. It follows from § 7.3.8 that to the modules $M_1, \ldots, M_n$ we can then assign the vector bundle $\mathcal{M}_1 \boxtimes \cdots \boxtimes \mathcal{M}_n$ with a flat connection on X^n. By construction, the space of conformal blocks $C_V(X, (x_i), (M_i))_{i=1}^n$ is a subspace of the fiber of $(\mathcal{M}_1 \boxtimes \cdots \boxtimes \mathcal{M}_n)^*$ at $(x_1, \ldots, x_n)$. One can show that for varying points $x_1, \ldots, x_n$ the spaces of conformal blocks may be organized into a sheaf of conformal blocks on $\overset{\circ}{X}{}^n = X^n \backslash \Delta$, which is a subsheaf of $(\mathcal{M}_1 \boxtimes \cdots \boxtimes \mathcal{M}_n)^*$ restricted to $\overset{\circ}{X}{}^n$ (if the eigenvalues of $L_0^{M_i}$ are not integers, we obtain instead a sheaf on the space of n-tuples $(x_i, \tau_i)_{i=1,\ldots,n}$, where $x_i \in X$ and τ_i is a non-zero tangent vector at x_i). Moreover, the flat connection on $(\mathcal{M}_1 \boxtimes \cdots \boxtimes \mathcal{M}_n)^*$ preserves this subsheaf. In other words, we can identify the spaces of conformal blocks at nearby points on $\overset{\circ}{X}{}^n$. We will not prove these results here, but they follow from the general formalism that will be developed in Chapter 17.

Let us choose $\varphi \in C_V(X, (x_i), (M_i))_{i=1}^n$ and let $\varphi_{y_1,\ldots,y_n}$ be the corresponding horizontal section of the sheaf of conformal blocks in the neighborhood of $(x_1, \ldots, x_n) \in \overset{\circ}{X}{}^n$ (depending on $y_1, \ldots, y_n$). Choosing local sections $A_i(y_i)$ of the bundles $\mathcal{M}_i$ near x_i for $i = 1, \ldots, n$, and evaluating $\varphi_{y_1,\ldots,y_n}$ on them, we obtain functions $\langle \varphi_{y_1,\ldots,y_n}, A_1 \otimes \ldots \otimes A_n \rangle$, which physicists call *chiral correlation functions* and denote by

$$\langle A_1(y_1) \cdots A_n(y_n) \rangle.$$

The most interesting problem is understanding the behavior of these sections near the diagonals. We will now study this question in the simplest situation when all V–modules M_i are chosen to be V.

First we need to describe the space $C_V(X, (x_i), (V))_{i=1}^n$.

10.3.1. Theorem. *There is a canonical isomorphism*

$$C_V(X, (x_1, \ldots, x_n, y), (M_1, \ldots, M_n, V)) \simeq C_V(X, (x_1, \ldots, x_n), (M_1, \ldots, M_n)),$$

given by restricting φ to $\mathcal{M}_{1,x_1} \otimes \ldots \otimes \mathcal{M}_{n,x_n} \otimes |0\rangle$.

10.3.2. Proof. We will prove the theorem in the case $n = 1$, $M_1 = V$. The extension of our argument to the general case is straightforward.

We will construct a linear map

$$C(X, x; V) \to C_V(X, (x, y), (V, V))$$

which is the inverse of the insertion of $|0\rangle$ at y. Let $\varphi \in C(X, x; V) \subset \mathcal{V}_x^*$ be a one–point conformal block. Then for any $A \in \mathcal{V}_x$, $\langle \varphi, \mathcal{Y}_x \cdot A \rangle$ extends to a section φ_A of $\mathcal{V}^*$ on $X \backslash x$. Define $\widetilde{\varphi}_y \in (\mathcal{V}_x \otimes \mathcal{V}_y)^*$ by the formula

$$\langle \widetilde{\varphi}_y, A \otimes B \rangle = \varphi_A(B), \qquad B \in \mathcal{V}_y.$$

Note that for y near x, we have, by the definition of φ_A,

$$\langle \widetilde{\varphi}_y, A \otimes |0\rangle \rangle = \langle \varphi, \mathcal{Y}_x(|0\rangle) \rangle \cdot A = \langle \varphi, Y(|0\rangle, y - x) \cdot A \rangle = \langle \varphi, A \rangle.$$

Hence $\langle \widetilde{\varphi}_y, A \otimes |0\rangle \rangle = \langle \varphi, A \rangle$.

We claim that this $\widetilde{\varphi}_y$ is the desired two-point conformal block in

$$C_V(X, (x, y), (V, V)).$$

In other words, we "create" an extra point by using the sections φ_A of $\mathcal{V}^*$, which are just the sections $\langle \varphi, \mathcal{Y}_x \cdot A \rangle$ of $\mathcal{V}^*|_{D_x^\times}$ extended to $X \backslash x$.

To show that $\widetilde{\varphi}_y \in C_V(X, (x, y), (V, V))$, we must produce, for every $A \in \mathcal{V}_x$ and $B \in \mathcal{V}_y$, a section $\varphi_{A,B}$ of $\mathcal{V}^*$ on $X \backslash \{x, y\}$ whose restrictions to $D_x^\times$ and $D_y^\times$ are equal to $\langle \widetilde{\varphi}_y, (\mathcal{Y}_x \cdot A) \otimes B \rangle$ and $\langle \widetilde{\varphi}_y, A \otimes (\mathcal{Y}_y \cdot B) \rangle$, respectively.

Suppose that y is near x (in what follows we will use analytic topology, but the same argument applies if x and y are infinitesimally nearby points in the sense of § A.2.3). Let t be a coordinate near x, inducing coordinates $t_x = t - x$ and $t_y = t - y$ at x, y. We use t to trivialize $\mathcal{V}$ in a neighborhood of x containing y. By the definitions of φ_A and $\widetilde{\varphi}_y$, we have, in this trivialization,

$$\langle \widetilde{\varphi}_y, A \otimes B \rangle = \langle \varphi, Y(B, y - x) \cdot A \rangle.$$

Thus for any $C \in V$, we have

$$\begin{aligned}
\langle \widetilde{\varphi}_y, (\mathcal{Y}_x(C) \cdot A) \otimes B \rangle &= \langle \varphi, Y(B, y - x)Y(C, t_x) \cdot A \rangle, \\
\langle \widetilde{\varphi}_y, A \otimes (\mathcal{Y}_y(C) \cdot B) \rangle &= \langle \varphi, Y(Y(C, t_y) \cdot B, y - x) \cdot A \rangle,
\end{aligned}$$

(where by the right hand sides we understand the corresponding analytic continuations). By Theorem 7.3.2, as we fix A and vary B, C, these two expressions give rise to sections of $(\mathcal{V} \boxtimes \mathcal{V})^*$ over D_x^2 with poles at x and on the diagonal. Moreover, these two sections agree by the locality and associativity properties of the vertex algebra V:

$$Y(B, y - x)Y(C, t_x) = Y(C, t_x)Y(B, y - x) = Y(Y(C, t_y) \cdot B, y - x)$$

(since $t_x - (y - x) = t_y$).

We now utilize the fact that φ is a conformal block. We first expand

$$\langle \varphi, Y(B, y - x)Y(C, t_x) \cdot A \rangle = \sum_{n \leq N} \langle \varphi, Y(B, y - x)C_{(n)} \cdot A \rangle \, t_x^{-n-1}.$$

Since φ is a conformal block, each coefficient in this expansion extends (as a function of $y - x$) to all of $X \backslash x$. For the same reason, each coefficient of the series

$$\langle \varphi, Y(C, t_x)Y(B, y - x) \cdot A \rangle = \sum_{m \leq M} \langle \varphi, Y(C, t_x)B_{(m)} \cdot A \rangle \, (y - x)^{-m-1}$$

also extends (as a function of t_x) to $X \backslash x$.

Thus, we see that our meromorphic section of $(\mathcal{V} \boxtimes \mathcal{V})^*$ over D_x^2 extends to a meromorphic section on the formal completion of X^2 along

(10.3.1) $$Z = \{x \times X\} \cup \Delta \subset X^2$$

(see the picture). We denote this completion by $\widehat{Z}$.

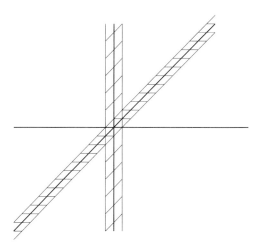

We need to prove that this section extends from $\widehat{Z}$ to all of X^2, with poles only along the divisors $x \times X, X \times x$, and Δ.

Recall that $(\mathcal{V} \boxtimes \mathcal{V})^*$ is the projective limit of vector bundles of finite rank on X^2, which we will denote by $\mathcal{F}_i$. Showing that a section of $(\mathcal{V} \boxtimes \mathcal{V})^*$ extends is equivalent to showing that its projections onto $\mathcal{F}_i$ may all be extended from $\widehat{Z}$ to X^2.

We may consider the meromorphic sections of $\mathcal{F}_i$ as holomorphic sections of their modifications. Hence without loss of generality we can switch to holomorphic (or regular) sections. A remarkable geometric fact, Theorem 10.3.3 below, implies that any regular section of a vector bundle over $\widehat{Z}$ necessarily extends to a regular section on all of X^2. Therefore our section of $(\mathcal{V} \boxtimes \mathcal{V})^*$ over $\widehat{Z}$ also extends to X^2 (with poles only along the divisors $x \times X, X \times x$ and Δ). If we restrict it to $y \times (X \backslash \{x, y\})$, we obtain a section of $(\mathcal{V} \otimes \mathcal{V}_y)^*$. Evaluating it further on $B \in \mathcal{V}_y$, we finally obtain the desired section $\varphi_{A,B}$ of $\mathcal{V}^*$ on $X \backslash \{x, y\}$, whose restriction to $D_x^\times$ and $D_y^\times$ equals $\langle \widetilde{\varphi}_y, (\mathcal{Y}_x \cdot A) \otimes B \rangle$ and $\langle \widetilde{\varphi}_y, A \otimes (\mathcal{Y}_y \cdot B) \rangle$, respectively. This proves Theorem 10.3.1.

10.3.3. Theorem. *Let $\mathcal{E}$ be a vector bundle on X^2, and let s be a section of $\mathcal{E}$ over the formal completion $\widehat{Z}$ of the divisor Z in X^2 given by (10.3.1). Then s extends to a regular section of $\mathcal{E}$ over all of X^2.*

10.3.4. Proof. Denote by p the projection $X \times X \to X$ onto the second factor. Let Δ_n be the nth formal neighborhood of the diagonal $\Delta \subset X \times X$, and p_n the projection of Δ_n onto the second factor of $X \times X$. Denote by $\mathcal{E}_n$ the restriction of $\mathcal{E}$ to Δ_n. It is not difficult to see that for large enough n, the natural map $p_* \mathcal{E} \to (p_n)_* \mathcal{E}$ is an injective map of vector bundles. Our section s gives rise to a section of $(p_n)_* \mathcal{E}$. By our assumption, the restriction of this section to the formal neighborhood of $x \in X$ lies in the subbundle $p_* \mathcal{E}$ (for large n). Therefore it lies in the subbundle $p_* \mathcal{E}$ everywhere. Hence s extends to a regular section of $\mathcal{E}$ over X^2.

10.3.5. Two–point functions. We now return to the question of describing the behavior of conformal blocks as we vary the points. According to Theorem 10.3.1, the spaces of conformal blocks $C_V(X, (x_i), (V))_{i=1}^n, n = 1, 2, \ldots,$ are canonically isomorphic to each other. For a given $\varphi \in C(X, x, V)$, let $\varphi_{x_1,\ldots,x_n}$ be the corresponding element of $C_V(X, (x_i), (V))_{i=1}^n$.

Let us look first at the case $n = 2$. According to the construction presented in the proof of Theorem 10.3.1, there is a regular section φ_2 of the bundle $(V \boxtimes V)^*$ over $X^2 \backslash \Delta$ such that for any $A_1 \in V_{x_1}$ and $A_2 \in V_{x_2}$, we have $\langle \varphi_2, A_1 \otimes A_2 \rangle = \langle \varphi_{x_1,x_2}, A_1 \otimes A_2 \rangle$. In other words, the conformal blocks φ_{x_1,x_2} with varying x_1, x_2 may be "glued" into a section of $(V \boxtimes V)^*$ on $X^2 \backslash \Delta$.

The section φ_2 satisfies the following property. Suppose that the points x_1 and x_2 are close to each other and to another, fixed, point x of X. Choosing a local coordinate z at x and denoting the corresponding coordinates of the points x_1, x_2 by z_1, z_2, we obtain from the proof of Theorem 10.3.1 that

$$(10.3.2) \qquad \langle \varphi_2, A_1 \otimes A_2 \rangle = \langle \varphi, Y(A_1, z_1)Y(A_2, z_2)|0\rangle.$$

The right hand side of this formula is a two–point function of the kind studied in § 4.5. In particular, by Theorem 4.5.1, it is an element of $\mathbb{C}[[z_1, z_2]][(z_1 - z_2)^{-1}]$.

According to Theorem 7.3.2 and Proposition 7.3.4, for any $\varphi \in V_x^*$ (not necessarily a conformal block) the matrix elements given by the right hand side of formula (10.3.2) give rise to a horizontal section of $(V \boxtimes V)^*$ on D_x^2 with poles on the diagonal $\Delta \subset D_x^2$. Therefore we obtain an embedding

$$(10.3.3) \qquad V_x^* \hookrightarrow \Gamma_\nabla(D_x^2, (V \boxtimes V)^*(\infty\Delta)) = \mathrm{Hom}_{D_x^2, \nabla}(V \boxtimes V, \mathcal{O}_{X^2}(\infty\Delta)),$$

where $\mathcal{O}_{X^2}(\infty\Delta)$ is the sheaf on X^2 whose sections are functions that are allowed to have poles on the diagonal. Here Γ_∇ (resp., $\mathrm{Hom}_{D_x^2, \nabla}$) denotes the space of horizontal sections (resp., homomorphisms of $\mathcal{O}$–modules on D_x^2 commuting with connections, where $\mathcal{O}_{X^2}(\infty\Delta)$ is equipped with the trivial connection).

Formula (10.3.2) shows that if in addition $\varphi \in V_x^*$ is a conformal block, then the image of φ in $\Gamma_\nabla(D_x^2, (V \boxtimes V)^*(\infty\Delta))$ under the embedding (10.3.3) extends to a horizontal section φ_2 of $(V \boxtimes V)^*$ over X^2 with poles on the diagonal Δ. Thus, we obtain a "global" analogue of the "local" embedding (10.3.3):

$$(10.3.4) \quad C(X, x, V) \hookrightarrow \Gamma_\nabla(X^2, (V \boxtimes V)^*(\infty\Delta)) = \mathrm{Hom}_{X^2, \nabla}(V \boxtimes V, \mathcal{O}_{X^2}(\infty\Delta)).$$

This is the analogue for two-point functions of the fact that the section $\langle \varphi, \mathcal{Y}_x \cdot |0\rangle\rangle$ of $V^*|_{D_x}$ extends to a global horizontal section of V^* on X, and hence we have a map

$$C(X, x, V) \hookrightarrow \Gamma_\nabla(X, V^*) = \mathrm{Hom}_{X, \nabla}(V, \mathcal{O}_X).$$

We want to describe the images of the maps (10.3.3) and (10.3.4). So we look for an appropriate quotient V_2^* of $(V \boxtimes V)^*(\infty\Delta)$ such that the resulting maps

$$(10.3.5) \qquad V_x^* \to \Gamma_\nabla(D_x^2, V_2^*), \qquad C(X, x, V) \to \Gamma_\nabla(X^2, V_2^*)$$

are isomorphisms.

According to Theorem 4.5.1, the functions appearing in the right hand side of formula (10.3.2) satisfy the "bootstrap" condition:

$$\langle \varphi, Y(A_1, z_1)Y(A_2, z_2)|0\rangle = \langle \varphi, Y(Y(A_1, z_1 - z_2) \cdot A_2, z_2)|0\rangle.$$

The desired quotient V_2^* is obtained from $(V \boxtimes V)^*(\infty\Delta)$ by imposing these conditions near the diagonal. The precise definition of V_2^* (due to Beilinson and Drinfeld) is given in § 20.2.4 using the formalism of chiral algebras.

Theorem 4.5.1 then implies that the first map in (10.3.5) is an isomorphism. Beilinson and Drinfeld have proved that the second map in (10.3.5) is also an isomorphism.

10.3.6. The n–point functions. The above picture may be generalized to the case of n–point functions. Denote by $\overset{\circ}{X}{}^n$ the complement of the union of all pairwise diagonals in X^n. In the same way as in the proof of Theorem 10.3.1, we show by induction on n that the there is a regular section φ_n of the bundle $(\mathcal{V}^{\boxtimes n})^*$ over $\overset{\circ}{X}{}^n$ whose value at $x_1, \ldots, x_n$ is equal to the conformal block $\varphi_{x_1, \ldots, x_n}$.

Let $A_i(x_i)$ be local sections of $\mathcal{V}$. Evaluating our conformal blocks on them, we obtain the chiral correlation functions

$$(10.3.6) \qquad \langle \varphi_n, A_1(x_1) \otimes \ldots \otimes A_n(x_n) \rangle.$$

Moreover, we can explicitly describe these functions when all x_i's are near each other in the neighborhood of a point x (this may be done in an analytic neighborhood of x as well as in a formal neighborhood). In that case, if we choose a local coordinate z at x and denote the coordinate of the point x_i by z_i, we obtain

$$(10.3.7) \qquad \langle \varphi_n, A_1(z_1) \otimes \ldots \otimes A_n(z_n) \rangle = \langle \varphi, Y(A_1, z_1) \ldots Y(A_n, z_n) |0\rangle \rangle,$$

where the right hand side is considered as an element of $\mathbb{C}[[z_1, \ldots, z_n]][(z_i - z_j)^{-1}]_{i \neq j}$ (see Theorem 4.5.1).

Denote by Δ the union of all pairwise diagonals in X^n, and by $(\mathcal{V}^{\boxtimes n})^*(\infty\Delta)$ the corresponding sheaf on X^n. According to Theorem 7.3.2, for any $\varphi \in \mathcal{V}_x^*$ the matrix elements (10.3.7) give rise to a horizontal section of $(\mathcal{V}^{\boxtimes n})^*$ on $D_x^n = \operatorname{Spec} \mathbb{C}[[z_1, \ldots, z_n]]$ with poles on the pairwise diagonals $z_i = z_j, i \neq j$. Then we obtain an embedding

$$(10.3.8) \qquad \mathcal{V}_x^* \hookrightarrow \Gamma_\nabla(D_x^n, (\mathcal{V}^{\boxtimes n})^*(\infty\Delta)) = \operatorname{Hom}_{D_x^n, \nabla}(\mathcal{V}^{\boxtimes n}, \mathcal{O}_{X^n}(\infty\Delta)).$$

(the meaning of a section of $(\mathcal{V}^{\boxtimes n})^*(\infty\Delta)$ over D_x^n is spelled out in § 7.3.1). If φ is a conformal block, then the corresponding section on D_x^n extends to a section of $(\mathcal{V}^{\boxtimes n})^*(\infty\Delta)$ on all of X^n, so we obtain a "global" analogue of the embedding (10.3.8):

$$(10.3.9) \qquad C(X, x, V) \hookrightarrow \Gamma_\nabla(X^n, (\mathcal{V}^{\boxtimes n})^*(\infty\Delta)) = \operatorname{Hom}_{X^n, \nabla}(\mathcal{V}^{\boxtimes n}, \mathcal{O}_{X^n}(\infty\Delta)).$$

By Theorem 4.5.1, the functions appearing in (10.3.7) satisfy "bootstrap" conditions relating n–point and $(n-1)$–point correlation functions:

$$(10.3.10) \quad \langle \varphi_n, A_1(z_1) \otimes \ldots \otimes A_n(z_n) \rangle$$
$$= \langle \varphi_{n-1}, A_1(z) \otimes \ldots \otimes (Y(A_i, z_i - z_j)A_j)(z_j) \otimes \ldots \otimes A_n(z_n) \rangle,$$

near the diagonal $z_i = z_j, i \neq j$. Using these conditions, it is possible to define a quotient $\mathcal{V}_n^*$ of the sheaf $(\mathcal{V}^{\boxtimes n})^*(\infty\Delta)$ in such a way that we have a commutative diagram

$$
\begin{array}{ccc}
C(X, x, V) & \xrightarrow{\;\sim\;} & \Gamma_\nabla(X^n, \mathcal{V}_n^*) \\
\downarrow & & \downarrow \\
\mathcal{V}_x^* & \xrightarrow{\;\sim\;} & \Gamma_\nabla(D_x^n, \mathcal{V}_n^*)
\end{array}
$$

Thus, we realize $\mathcal{V}_x^*$ and $C(X, x, V)$ in terms of sections of sheaves on powers of the disc D_x and the curve X, respectively, with poles along the pairwise diagonals.

Therefore we refer to the above isomorphisms as the *functional realizations*. In the next two sections we will consider two special cases of this construction: first when the curve X has genus 0, and then when X is arbitrary and V is the Heisenberg vertex algebra π.

10.3.7. Remark. Note that neither $(V^{\boxtimes n})^*$ nor V_n^* is quasi-coherent as an $\mathcal{O}_{X^n}$–module. It is more convenient to work with the dual sheaves $V^{\boxtimes n}$ and $\mathcal{V}_n$, which are quasi-coherent. The sheaf $\mathcal{V}_n$ on X^n is defined by Beilinson and Drinfeld in their construction of the factorization algebra corresponding to $\mathcal{V}$ (see § 20.2.4). Note that the space $\Gamma_\nabla(X^n, \mathcal{V}_n^*)$ is dual to the top de Rham cohomology $H^{2n}_{\mathrm{dR}}(X, \mathcal{V}_n^r)$ (which is therefore isomorphic to the space of coinvariants $H(X, x, V)$).

10.3.8. Remark. If we consider conformal blocks with general V–module insertions, then the horizontal sections of the corresponding flat bundle of conformal blocks over $\overset{\circ}{X}{}^n$ acquire non-trivial monodromy around the diagonals. For example, in the case of the Kac-Moody vertex algebra $V_k(\mathfrak{g})$ and $X = \mathbb{P}^1$, these sections are solutions of the Knizhnik–Zamolodchikov equations which we will study below in Chapter 13, and the monodromy matrices are given by the R–matrices of the quantum group $U_q(\mathfrak{g})$, see [**TK, Law, SV2**].

10.4. Conformal blocks in genus zero

In this section we describe the functional realization of the spaces of conformal blocks of an arbitrary conformal vertex algebra in the case when the curve X has genus zero, i.e., $X \simeq \mathbb{P}^1$. This case is simpler than the general case, because we may choose a global coordinate on $\mathbb{P}^1 \backslash \infty$.

We will use this functional realization to give a new interpretation of the vertex operation Y, the locality axiom and the associativity property, in terms of conformal blocks on $\mathbb{P}^1$ with three–point insertions.

10.4.1. The space of conformal blocks. Let V be a conformal vertex algebra. We will assume throughout this section that with respect to the $\mathbb{Z}$–grading on V induced by the operator L_0 we have the following decomposition:

$$V = \bigoplus_{n=0}^{\infty} V_n,$$

where each V_n is a finite-dimensional vector space.

Consider the projective line $\mathbb{P}^1$, on which we fix once and for all two distinct points, denoted by 0 and ∞. The discs around these points are denoted by D_0 and D_∞, respectively. We choose a global coordinate t on $\mathbb{P}^1 \backslash \infty$, so that the point 0 corresponds to $t = 0$, and the point ∞ corresponds to $t = \infty$. Such a coordinate is unique up to a non-zero scalar multiple.

By our general construction, we attach to V a vector bundle $\mathcal{V}$ on $\mathbb{P}^1$. Using the coordinate t, we trivialize the restriction of this bundle to $\mathbb{P}^1 \backslash \infty$. Let $\mathcal{V}_0$ be the fiber of $\mathcal{V}$ at 0. Recall that $U(\mathcal{V}_0) \simeq U(V)$ is a completion of the span of all Fourier coefficients $A_{(n)}, A \in V, n \in \mathbb{Z}$. Its Lie subalgebra $U_{\mathbb{P}^1 \backslash 0}(\mathcal{V}_0)$, as defined in § 9.2.5, is spanned by $A_{(n)}, A \in V, n < 0$. Therefore, by the vacuum axiom of vertex algebras, the subspace $U_{\mathbb{P}^1 \backslash 0}(\mathcal{V}_0) \cdot |0\rangle$ of $\mathcal{V}_0 \simeq V$ equals $\bigoplus_{n>0} V_n$. Hence the space of coinvariants $H(\mathbb{P}^1, 0, V) = \mathcal{V}_0 / U_{\mathbb{P}^1 \backslash 0}(\mathcal{V}_0) \cdot \mathcal{V}_0$ is one-dimensional and is canonically isomorphic to $\mathbb{C} \cdot |0\rangle$.

Let $\varphi_0 : V \to \mathbb{C}$ be the linear functional such that $\varphi_0(|0\rangle) = 1$ and $\varphi_0(A) = 0$ for all $A \in \bigoplus_{n>0} V_n$. Then the space of conformal blocks $C(\mathbb{P}^1, 0, V)$ is spanned by φ_0.

10.4.2. Functional realization. Now suppose we are given n distinct points $x_1, \ldots, x_n$ on $\mathbb{P}^1 \backslash \infty$, with coordinates $z_1, \ldots, z_n$. By Theorem 10.3.1, we have an isomorphism

$$(10.4.1) \qquad C(\mathbb{P}^1, (x_i), (V))_{i=1}^n \simeq C(\mathbb{P}^1, 0, V) = \mathbb{C}\varphi_0.$$

Denote by $\varphi_{z_1,\ldots,z_n} : V^{\otimes n} \to \mathbb{C}$ the element of $C(\mathbb{P}^1, (x_i), (V))_{i=1}^n$ corresponding to φ_0. We claim that $\varphi_{z_1,\ldots,z_n}$ is given by the following formula (compare with formula (10.3.7)):

$$(10.4.2) \qquad \langle \varphi_{z_1,\ldots,z_n}, A_1 \otimes \ldots \otimes A_n \rangle = \langle \varphi_0, Y(A_1, z_1) \ldots Y(A_n, z_n) |0\rangle \rangle,$$

where by the right hand side we understand the corresponding rational function in the z_i's with poles on the diagonals (see Theorem 4.5.1).

Indeed, by Definition 10.1.1, showing that formula (10.4.2) defines a conformal block is equivalent to showing that

$$\langle \varphi_0, Y(A_1, z_1) \ldots Y(Y(B, z - z_i)A_i, z_i) \ldots Y(A_n, z_n)|0\rangle \rangle$$

may be analytically continued to a rational function in $z, z_1, \ldots, z_n$, and all of these functions are equal for $i = 1, \ldots, n$. But by Theorem 4.5.1, these functions are expansions of one and the same element of $\mathbb{C}[[z_1, \ldots, z_n]][(z_i - z_j)^{-1}]_{i \neq j}$. Moreover, since φ_0 vanishes on $\bigoplus_{n>0} V_n$, we obtain that this element actually belongs to $\mathbb{C}[z_1, \ldots, z_n][(z_i - z_j)^{-1}]_{i \neq j}$.

To see that this conformal block corresponds to φ_0 under the isomorphism (10.4.1), we simply observe that if all $A_i, i = 1, \ldots, n$, are set equal to $|0\rangle$ in formula (10.4.2), we obtain 1.

Now we want to interpret in a similar way the right hand side of formula (10.4.2) when φ_0 is replaced by a more general functional on V. For that we need to introduce the notion of contragredient module.

10.4.3. Extension of $\mathcal{Y}$ to $\mathbb{P}^1$. By Theorem 6.5.4, we have an End $\mathcal{V}_0$–valued section $\mathcal{Y}_0$ of $\mathcal{V}^*|_{D_0^\times}$. By definition, in the trivialization $\imath_t$ of $\mathcal{V}|_{D_0}$ corresponding to the coordinate t, the section $\mathcal{Y}_0$ satisfies

$$\langle \varphi, \mathcal{Y}_0(\imath_t(A)) \cdot C \rangle = \langle \varphi, Y(A, t) \cdot C \rangle, \qquad A, C \in V; \varphi \in V^*.$$

Consider the restricted dual vector space

$$V^\vee = \bigoplus_{n \geq 0} V_n^*$$

of V. Then if $\varphi \in V^\vee$, we obtain by degree considerations that $\langle \varphi, Y(A, t) \cdot C \rangle \in \mathbb{C}[t, t^{-1}]$.

Recall that a global coordinate on $\mathbb{P}^1 \backslash \{0, \infty\}$ is unique up to a non-zero scalar multiple. Thus, the group $\operatorname{Aut} \mathcal{O}$ gets reduced to the multiplicative group $\mathbb{C}^\times$. Note that $V^\vee$ is preserved by the action of $\mathbb{C}^\times$. Let $\mathcal{V}_0$ and $\mathcal{V}_0^\vee$ be the twists of V and $V^\vee$ by the $\mathbb{C}^\times$–torsor of non-zero tangent vectors at $0 \in \mathbb{P}^1$. Then we obtain

10.4.4. Lemma. *For each $C \in \mathcal{V}_0$ and $\varphi \in \mathcal{V}_0^\vee$, the section $\langle \varphi, \mathcal{Y}_0 \cdot C \rangle$ of $\mathcal{V}^*|_{D_0^\times}$ extends to a meromorphic section of $\mathcal{V}^*$ on $\mathbb{P}^1$ with poles only at the points 0 and ∞.*

We denote the resulting section by $\langle \varphi, \mathcal{Y}_{\mathbb{P}^1} \cdot C \rangle$.

10.4.5. Expansion near the point ∞. Now we want to evaluate this section near the point $\infty \in \mathbb{P}^1$. We have a local coordinate $u = t^{-1}$ at ∞. Let x be a point of $\mathbb{P}^1$ near ∞ with u–coordinate z (and t–coordinate z^{-1}). Then it acquires two local coordinates: $u_z = u - z$ and $\widetilde{u}_z = u^{-1} - z^{-1}$. Let μ_z be the corresponding change of variables: $u_z = \mu_z(\widetilde{u}_z)$, i.e.,

$$\mu_z(y) = (y + z^{-1})^{-1} - z = -\frac{z^2 y}{zy + 1} = -z^2 y + z^3 y^2 - \cdots.$$

Realizing $\mu_z(y)$ as the composition of the Möbius transformations $y \mapsto -z^2 y$ and $y \mapsto y/(zy + 1)$, we obtain

$$(10.4.3) \qquad \mu_z(y) = e^{-zy^2 \partial_y} \left((-z^{-2})^{-y\partial_y} \cdot y \right).$$

Let us compute the restriction of the section $\langle \varphi, \mathcal{Y}_{\mathbb{P}^1} \cdot C \rangle$ of $\mathcal{V}^*$ to $D_\infty^\times$ with respect to the trivialization of $\mathcal{V}|_{D_\infty}$ induced by the coordinate u. It follows from the definitions of $\mathcal{V}$ and $\mathcal{Y}_{\mathbb{P}^1}$ that the value of $\langle \varphi, \mathcal{Y}_{\mathbb{P}^1} \cdot C \rangle|_{D_\infty}$ on the constant section of $\mathcal{V}|_{D_\infty}$ taking the value $A \in V$ in this trivialization is given by the formula

$$(10.4.4) \qquad \langle \varphi, Y(R(\mu_u)A, u^{-1}) \cdot C \rangle,$$

where

$$(10.4.5) \qquad R(\mu_u) = e^{uL_1}(-u^{-2})^{L_0},$$

by formula (10.4.3).

We will now use the above formula for the restriction of $\mathcal{Y}_{\mathbb{P}^1}$ to $D_\infty^\times$ to define a V–module structure on $V^\vee$.

10.4.6. Proposition. *Define a linear map $\widetilde{Y} : V \to \operatorname{End} V^\vee[[z, z^{-1}]]$ by the formula*

$$(10.4.6) \qquad \langle \widetilde{Y}(A, z) \cdot \varphi, C \rangle = \langle \varphi, Y(e^{zL_1}(-z^{-2})^{L_0} \cdot A, z^{-1}) \cdot C \rangle.$$

Then this map, together with the $\mathbb{Z}$–gradation $V^\vee = \bigoplus_{n \geq 0} V_n^*$, defines the structure of a V–module on $V^\vee$, which is called the contragredient module to V.

10.4.7. Proof. We need to check the axioms from Definition 5.1.1. The fact that $\widetilde{Y}(A, z)$ is a field of conformal dimension n on $V^\vee$ for each $A \in V_n$ and the formula $\widetilde{Y}(|0\rangle, z) = \operatorname{Id}_{V^\vee}$ follow directly from the definition. It remains to check the identity

$$(10.4.7) \qquad \widetilde{Y}(A, z)\widetilde{Y}(B, w) = \widetilde{Y}(Y(A, z - w)B, w)$$

(here and below $(z - w)^{-1}$ is expanded in positive powers of w/z). Using the definition of $\widetilde{Y}$ given in formula (10.4.6), we rewrite this identity as follows:

$$Y(R(\mu_w)B, w^{-1})Y(R(\mu_z)A, z^{-1}) = Y(R(\mu_w)Y(A, z - w)B, w^{-1})$$

(here we use the notation of formula (10.4.5)). Substituting $A = R(\mu_z)^{-1}A', B = R(\mu_w)^{-1}B'$, we obtain

$$Y(B', w^{-1})Y(A', z^{-1}) = Y(R(\mu_w)Y(R(\mu_z)^{-1}A', z - w)R(\mu_w)^{-1}B', w^{-1}).$$

This formula follows from the locality and associativity properties of Y and the identity
$$Y(A', z^{-1} - w^{-1}) = R(\mu_w)Y(R(\mu_z)^{-1}A', z - w)R(\mu_w)^{-1},$$
which is obtained from formula (6.5.1) by substituting $z^{-1} - w^{-1}$ instead of z and μ_w instead of ρ. This completes the proof.

10.4.8. Another proof. Alternatively, the proposition may be proved as follows. Recall that in § 9.2.2 we attached a Lie algebra $h(\mathcal{V} \otimes \Omega)(\Sigma)$ to any curve Σ. It follows from the construction that any automorphism of Σ induces an automorphism of the Lie algebra $h(\mathcal{V} \otimes \Omega)(\Sigma)$.

Consider the case $\Sigma = \mathbb{P}^1 \backslash \{0, \infty\}$. The Lie algebra $h(\mathcal{V} \otimes \Omega)(\mathbb{P}^1 \backslash \{0, \infty\})$ is the quotient of $\Gamma(\mathbb{P}^1 \backslash \{0, \infty\}, \mathcal{V} \otimes \Omega)$ by the subspace of total derivatives. Choosing a coordinate t on $\mathbb{P}^1 \backslash \{0, \infty\}$, we identify $h(\mathcal{V} \otimes \Omega)(\mathbb{P}^1 \backslash \{0, \infty\})$ with the Lie algebra $U'(V)$ introduced in § 4.1.1. The automorphism $t \mapsto t^{-1}$ of $\mathbb{P}^1 \backslash \{0, \infty\}$ induces an involution ω on $h(\mathcal{V} \otimes \Omega)(\mathbb{P}^1 \backslash \{0, \infty\}) \simeq U'(V)$. Recall that the Lie algebra $U'(V)$ is spanned by elements $A_{[n]}$, where $A \in V, n \in \mathbb{Z}$, which are the projections of the sections of $\mathcal{V} \otimes \Omega$ equal to $A \otimes z^n dz$ in the t–trivialization of $\mathcal{V}$. The involution ω maps $A_{[n]}$ to the projection onto $U'(V)$ of the same section, written in the t^{-1}–trivialization of $\mathcal{V} \otimes \Omega$. A short calculation similar to that of § 10.4.5 shows that this section equals $-R(\mu_{z^{-1}})A \otimes z^{-n-2}dz$ in the t^{-1}–trivialization (in the notation of formula (10.4.5)). Thus, if $\deg A = \Delta$ we have

$$(10.4.8) \qquad \omega(A_{[n]}) = (-1)^{\Delta+1} \sum_{m \geq 0} \frac{1}{m!}(L_1^m A)_{[2\Delta-n-m-2]}.$$

Now observe that the coefficient of z^{-n-1} in $\widetilde{Y}(A, z)$ given by formula (10.4.6) is a linear operator on $V^\vee$, which is nothing but the transpose of $-\omega(A_{[n]})$. Since ω is an involution, $\widetilde{Y}$ defines an action of the Lie algebra $U'(V)$ on $V^\vee$. Moreover, it is easy to see that with respect to this action $V^\vee$ is a bounded $U'(V)$–module (see § 5.1.5). Proposition 5.1.6 then implies that $V^\vee$ is a V–module.

10.4.9. Remark. Let M be a conformal V–module. Formula (10.4.6), with Y replaced by Y_M, defines a structure of V–module on $M^\vee$. It is called the contragredient module to M.

10.4.10. Conformal blocks. Consider the space of two-point conformal blocks
$$C(\mathbb{P}^1, (0, \infty), (V, V^\vee)).$$
Elements of this space are functionals $\Phi : V \otimes V^\vee \to \mathbb{C}$. Let $\Phi_0 : V \otimes V^\vee \to \mathbb{C}$ be the natural pairing. Recall that the action of V on $V^\vee$ is given by the restriction of the section $\mathcal{Y}_{\mathbb{P}^1}$ to $D_\infty^\times$ (compare formulas (10.4.6) and (10.4.4)). This implies that Φ_0 is an element of $C(\mathbb{P}^1, (0, \infty), (V, V^\vee))$.

Now choose n distinct points $x_1, \ldots, x_n$ on $\mathbb{P}^1 \backslash \infty$, with coordinates $z_1, \ldots, z_n$. By Theorem 10.3.1, we have an isomorphism
$$C(\mathbb{P}^1, ((x_i), \infty), ((V), V^\vee))_{i=1}^n \simeq C(\mathbb{P}^1, (0, \infty), (V, V^\vee)).$$

Let $\Phi_{z_1,\ldots,z_n,\infty}$ be the element of the space in the left hand side corresponding to Φ_0. Then we claim that
$$(10.4.9) \qquad \langle \Phi_{z_1,\ldots,z_n,\infty}, A_1 \otimes \ldots \otimes A_n \otimes \varphi \rangle = \langle \varphi, Y(A_1, z_1) \ldots Y(A_n, z_n)|0\rangle \rangle.$$

The proof of this statement follows from Theorem 4.5.1, by the same argument as we used in § 10.4.2. Note that the functions appearing in Theorem 4.5.1 belong to the space $\mathbb{C}[[z_1,\ldots,z_n]][(z_i - z_j)^{-1}]_{i \neq j}$, because φ is taken there to be an arbitrary functional on V. But here we consider only $\varphi \in V^\vee$. Therefore the functions given by the right hand side of (10.4.9) belong to $\mathbb{C}[z_1,\ldots,z_n][(z_i - z_j)^{-1}]_{i \neq j}$.

Thus we obtain a functional realization of the conformal blocks $\Phi_{z_1,\ldots,z_n,\infty}$.

10.4.11. A new interpretation of locality and associativity. In the special case $n = 3$ we obtain from formula (10.4.9)

$$(10.4.10) \qquad \langle \Phi_{0,z,\infty}, C \otimes A \otimes \varphi \rangle = \langle \varphi, Y(A,z)C \rangle.$$

Thus we obtain:

The vertex operation Y is a special case of conformal block.

Similarly, we obtain that if we insert M at 0, V at z, and $M^\vee$ at ∞, then the corresponding three-point conformal block gives us the module operation Y_M.

Schematically, this may be expressed as follows:

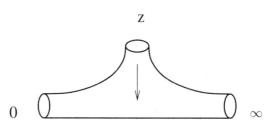

When $A \in V$, "inserted" at a point $z \in \mathbb{P}^1$, "acts" on $C \in M$, "inserted" at $0 \in \mathbb{P}^1$, we obtain an element of $M((z))$ (at ∞), whose pairing with $\varphi \in M^\vee$ is given by the left hand side of (10.4.10).

Now let us examine the implications of $\Phi_{0,z,\infty}$ being a conformal block. According to Definition 10.1.1, this means that the three power series

$$\langle \varphi, Y(B,w)Y(A,z)C \rangle, \qquad \langle \varphi, Y(Y(B,w-z)A,w)C \rangle, \qquad \langle \varphi, Y(A,z)Y(B,w)C \rangle$$

in $\mathbb{C}((w))((z))$, $\mathbb{C}((w-z))((w))$, and $\mathbb{C}((z))((w))$, respectively, are analytic continuations of the same rational function in z, w, namely, $\langle \Phi_{0,z,w,\infty}, C \otimes B \otimes A \otimes \varphi \rangle$. Indeed, the power series above correspond to

$$\langle \Phi_{0,z,\infty}, (\mathcal{Y}_0 \cdot C) \otimes B \otimes \varphi \rangle, \qquad \langle \Phi_{0,z,\infty}, C \otimes (\mathcal{Y}_z \cdot B) \otimes \varphi \rangle, \qquad \langle \Phi_{0,z,\infty}, C \otimes B \otimes (\mathcal{Y}_\infty \cdot \varphi) \rangle,$$

respectively. Since $\Phi_{0,z,\infty}$ is a conformal block, we should be able to "glue" these sections into a single section of $\mathcal{V}^*$ on $\mathbb{P}^1 \backslash \{0, z, \infty\}$. Following the proof of Theorem 10.3.1, we obtain that this section of $\mathcal{V}^*$ is nothing but

$$\Phi_{0,z,w,\infty} \in C(\mathbb{P}^1, (0, z, w, \infty), (V, V, V, V^\vee))$$

(with respect to the variable w). Therefore the four-point conformal block $\Phi_{0,z,w,\infty}$ is the object which unifies the above three power series. Thus, we obtain a new interpretation of locality, and associativity of vertex algebras. This may be illustrated by the following picture:

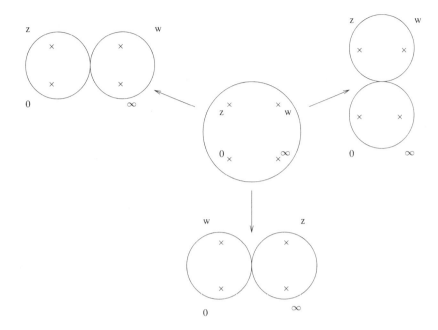

The three degenerations of $\mathbb{P}^1$ correspond to the three formal power series above. For example, in the degeneration represented in the upper left corner z is close to 0 and w is close to ∞, so we have $|w| \gg |z|$. Hence it corresponds to $\langle \varphi, Y(B,w)Y(A,z)C \rangle$, which is an element of $\mathbb{C}((w))((z))$. The three power series *a priori* make sense only in those three limits. The fact that $\Phi_{0,z,\infty}$ is a conformal block means that there is a rational function which "interpolates" between them. This function comes from the four-point conformal block $\Phi_{0,z,w,\infty}$, which corresponds to the central part of the above picture.

Likewise, if we start with a four-point conformal block and "act" by Y at each of the four points, we will obtain four power series in different domains. They correspond to four degenerations of $\mathbb{P}^1$ with five marked points. All of them are limits of a single rational function which comes from a five-point conformal block, and so on.

10.4.12. Remark. We have seen above that an element of the space of conformal blocks $C(\mathbb{P}^1, (0,z,\infty), (V,V,V^\vee))$ gives rise to the vertex operation $Y : V \to \operatorname{End} V[[z^{\pm 1}]]$. More generally, if M_1, M_2, and M_3 are conformal V–modules, then elements of the space of conformal blocks $C(\mathbb{P}^1, (0,z,\infty), (M_1, M_2, M_3^\vee))$ give rise to *intertwining operators* of type $\begin{pmatrix} M_3 \\ M_2\, M_1 \end{pmatrix}$ in the sense of [**FHL**]. These are linear maps from M_2 to formal power series in $z^{\pm 1}$ (possibly multiplied by $z^\alpha, \alpha \in \mathbb{C}$) with coefficients in $\operatorname{Hom}(M_1, M_3)$. In fact, the above spaces of conformal blocks and intertwining operators are isomorphic to each other.

For example, the bosonic vertex operator $V_\lambda(z) : \pi_\mu \to \pi_{\lambda+\mu}$ introduced in § 5.2 is the intertwining operator corresponding to a generator of the one-dimensional space of conformal blocks $C_\pi(\mathbb{P}^1, (0,z,\infty), (\pi_\mu, \pi_\lambda, \pi_{-\lambda-\mu}))$; note that $\pi_{\lambda+\mu}^\vee \simeq \pi_{-\lambda-\mu}$.

10.5. Functional realization of Heisenberg conformal blocks

In the case of the Heisenberg vertex algebra π, the dual space to Π_x and the space of conformal blocks $C(X, x, \pi)$ can be described in simpler terms than in the general case. Since the Heisenberg vertex algebra is generated by the field $b(z)$, and we have the projection $\Pi^* \twoheadrightarrow \Omega$, it is sufficient to consider sections of the exterior products of the canonical bundle Ω (with certain "bootstrap conditions" on the diagonals) instead of the "big" bundle Π^*. We will restrict now to the case where the parameter $\lambda = 0$, so that the conformal vector is $\omega_0 = \frac{1}{2} b_{-1}^2$.

We will describe a universal construction (due to Beilinson and Drinfeld) which when applied to the disc D_x gives Π_x^*, and when applied to a compact curve X gives $C(X, x, \pi)$.

In a nutshell, the construction goes as follows. Given a functional $\varphi \in \Pi_x^*$, we consider the collection of correlation functions

$$\langle \varphi, b(z_1) \dots b(z_n) |0\rangle \rangle dz_1 \dots dz_n.$$

Since $b(z)dz$ is an $\operatorname{End}\Pi_x$–valued one-form on $D_x^\times$, these are naturally sections of $\Omega^{\boxtimes n}$ over D_x^n with poles on the diagonals. Locality implies that these sections are symmetric. Using the OPE

$$b(z)b(w) = \frac{1}{(z-w)^2} + \operatorname{reg}.,$$

we can describe explicitly the bootstrap conditions that they satisfy near the diagonals. As before, denote by Δ the union of all pairwise diagonals. We will show that the vector space of all collections of symmetric sections of $\Omega^{\boxtimes n}(2\Delta)$ over D_x^n satisfying these conditions is isomorphic to Π_x^*. Furthermore, the analogous space of sections with D_x replaced by X turns out to be isomorphic to the space of conformal blocks $C(X, x, \pi)$.

10.5.1. Definition. We define the sheaf Ω_r of polydifferentials on X: its sections $\Omega_r(U)$ over an open subset $U \subset X$ are $(r+1)$–tuples

$$\omega = (\omega_0, \omega_1, \dots, \omega_r),$$

where $\omega_0 \in \mathbb{C}$ and $\omega_n \in \Gamma(U^n, \Omega^{\boxtimes n}(2\Delta)), n > 0$, satisfy the following conditions:

(1) ω_n is invariant under the symmetric group S_n acting by permutation of variables;

(2) the projection of the form ω_n to the sheaf

$$(10.5.1) \qquad \Omega^{\boxtimes n}(2\Delta)/\Omega^{\boxtimes n}(2\Delta - \Delta_{ij})|_{\Delta_{ij}} \simeq \Omega^{\boxtimes(n-2)}(2\Delta) \boxtimes \mathcal{O}_{\Delta_{ij}}$$

equals $\omega_{n-2} \boxtimes 1$.

The sheaves Ω_r form an inverse system. We denote by Ω_∞ the limit $\varprojlim \Omega_r$.

10.5.2. Remark. The isomorphism (10.5.1) is due to the isomorphism

$$\mathcal{O} \simeq \Omega^{\boxtimes 2}(2\Delta)/\Omega^{\boxtimes 2}(\Delta)|_\Delta,$$

which follows from Grothendieck's definition of Ω given in § 8.3.1.

If we choose local coordinates $z_1, \dots, z_n$, then the second condition can be rewritten as follows: near the diagonal $z_i = z_j$,

$$(10.5.2) \quad \omega_n(z_1, \dots, z_n) = \left(\frac{\omega_{n-2}(z_1, \dots, \widehat{z_i}, \dots, \widehat{z_j}, \dots, z_n)}{(z_i - z_j)^2} + \text{regular} \right) dz_i dz_j.$$

This is precisely the "bootstrap condition" (4.5.6) from § 4.5.

10.5.3. Proposition. *There is a canonical isomorphism $\Pi_x^* \simeq \Omega_\infty(D_x)$.*

10.5.4. Proof. Let us first construct a linear map $\Pi_x^* \to \Omega_\infty(D_x)$. This map sends a linear functional $\varphi \in \Pi_x^*$ to $(\omega_n)_{n\geq 0}$, where

$$(10.5.3) \qquad \omega_n = \langle \varphi, b(z_1) \ldots b(z_n)|0\rangle\rangle dz_1 \ldots dz_n.$$

By rewriting this formula as

$$(10.5.4) \qquad \omega_n = \sum_{i_1,\ldots,i_n \in \mathbb{Z}} \langle \varphi, b_{i_1} \cdots b_{i_n}|0\rangle\rangle z_1^{-i_1-1} \cdots z_n^{-i_n-1} dz_1 \ldots dz_n,$$

we see that ω_n lies in the space

$$\mathbb{C}((z_1)) \ldots \mathbb{C}((z_{n-1}))[[z_n]]dz_1 \ldots dz_n.$$

By Theorem 4.5.1, this power series is symmetric and is the expansion of an element of

$$\mathbb{C}[[z_1,\ldots,z_n]][(z_i - z_j)^{-1}]_{i\neq j}dz_1 \ldots dz_n.$$

Therefore it satisfies the first condition of Definition 10.5.1. By formula (4.5.6), it also satisfies the second condition. Since $b(z)dz$ does not depend on the choice of coordinate z on D_x as an $\mathrm{End}\,\Pi_x$–valued one-form on D_x, we see that ω_n is coordinate-independent (cf. Theorem 7.3.2). Thus, $(\omega_n)_{n\geq 0}$ is an element of $\Omega_\infty(D_x)$, and we obtain a linear map $\Pi_x^* \to \Omega_\infty(D_x)$.

This map is injective. Indeed, by formula (10.5.4), if the ω_n's vanish for all n, then $\langle \varphi, b_{i_1} \ldots b_{i_n}|0\rangle\rangle = 0, \forall i_j < 0, n \geq 0$, and so φ is identically equal to 0.

Now we want to construct the inverse map $\Omega_\infty(D_x) \to \Pi_x^*$. This map is induced from a pairing $\Pi_x \times \Omega_\infty(D_x)$ which we are about to construct. First we discuss a description of Π_x as the quotient of a Weyl algebra.

10.5.5. Π_x and the Weyl algebra. Recall the Heisenberg Lie algebra $\mathcal{H}$ and the Weyl algebra $\widetilde{\mathcal{H}}$ from § 2.1.2. The Lie algebra $\mathcal{H}$ was defined as the central extension of the commutative Lie algebra $\mathbb{C}((t))$. The algebra $\widetilde{\mathcal{H}}$ was defined as a completion of $U(\mathcal{H})/(\mathbf{1} - 1)$. The group $\mathrm{Aut}\,\mathcal{O}$ acts on $\mathcal{H}$ and on $\widetilde{\mathcal{H}}$. Let $\mathcal{H}_x$ and $\widetilde{\mathcal{H}}_x$ be the twists of $\mathcal{H}$ and $\widetilde{\mathcal{H}}$, respectively, by the torsor $\mathcal{A}ut_x$ of formal coordinates at x. Then $\mathcal{H}_x$ is simply the central extension of the commutative Lie algebra $\mathcal{K}_x$ (the topological ring of functions on $D_x^\times$) introduced in § 9.1.8. The algebra $\widetilde{\mathcal{H}}_x$ is a completion of $U(\mathcal{H}_x)/(\mathbf{1}-1)$ with respect to the topology, in which the basis of open neighborhoods of zero is formed by the left ideals generated by $\mathfrak{m}_x^N \subset \mathcal{K}_x, N \in \mathbb{Z}$ (here $\mathfrak{m}_x$ denotes the maximal ideal of $\mathcal{O}_x \subset \mathcal{K}_x$).

The space Π_x is naturally identified with the quotient of $\widetilde{\mathcal{H}}_x$ by the left ideal generated by $\mathcal{O}_x \subset \mathcal{K}_x$

Consider the filtration on $\widetilde{\mathcal{H}}$ whose rth term $\widetilde{\mathcal{H}}^{(r)}$ is the completion of

$$\mathrm{span}\{b_{i_1} \ldots b_{i_k}|i_j \in \mathbb{Z}, k \leq r\}.$$

This filtration is preserved by the action of $\mathrm{Aut}\,\mathcal{O}$ and hence induces a filtration on $\widetilde{\mathcal{H}}_x$ whose rth term $\widetilde{\mathcal{H}}_x^{(r)}$ is the completion of

$$\mathrm{span}\{f_1 \ldots f_k|f_i \in \mathcal{K}_x, k \leq r\}.$$

Likewise, π has a filtration (introduced in the proof of Proposition 9.3.2) with the terms

$$\pi^{(r)} = \mathrm{span}\{b_{i_1} \ldots b_{i_k}|0\rangle | i_j < 0, k \leq r\}.$$

The action of $\mathrm{Aut}\,\mathcal{O}$ preserves this filtration, and so we obtain a filtration of Π_x, whose rth term is

$$\Pi_x^{(r)} = \mathrm{span}\{f_1 \ldots f_k | f_i \in \mathcal{K}_x, k \leq r\}/\,\mathrm{span}\{f_1 \ldots f_k | f_i \in \mathcal{K}_x, f_k \in \mathcal{O}_x, k \leq r\}.$$

Given $f_1 \ldots f_n \in \widetilde{\mathcal{H}}_x^{(r)}$ and $\omega = (\omega_0, \ldots, \omega_r) \in \Omega_r(D_x^\times)$, we set

(10.5.5) $\langle \omega, f_1 \ldots f_n \rangle = \mathrm{Res}_{z_1=0} \ldots \mathrm{Res}_{z_n=0} \, f_1(z_1) \ldots f_n(z_n) \omega_n(z_1, \ldots, z_n).$

Here we consider the expansion of $f_1(z_1) \cdots f_r(z_r)\omega_r(z_1, \ldots, z_r)$ in $\mathbb{C}((z_1)) \cdots ((z_n))$, and take the residues first with respect to z_n, then with respect to z_{n-1}, etc. Clearly, the right hand side does not depend on the choice of the coordinate z on D_x.

10.5.6. Lemma. *Formula* (10.5.5) *gives rise to well-defined pairings* $\widetilde{\mathcal{H}}_x^{(r)} \times \Omega_r(D_x^\times) \to \mathbb{C}$ *and* $\Pi_x^{(r)} \times \Omega_r(D_x) \to \mathbb{C}.$

10.5.7. Proof. Formula (10.5.5) certainly gives a well-defined pairing between $\bigoplus_{n=0}^r \mathcal{K}_x^{\otimes n}$ and $\Omega_r(D_x)$. In order to prove that we have a well-defined pairing $\widetilde{\mathcal{H}}_x^{(r)} \times \Omega_r(D_x) \to \mathbb{C}$, we need to show that (10.5.5) vanishes on the ideal generated by the elements of the form $fg - gf + \mathrm{Res}\,fdg$.

Let us give a proof in the case $r = 2$. The generalization to $r > 2$ is straightforward.

We need to check that for all $\omega = (\omega_0, \omega_1, \omega_2) \in \Omega_2(D_x)$ and $f, g \in \widetilde{\mathcal{H}}_x$,

$$\langle \omega, (fg - gf) \rangle = -\langle \omega, \mathrm{Res}\,fdg \rangle = -\omega_0 \,\mathrm{Res}\,fdg$$

(recall that ω_0 is a scalar).

We find from formula (10.5.5) that

$$\mathrm{Res}_{z_1=0} \mathrm{Res}_{z_2=0} f(z_1)g(z_2)\omega_2(z_1, z_2) - \mathrm{Res}_{z_1=0} \mathrm{Res}_{z_2=0} g(z_1)f(z_2)\omega_2(z_1, z_2)$$

$$= \mathrm{Res}_{z_1=0} \mathrm{Res}_{z_2=0} f(z_1)g(z_2)\omega_2(z_1, z_2) - \mathrm{Res}_{z_2=0} \mathrm{Res}_{z_1=0} g(z_2)f(z_1)\omega_2(z_1, z_2),$$

where we used the symmetry of ω_2 and exchanged the variables in the second summand. Next, we find that

$$(\mathrm{Res}_{z_1=0} \mathrm{Res}_{z_2=0} - \mathrm{Res}_{z_2=0} \mathrm{Res}_{z_1=0}) f(z_1)f(z_2)\omega_2(z_1, z_2)$$

$$= \mathrm{Res}_{z_2=0} \mathrm{Res}_{z_1=z_2} f_1(z_1)f_2(z_2)\omega_2,$$

since $f(z_1)f(z_2)\omega_2(z_1, z_2)$ only has poles at $z_1 = 0, z_2 = 0$ and $z_1 = z_2$.

Finally, since

$$\omega_2 = \frac{\omega_0}{(z_1 - z_2)^2} + \text{regular terms},$$

we may evaluate the first residue to obtain

$$\mathrm{Res}_{z_2=0} g(z_2)\omega_0 \frac{\partial f(z_2)}{\partial z_2} dz_2 = -\omega_0 \,\mathrm{Res}\,fdg,$$

as desired.

Thus, formula (10.5.5) defines a pairing $\widetilde{\mathcal{H}}_x^{(r)} \times \Omega_r(D_x^\times) \to \mathbb{C}$.

Consider now the induced pairing $\widetilde{\mathcal{H}}_x^{(r)} \times \Omega_r(D_x) \to \mathbb{C}$. If $f_n \in \mathcal{O}_x$, i.e., $f_n(z_n)$ is regular at $z_n = 0$, then the first residue, and hence the entire pairing, will vanish. It thus follows that the pairing vanishes on the subspace $\mathrm{span}\{f_1 \ldots f_k | f_i \in \mathcal{K}_x, f_k \in$

$\mathcal{O}_x, k \leq r\}$ of $\widetilde{\mathcal{H}}_x^{(r)}$, and therefore the above pairing factors through the pairing $\Pi_x^{(r)} \times \Omega_r(D_x) \to \mathbb{C}$. This proves the lemma.

Hence we obtain a linear map $\Omega_r(D_x) \to (\Pi_x^{(r)})^*$. Now recall that we have constructed an injective map $\Pi_x^* \to \Omega_\infty(D_x)$. It is clear from the construction that this map comes from a compatible system of injective maps $(\Pi_x^{(r)})^* \to \Omega_r(D_x)$. Hence to complete the proof of Proposition 10.5.3 it suffices to show that the composition $\Omega_r(D_x) \to (\Pi_x^{(r)})^* \to \Omega_r(D_x)$ is the identity. This follows immediately from formulas (10.5.5) and (10.5.4). Therefore we obtain isomorphisms $(\Pi_x^{(r)})^* \simeq \Omega_r(D_x)$ and $\Pi_x^* \simeq \Omega_\infty(D_x)$. This completes the proof of Proposition 10.5.3.

10.5.8. Generalization. Denote by $\widetilde{\mathcal{H}}_N, N \geq 0$, the quotient of $\widetilde{\mathcal{H}}$ by the left ideal generated by $b_n, n \geq N$ (so that $\widetilde{\mathcal{H}}_0 = \pi$). Then its twist $\widetilde{\mathcal{H}}_{N,x}$ by the torsor of local coordinates at x is the quotient of $\widetilde{\mathcal{H}}_x$ by the left ideal generated by $\mathfrak{m}_x^N$, where $\mathfrak{m}_x$ is the maximal ideal of $\mathcal{O}_x$. Further, $\widetilde{\mathcal{H}}_{N,x}$ has a filtration with the terms

$$\widetilde{\mathcal{H}}_{N,x}^{(r)} = \operatorname{span}\{f_1 \ldots f_k | f_i \in \mathcal{K}_x, k \leq r\} / \operatorname{span}\{f_1 \ldots f_k | f_i \in \mathfrak{m}_x^N, k \leq r\}.$$

In the same way as above, one shows that formula (10.5.5) defines an isomorphism between $(\widetilde{\mathcal{H}}_{N,x}^{(r)})^*$ and the subspace $\Omega_r^N(D_x^\times)$ of $\Omega_r(D_x^\times)$ which consists of sections having poles of order less than or equal to N along the divisors $z_i = 0$.

Finally, define the following filtered subalgebra $\overline{\mathcal{H}}_x$ of the Weyl algebra $\widetilde{\mathcal{H}}_x$:

$$\overline{\mathcal{H}}_x = \bigcup_{N \geq 0} \bigcup_{r \geq 0} \lim \widetilde{\mathcal{H}}_{N,x}^{(r)}.$$

Then we obtain that $(\overline{\mathcal{H}}_x)^*$ is naturally isomorphic to $\Omega_\infty(D_x^\times)$.

We now turn to the functional realization of the spaces of conformal blocks.

10.5.9. Proposition. $C(X, x, \pi) \simeq \Omega_\infty(X)$.

10.5.10. Proof. Let $(\omega_n)_{n \geq 0}$ be the image of $\varphi \in C(X, x, \pi) \subset \Pi_x^*$ in $\Omega_\infty(D_x)$ under the isomorphism of Proposition 10.5.3. It follows from the discussion of § 10.3.6 that each ω_n extends to a section of $\Omega^{\boxtimes n}(2\Delta)$ over X^n given by $\langle \varphi_n, b_{-1} \otimes \ldots \otimes b_{-1} \rangle$ in the notation of § 10.3.6. Moreover, this section necessarily satisfies the conditions of Definition 10.5.1. Hence the image of $C(X, x, \pi)$ in $\Omega_\infty(D_x)$ belongs to its subspace $\Omega_\infty(X)$.

Now let us show that under the isomorphism $\Pi_x^* \to \Omega_\infty(D_x)$, every element $(\omega_n)_{n \geq 0}$ of $\Omega_\infty(X) \subset \Omega_\infty(D_x)$ corresponds to a conformal block. Indeed, let $\varphi \in \Pi_x^*$ be the corresponding element of Π_x^*. Then we need to show that for any $A \in \Pi_x$, the one-form $\langle \varphi, b(z) \cdot A \rangle dz$ on D_x extends to a regular one-form on $X \backslash x$ (here we use the definition of conformal blocks from § 9.2.11). Let $A = f_1 \ldots f_n$, where $f_i \in \mathcal{K}_x, i = 1, \ldots, n-1$, and $f_n \in \mathcal{K}_x/\mathcal{O}_x$ (see § 10.5.5). By our assumption,

$$\omega_{n+1} = \langle \varphi, b(z)b(z_1) \ldots b(z_n)|0\rangle \rangle dz dz_1 \ldots dz_n$$

extends to a section of $\Omega^{\boxtimes(n+1)}(2\Delta)$ over X^{n+1}. Hence the one-form

$$\operatorname{Res}_{z_1=0} \ldots \operatorname{Res}_{z_n=0} f_1(z_1) \ldots f_n(z_n)\omega_{n+1} = \langle \varphi, b(z) \cdot A \rangle dz$$

on D_x extends to a regular one-form on $X \backslash x$, which is what we wanted to prove.

Thus, we see that the sheaf Ω_∞ is a universal geometric object associated to the vertex algebra π. Its sections over the disc D_x give us the dual space to Π_x, over $D_x^\times$ – the dual to the completed Weyl algebra $\overline{\mathcal{H}}_x$, and over a complete curve X – the space of conformal blocks $C(X, x, \pi)$. These spaces form a commutative diagram:

$$
\begin{array}{ccc}
C(X, x, \pi) & \xrightarrow{\;\sim\;} & \Omega_\infty(X) \\
\downarrow & & \downarrow \\
\Pi_x^* & \xrightarrow{\;\sim\;} & \Omega_\infty(D_x) \\
\downarrow & & \downarrow \\
\overline{\mathcal{H}}_x^* & \xrightarrow{\;\sim\;} & \Omega_\infty(D_x^\times)
\end{array}
$$

10.5.11. The case of zero level. The above functional realization may also be applied to the Heisenberg vertex algebra π^0 of level 0 (see Remark 2.4.3). In § 9.4.3 we already found that $H(X, x, \pi^0) \simeq \operatorname{Sym} H^1(X, \mathcal{O}_X)$. To obtain this result using the functional realization, observe that the OPE reads

$$
b(z)b(w) = \frac{\kappa}{(z-w)^2} + \operatorname{reg}.
$$

Therefore there are no poles allowed when $\kappa = 0$, and our space of polydifferentials at $\kappa = 0$ is simply

$$
\Omega_{r, \kappa=0}(X) = \{(\omega_0, \ldots, \omega_r) | \omega_n \in \Gamma(X^n, \Omega^{\boxtimes n})^{S_n}\}.
$$

Hence there are no "bootstrap conditions" imposed on the components ω_n, and we obtain

$$
\begin{aligned}
C(X, x, \pi^0) &= \varprojlim \Omega_{k, \kappa=0}(X) \\
&= \widehat{\operatorname{Sym}} H^0(X, \Omega),
\end{aligned}
$$

the completed symmetric algebra of the space of one-forms on X. This agrees with our previous answer by Serre duality.

10.6. Bibliographical notes

Definition 10.1.1 and Theorem 10.3.1 are due to E. Frenkel. Various proofs of Theorem 10.3.3 were communicated to us by A. Beilinson, A. Hirschowitz and M. Schneider. Here we reproduced Hirschowitz's proof.

The definition of contragredient module over a vertex algebra is due to I. Frenkel, Y.-Z. Huang and J. Lepowsky [**FHL**], Sect. 5.2 (formula (10.4.8) for the involution ω appeared in [**B1**]). The relationship between conformal blocks and chiral correlation functions in genus zero is also discussed in [**Z2, MS**]. For a genus one analogue, see [**Z1, Z2**].

The construction of § 10.5 parallels the construction of A. Beilinson and V. Drinfeld [**BD1**] in the case of affine Kac–Moody algebras (see also [**BG, Gi2**] for related work).

A similar description of the chiral correlation functions for the fermionic vertex superalgebra $\bigwedge$ has been obtained by A.K. Raina [**Ra**]. He also gave in [**Ra**] an interpretation of various identifies for analytic functions on Riemann surfaces in terms of these correlation functions.

CHAPTER 11

Free Field Realization I

In this chapter and the next we develop the powerful technique of free field realization of the affine vertex algebra $V_k(\mathfrak{g})$. This involves an interplay of ideas from representation theory and "semi–infinite geometry", with the machinery of vertex algebras acting as an intermediary. An important application of this technique is the complete description of the Kac–Moody conformal blocks in genus zero, by reduction to the case of a Heisenberg vertex algebra. This application is the subject of Chapters 13 and 14, where we present a solution to the Knizhnik–Zamolodchikov equations describing the spaces of Kac–Moody conformal blocks when $X = \mathbb{P}^1$ and their dependence on the marked points.

A free field realization may be constructed from an embedding of a Lie algebra into a Weyl algebra. Such embeddings may be obtained from an action of the Lie algebra on an affine space. We start by considering examples of such actions in the finite-dimensional case. This naturally leads us to the study of flag manifolds and various representations associated to them. We then investigate the analogous construction for affine Kac–Moody algebras.

11.1. The idea

11.1.1. Conformal embeddings. Let V, W be conformal vertex algebras, and $V \to W$ a homomorphism which preserves conformal structures (i.e., the conformal vector in W is the image of the conformal vector in V). Then according to Lemma 10.2.1, for any set of W–modules $M_1, \ldots, M_N$ (which are then automatically V–modules) there is a natural embedding of the spaces of conformal blocks

$$(11.1.1) \qquad C_W(X, (x_i), (M_i))_{i=1}^N \hookrightarrow C_V(X, (x_i), (M_i))_{i=1}^N.$$

We will see in Chapter 17 that when we vary the data $(X; x_1, \ldots, x_N)$, the spaces of conformal blocks are organized into sheaves of $\mathcal{D}$–modules over the moduli space of N–pointed curves of a given genus. The above embeddings then give rise to a map from the sheaf on the moduli space corresponding to W to the sheaf corresponding to V.

In the case of curves of genus 0 (and more generally, in the case when we fix the curve X, and vary only the positions of the points) these sheaves are especially simple: they are sheaves of sections of vector bundles on the complement of pairwise diagonals in the Nth power of the curve X, equipped with flat connections. The flat connections are defined entirely in terms of the action of the Virasoro algebra on the modules M_i, considered as W–modules and as V–modules. If the homomorphism $V \to W$ preserves conformal structures, then these two actions coincide. Therefore the embedding of the bundles of conformal blocks is compatible with the flat connections. In particular, if we have a horizontal section of the sheaf of conformal blocks for W, its image in the sheaf of conformal blocks for V is also a horizontal section.

This observation, which follows tautologically from our general formalism, can be used to find explicit formulas for horizontal sections of the bundle of conformal blocks of a complicated vertex algebra V. Namely, suppose we can map V to a simpler vertex algebra W, for which the horizontal sections are already known. Then we can take their images under the embedding (11.1.1) and obtain horizontal sections of the bundle of V–conformal blocks.

As our main example, we will take as V the vertex algebra $V_k(\mathfrak{g})$ associated to an affine Lie algebra. In this case the horizontal sections of the corresponding bundle of conformal blocks are solutions of the Knizhnik–Zamolodchikov equations. We will show how to embed $V_k(\mathfrak{g})$ into a certain Heisenberg vertex algebra. The horizontal sections of the bundle of conformal blocks of the Heisenberg vertex algebra are easy to find (at least in genus 0). Using them we obtain explicit formulas for the horizontal sections of the bundle of conformal blocks for $\widehat{\mathfrak{g}}$, and hence explicit solutions of the Knizhnik–Zamolodchikov equations (see Chapter 13).

The above argument is the type of application that justifies developing the general machinery of vertex algebras and their conformal blocks, even if we are only interested in Kac–Moody algebras. Using this machinery, we can reduce the computation of conformal blocks for $\widehat{\mathfrak{g}}$ to a much simpler computation for a Heisenberg Lie algebra. While both of these objects are Lie algebras, the embedding we use does *not* come from an embedding of these Lie algebras, but only an embedding of the corresponding vertex algebras (thus, it is non-linear in the Lie algebra generators). Therefore in order to see that the corresponding spaces of conformal blocks are related, we must develop the formalism of conformal blocks for general vertex algebras (cf. § 10.2.2). Then the above embedding follows tautologically from the definitions.

11.1.2. Remark. From the physics point of view, the Heisenberg vertex algebra corresponds to a theory of free bosonic fields, which is the simplest conformal field theory, while the affine vertex algebra $V_k(\mathfrak{g})$ underlies the much more complicated WZW model. The formalism that we are about to describe is called the free field realization, as it corresponds to representing a complicated model in terms of a free field theory.

11.1.3. Vertex algebras and differential operators. Let us discuss how we can construct a homomorphism from $V_k(\mathfrak{g})$ to a vertex algebra W. Recall from § 2.4.4 that the vertex algebra $V_k(\mathfrak{g})$ is generated by fields $J^a(z)$, where $\{J^a\}$ is a basis of $\mathfrak{g}$. In particular, $V_k(\mathfrak{g})$ has a basis of lexicographically ordered monomials $J^{a_1}_{n_1} \ldots J^{a_k}_{n_k} v_k$.

11.1.4. Lemma. *Defining a homomorphism of vertex algebras $V_k(\mathfrak{g}) \to W$ is equivalent to choosing vertex operators $\widetilde{J}^a(z), a = 1, \ldots, \dim \mathfrak{g}$, of conformal dimension 1 in the vertex algebra W, whose Fourier coefficients satisfy the relations of $\widehat{\mathfrak{g}}$ of level k.*

11.1.5. Proof. Given a homomorphism $\rho : V_k(\mathfrak{g}) \to W$, take the images of the generating fields $J^a(z)$ of $V_k(\mathfrak{g})$ under ρ. The fact that ρ_k is a homomorphism of vertex algebras implies that the OPEs between these fields, and hence the commutation relations between their Fourier coefficients, are preserved.

Conversely, suppose that we are given vertex operators $\widetilde{J}^a(z)$ satisfying the condition of the lemma. Denote by $\widetilde{J}^a_n$ the corresponding Fourier coefficients. Define a linear map $\rho : V_k(\mathfrak{g}) \to W$ by the formula

$$J^{a_1}_{n_1} \ldots J^{a_m}_{n_m} v_k \mapsto \widetilde{J}^{a_1}_{n_1} \ldots \widetilde{J}^{a_m}_{n_m} |0\rangle.$$

We leave it to the reader to check that this map is a vertex algebra homomorphism.

11.1.6. We have attached to each vertex algebra W a Lie algebra $U(W)$, which is spanned by the Fourier coefficients of the vertex operators from W (see § 4.1.1).

The above lemma shows that in order to define a homomorphism of vertex algebras we need to define a Lie algebra homomorphism $\widehat{\mathfrak{g}} \to U(W)$.

If W is a Heisenberg vertex algebra, then $U(W)$ belongs to a completion of the universal enveloping algebra of a Heisenberg Lie algebra, in which we identify the central element with the unit element. In other words, $U(W)$ belongs to a completed Weyl algebra. But a Weyl algebra can be thought of as the ring of (algebraic) differential operators on an affine space. Therefore, roughly speaking, our task is to embed our Lie algebra $\widehat{\mathfrak{g}}$ into an algebra of differential operators. We know that a Lie algebra can be mapped to differential operators whenever we have an infinitesimal action of this Lie algebra on a manifold by vector fields. Thus, homogeneous manifolds of Lie groups provide a source for such embeddings. In the next section we consider finite-dimensional examples of such embeddings, which will serve as prototypes for the embeddings we are looking for in the infinite-dimensional case.

11.2. Finite–dimensional setting

Suppose S is a homogeneous space for a Lie group G. Then the Lie algebra $\mathfrak{g}$ of G maps to the Lie algebra $\operatorname{Vect} S$ of vector fields on S. Suppose that $U \subset S$ is a Zariski open subset isomorphic to the affine space $\mathbb{A}^m$ with coordinates $y_1, \ldots, y_m$. In general, it is impossible to restrict the action of G to U, but one may always resrict the action of the corresponding Lie algebra,

$$\mathfrak{g} \to \operatorname{Vect} S \hookrightarrow \operatorname{Vect} U.$$

Thus we obtain a map from $\mathfrak{g}$ to vector fields on U.

Denote by $\mathcal{D}(U)$ the algebra of differential operators on the affine space $U = \mathbb{A}^m$. It has generators $y_i, \dfrac{\partial}{\partial y_i}, i = 1, \ldots, m$, and relations

$$\left[\frac{\partial}{\partial y_i}, y_j \right] = \delta_{ij}.$$

This algebra is known as the *Weyl algebra*. This is nothing but the enveloping algebra of a Heisenberg Lie algebra, in which the central element is identified with 1. An example of a (completed) Weyl algebra is the algebra $\widehat{\mathcal{H}}$ introduced in § 2.1.2. Note that the representations of a Heisenberg Lie algebra on which the central element acts as the identity are the same as representations of the corresponding Weyl algebra.

The algebra $\mathcal{D}(U)$ has a natural filtration $\{\mathcal{D}_{\leq i}(U)\}$ by the order of the differential operator. Functions and vector fields on U fit into the short exact sequence

(11.2.1) $0 \to \operatorname{Fun} U \to \mathcal{D}_{\leq 1}(U) \to \operatorname{Vect} U \to 0,$

which has a canonical splitting: namely, we lift $\xi \in \operatorname{Vect} U$ to the unique first order differential operator D_ξ whose symbol equals ξ and which kills the constant functions $D_\xi \cdot 1 = 0$. In fact, we usually take this for granted and do not distinguish between vector fields and the corresponding differential operators.

Now we see that we may lift the map $\mathfrak{g} \to \operatorname{Vect}(U)$ to a map $\mathfrak{g} \to \mathcal{D}(U)$. In summary, a map from a given Lie algebra $\mathfrak{g}$ to a Weyl algebra can be obtained by considering the action of $\mathfrak{g}$ on an open subset of a homogeneous space. We will use this method in the infinite-dimensional context to obtain free field realizations.

11.2.1. Example. Let $\mathfrak{g} = \mathfrak{sl}_2$ and $Y = \mathbb{P}^1 = SL_2/B_-$, where B_- is the lower Borel subgroup of SL_2,

$$B_- = \begin{pmatrix} * & 0 \\ * & * \end{pmatrix}.$$

As an open subset of $\mathbb{P}^1$ we take

$$U = \left\{ \mathbb{C} \begin{pmatrix} y \\ -1 \end{pmatrix} \right\} \subset \mathbb{P}^1$$

(the minus sign is chosen here for notational convenience). We obtain an embedding $\mathfrak{sl}_2 \hookrightarrow \operatorname{Vect} U \hookrightarrow \mathcal{D}(U)$ sending a to v_a which can be calculated explicitly by the formula

$$(v_a \cdot f)(y) = \frac{\partial}{\partial \epsilon} f(\exp(-\epsilon a)y)|_{\epsilon=0}.$$

11.2.2. Exercise. The standard basis for $\mathfrak{sl}_2$ is

$$e = \begin{pmatrix} 0 & 1 \\ 0 & 0 \end{pmatrix}, \quad h = \begin{pmatrix} 1 & 0 \\ 0 & -1 \end{pmatrix}, \quad f = \begin{pmatrix} 0 & 0 \\ 1 & 0 \end{pmatrix}.$$

Show that under this embedding,

$$(11.2.2) \qquad e \mapsto \frac{\partial}{\partial y}, \quad h \mapsto -2y\frac{\partial}{\partial y}, \quad f \mapsto -y^2\frac{\partial}{\partial y}.$$

11.2.3. Flag manifold. Now consider an arbitrary simple Lie algebra $\mathfrak{g}$. It has a Cartan decomposition (as a vector space)

$$(11.2.3) \qquad\qquad \mathfrak{g} = \mathfrak{n}_+ \oplus \mathfrak{h} \oplus \mathfrak{n}_-,$$

where $\mathfrak{h}$ is the Cartan subalgebra and $\mathfrak{n}_\pm$ are the upper and lower nilpotent subalgebras. Introduce the upper and lower Borel subalgebras:

$$\mathfrak{b}_\pm = \mathfrak{h} \oplus \mathfrak{n}_\pm.$$

Let G be the connected simply-connected Lie group corresponding to $\mathfrak{g}$, and $N_\pm$ (respectively, $B_\pm$) the upper and lower unipotent subgroups (respectively, Borel subgroups) of G corresponding to $\mathfrak{n}_\pm$ (respectively, $\mathfrak{b}_\pm$).

A natural candidate for a homogeneous space on which we can realize $\mathfrak{g}$ by differential operators is the flag manifold G/B_-. The flag manifold has an open subset, the so-called big cell $U = N_+ \cdot [B_-] \subset G/B_-$, which is isomorphic to N_+. Note that, in the case of $\mathfrak{sl}_2$ with the above conventions,

$$N_+ = \begin{pmatrix} 1 & * \\ 0 & 1 \end{pmatrix} \simeq \mathbb{A}^1.$$

The action of G on G/B_- gives us an embedding of $\mathfrak{g}$ into the Lie algebra of first order differential operators on N_+, via its identification with the big cell U. Since N_+ is a unipotent Lie group, the exponential map $\mathfrak{n}_+ \to N_+$ is an isomorphism.

Therefore N_+ is isomorphic to the vector space $\mathfrak{n}_+$. Choose a root basis of $\mathfrak{n}_+$, $\{e^\alpha\}_{\alpha \in \Delta_+}$, where Δ_+ is the set of positive roots of $\mathfrak{g}$, so that $[h, e^\alpha] = \alpha(h)e^\alpha$ for all $h \in \mathfrak{h}$. This basis gives us a system of coordinates $\{y_\alpha\}_{\alpha \in \Delta_+}$ on N_+. Thus, N_+ is isomorphic to the affine space $\mathbb{A}^{|\Delta_+|}$, and the space $\mathbb{C}[N_+]$ of regular functions on N_+ is isomorphic to the free polynomial algebra $\mathbb{C}[y_\alpha]_{\alpha \in \Delta_+}$. Hence the algebra of differential operators on N_+ is the Weyl algebra with generators $\{y_\alpha, \partial/\partial y_\alpha\}_{\alpha \in \Delta_+}$, and the standard relations.

Recall the exact sequence (11.2.1). As explained above, this sequence has a canonical splitting, which gives us an embedding $\mathfrak{g} \to \mathcal{D}_{\leq 1}(N_+)$. Given this embedding, we obtain a structure of $\mathfrak{g}$–module on the space of functions $\mathbb{C}[N_+]$ on N_+. We will identify this module with the contragredient Verma module.

11.2.4. Verma modules and contragredient Verma modules. If a vector space $V = \bigoplus_{\alpha \in A} V_\alpha$ is graded by an abelian group A, with finite-dimensional homogeneous components V_α, then we define its restricted dual vector space as $V^\vee = \bigoplus_{\alpha \in A} V_\alpha^*$. Note that if V is a module over a Lie algebra $\mathfrak{g}$, then its restricted dual is also a $\mathfrak{g}$–module. For example, the universal enveloping algebra $U(\mathfrak{n}_+)$ is graded by the root lattice of $\mathfrak{g}$, with finite dimensional homogeneous components. We define the restricted dual $U(\mathfrak{n}_+)^\vee$ using this gradation.

By construction, the action of $\mathfrak{n}_+$ on $\mathbb{C}[N_+]$ satisfies $e^\alpha \cdot y_\alpha = 1$, and $e^\alpha \cdot y_\beta \neq 0$ only if $\beta = \alpha + \gamma$ where $\gamma \in \Delta_+$. Therefore, for any $A \in \mathbb{C}[N_+]$ there exists $P \in U(\mathfrak{n}_+)$ such that $P \cdot A = 1$. Consider the pairing $U(\mathfrak{n}_+) \times \mathbb{C}[N_+] \to \mathbb{C}$ which maps (P, A) to the value of the function $P \cdot A$ at the identity element of N_+. This pairing is $\mathfrak{n}_+$–invariant. The Poincaré-Birkhoff-Witt basis $\{e^{\alpha(1)} \dots e^{\alpha(k)}\}$ (with respect to some ordering on Δ_+) of $U(\mathfrak{n}_+)$ and the monomial basis $\{y_{\alpha(1)} \cdots y_{\alpha(k)}\}$ of $\mathbb{C}[N_+]$ are dual to each other with respect to this pairing. Moreover, both $U(\mathfrak{n}_+)$ and $\mathbb{C}[N_+]$ are graded by the root lattice of $\mathfrak{g}$ (under the action of the Cartan subalgebra $\mathfrak{h}$), and this pairing identifies each graded component of $\mathbb{C}[N_+]$ with the dual of the corresponding graded component of $U(\mathfrak{n}_+)$. Therefore the $\mathfrak{n}_+$–module $\mathbb{C}[N_+]$ is isomorphic to the restricted dual $U(\mathfrak{n}_+)^\vee$ of $U(\mathfrak{n}_+)$. Since $U(\mathfrak{n}_+)$ is a free $\mathfrak{n}_+$–module (with one generator), it is natural to call $\mathbb{C}[N_+]$ a "co-free" $\mathfrak{n}_+$–module.

Now we show that, as a $\mathfrak{g}$–module, $\mathbb{C}[N_+]$ is isomorphic to the module M_0^* contragredient to the Verma module M_0 over $\mathfrak{g}$ with highest weight 0.

First we recall the definition of Verma modules and contragredient Verma modules. For each $\chi \in \mathfrak{h}^*$, consider the one-dimensional representation of $\mathfrak{b}_\pm$ on which $\mathfrak{h}$ acts according to χ, and $\mathfrak{n}_\pm$ acts by 0. The *Verma module* M_χ with highest weight $\chi \in \mathfrak{h}^*$ is the induced module:

$$M_\chi \overset{\text{def}}{=} U(\mathfrak{g}) \otimes_{U(\mathfrak{b}_+)} \mathbb{C}_\chi.$$

The Cartan decomposition (11.2.3) gives us the isomorphisms

$$U(\mathfrak{g}) \simeq U(\mathfrak{n}_-) \otimes U(\mathfrak{b}_+), \qquad U(\mathfrak{g}) \simeq U(\mathfrak{n}_+) \otimes U(\mathfrak{b}_-).$$

Therefore, as an $\mathfrak{n}_-$–module, $M_\chi \simeq U(\mathfrak{n}_-)$.

The *contragredient Verma module* M_χ^* with highest weight $\chi \in \mathfrak{h}^*$ is defined as the coinduced module

$$M_\chi^* = \mathrm{Hom}_{U(\mathfrak{b}_-)}^{\mathrm{res}}(U(\mathfrak{g}), \mathbb{C}_\chi).$$

Here $U(\mathfrak{g})$ is considered as a $U(\mathfrak{b}_-)$–module with respect to the right action, and we consider only those homomorphisms $U(\mathfrak{g}) \to \mathbb{C}_\chi$ which belong to $U(\mathfrak{b}_-)^* \otimes U(\mathfrak{n}_+)^\vee \subset U(\mathfrak{g})^*$. Thus, as an $\mathfrak{n}_+$–module, $M_\chi^* \simeq U(\mathfrak{n}_+)^\vee$.

11.2.5. Remark. The module M_χ^* is contragredient to the Verma module M_χ in the following sense. There is an anti-involution ω on $\mathfrak{g}$, which preserves $\mathfrak{h}$ and sends the standard generators f_i of $\mathfrak{n}_+$ to the standard generators e_i of $\mathfrak{n}_+$, and vice versa. Let M be a $\mathfrak{g}$–module such that the action of $\mathfrak{h}$ on M is diagonal, and the eigenspaces corresponding to various eigenvalues $\lambda : \mathfrak{h} \to \mathbb{C}$ (called weights) are all finite–dimensional. Then we define its contragredient module M^* as the restricted dual of M with respect to the weight decomposition induced by the $\mathfrak{h}$–action, with the action of $\mathfrak{g}$ given by the formula

$$(x \cdot f)(v) = f(\omega(x) \cdot v).$$

It follows from the above definitions that M_χ^* is the contragredient module to M_χ.

The advantage of using the anti-involution ω (as opposed to the anti-involution $x \mapsto -x$, which leads to the ordinary restricted dual modules) is the following. If $\mathfrak{n}_+$ acts locally nilpotently on M, i.e., M belongs to the so-called category $\mathcal{O}$, then the contragredient module to M will also have this property, whereas the dual to M will carry a locally nilpotent action of $\mathfrak{n}_-$, so will belong to a different category.

11.2.6. Identification of $\mathbb{C}[N_+]$ with M_χ^*. Let us choose a generator v_χ of M_χ and the corresponding vector v_χ^* in M_χ^*.

Now observe that, given any $\mathfrak{g}$–module M with a vector v such that

$$\mathfrak{n}_+ \cdot v = 0, \qquad h \cdot v = \chi(h)v, \quad h \in \mathfrak{h},$$

there is a unique homomorphism $M_\chi \to M$ sending v_χ to v and a unique homomorphism $M \to M_\chi^*$ sending v to v_χ^*.

In particular, the constant function 1 is such a vector in the $\mathfrak{g}$–module $\mathbb{C}[N_+]$, with $\chi = 0$. Thus, we obtain a homomorphism $\mathbb{C}[N_+] \to M_0^*$ sending $1 \in \mathbb{C}[N_+]$ to $v_0^* \in M_0^*$. Since both $\mathbb{C}[N_+]$ and M_0^*, considered as $\mathfrak{n}_+$–modules, are isomorphic to $U(\mathfrak{n}_+)^\vee$, we obtain that this map is necessarily an isomorphism. Hence we have an isomorphism of $\mathfrak{g}$–modules: $\mathbb{C}[N_+] \simeq M_0^*$.

Furthermore, the module M_χ^* with an arbitrary weight χ may also be identified with $\mathbb{C}[N_+]$, where the latter is equipped with a modified action of $\mathfrak{g}$.

To see that, recall that we have a canonical lifting of $\mathfrak{g}$ to $\mathcal{D}_{\leq 1}(N_+)$, $a \mapsto v_a$. But this lifting is not unique. We can modify it by adding to each v_a a function $\phi_a \in \mathbb{C}[N_+]$ so that $\phi_{a+b} = \phi_a + \phi_b$. One readily checks that the modified differential operators $v_a + \phi_a$ satisfy the commutation relations of $\mathfrak{g}$ if and only if

$$v_a \cdot \phi_b - v_b \cdot \phi_a = \phi_{[a,b]}.$$

In other words, the linear map $\mathfrak{g} \to \mathbb{C}[N_+]$ given by $a \mapsto \phi_a$ should be a one-cocycle of $\mathfrak{g}$ with coefficients in $\mathbb{C}[N_+]$ (see § A.4).

Each such lifting gives $\mathbb{C}[N_+]$ the structure of a $\mathfrak{g}$–module. But we impose the extra condition that the modified action of $\mathfrak{h}$ on V remains diagonalizable. This means that ϕ_h is a constant function on N_+ for each $h \in \mathfrak{h}$, and therefore our one-cocycle should be $\mathfrak{h}$–invariant: $\phi_{[h,a]} = v_h \cdot \phi_a$, for all $h \in \mathfrak{h}, a \in \mathfrak{g}$. We claim that the space of $\mathfrak{h}$–invariant one-cocycles of $\mathfrak{g}$ with coefficients in $\mathbb{C}[N_+]$ is canonically isomorphic to the first cohomology of $\mathfrak{g}$ with coefficients in $\mathbb{C}[N_+]$, i.e., $H^1(\mathfrak{g}, \mathbb{C}[N_+])$ (see § A.4).

Indeed, it is well-known (see, e.g., [**Fu**]) that if a Lie subalgebra $\mathfrak{h}$ of $\mathfrak{g}$ acts diagonally on $\mathfrak{g}$ and on a $\mathfrak{g}$–module M, then $\mathfrak{h}$ must act by 0 on $H^1(\mathfrak{g}, M)$. Hence $H^1(\mathfrak{g}, \mathbb{C}[N_+])$ is equal to the quotient of the space of $\mathfrak{h}$–invariant one-cocycles by its subspace of $\mathfrak{h}$–invariant coboundaries (i.e., those cocycles which have the form

$\phi_b = v_b \cdot f$ for some $f \in \mathbb{C}[N_+]$). But it is clear that the space of $\mathfrak{h}$–invariant coboundaries is equal to 0 in our case, hence the result.

Thus, the set of liftings of $\mathfrak{g}$ to $\mathcal{D}_{\leq 1}(N_+)$ making $\mathbb{C}[N_+]$ into a $\mathfrak{g}$–module with diagonal action of $\mathfrak{h}$ is naturally isomorphic to $H^1(\mathfrak{g}, \mathbb{C}[N_+])$. By Shapiro's lemma (see § A.4),

$$H^1(\mathfrak{g}, \mathbb{C}[N_+]) = H^1(\mathfrak{g}, M_0^*) \simeq H^1(\mathfrak{b}_-, \mathbb{C}_0) = (\mathfrak{b}_-/[\mathfrak{b}_-, \mathfrak{b}_-])^* \simeq \mathfrak{h}^*.$$

Thus, for each $\chi \in \mathfrak{h}^*$ we obtain an embedding $l_\chi : \mathfrak{g} \hookrightarrow \mathcal{D}_{\leq 1}(N_+)$ and hence the structure of an $\mathfrak{h}^*$–graded $\mathfrak{g}$–module on $\mathbb{C}[N_+]$. Let us analyze this $\mathfrak{g}$–module in more detail.

We have: $v_h \cdot y_\alpha = -\alpha(h)y_\alpha, \alpha \in \Delta_+$, so the weight of any monomial in $\mathbb{C}[N_+]$ is equal to a sum of negative roots. Since our one-cocycle is $\mathfrak{h}$–invariant, we obtain $v_h \cdot \phi_{e^\alpha} = \phi_{[h,e^\alpha]} = \alpha(h)\phi_{e^\alpha}, \alpha \in \Delta_+$, so the weight of ϕ_{e^α} has to be equal to the positive root α. Therefore $\phi_{e^\alpha} = 0$ for all $\alpha \in \Delta_+$. Thus, the action of $\mathfrak{n}_+$ on $\mathbb{C}[N_+]$ is not modified, and remains co-free. On the other hand, by construction, the action of $h \in \mathfrak{h}$ is modified by $h \mapsto h + \chi(h)$. Therefore the vector $1 \in \mathbb{C}[N_+]$ is still annihilated by $\mathfrak{n}_+$, but now it has weight χ with respect to h. Hence there is a homomorphism $\mathbb{C}[N_+] \to M_\chi^*$ sending $1 \in \mathbb{C}[N_+]$ to $v_\chi^* \in M_\chi^*$. In the same way as in the case $\chi = 0$ we show that this map is an isomorphism. Thus, under the modified action obtained via the lifting l_χ, the $\mathfrak{g}$–module $\mathbb{C}[N_+]$ is isomorphic to the contragredient Verma module M_χ^*.

11.2.7. Remark. If χ is an integral weight, then the above lifting $l_\chi : \mathfrak{g} \to \mathcal{D}_{\leq 1}(N_+)$ can be obtained as follows. There is a $\mathfrak{g}$–equivariant line bundle $\mathcal{L}_\chi$ on G/B_-, and the space of sections of this line bundle over the big cell U is naturally isomorphic to M_χ^*. Since U is an affine space, the bundle $\mathcal{L}_\chi|_U$ can be trivialized, and the space of its sections over U can be identified with $\mathbb{C}[N_+]$. The action of $\mathfrak{g}$ on the space of sections of $\mathcal{L}_\chi|_U$ can then be written in terms of first order differential operators. These are exactly the differential operators that we obtain when we apply the lifting l_χ corresponding to χ.

For general χ there is no line bundle $\mathcal{L}_\chi$, but there exists a sheaf $\mathcal{D}_\chi$ of twisted differential operators (see § A.3.3), which for integral χ is the sheaf $\mathcal{D}_{\mathcal{L}_\chi}$ of differential operators acting on sections of $\mathcal{L}_\chi$. The module M_χ^* is naturally the space of sections over $U \subset G/B$ of a $\mathcal{D}_\chi$–module, and the action of $\mathfrak{g}$ on it is obtained via an embedding $\mathfrak{g} \hookrightarrow \Gamma(G/B_-, \mathcal{D}_\chi)$. (In fact, it is known that $\Gamma(G/B_-, \mathcal{D}_\chi) \simeq U(\mathfrak{g})/I_\chi$, where I_χ is the ideal determined by the character of the center of $U(\mathfrak{g})$ acting on M_χ^*.) Over U, the sheaf $\mathcal{D}_\chi$ can be identified with the sheaf of ordinary differential operators, and so we obtain our embedding l_χ of $\mathfrak{g}$ into differential operators.

11.2.8. Exercise. Show that in the case of $\mathfrak{sl}_2$ the formulas for the liftings $\mathfrak{sl}_2 \to \mathcal{D}_{\leq 1}(\mathbb{A}^1)$ corresponding to $\nu \in \mathfrak{h}^*$ read as follows:

$$(11.2.4) \qquad e \mapsto \frac{\partial}{\partial y}, \qquad h \mapsto -2y\frac{\partial}{\partial y} + \nu, \qquad f \mapsto -y^2\frac{\partial}{\partial y} + \nu y.$$

11.2.9. Representations of Weyl algebras. Up to now, we have considered representations of $\mathfrak{g}$ (such as contragredient Verma modules) obtained from the action of the algebra of differential operators on an affine space U (i.e., a Weyl algebra) on the space of polynomial functions on U. There are however other choices for representations of the Weyl algebras, which lead to other classes of representations of $\mathfrak{g}$.

Let S be an even–dimensional vector space with an anti–symmetric non–degenerate inner product $(\cdot, \cdot)$. The corresponding *Weyl algebra* of S is by definition

$$(11.2.5) \qquad\qquad A = A(S) = T^\bullet S / I,$$

where I denotes the two-sided ideal of $T^\bullet S$ generated by elements of the form $\{a \otimes b - b \otimes a - (a, b) \cdot 1\}$. Thus, A has generators $a \in S$ and relations

$$(11.2.6) \qquad\qquad [a, b] = (a, b) \cdot 1.$$

How can one construct an irreducible representation of A? Pick a Lagrangian (i.e., maximal isotropic) subspace $L \subset S$ (so we have $\dim L = \frac{1}{2} \dim S$). Then the symmetric algebra $\mathrm{Sym}^\bullet L$ is a subalgebra of A, since the elements of L commute in A. This is a maximal commutative subalgebra in A; in fact, all maximal commutative subalgebras in A arise this way. Consider $\mathbb{C}$ as a trivial representation of $\mathrm{Sym}^\bullet L$ and take the induced A–module

$$(11.2.7) \qquad\qquad M_L = \mathrm{Ind}_{\mathrm{Sym}^\bullet L}^A \mathbb{C}.$$

In order to see the structure of this A–module, pick a Lagrangian subspace $L' \subset S$ which is complementary to L, so that $S \simeq L \oplus L'$. Since $(\cdot, \cdot)$ is non–degenerate, $L' \simeq L^*$, and thus S is isomorphic to the cotangent bundle of L. Therefore the module M_L is isomorphic to the algebra $\mathrm{Sym}^\bullet L^*$ as a $\mathrm{Sym}^\bullet L^*$–module, and L acts on M_L by derivations (contractions).

11.2.10. Examples. Let $V = \mathbb{C}^2$, with coordinates x, y. Set $S = T^*V \simeq V \oplus V^*$ with its standard skew-symmetric form: $(a, b) = 0$ if a and b both belong to V or V^*, and

$$(11.2.8) \qquad (a, b) = -(b, a) = \langle a, b \rangle, \qquad a \in V, b \in V^*.$$

We consider several cases.

1. $L = V$. Then $M_L = \mathbb{C}[V] \simeq \mathbb{C}[x, y]$ is generated by a vector v satisfying
$$\partial_x v = \partial_y v = 0.$$

2. $L = V^*$. Then M_L is generated by a vector v satisfying
$$x \cdot v = y \cdot v = 0.$$

   These are the relations satisfied by the delta–function supported at the origin $x = y = 0$. Hence
$$M_L = \mathbb{C}[\partial_x, \partial_y] v$$
   can informally be thought of as a space of generalized functions (or "delta–functions") supported at the origin in $\mathbb{C}^2$.

3. $L = \mathrm{span}\{\frac{\partial}{\partial x}, y\}$. M_L is generated by a vector v satisfying
$$\partial_x v = y \cdot v = 0.$$

   Thus
$$M_L = \mathbb{C}[x, \partial_y] v,$$
   so M_L may be informally thought of as the space of "delta–functions" on $\mathbb{C}^2$ supported at the line $y = 0$.

The above pattern is replicated in higher dimensions: we may identify S with the cotangent bundle of any given Lagrangian subspace L, and the module $M_{L'}$, where L' is another Lagrangian subspace, can then be thought of as the space of delta–functions on L supported on $L \cap L'$.

11.2.11. Twisted Verma modules. We have constructed above a family l_χ of embeddings of $\mathfrak{g}$ into the algebra of differential operators on the big cell $U \subset G/B_-$. Choosing different representations of this Weyl algebra, we obtain different $\mathfrak{g}$–modules. We have seen that the module of functions gives us the contragredient Verma module. Now consider the module of delta–functions supported at the origin of the big cell U, i.e., $1 \cdot [B_-] \subset G/B_-$.

It is well-known that the space of delta–functions on a finite-dimensional Lie group G supported at $1 \in G$ is canonically identified with the universal enveloping algebra $U(\mathfrak{g})$. This implies that the corresponding $\mathfrak{g}$–module of delta-functions supported at the origin of U is isomorphic to $U(\mathfrak{g}) \otimes_{U(\mathfrak{b}_-)} \mathbb{C}_\chi$, where χ reflects the choice of the embedding $\mathfrak{g} \to \mathcal{D}(U)$ as before. This is the "upside-down" Verma module, which is free as an $\mathfrak{n}_+$–module rather than as an $\mathfrak{n}_-$–module.

To construct the Verma module M_χ, we should take the space of delta–functions supported not at the origin of the big cell $U \simeq N_+$, but at the origin of the opposite big cell, i.e., the orbit of N_- on G/B_-. This point is actually the unique one-point N_+–orbit on the flag manifold G/B_-.

More generally, suppose that the Lie algebra $\mathfrak{g}$ acts on a variety S. Then we may restrict the action of $\mathfrak{g}$ not only to open subsets U, but to the formal completion $\widehat{S}_Z$ of S along a subvariety $Z \subset S$. The differential operators on $\widehat{S}_Z$ act on the space of delta–functions supported on Z (in the sense of § 11.2.10). In the case of the flag manifold, if we choose Z to be the open N_+–orbit on the flag manifold, we obtain the contragredient Verma modules. If we choose Z to be the one-point N_+–orbit, then we obtain the Verma modules.

Both M_χ and M_χ^* belong to the so-called category $\mathcal{O}$, whose objects are $\mathfrak{g}$–modules with diagonal action of $\mathfrak{h}$ and locally nilpotent action of $\mathfrak{n}_+$ (analogous categories may be defined for all conjugates of $\mathfrak{n}_+$, such as $\mathfrak{n}_-$, and they are all equivalent to each other). It is easy to see that a $\mathfrak{g}$–module of delta-functions supported on a subvariety Z of the flag manifold will belong to the category $\mathcal{O}$ if and only if Z is a union of N_+–orbits. In fact, the space of global sections of any N_+–equivariant $\mathcal{D}$–module on the flag manifold (such as the sheaf of delta-functions supported on an N_+–orbit) gives us a $\mathfrak{g}$–module from the category $\mathcal{O}$.

The N_+–orbits, known as the Schubert cells, in the flag manifold G/B_- are parameterized by the Weyl group $W = N(H)/H$, where H is the Cartan subgroup of G, and $N(H)$ is its normalizer. The orbit corresponding to $w \in W$ is $N_+ \cdot [w] \subset G/B_-$. In particular, the identity element corresponds to the big cell U, and the longest element corresponds to the one-point N_+–orbit. As we have seen above, the $\mathfrak{g}$–modules associated to these orbits are the contragredient Verma modules and the Verma modules, respectively. To other N_+–orbits we may also attach families of $\mathfrak{g}$–modules, called twisted Verma modules (see [**FF2**]), which are "intermediate" between the Verma modules and the contragredient Verma modules.

We will see below that this flexible construction of representations from delta–functions is particularly well suited for our infinite–dimensional applications.

11.3. Infinite–dimensional setting

11.3.1. Generalized flag manifolds. Now let us try to generalize the constructions of the previous section to the affine Kac–Moody algebras.

It is possible to define a flag manifold for an arbitrary Kac–Moody algebra, and in particular for affine Kac–Moody algebras. Recall that the flag manifold was

defined in the finite-dimensional setting as the quotient of the Lie group G by its
Borel subgroup. However in the case of affine algebras there are several candidates
for the "Borel subgroup", and hence several choices for the flag manifold.

By its very definition, each Kac–Moody algebra $\mathfrak{g}$ comes equipped with a Cartan
decomposition (11.2.3). Therefore each Kac–Moody algebra comes equipped with
two obvious choices for a Borel subalgebra: $\mathfrak{b}_- = \mathfrak{h} \oplus \mathfrak{n}_-$, and $\mathfrak{b}_+ = \mathfrak{h} \oplus \mathfrak{n}_+$.
However, there are other choices as well. Consider the anti-involution ω on $\mathfrak{g}$ which
preserves $\mathfrak{h}$ and sends the standard generators of $\mathfrak{n}_+$ to those of $\mathfrak{n}_-$ and vice versa
(cf. Remark 11.2.5). It is natural to call a Lie subalgebra $\mathfrak{b}$ of $\mathfrak{g}$ a generalized Borel
subalgebra if $\mathfrak{b} = \mathfrak{h} \oplus \mathfrak{n}$, where $\mathfrak{n}$ is a Lie subalgebra of $\mathfrak{g}$ such that

$$\mathfrak{g} = \mathfrak{n} \oplus \mathfrak{h} \oplus \omega(\mathfrak{n}).$$

Now, given a generalized Borel subalgebra, we consider the corresponding subgroup
B of the Kac–Moody group, and define the corresponding flag manifold as G/B.
The flag manifolds corresponding to two different choices of Borel subgroups, B and
B', are isomorphic if and only if B and B' are conjugate by an element of G. For
example, if $\mathfrak{g}$ is a finite-dimensional simple Lie algebra, then all Borel subalgebras
are conjugate to each other, and therefore there is only one flag manifold up to an
isomorphism (e.g., we have $G/B_- \simeq G/B_+$). But if $\mathfrak{g}$ is infinite-dimensional, this
is no longer true.

11.3.2. The case of affine algebras. Let us examine more closely generalized Cartan decompositions of affine Kac–Moody algebras. Consider for a moment
the "polynomial" version of the affine algebra $\widehat{\mathfrak{g}}$, which is the central extension of
the Lie algebra $\mathfrak{g}[t, t^{-1}]$ (rather than $\mathfrak{g}((t))$). By abuse of notation, we will again
denote it by $\widehat{\mathfrak{g}}$. This Lie algebra has the following Cartan decomposition:

(11.3.1) $$\widehat{\mathfrak{g}} = \widehat{\mathfrak{n}}_- \oplus \widehat{\mathfrak{h}} \oplus \widehat{\mathfrak{n}}_+,$$

where

$$\widehat{\mathfrak{n}}_- = (\mathfrak{n}_- \otimes 1) \oplus (\mathfrak{g} \otimes t^{-1}\mathbb{C}[t^{-1}]),$$
$$\widehat{\mathfrak{h}} = (\mathfrak{h} \otimes 1) \oplus \mathbb{C}K,$$
$$\widehat{\mathfrak{n}}_+ = (\mathfrak{n}_+ \otimes 1) \oplus (\mathfrak{g} \otimes t\mathbb{C}[t]).$$

The obvious candidates for Borel subalgebras are $\widehat{\mathfrak{b}}_- = (\widehat{\mathfrak{n}}_- \oplus \widehat{\mathfrak{h}})$ and $\widehat{\mathfrak{b}}_+ = (\widehat{\mathfrak{n}}_+ \oplus \widehat{\mathfrak{h}})$. But unlike the finite-dimensional case, they are not conjugate to each
other, and hence give rise to different flag manifolds. In fact, the picture is much
richer, because in addition to the above two obvious choices of Borel subalgebras
there are other "intermediate" choices. The most natural among them are the Lie
subalgebras

(11.3.2) $$\mathfrak{n}_\pm[t, t^{-1}] \oplus \mathfrak{h}[t^{\mp 1}].$$

Now let us consider the corresponding groups. At this point we return to the
formal power series versions of our Lie algebras. Note that as far as flag manifolds
are concerned, the central extension is irrelevant, since any Borel subgroup will
contain the center, and so when we divide by a Borel subgroup, we eliminate the
center. Hence we can look at the loop groups instead. The Lie group of $\mathfrak{g}((t))$
is the ind-group $G((t))$. The Lie groups $\widehat{B}_+$ and $\widehat{N}_+$ of $\widehat{\mathfrak{b}}_+$ and $\widehat{\mathfrak{n}}_+$, respectively,
are subgroups of the group $G[[t]]$. The Lie groups $\widehat{B}_-$ and $\widehat{N}_-$ corresponding to

$\widehat{\mathfrak{b}}_-$ and $\widehat{\mathfrak{n}}_-$, respectively, are subgroups of the ind-group $G[t^{-1}]$. Now let us look at the corresponding flag manifolds and their $\widehat{N}_+$–orbits. The reason why we are interested in $\widehat{N}_+$–orbits is the same as in the finite-dimensional case (see § 11.2.11). Namely, there is an analogue of the category $\mathcal{O}$ in the affine case, whose objects are $\widehat{\mathfrak{g}}$–modules with diagonal action of $\widehat{\mathfrak{h}}$ and locally nilpotent action of $\widehat{\mathfrak{n}}_+$. The spaces of delta-functions supported on $\widehat{N}_+$–orbits (and spaces of global sections of more general $\widehat{N}_+$–equivariant sheaves) on any of the flag manifolds give us $\widehat{\mathfrak{g}}$–modules from the category $\mathcal{O}$.

One can show that the flag manifold $G((t))/\widehat{B}_-$ may be given the structure of a scheme of infinite type. Furthermore, it is stratified by $\widehat{N}_+$–orbits of finite codimension, parameterized by the affine Weyl group. In particular, there is an open orbit $\widehat{U}$ which is the analogue of the big cell $U \subset G/B_-$. The orbit $\widehat{U}$ is isomorphic to the projective limit of affine spaces, and so the space of functions on $\widehat{U}$ is the ring of polynomials in infinitely many variables. The Lie algebra $\widehat{\mathfrak{g}}$ acts on $\widehat{U}$, and the space of functions on $\widehat{U}$, considered as $\widehat{\mathfrak{g}}$–modules via different liftings of $\widehat{\mathfrak{g}}$ to the algebra of differential operators, are just the contragredient Verma modules over $\widehat{\mathfrak{g}}$.

On the other hand, the "opposite" flag manifold $G((t))/\widehat{B}_+$ may be given a structure of an ind-scheme, stratified by finite-dimensional $\widehat{N}_+$–orbits (moreover, the ind-structure may be obtained from the closures of these orbits, which are finite-dimensional algebraic varieties). The space of delta-functions supported at the one-point $\widehat{N}_+$–orbit gives rise to the Verma modules over $\widehat{\mathfrak{g}}$. Other $\widehat{N}_+$–orbits in $G((t))/\widehat{B}_+$ and $G((t))/\widehat{B}_-$ give rise to various twisted Verma modules and twisted contragredient Verma modules. Note that Verma modules and contragredient Verma modules over $\widehat{\mathfrak{g}}$ now "live" on different flag manifolds.

11.3.3. Semi-infinite flag manifold. Now let us look at the generalized Borel subgroup corresponding to the Lie algebra $\widetilde{\mathfrak{b}}_-$ given by formula (11.3.2), or more precisely, its completion in $\mathfrak{g}((t))$. It is equal to the product of the ind-groups $N_-((t))$ and $H[[t]]$. The flag manifold attached to it should therefore be a quotient of $G((t))$ by this subgroup. However, in general a quotient of an ind-scheme (such as $G((t))$) by an ind-group cannot be given the structure of a scheme or an ind-scheme (it exists as a sheaf on the category of schemes).

Note that $N_-((t)) \cdot H[[t]]$ is the connected component $B_-((t))_0$ of the ind-group $B_-((t))$. The corresponding quotient $G((t))/B_-((t))_0$ is called the semi-infinite flag manifold. Morally, it should be viewed as the space of maps from the punctured disc $D^\times = \operatorname{Spec} \mathbb{C}((t))$ to the finite-dimensional flag manifold G/B_-. It is natural to call it the "formal loop space" of G/B_-.

The definition of formal loop spaces for general algebraic varieties is a subtle issue in algebraic geometry. Consider first the simpler question of constructing the space of maps from the disc D to a variety Z. In § 9.4.4 we associated to an affine algebraic variety Z its jet scheme JZ, which represents the space of maps $D \to Z$. If S is a general scheme equipped with an affine cover Z_i, the jet schemes JZ_i can be naturally "glued" together to form the jet scheme JS of S. This follows from the fact that rational changes of coordinates $x_i, i = 1, \ldots, N$, lead to well-defined changes of the coordinates $x_{i,n}$ of the jet scheme, which are given by functions that are rational in $x_{i,0}, i = 1, \ldots, N$, and polynomial in $x_{i,n}, i = 1, \ldots, N, n < 0$ (here we use the notation introduced in § 9.4.4).

For instance, suppose that $S = \mathbb{P}^1$, covered by two affine lines. Then we need to implement the change of variables $x \mapsto x^{-1}$. At the level of jet schemes, it results in the following change of variables:

$$x_0 \mapsto x_0^{-1}, \qquad x_1 \mapsto -x_1 x_0^{-2}, \qquad x_2 \mapsto -x_2 x_0^{-2} + x_1^2 x_0^{-3},$$

etc., obtained by equating the coefficients in $x(t) \mapsto x(t)^{-1}$ and viewing the variables $x_n, n < 0$, as "small" compared to x_0.

Now consider the question of constructing the formal loop space LZ of maps $D^\times \to Z$. For an affine variety $Z = \operatorname{Spec} A$, where $A = \mathbb{C}[x_1, \dots, x_N]/(P_1, \dots, P_k)$, it is represented by an ind-scheme, which is constructed as follows. For each $m \geq 0$ we introduce the ring

$$A_\infty^{(m)} = \mathbb{C}[x_{1,n}, \dots, x_{N,n}]_{n \leq m}/(P_{1,n}, \dots, P_{k,n})_{n \leq m},$$

where $P_{i,n}$ is the coefficient of t^n in the series $P_i(x_1(t), \dots, x_N(t))$, where the $x_i(t) = \sum_{n \leq m} x_{i,n} t^{-n}$ (so $A_\infty^{(0)}$ is the ring A_∞ introduced in § 9.4.4). We have natural maps $A_\infty^{(m+1)} \to A_\infty^{(m)}$, and we define an ind-scheme LZ as $\varinjlim \operatorname{Spec} A_\infty^{(m)}$. Then we have $\operatorname{Hom}(\operatorname{Spec} R((t)), Z) = \operatorname{Hom}(\operatorname{Spec} R, LZ)$, so LZ represents the formal loop space of Z.

However, unlike jet schemes, these formal loop spaces do not glue together in any obvious way, because now changes of variables x_i do not result in well-defined changes of the coefficients of the series $x_i(t)$ (for instance, it is not clear what to make of the series $x(t)^{-1}$). Therefore it is not clear how to define the formal loop spaces for a general algebraic variety, such as G/B_-. Nevertheless, we can define a "formal neighborhood" of JS in LS for any algebraic variety S by requiring that all coefficients $x_{i,m}, m > 0$, of the series $x_i(t)$ (those in front of the negative powers of t) be nilpotent. Then any rational changes of coordinates can be implemented (it is again instructive to look at the case $x(t) \mapsto x(t)^{-1}$), and so we can "glue" together different affine pieces.

Recall that our goal at the moment is to attach $\widehat{\mathfrak{g}}$-modules to $\widehat{N}_+$-orbits on various flag manifolds. Even if the flag manifold in question is not well-defined, its $\widehat{N}_+$-orbits and even their "formal neighborhoods" may be perfectly well-defined. Since we are interested in the modules of "delta-functions" supported on a given orbit, these data are sufficient for our purposes.

In the case of the semi-infinite flag manifold, the $\widehat{N}_+$-orbit of the unit coset is isomorphic to $N_+[[t]]$, and its "open neighborhood" is isomorphic to $N_+((t))$, the formal loop space of the big cell $U \simeq N_+$ of G/B_-. Hence we can try to construct the corresponding $\widehat{\mathfrak{g}}$-modules using these data. This leads us to the construction of the Wakimoto modules introduced in [**Wak, FF1**]. In the remainder of this chapter and in the next one we will discuss this construction, following [**FF2**]. Along the way we obtain some results which may be viewed as the ingredients of a new theory of "semi–infinite varieties".

11.3.4. Vector fields on an infinite-dimensional big cell.
Consider the simplest case of the loop algebra $L\mathfrak{sl}_2 = \mathfrak{sl}_2((t))$. The corresponding semi-infinite flag manifold should be thought of as the formal loop space of $\mathbb{P}^1$. For our purposes, we only need a neighborhood of a single $\widehat{N}_+$-orbit there. For this purpose we can

simply take the formal loop space of the big cell $\mathbb{A}^1 \subset \mathbb{P}^1$:

$$L\mathbb{A}^1 = \left\{ \begin{pmatrix} x(t) \\ -1 \end{pmatrix} \right\}, \qquad x(t) = \sum_{n \in \mathbb{Z}} x_n t^n \in \mathbb{C}((t)).$$

We identify $L\mathbb{A}^1$ with the space $\mathcal{K}$ of functions on the punctured disc. Our $\widehat{N}_+$–orbit is then identified with the subspace $\mathcal{O} \subset \mathcal{K}$ of functions on the disc. To simplify the exposition, in what follows we choose a coordinate t on the disc, and identify $\mathcal{K}$ with $\mathbb{C}((t))$, and $\mathcal{O}$ with $\mathbb{C}[[t]]$. However, our constructions below do not depend on the choice of t.

Let us define the spaces of functions and vector fields on $\mathcal{K}$. We view $\mathcal{K}$ as a complete topological vector space with the basis of open neighborhoods of zero formed by the subspaces $\mathfrak{m}^N \subset \mathcal{K}, N \in \mathbb{Z}$, where $\mathfrak{m}$ is the maximal ideal of $\mathcal{O}$. From the algebraic point of view, $\mathcal{K}$ is an affine ind-scheme

$$\mathcal{K} = \varinjlim \mathfrak{m}^N, \qquad N < 0.$$

We have

$$\mathfrak{m}^N \simeq t^N \mathbb{C}[[t]] = \left\{ \sum_{n \geq N} x_n t^n \right\} = \operatorname{Spec} \mathbb{C}[x_m]_{m \geq N}.$$

Therefore we obtain that the ring of functions on $\mathcal{K}$, denoted by $\operatorname{Fun} \mathcal{K}$, is the inverse limit of the rings $\mathbb{C}[x_m]_{m \geq N}, N < 0$, with respect to the natural surjective homomorphisms $s_{N,M} : \mathbb{C}[x_m]_{m \geq N} \to \mathbb{C}[x_m]_{m \geq M}$, for $N < M$, such that $x_m \mapsto 0$ for $N \leq m < M$ and $x_m \mapsto x_m, m \geq M$. This is a complete topological ring, with the basis of open neighborhoods of 0 given by the ideals generated by $x_k, k < N$, i.e., the kernels of the homomorphisms $s_{\infty,M} : \operatorname{Fun} \mathcal{K} \to \operatorname{Fun} \mathfrak{m}^N$.

A vector field on $\mathcal{K}$ is by definition a continuous linear endomorphism ξ of $\operatorname{Fun} \mathcal{K}$ which satisfies the Leibniz rule: $\xi(fg) = \xi(f)g + f\xi(g)$.

In other words, a vector field is a linear endomorphism ξ of $\operatorname{Fun} \mathcal{K}$, such that for any $M < 0$ there exist $N \leq M$ and a derivation $\xi_{N,M} : \mathbb{C}[x_m]_{m \geq N} \to \mathbb{C}[x_m]_{m \geq M}$ (i.e., a linear map satisfying $\xi_{N,M}(ab) = \xi_{N,M}(a)s_{N,M}(b) + s_{N,M}(a)\xi_{N,M}(b)$) which satisfies

$$s_{\infty,M}(\xi \cdot f) = \xi_{N,M} \cdot s_{\infty,N}(f)$$

for all $f \in \operatorname{Fun} \mathcal{K}$. The space of vector fields is naturally a Lie algebra, which we denote by $\operatorname{Vect} \mathcal{K}$.

11.3.5. Remark. More concretely, an element of $\operatorname{Fun} \mathcal{K}$ may be represented as a (possibly infinite) series

$$P_0 + \sum_{n < 0} P_n x_n,$$

where $P_0 \in \mathbb{C}[x_m]_{m \geq 0}$, and the P_n's are arbitrary (finite) polynomials in $x_m, m \in \mathbb{Z}$.

The Lie algebra $\operatorname{Vect} \mathcal{K}$ may also be described as follows. Identify the tangent space $T_0 \mathcal{K}$ to the origin in $\mathcal{K}$ with $\mathcal{K}$, equipped with the structure of a complete topological vector space. Then $\operatorname{Vect} \mathcal{K}$ is isomorphic to the completed tensor product of $\operatorname{Fun} \mathcal{K}$ and $\mathcal{K}$. This means that vector fields on $\mathcal{K}$ can be described more concretely as series

$$\sum_{n \in \mathbb{Z}} P_n \cdot \frac{\partial}{\partial x_n},$$

where $P_n \in \text{Fun}\,\mathcal{K}$ satisfies the following property: for each $M \geq 0$, there exists $N \leq M$ such that each $P_n, n \leq N$, lies in the ideal generated by the $x_m, m \leq M$. The commutator between two such series is computed in the standard way (term by term), and it is again a series of the above form.

Note that in the above definitions of functions and vector fields on $\mathcal{K}$ we used an isomorphism $\mathcal{K} \simeq \mathbb{C}((t))$ (i.e., a formal coordinate on our disc). But it is easy to see that these spaces do not depend on the choice of the isomorphism $\mathcal{K} \simeq \mathbb{C}((t))$.

11.3.6. The action of $L\mathfrak{sl}_2$. Since we have identified $\mathcal{K}$ with the neighborhood of an $\widehat{N}_+$–orbit in the semi-infinite flag manifold, we obtain that the Lie algebra $L\mathfrak{sl}_2$ naturally acts on $\mathcal{K}$ by vector fields. More precisely, we have a homomorphism of Lie algebras $L\mathfrak{sl}_2 \to \text{Vect}\,\mathcal{K}$. We can easily calculate the image of the (topological) basis elements of $L\mathfrak{sl}_2$,

$$e_n = e \otimes t^n, \qquad h_n = h \otimes t^n, \qquad f_n = f \otimes t^n.$$

For example, we calculate the action of $e_n = \begin{pmatrix} 0 & t^n \\ 0 & 0 \end{pmatrix}$,

$$\exp(-\epsilon e_n) \begin{pmatrix} \sum_m x_m t^m \\ -1 \end{pmatrix} = \begin{pmatrix} \sum_{m \neq n} x_m t^m + (x_n + \epsilon)t^n \\ -1 \end{pmatrix},$$

and so, differentiating at $\epsilon = 0$, we obtain that e_n acts as $\dfrac{\partial}{\partial x_n}$. Similarly one finds the actions of h_n and f_n. The result is

$$e_n \;\mapsto\; \frac{\partial}{\partial x_n},$$

(11.3.3)
$$h_n \;\mapsto\; -2 \sum_{-i+j=n} x_i \frac{\partial}{\partial x_j},$$

$$f_n \;\mapsto\; - \sum_{-i-j+k=n} x_i x_j \frac{\partial}{\partial x_k}.$$

These formulas may be simplified by the introduction of generating functions. Thus we set

$$e(z) = \sum_{n \in \mathbb{Z}} e_n z^{-n-1},$$

and similarly for h and f. Note that $e(z)$ looks like the vertex operator $Y(e_{-1}v_k, z)$ from the vertex algebra $V_k(\widehat{\mathfrak{sl}}_2)$, as do $h(z)$ and $f(z)$.

Now introduce the notation $a_n = \dfrac{\partial}{\partial x_n}$ and $a_n^* = x_{-n}$, and set

(11.3.4)
$$a(z) \;=\; \sum_{n \in \mathbb{Z}} a_n z^{-n-1} = \sum_{n \in \mathbb{Z}} \frac{\partial}{\partial x_n} z^{-n-1},$$

(11.3.5)
$$a^*(z) \;=\; \sum_{n \in \mathbb{Z}} a_n^* z^{-n} = \sum_{n \in \mathbb{Z}} x_{-n} z^{-n}.$$

In terms of these series, equations (11.3.3) become

$$e(z) \;\mapsto\; a(z),$$

(11.3.6)
$$h(z) \;\mapsto\; -2a^*(z)a(z),$$

$$f(z) \;\mapsto\; -a^*(z)^2 a(z)$$

(compare with formulas (11.2.2)).

The relevant Weyl algebra has generators a_n and a_n^*, $n \in \mathbb{Z}$, and relations

$$(11.3.7) \qquad [a_n, a_m^*] = \delta_{n,-m}, \qquad [a_n, a_m] = [a_n^*, a_m^*] = 0.$$

We denote this algebra by A. Note the difference between A and the Weyl algebra $\widetilde{\mathcal{H}}$ introduced in § 2.1.2: the latter has only one series of generators b_n, and the element b_0 is central.

11.3.7. Heisenberg vertex algebra. Our goal is to interpret the formulas in (11.3.6) as coming from a map of $V_k(\mathfrak{sl}_2)$ to a vertex algebra associated with the above Weyl algebra A.

Using the experience we have gained from analyzing previous examples, it is straightforward to define the corresponding vertex algebra. Indeed, the first formula in (11.3.6) shows that $a(z)$ has to be of conformal dimension 1, and the second then shows that $a^*(z)$ should have conformal dimension 0.

Therefore the underlying vector space of the vertex algebra we are looking for should be the irreducible representation M of A generated by a vector $|0\rangle$ which satisfies

$$(11.3.8) \qquad a_n|0\rangle = 0, \quad n \geq 0; \qquad a_n^*|0\rangle = 0, \quad n > 0.$$

Thus,

$$M \simeq \mathbb{C}[a_n]_{n<0} \otimes \mathbb{C}[a_m^*]_{m \leq 0}.$$

The Reconstruction Theorem 2.3.11 then leads us to the following result.

11.3.8. Lemma. *Define a linear operator T on M by the formulas*

$$T|0\rangle = 0, \qquad [T, a_n] = -na_{n-1}, \qquad [T, a_n^*] = -(n-1)a_{n-1}^*.$$

Define $Y(|0\rangle, z) = \mathrm{Id}$,

$$Y(a_{-1}|0\rangle, z) = a(z), \qquad Y(a_0^*|0\rangle, z) = a^*(z),$$

$$(11.3.9) \quad Y(a_{n_1} \ldots a_{n_k} a_{m_1}^* \ldots a_{m_l}^* |0\rangle, z) = \prod_{i=1}^{k} \frac{1}{(-n_i - 1)!} \prod_{j=1}^{l} \frac{1}{(-m_j)!}$$

$$\cdot :\partial_z^{-n_1 - 1} a(z) \ldots \partial_z^{-n_k - 1} a(z) \partial_z^{-m_1} a^*(z) \ldots \partial_z^{-m_l} a^*(z): .$$

Then these structures satisfy the axioms of a vertex algebra.

Assigning to a_n and a_n^* degree $-n$, and to $|0\rangle$ degree zero, we obtain a $\mathbb{Z}_+$–gradation on M which makes it into a $\mathbb{Z}_+$–graded vertex algebra. Note that it has infinite–dimensional homogeneous components. If however we introduce in addition a "charge" gradation, with the charge of a_n being 1, the charge of a_n^* being -1 and the charge of $|0\rangle$ being zero, then the bigraded pieces become finite–dimensional.

11.3.9. Remark. The vertex algebra M may be viewed as a "bosonic" counterpart of the fermionic vertex superalgebra $\bigwedge$ introduced in § 5.3.1. Indeed, $\bigwedge$ is a module over the Clifford algebra $\mathcal{C}$ associated to the vector space $\mathcal{K} \oplus \Omega_{\mathcal{K}}$ equipped with a natural symmetric form, while M is a module over the Weyl algebra associated to the same vector space, but with a skew-symmetric form (see § 12.1.1).

In physics literature the fields $a(z)$ and $a^*(z)$ are traditionally denoted by $\beta(z)$ and $\gamma(z)$, and the corresponding conformal field theory is called the $\beta\gamma$–system.

11.3.10. Normal ordering is needed. According to Lemma 11.1.4, constructing a map of vertex algebras $V_k(\mathfrak{sl}_2) \to M$ means choosing three vertex operators $\widetilde{e}(z), \widetilde{h}(z)$, and $\widetilde{f}(z)$ in M such that the corresponding Fourier coefficients satisfy the commutation relations of $\widehat{\mathfrak{sl}}_2$. So at this point it looks like formulas (11.3.6) give rise to a homomorphism $V_k(\mathfrak{sl}_2) \to M$. However, there is a serious obstacle here: the series appearing in the right hand sides of formulas (11.3.6) are *not* normally ordered, and hence they are not well-defined fields on M (let alone vertex operators).

For example, the (-1)st Fourier coefficient of the series $-2a^*(z)a(z)$ equals

$$-2 \sum_{n \in \mathbb{Z}} a_n^* a_{-n}.$$

This infinite sum is not well-defined on M, because it contains infinitely many terms in which the annihilation operator follows a creation operator. This problem is familiar to us from the study of our first example of a vertex algebra, the Fock representation π, in § 2.2.1. It can be solved by considering normally ordered products of fields. Indeed, according to formula (11.3.9), $Y(a_0^* a_{-1}|0\rangle, z)$ is nothing but $:a^*(z)a(z):$, the normally ordered version of the ill-fated expression $a^*(z)a(z)$. Unlike $a^*(z)a(z)$, this normally ordered product is a well-defined field on M.

In the next chapter we will see how the general structure of modules over Weyl algebras (see § 11.2.9), when transported to the "semi–infinite" setting, helps explain the need to introduce normal ordering.

11.4. Bibliographical notes

For details on the structure theory of Kac–Moody algebras and their representations the reader may consult [**Kac2, KR, FZe, FF2, EFK**].

Flag manifolds for affine Kac–Moody algebras are discussed in detail in [**BL1, LS, KNR, Ku2**]. For the definition of the flag manifolds for general Kac–Moody algebras, see [**Ku1, Ma, Ka**].

Semi-infinite flag manifolds were introduced by B. Feigin and E. Frenkel in [**FF2**]. The general picture outlined in §§ 11.3.2 and 11.3.3 is borrowed from [**FF2**]. The algebro-geometric structure of semi-infinite flag manifolds is elucidated in [**FM, FFKM**].

CHAPTER 12

Free Field Realization II

In this chapter we take up the question of regularization of the first order differential operators corresponding to the action of $L\mathfrak{sl}_2$ on $\mathcal{K}$. We start out by describing various classes of representations and completions of the corresponding Weyl algebra with infinitely many generators. After that we explain how the operation of normal ordering allows us to make affine Kac–Moody algebras act on the Fock representation M introduced in the previous chapter. Thus, we obtain a new series of representations of $\widehat{\mathfrak{sl}}_2$, called Wakimoto modules, and the free field realization we were looking for.

12.1. Weyl algebras in the infinite-dimensional case

12.1.1. Delta-functions with semi-infinite support. Recall the Weyl algebra A and the A–module M introduced in §§ 11.3.6 and 11.3.7. We now want to interpret A and M along the lines of the general theory of Weyl algebras and their modules, as described in §§ 11.2.9 and 11.2.10.

In the setting of § 11.2.10, we take as V the space $\mathcal{K} \simeq \mathbb{C}((t))$ of functions on the punctured disc. Using the residue pairing, we identify the restricted dual space to V with the space $\Omega_{\mathcal{K}} \simeq \mathbb{C}((t))dt$ of differentials on the punctured disc. The residue pairing then gives us a skew-symmetric non-degenerate inner product on $S = \mathcal{K} \oplus \Omega_{\mathcal{K}}$ (see formula (11.2.8)), and we define the corresponding Weyl algebra $A(\mathcal{K} \oplus \Omega_{\mathcal{K}})$ as in § 11.2.9. By abuse of notation, we will denote this Weyl algebra also by A. By construction, it is topologically generated by the elements

$$(12.1.1) \qquad a_n = t^n, \qquad a_n^* = t^{n-1}dt, \qquad n \in \mathbb{Z}.$$

With respect to the inner product on $\mathcal{K} \oplus \Omega_{\mathcal{K}}$, we have

$$\left(t^n, t^{m-1}dt\right) = \mathrm{Res}_{t=0}\, t^{n+m-1}dt = \delta_{n,-m}, \quad (t^n, t^m) = \left(t^{n-1}dt, t^{m-1}dt\right) = 0,$$

so formulas (11.2.6) become precisely the commutation relations (11.3.7).

The generating space $\mathcal{K} \oplus \Omega_{\mathcal{K}}$ of A is a complete topological vector space, in which the basis of open neighborhoods of 0 is formed by the subspaces $t^N \mathcal{O} \oplus t^M \Omega_{\mathcal{O}}, N, M \in \mathbb{Z}$ (here as usual $\mathcal{O} \simeq \mathbb{C}[[t]]$). Introduce a topology on A in which the basis of open neighborhoods of 0 is formed by the left ideals generated by the subspaces $t^N \mathcal{O} \oplus t^M \Omega_{\mathcal{O}}, N, M \in \mathbb{Z}$. The *completed Weyl algebra* $\widetilde{A}$ is by definition the completion of A with respect to this topology. The algebra $\widetilde{A}$ is topologically generated by the elements (12.1.1).

The module M defined by formulas (11.3.8) is nothing but the A–module M_L defined by formula (11.2.7) with $L = \mathcal{O} \oplus \Omega_{\mathcal{O}}$, i.e., L is the subspace of $\mathcal{K} \oplus \Omega_{\mathcal{K}}$ consisting of elements regular at the origin. Therefore we may view M as the module of delta-functions on $\mathcal{K}$ supported at the subspace $\mathcal{O} \subset \mathcal{K}$. This subspace is an example of a "semi-infinite" subspace, as it has both infinite dimension and infinite codimension in $\mathcal{K}$. The action of A on M extends to the action of the completion $\widetilde{A}$, which may be viewed as a completed algebra of differential operators on $\mathcal{K}$.

In fact, M can be defined in terms of $\widetilde{A}$. Let $\widetilde{\mathrm{Sym}}(\mathcal{O} \oplus \Omega_\mathcal{O})$ be the commutative subalgebra of $\widetilde{A}$ topologically generated by the Lagrangian subspace $(\mathcal{O} \oplus \Omega_\mathcal{O}) \subset (\mathcal{K} \oplus \Omega_\mathcal{K})$. Then $M = M_{\mathcal{O} \oplus \Omega_\mathcal{O}}$ is the $\widetilde{A}$–module induced from the trivial representation of $\widetilde{\mathrm{Sym}}(\mathcal{O} \oplus \Omega_\mathcal{O})$. Note that by construction this module has discrete topology, but the action of $\widetilde{A}$ on it is continuous.

12.1.2. Remark. To clarify the difference between the algebras A and $\widetilde{A}$ introduced above, observe that elements of A are finite polynomials in the series of the form

$$\sum_{n \geq N} \alpha_n a_n, \qquad \sum_{m \geq M} \beta_m a_m^*,$$

with arbitrary $\alpha_n, \beta_m \in \mathbb{C}$, while elements of $\widetilde{A}$ are arbitrary series of the form

$$(12.1.2) \qquad \sum_{n > 0} P_n a_n + \sum_{m > 0} Q_m a_m^* + R, \qquad P_n, Q_m, R \in A.$$

12.1.3. Another completion: naive differential operators. In addition to $\widetilde{A}$, there are other completions of the Weyl algebra $A = A(\mathcal{K} \oplus \Omega_\mathcal{K})$. Each of these completions acts on a particular induced module. For instance, let us define the completion which acts on the module $\mathrm{Fun}\,\mathcal{K}$ of functions on $\mathcal{K}$, defined § 11.3.4.

Introduce a new topology on A in which the basis of open neighborhoods of 0 the subspaces $U + V$, where U is the left ideal generated by $t^N \mathcal{O}$ for some $N \in \mathbb{Z}$ and V is the right ideal generated by $t^M \Omega_\mathcal{O}$ for some $M \in \mathbb{Z}$. Denote the completion of A with respect to this topology by $A^\natural$.

Let $\widetilde{\mathrm{Sym}}(\mathcal{K})$ (resp., $\widetilde{\mathrm{Sym}}(\Omega_\mathcal{K})$) be the commutative subalgebra of $A^\natural$ topologically generated by the Lagrangian subspace $\mathcal{K}$ (resp., $\Omega_\mathcal{K}$) of the space $\mathcal{K} \oplus \Omega_\mathcal{K}$ of generators of A. Note that $\widetilde{\mathrm{Sym}}(\Omega_\mathcal{K})$ is nothing but the topological space $\mathrm{Fun}\,\mathcal{K}$ of functions on $\mathcal{K}$, defined in § 11.3.4. Let $M_\mathcal{K}$ be the $A^\natural$–module induced from the trivial representation of $\widetilde{\mathrm{Sym}}(\mathcal{K})$. Then $M_\mathcal{K}$ is naturally identified with $\widetilde{\mathrm{Sym}}(\Omega_\mathcal{K})$, and hence with $\mathrm{Fun}\,\mathcal{K}$. Note that the action of $A^\natural$ on $M_\mathcal{K} \simeq \mathrm{Fun}\,\mathcal{K}$ is continuous.

More generally, each Lagrangian subspace L of $\mathcal{K} \oplus \Omega_\mathcal{K}$ gives rise to a commutative subalgebra of $A = A(\mathcal{K} \oplus \Omega_\mathcal{K})$ and the corresponding induced A–module M_L. This module carries a continuous action of a suitable completion of A.

If the Lagrangian subspaces L_1 and L_2 are not commensurable, i.e., if the spaces $L_1/(L_1 \cap L_2)$ and $L_2/(L_1 \cap L_2)$ are infinite–dimensional, then the completions of A corresponding to the modules M_{L_1} and M_{L_2} are different. Moreover, the completion corresponding to M_{L_2} does not act on M_{L_1}.

For example, the Lagrangian subspaces $L_1 = \mathcal{O} \oplus \Omega_\mathcal{O}$ and $L_2 = \mathcal{K}$ are not commensurable. The series

$$h_0 = -2 \sum_{n \in \mathbb{Z}} a_n^* a_{-n}$$

belongs to the completion $A^\natural$ corresponding to L_2, and its action is well-defined on $M_{L_2} = \mathrm{Fun}\,\mathcal{K}$. But it does not belong to the completion $\widetilde{A}$ corresponding to L_1, and its action is not well-defined on the module $M_{L_1} = M$.

12.1.4. Lifting to differential operators. In § 11.3.4 we defined the Lie algebra Vect $\mathcal{K}$ of vector fields on $\mathcal{K}$. The vector fields on $\mathcal{K}$ corresponding to the infinitesimal action of $L\mathfrak{sl}_2$ are given by the right hand sides of formulas (11.3.3). We want to lift these vector fields to first order differential operators on $\mathcal{K}$. This is a non-trivial task, since as we have seen above there are different candidates for the algebra of differential operators on $\mathcal{K}$, and they act on different modules.

One of the candidates, the algebra $A^\natural$ defined in § 12.1.3, contains the space of functions $A_0^\natural = \operatorname{Fun}\mathcal{K}$, identified with $\widetilde{\operatorname{Sym}}(\Omega_\mathcal{K})$, and the space of vector fields Vect $\mathcal{K}$, identified with the completed tensor product of $\widetilde{\operatorname{Sym}}(\Omega_\mathcal{K})$ and $\mathcal{K}$ (see § 11.3.4). Let us define the space $A_{\leq 1}^\natural \subset A^\natural$ of differential operators of order ≤ 1 on $\mathcal{K}$ as the span of Fun $\mathcal{K}$ and Vect $\mathcal{K}$ in $A^\natural$. Then $A_{\leq 1}^\natural$ is a Lie subalgebra of $A^\natural$, and we have an exact sequence of Lie algebras

$$(12.1.3) \qquad\qquad 0 \to \operatorname{Fun}\mathcal{K} \to A_{\leq 1}^\natural \to \operatorname{Vect}\mathcal{K} \to 0.$$

The action of the algebra $A^\natural$ on the space of functions Fun $\mathcal{K}$ (which is nothing but the Weyl module $M_\mathcal{K}$ corresponding to the Lagrangian subspace $\mathcal{K}$ of $\mathcal{K} \oplus \Omega_K$) provides us with a canonical splitting of this sequence. Namely, we map vector fields to the first order differential operators which annihilate the constant function $1 \in \operatorname{Fun}\mathcal{K}$. This sequence is therefore similar to the sequence (11.2.1) in the finite-dimensional case.

As we know, the Lie algebra $L\mathfrak{sl}_2$ maps to Vect $\mathcal{K}$ by formulas (11.3.6). Since the exact sequence (12.1.3) is split, we obtain a map $L\mathfrak{sl}_2 \to A_{\leq 1}^\natural$ and hence an action of $L\mathfrak{sl}_2$ on Fun $\mathcal{K}$. However, this is not what we need. We want to make $L\mathfrak{sl}_2$ act on the Weyl module $M_{\mathcal{O} \oplus \Omega_\mathcal{O}}$ corresponding to the Lagrangian subspace $\mathcal{O} \oplus \Omega_\mathcal{O}$ of $\mathcal{K} \oplus \Omega_\mathcal{K}$. Indeed, $M_{\mathcal{O} \oplus \Omega_\mathcal{O}}$ may be thought of as the space of delta-functions on an $\widehat{N}_+$–orbit of the semi-infinite flag manifold of $L\mathfrak{sl}_2$ (see the end of § 11.3.3), and our objective is to make this space into an $L\mathfrak{sl}_2$–module. This module, which we have denoted by M before, is a vertex algebra, and we need $L\mathfrak{sl}_2$ to act on M in order to construct a vertex algebra homomorphism $V_k(\mathfrak{sl}_2) \to M$.

But $A^\natural$, and hence in particular $A_{\leq 1}^\natural$, does not act on M. It acts on its own module $M_\mathcal{K} = \operatorname{Fun}\mathcal{K}$. Hence we cannot use the map $L\mathfrak{sl}_2 \to A_{\leq 1}^\natural$ to obtain an action of $L\mathfrak{sl}_2$ on M.

Thus, the "obvious" lifting of the vector fields of the action of $L\mathfrak{sl}_2$ on $\mathcal{K}$ given by formulas (11.3.3) to $A_{\leq 1}^\natural$ lies in the "wrong" completion of the Weyl algebra A, whose action is not well-defined on the module M. The "right" completion, which acts on the module M, is the completion $\widetilde{A}$ of A defined in § 12.1.1, and we should try to lift our vector fields to $\widetilde{A}$.

12.1.5. Analogues of functions and vector fields in $\widetilde{A}$. We need an analogue of the exact sequence (12.1.3) for the algebra $\widetilde{A}$. First, we define the analogue of $A_0^\natural = \operatorname{Fun}\mathcal{K}$. Let A_0 be the subalgebra of A generated by $\Omega_\mathcal{K}$, and $\widetilde{A}_0$ its completion in $\widetilde{A}$ (i.e., $\widetilde{A}_0$ is topologically generated by $\Omega_\mathcal{K}$). Next, we define the analogue of $A_{\leq 1}^\natural$. Let $A_{\leq 1}$ be the subspace of A spanned by the products of elements of A_0 and elements of $\mathcal{K}$, and $\widetilde{A}_{\leq 1}$ its completion in $\widetilde{A}$. Thus, $\widetilde{A}_{\leq 1}$ consists of all elements P of $\widetilde{A}$ with the property that $P \bmod I_{N,M} \in A_{\leq 1} \bmod I_{N,M}$, for all $N, M \in \mathbb{Z}$, where $I_{N,M}$ is the left ideal of $\widetilde{A}$ topologically generated by $t^N \mathcal{O} \oplus t^M \Omega_\mathcal{O}$.

Here is a more concrete description of $\widetilde{A}_0$ and $\widetilde{A}_{\leq 1}$ using the realization of $\widetilde{A}$ by series of the form (12.1.2). The space $\widetilde{A}_0$ consists of series of the form (12.1.2), where all $P_n = 0$, and $Q_m, R \in A_0$, where $A_0 \subset A$ is the subalgebra generated by the $a_m^*, m \in \mathbb{Z}$. In other words, these are the series which do not contain $a_n, n \in \mathbb{Z}$. The space $\widetilde{A}_{\leq 1}$ consists of elements of the form (12.1.2), where $P_n \in A_0$, and $Q_m, R \in A_{\leq 1}$, where $A_{\leq 1}$ is the subspace of A spanned by monomials of degree at most 1 in the $a_n, n \in \mathbb{Z}$.

12.1.6. Proposition.

(1) $\widetilde{A}_{\leq 1}$ is a Lie algebra, and $\widetilde{A}_0$ is its ideal.

(2) $\widetilde{A}_0 \simeq \mathrm{Fun}\, \mathcal{K}$.

(3) $\widetilde{A}_{\leq 1}/\widetilde{A}_0 \simeq \mathrm{Vect}\, \mathcal{K}$ as Lie algebras.

12.1.7. Proof.

(1) follows from the definition of $\widetilde{A}_{\leq 1}$ and $\widetilde{A}_0$ as completions of $A_{\leq 1}$ and A_0, and the fact that $A_{\leq 1}$ is a Lie algebra containing A_0 as an ideal.

(2) follows from the definition of $\widetilde{A}_0$ and the definition of $\mathrm{Fun}\, \mathcal{K}$ given in § 11.3.4.

(3) We need to define a map from $\widetilde{A}_{\leq 1}$ to $\mathrm{Vect}\, \mathcal{K}$. By definition of the completions $\widetilde{A}$ and $A^\natural$, for any $n, m \in \mathbb{Z}$, the quotient of $\widetilde{A}$ by its left ideal topologically generated by $t^n \mathcal{O} \oplus t^m \Omega_\mathcal{O}$ coincides with the quotient of $A^\natural$ by the sum of the left ideal generated by $t^n \mathcal{O}$ and the right ideal generated by $t^m \Omega_\mathcal{O}$. Denote by $\widetilde{A}_{\leq 1}^{n,m}$, $\widetilde{A}_0^{n,m}$, $A_{\leq 1}^{\natural n,m}$, $A_0^{\natural n,m}$ the corresponding subspaces in these quotients. We have $\widetilde{A}_{\leq 1}^{n,m} = A_{\leq 1}^{\natural n,m}$, $\widetilde{A}_0^{n,m} = A_0^{\natural n,m}$, etc. Denote $A_{\leq 1}^{\natural n,m}/A_0^{\natural n,m}$ by $\mathrm{Vect}^{n,m}\, \mathcal{K}$.

For $n' > n, m' > m$, we have the following surjections:

$$
\begin{aligned}
p_{n,m}^{n',m'} &: \quad \widetilde{A}_{\leq 1}^{n',m'} &\rightarrow \quad \widetilde{A}_{\leq 1}^{n,m}, \\
q_{n,m}^{n',m'} &: \quad A_{\leq 1}^{\natural n',m'} &\rightarrow \quad A_{\leq 1}^{\natural n,m}, \\
r_{n,m}^{n',m'} &: \quad \mathrm{Vect}^{n',m'}\, \mathcal{K} &\rightarrow \quad \mathrm{Vect}^{n,m}\, \mathcal{K}.
\end{aligned}
$$

Let P be an element of $\widetilde{A}_{\leq 1}^{n',m'}$ and $\overline{P}$ the corresponding element of $A_{\leq 1}^{\natural n',m'}$. From the fact that the commutator of a first order differential operator and a function is a function we find that $p_{n,m}^{n',m'}(P)$ and $q_{n,m}^{n',m'}(\overline{P})$ (considered as elements of $\widetilde{A}_{\leq 1}^{n,m} = A_{\leq 1}^{\natural n,m}$) may only differ by an element of $\widetilde{A}_0^{n,m} = A_0^{\natural n,m}$. Define a map $s_{n,m} : \widetilde{A}_{\leq 1}^{n,m} \rightarrow \mathrm{Vect}^{n,m}\, \mathcal{K}$ as the composition of the isomorphism $\widetilde{A}_{\leq 1}^{n,m} \xrightarrow{\sim} A_{\leq 1}^{\natural n,m}$ and the projection $A_{\leq 1}^{\natural n,m} \twoheadrightarrow \mathrm{Vect}^{n,m}\, \mathcal{K}$. This map satisfies $s_{n,m} \circ p_{n,m}^{n',m'} = r_{n,m}^{n',m'} \circ s_{n',m'}$. Therefore we obtain a well-defined linear map $\widetilde{A}_{\leq 1} \rightarrow \mathrm{Vect}\, \mathcal{K}$, whose kernel is $\widetilde{A}_0$. The statement that it is a homomorphism of Lie algebras follows from the fact that the Lie brackets on both $\widetilde{A}_{\leq 1}$ and $\mathrm{Vect}\, \mathcal{K}$ are continuous. This completes the proof.

In more concrete terms, the homomorphism $\widetilde{A}_{\leq 1} \rightarrow \mathrm{Vect}\, \mathcal{K}$ may be described as follows. Represent an element of $\widetilde{A}_{\leq 1}$ as a series of the form (12.1.2). Then replace each a_n by $\partial/\partial x_n$ and each a_n^* by x_{-n}. Finally, remove those summands of the resulting series which do not contain the factors $\partial/\partial x_n, n \in \mathbb{Z}$, and move those factors to the right in the remaining summands. It is straightforward to check that the resulting series satisfies the conditions described at the end of § 11.3.4. Further, it is clear that under this map all elements of $\widetilde{A}_0$ go to 0. Hence we obtain a well-defined linear map $\widetilde{A}_{\leq 1}/\widetilde{A}_0 \rightarrow \mathrm{Vect}\, \mathcal{K}$. To see that this map is a homomorphism of

Lie algebras, observe that in both Lie algebras the commutator between two series may be computed term-by-term. Hence it suffices to check the homomorphism property for any two monomials in $A_{\leq 1}$, which is obvious.

12.1.8. Non–splitting. According to Proposition 12.1.6, we have an exact sequence of Lie algebras

$$(12.1.4) \qquad\qquad 0 \to \mathrm{Fun}\,\mathcal{K} \to \widetilde{A}_{\leq 1} \to \mathrm{Vect}\,\mathcal{K} \to 0,$$

which is *a priori* different from the "naive" sequence (12.1.3), i.e., we obtain a different extension of the Lie algebra of vector fields on $\mathcal{K}$ by its module of functions on $\mathcal{K}$. In contrast to the sequence (12.1.3), there is no reason for the sequence (12.1.4) to be split. Indeed, the natural action of $\mathrm{Fun}\,\mathcal{K}$ on itself by multiplication cannot be extended to an action of $\widetilde{A}_{\leq 1}$, because the latter is part of the completion $\widetilde{A}$ of the Weyl algebra, which does not act on the module $M_{\mathcal{K}} = \mathrm{Fun}\,\mathcal{K}$. Hence we cannot use the "annihilation of constant functions" condition to split the sequence as we did in the case of the sequence (12.1.3). The completion $\widetilde{A}_{\leq 1}$ acts on the module $M_{\mathcal{O}\oplus\Omega_{\mathcal{O}}} = M$, but this is a module of "delta-functions supported at $\mathcal{O} \subset \mathcal{K}$", and so M does not contain "constant functions", i.e., vectors annihilated by all $a_n, n \in \mathbb{Z}$. In fact, we have the following result.

12.1.9. Lemma. *The sequence* (12.1.4) *does not split.*

12.1.10. Proof. It is easy to see that the following series belongs to $\widetilde{A}_{\leq 1}$:

$$(12.1.5) \qquad\qquad z_m = \sum_{k+l=m} :a_k^* a_l: , \qquad m \in \mathbb{Z}.$$

By the above definition of the homomorphism $\widetilde{A}_{\leq 1} \to \mathrm{Vect}\,\mathcal{K}$, the image of z_m in $\mathrm{Vect}\,\mathcal{K}$ equals

$$(12.1.6) \qquad\qquad \overline{z}_m = \sum_{k\in\mathbb{Z}} x_k \frac{\partial}{\partial x_{k+m}}.$$

We have $[\overline{z}_n, \overline{z}_m] = 0$ for all $n, m \in \mathbb{Z}$. But

$$[z_n, z_m] = -\delta_{n,-m},$$

as we will see below in § 12.2.11. If the sequence (12.1.4) were split, then we would be able to find "correction terms" $y_m \in \mathrm{Fun}\,\mathcal{K}$ such that

$$[z_n + y_n, z_m + y_m] = 0, \qquad \forall n, m \in \mathbb{Z}.$$

This is equivalent to the formula

$$\overline{z}_n \cdot y_m - \overline{z}_m \cdot y_n = -\delta_{n,-m}.$$

But since $\overline{z}_n$ is a linear vector field, the left hand side of this formula cannot be a non-zero constant for any choice of $y_n, n \in \mathbb{Z}$. Therefore the sequence (12.1.4) does not split.

12.1.11. Remark. Let $\mathcal{E}$ be the completed tensor product of $\Omega_{\mathcal{K}}$ and $\mathcal{K}$, considered as a subspace of linear vector fields in $\text{Vect}\,\mathcal{K}$. This is in fact a Lie subalgebra of $\text{Vect}\,\mathcal{K}$ isomorphic to the Lie algebra of continuous linear endomorphisms of $\mathcal{K} \simeq \mathbb{C}((t))$. Thus, $\mathcal{E}$ is a completion of the Lie algebra $\mathfrak{gl}_\infty$ of infinite matrices with finitely many non-zero entries. The corresponding subspace $\widehat{\mathcal{E}}$ of $\widetilde{A}_{\leq 1}$ is defined as the span of the constants and the elements linear in the generators from $\mathcal{K}$ and $\Omega_{\mathcal{K}}$ (in other words, the a_n's and a_n^*'s). The reader should check that $\widehat{\mathcal{E}}$ is a Lie subalgebra of $\widetilde{A}_{\leq 1}$. In other words, the restriction of the extension (12.1.4) to $\mathcal{E}$ gives rise to the central extension $\widehat{\mathcal{E}}$ of $\mathcal{E}$ by $\mathbb{C}$. It is known that this central extension is universal, i.e. $H^2(\mathcal{E}, \mathbb{C})$ is one-dimensional and is generated by the class of this extension.

Observe that each $f \in \mathcal{K}$ defines a linear operator of multiplication by f on $\mathcal{K}$ which is continuous. Therefore the abelian Lie algebra $\mathcal{K}$ naturally embeds into $\mathcal{E}$. Under this embedding, the element $t^m \in \mathbb{C}((t)) \simeq \mathcal{K}$ maps to $\overline{z}_{-m} \in \mathcal{E}$ given by formula (12.1.6). The pull-back of the universal central extension of $\mathcal{E}$ to $\mathcal{K}$ defines a central extension of $\mathcal{K}$. The computation made in the proof of Lemma 12.1.9 shows that this central extension is nothing but the Heisenberg Lie algebra $\mathcal{H}$ introduced in § 2.1.2.

In fact, we can obtain other important central extensions, such as the affine Lie algebra $\widehat{\mathfrak{g}}$ and the Virasoro algebra Vir from the above universal central extension of $\mathcal{E}$ by using the natural embeddings into $\mathcal{E}$ of the loop algebra $\mathfrak{g}(\mathcal{K})$ and the Lie algebra $\text{Der}\,\mathcal{K}$. In this sense, $\widehat{\mathcal{E}}$ may be viewed as a "master Lie algebra" which contains all of our favorite centrally extended Lie algebras that are related to the punctured disc.

12.2. Local completion

We are now in a position to state our original question more precisely. We are given a homomorphism of Lie algebras $L\mathfrak{sl}_2 \to \text{Vect}\,\mathcal{K}$, and we want to know whether it is possible to lift this homomorphism to $\widetilde{A}_{\leq 1}$. If we could do that, then we could make $L\mathfrak{sl}_2$ act on the $\widetilde{A}$–module M. Since the sequence (12.1.4) does not split, it is not clear *a priori* whether such a lifting exists or not.

In order to determine this, we examine the extension (12.1.4) in more detail.

12.2.1. Defining the local completion. First of all, we observe that the image of $L\mathfrak{sl}_2$ in $\text{Vect}\,\mathcal{K}$ consists of vector fields of a special kind, which may be called "local" by analogy with the Fourier coefficients of vertex operators.

Consider the module M over the Heisenberg algebra A (and over its completion $\widetilde{A}$) defined by formulas (11.3.8), with its vertex algebra structure. Recall that the completion of the span of all Fourier coefficients of vertex operators from M forms a Lie algebra $U(M)$ defined in § 4.1.1. More precisely,

$$U(M) \overset{\text{def}}{=} (M \otimes \mathbb{C}((t)))/\text{Im}(T \otimes 1 + \text{Id} \otimes \partial_t).$$

Define a linear map $M \otimes \mathbb{C}((t)) \to \widetilde{A}$ sending $A \otimes f(t)$ to $\text{Res}\,Y(A, z)f(z)dz$. According to Theorem 4.1.2, $U(M)$ is a Lie algebra, and the above map is a homomorphism of Lie algebras (actually, an embedding in this case).

Note that $U(M)$ is not an algebra. For instance, it contains the generators a_n, a_n^* of the Heisenberg algebra, but does not contain monomials in these generators of degrees greater than one. However, we will only need the Lie algebra structure on $U(M)$.

The elements of $\widetilde{A}_0 = \operatorname{Fun}\mathbb{C}((t))$ which lie in the image of $U(M)$ are called "local functionals" (see § 4.2.7). The elements of $\widetilde{A}$ which belong to $U(M)$ are given by (possibly infinite) linear combinations of Fourier coefficients of normally ordered polynomials in $a(z), a^*(z)$ and their derivatives. We refer to them as "local" elements of $\widetilde{A}$.

12.2.2. Local extension. For our purposes we can replace $\widetilde{A}$, which is a very large topological algebra, by a relatively small "local part" $U(M)$. Thus, we replace $\widetilde{A}_0$ and $\widetilde{A}_{\leq 1}$ by their local versions $A_{0,\mathrm{loc}} = \widetilde{A}_0 \cap U(M)$ and $A_{\leq 1,\mathrm{loc}} = \widetilde{A}_{\leq 1} \cap U(M)$.

Let us describe $A_{0,\mathrm{loc}}$ and $A_{\leq 1,\mathrm{loc}}$ more explicitly. The space $A_{0,\mathrm{loc}}$ is spanned (topologically) by the Fourier coefficients of all polynomials in the $\partial_z^n a^*(z), n \geq 0$. Note that because the a_n^*'s commute among themselves, these polynomials are automatically normally ordered. The space $A_{\leq 1,\mathrm{loc}}$ is spanned by the Fourier coefficients of the fields of the form $:P(a^*(z), \partial_z a^*(z), \ldots)a(z):$ (the normally ordered product of $P(a^*(z), \partial_z a^*(z), \ldots)$ and $a(z)$). Observe that the Fourier coefficients of all fields of the form

$$:P(a^*(z), \partial_z a^*(z), \ldots)\partial_z^m a(z): , \qquad m > 0,$$

can be expressed as linear combinations of the fields of the form

$$:P(a^*(z), \partial_z a^*(z), \ldots)a(z): .$$

Further, we define a local version $\mathfrak{T}_{\mathrm{loc}}$ of $\operatorname{Vect}\mathcal{K}$ as the subspace, which consists of finite linear combinations of Fourier coefficients of the formal power series

$$P(a^*(z), \partial_z a^*(z), \ldots)a(z),$$

where $a(z)$ and $a^*(z)$ are given by formulas (11.3.4), (11.3.5).

Since $A_{\leq 1,\mathrm{loc}}$ is the intersection of Lie subalgebras of $\widetilde{A}$, it is also a Lie subalgebra of $\widetilde{A}$. By construction, its image in $\operatorname{Vect}\mathcal{K}$ under the homomorphism $\widetilde{A}_{\leq 1} \to \operatorname{Vect}\mathcal{K}$ equals $\mathfrak{T}_{\mathrm{loc}}$. Finally, the kernel of the resulting surjective Lie algebra homomorphism $A_{\leq 1,\mathrm{loc}} \to \mathfrak{T}_{\mathrm{loc}}$ equals $A_{0,\mathrm{loc}}$. Hence we obtain that the extension (12.1.4) restricts to the "local" extension

$$(12.2.1) \qquad\qquad 0 \to A_{0,\mathrm{loc}} \to A_{\leq 1,\mathrm{loc}} \to \mathfrak{T}_{\mathrm{loc}} \to 0.$$

The argument we used in the proof of Lemma 12.1.9 shows that this sequence does not split.

By looking at formulas (11.3.6), we find that the image of $L\mathfrak{sl}_2$ in $\operatorname{Vect}\mathcal{K}$ belongs to $\mathfrak{T}_{\mathrm{loc}}$. Therefore we can now reformulate our question as to whether the image of $L\mathfrak{sl}_2$ in $\mathfrak{T}_{\mathrm{loc}}$ can be lifted to $A_{\leq 1,\mathrm{loc}}$.

12.2.3. The two–cocycle. Before answering this question, we compute the two-cocycle representing the extension (12.2.1). Recall that an exact sequence of Lie algebras

$$0 \to C \to \widehat{\mathfrak{g}} \to \mathfrak{g} \to 0,$$

where C is an abelian ideal, with prescribed $\mathfrak{g}$–module structure, gives rise to a two-cocycle of $\mathfrak{g}$ with coefficients in C. It is constructed as follows (compare with the discussion of § 2.1.1). Choose a splitting $\imath : \mathfrak{g} \to \widehat{\mathfrak{g}}$ of this sequence (considered as a vector space), and define $\gamma : \bigwedge^2 \mathfrak{g} \to C$ by the formula

$$\gamma(a, b) = \imath([a, b]) - [\imath(a), \imath(b)].$$

One checks that γ is a two-cocycle in the Chevalley complex of $\mathfrak{g}$ with coefficients in C (see § A.4), and that changing the splitting $\imath$ amounts to changing γ by a coboundary. Therefore we obtain a bijection between the set of isomorphism classes of extensions of $\mathfrak{g}$ by C and the cohomology group $H^2(\mathfrak{g}, C)$.

In the case of the sequence (12.2.1), the operation of normal ordering provides us with a splitting.

Namely, given an element ξ of $\mathcal{T}_{\mathrm{loc}}$ of the form

$$(12.2.2) \qquad \xi = \int P(a^*(z), \partial_z a^*(z), \ldots) a(z) z^n dz,$$

we define

$$(12.2.3) \qquad \imath(\xi) = \int :P(a^*(z), \partial_z a^*(z), \ldots) a(z): z^n dz,$$

and extend this to a map $\imath : \mathcal{T}_{\mathrm{loc}} \to A_{\leq 1, \mathrm{loc}}$ by linearity. In order to compute the corresponding cocycle we have to learn how to compute commutators of elements of the form (12.2.2) and (12.2.3). Those may be computed from the OPE of the corresponding vertex operators. We now explain how to compute the OPEs of vertex operators using the Wick formula. First we discuss the operation of normal ordering in more detail.

12.2.4. Normal ordering. Let us describe the operation of normal ordering in our Heisenberg vertex algebra M more explicitly. This description is similar to the one in the case of the vertex algebra π given in Remark 2.2.6.

Let us call the generators $a_n, n \geq 0$, and $a_m^*, m > 0$, annihilation operators, and the generators $a_n, n < 0$, and $a_m^*, m \leq 0$, creation operators. A monomial P in A is called normally ordered if all factors of P which are annihilation operators stand to the right of all factors of P which are creation operators. Given any monomial P, we define the normally ordered monomial $:P:$ as the monomial obtained by moving all factors of P which are annihilation operators to the right, and all factors of P which are creation operators to the left. Note that since the annihilation operators commute among themselves, it does not matter how we order them among themselves. The same is true for the creation operators. This shows that $:P:$ is well-defined by the above conditions.

For example,

$$:a_5 a_{-4}^* a_{-2} a_{-5}^*: \; = a_{-4}^* a_{-5}^* a_{-2} a_5.$$

Given two monomials P and Q, their normally ordered product is by definition the normally ordered monomial $:PQ:$. By linearity, we define the normally ordered product of any two elements of A, and of any two elements of $U(M)$, termwise. Given two vertex operators from the vertex algebra M, we define their normally ordered product by applying the above definition to each of their Fourier coefficients.

It is easy to check by induction that this definition of normally ordered product of two vertex operators coincides with the general Definition 2.2.2. Recall that all vertex operators from M are given by finite linear combinations of normally ordered products of the fields $a(z), a^*(z)$ and their derivatives. This implies that each Fourier coefficient of any vertex operator from M is a series of normally ordered monomials.

12.2.5. Contractions of fields. In order to state the Wick formula, we have to introduce the notion of contraction of two fields.

From the commutation relations, we obtain the following OPEs:

$$a(z)a^*(w) = \frac{1}{z-w} + :a(z)a^*(w): \,,$$

$$a^*(z)a(w) = -\frac{1}{z-w} + :a^*(z)a(w): \,.$$

We view them now as identities on formal power series, in which by $1/(z-w)$ we understand its expansion in positive powers of w/z. Differentiating several times, we obtain

$$(12.2.4) \qquad \partial_z^n a(z)\partial_w^m a^*(w) = (-1)^n \frac{(n+m)!}{(z-w)^{n+m+1}} + :\partial_z^n a(z)\partial_w^m a^*(w): \,,$$

$$(12.2.5) \qquad \partial_z^m a^*(z)\partial_w^n a(w) = (-1)^{m+1} \frac{(n+m)!}{(z-w)^{n+m+1}} + :\partial_z^m a^*(z)\partial_w^n a(w):$$

(here again by $1/(z-w)^n$ we understand its expansion in positive powers of w/z).

Suppose that we are given two normally ordered monomials in $a(z), a^*(z)$ and their derivatives. Denote them by $P(z)$ and $Q(z)$. A single *pairing* between $P(z)$ and $Q(w)$ is by definition either the pairing $(\partial_z^n a(z), \partial_w^m a^*(w))$ of $\partial_z^n a(z)$ occuring in $P(z)$ and $\partial_w^m a^*(w)$ occuring in $Q(w)$, or $(\partial_z^m a^*(z), \partial_w^n a(w))$ of $\partial_z^m a^*(z)$ occuring in $P(z)$ and $\partial_w^n a(w)$ occuring in $Q(w)$. We attach to it the functions

$$(-1)^n \frac{(n+m)!}{(z-w)^{n+m+1}} \qquad \text{and} \qquad (-1)^{m+1} \frac{(n+m)!}{(z-w)^{n+m+1}},$$

respectively. A *multiple pairing* B is by definition a disjoint union of single pairings. We attach to it the function $f_B(z, w)$, which is the product of the functions corresponding to the single pairings in B (considered as power series in w/z).

Note that the monomials $P(z)$ and $Q(z)$ may well have multiple pairings of the same type. For example, the monomials $:a^*(z)^2\partial_z a(z):$ and $:a(w)\partial_z^2 a^*(w):$ have two different pairings of type $(a^*(z), a(w))$; the corresponding function is $-1/(z-w)$. In such a case we say that the multiplicity of this pairing is 2. Note that these two monomials also have a unique pairing $(\partial_z a(z), \partial_w^2 a^*(w))$, and the corresponding function is $-6/(z-w)^4$.

Given a multiple pairing B between $P(z)$ and $Q(w)$, we define $(P(z)Q(w))_B$ as the product of all factors of $P(z)$ and $Q(w)$ which do not belong to the pairing (if there are no factors left, we set $(P(z)Q(w))_B = 1$). The *contraction* of $P(z)Q(w)$ with respect to the pairing B, denoted $:P(z)Q(w):_B$, is by definition the normally ordered formal power series $:(P(z)Q(w)_B):$ multiplied by the function $f_B(z, w)$. We extend this definition to the case when B is the empty set by stipulating that

$$:P(z)Q(w):_\emptyset = :P(z)Q(w): \,.$$

Now we are in a position to state the *Wick formula*, which gives the OPE of two arbitrary normally ordered monomial vertex operators. The proof of this formula is straightforward and is left to the reader.

12.2.6. Lemma (Wick formula). *Let $P(z)$ and $Q(w)$ be two monomials as above. Then $P(z)Q(w)$ equals the sum of terms $:P(z)Q(w):_B$ over all pairings B between P and Q including the empty one, counted with multiplicity.*

12.2.7. Example.

$$:a^*(z)^2\partial_z a(z):\, :a(w)\partial_w^2 a^*(w): = \, :a^*(z)^2\partial_z a(z)a(w)\partial_w^2 a^*(w):$$

$$-\frac{2}{z-w}:a^*(z)\partial_z a(z)\partial_w^2 a^*(w):$$

$$-\frac{6}{(z-w)^4}:a^*(z)^2 a(w): \, .$$

Observe that the terms appearing in the right hand side of this formula contain $a(z), a^*(z)$, as well as $a(w), a^*(w)$. In order to obtain the OPE formula, we need to replace $a(z), a^*(z)$ by their Taylor expansions at w.

12.2.8. Computation of the two–cocycle.
Now we can compute our two–cocycle. For that, we need to apply the Wick formula to the fields of the form

$$:R(a^*(z), \partial_z a^*(z), \dots)a(z): \, ,$$

whose Fourier coefficients span the pre-image of $\mathcal{T}_{\text{loc}}$ in $A_{\leq 1,\text{loc}}$ under our splitting $\imath$. Two fields of this form may have only single or double pairings, and therefore their OPE can be written quite explicitly.

A field of the above form may be written as $Y(P(a_n^*)a_{-1}, z)$ (or $Y(Pa_{-1}, z)$ for short), where P is a polynomial in the $a_n^*, n \leq 0$ (recall that $a_n^*, n \leq 0$, corresponds to $\partial_z^{-n} a^*(z)/(-n)!$). Applying the Wick formula, we obtain

12.2.9. Lemma.

$$Y(Pa_{-1}, z)Y(Qa_{-1}, w) = \, :Y(Pa_{-1}, z)Y(Qa_{-1}, w):$$

$$+ \sum_{n \geq 0} \frac{1}{(z-w)^{n+1}} :Y(P, z)Y\left(\frac{\partial Q}{\partial a_{-n}^*}a_{-1}, w\right):$$

$$- \sum_{n \geq 0} \frac{(-1)^m}{(z-w)^{n+1}} :Y\left(\frac{\partial P}{\partial a_{-n}^*}a_{-1}, z\right)Y(Q, w):$$

$$- \sum_{n \geq 0, m \geq 0} \frac{(-1)^m}{(z-w)^{n+m+2}} :Y\left(\frac{\partial P}{\partial a_{-n}^*}, z\right)Y\left(\frac{\partial Q}{\partial a_{-m}^*}, w\right): \, .$$

12.2.10. Double contractions.
Using this formula, we can now easily obtain the commutators of the Fourier coefficients of the fields $Y(Pa_{-1}, z)$ and $Y(Qa_{-1}, w)$, using the residue calculus (see § 3.3).

The first two terms in the right hand side of the formula in Lemma 12.2.9 correspond to single contractions between $Y(Pa_{-1}, z)$ and $Y(Qa_{-1}, w)$. The part in the commutator of the Fourier coefficients induced by these terms will be exactly the same as the commutator of the corresponding vector fields, computed in $\mathcal{T}_{\text{loc}}$. Thus, we see that the discrepancy between the commutators in $A_{\leq 1,\text{loc}}$ and in $\mathcal{T}_{\text{loc}}$ (as measured by our two–cocycle) is due to the last term in the formula from Lemma 12.2.9, which comes from the *double contractions* between $Y(Pa_{-1}, z)$ and $Y(Qa_{-1}, w)$.

12.2.11. Back to the embedding.
Recall that our intention is to embed $L\mathfrak{sl}_2$ into $A_{\leq 1,\text{loc}}$ using its embedding into $\mathcal{T}_{\text{loc}}$. The sequence (12.2.1) does not split, but it is still possibile that this can be done: even if we cannot lift the entire $\mathcal{T}_{\text{loc}}$ to $A_{\leq 1,\text{loc}}$, we may still be able to lift its Lie subalgebra $L\mathfrak{sl}_2$.

The first thing that comes to mind is to apply our lifting $\imath : \mathcal{T}_{\text{loc}} \to A_{\leq 1, \text{loc}}$ to the vector fields (11.3.3), which amounts to imposing normal ordering. Then we should check if these lifts still satisfy the commutation relations in $L\mathfrak{sl}_2$, i.e., that the corresponding fields have the right OPEs.

Since $e(z)$ is linear, we do not need any ordering there, and its lift is $\widetilde{e}(z) = a(z)$. For $h(z)$ we obtain

$$\imath(h(z)) = \widetilde{h}(z) = -2{:}a^*(z)a(z){:} = Y(-2a_0^* a_{-1} \cdot v, z).$$

However, by doing so we distort the commutation relations. Originally, the h_n's commute among themselves. But their lifts, which we denote by $\widetilde{h}_n$, do not commute. By applying Lemma 12.2.9 in this case, we obtain

$$\widetilde{h}(z)\widetilde{h}(w) = -\frac{4}{(z-w)^2} + {:}\widetilde{h}(z)\widetilde{h}(w){:} \, .$$

Note that the first term in the right hand side came from the double contraction of the fields $\widetilde{h}(z)$ and $\widetilde{h}(w)$. This formula gives us the following commutation relations:

$$(12.2.6) \qquad\qquad [h_n, h_m] = -4n\delta_{n-m}.$$

Thus, we see that the commutation relations between h_n's are distorted, although only by a constant term. The situation is worse for the field $f(z)$. Its naive lift is

$$(12.2.7) \qquad\qquad \imath(f(z)) = \widetilde{f}(z) = -{:}a^*(z)^2 a(z){:} \, .$$

Unfortunately, one readily sees that this field does not have the right OPE with the other two fields, and that its Fourier coefficients do not commute with each other as they should.

Formula (12.2.6) shows that after the lifting the abelian Lie algebra spanned by the generators $h_n, n \in \mathbb{Z}$, becomes a Heisenberg Lie algebra. This suggests that while we cannot lift $L\mathfrak{sl}_2$ to $A_{\leq 1, \text{loc}}$, we may be able to lift it *projectively* – in other words, to obtain a map $\widehat{\mathfrak{sl}}_2 \to A_{\leq 1, \text{loc}}$ instead (necessarily of level -2, since $(h, h) = 2$). Note that this is not automatic at all, since the extension (12.1.4) is *not* an extension by constants, but by a module $A_{0, \text{loc}}$, on which vector fields act in a non-trivial way.

We still have some leverage, since we can modify our liftings by elements of $A_{0, \text{loc}}$. If we compute the OPE between $\widetilde{e}(z)$ and the field (12.2.7), we obtain

$$\widetilde{e}(z)\widetilde{f}(w) = \frac{\widetilde{h}(w)}{z-w} + \text{regular terms.}$$

The term $-2/(z-w)^2$ corresponding to the central extension of level -2 (since $(e, f) = 1$) is lacking. But we can introduce it if we add $-2\partial_w a^*(w)$ to the field (12.2.7), since

$$a(z)\partial_w a^*(w) = \frac{1}{(z-w)^2} + \text{regular terms.}$$

This suggests that we should redefine our field as follows:

$$(12.2.8) \qquad\qquad \widetilde{f}(z) = -{:}a^*(z)^2 a(z){:} - 2\partial_z a^*(z).$$

12.2.12. Exercise. Check that the fields $\widetilde{e}(z) = a(z), \widetilde{h}(z) = -2{:}a(z)a^*(z){:}$, and $\widetilde{f}(z)$ given by formula (12.2.8) have the OPEs of $\widehat{\mathfrak{sl}}_2$ of level -2, and therefore their Fourier coefficients span a Lie subalgebra of $A_{\leq 1,\mathrm{loc}}$ isomorphic to $\widehat{\mathfrak{sl}}_2$ of level -2. Thus we obtain

12.2.13. Proposition. *The restriction of the class of the extension* (12.2.1) *to $L\mathfrak{sl}_2 \subset \mathcal{T}_{\mathrm{loc}}$ coincides, as an element of $H^2(L\mathfrak{sl}_2, \mathcal{T}_{\mathrm{loc}})$, with the image of the central extension class from $H^2(L\mathfrak{sl}_2, \mathbb{C})$ corresponding to level -2.*

Using Lemma 11.1.4, we now obtain a homomorphism of vertex algebras:

12.2.14. Corollary. *There is a Lie algebra homomorphism $\widehat{\mathfrak{sl}}_2 \to U(M)$ sending the central element K to -2, and it gies rise to a homomorphism of vertex algebras $V_{-2}(\mathfrak{sl}_2) \to M$.*

12.3. Wakimoto realization

The above homomorphism $V_{-2}(\mathfrak{sl}_2) \to M$ is not an embedding, as can be easily seen by comparing the characters of both spaces. It is also defined only at the *critical level $k = -2$* of $\widehat{\mathfrak{sl}}_2$, where the vertex algebra $V_{-2}(\mathfrak{sl}_2)$ does not have a conformal vector. We would like to deform this homomorphism to other levels k.

Consider the Heisenberg vertex algebra $\pi^{2(k+2)}$ defined as in Remark 2.3.9. It is generated by the vertex operator $b(z)$ (note that this vertex algebra becomes commutative when $k = -2$). Introduce the vertex algebra

$$W_k = M \otimes_{\mathbb{C}} \pi^{2(k+2)}.$$

The following theorem, which is proved by a direct computation, shows that the action of $\widehat{\mathfrak{sl}}_2$ on M with level -2 may be deformed to an action of $\widehat{\mathfrak{sl}}_2$ on W_k with level k (here we again use Lemma 11.1.4).

12.3.1. Theorem.

(1) *The assignments*

$$
\begin{aligned}
e(z) &\mapsto a(z) \\
(12.3.1) \qquad h(z) &\mapsto -2{:}a^*(z)a(z){:} + b(z) \\
f(z) &\mapsto -{:}a^*(z)^2 a(z){:} + k\partial_z a^*(z) + a^*(z)b(z)
\end{aligned}
$$

give rise to homomorphisms $w_k : V_k(\mathfrak{sl}_2) \to W_k$ and $\widehat{\mathfrak{sl}}_2 \to U(W_k)$ with $K \mapsto k$.

(2) *The Segal-Sugawara vector*

$$S = \frac{1}{2}\left(e_{-1}f_{-1} + f_{-1}e_{-1} + \frac{1}{2}h_{-1}^2\right)v_k$$

of $V_k(\mathfrak{sl}_2)$ is mapped by w_k to the following vector in W_k:

$$(12.3.2) \qquad S \mapsto \left((k+2)a^*_{-1}a_{-1} + \frac{1}{2}b_{-1}^2 + \frac{1}{2}b_{-2}\right)|0\rangle.$$

Thus, for $k \neq -2$, w_k is a homomorphism of conformal vertex algebras, where we choose $\frac{1}{k+2}w_k(S)$ as the conformal vector in W_k.

12.3.2. Remarks. To see that we indeed get the correct level k, note that in $\pi^{2(k+2)}$,

$$b(z)b(w) = \frac{2(k+2)}{(z-w)^2} + :b(z)b(w):,$$

so that

$$h(z)h(w) = -\frac{4}{(z-w)^2} + \frac{2(k+2)}{(z-w)^2} + \text{reg.} = \frac{k(h,h)}{(z-w)^2} + \text{reg.}$$

(here we use the fact that $(h,h) = 2$). This formula coincides with the OPE formula (3.4.1) in the case when $J^a = J^b = h$.

Note that if $k = -2$, then the Heisenberg vertex algebra $\pi^{2(k+2)}$ (and the Weyl algebra $\widetilde{\mathcal{H}}_{2(k+2)}$) becomes commutative. Therefore it has a trivial one-dimensional representation. Hence at level -2 we obtain a homomorphism $V_{-2}(\mathfrak{sl}_2) \to M$. However, when $k \neq -2$, the smallest Heisenberg modules are the Fock representations $\pi_\nu^{2(k+2)}$. Therefore the smallest vertex algebra to which we can map $V_k(\mathfrak{sl}_2)$ this way is W_k.

If k is not equal to -2 plus a non-negative rational number, then $V_k(\mathfrak{sl}_2)$ is irreducible. Since $w_k(v_k) \neq 0$ by definition, we find that w_k is injective in this case. On can also show that w_k remains injective for all values of k, even when $V_k(\mathfrak{sl}_2)$ is reducible.

12.3.3. Wakimoto modules. Using the homomorphism $V_k(\mathfrak{sl}_2) \to W_k$ of Theorem 12.3.1, we can now construct a family of modules over $V_k(\mathfrak{sl}_2)$ (and hence over $\widehat{\mathfrak{sl}}_2$).

For $\nu \in \mathbb{C}$, denote by $\pi_\nu^{2(k+2)}$ the Fock representation with highest weight ν defined as in § 5.2.1. For each $\nu \in \mathbb{C}$, the tensor product

$$W_{\nu,k} = M \otimes \pi_\nu^{2(k+2)}$$

is a W_k–module. In particular, $W_{0,k} = W_k$. Via the homomorphism w_k, each $W_{\nu,k}$ becomes a $V_k(\mathfrak{sl}_2)$–module and an $\widehat{\mathfrak{sl}}_2$–module of level k. These modules are called the *Wakimoto modules* over $\widehat{\mathfrak{sl}}_2$. Note that they are $\mathbb{Z}$–bigraded, with respect to the degree and charge, with finite–dimensional homogeneous components.

12.3.4. The top component. Let us find out what the top degree homogeneous component $\overline{W}_{\nu,k} \subset W_{\nu,k}$ looks like. If we denote the highest–weight vector $v \otimes |\nu\rangle$ of $W_{\nu,k}$ by $v_{\nu,k}$, then

$$(12.3.3) \qquad\qquad \overline{W}_{\nu,k} = \mathbb{C}[a_0^*] \cdot v_{\nu,k}.$$

The Lie algebra $\mathfrak{sl}_2 = \mathfrak{sl}_2 \otimes 1 \subset \widehat{\mathfrak{sl}}_2$ preserves $\overline{W}_{\nu,k}$. If we denote the variable a_0^* by y, then $\overline{W}_{\nu,k}$ becomes isomorphic to $\mathbb{C}[y]$, with the action of $\mathfrak{sl}_2$ on $\overline{W}_{\nu,k}$ given by formulas (11.2.4) defining the action of $\mathfrak{sl}_2$ on $\mathbb{C}[y] = \mathbb{C}[N_+]$ corresponding to the highest weight ν. Therefore $\overline{W}_{\nu,k}$ is isomorphic, as an $\mathfrak{sl}_2$–module, to the contragredient Verma module M_ν^*.

12.3.5. The case of arbitrary $\mathfrak{g}$. For a general simple Lie algebra $\mathfrak{g}$, consider the corresponding flag manifold G/B_- and its open N_+–orbit $U \subset G/B_-$. Then U is isomorphic to the vector space $\mathbb{C}^d$, $d = \dim N_+$. We define the completion $\widetilde{A}^{\mathfrak{g}}$ of the corresponding Weyl algebra $A((U \otimes \mathcal{K}) \oplus (U \otimes \Omega_{\mathcal{K}}))$ as in § 12.1.1. This algebra is topologically generated by elements $a_{\alpha,n}, a^*_{\alpha,n}, \alpha \in \Delta_+, n \in \mathbb{Z}$, satisfying the relations

$$[a_{\alpha,n}, a^*_{\beta,m}] = \delta_{\alpha,\beta}\delta_{n,-m}, \qquad [a_{\alpha,n}, a_{\beta,m}] = [a^*_{\alpha,n}, a^*_{\beta,m}] = 0.$$

Let $M_{\mathfrak{g}}$ be the Fock representation of $\widetilde{A}^{\mathfrak{g}}$ generated by a vector $|0\rangle$ such that

$$a_{\alpha,n}|0\rangle = 0, \quad n \geq 0; \qquad a^*_{\alpha,n}|0\rangle = 0, \quad n > 0.$$

Then $M_{\mathfrak{g}}$ carries a vertex algebra structure, which is defined in the same way as in Lemma 11.3.8 (in fact, $M_{\mathfrak{g}}$ is the tensor product of d copies of M). Then we obtain a Lie algebra $A^{\mathfrak{g}}_{\leq 1,\mathrm{loc}} = \widetilde{A}^{\mathfrak{g}}_{\leq 1} \cap U(M_{\mathfrak{g}})$ in the same way as in § 12.2.2 and an analogue of the extension (12.2.1):

$$0 \to A^{\mathfrak{g}}_{0,\mathrm{loc}} \to A^{\mathfrak{g}}_{\leq 1,\mathrm{loc}} \to \mathcal{T}^{\mathfrak{g}}_{\mathrm{loc}} \to 0,$$

where $\mathcal{T}^{\mathfrak{g}}_{\mathrm{loc}}$ is a Lie subalgebra of the Lie algebra $\mathrm{Vect}(U \otimes \mathcal{K})$ of vector fields on $U \otimes \mathcal{K}$.

The infinitesimal action of $L\mathfrak{g}$ on $U \otimes \mathcal{K}$ gives us a Lie algebra homomorphism $L\mathfrak{g} \to \mathcal{T}^{\mathfrak{g}}_{\mathrm{loc}}$. Then we have (recall that $h^\vee$ denotes the dual Coxeter number of $\mathfrak{g}$)

12.3.6. Theorem ([FF1, FF2, F6]). *The homomorphism $L\mathfrak{g} \to \mathcal{T}^{\mathfrak{g}}_{\mathrm{loc}}$ may be lifted to a homomorphism $\widehat{\mathfrak{g}} \to A^{\mathfrak{g}}_{\leq 1,\mathrm{loc}}$, sending the central element K to $-h^\vee$. Moreover, there is a homomorphism of vertex algebras $V_{-h^\vee}(\mathfrak{g}) \to M_{\mathfrak{g}}$.*

The proof given in [**FF2, F6**] is based on a cohomological argument. Explicit formulas for the homomorphism $\widehat{\mathfrak{g}} \to A^{\mathfrak{g}}_{\leq 1,\mathrm{loc}}$ for $\mathfrak{g} = \mathfrak{sl}_n$ may be found in [**FF1, FF2**]. In general, they are rather complicated (see [**dBF**]).

12.3.7. Critical level. To see where the dual Coxeter number is coming from, we compute the restriction of the homomorphism $\widehat{\mathfrak{g}} \to A^{\mathfrak{g}}_{\leq 1,\mathrm{loc}}$ to $\widehat{\mathfrak{h}}$, the Heisenberg Lie subalgebra of $\widehat{\mathfrak{g}}$. It is spanned by the $u_n, u \in \mathfrak{h}, n \in \mathbb{Z}$, and K, with the relations

$$(12.3.4) \qquad\qquad [u_n, u'_m] = n(u, u')K\delta_{n,-m}.$$

The formulas for the images of u_n in $\mathcal{T}_{\mathrm{loc}}$ are easy to find. Applying the lifting $\imath$ (see formula (12.2.3)) to the corresponding generating functions $u(z)$, we obtain the following formula:

$$u(z) \mapsto \widetilde{u}(z) = -\sum_{\alpha \in \Delta_+} \alpha(u){:}a^*_\alpha(z)a_\alpha(z){:}\,.$$

Using Lemma 12.2.9, we find the OPE

$$u(z)u'(w) = -\frac{\sum_{\alpha \in \Delta_+} \alpha(u)\alpha(u')}{(z-w)^2} + {:}u(z)u'(w){:}\,,$$

which gives the commutation relations (12.3.4) with

$$K = -\frac{1}{(u, u')}\sum_{\alpha \in \Delta_+} \alpha(u)\alpha(u').$$

The sum in the numerator is equal to

$$\mathrm{Tr}_{\mathfrak{n}_+}(\mathrm{ad}\,u\,\mathrm{ad}\,u') = \frac{1}{2}\,\mathrm{Tr}_{\mathfrak{g}}(\mathrm{ad}\,u\,\mathrm{ad}\,u'),$$

i.e., one half of the Killing form, and hence to $h^\vee(u, u')$ (see § 2.4.1). Therefore we obtain that K maps to $-h^\vee$, which is called the *critical level*.

12.3.8. Remark. Having established the existence of a lifting $\widehat{\mathfrak{g}} \to A^{\mathfrak{g}}_{\leq 1, \mathrm{loc}}$, it is natural to ask how to describe all possible liftings. It is easy to see that, if non-empty, the set of splittings of the exact sequence

$$0 \to C \to \widehat{\mathfrak{g}} \to \mathfrak{g} \to 0$$

is a torsor over the group $H^1(\mathfrak{g}, C)$. In our case we obtain a torsor over the group $H^1(L\mathfrak{g}, A^{\mathfrak{g}}_{0,\mathrm{loc}})$. One can show that this group is isomorphic to the dual space $(L\mathfrak{h})^* \simeq \mathfrak{h}^* \otimes \Omega_{\mathcal{K}}$. Each lifting gives rise to a different $\widehat{\mathfrak{g}}$–module structure on M, and hence we obtain a family of $\widehat{\mathfrak{g}}$–modules parametrized by such liftings. These are the Wakimoto modules of critical level. One can show that they are irreducible for a generic lifting (see [**F1, F6**]).

Moreover, one can describe the torsor of liftings explicitly. For instance, in the case of $\mathfrak{sl}_2$ we obtain a torsor over the space $\Omega_{\mathcal{K}}$ of one-forms on the punctured disc. This torsor may be canonically identified with the space of affine connections on the punctured disc (see § 8.1). One can show this by computing the action of the group $\mathrm{Aut}\, \mathcal{O}$ of changes of coordinates on these liftings and verifying that this action coincides with the action of $\mathrm{Aut}\, \mathcal{O}$ on affine connections. As explained in [**F6**], for general $\mathfrak{g}$, the torsor of liftings may be identified with the connections on a certain ${}^L H$–bundle on the punctured disc, where ${}^L H$ is the Langlands dual group of the Cartan subgroup H of G (see § 18.4.9).

12.3.9. Other levels. In order to obtain a free field realization of $\widehat{\mathfrak{g}}$ at levels other than critical, we consider another copy of the Heisenberg Lie algebra $\widehat{\mathfrak{h}}$. Recall that $\widehat{\mathfrak{h}}$ is just a central extension of $L\mathfrak{h}$, where $\mathfrak{h}$ is the Cartan subalgebra of $\mathfrak{g}$ (see § 5.4.1). Consider the corresponding vertex algebra $\pi^{(k+h^\vee)}$, defined as in Remark 2.3.9. Form the tensor product $W_k = M_{\mathfrak{g}} \otimes \pi^{(k+h^\vee)}$. Then we have the following generalization of Theorem 12.3.1 (see [**FF1, FF2, F6**]):

12.3.10. Theorem. *For each $k \in \mathbb{C}$, there is an injective homomorphism of vertex algebras $V_k(\mathfrak{g}) \to W_k$, and an injective homomorphism of Lie algebras $\widehat{\mathfrak{g}} \to U(W_k)$ sending K to k.*

Under the above homomorphism, for each $\nu \in \mathfrak{h}^*$, the tensor product

$$W_{\nu,k} = M_{\mathfrak{g}} \otimes \pi^{2(k+h^\vee)}_\nu$$

becomes a $V_k(\mathfrak{g})$–module and a $\widehat{\mathfrak{g}}$–module of level k. These modules are called the *Wakimoto modules* over $\widehat{\mathfrak{g}}$ of level k.

12.3.11. Semi–infinite induction. The above construction of Wakimoto modules can be cast as the application of a "semi-infinite induction" functor.

Recall the Cartan decomposition (11.2.3). In § 11.2.4 we defined the Verma module M_χ over $\mathfrak{g}$ corresponding to $\chi \in \mathfrak{h}^*$ as the representation of $\mathfrak{g}$ induced from the one–dimensional $\mathfrak{b}_+$–module $\mathbb{C}_\chi$:

$$M_\chi = \mathrm{Ind}^{\mathfrak{g}}_{\mathfrak{b}_+} \mathbb{C}_\chi = U(\mathfrak{g}) \otimes_{U(\mathfrak{b}_+)} \mathbb{C}_\chi.$$

The induction functor is a guise of the coinvariants functor, or the zeroth Lie algebra homology:

$$M_\chi = H_0(\mathfrak{b}_+, U(\mathfrak{g}) \otimes_{\mathbb{C}} \mathbb{C}_\chi).$$

In § 11.2.4 we also defined the contragredient Verma module M_χ^* as the module coinduced from the one–dimensional $\mathfrak{b}_-$–module $\mathbb{C}_\chi$,

$$M_\chi^* = \mathrm{Coind}_{\mathfrak{b}_-}^{\mathfrak{g}} \, \mathbb{C}_\chi = \mathrm{Hom}_{U(\mathfrak{b}_-)}^{\mathrm{res}}(U(\mathfrak{g}), \mathbb{C}_\chi),$$

where $\mathbb{C}_\chi$ is a $\mathfrak{b}_-$–module defined as above after replacing $\mathfrak{n}_+$ by $\mathfrak{n}_-$. Thus, coinduced modules arise from the invariants functor, or the zeroth Lie algebra cohomology:

$$M_\chi^* = H^0(\mathfrak{b}_-, \mathrm{Hom}_{\mathbb{C}}^{\mathrm{res}}(U(\mathfrak{g}), \mathbb{C}_\chi)).$$

More generally, let $\mathfrak{g}_1$ and $\mathfrak{g}_2$ be Lie algebras, and B a module over $\mathfrak{g}_1 \oplus \mathfrak{g}_2$. Then we can define the functors

$$(\mathfrak{g}_2\text{–mod}) \;\; \rightarrow \;\; (\mathfrak{g}_1\text{–mod}),$$
$$F_i' : M \;\; \mapsto \;\; H_i(\mathfrak{g}_2, B \otimes M),$$
$$F^i : M \;\; \mapsto \;\; H^i(\mathfrak{g}_2, \mathrm{Hom}(B, M)).$$

The functor F_0' specializes to the previous example of induced modules when $\mathfrak{g}_1 = \mathfrak{g}$, $\mathfrak{g}_2 = \mathfrak{b}_+$ and $B = U(\mathfrak{g})$, considered as a $(\mathfrak{g}, \mathfrak{b}_+)$–bimodule (with $\mathfrak{g}$ acting from the left and $\mathfrak{b}_+$ acting from the right). The functor F^0 specializes to the example of coinduced modules with $\mathfrak{g}_1 = \mathfrak{g}$, $\mathfrak{g}_2 = \mathfrak{b}_-$.

Now consider the affine algebra $\widehat{\mathfrak{g}}$. Its standard Cartan decomposition is given by formula (11.3.1). However, we may choose instead the following decomposition (cf. § 11.3.2):

$$\widehat{\mathfrak{g}} = \mathfrak{n}_-((t)) \oplus \widehat{\mathfrak{h}} \oplus \mathfrak{n}_+((t)),$$

where $\widehat{\mathfrak{h}} = \mathfrak{h}((t)) \oplus \mathbb{C}K$. As in the finite–dimensional case, we may define representations of $\widehat{\mathfrak{g}}$ by picking a character $\chi(t)dt \in \mathfrak{h}((t))^*dt$ of $\mathfrak{h}((t))$, and extending it to a character of $\mathfrak{b}_-((t))$ by letting $\mathfrak{n}_-((t))$ and K act by zero. We may then take the coinduced representation of $\widehat{\mathfrak{g}}$ from this one-dimensional representation of $\mathfrak{b}_-((t))$. The resulting $\widehat{\mathfrak{g}}$–module $M_{\chi(t)}^*$ is realized in the space of functions on the big cell $U \otimes \mathcal{K}$ (which we discussed above in detail in the case of $\widehat{\mathfrak{sl}}_2$). However, this representation does not belong to the category $\mathcal{O}$ (see § 11.3.2); for example, the $\mathbb{Z}$–gradation on it is not bounded from below. As we found out above, a suitable $\widehat{\mathfrak{g}}$–module is the module M defined by formula (11.3.8). It is the space not of functions on $U \otimes \mathcal{K}$, but of delta–functions supported at the subspace $U \otimes \mathcal{O}$.

The module M arises neither out of induction (that would be the space of delta–functions supported at a point) nor out of coinduction (space of functions). Rather, M may be constructed by a process that it is natural to call "semi–infinite induction", which appears neither as homology nor as cohomology but as an intermediate functor, *semi–infinite cohomology*.

First one needs to construct a $\widehat{\mathfrak{g}}$–bimodule $U^{\infty/2}(\widehat{\mathfrak{g}})$. It may be defined in a similar way to the above definition of the Wakimoto modules, as the space of delta–functions on a neighborhood of $1 \in G((t))$, supported on the intersection of this neighborhood with $G[[t]]$. An interesting feature of the module $U^{\infty/2}(\widehat{\mathfrak{g}})$ is that it carries two commuting actions of $\widehat{\mathfrak{g}}$ of levels k_1, k_2, where $k_1 + k_2 = -2h^\vee$, rather than 0, as is the case for the ordinary $\widehat{\mathfrak{g}}$–module $U(\widehat{\mathfrak{g}})$.

Having constructed the bimodule $U^{\infty/2}(\widehat{\mathfrak{g}})$, we can use it to obtain Wakimoto modules by analogy with the construction of the Verma modules. Namely, we again pick a character $\chi(t)dt$ of $\mathfrak{h}((t))$ and extend it to a one–dimensional representation $\mathbb{C}_{\chi(t)}$ of $\mathfrak{b}_-((t))$. But now we take the semi–infinite cohomology of $\mathfrak{b}_-((t))$

with coefficients in $U^{\infty/2}(\widehat{\mathfrak{g}}) \otimes \mathbb{C}_{\chi(t)}$ (see § A.4), where $U^{\infty/2}(\widehat{\mathfrak{g}})$ is considered as a $\widehat{\mathfrak{g}}$–bimodule with levels $(-h^\vee, -h^\vee)$; this is necessary for the semi-infinite cohomology to be well-defined. The result is the Wakimoto module $W_{\chi(t)}$ of level $-h^\vee$. However, because of the non-trivial action of the group $\mathrm{Aut}\,\mathcal{O}$ on the semi-infinite cohomology complex, we obtain that under changes of the coordinate t, $\chi(t)dt$ does not transform as a one-form, but as a connection – see Remark 12.3.8.

If we apply this construction to the $\widehat{\mathfrak{h}}$–module $\pi_\nu^{(k+h^\vee)}$ of level $(k + h^\vee)$, we end up with the Wakimoto module $W_{\nu,k}$ of level k (so the level always shifts by $-h^\vee$).

12.4. Bibliographical notes

The Wakimoto modules were defined by M. Wakimoto [**Wak**] for $\widehat{\mathfrak{sl}}_2$ and by B. Feigin and E. Frenkel [**FF1, FF2, F6**] for an arbitrary affine algebra. The exposition in this chapter essentially follows [**FF2**] (see also [**F3, F6**]).

Semi-infinite cohomology was introduced by B. Feigin [**Fe**] (see also [**FGZ, FF6, Vo1**]). The construction of the Wakimoto modules by semi-infinite induction is discussed in [**Vo2**]. Various definitions of the bimodule $U^{\infty/2}(\widehat{\mathfrak{g}})$ are given in [**Vo2, Ar, Soe**].

CHAPTER 13

The Knizhnik–Zamolodchikov Equations

In this chapter and its sequel, we apply the general results on conformal blocks and the technique of free field realization to the study of conformal blocks associated to the Kac–Moody vertex algebra $V_k(\mathfrak{g})$ and the curve $X = \mathbb{P}^1$. Our goal is to find explicit formulas for the horizontal sections of the bundle of conformal blocks over the configuration space of points on $\mathbb{P}^1$. When we trivialize this bundle, the horizontality condition becomes equivalent to a remarkable holonomic system of differential equations with regular singularities, called the Knizhnik–Zamolodchikov (KZ) equations. This example should serve as an illustration of the general construction of flat connections (or more generally, $\mathcal{D}$–module structures) on the sheaves of conformal blocks and coinvariants that we present below in Chapter 17.

We start with a "baby version" of the Knizhnik–Zamolodchikov system, coming from the bundle of conformal blocks associated to the Heisenberg vertex algebra. By the free field realization technique developed in Chapters 11 and 12, we know how to embed the Kac–Moody vertex algebra into a Heisenberg vertex algebra. Using the functoriality of conformal blocks (see § 10.2) we may then apply the results of our calculations of the Heisenberg conformal blocks to the computation of the Kac–Moody conformal blocks. This way we find all solutions of the Knizhnik–Zamolodchikov equations, at least for generic values of the parameters, as we will see in Chapter 14.

13.1. Conformal blocks in the Heisenberg case

13.1.1. Fock representations. Recall the Heisenberg Lie algebra $\mathcal{H}$ introduced in § 2.1.2. Let $\mathbb{C}_\nu^\kappa$ be the one-dimensional representation of the Lie subalgebra $(\mathbb{C}[[t]] \oplus \mathbf{1}) \subset \mathcal{H}$, on which $\mathbf{1}$ acts by multiplication by κ, the element $1 \in \mathbb{C}[[t]]$ (i.e, b_0) acts by ν, and elements of $t\mathbb{C}[[t]]$ (i.e., $b_n, n > 0$) act by 0. We define the Fock representation π_ν^κ of $\mathcal{H}$, of level $\kappa \in \mathbb{C}^\times$ (not necessarily equal to 1) and highest weight $\nu \in \mathbb{C}$ as the induced representation

$$(13.1.1) \qquad \pi_\nu^\kappa = \mathrm{Ind}_{\mathbb{C}[[t]] \oplus \mathbf{1}}^{\mathcal{H}} \mathbb{C}_\nu^\kappa.$$

In other words, the module π_ν^κ is generated by a vector $|\nu\rangle$ such that

$$b_n|\nu\rangle = \delta_{n,0}\nu|\nu\rangle, \qquad n \geq 0.$$

Thus,

$$\pi_\nu^\kappa = \mathbb{C}[b_{-1}, b_{-2}, \ldots] \cdot |\nu\rangle.$$

The Fock space $\pi^\kappa = \pi_0^\kappa$ is a conformal vertex algebra for any $\kappa \in \mathbb{C}^\times$ with the same operation Y as for $\kappa = 1$, as defined in § 2.3.6, and with the conformal vector

$$(13.1.2) \qquad \omega = \frac{1}{2\kappa}b_{-1}^2|0\rangle.$$

The corresponding Virasoro vertex operator reads:

$$(13.1.3) \qquad\qquad Y(\omega, z) = \frac{1}{2\kappa} {:}b(z)^2{:} .$$

13.1.2. Conformal blocks. From now on, we will concentrate on the genus zero case, setting $X = \mathbb{P}^1$, fixing a point $\infty \in \mathbb{P}^1$ and fixing a global coordinate t on $\mathbb{P}^1\backslash\infty$ (these choices are unique up to a Möbius transformation). Let $\vec{x} = \{x_1, \ldots, x_N\}$ be a collection of N distinct points on $\mathbb{P}^1\backslash\infty$, to which we attach the Fock representations $\pi^\kappa_{\nu_1}, \ldots, \pi^\kappa_{\nu_N}$, where $\nu_i \in \mathbb{C}$.

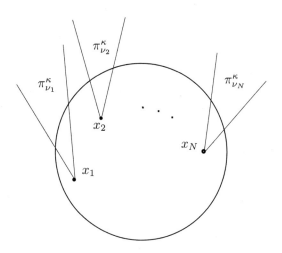

Let $\mathcal{H}_{x_i}$ be the Heisenberg Lie algebra which is the central extension of the commutative Lie algebra $\mathbb{C}((t - x_i))$. Denote by $\mathcal{H}^+_{x_i}$ its positive half $\mathbb{C}[[t - x_i]]$.

Consider the central extension

$$0 \to \mathbb{C}\mathbf{1} \to \mathcal{H}(\vec{x}) \to \bigoplus_{i=1}^N \mathbb{C}((t - x_i)) \to 0$$

defined by the two-cocycle which is the sum of the two-cocycles coming from each summand. In other words, $\mathcal{H}(\vec{x})$ is the direct sum of the Heisenberg Lie algebras $\mathcal{H}_{x_i}, i = 1, \ldots, N$, in which we identify the central elements. We have a Lie subalgebra

$$\mathcal{H}_{\vec{x}} \overset{\text{def}}{=} \mathbb{C}[\mathbb{P}^1\backslash\{x_1, \ldots, x_N\}] \subset \bigoplus_{i=1}^N \mathbb{C}((t - x_i)).$$

By the residue theorem, the restriction of the central extension to $\mathcal{H}_{\vec{x}}$ is equal to 0 (cf. Lemma 9.1.5). Therefore $\mathcal{H}_{\vec{x}}$ is naturally a Lie subalgebra of $\mathcal{H}(\vec{x})$.

Consider the $\mathcal{H}(\vec{x})$–module

$$\pi^\kappa_{\vec{\nu}} = \pi^\kappa_{\nu_1, \ldots, \nu_N} \overset{\text{def}}{=} \bigotimes_{i=1}^N \pi^\kappa_{\nu_i},$$

where $\vec{\nu} = (\nu_1, \ldots, \nu_N)$, and the corresponding space of conformal blocks

$$C_{\pi^\kappa}(\mathbb{P}^1, (x_i), (\pi^\kappa_{\nu_i}))_{i=1}^N$$

as defined in Definition 10.1.1. Generalizing the argument of Theorem 9.3.3, we obtain that

$$(13.1.4) \qquad C_{\pi^\kappa}(\mathbb{P}^1, (x_i), (\pi^\kappa_{\nu_i}))_{i=1}^N = \mathrm{Hom}_{\mathcal{H}_{\vec{x}}}(\pi^\kappa_{\vec{\nu}}, \mathbb{C}).$$

Let us describe this space.

13.1.3. Lemma. *We have the decomposition*

$$(13.1.5) \qquad \bigoplus_{i=1}^N \mathbb{C}((t - x_i)) = \bigoplus_{i=1}^N \mathbb{C}[[t - x_i]] + \mathbb{C}[\mathbb{P}^1 \backslash \{x_1, \ldots, x_N\}],$$

and

$$(13.1.6) \qquad \bigoplus_{i=1}^N \mathbb{C}[[t - x_i]] \cap \mathbb{C}[\mathbb{P}^1 \backslash \{x_1, \ldots, x_N\}] = \mathbb{C}.$$

13.1.4. Proof. The lemma may be interpreted as a calculation of the cohomology of $\mathbb{P}^1$ using a formal version of the Čech complex with respect to the "cover" $\mathbb{P}^1 = \bigcup_{i=1}^N D_{x_i} \cup (\mathbb{P}^1 \backslash \{x_1, \ldots, x_N\})$. This complex computes the cohomologies of $\mathbb{P}^1$ with coefficients in $\mathcal{O}_{\mathbb{P}^1}$ (the proof is analogous to the proof of Lemma 17.3.6 below). The intersection (13.1.6) is isomorphic to $H^0(\mathbb{P}^1, \mathcal{O})$, which consists of constant functions. On the other hand, the quotient of the left hand side of (13.1.5) by the right hand side of (13.1.5) is isomorphic to $H^1(\mathbb{P}^1, \mathcal{O}) = 0$.

13.1.5. Corollary.

$$C_{\pi^\kappa}(\mathbb{P}^1, (x_i), (\pi^\kappa_{\nu_i}))_{i=1}^N = \begin{cases} \mathbb{C}, & if \ \sum_{i=1}^N \nu_i = 0, \\ 0, & otherwise. \end{cases}$$

13.1.6. Proof. Denote by $\mathbb{C}[\mathbb{P}^1 \backslash \{x_1, \ldots, x_N\}]_0$ the space of functions on $\mathbb{P}^1$ which are regular away from $x_1, \ldots, x_N$ and vanish at ∞. Let $\mathcal{H}^0_{\vec{x}}$ be the corresponding Lie subalgebra of $\mathcal{H}(\vec{x})$.

By Lemma 13.1.3, we have a decomposition of $\mathcal{H}(\vec{x})$, considered as a vector space, into a direct sum of its Lie subalgebras,

$$(13.1.7) \qquad \mathcal{H}(\vec{x}) = \mathcal{H}^0_{\vec{x}} \oplus \mathcal{H}(\vec{x})_+,$$

where

$$\mathcal{H}(\vec{x})_+ = \bigoplus_{i=1}^N \mathcal{H}^+_{x_i} \oplus \mathbb{C}\mathbf{1}.$$

From the definition (13.1.1) of the module π^κ_ν we conclude that

$$\pi^\kappa_{\vec{\nu}} \simeq \mathrm{Ind}_{\mathcal{H}(\vec{x})_+}^{\mathcal{H}(\vec{x})} \mathbb{C}^\kappa_{\vec{\nu}},$$

where

$$\mathbb{C}^\kappa_{\vec{\nu}} = \mathbb{C} \cdot |\nu_1\rangle \otimes \ldots \otimes |\nu_N\rangle,$$

with its natural $\mathcal{H}(\vec{x})_+$–action. According to the decomposition (13.1.7), $\pi^\kappa_{\vec{\nu}}$ is a free $\mathcal{H}^0_{\vec{x}}$–module with one generator. Therefore the space of $\mathcal{H}^0_{\vec{x}}$–invariant functionals on $\pi^\kappa_{\vec{\nu}}$ is one-dimensional and is isomorphic to the space dual to the one-dimensional space $\mathbb{C}^\kappa_{\vec{\nu}}$.

Now by formula (13.1.4) the space of conformal blocks we are trying to compute is the space of $\mathcal{H}_{\vec{x}}$–invariant functionals on $\pi^\kappa_{\vec{\nu}}$. Since $\mathcal{H}_{\vec{x}} = \mathcal{H}^0_{\vec{x}} \oplus \mathbb{C}$, this space is equal to the subspace of the space of $\mathcal{H}^0_{\vec{x}}$–invariant functionals consisting of functionals invariant under the action of the one-dimensional Lie algebra of constant functions on $\mathbb{P}^1$. This Lie algebra acts on $\mathbb{C}_{\vec{\nu}}^\kappa$ by multiplication by $\sum_{i=1}^N \nu_i$. This proves the corollary.

13.1.7. Modified space of conformal blocks. In view of Corollary 13.1.5, it makes sense to modify the definition of the conformal blocks, and consider instead the space of $\mathcal{H}_{\vec{x}}^0$–invariant functionals on $\pi_{\vec{\nu}}^\kappa$, denoted by $C_{\pi^\kappa}^0(\mathbb{P}^1, (x_i), (\pi_{\nu_i}^\kappa))_{i=1}^N$, or $C^0(\vec{x}, \pi_{\vec{\nu}}^\kappa)$ for short. From now on by the space of Heisenberg conformal blocks we will understand $C^0(\vec{x}, \pi_{\vec{\nu}}^\kappa)$.

According to the proof of Corollary 13.1.5, $C^0(\vec{x}, \pi_{\vec{\nu}}^\kappa)$ is always one–dimensional. Moreover, it follows that a functional $\tau \in C^0(\vec{x}, \pi_{\vec{\nu}}^\kappa)$ is uniquely determined by its value on $|\nu_1\rangle \otimes \ldots \otimes |\nu_N\rangle$.

It is possible to define $C^0(\vec{x}, \pi_{\vec{\nu}}^\kappa)$ along the lines of § 10.1.2. Recall the Lie algebras $U(\Pi_x^\kappa)$ and $U_\Sigma(\pi^\kappa)$ defined in § 9.2.2 and § 10.1.2, and the homomorphism $U_\Sigma(\pi^\kappa) \to \bigoplus_{i=1}^N U(\Pi_{x_i}^\kappa)$, where $\Sigma = \mathbb{P}^1 \backslash \{x_1, \ldots, x_N\}$. Via this homomorphism the Lie algebra $U_\Sigma(\pi^\kappa)$ acts on $\pi_{\vec{\nu}}^\kappa$, and the ordinary space of conformal blocks $C(\mathbb{P}^1; \vec{x}; \pi_{\vec{\nu}}^\kappa)$ is by definition the space of $U_\Sigma(\pi^\kappa)$–invariant functionals on $\pi_{\vec{\nu}}^\kappa$.

Since we have chosen a global coordinate t on $\mathbb{P}^1$, we identify Π_x^κ with π^κ and $U(\Pi_x^\kappa)$ with $U(\pi^\kappa)$. The latter is just the completion of the span of all Fourier coefficients of vertex operators from π^κ. Using this identification, the image of $U_\Sigma(\pi^\kappa)$ in $\bigoplus_{i=1}^N U(\Pi_{x_i}^\kappa)$ may be described as the span of N–tuples $(R_1, \ldots, R_N) \in U(\pi^\kappa)^{\oplus N}$, where

$$(13.1.8) \qquad R_i = \mathrm{Res}_{t=x_i} f(t) Y(A, t) dt, \qquad f(t) \in \mathbb{C}[\mathbb{P}^1 \backslash \{x_1, \ldots, x_N\}], A \in \pi^\kappa.$$

Now consider its subspace $U_\Sigma^0(\pi^\kappa)$ spanned by N–tuples as above, where $f(t)$ satisfies the extra condition that it vanishes at $\infty \in \mathbb{P}^1$ up to order $2 \deg A - 1$ (for homogeneous $A \in \pi^\kappa$). It is straightforward to check that such elements form a Lie subalgebra of $U(\pi^\kappa)^{\oplus N}$. Moreover, $U_\Sigma^0(\pi^\kappa)$ contains $\mathcal{H}_{\vec{x}}^0$ as a Lie subalgebra. Following the same argument as in the proof of Theorem 9.3.3, one shows that the space of $U_\Sigma^0(\pi^\kappa)$–invariant functionals on $\pi_{\vec{\nu}}^\kappa$ coincides with the space $C^0(\vec{x}, \pi_{\vec{\nu}}^\kappa)$ of $\mathcal{H}_{\vec{x}}^0$–invariant functionals on $\pi_{\vec{\nu}}^\kappa$.

13.1.8. One-forms vs. affine connections. One of the advantages of the modified space of conformal blocks is that it does not depend on the choice of conformal vector in π^κ (and on the resulting action of $\mathrm{Aut}\,\mathcal{O}$ on the Fock representations). To explain this point, consider first the space of ordinary conformal blocks $C_{\pi^\kappa}(\mathbb{P}^1, (x_i), (\pi_{\nu_i}^\kappa))_{i=1}^N$. In the case of the conformal vector (13.1.2) (without the linear term), the expression $b(t - x_i) dt$ is a well-defined one–form with values in $\mathrm{End}\,\Pi_{x_i}^\kappa$. By definition, a functional φ on $\bigotimes_{i=1}^N \Pi_{x_i}^\kappa \simeq \pi_{\vec{\nu}}^\kappa$ is in $C_{\pi^\kappa}(\mathbb{P}^1, (x_i), \pi_{\nu_i}^\kappa)_{i=1}^N$ if and only if for any $A \in \pi_{\vec{\nu}}^\kappa$ the one-forms

$$(13.1.9) \qquad \varphi(b(t - x_i)^{(i)} \cdot A) dt, \qquad i = 1, \ldots, N,$$

on $D_{x_i}^\times$ extend to a regular one-form on $\mathbb{P}^1 \backslash \{x_1, \ldots, x_N\}$. For example, since $b_0^{(i)}$ acts on $\pi_{\vec{\nu}}^\kappa$ by multiplication by ν_i, in the case when $A = |\nu_1\rangle \otimes \ldots \otimes |\nu_N\rangle$ the one-form (13.1.9) reads $\dfrac{\nu_i}{t - x_i} dt +$ regular terms. Hence this one-form must have poles of order 1 at $x_1, \ldots, x_N$ with residues $\nu_1, \ldots, \nu_N$, respectively, and be regular at all other points (including ∞). But such a one-form exists if and only if $\sum_{i=1}^N \nu_i = 0$ (otherwise, there is a pole at ∞), so we obtain the condition from Corollary 13.1.5.

If on the other hand we consider the conformal vector with a linear term

(13.1.10)
$$\left(\frac{1}{2\kappa}b_{-1}^2 + \frac{\lambda}{\kappa}b_{-2}\right)|0\rangle,$$

then the expressions (13.1.9) transform as affine λ–connections on D_{x_i}. The condition on conformal blocks is that these can be extended to a global affine λ–connection on $\mathbb{P}^1\backslash\{x_1,\ldots,x_N\}$. Because of the term $\lambda\frac{\rho''(t)}{\rho'(t)}$ in the transformation formula (8.1.5) for $b(t)$, we obtain an extra term $\frac{2\lambda}{t}$ when we make the transformation $\rho(t) = \frac{1}{t}$ to calculate the expansion of the affine connection at $\infty \in \mathbb{P}^1$. For this reason the condition that our affine connection be regular at $\infty \in \mathbb{P}^1$ is $\sum_{i=1}^N \nu_i = 2\lambda$. The space of conformal blocks is one-dimensional if this condition is satisfied, and is zero otherwise.

Thus, we see that the spaces of ordinary conformal blocks are different for different choices of conformal vectors in the vertex algebra π^κ.

Now consider the modified space of conformal blocks $C^0(\vec{x}, \pi_\nu^\kappa)$. In the case of the conformal vector (13.1.2), we find that $\varphi \in C^0(\vec{x}, \pi_\nu^\kappa)$ if and only if the one-forms (13.1.9) extend to a regular one-form on $\mathbb{P}^1\backslash\{x_1,\ldots,x_N,\infty\}$ having a pole of order at most 1 at ∞. Hence the condition $\sum_{i=1}^N \nu_i = 0$ is not relevant any more, and we find that the space $C^0(\vec{x}, \pi_\nu^\kappa)$ is one-dimensional for all $\nu_1,\ldots,\nu_N$. Likewise, in the case of the conformal vector (13.1.10), the λ–affine connection is allowed to have a pole of order at most 1 at ∞, so the condition $\sum_{i=1}^N \nu_i = 2\lambda$ also becomes irrelevant. Therefore we obtain that the space of conformal blocks is again one-dimensional. Thus, the two spaces coincide for all values of λ, i.e., the modified space of conformal blocks is independent of the choice of conformal vector.

13.2. Moving the points

We wish to study the variation of the Heisenberg conformal blocks as we move the points $x_1,\ldots,x_N$ along $\mathbb{A}^1 = \mathbb{P}^1\backslash\infty$. Thus we work over the configuration space

$$\mathcal{C}_N = \mathbb{A}^N\backslash\bigcup\{x_i = x_j\},$$

with the coordinate ring

$$\mathcal{B}_N = \mathbb{C}[\mathcal{C}_N] = \mathbb{C}[x_i, (x_i - x_j)^{-1}]_{1\leq i,j\leq N, i\neq j}.$$

The characteristic feature of $\mathcal{C}_N$ is that the trivial $\mathbb{P}^1$–bundle $\mathbb{P}^1 \times \mathcal{C}_N$ over $\mathcal{C}_N$ has N sections (disjoint from the constant section ∞), again denoted by $x_1,\ldots,x_N$, and any parameterized family of N–tuples of distinct points in $\mathbb{P}^1\backslash\infty$ is obtained from this universal family.

13.2.1. Bundle of conformal blocks. Allowing the N–tuple $\vec{x}$ to vary simply means repeating the above discussion over the ground ring $\mathcal{B}_N$, instead of over $\mathbb{C}$, i.e., we consider $x_1,\ldots,x_N$ as $\mathcal{B}_N$–valued points of $\mathbb{A}^1$, rather than as $\mathbb{C}$–points.

More precisely, let $\mathcal{H}(\mathbf{x})$ be the Lie algebra over $\mathcal{B}_N$ which is the central extension

$$0 \to \mathcal{B}_N \cdot \mathbf{1} \to \mathcal{H}(\mathbf{x}) \to \bigoplus_{i=1}^N \mathcal{B}_N((t - x_i)) \to 0$$

defined by the same two-cocycle as before. Each set of distinct complex numbers $a_1,\ldots,a_N$ defines a homomorphism $\mathcal{B}_N \to \mathbb{C}$ (specialization) sending x_i to a_i, and a homomorphism of Lie algebras $\mathcal{H}(\mathbf{x}) \to \mathcal{H}(\vec{a})$.

Next, denote by $\mathcal{B}'_N$ the ring of functions on $(\mathbb{P}^1 \times \mathcal{C}_N)$ minus the diagonals which vanish at $\infty \times \mathcal{C}_N$. Thus, $\mathcal{B}'_N$ is a subring of

$$\mathbb{C}[t, x_i, (x_i - x_j)^{-1}, (t - x_i)^{-1}]_{1 \leq i,j \leq N, i \neq j},$$

consisting of those elements $f(t, x_1, \ldots, x_N)$ which satisfy $f(\infty, x_1, \ldots, x_N) = 0$. We denote by $\mathcal{H}^0_{\mathbf{x}}$ the space $\mathcal{B}'_N$ considered as a commutative Lie algebra. Each set of distinct complex numbers $a_1, \ldots, a_N$ defines a natural surjective Lie algebra homomorphism $\mathcal{H}^0_{\mathbf{x}} \to \mathcal{H}^0_{\vec{a}}$.

We have a natural embedding of Lie algebras $\mathcal{H}^0_{\mathbf{x}} \hookrightarrow \mathcal{H}(\mathbf{x})$, which specializes to the embedding $\mathcal{H}^0_{\vec{a}} \hookrightarrow \mathcal{H}(\vec{a})$ for each $(a_1, \ldots, a_N) \in \operatorname{Spec} \mathcal{B}_N$. The $\mathcal{B}_N$–module $\operatorname{Hom}_{\mathcal{H}^0_{\mathbf{x}}}(\pi^\kappa_{\vec{\nu}}, \mathcal{B}_N)$ gives rise to a coherent sheaf $C^0(\mathbf{x}, \pi^\kappa_{\vec{\nu}})$ on the configuration space $\mathcal{C}_N = \operatorname{Spec} \mathcal{B}_N$ whose fiber at $(x_1, \cdots, x_N) \in \mathcal{C}_N$ equals $C^0(\vec{x}, \pi^\kappa_{\vec{\nu}})$. This is the sheaf of conformal blocks on our configuration space. Since we know from Corollary 13.1.5 that each of these spaces is canonically isomorphic to $\mathbb{C}$, we find that this sheaf is actually the sheaf of sections of a line bundle $\mathcal{L}$ over $\mathcal{C}_N$, which comes with a canonical trivialization. In other words, we have a canonical isomorphism

$$\operatorname{Hom}_{\mathcal{H}^0_{\mathbf{x}}}(\pi^\kappa_{\vec{\nu}}, \mathcal{B}_N) \simeq \mathcal{B}_N.$$

13.2.2. Flat connection. It follows from the general results on conformal blocks presented in Chapter 17 that $\mathcal{L}$ carries a natural flat connection over $\mathcal{C}_N$. With respect to our canonical trivialization, this connection can be written as a set of commuting first order differential operators $\nabla_i, i = 1, \ldots, N$. Let us find explicit formulas for these operators.

First we construct a connection on the bigger $\mathcal{B}_N$–module $\operatorname{Hom}_{\mathbb{C}}(\pi^\kappa_{\vec{\nu}}, \mathcal{B}_N)$, which represents a bundle over $\mathcal{C}_N$ with the fiber $(\pi^\kappa_{\vec{\nu}})^*$. Note that this bundle is the restriction of the vector bundle $(\boxtimes^N_{i=1} \Pi^\kappa_{\nu_i})^*$ on $\mathbb{A}^N$ to $\mathcal{C}_N \subset \mathbb{A}^N$.

We adopt the notational convention that if P is an operator on $\Pi^\kappa_{\vec{\nu}}$, then P^* denotes the adjoint operator on $(\Pi^\kappa_{\vec{\nu}})^*$. Further, $P^{(i)}$ will denote the operator acting as P on the ith factor and as the identity on all other factors of a tensor product. Introduce operators ∇_i $(i = 1, \ldots, N)$ on F by the formula

$$\nabla_i = \frac{\partial}{\partial x_i} - L^{(i)*}_{-1}.$$

According to Proposition 7.3.4, these operators define a flat connection on the bundle $(\boxtimes^N_{i=1} \Pi^\kappa_{\nu_i})^*$. We now show that this connection preserves its subbundle of conformal blocks. Actually, both statements of the following lemma are consequences of the general results of Chapter 17. However, we will give here an alternative proof by explicit computation.

13.2.3. Lemma.

(1) $[\nabla_i, \nabla_j] = 0$.
(2) *The operators* ∇_i *normalize* $\mathcal{H}^0_{\mathbf{x}}$ – *in other words,* $[\mathcal{H}^0_{\mathbf{x}}, \nabla_i] \subset \mathcal{H}^0_{\mathbf{x}}$.

13.2.4. Proof. First of all, observe that

$$\left[\frac{\partial}{\partial x_i} - L^{(i)*}_{-1}, \frac{\partial}{\partial x_j} - L^{(j)*}_{-1} \right] = 0,$$

since the two L_{-1} operators act on different factors, and are independent of the parameters, hence killing the cross–terms. This proves part (1).

In order to establish part (2), note that $\mathcal{H}_{\mathbf{x}}^0$ is spanned by elements of the form

$$A_n^i = \frac{1}{(t - x_i)^n}, \qquad n > 0, \ i = 1, \ldots, N.$$

Thus it is sufficient to check that $[\nabla_j, A_n^i]$ is a linear combination of the A_m^k. To see this, we need to express A_n^i as an element of the Lie algebra $\mathcal{H}(\mathbf{x})$, i.e., find the expansions of A_n^i at the points $x_1, \ldots, x_N$.

Clearly, the expansion of A_n^i at x_i gives us $b_{-n}^{(i)}$. Next, we find the following formula for the expansion of A_1^i at the point $x_j \neq x_i$:

$$
\begin{aligned}
\frac{1}{t - x_i} &= \frac{1}{(t - x_j) - (x_i - x_j)} \\
(13.2.1) \qquad &= -\frac{1}{x_i - x_j} \cdot \frac{1}{1 - \frac{t - x_j}{x_i - x_j}} \\
&= -\sum_{m=0}^{\infty} \frac{1}{(x_i - x_j)^{m+1}} (t - x_j)^m,
\end{aligned}
$$

Differentiating this formula $n - 1$ times with respect to x_i and combining these formulas for all j, we obtain the following expression of A_n^i as an element of $\mathcal{H}(\mathbf{x})$:

$$(13.2.2) \qquad A_n^i = b_{-n}^{(i)} - \frac{1}{(n-1)!} \frac{\partial^{n-1}}{\partial x_i^{n-1}} \sum_{j \neq i} \sum_{m=0}^{\infty} \frac{b_m^{(j)}}{(x_i - x_j)^{m+1}}.$$

When $i \neq j$, we find that

$$
\begin{aligned}
\left[\frac{\partial}{\partial x_j}, A_n^i\right] &= -\frac{1}{(n-1)!} \frac{\partial^{n-1}}{\partial x_i^{n-1}} \sum_{m=0}^{\infty} \frac{(m+1) b_m^{(j)}}{(x_i - x_j)^{m+2}} \\
(13.2.3) \qquad &= -[L_{-1}^{(j)}, A_n^i],
\end{aligned}
$$

since

$$[L_{-1}, b_{m+1}] = -(m+1) b_m.$$

When $i = j$, we find that

$$(13.2.4) \qquad \left[\frac{\partial}{\partial x_i} + L_{-1}^{(i)}, A_n^i\right] = n b_{-n-1}^{(i)} - \frac{1}{(n-1)!} \frac{\partial^n}{\partial x_i^n} \sum_{j \neq i} \sum_{m=0}^{\infty} \frac{b_m^{(j)}}{(x_i - x_j)^{m+1}}.$$

Combining these formulas, we obtain

$$\left[\frac{\partial}{\partial x_j}, A_n^i\right] = -[L_{-1}^{(j)}, A_n^i] + \delta_{ij} n A_{n+1}^i.$$

Finally, passing to the dual operators, and noting that $[A, B]^* = -[A^*, B^*]$, we obtain

$$\left[\frac{\partial}{\partial x_j} - L_{-1}^{(j)*}, A_n^{i*}\right] = \delta_{ij} n A_{n+1}^{i*},$$

completing the proof of the lemma.

13.2.5. Derivation of the baby KZ equations. The above lemma implies that the operators ∇_i preserve the subbundle of conformal blocks $\mathcal{L} = C^0(\mathbf{x}, \pi_{\vec{\nu}}^\kappa)$ inside the trivial vector bundle with fibers $\pi_{\vec{\nu}}^\kappa$. This subbundle also has a canonical trivialization. Namely, every $\eta \in \mathcal{B}_N$ can be lifted uniquely to

$$\widetilde{\eta} \in \mathrm{Hom}_{\mathcal{H}_{\mathbf{x}}^0}\left(\pi_{\vec{\nu}}^\kappa, \mathcal{B}_N\right)$$

in such a way that

$$\widetilde{\eta}(\mathbf{v}) = \eta,$$

where $\mathbf{v} = \bigotimes_{i=1}^N |\nu_i\rangle$.

Therefore to find $\nabla_i \cdot \eta$ we should evaluate ∇_i on $\widetilde{\eta}$ upstairs, obtaining a new $\mathcal{H}_{\mathbf{x}}^0$–invariant functional $\nabla_i \widetilde{\eta}$, and then evaluate it on the vector $\mathbf{v}$. We find from formula (13.1.3) that

$$L_{-1} = \frac{1}{2\kappa} \sum_{m \in \mathbb{Z}} :b_m b_{-m-1}:.$$

But

(13.2.5) $$b_m^{(i)} \cdot \mathbf{v} = \delta_{m,0}\nu_i \mathbf{v}, \qquad m \geq 0.$$

Therefore the only term that will survive in $L_{-1}^{(i)} \cdot \mathbf{v}$ is $\frac{1}{\kappa}b_{-1}^{(i)}b_0^{(i)}$, since in all other terms either m or $-m-1$ will be positive and hence they will annihilate $\mathbf{v}$.

Hence

$$\left(L_{-1}^{(i)*} \cdot \widetilde{\eta}\right)(\mathbf{v}) = \widetilde{\eta}\left(L_{-1}^{(i)} \cdot \mathbf{v}\right) = \frac{1}{\kappa}\widetilde{\eta}(b_{-1}^{(i)}\nu_i \mathbf{v}).$$

By $\mathcal{H}_{\mathbf{x}}^0$–invariance of $\widetilde{\eta}$, we have

$$\widetilde{\eta}\left(\frac{1}{t - x_i} \cdot C\right) = 0$$

for all $C \in \pi_{\vec{\nu}}^\kappa$. Here $1/(t - x_i)$ stands for the corresponding element of $\mathcal{H}(\mathbf{x})$ which acts on $\pi_{\vec{\nu}}^\kappa$ as the sum of its expansions around the points $x_j, j = 1, \ldots, N$. Thus, using formula (13.2.1), we find that

$$\widetilde{\eta}(b_{-1}^{(i)} \cdot C) = \sum_{j \neq i} \sum_{m \geq 0} \frac{\widetilde{\eta}(b_m^{(j)} \cdot C)}{(x_i - x_j)^{m+1}}.$$

Recalling formula (13.2.5), we obtain

(13.2.6) $$\widetilde{\eta}\left(L_{-1}^{(i)} \cdot \mathbf{v}\right) = \sum_{j \neq i} \frac{\nu_i \nu_j / \kappa}{x_i - x_j}\widetilde{\eta}(\mathbf{v}) = \sum_{j \neq i} \frac{\nu_i \nu_j / \kappa}{x_i - x_j}\eta.$$

Thus the formula for the action of ∇_i on η is

$$\nabla_i = \frac{\partial}{\partial x_i} - \frac{1}{\kappa} \sum_{j \neq i} \frac{\nu_i \nu_j}{x_i - x_j}.$$

We have proved the following result.

13.2.6. Lemma. *The connection operators on the line bundle $\mathcal{L}$ of Heisenberg conformal blocks have the form*

(13.2.7) $$\nabla_i = \frac{\partial}{\partial x_i} - \frac{1}{\kappa} \sum_{j \neq i} \frac{\nu_i \nu_j}{x_i - x_j}.$$

13.2.7. Solutions. A horizontal section of the bundle of conformal blocks is, with respect to our trivialization, a function $\psi(x_1, \ldots, x_n)$ which is a solution of the system

$$(13.2.8) \qquad \kappa \frac{\partial}{\partial x_i} \psi = \sum_{j \neq i} \frac{\nu_i \nu_j}{x_i - x_j} \psi, \qquad i = 1, \ldots, N.$$

The space of solutions of this system is one-dimensional, and it is spanned by the multivalued function

$$(13.2.9) \qquad \psi(x_1, \ldots, x_N) = \prod_{i<j} (x_i - x_j)^{\nu_i \nu_j / \kappa}.$$

Thus, we have found the horizontal sections of the bundle of conformal blocks in genus 0 in the case of the Heisenberg vertex algebra.

The above construction can be easily generalized to the case of the Heisenberg Lie algebra $\widehat{\mathfrak{h}}$, the central extension of $\mathfrak{h}((t))$, where $\mathfrak{h}$ is a finite-dimensional abelian Lie algebra with a non-degenerate inner product (see § 5.4.1). The modules π_ν^κ are constructed as before, but with $\nu \in \mathfrak{h}^*$ as a highest weight. The spaces of conformal blocks are again fibers of a trivial line bundle over the configuration space, with a horizontal section given by formula (13.2.9) with $\nu_i \nu_j$ replaced by (ν_i, ν_j) (here we use the inner product on $\mathfrak{h}^*$ induced by that on $\mathfrak{h}$).

Our next step is to generalize this analysis to the case of affine Kac-Moody algebras.

13.3. Conformal blocks for affine Kac-Moody algebras

Let $\mathfrak{g}$ denote a finite–dimensional simple Lie algebra, and $\widehat{\mathfrak{g}}$ the corresponding affine Kac–Moody algebra. In this section we study the space of conformal blocks $C^0_{V_k(\mathfrak{g})}(\mathbb{P}^1, (x_i), (M_i))_{i=1}^N$ defined in the same way as in the Heisenberg case. We again choose a global coordinate t on $\mathbb{P}^1$, and a set of distinct points $x_1, \ldots, x_N$ on $\mathbb{P}^1 \backslash \infty$.

13.3.1. Definition of the space of conformal blocks. Set

$$L\mathfrak{g}(x_i) = \mathfrak{g}((t - x_i)).$$

Consider the diagonal central extension of $\bigoplus_{i=1}^N L\mathfrak{g}(x_i)$,

$$0 \to \mathbb{C} \to \widehat{\mathfrak{g}}(\vec{x}) \to \bigoplus_{i=1}^N L\mathfrak{g}(x_i) \to 0,$$

defined in the same way as in the Heisenberg case. Let $\mathfrak{g}_{\vec{x}}^0 = \mathfrak{g}_{x_1, \ldots, x_N}^0$ be the Lie algebra $\mathfrak{g} \otimes \mathbb{C}[\mathbb{P}^1 \backslash \{x_1, \ldots, x_N\}]_0$, where the zero subscript indicates the subspace of those functions vanishing at ∞. By the residue theorem, the embedding $\mathfrak{g}_{\vec{x}}^0 \hookrightarrow \bigoplus_{i=1}^N L\mathfrak{g}(x_i)$ can be lifted to a Lie algebra embedding $\mathfrak{g}_{\vec{x}}^0 \hookrightarrow \widehat{\mathfrak{g}}(\vec{x})$. Given a set of $\widehat{\mathfrak{g}}$–modules $M_1, \ldots, M_N$ of equal level, we define $C^0(\mathbb{P}^1, (x_i), (M_i))_{i=1}^N$ as the space of $\mathfrak{g}_{\vec{x}}^0$–invariant functionals on $\bigotimes_{i=1}^N M_i$. In this section we will restrict ourselves to a special class of representations that we will now describe.

13.3.2. Induced representations. Given a $\mathfrak{g}$–module M, we construct a representation $\mathbb{M}^k$ of $\widehat{\mathfrak{g}}$ of level k as follows: we extend the action of $\mathfrak{g} = \mathfrak{g} \otimes 1$ on M to the Lie subalgebra $\mathfrak{g}[[t]]$ of $\widehat{\mathfrak{g}}$, so that $\mathfrak{g} \otimes t\mathbb{C}[[t]]$ acts by 0, and we make K act on M by the scalar k. Then we define the induced $\widehat{\mathfrak{g}}$–module:

$$\mathbb{M}^k = \operatorname{Ind}_{\mathfrak{g}[[t]] \oplus \mathbb{C}K}^{\widehat{\mathfrak{g}}} M.$$

As a vector space,

$$\mathbb{M}^k \simeq U(\mathfrak{g} \otimes t^{-1}\mathbb{C}[t^{-1}]) \otimes_{\mathbb{C}} M.$$

In particular, M is embedded into $\mathbb{M}^k$ as the subspace of $\mathfrak{g} \otimes t\mathbb{C}[[t]]$–invariants.

Note that if M is the trivial one-dimensional module, $\mathbb{M}^k$ is precisely the vacuum $\widehat{\mathfrak{g}}$–module underlying the vertex algebra $V_k(\mathfrak{g})$.

Now let $M_1, \ldots, M_N$ be representations of $\mathfrak{g}$ and $\mathbb{M}_1^k, \ldots, \mathbb{M}_N^k$ the corresponding induced $\widehat{\mathfrak{g}}$–modules of the same level k. To simplify notation, from this point on we will suppress the upper index k in $\mathbb{M}_i^k$, and also denote the space of conformal blocks $C_{V_k(\mathfrak{g})}^0(\mathbb{P}^1, (x_i), (\mathbb{M}_i^k))_{i=1}^N$ by $C_k^0\left(\vec{x}, \bigotimes_{i=1}^N \mathbb{M}_i\right)$. Denote

$$\mathbf{M} = \bigotimes_{i=1}^N M_i.$$

The following result is proved in the same way as Corollary 13.1.5.

13.3.3. Lemma. *The restriction map* $\tau \mapsto \tau|_{\mathbf{M}}$ *gives rise to an isomorphism*

$$C_k^0\left(\vec{x}, \bigotimes_{i=1}^N \mathbb{M}_i\right) \simeq \mathbf{M}^*.$$

13.3.4. Remark. Consider the Lie algebra $\mathfrak{g}_{\vec{x}} = \mathfrak{g} \otimes \mathbb{C}[\mathbb{P}^1 \backslash \{x_1, \ldots, x_N\}]$, i.e., we drop the requirement of vanishing at ∞. The corresponding ("true") space of conformal blocks $\operatorname{Hom}_{\mathfrak{g}_{\mathbf{x}}}\left(\bigotimes_{i=1}^N \mathbb{M}_i, \mathbb{C}\right)$ is isomorphic to the subspace of $\mathfrak{g}$–invariants in $\mathbf{M}^*$.

13.3.5. Varying the points. We now allow the points $x_1, \ldots, x_N$ to vary over the configuration space $\mathcal{C}_N = \mathbb{A}^N \backslash \Delta$ in the same way as we did in the Heisenberg case (see § 13.2.1). For that we consider all objects involved in the definition of the space of conformal blocks as $\mathcal{B}_N$–modules, where $\mathcal{B}_N = \mathbb{C}[\mathcal{C}_N]$. We define the Lie algebra $\widehat{\mathfrak{g}}(\mathbf{x})$ as the central extension

$$0 \to \mathcal{B}_N \to \widehat{\mathfrak{g}}(\mathbf{x}) \to \bigoplus_{i=1}^N L\mathfrak{g}(x_i) \to 0.$$

Then the Lie algebra $\mathfrak{g}_{\mathbf{x}}^0 = \mathfrak{g} \otimes \mathcal{B}_N'$ (with $\mathcal{B}_N'$ defined in § 13.2.1) naturally embeds into $\widehat{\mathfrak{g}}(\mathbf{x})$.

The $\mathcal{B}_N$–module $\operatorname{Hom}_{\mathfrak{g}_{\mathbf{x}}^0}\left(\bigotimes_{i=1}^N \mathbb{M}_i, \mathcal{B}_N\right)$ gives rise to a coherent sheaf, denoted by $C_k^0\left(\mathbf{x}, \bigotimes_{i=1}^N \mathbb{M}_i\right)$, on $\mathcal{C}_N$, whose fiber at $(x_1, \ldots, x_N) \in \mathcal{C}_N$ is the space of conformal blocks $C_k^0\left(\vec{x}, \bigotimes_{i=1}^N \mathbb{M}_i\right)$. But according to Lemma 13.3.3 each fiber of this sheaf is canonically isomorphic to $\mathbf{M}^*$. Therefore $C_k^0(\mathbf{x}, \bigotimes_{i=1}^N \mathbb{M}_i)$ is the sheaf of sections of a vector bundle which is canonically isomorphic to the trivial bundle with fiber $\mathbf{M}^*$.

13.3.6. Derivation of the KZ equations. Our goal now is to study the flat connection on the bundle $C_k^0(\mathbf{x}, \bigotimes_{i=1}^N \mathbb{M}_i)$. To obtain explicit formulas for the connection operators with respect to our trivialization of $C_k^0(\mathbf{x}, \bigotimes_{i=1}^N \mathbb{M}_i)$, we follow the same steps as in the Heisenberg case.

First we define the connection on a larger $\mathcal{B}_N$–module $\operatorname{Hom}_\mathbb{C}\left(\bigotimes_{i=1}^N \mathbb{M}_i, \mathcal{B}_N\right)$, which represents the trivial vector bundle over $\mathcal{C}_N$ with fibers $\left(\bigotimes_{i=1}^N \mathbb{M}_i\right)^*$. Following the general definition of the connection on bundles of vertex algebra modules given in § 7.3.8, we define the connection operators ∇_i $(i = 1, \ldots, N)$ on this vector bundle by the formulas

$$\nabla_i = \frac{\partial}{\partial x_i} - L_{-1}^{(i)*}.$$

Here L_{-1}^* is the adjoint of the Virasoro operator L_{-1} coming from the Sugawara conformal vector (see § 2.5.10)

$$\omega = \frac{S}{k + h^\vee} = \frac{1}{2(k + h^\vee)} \sum_{a=1}^d J_{a,-1} J_{-1}^a v_k \in V_k(\mathfrak{g}),$$

where $\{J^a\}$ and $\{J_a\}$ are dual bases of $\mathfrak{g}$ with respect to the normalized invariant inner product, and we use the notation A_m for $A \otimes t^m \in \widehat{\mathfrak{g}}$, as before. From now on we assume that $k \neq -h^\vee$, for otherwise $V_k(\mathfrak{g})$ does not have a confomal vector.

Then we find that

(13.3.1) $$L_{-1} = \frac{1}{2(k + h^\vee)} \sum_{m \in \mathbb{Z}} \sum_{a=1}^d {:}J_{a,m} J_{-m-1}^a{:} .$$

The proof of the following lemma is the same as the proof of Lemma 13.2.3.

13.3.7. Lemma.

   (1) $[\nabla_i, \nabla_j] = 0$.

   (2) *The operators ∇_i normalize $\mathfrak{g}_\mathbf{x}^0$ – in other words, $[\mathfrak{g}_\mathbf{x}^0, \nabla_i] \subset \mathfrak{g}_\mathbf{x}^0$.*

13.3.8. Connection operators. Lemma 13.3.7 implies that the operators ∇_i preserve the subbundle $C_k^0(\mathbf{x}, \bigotimes_{i=1}^N \mathbb{M}_i)$ (with the fiber $\mathbf{M}^*$) of the trivial bundle with the fiber $\left(\bigotimes_{i=1}^N \mathbb{M}_i\right)^*$. Let us compute the restriction of our connection to $C_k^0(\mathbf{x}, \bigotimes_{i=1}^N \mathbb{M}_i)$ with respect to our trivialization.

With respect to this trivialization, a section of $C_k^0(\mathbf{x}, \bigotimes_{i=1}^N \mathbb{M}_i)$ is a function

$$\eta \in \operatorname{Hom}_\mathbb{C}\left(\mathbf{M}, \mathcal{B}_N\right).$$

According to Lemma 13.3.3, it can be extended uniquely to

$$\widetilde{\eta} \in \operatorname{Hom}_{\mathfrak{g}_\mathbf{x}}\left(\bigotimes_{i=1}^N \mathbb{M}_i, \mathcal{B}_N\right).$$

The action of ∇_i on η can be evaluated by acting with ∇_i on $\widetilde{\eta}$ upstairs, obtaining a new $\mathfrak{g}_\mathbf{x}^0$–invariant functional $\nabla_i\widetilde{\eta}$, and restricting it back to $C_k^0(\mathbf{x}, \bigotimes_{i=1}^N \mathbb{M}_i)$ (cf. § 13.2.5).

Thus, given $v \in \mathbf{M}^*$, we need to calculate

$$\left(L_{-1}^{(i)*} \cdot \widetilde{\eta}\right)(v) = \widetilde{\eta}\left(L_{-1}^i \cdot v\right).$$

Since $v \in \mathbf{M}$ is in the top component of $\bigotimes_{i=1}^{N} \mathbb{M}_i$, it follows that

$$J_m^a \cdot v = 0, \qquad m > 0,$$
$$J_0^a \cdot v = J^a \cdot v.$$

Using the explicit formula (13.3.1) for L_{-1}, we find that

$$L_{-1} \cdot v = 2 J_{-1}^a J_{a,0} \cdot v.$$

Therefore

$$\widetilde{\eta}\left(L_{-1}^{(i)} \cdot v\right) = \frac{\widetilde{\eta}}{k + h^\vee} \sum_{a=1}^{\dim \mathfrak{g}} (J_{-1}^a)^{(i)} (J_a)^{(i)} v.$$

By the $\mathfrak{g}_{\mathbf{x}}^0$–invariance of $\widetilde{\eta}$, we obtain, in the same way as in § 13.2.5,

$$\widetilde{\eta}\left(L_{-1}^{(i)} \cdot v\right) = \left[\frac{\Xi_i}{k + h^\vee}\eta\right](v),$$

where

(13.3.2) $$\Xi_i = \sum_{j \neq i} \frac{\sum_a J_a^{(i)} J^{a(j)}}{x_i - x_j} = \sum_{j \neq i} \frac{\Omega_{ij}}{x_i - x_j},$$

$\Omega = \sum_a J_a \otimes J^a$ being the Casimir tensor in $\mathfrak{g} \otimes \mathfrak{g}$.

Thus the connection operator ∇_i on $C_k^0(\mathbf{x}, \bigotimes_{i=1}^{N} \mathbb{M}_i)$ is given by the formula

$$\nabla_i = \frac{\partial}{\partial x_i} - \frac{1}{k + h^\vee}\Xi_i.$$

The horizontal sections of $C_k^0(\mathbf{x}, \bigotimes_{i=1}^{N} \mathbb{M}_i)$ are therefore functions

$$\Phi : \mathcal{C}_N \to \mathbf{M}^*$$

that satisfy the system of differential equations

(13.3.3) $$(k + h^\vee)\frac{\partial}{\partial x_i}\Phi = \Xi_i \Phi, \qquad i = 1, \ldots, N.$$

These are the *Knizhnik–Zamolodchikov equations* (or KZ equations for short). The KZ system is holonomic with regular singularities along the diagonals $x_i = x_j$. In the next chapter we will find explicit solutions of this system, using the free field realization of $\widehat{\mathfrak{g}}$ that we have developed in Chapters 11, 12.

13.4. Bibliographical notes

The KZ equations were defined by V. Knizhnik and A. Zamolodchikov in [**KZ**]. Our presentation follows [**FFR**]. The mathematical structure of the KZ equations has been extensively studied in the literature; see [**TK, Law, SV1, SV2, D, KL, Fel, EFK**].

CHAPTER 14

Solving the KZ Equations

In this chapter we use the free field realization of $\widehat{\mathfrak{sl}}_2$ constructed in Chapter 12 to obtain explicit solutions of the KZ equations.

14.1. Conformal blocks from the point of view of free field realization

14.1.1. Conformal blocks for the Heisenberg Lie algebra.
We start by considering the space of conformal blocks of the tensor product $\bigotimes_{i=1}^{N} W_{\nu_i,k}$, where $W_{\nu,k} = M \otimes \pi_\nu^{2(k+2)}$ is a module over the vertex algebra $W_k = M \otimes \pi^{2(k+2)}$.

According to the definition of the Fock representation M given in § 12.1.1, M is a representation of the Heisenberg Lie algebra $\widehat{\mathcal{F}}$, which is the extension

$$0 \to \mathbb{C} \to \widehat{\mathcal{F}} \to \mathcal{F} \to 0,$$

where $\mathcal{F} = \mathcal{K} \oplus \Omega_{\mathcal{K}}$, with the cocycle induced by the non–degenerate residue pairing between $\mathcal{K}$ and $\Omega_{\mathcal{K}}$.

On the other hand, $\pi^{2(k+2)}$ is a representation of the Heisenberg Lie algebra $\mathcal{H}$, which is the extension

$$0 \to \mathbb{C} \to \mathcal{H} \to \mathcal{K} \to 0$$

defined by the cocycle $(f, g) = -\operatorname{Res}_{t=0} f \, dg$. This cocycle has a one–dimensional kernel corresponding to the constant functions, which produces a central element b_0 in the resulting Lie algebra, and hence gives us some flexibility in our definition of $\mathcal{H}$–modules (namely, we can have b_0 act by any scalar ν).

We return to the geometric setup of the previous chapter, with the curve $X = \mathbb{P}^1$ and marked points $x_1, \dots, x_N$. Introduce the Lie algebra $\Gamma(\vec{x})$ as the diagonal central extension

$$0 \to \mathbb{C} \to \Gamma(\vec{x}) \to \bigoplus_{i=1}^{N} (\mathcal{F}(x_i) \oplus \mathcal{H}(x_i)) \to 0$$

of the direct sum of the Lie algebras $\mathcal{F}$ and $\mathcal{H}$ attached to each of the points x_i. Let $\Gamma_{\vec{x}}$ be the Lie subalgebra of $\bigoplus_{i=1}^{N} (\mathcal{F}(x_i) \oplus \mathcal{H}(x_i))$ consisting of all elements regular away from the points $x_1, \dots, x_N$. By the residue theorem there is a natural lifting of $\Gamma_{\vec{x}}$ to $\Gamma(\vec{x})$. The space of W_k–conformal blocks on $\bigotimes_{i=1}^{N} W_{\nu_i,k}$ can then be described as the space of $\Gamma_{\vec{x}}$–invariant functionals on $\bigotimes_{i=1}^{N} W_{\nu_i,k}$.

As before, we will consider a slight modification of this space. Let $\mathcal{H}_{\vec{x}}^0 = \mathcal{K}_{\vec{x}}^0$ denote the subspace of elements of $\mathcal{H}_{\vec{x}}$ which vanish at ∞, and let

$$\mathcal{F}_{\vec{x}}^0 = \mathcal{K}_{\vec{x}}^0 \oplus \Omega_{\vec{x}}^0,$$

where $\Omega_{\vec{x}}^0$ consists of the one-forms on $\mathbb{P}^1 \backslash \{x_1, \dots, x_N, \infty\}$ with regular singularity (a pole of order ≤ 1) at ∞. Let $\Gamma_{\vec{x}}^0 = \mathcal{H}_{\vec{x}}^0 \oplus \mathcal{F}_{\vec{x}}^0$. This is a Lie subalgebra of $\Gamma(\vec{x})$. The *modified* space of W_k–conformal blocks $C_{W_k}^0(\mathbb{P}^1, (x_i), (W_{\nu_i,k}))_{i=1}^{N}$ is by definition the space of linear functionals on $\bigotimes_{i=1}^{N} W_{\nu_i,k}$ which are invariant under $\Gamma_{\vec{x}}^0$. Using the decompositions $W_{\nu_i,k} = M \otimes \pi_{\nu_i}^{2(k+2)}$ and $\Gamma_{\vec{x}}^0 = \mathcal{H}_{\vec{x}}^0 \oplus \mathcal{F}_{\vec{x}}^0$, we obtain that this space is isomorphic to

$$\mathrm{Hom}_{\mathcal{F}_{\vec{x}}^0}(M^{\otimes N}, \mathbb{C}) \otimes \mathrm{Hom}_{\mathcal{H}_{\vec{x}}^0}(\pi_{\nu_1,\ldots,\nu_N}^{2(k+2)}, \mathbb{C}).$$

We have already computed in § 13.1.7 the second factor: it is one–dimensional and the evaluation map $\tau \mapsto \tau\left(\bigotimes_{i=1}^{N} |\nu_i\rangle\right)$ induces an isomorphism

$$\mathrm{Hom}_{\mathcal{H}_{\vec{x}}^0}(\pi_{\nu_1,\ldots,\nu_N}^{2(k+2)}, \mathbb{C}) \simeq \left(\mathbb{C}\bigotimes_{i=1}^{N} |\nu_i\rangle\right)^*.$$

The following statement is proved along the same lines. Recall that v denotes the generating vector of M, and $v_{\nu_i,k} = v \otimes |\nu_i\rangle$ is the highest weight vector of $W_{\nu_i,k}$

14.1.2. Lemma. *The restriction map* $\tau \mapsto \tau(v^{\otimes N})$ *induces an isomorphism* $\mathrm{Hom}_{\mathcal{F}_{\vec{x}}^0}(M^{\otimes N}, \mathbb{C}) \simeq (\mathbb{C}v^{\otimes N})^*$. *Thus*

$$(14.1.1) \qquad C_{W_k}^0(\mathbb{P}^1, (x_i), (W_{\nu_i,k}))_{i=1}^{N} \simeq (\mathbb{C}\mathbf{v})^*,$$

where $\mathbf{v} = \bigotimes_{i=1}^{N} v_{\nu_i,k}$.

14.1.3. Application of the functoriality of conformal blocks. Now we can use the functoriality property of conformal blocks established in § 10.2. Namely, we have constructed in Theorem 12.3.1 a homomorphism of vertex algebras $w_k : V_k(\mathfrak{g}) \to W_k$. Therefore, according to Lemma 10.2.1 (adapted to the case of modified spaces of conformal blocks) we have an embedding

$$(14.1.2) \qquad C_{W_k}^0(\mathbb{P}^1, (x_i), (W_{\nu_i,k}))_{i=1}^{N} \hookrightarrow C_{V_k(\mathfrak{sl}_2)}^0(\mathbb{P}^1, (x_i), (W_{\nu_i,k}))_{i=1}^{N}.$$

Using the description of the spaces of conformal blocks of $\bigotimes_{i=1}^{N} W_{\nu_i,k}$ with respect to W_k (resp., $V_k(\mathfrak{sl}_2)$) as the space of $\Gamma_{\vec{x}}^0$–invariant (resp., $\mathfrak{g}_{\vec{x}}^0$–invariant) functionals on $\bigotimes_{i=1}^{N} W_{\nu_i,k}$, we obtain that any $\Gamma_{\vec{x}}^0$–invariant functional is automatically $\mathfrak{g}_{\vec{x}}^0$–invariant.

This is a non-trivial statement (compare with Example 10.2.2). Indeed, $\mathfrak{g}_{\vec{x}}^0$ is not embedded either into $\Gamma_{\vec{x}}^0$ or into its universal enveloping algebra. We have constructed an embedding of $\widehat{\mathfrak{g}}$ into $U(W_k)$, a local completion of the universal enveloping algebra of the Lie algebra Γ. But the formulas are non-linear, and hence do not give us an embedding of $\mathfrak{g}_{\vec{x}}^0$ into $\Gamma_{\vec{x}}^0$.

The reason why we have an embedding of conformal blocks is that the space of Heisenberg conformal blocks $C_{W_k}^0(\mathbb{P}^1, (x_i), (W_{\nu_i,k}))_{i=1}^{N}$ has two equivalent descriptions: as the space of $\Gamma_{\vec{x}}^0$–invariants in $\bigotimes_{i=1}^{N} W_{\nu_i,k}$ and as the space the invariants with respect to the big Lie algebra $U_\Sigma^0(W_k)$, where $\Sigma = \mathbb{P}^1 \backslash \{x_1, \ldots, x_n\}$, corresponding to all fields of the vertex algebra $W_k = M \otimes \pi^{2(k+2)}$ (see § 13.1.7). This big Lie algebra contains $\mathfrak{g}_{\vec{x}}$, and therefore we obtain the embedding (14.1.2).

14.1.4. Varying the points. As the points $x_1, \ldots, x_N$ vary over $\mathcal{C}_N$, we obtain a line bundle with the fibers (14.1.1), which we denote by $\mathcal{L}$. This line bundle is trivialized via the identification of the fibers with $(\mathbb{C}\mathbf{v})^*$.

We also have a vector bundle over $\mathcal{C}_N$ with the fibers

$$C_{V_k(\mathfrak{sl}_2)}^0(\mathbb{P}^1, (x_i), (W_{\nu_i,k}))_{i=1}^{N},$$

which we denote by $\mathcal{G}$. The maps (14.1.2) give rise to an embedding of vector bundles $\mathcal{L} \hookrightarrow \mathcal{G}$. Moreover, the bundle $\mathcal{G}$ carries a flat connection, which is determined by the action of the operators $L_{-1}^{(i)}$ coming from the Sugawara conformal vector $S/(k + h^\vee)$ in $V_k(\mathfrak{sl}_2)$. The image of this vector under the homomorphism $w_k : V_k(\mathfrak{sl}_2) \to W_k$ is a conformal vector of W_k described in Theorem 12.3.1. The corresponding operators $L_{-1}^{(i)}$ give rise to a flat connection on the line bundle $\mathcal{L}$. By construction, the embedding $\mathcal{L} \hookrightarrow \mathcal{G}$ respects the flat connections. Therefore the image of a horizontal section of $\mathcal{L}$ in $\mathcal{G}$ is a horizontal section of $\mathcal{G}$. It is much easier to find horizontal sections of the bundle $\mathcal{L}$ of Heisenberg conformal blocks than those of the bundle $\mathcal{G}$ of Kac–Moody conformal blocks. We will now use the above embedding of flat bundles of conformal blocks to construct horizontal sections of $\mathcal{G}$ using those of $\mathcal{L}$, and in particular to find solutions of the KZ equations.

The first step is the calculation of the flat connection on $\mathcal{L}$ which is compatible with the KZ connection on $\mathcal{G}$.

Using formula (12.3.2) for the image of the Sugawara conformal vector under the map $w_k : V_k(\mathfrak{sl}_2) \to W_k$, we compute the connection operators on $\mathcal{L}$ in the same way as in § 13.2.5. We find that the first and the last terms in formula (12.3.2) do not make contributions to the connection operators, so the computation turns out to be identical to that of § 13.2.5. Thus, with respect to our trivialization of $\mathcal{L}$, the connection operators read

$$\nabla_i = \frac{\partial}{\partial x_i} - \frac{1}{2(k+2)} \sum_{j \neq i} \frac{\nu_i \nu_j}{x_i - x_j}, \qquad i = 1, \ldots, N.$$

Hence we obtain

14.1.5. Lemma. *Let $\tau(\mathbf{x}) \in \mathrm{Hom}_{\Gamma_\mathbf{x}^0} \left(\bigotimes_{i=1}^N W_{\nu_i,k}, \mathcal{B}_N \right)$ be the unique functional whose value on $\mathbf{v}$ equals*

$$\prod_{i<j} (x_i - x_j)^{\nu_i \nu_j / 2(k+2)}.$$

Then $\tau(\mathbf{x})$ belongs to $\mathrm{Hom}_{\mathfrak{g}_\mathbf{x}^0} \left(\bigotimes_{i=1}^N W_{\nu_i,k}, \mathcal{B}_N \right)$ and is horizontal with respect to the connection on the bundle $\mathcal{G}$.

14.1.6. Restricting $\widehat{\mathfrak{sl}}_2$–conformal blocks. Recall that in § 13.3 we have defined, for each $\mathfrak{g}$–module M, a $\widehat{\mathfrak{g}}$–module $\mathbb{M} = \mathbb{M}^k$ of level k, and established an isomorphism

$$(14.1.3) \qquad \mathrm{Hom}_{\mathfrak{g}_\mathbf{x}^0} \left(\bigotimes_{i=1}^N \mathbb{M}_i, \mathcal{B}_N \right) \simeq \mathrm{Hom}_{\mathbb{C}} \left(\bigotimes_{i=1}^N M_i, \mathcal{B}_N \right).$$

We have also shown that under this isomorphism the flat connection on the bundle of conformal blocks becomes the Knizhnik–Zamolodchikov connection. We have now constructed a flat section of the bundle of conformal blocks, but with the insertions of the $\widehat{\mathfrak{sl}}_2$–modules $W_{\nu_i,k}$ rather than modules of type $\mathbb{M}^k$.

One can show that for generic ν and k, $W_{\nu,k}$ is isomorphic to the module induced from the contragredient Verma module M_ν^* of $\mathfrak{g}$. Therefore we can use this flat section to obtain a solution of the KZ system, at least for generic ν, k.

As far as solving the KZ equations is concerned, however, we do not necessarily need our modules to be of type $\mathbb{M}^k$. What matters is the top homogeneous component of the $\widehat{\mathfrak{sl}}_2$–module. But we have shown in § 12.3.4 that the top component $\overline{W}_{\nu_i,k}$ of the Wakimoto module $W_{\nu_i,k}$ is isomorphic to the contragredient Verma module $M^*_{\nu_i}$ for any ν, k. Therefore, given

$$\tau(\mathbf{x}) \in \mathrm{Hom}_{\mathfrak{g}^0_{\mathbf{x}}}\left(\bigotimes_{i=1}^N W_{\nu_i,k}, \mathcal{B}_N\right),$$

its restriction

(14.1.4) $$\Phi(\mathbf{x}) = \tau(\mathbf{x})|_{\bigotimes_{i=1}^N \overline{W}_{\nu_i,k}}.$$

can be viewed as an element of $\mathrm{Hom}_{\mathbb{C}}\left(\bigotimes_{i=1}^N M^*_{\nu_i}, \mathcal{B}_N\right)$. In fact, more is true.

14.1.7. Lemma. *If $\widetilde{\tau}(\mathbf{x})$ is a horizontal section of the bundle $\mathcal{G}$, i.e., an element of $\mathrm{Hom}_{\mathfrak{g}^0_{\mathbf{x}}}\left(\bigotimes_{i=1}^N W_{\nu_i,k}, \mathcal{B}_N\right)$ satisfying*

(14.1.5) $$\left(\frac{\partial}{\partial x_i} - L^{(i)*}_{-1}\right) \cdot \widetilde{\tau}(\mathbf{x}) = 0,$$

then $\Phi(\mathbf{x})$ satisfies the KZ equations with values in $\left(\bigotimes_{i=1}^N M^*_{\nu_i}\right)^*$.

14.1.8. Proof. Note that in our original derivation of the KZ equations in § 13.3.8 we only used the fact that $M^*_{\nu_i}$ is the $\mathfrak{g}$–module in the top homogeneous component of the corresponding induced $\widehat{\mathfrak{g}}$–module, and that

$$(\mathfrak{g} \otimes t^n) \cdot M^*_{\nu_i} = 0, \qquad n > 0.$$

Therefore the same argument will go through with the $\widehat{\mathfrak{g}}$–module $W_{\nu_i,k}$.

14.1.9. Explicit solution. In Lemma 14.1.5 we have found a horizontal section $\tau(\mathbf{x})$ of $\mathcal{G}$. According to Lemma 14.1.7, the restriction of this section to the subspace $\bigotimes_{i=1}^N \overline{W}_{\nu_i,k}$ is a solution of the KZ equations. Let us compute the value of $\tau(\mathbf{x})$ on

$$A_1 \otimes \cdots \otimes A_N \in \bigotimes_{i=1}^N \overline{W}_{\nu_i,k} = \bigotimes_{i=1}^N M^*_{\nu_i}.$$

By our construction,

(14.1.6) $$(\tau(\mathbf{x}))(\mathbf{v}) = (\Phi(\mathbf{x}))(\mathbf{v}) = \prod_{i<j}(x_i - x_j)^{\nu_i\nu_j/2(k+2)}.$$

Note that $\bigotimes_{i=1}^N M^*_{\nu_i}$ has a gradation by weights, with finite-dimensional homogeneous components. The KZ equations preserve this gradation; hence each solution is a sum of solutions with values in the homogeneous components.

14.1.10. Lemma. *The solution $\Phi(\mathbf{x})$ vanishes on all homogeneous components of $\bigotimes_{i=1}^N M^*_{\nu_i}$ except for the highest weight subspace spanned by the highest weight vector $\bigotimes_{i=1}^N v_{\nu_i}$, on which the value of $\Phi(\mathbf{x})$ is given by formula (14.1.6).*

14.1.11. Proof. Recall that

$$\bigotimes_{i=1}^{N} \overline{W}_{\nu_i,k} = \bigotimes_{i=1}^{N} \left(\mathbb{C}[(a_0^*)^{(i)}]_{i=1,\dots,N} \right) \cdot v_{\nu_i}.$$

Suppose we know the value of $\Phi(\mathbf{x})$ on some $A \in \bigotimes_{i=1}^{N} \overline{W}_{\nu_i,k}$. Let us calculate the value of $\Phi(\mathbf{x})$ on $(a_0^*)^{(i)} \cdot A$. Consider the 1–form

$$\frac{1}{t-x_i} dt \in \Omega_{\mathbf{x}}^0 \subset \Gamma_{\mathbf{x}}^0.$$

At the point x_i this operator acts as a_0^* (see formula (12.1.1)), while at $x_j, j \neq i$, it acts as

$$-\sum_{m=1}^{\infty} \frac{(a_m^*)^{(j)}}{(x_i - x_j)^m}$$

(see (13.2.1)). Combining these equations, we obtain

$$\tau(A_1 \otimes \cdots a_0^{*(i)} A_i \dots \otimes A_N) = \sum_{j \neq i} \sum_{m=1}^{\infty} \tau(A_1 \otimes \cdots a_m^* \cdot A_j \cdots \otimes A_N) \frac{1}{(x_j - x_i)^m}.$$

But $a_m^* \cdot A_j = 0$ for $A_j \in \overline{W}_{\nu_j,k}$ and $m > 0$. By induction we obtain the statement of the lemma.

Note that the weight of the highest weight vector $\bigotimes_{i=1}^{N} v_{\nu_i}$ is $\sum_{i=1}^{N} \nu_i$, while each application of $a_0^{(i)*}$ reduces the weight by 2.

14.2. Generalization: singular vectors

14.2.1. Throwing in extra points. Thus far, we have only been able to find one solution of the KZ equation, which is concentrated on the highest weight component of $\bigotimes_{i=1}^{N} M_{\nu_i}^*$. The vector space of all solutions is isomorphic to the fiber of our bundle at any point, i.e., to the dual space of $\bigotimes_{i=1}^{N} M_{\nu_i}^*$. In order to find explicit formulas for other solutions we use the following trick.

Suppose we are given a module $W_{\mu,k}$ and a vector B in it satisfying

(14.2.1) $(\mathfrak{g} \otimes t^r) \cdot B = 0, \qquad r \geq 0.$

Thus, it has the same properties as the vacuum vector of $V_k(\mathfrak{sl}_2)$. Such a vector is known as a *singular vector* (compare with the notion of a primary vector introduced in Definition 6.4.1). We may construct new solutions of KZ equations using these vectors. Write $N = n + m$ and relabel the coordinates

$$x_1, \dots, x_n \mapsto z_1, \dots, z_n; \qquad x_{n+1}, \dots, x_N \mapsto w_1, \dots, w_m$$

and the weight parameters

$$\nu_1, \dots, \nu_n \mapsto \lambda_1, \dots, \lambda_n; \qquad \nu_{n+1}, \dots, \nu_N \mapsto \mu, \dots, \mu.$$

In other words, we insert the same representation $W_{\mu,k}$ at all points w_j. Consider the horizontal section of the bundle of conformal blocks $\tau(\mathbf{z}, \mathbf{w})$ introduced above. By construction,

$$\tau(\mathbf{z}, \mathbf{w}) \in \mathrm{Hom}_{\Gamma_{\mathbf{z},\mathbf{w}}^0} \left(\bigotimes_{i=1}^{n} W_{\lambda_i,k} \otimes (W_{\mu,k})^{\otimes m}, \mathcal{B}_{n+m} \right).$$

Now let

$$(14.2.2) \qquad \Phi(\mathbf{z}; \mathbf{w}) = \tau(\mathbf{z}, \mathbf{w})|_{W_B}, \qquad W_B \overset{\text{def}}{=} \bigotimes_{i=1}^{n} \overline{W}_{\lambda_i, k} \otimes B \otimes \cdots \otimes B.$$

Following the derivation of the KZ equations in § 13.3.8 and Lemma 14.1.7, we find that the terms of the type

$$\sum_{m \geq 0} \frac{(J_m^a)^{(j)} \cdot B}{(z_i - w_j)^{m+1}}$$

vanish. Therefore $\Phi(\mathbf{z}; \mathbf{w})$ is a solution of the KZ system with values in the dual space to $\bigotimes_{i=1}^{n} M_{\lambda_i}^*$. In other words, the vectors B inserted at the auxiliary points w_j are "invisible" to our connection, as if we had simply placed the vacuum vector at those points.

Clearly, the vector $v_{0,k} \in W_{0,k}$ satisfies the conditions (14.2.1). But the corresponding solution $\Phi(\mathbf{z}; \mathbf{w})$ coincides with the previously constructed solutions $\Phi(\mathbf{z})$. Thus, to be able to produce new solutions we need to find vectors satisfying conditions (14.2.1) other than $v_{0,k}$. For non–critical level ($k \neq -2$) the Wakimoto modules $W_{\mu,k}$ do not contain any such vectors. But it turns out that there exist vectors which satisfy the above conditions "up to total derivatives". These total derivatives can then be killed by integration, and this way we can obtain new solutions. The following lemma is proved by direct computation using formulas (12.3.1).

14.2.2. Lemma. *Let $B = a_{-1} \cdot v_{-2,k} \in W_{-2,k}$, and assume that $k \neq -2$. Then*

$$e_n \cdot B = h_n \cdot B = 0, \quad n \geq 0; \qquad f_n \cdot B = 0, \quad n > 1,$$
$$f_1 \cdot B = -(k+2)v_{-2,k}, \qquad f_0 \cdot B = -(k+2)L_{-1} \cdot v_{-2,k}.$$

Thus, the OPEs $e(z)Y(B, w)$ and $h(z)Y(B, w)$ have no singular terms at the diagonal, while

$$\begin{aligned} f(z)Y(B, w) &= -(k+2)\frac{Y(v_{-2,k}, w)}{(z-w)^2} + \frac{Y(L_{-1}v_{-2,k}, w)}{z-w} + \text{reg.} \\ &= -(k+2)\partial_w \left(\frac{Y(v_{-2,k}, w)}{z-w} \right) + \text{reg.} \end{aligned}$$

is regular up to a total derivative.

This immediately leads us to the following result.

14.2.3. Proposition. *The functional $\Phi(\mathbf{z}; \mathbf{w})$ defined by (14.2.2) with $\mu = -2$ and $B = a_{-1}v_{-2,k}$, satisfies the KZ equations in the z variables, with values in the dual space to $\bigotimes_{i=1}^{n} M_{\lambda_i}^*$, up to total derivatives in the w variables.*

14.2.4. Proof. By our construction, τ satisfies

$$\frac{\partial}{\partial z_i} \tau(\mathbf{z}, \mathbf{w}) = L_{-1}^{(z_i)*} \cdot \tau(\mathbf{z}, \mathbf{w}).$$

Restrict this equation to the subspace W_B of (14.2.2). Then for each

$$\omega \in \bigotimes_{i=1}^{n} \overline{W}_{\lambda_i, k} \simeq \bigotimes_{i=1}^{n} M_{\lambda_i}^*,$$

we have

$$\frac{\partial}{\partial z_i}\tau(\omega \otimes B \otimes \cdots \otimes B) = \tau((L_{-1}^{(z_i)} \cdot \omega) \otimes B \otimes \cdots \otimes B).$$

Since

$$L_{-1} = \frac{1}{2(k+2)}\sum_{n \in \mathbb{Z}}\left(:e_n f_{-1-n}: + :f_n e_{-1-n}: + \frac{1}{2}:h_n h_{-1-n}:\right),$$

we obtain

(14.2.3) $$L_{-1}^{(z_i)} \cdot \omega = \frac{1}{k+2}\left(e_{-1}^{(z_i)} f_0^{(z_i)} + f_{-1}^{(z_i)} e_0^{(z_i)} + \frac{1}{2}h_{-1}^{(z_i)} h_0^{(z_i)}\right) \cdot \omega.$$

We wish now to manipulate this expression, term by term, into the operator Ξ_i of (13.3.2) plus some extra terms. We begin with the first summand. Consider $\frac{e}{t - z_i} \in \mathfrak{g}_{\mathbf{z},\mathbf{w}}^0$, which acts at z_i as $e_{-1}^{(z_i)}$, and at $z_j, j \neq i$ (resp., at w_j) as

$$-\sum_{n=0}^{\infty}\frac{e_n^{(z_j)}}{(z_i - z_j)^{n+1}} \qquad \left(\text{resp.,} \quad -\sum_{n=0}^{\infty}\frac{e_n^{(w_j)}}{(z_i - w_j)^{n+1}}\right).$$

Utilizing the $\mathfrak{g}_{\mathbf{z},\mathbf{w}}^0$–invariance of τ, we obtain that the first term of (14.2.3) equals

$$\tau\left(e_{-1}^{(z_i)} f_0^{(z_i)} \cdot \omega \otimes B \cdots B\right) = \sum_{j \neq i}\sum_{n=0}^{\infty}\frac{\tau(e_n^{(z_j)} f_0^{(z_i)} \cdot \omega \otimes B \cdots B)}{(z_i - z_j)^n}$$

$$+ \sum_{j=1}^{m}\sum_{n=0}^{\infty}\frac{\tau\left(f_0^{(z_j)} \cdot \omega \otimes B \cdots e_n \cdot B \cdots B\right)}{(z_i - w_j)^{m+1}} \underset{j}{}.$$

By degree considerations, for $m > 0$, $e_m^{(z_j)} f_0^{(z_i)} \cdot \omega = 0$. Also, by Lemma 14.2.2, for $n \geq 0$, $e_n \cdot B = 0$. Hence we are left with

$$\sum_{j \neq i}\frac{\Phi\left(e_0^{(z_j)} f_0^{(z_i)} \cdot \omega \otimes B \cdots B\right)}{z_i - z_j},$$

which is a summand of $\Xi_i \cdot \Phi(\mathbf{z}; \mathbf{w})$ (corresponding to the basis element $J^a = e$ and its dual $J_a = f$).

Since $h_n \cdot B = 0$ for $n \geq 0$ as well, by Lemma 14.2.2, the analysis of the third term in (14.2.3) proceeds identically. The problematic term is the second term, for which we have to use more complicated relations from Lemma 14.2.2. We find that

$$\tau(f_{-1}^{(z_i)} e_0^{(z_i)} \cdot \omega \otimes B \cdots B) = \sum_{j \neq i}\frac{\tau(f_0^{(z_j)} e_0^{(z_i)} \cdot \omega \otimes B \cdots B)}{z_i - z_j}$$

$$+ \sum_{j=1}^{m}\frac{\tau\left(e_0^{(z_i)} \cdot \omega \otimes B \cdots f_0 \cdot B \cdots B\right)}{z_i - w_j} \underset{j}{} + \sum_{j=1}^{m}\frac{\tau\left(e_0^{(z_i)} \cdots \omega \otimes B \cdots f_1 \cdot B \cdots B\right)}{(z_i - w_j)^2} \underset{j}{}.$$

By Lemma 14.2.2, the last two terms sum up to $-(k+2)$ times

$$\sum_{j=1}^{m}\frac{\tau\left(e_0^{(z_i)} \cdot \omega \otimes B \cdots L_{-1} v_{-2,k} \cdots B\right)}{z_i - w_j} \underset{j}{} + \sum_{j=1}^{m}\frac{\tau\left(e_0^{(z_i)} \cdots \omega \otimes B \cdots v_{-2,k} \cdots B\right)}{(z_i - w_j)^2} \underset{j}{}.$$

But τ satisfies

$$L_{-1}^{(w_j)*}\tau = \frac{\partial}{\partial w_j}\tau,$$

so that we may replace $\tau(\cdots \otimes L_{-1}v_{-2,k} \otimes \cdots)$ by $\frac{\partial}{\partial w_j}\tau(\cdots \otimes v_{-2,k} \otimes \cdots)$. Thus the last two terms of the above formula combine into a total derivative

$$-(k+2)\frac{\partial}{\partial w_j} \cdot \frac{\tau\left(e_0^{(z_i)} \cdot \omega \otimes \underset{j}{B \cdots v_{-2,k} \cdots B}\right)}{z_i - w_j}.$$

Therefore $\Phi(\mathbf{z};\mathbf{w})$ indeed satisfies the KZ equations up to the above total derivative terms. This proves the proposition.

14.3. Finding solutions

14.3.1. Integrating out the auxiliary variables.
In order to obtain actual solutions of the KZ system (in the z_i variables), we now need to integrate out the w_j variables, thus killing the total derivatives. If $\Phi(\mathbf{z};\mathbf{w})$ were a function, then for fixed z_i we would consider the closed holomorphic m–form

$$\Phi(\mathbf{z};\mathbf{w})\, dw_1 \cdots dw_m$$

on the space

$$\mathcal{C}_{m,\mathbf{z}} = \mathbb{A}^m \backslash \{w_i = w_j, w_j = z_i\}.$$

If $\partial \psi / \partial w_i$ is a total derivative, then

$$\left(\frac{\partial}{\partial w_i}\psi\right) dw_1 \cdots dw_m = d\left(\psi\, dw_1 \cdots \widehat{dw_j} \cdots dw_m\right)$$

is an exact form. Hence it vanishes after taking the integral over any m–cycle in $\mathcal{C}_{m,\mathbf{z}}$. Therefore the integral of the above m–form over a cycle would be a solution of the KZ system.

In fact, $\eta = \Phi(\mathbf{z};\mathbf{w})\, dw_1 \cdots dw_m$ is a closed m–form, but with coefficients in the tensor product of the dual space to $\bigotimes_{i=1}^{n} M_{\lambda_i}^*$ and a rank one local system $\mathcal{L}$ on $\mathcal{C}_{m,\mathbf{z}}$ defined by the multivalued function

$$(14.3.1) \quad \ell = \tau(\mathbf{z},\mathbf{w})\left(\bigotimes_{i=1}^{n} v_{\lambda_i,k} \otimes (v_{-2,k})^{\otimes m}\right)$$

$$= \prod_{i<j}(z_i - z_j)^{\lambda_i \lambda_j / 2(k+2)} \cdot \prod(z_i - w_j)^{-\lambda_i/(k+2)} \cdot \prod_{i<j}(w_i - w_j)^{2/(k+2)}.$$

Hence η defines a class in the mth de Rham cohomology of $\mathcal{C}_{m,\mathbf{z}}$ with coefficients in the local system $\mathcal{L}$ (tensored with the dual space to $\bigotimes_{i=1}^{n} M_{\lambda_i}^*$). This form may be integrated against m–cycles with coefficients in the dual local system $\mathcal{L}^*$.

Here by a local system on $\mathcal{C}_{m,\mathbf{z}}$ we understand a locally constant sheaf, i.e., a sheaf on $\mathcal{C}_{m,\mathbf{z}}$ such that for a sufficiently fine open covering of $\mathcal{C}_{m,\mathbf{z}}$, the restriction of $\mathcal{C}_{m,\mathbf{z}}$ to each open set in the covering is a constant sheaf. Local systems may be obtained from vector bundles equipped with a flat connection with regular singularities (see § 6.6.1) by taking the sheaf of horizontal sections. For instance, $\mathcal{L}$ and $\mathcal{L}^*$ come from the trivial rank one bundles on $\mathcal{C}_{m,\mathbf{z}}$ with non-trivial connections given by the formulas $d + d\log \ell$ and $d - d\log \ell$, respectively. Local horizontal sections of $\mathcal{L}$ and $\mathcal{L}^*$ are constant multiples of a particular branch of the multivalued functions

ℓ^{-1} and ℓ, respectively. One can cover $\mathcal{C}_{m,\mathbf{z}}$ by sufficiently small open analytic subsets, over which the sheaf of horizontal sections of $\mathcal{L}$ (or $\mathcal{L}^*$) is isomorphic to the constant sheaf $\mathbb{C}$. Hence the sheaf of horizontal sections of $\mathcal{L}$ (or $\mathcal{L}^*$) is indeed a locally constant sheaf.

Local systems, such as $\mathcal{L}$ or $\mathcal{L}^*$, may be used as systems of coefficients for singular homology or cohomology, after a minor modification of the ordinary definitions. In the case at hand, an $\mathcal{L}^*$–valued m–chain is represented as a linear combination of m–cells in $\mathcal{C}_{m,\mathbf{z}}$ (with respect to a sufficiently fine cell decomposition such that the restriction of $\mathcal{L}^*$ to each cell is isomorphic to the constant sheaf), together with the choice of a branch of the function ℓ over each cell. There is an obvious boundary map from the space of m–chains to the space of $(m-1)$–chains. The resulting chain complex computes the homology groups $H_m(\mathcal{C}_{m,\mathbf{z}}, \mathcal{L}^*)$ of $\mathcal{C}_{m,\mathbf{z}}$ with coefficients in $\mathcal{L}^*$.

On the other hand, to $\mathcal{L}$ we associate the de Rham complex, in which the mth group consists of the differential forms $\ell \cdot \omega$, where ω is an ordinary m–form, and the differential is the standard de Rham differential d. (Equivalently, we can take as the mth group the space of ordinary differential m–forms ω, but take the connection $d + d\log\ell$ as the differential.) This complex computes the cohomology groups $H^m(\mathcal{C}_{m,\mathbf{z}}, \mathcal{L})$ of $\mathcal{C}_{m,\mathbf{z}}$ with coefficients in $\mathcal{L}$.

Any element $\ell \cdot \omega$ of the mth group of the de Rham complex may be integrated over any m–chain with coefficients in $\mathcal{L}^*$ (this explains why we need the data of branches of ℓ over m–cells). Moreover, an analogue of the Stokes formula holds (so in particular the integral of an exact form over a cycle is equal to zero). Hence we obtain a non–degenerate pairing

$$H_m(\mathcal{C}_{m,\mathbf{z}}, \mathcal{L}^*) \times H^m(\mathcal{C}_{m,\mathbf{z}}, \mathcal{L}) \to \mathbb{C}, \qquad \Delta \times \eta \mapsto \int_\Delta \eta.$$

Proposition 14.2.3 now implies:

14.3.2. Theorem. *Let $\Delta_\mathbf{z}$ be a (local) family of cycles on $\mathcal{C}_{m,\mathbf{z}}$ with coefficients in $\mathcal{L}^*$. Then*

$$(14.3.2) \qquad \int_{\Delta_\mathbf{z}} \Phi(\mathbf{z}; \mathbf{w}) \, dw_1 \cdots dw_m$$

is a solution of the Knizhnik–Zamolodchikov equations with values in the restricted dual space to $\bigotimes_{i=1}^n M_{\lambda_i}^*$, i.e., in $\bigotimes_{i=1}^n M_{\lambda_i}$.

14.3.3. Remark. The monodromies arising from exchanging the z variables in the above solution are given by the R–matrices for the quantum group $U_q\mathfrak{sl}_2$, where $q = \exp \pi i/(k+2)$ (see [**Law, SV1**]).

14.3.4. Explicit formula. We can write down the above solutions explicitly. Recall that $M_\lambda^* \simeq \mathbb{C}[y]$, and so

$$\bigotimes_{i=1}^n M_{\lambda_i}^* \simeq \mathbb{C}[y_1, \ldots, y_n].$$

We need to find the value of our solution on a basis element $y_1^{p_1} \cdots y_n^{p_n}$ of $\bigotimes_{i=1}^n M_{\lambda_i}^*$. We find that

$$\Phi(\mathbf{z}; \mathbf{w})(y_1^{p_1} \cdots y_n^{p_n}) = \tau((a_0^*)^{p_1} v_{\lambda_1, k} \cdots (a_0^*)^{p_n} v_{\lambda_n, k} \otimes a_{-1} v_{-2, k} \cdots a_{-1} v_{-2, k}).$$

Now we can use the $\mathcal{F}^0_{\mathbf{z},\mathbf{w}}$–invariance of τ to switch the a_{-1} at the point w_j to the other points:

$$a_{-1}^{(w_j)} \mapsto \sum_{i=1}^{n} \frac{a_0^{(z_i)}}{w_j - z_i} + \sum_{s \neq j} \frac{a_0^{(w_s)}}{w_j - w_s}$$

(all other terms drop out by degree considerations.) But $a_0 \cdot a_{-1} v_{-2,k} = 0$. Therefore the second term drops out.

Finally, recall that a_0 acts as $\partial/\partial a_0^*$. Therefore we obtain

$$\Phi(\mathbf{z};\mathbf{w})\,(y_1^{p_1} \cdots y_n^{p_n}) = \prod_{j=1}^{m} \left(\sum_{i=1}^{n} \frac{\partial_{y_i}}{w_j - z_i} \right) y_1^{p_1} \cdots y_n^{p_n} |_{y_1 = \cdots = y_n = 0} \cdot \ell.$$

In particular, we see that this solution is non–zero only if $\sum_{i=1}^{n} p_i = m$. Thus we have obtained a solution which is nontrivial precisely on the component of weight $\sum_{i=1}^{n} \lambda_i - 2m$. Since all homogeneous components in $\bigotimes_{i=1}^{n} M_{\lambda_i}^*$ are finite-dimensional, we obtain solutions of the KZ system with values in the restricted dual space of $\bigotimes_{i=1}^{n} M_{\lambda_i}^*$, which is the same as $\bigotimes_{i=1}^{n} M_{\lambda_i}$.

One can check that each of the above solutions actually belongs to the subspace $(\bigotimes_{i=1}^{n} M_{\lambda_i})^{\mathfrak{n}_+}$ of vectors invariant under the diagonal action of $\mathfrak{n}_+ \subset \mathfrak{g}$. In fact, the KZ operators preserve this subspace, so we can consider the KZ system with values in $(\bigotimes_{i=1}^{n} M_{\lambda_i})^{\mathfrak{n}_+}$, or in any of its weight components.

Let us summarize the above calculation.

14.3.5. Corollary. *The following formula gives a solution of the KZ system with values in the component of weight $\sum_{i=1}^{n} \lambda_i - 2m$ in $(\bigotimes_{i=1}^{n} M_{\lambda_i})^{\mathfrak{n}_+}$. The value of this solution on $P \in \mathbb{C}[y_1, \ldots, y_n] \simeq \bigotimes_{i=1}^{n} M_{\lambda_i}^*$ is given by*

$$\int_{\Delta_{\mathbf{z}}} \prod_{j=1}^{m} \left(\sum_{i=1}^{n} \frac{\partial_{y_i}}{w_j - z_i} \right) P|_{y_1 = \cdots = y_n = 0} \cdot \ell \; dw_1 \cdots dw_m.$$

14.3.6. Remark. A dimension count of the space of cycles $\Delta_{\mathbf{z}}$ reveals that generically (with fixed m) the dimension of $H_m(\mathcal{C}_{m,\mathbf{z}}, \mathcal{L}^*)$ equals the dimension of the component of weight $\sum_{i=1}^{n} \lambda_i - 2m$ in $(\bigotimes_{i=1}^{n} M_{\lambda_i})^{\mathfrak{n}_+}$. Thus, for generic values of the λ_i's and k we obtain *all* solutions of the KZ system with values in $(\bigotimes_{i=1}^{n} M_{\lambda_i})^{\mathfrak{n}_+}$.

14.3.7. Interpretation in terms of the Gauss-Manin connection. The existence of the integral formulas for solutions of the KZ equations means that the KZ connection may be identified with a Gauss-Manin connection.

The basic topological idea behind the Gauss-Manin connection is very simple. Let X and Y be two smooth manifolds, and $\pi : X \to Y$ a locally trivial bundle, i.e., for any point $y \in Y$ there is an open neighborhood $N_y \subset Y$ such that $\pi^{-1}(N_y) \simeq N_y \times X_y$, where X_y denotes the fiber at y. Then the nth cohomologies (with complex coefficients, say) of the fibers X_y form a (locally trivial) vector bundle over Y, which we denote by $\mathcal{H}^n(X)$. Moreover, the bundle $\mathcal{H}^n(X)$ carries a canonical flat connection, which is called the *Gauss-Manin connection*. In order to define this connection, we need to identify the fibers of $\mathcal{H}^n(X)$ at nearby points $y_1, y_2 \in Y$, i.e., identify the cohomology spaces $H^n(X_{y_1})$ and $H^n(X_{y_2})$. Without loss of generality, we may assume that there exists an open subset $N \subset Y$ containing y_1 and y_2, and such that $\pi^{-1}(N)$ is a trivial bundle. But then the embeddings $X_{y_1} \hookrightarrow \pi^{-1}(N)$

and $X_{y_2} \hookrightarrow \pi^{-1}(N)$ are homotopy equivalences, and hence we obtain a canonical identification $H^n(X_{y_1}) \simeq H^n(\pi^{-1}(N)) \simeq H^n(X_{y_2})$.

Now suppose that we are given a local system $\mathcal{L}$ on X such that the restrictions of $\mathcal{L}$ to nearby fibers are isomorphic. Then one constructs in the same way as above flat connections on the (higher) direct images $R^n\pi_*\mathcal{L}$ of $\mathcal{L}$ to Y (or direct images with compact support $R^n\pi_!\mathcal{L}$). Note that the fibers of $R^n\pi_*\mathcal{L}$ are the cohomologies $H^n(X_y, \mathcal{L}|_{X_y})$ (resp., cohomologies with compact support $H^n_c(X_y, \mathcal{L}|_{X_y})$).

In our case, Y is the configuration space $\mathcal{C}_n = \mathbb{A}^n \setminus \{\mathrm{diag}\}$ of n distinct points on the affine line (with coordinates $z_1, \ldots, z_n$), X is the configuration space $\mathcal{C}_{n+m} = \mathbb{A}^{n+m} \setminus \{\mathrm{diag}\}$ of $n+m$ distinct points on the affine line (with coordinates $z_1, \ldots, z_n$, $w_1, \ldots, w_m$), and π is the natural projection sending $(z_1, \ldots, z_n, w_1, \ldots, w_m)$ to $(z_1, \ldots, z_n)$. As the local system $\mathcal{L}$, we take the rank one local system on $\mathcal{C}_{n+m}$ defined by the multivalued function ℓ given by formula (14.3.1).

Consider the flat bundle on $\mathcal{C}_{n+m}$ with fibers isomorphic to the component of $(\bigotimes_{i=1}^n M_{\lambda_i})^{\mathfrak{n}_+}$ of weight $\sum_{i=1}^n \lambda_i - 2m$ and with the connection given by the KZ operators. Corollary 14.3.5 and Remark 14.3.6 imply that for generic values of the parameters it is isomorphic to the dual of the flat bundle $R^m\pi_*\mathcal{L}$. Indeed, for fixed values of $z_1, \ldots, z_n$, the integrand in formula (14.3.2) for the horizontal sections of the KZ bundle are de Rham representatives of the mth cohomology groups of the fibers $\mathcal{C}_{m,\mathbf{z}}$ of π with coefficients in the restriction of the local system $\mathcal{L}$. Hence formula (14.3.2) defines a section of the dual bundle to $R^m\pi_*\mathcal{L}$. The fact that this section satisfies the KZ system means precisely that the horizontal sections of the KZ bundle are the same as the horizontal sections of the dual bundle to $R^m\pi_*\mathcal{L}$ (with respect to the Gauss-Manin connection). Therefore we obtain that the KZ connection is dual to the Gauss-Manin connection coming from a rank one local system on a larger configuration space.

Note that there is no *a priori* reason for the KZ connection to be related to the Gauss-Manin connection coming from such a simple rank one local system. We obtain this fact *a posteriori*, since we have found integral formulas for solutions of the KZ system; the existence of such solutions is essentially equivalent to the identification of the KZ connection with a Gauss-Manin connection. The free field realization has enabled us to find these integral solutions, and therefore it may be deemed "responsible" for the identification with the Gauss-Manin connection.

14.3.8. Generalization to an arbitrary $\mathfrak{g}$. For a general simple Lie algebra $\mathfrak{g}$, solutions of the KZ system may be derived in the same way as in the case of $\mathfrak{sl}_2$. The role of the module $W_{-2,k}$ is now played by the Wakimoto modules $W_{-\alpha_i,k}$, where the α_i's are the simple roots of $\mathfrak{g}$. The module $W_{-\alpha_i,k}$ contains a vector B_i which satisfies the relations similar to those satisfied by vector $B = a_{-1}v_{-2,k} \in W_{-2,k}$ in the case of $\mathfrak{sl}_2$ (see Lemma 14.2.2). Therefore we may, as in the case of $\mathfrak{sl}_2$, "throw in" extra points $w_1, \ldots, w_m$ with the modules $W_{-\alpha_{i(1)},k}, \ldots, W_{-\alpha_{i(m)},k}$ attached to them.

Next, as in the case of $\mathfrak{sl}_2$, one shows that the top degree component $\overline{W}_{\lambda,k}$ of the Wakimoto module $W_{\lambda,k}$ is isomorphic, as a $\mathfrak{g}$–module, to M_λ^*. We may therefore follow the procedure explained above in the case of $\mathfrak{sl}_2$. Namely, we restrict the Heisenberg conformal block

$$\tau(\mathbf{z}, \mathbf{w}) \in \mathrm{Hom}_{\Gamma^0_{\mathbf{x}}}\left(\bigotimes_{i=1}^n W_{\lambda_i,k} \otimes \bigotimes_{j=1}^m W_{-\alpha_{i(j)},k}, \mathcal{B}_{n+m}\right)$$

to the subspace $\bigotimes_{i=1}^{n} \overline{W}_{\lambda_i,k} \otimes B_{i(1)} \otimes \ldots \otimes B_{i(m)}$. Then we obtain a (multivalued) function $\Phi(\mathbf{z};\mathbf{w})$ with values in $\bigotimes_{i=1}^{n} M_{\lambda_i}$ which satisfies the KZ system in the z variables, up to total derivatives in the w variables.

The m–form $\Phi(\mathbf{z};\mathbf{w})\, dw_1 \ldots dw_m$ may then be interpreted as a $\bigotimes_{i=1}^{n} M_{\lambda_i}$–valued section of a local system $\mathcal{L}$ on $\mathcal{C}_{m,\mathbf{z}}$ defined by the multivalued function

$$\prod_{i<j}(z_i-z_j)^{(\lambda_i,\lambda_j)/(k+h^\vee)} \cdot \prod_{s,j}(z_s-w_j)^{-(\lambda_s,\alpha_{i(j)})/(k+h^\vee)} \cdot \prod_{s<j}(w_i-w_j)^{(\alpha_{i(s)},\alpha_{i(j)})/(k+h^\vee)}.$$

Integrating this m–form against m–cycles with coefficients in the dual local system $\mathcal{L}^*$, we obtain solutions of the KZ system with values in $\bigotimes_{i=1}^{n} M_{\lambda_i}$. In fact, they take values in the component of $(\bigotimes_{i=1}^{n} M_{\lambda_i})^{\mathfrak{n}+}$ of weight $\sum_{i=1}^{n} \lambda_i - \sum_{j=1}^{m} \alpha_{i(j)}$, and for generic values of parameters one obtains in this way all solutions with values in this space. One can compute these solutions explicitly. For more details, the reader is referred to [**FFR**] and [**ATY**].

It is shown in [**SV2, Va**] that the monodromies of these solutions are given by the R–matrices of the quantum group $U_q(\mathfrak{g})$ with $q = \exp(\pi i/r^\vee(k+h^\vee))$, where $r^\vee$ is the lacing number of $\mathfrak{g}$ (see Remark 15.4.13).

14.3.9. Remark. In § 10.4 we showed that conformal blocks in genus 0 may be expressed in terms of N–point functions of vertex operators. To be precise, we discussed only the case when all insertions are equal to the vertex algebra itself. However, this construction may be generalized to arbitrary module insertions. In that case conformal blocks are expressed in terms of matrix elements of products of the so-called intertwining operators. These act between different modules, and are labeled by the corresponding three–point conformal blocks (see Remark 10.4.12).

The solutions of the KZ equations which correspond to the genus 0 conformal blocks may therefore be written as matrix elements of operators of this type (integrated over the w variables), see, e.g., [**EFK**]. However, in our view, the approach presented above is more conceptual, and it may be generalized to higher genera.

14.4. Bibliographical notes

The integral solutions of the KZ equations in the case of $\mathfrak{sl}_2$ were first written down by P. Christe and R. Flume [**CF**] in some special cases. The general solutions for arbitrary $\mathfrak{g}$ were obtained by V. Schechtman and A. Varchenko [**SV1, SV2, Va**]. In the case of $\mathfrak{sl}_2$ the general solutions were independently found by R. Lawrence [**Law**]. See also the related works [**DJMM, Mat, Ch**]. In this section we derive the solutions using the free field realization following [**FFR**] (see also [**ATY, Kur**]). These solutions coincide with the Schechtman–Varchenko solutions.

The homology and cohomology with coefficients in local systems and the construction of the integration cycles $\Delta_{\mathbf{z}}$ are discussed in detail in [**SV2, Va**].

At the critical level, the vectors $B_i \subset W_{-\alpha_i,-h^\vee}$ may be used to obtain the eigenvectors of the Gaudin hamiltonians (the operators appearing in the right hand side of the KZ equations) acting on $\bigotimes_{i=1}^{n} M_{\lambda_i}$, see [**FFR**].

CHAPTER 15

Quantum Drinfeld–Sokolov Reduction and $\mathcal{W}$–algebras

In this chapter we introduce a class of vertex algebras called the $\mathcal{W}$–algebras. These are one-parameter families of vertex algebras associated to simple Lie algebras. The $\mathcal{W}$–algebra associated to $\mathfrak{sl}_2$ is the Virasoro vertex algebra Vir_c introduced in § 2.5 (here the parameter is the central charge c). However, unlike the Virasoro vertex algebra, the $\mathcal{W}$–algebras associated to other simple Lie algebras are not generated by Lie algebras. Hence the vertex algebra formalism cannot be avoided in this case.

The $\mathcal{W}$–algebras may be defined in several ways. In the main body of this chapter we define them using the quantum Drinfeld–Sokolov reduction. The relation to the classical Drinfeld–Sokolov reduction will be discussed later in § 16.8. Another approach, reviewed at the end of this chapter, is to define the $\mathcal{W}$–algebra as the intersection of kernels of certain operators (called screening operators) acting on a Heisenberg vertex algebra. Finally, a quasi-classical (i.e., Poisson) version of $\mathcal{W}$–algebra can be defined as the center of the Kac–Moody vertex algebra $V_{-h^\vee}(\mathfrak{g})$ at the critical level (see §§ 18.4.3 and 18.4.6).

15.1. The BRST complex

15.1.1. Fermionic vertex superalgebra. Let U be a finite-dimensional vector space. The complete topological vector space $U((t)) \oplus U^*((t))dt$ is equipped with the natural non-degenerate symmetric inner product induced by the residue pairing. Let $\mathcal{C}l_U$ be the corresponding (complete topological) Clifford algebra. If $\{v_i\}_{i \in I}$ is a basis of U and $\{v_i^*\}$ is the dual basis of U^*, then $\mathcal{C}l_U$ has topological generators $\psi_{i,n} = v_i \otimes t^n, \psi_{i,n}^* = v_i^* \otimes t^{n-1}dt, i \in I, n \in \mathbb{Z}$, satisfying the anti-commutation relations

$$(15.1.1) \qquad [\psi_{i,n}, \psi_{j,m}^*]_+ = \delta_{i,j}\delta_{n,-m}, \qquad [\psi_{i,n}, \psi_{j,m}]_+ = [\psi_{i,n}^*, \psi_{j,m}^*]_+ = 0.$$

Let $\bigwedge_U$ be the Fock representation of $\mathcal{C}l_U$, generated by a vector $|0\rangle$, such that

$$\psi_{i,n}|0\rangle = 0, \quad n \geq 0, \qquad \psi_{i,n}^*|0\rangle = 0, \quad n > 0.$$

Then $\bigwedge_U$ is a vertex superalgebra isomorphic to the tensor product of $|I|$ copies of the free fermionic vertex superalgebra $\bigwedge$ from § 5.3.1. In particular, we have

$$Y(\psi_{i,-1}|0\rangle, z) = \psi_i(z) \overset{\mathrm{def}}{=} \sum_{n \in \mathbb{Z}} \psi_{i,n}z^{-n-1}, \qquad Y(\psi_{i,0}^*|0\rangle, z) = \psi_i^*(z) \overset{\mathrm{def}}{=} \sum_{n \in \mathbb{Z}} \psi_{i,n}^*z^{-n}.$$

Introduce an additional $\mathbb{Z}$–gradation on $\mathcal{C}l_U$ and $\bigwedge_U$, called the *charge gradation*, by setting char $\psi_{i,n}^* = -$ char $\psi_{i,n} = 1$, char $|0\rangle = 0$. Note that parity on $\bigwedge_U$ equals charge modulo 2.

15.1.2. Representations of Lie algebras on $\bigwedge_U$. Suppose that a Lie algebra $\mathfrak{h}$ acts on the vector space U via a representation ρ. Let $\{h^a\}$ be a basis of $\mathfrak{h}$. Denote by (f^a_{ij}) the matrix of $\rho(h^a)$ in the basis $\{v_i\}$ of U, so that $\rho(h^a) \cdot v_j = \sum_i f^a_{ij} v_i$. The fields

$$(15.1.2) \qquad \widetilde{h}^a(z) = \sum_{n \in \mathbb{Z}} \widetilde{h}^a_n z^{-n-1} = \sum_i f^a_{ij} \, {:}\psi_i(z)\psi^*_j(z){:}$$

are vertex operators from the vertex algebra $\bigwedge_U$.

Define an invariant inner product $\langle \cdot, \cdot \rangle$ on $\mathfrak{h}$ (possibly degenerate) by the formula $\langle x, y \rangle = \mathrm{Tr}_U \, \rho(x)\rho(y)$. Let $\widehat{\mathfrak{h}}$ be the one-dimensional central extension of $\mathfrak{h}((t))$ with the commutation relations

$$[A \otimes f(t), B \otimes g(t)] = [A, B] \otimes f(t)g(t) - \langle A, B \rangle (\mathrm{Res}_{t=0} \, f dg) K,$$

where K denotes a generator of the one-dimensional center (compare with the definition of the affine Kac–Moody algebras in § 2.4).

Let $\mathbb{C}_1$ denote the one-dimensional representation of $\mathfrak{h}[[t]] \oplus \mathbb{C}K$, on which $\mathfrak{h}[[t]]$ acts by 0 and K acts by multiplication by 1. The induced representation

$$V(\mathfrak{h}, U) = \mathrm{Ind}^{\widehat{\mathfrak{h}}}_{\mathfrak{h}[[t]] \oplus \mathbb{C}K} \mathbb{C}_1$$

of $\widehat{\mathfrak{h}}$ carries a vertex algebra structure, defined in the same way as in § 2.4.4. The proof of the following lemma is straightforward and is left to the reader.

15.1.3. Lemma. *The assignment $h^a \otimes t^n \mapsto \widetilde{h}^a_n$ (given by formula (15.1.2)) and $K \mapsto \mathrm{Id}$ defines a representation of $\widehat{\mathfrak{h}}$ on $\bigwedge_U$. The corresponding map of $\widehat{\mathfrak{h}}$–modules $V(\mathfrak{h}, U) \to \bigwedge_U$, sending the vacuum vector to the vacuum vector, is a homomorphism of vertex (super)algebras.*

15.1.4. The complex. Let $\mathfrak{g}$ be a simple Lie algebra of rank ℓ with the Cartan decomposition

$$\mathfrak{g} = \mathfrak{n}_+ \oplus \mathfrak{h} \oplus \mathfrak{n}_-,$$

where $\mathfrak{n}_+$ (resp., $\mathfrak{n}_-$) is the upper (resp., lower) nilpotent subalgebra and $\mathfrak{h}$ is the Cartan subalgebra. Denote by Δ_+ the set of positive roots of $\mathfrak{g}$ and by $\{\alpha_1, \ldots, \alpha_\ell\}$ the subset of simple roots. Consider the Clifford algebra $\mathcal{C}l_{\mathfrak{n}_+}$ associated to the vector space $\mathfrak{n}_+((t)) \oplus \mathfrak{n}^*_+((t))dt$ and the corresponding vertex superalgebra $\bigwedge_{\mathfrak{n}_+}$. Choose a basis $\{e^\alpha\}_{\alpha \in \Delta_+}$ of root vectors in $\mathfrak{n}_+$. Then $\mathcal{C}l_{\mathfrak{n}_+}$ has topological generators $\psi_{\alpha,n}, \psi^*_{\alpha,n}, \alpha \in \Delta_+, n \in \mathbb{Z}$, satisfying the anti-commutation relations (15.1.1).

Recall the Kac–Moody vertex algebra $V_k(\mathfrak{g})$. Consider the vertex superalgebra

$$(15.1.3) \qquad C^\bullet_k(\mathfrak{g}) = V_k(\mathfrak{g}) \otimes \bigwedge{}^\bullet_{\mathfrak{n}_+},$$

where the upper dot indicates the charge gradation introduced above (note that the parity on $C^\bullet_k(\mathfrak{g})$ equals the charge modulo 2). Next we define a differential d_{st} of charge 1 on $C^\bullet_k(\mathfrak{g})$ (the "standard differential"). The complex $(C^\bullet_k(\mathfrak{g}), d_{\mathrm{st}})$ is the standard complex of *semi-infinite cohomology* of the Lie algebra $\mathfrak{n}_+((t))$ with coefficients in the module $V_k(\mathfrak{g})$, see [**Fe, FGZ, Vo1**] and § A.4.

To write down an explicit formula for the differential d_{st}, denote by $\{c^\gamma_{\alpha\beta}\}$ the structure constants in $\mathfrak{n}_+$, i.e.,

$$[e^\alpha, e^\beta] = \sum_{\gamma \in \Delta_+} c^{\alpha\beta}_\gamma e^\gamma.$$

Recall our notation $A_n \overset{\text{def}}{=} A \otimes t^n \in \widehat{\mathfrak{g}}$ and $A(z) \overset{\text{def}}{=} \sum_{n\in\mathbb{Z}} A_n z^{-n-1}$ for $A \in \mathfrak{g}$. Introduce a vector of degree 1 in $C_k^1(\mathfrak{g})$,

$$(15.1.4) \qquad Q = \sum_{\alpha\in\Delta_+} e_{-1}^{\alpha} v_k \otimes \psi_{\alpha,0}^* |0\rangle - \frac{1}{2} \sum_{\alpha,\beta,\gamma\in\Delta_+} c_{\gamma}^{\alpha\beta} v_k \otimes \psi_{\alpha,0}^* \psi_{\beta,0}^* \psi_{\gamma,-1} |0\rangle,$$

and the corresponding vertex operator

$$Q(z) = Y(Q,z) = \sum_{\alpha\in\Delta_+} e^{\alpha}(z)\psi_{\alpha}^*(z) - \frac{1}{2} \sum_{\alpha,\beta,\gamma\in\Delta_+} c_{\gamma}^{\alpha\beta} {:}\psi_{\alpha}^*(z)\psi_{\beta}^*(z)\psi_{\gamma}(z){:}$$

(here and below we suppress the tensor product sign).

Since the charge of Q is equal to 1, the field $Q(z)$ is odd. We set

$$(15.1.5) \qquad d_{\text{st}} \overset{\text{def}}{=} Q_{(0)} = \int Q(z)dz.$$

15.1.5. Exercise. Show that the OPE $Q(z)Q(w)$ is regular at $z = w$. Derive from it that $d_{\text{st}}^2 = 0$.

15.1.6. Remark. If we replace $\bigwedge_{\mathfrak{n}_+}$ by another $\mathcal{C}l_{\mathfrak{n}_+}$–module in (15.1.3), we obtain another cohomological complex of the Lie algebra $\mathfrak{n}_+((t))$. The representation theory of Clifford algebras is similar to that of Weyl algebras as described in Chapter 12, where we considered modules over the Weyl algebra A (which is the bosonic counterpart of our Clifford algebra). In particular, the module Fun $\mathcal{K}$ over A is analogous to the module $\bigwedge(\psi_{\alpha,n}^*)$ over $\mathcal{C}l_{\mathfrak{n}_+}$, while the module $M_{0\oplus\Omega_0}$ is analogous to $\bigwedge_{\mathfrak{n}_+}$. The tensor product of $\bigwedge(\psi_{\alpha,n}^*)$ with an $\mathfrak{n}_+((t))$–module M is the Chevalley complex computing the cohomology of $\mathfrak{n}_+((t))$ with coefficients in M (see § A.4), and the tensor product of the $\mathcal{C}l_{\mathfrak{n}_+}$–module $\bigwedge(\psi_{\alpha,n})$ with M is the Chevalley complex of homology of $\mathfrak{n}_+((t))$ with coefficients in M. The differentials in those complexes are given by the same formula as d_{st}.

15.1.7. Twist by the Drinfeld–Sokolov character. Thus, d_{st} is a differential on $C_k^\bullet(\mathfrak{g})$. Now consider a linear functional χ on $\mathfrak{n}_+((t))$ defined by the formula

$$(15.1.6) \qquad \chi(e_n^{\alpha}) = \begin{cases} 1, & \text{if } \alpha \text{ is simple}, n = -1, \\ 0, & \text{otherwise}. \end{cases}$$

This functional is a character of $\mathfrak{n}_+((t))$, i.e., $\chi([x,y]) = 0$ for all $x, y \in \mathfrak{n}_+((t))$. Let

$$\chi = \sum_{i=1}^{\ell} \psi_{\alpha_i,1}^* = \sum_{i=1}^{\ell} \int \psi_{\alpha_i}^*(z)dz$$

be the corresponding element of $\mathfrak{n}_+^*((t))dt \subset \mathcal{C}l_{\mathfrak{n}_+}$.

Clearly, $\chi^2 = 0$. The fact that χ is a character of $\mathfrak{n}_+((t))$ also implies that $[d_{\text{st}}, \chi]_+ = 0$. We obtain that

$$d = d_{\text{st}} + \chi$$

satisfies $d^2 = 0$. Thus, we have a complex $(C_k^\bullet(\mathfrak{g}), d)$, with the cohomological gradation being the charge gradation, called the *BRST complex of the quantum Drinfeld–Sokolov reduction*. We denote by $H_k^\bullet(\mathfrak{g})$ its cohomology. The complex $(C_k^\bullet(\mathfrak{g}), d)$ is nothing but the standard complex of semi-infinite cohomology of $\mathfrak{n}_+((t))$ with coefficients in the $\mathfrak{n}_+((t))$–module $V_k(\mathfrak{g}) \otimes \mathbb{C}_\chi$, where $\mathbb{C}_\chi$ is the one-dimensional module on which $\mathfrak{n}_+((t))$ acts according to the character χ. Tensoring by $\mathbb{C}_\chi$ does not change the complex as a vector space, but it changes the differential from d_{st} to d.

The motivation for considering the complex $(C_k^\bullet(\mathfrak{g}), d)$ comes from the general construction of quantization of hamiltonian reduction (see § 16.7.3). As shown by B. Kostant and S. Sternberg [**KS**], the Poisson algebra of functions on the reduced Poisson manifold obtained by hamiltonian reduction may be realized as the 0th cohomology of the so-called classical BRST complex. In § 16.8 we will see that our complex $(C_k^\bullet(\mathfrak{g}), d)$ is a quantum deformation of the classical BRST complex for the Drinfeld-Sokolov hamiltonian reduction (with k^{-1} playing the role of the deformation parameter), or more precisely, of its vertex Poisson algebra version. In the limit as $k \to \infty$, the 0th cohomology of our complex $(C_k^\bullet(\mathfrak{g}), d)$ becomes the reduced vertex Poisson algebra of the Drinfeld-Sokolov reduction (i.e., the classical $\mathcal{W}$–algebra); see [**FF5, FF7, F2, FKW**] and § 16.8 below for more details. This explains the terminology "quantum Drinfeld–Sokolov reduction" and "quantum BRST complex".

15.1.8. Vertex algebra structure on the cohomology. Since d is the residue of a vertex operator, namely,

$$(15.1.7) \qquad d(z) = \sum_{\alpha \in \Delta_+} e^\alpha(z) \psi_\alpha^*(z)$$

$$-\frac{1}{2} \sum_{\alpha, \beta, \gamma \in \Delta_+} c_\gamma^{\alpha\beta} {:} \psi_\alpha^*(z) \psi_\beta^*(z) \psi_\gamma(z){:} + \sum_{i=1}^\ell \psi_{\alpha_i}^*(z),$$

we obtain from § 5.7.3 that $H_k^\bullet(\mathfrak{g})$ is a vertex superalgebra (with the parity given, as before, by the charge gradation modulo 2).

However, we need to be careful when defining the vertex superalgebra $\mathbb{Z}$–gradation on $C_k^\bullet(\mathfrak{g})$ (not to be confused with the charge gradation reflected in the upper index). We want it to be such that the field $d(z)$ given by (15.1.7) has conformal dimension 1, so as to make the differential d homogeneous of degree 0. Since $d(z)$ contains $\psi_{\alpha_i}^*(z), i = 1, \ldots, \ell$, as summands, we have to assign to them conformal dimension 1. The condition that the other terms of $d(z)$ also have conformal dimension 1 then determines the conformal dimensions of the vertex operators $e^\alpha(z), \psi_\alpha(z), \psi_\alpha^*(z)$ for all $\alpha \in \Delta_+$. Finally, the condition that the commutation relations in $\widehat{\mathfrak{g}}$ are homogeneous determines the conformal dimensions of the remaining generating fields of $C_k^\bullet(\mathfrak{g})$.

To describe the result, recall that $\mathfrak{g}$ is generated by $e_i = e^{\alpha_i}, h_i$ and f_i, where $i = 1, \ldots, \ell$. Introduce the principal gradation on $\mathfrak{g}$ by setting

$$\mathrm{pr.\,gr.}\, e_i = -\,\mathrm{pr.\,gr.}\, f_i = 1, \qquad \mathrm{pr.\,gr.}\, h_i = 0.$$

In particular, $\mathrm{pr.\,gr.}\, e^\alpha = \mathrm{ht}\,\alpha = \sum_{i=1}^\ell a_i$, where $\alpha = \sum_{i=1}^\ell a_i \alpha_i$.

The following formulas define a $\mathbb{Z}$–gradation on $C_k^\bullet(\mathfrak{g})$, with respect to which the field $d(z)$ has conformal dimension 1 (and so the differential d has degree 0):

$$(15.1.8) \qquad \deg A_n = -n - \mathrm{pr.\,gr.}\, A, \qquad A \in \mathfrak{g},$$

$$\deg \psi_{\alpha,n} = -n - \mathrm{ht}\,\alpha, \qquad \deg \psi_{\alpha,n}^* = -n + \mathrm{ht}\,\alpha.$$

Since d preserves the $\mathbb{Z}$–gradation defined by these formulas, we obtain a $\mathbb{Z}$–gradation on the cohomology $H_k^\bullet(\mathfrak{g})$, which is part of the vertex superalgebra structure on $H_k^\bullet(\mathfrak{g})$.

In particular, $H_k^0(\mathfrak{g})$ is a $\mathbb{Z}$–graded vertex algebra, which is called the $\mathcal{W}$–algebra associated to $\widehat{\mathfrak{g}}$ at level k, and is denoted by $\mathcal{W}_k(\mathfrak{g})$. The following theorem describes the structure of this vertex algebra. Recall that to each simple Lie algebra one can attach the set of *exponents* $d_1, \ldots, d_\ell$ (for example, when $\mathfrak{g} = \mathfrak{sl}_n$, they are the integers between 1 and $n - 1$).

15.1.9. Theorem. *The $\mathbb{Z}$–graded vertex algebra $H_k^0(\mathfrak{g}) = \mathcal{W}_k(\mathfrak{g})$ is generated (in the sense of the Reconstruction Theorem) by elements W_i of degrees $d_i + 1, i = 1, \ldots, \ell$, where d_i is the ith exponent of $\mathfrak{g}$. Furthermore, $H_k^i(\mathfrak{g}) = 0, i \neq 0$.*

15.2. Proof of the main theorem

15.2.1. Decomposition of the complex $C_k^\bullet(\mathfrak{g})$. Let $\{J^a\}$ be a basis of $\mathfrak{g}$ which is the union of the basis $\{J^\alpha\}_{\alpha \in \Delta_+}$ (where $J^\alpha = e^\alpha$) of $\mathfrak{n}_+$ and a basis $\{J^{\bar{a}}\}_{\bar{a} \in \Delta_- \cup I}$ of $\mathfrak{b}_- = \mathfrak{n}_- \oplus \mathfrak{h}$ consisting of root vectors $f^\alpha, \alpha \in \Delta_+$, in $\mathfrak{n}_-$ and vectors $h^i, i \in I = \{1, \ldots, \ell\}$, in $\mathfrak{h}$. We will use latin upper indices to denote arbitrary basis elements, latin indices with a bar to denote elements of $\mathfrak{b}_-$, and greek indices to denote basis elements of $\mathfrak{n}_+$. We denote by c_d^{ab} the structure constants of $\mathfrak{g}$ with respect to the basis $\{J^a\}$.

Set

$$(15.2.1) \qquad \widehat{J}^a(z) = \sum_{n \in \mathbb{Z}} \widehat{J}_n^a z^{-n-1} = J^a(z) + \sum_{\beta, \gamma \in \Delta_+} c_\gamma^{a\beta} {:}\psi_\gamma(z)\psi_\beta^*(z){:} \, .$$

It is clear that the fields $\widehat{J}^a(z), \psi_\alpha(z), \psi_\alpha^*(z)$ generate the vertex superalgebra $C_k^\bullet(\mathfrak{g})$ (in the sense of the Reconstruction Theorem).

Let us define two subspaces of $C_k^\bullet(\mathfrak{g})$. The first, denoted by $C_k^\bullet(\mathfrak{g})_0$, is spanned by all monomials of the form

$$(15.2.2) \qquad \widehat{J}_{n_1}^{\bar{a}(1)} \ldots \widehat{J}_{n_r}^{\bar{a}(r)} \psi_{\alpha(1), m_1}^* \ldots \psi_{\alpha(s), m_s}^* |0\rangle$$

(recall that $J^{\bar{a}} \in \mathfrak{b}_-$). The second, denoted by $C_k^\bullet(\mathfrak{g})'$, is spanned by all monomials of the form

$$\widehat{J}_{n_1}^{\alpha(1)} \ldots \widehat{J}_{n_r}^{\alpha(r)} \psi_{\alpha(1), m_1} \ldots \psi_{\alpha(s), m_s} |0\rangle$$

(recall that $J^\alpha \in \mathfrak{n}_+$). We have a decomposition of vector spaces

$$(15.2.3) \qquad C_k^\bullet(\mathfrak{g}) \simeq C_k^\bullet(\mathfrak{g})_0 \otimes C_k^\bullet(\mathfrak{g})'.$$

15.2.2. Lemma. *$C_k^\bullet(\mathfrak{g})_0$ and $C_k^\bullet(\mathfrak{g})'$ are vertex subalgebras of $C_k^\bullet(\mathfrak{g})$.*

15.2.3. Proof. It suffices to check that the span of $\widehat{J}_n^\alpha, \psi_{\alpha,n}, \alpha \in \Delta_+, n \in \mathbb{Z}$, is stable under the commutator, as well as the span of $\widehat{J}_n^{\bar{a}}, \bar{a} \in \Delta_- \cup I, n \in \mathbb{Z}$, and $\psi_{\alpha,n}^*, \alpha \in \Delta_+, n \in \mathbb{Z}$.

According to formula (15.2.1), each $\widehat{J}_n^a$ is a sum of two terms which commute with each other. The first of them is J_n^a, and $\mathrm{span}\{J_n^\alpha\} = \mathfrak{n}_+[t, t^{-1}]$ (resp., $\mathrm{span}\{J_n^{\bar{a}}\} = \mathfrak{b}_-[t, t^{-1}]$) is certainly closed under the commutator. The second term is given by a formula of the type discussed in § 15.1.2, where $U = \mathfrak{n}_+$. Here $\mathfrak{n}_+$ acts on itself via the adjoint representation, and $\mathfrak{b}_-$ acts on $\mathfrak{n}_+$ via the identification $\mathfrak{n}_+ \simeq \mathfrak{g}/\mathfrak{b}_-$. By Lemma 15.1.3, these terms are also closed under the commutator. In fact, we obtain that the span of $\{J_n^\alpha\}$ is isomorphic to $\mathfrak{n}_+((t))$, while the span of $\{J_n^{\bar{a}}\}$ is isomorphic to a central extension of $\mathfrak{b}_-((t))$.

Finally, the commutator of $\widehat{J}_n^\alpha$ and $\psi_{\beta,m}$ is proportional to $\psi_{\alpha+\beta,n+m}$; the commutator of $\widehat{J}_n^{\bar{a}}$ and $\psi_{\alpha,m}^*$ is proportional to $\psi_{\gamma,n+m}^*$ for some γ. This completes the proof.

15.2.4. Commutation relations with the differential. Thus, as a vector space, $C_k^\bullet(\mathfrak{g})$ is isomorphic to the tensor product of its two vertex subalgebras: $C_k^\bullet(\mathfrak{g})_0$ and $C_k^\bullet(\mathfrak{g})'$.

Now we find the commutation relations between $\widehat{J}_n^a$ and the differential d. After a straightforward computation using the OPE formula (3.3.1) we obtain the following relations:

$$[\chi, \widehat{J}^a(z)] = \sum_{\beta,\gamma} c_\gamma^{a\beta} \chi(J^\gamma(z))\psi_\beta^*(z),$$

$$[\chi, \psi_\alpha^*(z)]_+ = 0,$$

$$[\chi, \psi_\alpha(z)]_+ = \chi(J^\alpha(z)),$$

$$[d_{\mathrm{st}}, \widehat{J}^a(z)] = \sum_{\bar{b},\alpha} c_{\bar{b}}^{\alpha a} :\widehat{J}^{\bar{b}}(z)\psi_\alpha^*(z): + k\sum_\alpha (J^a, J^\alpha)\partial_z \psi_\alpha^*(z) - \sum_{\alpha,\beta,b} c_\beta^{\alpha b} c_b^{\beta a} \partial_z \psi_\alpha^*(z),$$

$$[d_{\mathrm{st}}, \psi_\alpha^*(z)]_+ = -\frac{1}{2}\sum_{\beta,\gamma} c_\alpha^{\beta\gamma} \psi_\beta^*(z)\psi_\gamma^*(z),$$

$$[d_{\mathrm{st}}, \psi_\alpha(z)]_+ = \widehat{J}^\alpha(z).$$

These formulas follow from the corresponding OPEs, which are obtained using a fermionic analogue of the Wick formula from Lemma 12.2.6. We leave the details to the reader. Actually, only the fourth formula is difficult to compute. However, as will become clear below, the only fact that we will need about this formula is that the right hand side involves only the fields $\widehat{J}^a(z)$, $\psi_\alpha^*(z)$ and their derivatives, which is fairly easy to see.

The above formulas together with $\chi \cdot |0\rangle = d_{\mathrm{st}} \cdot |0\rangle = 0$ completely determine the action of χ and d_{st} on $C_k^\bullet(\mathfrak{g})$.

We find from these formulas that each vertex subalgebra, $C_k^\bullet(\mathfrak{g})_0$ and $C_k^\bullet(\mathfrak{g})'$, is preserved by $d = d_{\mathrm{st}} + \chi$. Moreover, if $A \in C_k^\bullet(\mathfrak{g})_0$, $B \in C_k^\bullet(\mathfrak{g})'$ and $A \otimes B \in C_k^\bullet(\mathfrak{g})$, then $d(A \otimes B) = dA \otimes B + (-1)^{p(A)}A \otimes dB$. Hence the decomposition (15.2.3) is true at the level of complexes, and we obtain

15.2.5. Lemma. *The cohomology of $C_k^\bullet(\mathfrak{g})$ is isomorphic (as a vector space) to the tensor product of the cohomologies of its subcomplexes $C_k^\bullet(\mathfrak{g})_0$ and $C_k^\bullet(\mathfrak{g})'$.*

15.2.6. Computation of the cohomology of $C_k^\bullet(\mathfrak{g})'$. To compute the cohomology of $C_k^\bullet(\mathfrak{g})'$, observe that

$$[d, \widehat{J}_n^\alpha] = 0, \qquad [d, \psi_n^\alpha] = \widehat{J}_n^\alpha + \chi(J_n^\alpha), \qquad \forall \alpha \in \Delta_+.$$

Therefore $C_k^\bullet(\mathfrak{g})'$ is isomorphic to the tensor product of infinitely many complexes of the form $\mathbb{C}[x] \otimes \bigwedge(\psi)$ with the differential $x \otimes \psi^*$ (or $(x+1) \otimes \psi^*$), where ψ^* is the contraction with ψ. Each of these complexes has one-dimensional zeroth cohomology and zero first cohomology. Hence the complex $C_k^\bullet(\mathfrak{g})'$ has no higher cohomologies and its zeroth cohomology is one-dimensional. Lemma 15.2.5 and the fact that $C_k^\bullet(\mathfrak{g})_0$ is a vertex subalgebra of $C_k^\bullet(\mathfrak{g})$ stable under the differential imply

15.2.7. Lemma. *The cohomology of $C_k^\bullet(\mathfrak{g})$ is isomorphic as a vertex superalgebra to the cohomology of its subcomplex $C_k^\bullet(\mathfrak{g})_0$.*

We have thus reduced the problem of finding $H_k^\bullet(\mathfrak{g})$ to the computation of the cohomology of $C_k^\bullet(\mathfrak{g})_0$. Note that the cohomological gradation (by charge) takes only non-negative values on $C_k^\bullet(\mathfrak{g})_0$.

15.2.8. Double complex. Let us refine the cohomological gradation on the complex $C_k^\bullet(\mathfrak{g})_0$ to a double gradation by setting

$$\text{bideg } A_n = (-k, k), \qquad \text{if} \quad \text{pr. gr. } A = k,$$

$$\text{bideg } \psi_{\alpha,n}^* = (\text{ht } \alpha, -\text{ht } \alpha + 1)$$

(see § 15.1.7 for the definition of pr. gr.). Then $\text{bideg } \chi = (1, 0)$ and $\text{bideg } d_{\text{st}} = (0, 1)$. According to the above formulas, we have $\chi^2 = 0$, $d_{\text{st}}^2 = 0$ and $[d_{\text{st}}, \chi]_+ = 0$. Therefore we obtain the structure of a bicomplex on $C_k^\bullet(\mathfrak{g})_0$. We associate to it a spectral sequence in which the zeroth differential is χ and the first differential is d_{st} (those readers who are not familiar with spectral sequences may learn basic facts about them from any textbook in homological algebra).

15.2.9. The spectral sequence. Now we compute the cohomology of $C_k^\bullet(\mathfrak{g})_0$ using this spectral sequence. We start by computing the cohomology of the differential χ. It follows from the formulas of § 15.2.4 that we can write

$$(15.2.4) \qquad [\chi, \widehat{J^{\bar{a}}}(z)] = \sum_\beta \chi([J^{\bar{a}}, J^\beta(z)])\psi_\beta^*(z) = \sum_\beta ([p_-, J^{\bar{a}}], J^\beta)\psi_\beta^*(z).$$

Here $(\cdot, \cdot)$ is the invariant inner product on $\mathfrak{g}$ normalized as in § 2.4.1, and

$$(15.2.5) \qquad p_- = \sum_{i=1}^\ell \frac{(\alpha_i, \alpha_i)}{2} f_i.$$

Since $(f_i, e_j) = \frac{2}{(\alpha_i, \alpha_i)}\delta_{ij}$, we have $(p_-, e^\alpha) = \chi(e^\alpha)$.

We recall the structure of $\mathfrak{g}$ with respect to the operator $\text{ad } p_-$. One can complete p_- in a unique way to an $\mathfrak{sl}_2$–triple $\{p_+, p_0, p_-\}$, so that $p_0 \in \mathfrak{h}$ and $p_+ \in \mathfrak{n}_+$. Under the adjoint action of this $\mathfrak{sl}_2$, the Lie algebra $\mathfrak{g}$ decomposes into a direct sum of $(2d_i + 1)$-dimensional representations R_i, $i = 1, \ldots, \ell$, where d_i denotes the ith exponent of $\mathfrak{g}$ (in particular, $R_1 = \text{span}\{p_+, p_0, p_-\}$). Let $p_-^{(i)}$ be a non-zero lowest weight vector in R_i (in particular, we can take $p_-^{(1)} = p_-$). Then pr. gr. $p_-^{(i)} = -d_i$. The vector space $\mathfrak{a}_- = \text{span}\{p_-^{(i)}\}_{i=1}^\ell$ is the kernel of $\text{ad } p_-$ in $\mathfrak{b}_-$. It is known that $\mathfrak{a}_-$ is a maximal abelian Lie subalgebra of $\mathfrak{g}$.

Denote by $P_{(n)}^{(i)}$ the linear combination of $\widehat{J_n^{\bar{a}}}$'s corresponding to $p_-^{(i)}$, and write

$$P^{(i)}(z) = \sum_{n \in \mathbb{Z}} P_{(n)}^{(i)} z^{-n-1}.$$

Let $V(\mathfrak{a}_-)$ be the subspace of $C_k^0(\mathfrak{g})_0$ spanned by the monomials $P_{(n_1)}^{(i_1)} \ldots P_{(n_m)}^{(i_m)}|0\rangle$. Since $\mathfrak{a}_-$ is an (abelian) Lie algebra, $V(\mathfrak{a}_-)$ is a (commutative) vertex subalgebra of $C_k^0(\mathfrak{g})_0$. According to formula (15.2.4), we have $[\chi, P^{(i)}(z)] = 0$. Therefore $V(\mathfrak{a}_-)$ belongs to the kernel of χ in $C_k^0(\mathfrak{g})_0$.

Moreover, it is clear that for any $\alpha \in \Delta_+$, there exists an element $I^\alpha \in \mathfrak{b}_-$ such that

$$([p^-, I^\alpha], J^\beta) = \delta_{\alpha, \beta}$$

for all $\beta \in \Delta_+$. Let $\widehat{I}^\alpha(z)$ be the corresponding vertex operator in $C_k^\bullet(\mathfrak{g})_0$, defined as in formula (15.2.1). We then have a vector space decomposition

$$(15.2.6) \qquad C_k^\bullet(\mathfrak{g})_0 \simeq V(\mathfrak{a}_-) \otimes \widetilde{C}_k^\bullet(\mathfrak{g})_0,$$

where $\widetilde{C}_k^\bullet(\mathfrak{g})_0$ is spanned by all monomials in $\widehat{I}_n^\alpha$, $n < 0$ and $\psi_{\alpha,n}^*$, $n \leq 0$, applied to $|0\rangle$. By formula (15.2.4), we have

$$[\chi, \widehat{I}_n^\alpha] = \psi_{\alpha,n+1}^*, \qquad [\chi, \psi_{\alpha,n}^*]_+ = 0.$$

Therefore the decomposition (15.2.6) holds at the level of complexes (with respect to the differential χ).

If the elements $\widehat{I}_n^\alpha$ and $\psi_{\beta,m}^*$ commuted with each other, we would be able to apply the argument used in § 15.2.6 to show that the zeroth cohomology of $\widetilde{C}_k^\bullet(\mathfrak{g})_0$ is $\mathbb{C}$ and all other cohomologies vanish. This would imply that the zeroth cohomology of $C_k^\bullet(\mathfrak{g})_0$ is $V(\mathfrak{a}_-)$, and all other cohomologies vanish. Since these elements do not commute, we cannot apply that argument directly. However, let us observe that the commutator of $\widehat{I}_n^\alpha$ and $\psi_{\beta,m}^*$ is a linear combination of the $\psi_{\alpha,n}^*$'s only. Therefore we introduce a filtration on the complex $\widetilde{C}_k^\bullet(\mathfrak{g})_0$ with the differential χ as follows.

We choose a lexicographycally ordered basis of monomials in the generators $\widehat{I}_n^\alpha$ and $\psi_{\beta,m}^*$ and define the pth term of the filtration as the span of those monomials in which the total number of the generators is less than or equal to p. In the corresponding spectral sequence the zeroth differential will then be just the Koszul differential (as in § 15.2.6). Indeed, the commutator terms will not survive in the leading part of the differential with respect to this filtration because each commutator will decrease the number of the $\widehat{I}_n^\alpha$'s while preserving the number of the $\psi_{\beta,m}^*$'s. Therefore the cohomology of this zeroth differential will be $\mathbb{C}$ in the cohomological degree zero and 0 in all other cohomological degrees. But then all higher differentials in this spectral sequence will vanish, and so this will be the cohomology of the complex $\widetilde{C}_k^\bullet(\mathfrak{g})_0$, as desired. Thus, we obtain

15.2.10. Lemma. *The zeroth cohomology of $C_k^\bullet(\mathfrak{g})_0$ with respect to χ is the commutative vertex subalgebra $V(\mathfrak{a}_-)$ of $C_k^0(\mathfrak{g})_0$, and all other cohomologies vanish.*

15.2.11. Completion of the proof of Theorem 15.1.9. Lemma 15.2.10 implies that the first term of our spectral sequence is equal to $V(\mathfrak{a}_-)$ placed in cohomological degree 0. Therefore the spectral sequence collapses in the first term, and we obtain that, as a vector space, $H_k^0(\mathfrak{g}) \simeq V(\mathfrak{a}_-)$ and $H_k^i(\mathfrak{g}) = 0$ for all $i \neq 0$.

However, elements of $V(\mathfrak{a}_-)$ do not necessarily represent cocycles in the complex $C_k^\bullet(\mathfrak{g})_0$ (and hence in $C_k^\bullet(\mathfrak{g})$). But from each element X_0 of $V(\mathfrak{a}_-)$ a cocycle can be reconstructed following the standard procedure. Namely, we first need to find $X_1 \in C^0(\mathfrak{g})_0$ such that $\chi \cdot X_1 = -d_{\mathrm{st}} \cdot X_0$ (such an element exists by Lemma 15.2.10). Then we find X_2 such that $\chi \cdot X_2 = -d_{\mathrm{st}} \cdot X_1$, etc. Eventually, we will have $d_{\mathrm{st}} \cdot X_n = 0$, since our double gradation satisfies $C_k^{p,q}(\mathfrak{g})_0 = 0$ for $p < 0$. Then $\sum_{i=0}^n X_i$ is a cocycle in $C_k^\bullet(\mathfrak{g})$ representing $X_0 \in V(\mathfrak{a}_-)$. In fact, it is uniquely determined by X_0 and the condition that it belongs to $C_k^0(\mathfrak{g})_0$, because $C_k^i(\mathfrak{g})_0 = 0$ for all $i < 0$.

Because of the correction terms, the vertex algebra structure on $\mathcal{W}_k(\mathfrak{g}) = H_k^0(\mathfrak{g})$ is not the same as on $V(\mathfrak{a}_-)$. However, since both χ and d_{st} commute with the gradation operator and the translation operator T, the $\mathbb{Z}$–gradations and the actions of T on $\mathcal{W}_k(\mathfrak{g})$ and $V(\mathfrak{a}_-)$ are the same. According to the definition of the gradation given in § 15.1.7, we have $\deg P_{(n)}^{(i)} = -n - d_i$.

Let W_i be the cocycle in $C_k^0(\mathfrak{g})_0$ corresponding to $P_{(-1)}^{(i)}|0\rangle \in V(\mathfrak{a}_-)$. It has the form $P_{(-1)}^{(i)}|0\rangle$ (which has bidegree $(d_i, -d_i)$) plus a sum of terms of bidegrees $(p, -p)$ with $0 \le p < d_i$. Let us write

$$Y(W_i, z) = \sum_{n \in \mathbb{Z}} W_{i,(n)} z^{-n-1}.$$

Then the monomials of the form $W_{i_1,(n_1)} \ldots W_{i_m,(n_m)}|0\rangle$ are the cocycles corresponding to the monomials $P_{(n_1)}^{(i_1)} \ldots P_{(n_m)}^{(i_m)}|0\rangle \in V(\mathfrak{a}_-)$. Hence they form a basis of $\mathcal{W}_k(\mathfrak{g})$. Therefore $W_i, i = 1, \ldots, \ell$, generate $\mathcal{W}_k(\mathfrak{g})$ in the sense of the Reconstruction Theorem 4.4.1. This completes the proof of Theorem 15.1.9.

15.3. Examples

15.3.1. Conformal vector. Recall that the smallest exponent d_1 is equal to 1 for any $\mathfrak{g}$. The statement of Theorem 15.1.9 implies that the homogeneous component of $\mathcal{W}_k(\mathfrak{g})$ of degree 1 is zero, and the component of degree 2 is spanned by the vector W_1. Hence we may expect that W_1 can be chosen to be a conformal vector of the vertex algebra $\mathcal{W}_k(\mathfrak{g})$. This is indeed the case if $k \ne -h^\vee$. In this case the complex $C_k^\bullet(\mathfrak{g}) = V_k(\mathfrak{g}) \otimes \bigwedge_{\mathfrak{n}_+}^\bullet$ is a conformal vertex algebra, but the choice of conformal vector is not unique. Namely, we can take the sum of one of the conformal vectors of $V_k(\mathfrak{g})$ described in § 8.5.3 and the sum of conformal vectors of the fermionic vertex superalgebra corresponding to each α (see § 5.3.4). But only one of these linear combinations may give rise to a conformal vector in $\mathcal{W}_k(\mathfrak{g})$, namely, the one which belongs to the kernel of the differential d.

This condition means in particular that the conformal dimensions of the generating fields of $C_k^\bullet(\mathfrak{g})$ with respect to this conformal structure should be as described in § 15.1.7, so that the operator L_0 coincides with the gradation operator. Using this condition, we determine the conformal vector and the corresponding field $T(z)$. Namely, let $\rho^\vee$ be the vector in $\mathfrak{h}$ which is the sum of the fundamental coroots (then the operator $\operatorname{ad} \rho^\vee$ equals the principal gradation operator on $\mathfrak{g}$), and $\rho^\vee(z)$ the corresponding vertex operator of $V_k(\mathfrak{g})$. Then the vertex operator

$$(15.3.1) \qquad T(z) = Y(\omega_k, z) = \frac{1}{k + h^\vee} S(z) + \partial_z \rho^\vee(z)$$

$$+ \sum_{\alpha \in \Delta_+} \left((1 - \operatorname{ht} \alpha){:}\psi_\alpha(z) \partial_z \psi_\alpha^*(z){:} + \operatorname{ht} \alpha{:}\psi_\alpha^*(z) \partial_z \psi_\alpha(z){:} \right),$$

where $S(z)$ is the Sugawara field, commutes with the differential d and provides the desired conformal structure on $C_k^\bullet(\mathfrak{g})$. The corresponding vector $\omega_k \in C_k^\bullet(\mathfrak{g})$ is a cocycle. It cannot be a coboundary, for otherwise the above field $T(z)$, and in particular its coefficient L_0, would act by 0 on the cohomology. Since L_0 coincides with the gradation operator, that would mean that the cohomology of $C_k^\bullet(\mathfrak{g})$ is concentrated in (conformal) degree 0. But by Theorem 15.1.9 this is not so. Hence ω_k represents a non-trivial cohomology class in $H_k^0(\mathfrak{g})$, and we may take W_1 to be equal to ω_k.

15.3.2. Special cases. In the case when $\mathfrak{g} = \mathfrak{sl}_2$, we obtain from Theorem 15.1.9 that $\mathcal{W}_k(\mathfrak{sl}_2)$ is generated by a vector W_1, which can be chosen to be the conformal vector if $k \neq -2$. Therefore $\mathcal{W}_k(\mathfrak{sl}_2)$ is isomorphic to the Virasoro vertex algebra $\mathrm{Vir}_{c(k)}$ with some value of the central charge. Formula (15.3.1) becomes in this case

$$T(z) = \frac{1}{k+2}S(z) + \frac{1}{2}\partial_z h(z) + {:}\psi^*(z)\partial_z\psi(z){:}\ .$$

From this formula we find the central charge

$$c(k) = \frac{3k}{k+2} - 6k - 2 = 1 - \frac{6(k+1)^2}{k+2}.$$

For $\mathfrak{g} = \mathfrak{sl}_3$, in addition to the Virasoro field $T(z) = W_1(z)$ there is a field $W_2(z)$ of conformal dimension 3. The central charge is

$$c(k) = 2 - \frac{24(k+2)^2}{k+3}.$$

For $c \neq -22/5$ the field $W_2(z)$ may be chosen, up to a scalar multiple, in a unique way so that it is a primary field (see Definition 6.4.1). Then the OPE of this field with itself reads

$$W_2(z)W_2(w) = \frac{c/3}{(z-w)^6} + \frac{2T(w)}{(z-w)^4} + \frac{\partial_w T(w)}{(z-w)^3}$$
$$+ \frac{1}{(z-w)^2}\left(2\Lambda(w) + \frac{3}{10}\partial_w^2 T(w)\right)$$
$$+ \frac{1}{z-w}\left(\partial_w \Lambda(w) + \frac{1}{15}\partial_w^3 T(w)\right),$$

where

$$\Lambda(w) = \frac{16}{22+5c}\left({:}T(w)^2{:} - \frac{3}{10}\partial_w^2 T(w)\right).$$

Note that the pole at $c = -22/5$ is inessential and can be removed if we multiply $W_2(z)$ by $22 + 5c$ (at the expense of making the above OPE formula even uglier). The most important property of the above formula is that it contains the quadratic term ${:}T(z)^2{:}$ in the right hand side. This means that the vertex algebra $\mathcal{W}_k(\mathfrak{sl}_3)$ cannot be generated by a "small" Lie algebra (in the way $\mathcal{W}_k(\mathfrak{sl}_2)$ is generated by the Virasoro Lie algebra). Indeed, such a Lie algebra would then necessarily contain the Fourier coefficients of the field ${:}T(z)^2{:}$. But then we also have to adjoin to this Lie algebra the Fourier coefficients of the OPEs of this field with the generating fields, and so on. It is easy to see that this way we obtain the entire Lie algebra $U(\mathcal{W}_k(\mathfrak{sl}_3))$, which is very large. Nevertheless, from the point of view of the theory of vertex algebras, $\mathcal{W}_k(\mathfrak{sl}_3)$, and $\mathcal{W}_k(\mathfrak{g})$ for general $\mathfrak{g}$, is finitely generated (by the fields $W_1(z), \ldots, W_\ell(z)$), so one can appreciate the convenience of the vertex algebra formalism here.

When $k \to \infty$, the $\mathcal{W}$–algebra $\mathcal{W}_k(\mathfrak{g})$ degenerates into a vertex Poisson algebra $\mathcal{W}_\infty(\mathfrak{g})$, as will be explained in § 16.8. The latter is called the *classical* $\mathcal{W}$–*algebra*. The corresponding Poisson algebra first appeared in the work of V. Drinfeld and V. Sokolov on KdV type hierarchies of soliton equations (in the case when $\mathfrak{g} = \mathfrak{sl}_n$ it appeared earlier in the works of M. Adler and I.M. Gelfand–L.D. Dickey).

Finally, we consider the case $k = -h^\vee$. In this case $T(z)$ given by formula (15.3.1) is not well-defined at $k = -h^\vee$, but $(k + h^\vee)T(z)$ is well-defined and becomes equal to $S(z)$ at $k = -h^\vee$. The Fourier coefficients of $S(z)$ are central elements of $U(V_{-h^\vee}(\mathfrak{g}))$. Therefore they automatically commute with the differential d in $C^\bullet_{-h^\vee}(\mathfrak{g})$. Thus, the space of polynomials in $S_n, n \leq -2$, maps to $\mathcal{W}_{-h^\vee}(\mathfrak{g})$. Moreover, this map is an embedding, since the complex $C^\bullet_{-h^\vee}(\mathfrak{g})_0$ has no terms with negative cohomological degrees. In particular, for $\mathfrak{g} = \mathfrak{sl}_2$ we obtain that $\mathcal{W}_{-2}(\mathfrak{sl}_2)$ is the commutative vertex algebra $\mathbb{C}[S_n]_{n \leq -2}$.

It is shown in [**FF7**] that for general $\mathfrak{g}$ the vertex algebra $\mathcal{W}_{-h^\vee}(\mathfrak{g})$ is also a commutative vertex algebra, which is isomorphic to the center of $V_{-h^\vee}(\mathfrak{g})$ (see also § 18.4).

15.3.3. Rational $\mathcal{W}$–algebras. The simple quotient of $\mathcal{W}_k(\mathfrak{g})$ for $k = -h^\vee + p/q$, where p, q are relatively prime integers greater than or equal to $h^\vee$, is believed to be a rational vertex algebra. For example, in the case when $\mathfrak{g} = \mathfrak{sl}_2$, we obtain the vertex algebra $L_{c(p,q)}$ from § 4.4.3. Moreover, it is conjectured in [**FKW**] that if $\mathfrak{g}$ is simply-laced and $q = p + 1$, then this vertex algebra is isomorphic to the coset vertex algebra of $(L_1(\mathfrak{g}) \otimes L_{p-2}(\mathfrak{g}), L_{p-1}(\mathfrak{g}))$, as defined in § 5.7.2.

Given a $V_k(\mathfrak{g})$–module M, the tensor product $C^\bullet_k(M) = M \otimes \bigwedge^\bullet_{\mathfrak{n}_+}$ is a module over the vertex superalgebra $C^\bullet_k(\mathfrak{g})$. The operator d is well-defined on $C^\bullet_k(M)$ and makes it into a complex. The cohomology $H^\bullet_k(\mathfrak{g})$ of this complex is the semi-infinite cohomology of $\mathfrak{n}_+((t))$ with coefficients in $M \otimes \chi$ (see § A.4). It is straightforward to verify that $\mathcal{W}_k(\mathfrak{g})$ naturally acts on $H^i_k(M)$ for any $i \in \mathbb{Z}$. Thus we obtain a functor from the category of graded $\widehat{\mathfrak{g}}$–modules of level k with gradation bounded from below (these are the same as $V_k(\mathfrak{g})$–modules) to the category of $\mathcal{W}_k(\mathfrak{g})$–modules. This functor was studied in [**FKW**] (see also [**FF4**] and [**BO**] in the case when $\mathfrak{g} = \mathfrak{sl}_2$).

In particular, in [**FKW**] this functor (and its analogue for the Lie algebra $\mathfrak{n}_-$) was applied to irreducible modules of $\widehat{\mathfrak{g}}$ of rational levels. It was conjectured in [**FKW**] that $H^i_k(M) = 0$ for all $i \neq 0$ if M belongs to a certain finite set of the so-called modular invariant representations. Recently, this conjecture was proved by T. Arakawa [**A**].

15.4. The second computation

Now we compute the cohomology of $C^\bullet_k(\mathfrak{g})_0$ in a different way in the case of generic k. We first define a limit of the complex $C^\bullet_k(\mathfrak{g})$ as $k \to \infty$. Then we show that all higher cohomologies of the limiting complex vanish. Since $C^\bullet_k(\mathfrak{g})_0$ is a direct sum of finite-dimensional complexes (namely, its homogeneous components with respect to the conformal $\mathbb{Z}$–gradation), this implies that the higher cohomologies of $C^\bullet_k(\mathfrak{g})_0$ vanish for generic k as well. Moreover, we will express the 0th cohomology (the $\mathcal{W}$–algebra) for generic k as the intersection of kernels of the so-called screening operators inside a Heisenberg vertex algebra. Thus, we will obtain a "free field realization" of $\mathcal{W}_k(\mathfrak{g})$.

In § 16.8 we will show that $\mathcal{W}_\infty(\mathfrak{g})$ is a vertex Poisson algebra of regular functions on the space of $\mathfrak{g}$–opers on the disc. Thus, $\mathcal{W}_k(\mathfrak{g})$ may be viewed as a quantum deformation of $\mathcal{W}_\infty(\mathfrak{g})$ with k^{-1} playing the role of the quantization parameter.

15.4.1. The limit as $k \to \infty$. We start by defining the limit of our complex $C^\bullet_k(\mathfrak{g})_0$ as $k \to \infty$. In order to do that we define a new complex, denoted by $C^\bullet_{\mathbf{k}}(\mathfrak{g})_0$, which is a free module over the ring $\mathbb{C}[\mathbf{k}^{-1}]$, such that $C^\bullet_k(\mathfrak{g})_0$ is the quotient of $C^\bullet_{\mathbf{k}}(\mathfrak{g})_0$ by the image of the ideal generated by $(\mathbf{k}^{-1} - k^{-1})$ (for more on this procedure, see § 16.3).

Recall that $C_k^\bullet(\mathfrak{g})_0$ has a basis of monomials (15.2.2). Let $\widetilde{C}_\mathbf{k}^\bullet(\mathfrak{g})_0$ be the free module over $\mathbb{C}[\mathbf{k}, \mathbf{k}^{-1}]$ with the basis (15.2.2), and the action of the differential d is given by the formulas of § 15.2.4, in which k is replaced by $\mathbf{k}$.

Next, we define $C_\mathbf{k}^\bullet(\mathfrak{g})_0$ as the $\mathbb{C}[\mathbf{k}^{-1}]$–submodule of $\widetilde{C}_\mathbf{k}^\bullet(\mathfrak{g})_0$ which is the $\mathbb{C}[\mathbf{k}^{-1}]$–span of the monomials

$$(15.4.1) \qquad \mathbf{k}^{-r} \widehat{J}_{n_1}^{\overline{a}(1)} \ldots \widehat{J}_{n_r}^{\overline{a}(r)} \psi_{\alpha(1),m_1}^* \ldots \psi_{\alpha(s),m_s}^* |0\rangle.$$

We obtain from the above formulas for the action of d_{st} and χ on $C_\mathbf{k}^\bullet(\mathfrak{g})_0$ that d_{st} and $\mathbf{k}\chi$ preserve $C_\mathbf{k}^\bullet(\mathfrak{g})_0$. Moreover, each of these operators is nilpotent and they anti-commute with each other. Consider the differential $d_\mathbf{k} = d_{\mathrm{st}} + \mathbf{k}\chi$. It is clear that for $\mathbf{k}^{-1} \neq 0$ the cohomology of this differential is the same as the cohomology of $d = d_{\mathrm{st}} + \chi$.

The complex $C_\infty^\bullet(\mathfrak{g})_0$ is by definition the specialization $C_\mathbf{k}^\bullet(\mathfrak{g})_0/(\mathbf{k}^{-1} \cdot C_\mathbf{k}^\bullet(\mathfrak{g})_0)$ at $\mathbf{k}^{-1} = 0$, with a differential d_∞ induced by $d_\mathbf{k}$. We denote by $H_\infty^\bullet(\mathfrak{g})$ the cohomology of the complex $(C_\infty^\bullet(\mathfrak{g})_0, d_\infty)$.

Since $\mathbf{k}^{-1} \widehat{J}_n^{\overline{a}}$ and $\mathbf{k}^{-1} \widehat{J}_m^{\overline{b}}$ commute modulo $\mathbf{k}^{-1}$, we can identify linear combinations of monomials (15.4.1) with polynomials in the commuting variables $\widehat{J}_n^{\overline{a}}, n < 0$, and anti-commuting variables $\psi_{\alpha,n}^*, n \leq 0$. Thus, we obtain an isomorphism

$$C_\infty^\bullet(\mathfrak{g})_0 \simeq \mathrm{Sym}\, \mathfrak{b}_-((t))/\mathfrak{b}_-[[t]] \otimes \bigwedge{}^\bullet \mathfrak{n}_+[[t]]^*.$$

Using the residue pairing and the isomorphism $\mathfrak{b}_- \simeq \mathfrak{b}_+^*$ induced by the normalized invariant inner product on $\mathfrak{g}$, we identify $\mathrm{Sym}\, \mathfrak{b}_-((t))/\mathfrak{b}_-[[t]]$ with the ring of polynomial functions on the space $\mathfrak{b}_+[[t]]dt \simeq \mathfrak{b}_+[[t]]$. Therefore $C_\infty^\bullet(\mathfrak{g})_0$ is isomorphic to

$$(15.4.2) \qquad \mathrm{Fun}\, \mathfrak{b}_+[[t]] \otimes \bigwedge{}^\bullet \mathfrak{n}_+[[t]]^*.$$

15.4.2. Action of the differential in the limit. Now we compute the action of the differentials d_{st} and $\widetilde{\chi} \overset{\mathrm{def}}{=} \mathbf{k}\chi$ when $\mathbf{k}^{-1} = 0$. They act as superderivations of the complex, i.e., the super-Leibniz rule holds: $d_{\mathrm{st}}(AB) = (d_{\mathrm{st}}A)B + (-1)^{p(A)}A(d_{\mathrm{st}}B)$, and similarly for $\widetilde{\chi}$. Hence it is sufficient to find the action of these operators on the generators. Setting $\mathbf{k}^{-1} = 0$ in the formulas obtained in § 15.2.4, we find that

$$\widetilde{\chi} \cdot \mathbf{k}^{-1}\widehat{J}^{\overline{a}}(z) = \sum_{i=1}^\ell \sum_\beta c_{\alpha_i}^{\overline{a}\beta} \psi_\beta^*(z),$$

$$\widetilde{\chi} \cdot \psi_\alpha^*(z) = 0,$$

$$d_{\mathrm{st}} \cdot \mathbf{k}^{-1}\widehat{J}^{\overline{a}}(z) = \sum_{\overline{b},\alpha} c_{\overline{b}}^{\alpha\overline{a}} \mathbf{k}^{-1}\widehat{J}^{\overline{b}}(z)\psi_\alpha^*(z) + \sum_\alpha (J^{\overline{a}}, J^\alpha)\partial_z \psi_\alpha^*(z),$$

$$d_{\mathrm{st}} \cdot \psi_\alpha^*(z) = -\frac{1}{2} \sum_{\beta,\gamma} c_\alpha^{\beta\gamma} \psi_\beta^*(z)\psi_\gamma^*(z).$$

Here all power series in z are understood to contain only terms with non-negative degrees in z, i.e., $\widehat{J}^{\overline{a}}(z) = \sum_{n<0} \widehat{J}_n^{\overline{a}} z^{-n-1}$, and so on.

15.4.3. Connections. On the other hand, consider the space $\mathcal{C}'(\mathfrak{g})$ of the first order differential operators

$$\nabla = \partial_t - p_- - A(t), \qquad A(t) \in \mathfrak{b}_+[[t]].$$

This space is isomorphic to $\mathfrak{b}_+[[t]]$. The group $N_+[[t]]$ acts on $\mathcal{C}'(\mathfrak{g})$ by gauge transformations:

$$X(t) \cdot (\partial_t - p_- - A(t)) = X(t)(\partial_t - p_- - A(t))X(t)^{-1}.$$

This action induces an action of the Lie algebra $\mathfrak{n}_+[[t]]$ on $\mathcal{C}'(\mathfrak{g})$ by infinitesimal gauge transformations:

(15.4.3) $$x(t) : A(t) \mapsto [x(t), p_- + A(t)] + \partial_t x(t).$$

Let $\operatorname{Fun} \mathcal{C}'(\mathfrak{g})$ be the space of polynomial functions on $\mathcal{C}'(\mathfrak{g})$. This is an $\mathfrak{n}_+[[t]]$–module. We identify it as a vector space with $\operatorname{Fun} \mathfrak{b}_+[[t]]$. Consider the Chevalley complex of the cohomology of $\mathfrak{n}_+[[t]]$ with coefficients in this module (see § A.4). As a vector space, this Chevalley complex coincides with the complex (15.4.2), i.e., it coincides with our complex $C_\infty^\bullet(\mathfrak{g})_0$. Moreover, using the above formulas for the action of d_{st} and $\widetilde{\chi}$ on $C_\infty^\bullet(\mathfrak{g})_0$, and for the $\mathfrak{n}_+((t))$–action on $\operatorname{Fun} \mathcal{C}'(\mathfrak{g})$, it is straightforward to check that the action of the Chevalley differential coincides with the action of $d_\infty = d_{\mathrm{st}} + \widetilde{\chi}$ on $C_\infty^\bullet(\mathfrak{g})_0$. Hence the cohomology $H_\infty^\bullet(\mathfrak{g})$ of $C_\infty^\bullet(\mathfrak{g})_0$ coincides with the cohomology of $\mathfrak{n}_+[[t]]$ with coefficients in $\operatorname{Fun} \mathcal{C}'(\mathfrak{g})$, i.e., $H^\bullet(\mathfrak{n}_+[[t]], \operatorname{Fun} \mathcal{C}'(\mathfrak{g}))$. We now compute this cohomology.

Recall the $\mathfrak{sl}_2$–triple $\{p_+, p_0, p_-\}$ and the decomposition $\mathfrak{g} = \bigoplus_{i=1}^\ell R_i$ from § 15.2.9. Let $p_+^{(i)}$ be a non-zero highest weight vector in R_i (in particular, we can take $p_+^{(1)} = p_+$). Then $\operatorname{pr.gr.} p_+^{(i)} = d_i$. The vector space $\mathfrak{a}_+ = \operatorname{span}\{p_+^{(i)}\}_{i=1}^\ell$ is the kernel of $\operatorname{ad} p_-$ in $\mathfrak{b}_+$. It is known that $\mathfrak{a}_+$ is a maximal abelian Lie subalgebra of $\mathfrak{g}$. Moreover, the image of $\operatorname{ad} p_-$ in $\mathfrak{b}_+$ projects isomorphically onto the quotient $\mathfrak{b}_+/\mathfrak{a}_+$, so that we can canonically decompose $\mathfrak{b}_+$ into a direct sum $\mathfrak{a}_+ \oplus \operatorname{Im} \operatorname{ad} p_+$.

15.4.4. Lemma. *The action of $N_+[[t]]$ on $\mathcal{C}'(\mathfrak{g})$ is free. Moreover, the quotient $\mathcal{C}'(\mathfrak{g})/N_+[[t]]$ is canonically isomorphic to the space $\mathcal{C}(\mathfrak{g})$ of operators*

$$\nabla = \partial_t - p_- - A(t), \qquad A(t) \in \mathfrak{a}_+[[t]],$$

and hence to $\mathfrak{a}_+[[t]]$.

15.4.5. Proof. Recall that we have $\mathfrak{b}_+ = \mathfrak{a}_+ \oplus \operatorname{Im} \operatorname{ad} p_-$, and $\mathfrak{a}_+ = \operatorname{Ker} \operatorname{ad} p_+$. Decompose $\mathfrak{b}_+$ and $\mathfrak{a}_+$ into a direct sum using the principal gradation: $\mathfrak{b}_+ = \bigoplus_{i \geq 0} \mathfrak{b}_{+,i}$ and $\mathfrak{a}_+ = \bigoplus_{j \in E} \mathfrak{a}_{+,j}$, where E is the set of exponents of $\mathfrak{g}$ counted with their multiplicities. It is known that the operator $\operatorname{ad} p_-$ acts from $\mathfrak{b}_{+,i}$ to $\mathfrak{b}_{+,i-1}$ injectively for all $i > 1$.

We claim that each element $\partial_t - p_- - A(t)$ of $\mathcal{C}'(\mathfrak{g})$ can be uniquely represented in the form

(15.4.4) $$\partial_t - p_- - A(t) = \exp(\operatorname{ad} U(t)) \cdot (\partial_t - p_- - B(t)),$$

where $U \in \mathfrak{n}_+[[t]]$ and $B \in \mathfrak{a}_+[[t]]$. To see this, we use the decomposition with respect to the principal gradation: $U = \sum_{j \geq 0} U_j$, $B = \sum_{j \in E} B_j$, $A = \sum_{j \geq 0} A_j$. Equating the homogeneous components of degree i in both sides of (15.4.4), we obtain that $B_i + [U_{i+1}, p_-]$ is expressed in terms of $A_i, B_1, \ldots, B_{i-1}, U_1, \ldots, U_i$. The injectivity of $\operatorname{ad} p_-$ then allows us to determine B_i and U_{i+1} uniquely. Hence

U and B satisfying equation (15.4.4) can be found uniquely by induction, and the assertion of the lemma follows.

Using this lemma, we describe the cohomology of the complex $C^\bullet_\infty(\mathfrak{g})_0$:

15.4.6. Corollary. $H^0_\infty(\mathfrak{g}) \simeq H^0(\mathfrak{n}_+[[t]], \operatorname{Fun} \mathcal{C}'(\mathfrak{g}))$ *is isomorphic to the space of functions on* $\mathcal{C}(\mathfrak{g}) \simeq \mathfrak{a}_+[[t]]$, *and* $H^i_\infty(\mathfrak{g}) \simeq H^i(\mathfrak{n}_+[[t]], \operatorname{Fun} \mathcal{C}'(\mathfrak{g})) = 0$ *for* $i > 0$.

15.4.7. Cohomology of $C^\bullet_k(\mathfrak{g})_0$ for generic k. If we have a complex which is a free $\mathbb{C}[\mathbf{h}]$–module of finite rank, then the dimensions of the cohomologies of its specializations at $\mathbf{h} = h \in \mathbb{C}$ are the same for generic values of h, and may only increase for special values of h (this is called the semi-continuity property of cohomology).

Recall that $C^\bullet_\mathbf{k}(\mathfrak{g})_0$ is a direct sum of its homogeneous components with respect to the $\mathbb{Z}$–gradation. Each of them forms a subcomplex, which is a free $\mathbb{C}[\mathbf{k}^{-1}]$–module of finite rank. By our construction, the specialization of $C^\bullet_\mathbf{k}(\mathfrak{g})_0$ at $\mathbf{k}^{-1} = k^{-1}$ is the complex $C^\bullet_k(\mathfrak{g})_0$ for any $k \in \mathbb{C}^\times$, and the specialization of $C^\bullet_\mathbf{k}(\mathfrak{g})_0$ at $\mathbf{k}^{-1} = 0$ is the complex $C^\bullet_\infty(\mathfrak{g})_0$. Therefore vanishing of the ith cohomology group of $C^\bullet_\infty(\mathfrak{g})_0$ implies vanishing of the ith cohomology group of $C^\bullet_k(\mathfrak{g})_0$ for generic values of k. Since the Euler characteristics of the homogeneous components $C^\bullet_k(\mathfrak{g})_0$ do not depend on k, vanishing of all but the 0th cohomology of $C^\bullet_\infty(\mathfrak{g})_0$ implies that the dimensions of all homogeneous components of the 0th cohomology of $C^\bullet_k(\mathfrak{g})_0$ are the same for generic k as for $k = \infty$.

Thus, we obtain that for generic k, the 0th cohomology of $C^\bullet_k(\mathfrak{g})_0$ (and hence of $C^\bullet_k(\mathfrak{g})$) is isomorphic to the ring of functions on $\mathcal{C}(\mathfrak{g}) \simeq \mathfrak{a}_+[[t]]$, and all other cohomologies vanish. This gives us another proof of Theorem 15.1.9 in the case of generic k.

15.4.8. The opposite spectral sequence. Now we want to describe the 0th cohomology $H^0_k(\mathfrak{g})$ of the complex $C^\bullet_k(\mathfrak{g})_0$ more explicitly. For that we consider the opposite spectral sequence on $C^\bullet_k(\mathfrak{g})_0$, in which the zeroth differential is d_{st} and the first differential is χ. Since $C^i_k(\mathfrak{g})_0 = 0$ for $i < 0$, we only need to compute the 0th and the 1st terms of this spectral sequence, and the differentials acting between them.

Let us compute them first in the limit as $k \to \infty$. Thus, we consider again the complex $C^\bullet_\infty(\mathfrak{g})_0$, but with the differential d_{st}, rather than $d_{\mathrm{st}} + \widetilde{\chi}$.

Let $\overline{\mathcal{C}}'(\mathfrak{g})$ be the space of operators

$$\nabla = \partial_t + A(t), \qquad A(t) \in \mathfrak{b}_+[[t]].$$

The group $N_+[[t]]$ acts on $\overline{\mathcal{C}}'(\mathfrak{g})$ by gauge transformations:

$$X(t) \cdot (\partial_t - A(t)) = X(t)(\partial_t - A(t))X(t)^{-1}.$$

This action induces an action of the Lie algebra $\mathfrak{n}_+[[t]]$ on $\overline{\mathcal{C}}'(\mathfrak{g})$ by infinitesimal gauge transformations:

$$x(t) : A(t) \mapsto [x(t), A(t)] + \partial_t x(t)$$

(compare with formula (15.4.3)).

This makes the space $\operatorname{Fun} \overline{\mathcal{C}}'(\mathfrak{g})$ of polynomial functions on $\overline{\mathcal{C}}'(\mathfrak{g})$ into an $\mathfrak{n}_+[[t]]$–module. We identify it as a vector space with $\operatorname{Fun} \mathfrak{b}_+[[t]]$. Recall that the standard Chevalley complex of the cohomology of $\mathfrak{n}_+[[t]]$ with coefficients in this module is isomorphic to the complex $C^\bullet_\infty(\mathfrak{g})_0$. It follows from the above formulas that the action of d_{st} on $C^\bullet_\infty(\mathfrak{g})_0$ coincides with the action of the Chevalley differential on this complex. Hence the cohomology of the complex $(C^\bullet_\infty(\mathfrak{g})_0, d_{\mathrm{st}})$ equals $H^\bullet(\mathfrak{n}_+[[t]], \mathrm{Fun}\, \overline{\mathcal{C}}'(\mathfrak{g}))$.

Denote by $\mathfrak{n}_+[[t]]_0$ the Lie algebra ideal $\mathfrak{n}_+ \otimes t\mathbb{C}[[t]]$ in $\mathfrak{n}_+[[t]]$ consisting of all elements vanishing at $t = 0$. Let $N_+[[t]]_0$ be the corresponding normal subgroup of $N_+[[t]]$. The proof of the following lemma is similar to the proof of Lemma 15.4.4.

15.4.9. Lemma. *The action of $N_+[[t]]_0$ on $\overline{\mathcal{C}}'(\mathfrak{g})$ is free, and the quotient $\mathcal{C}'(\mathfrak{g})/N_+[[t]]_0$ is canonically isomorphic to the space of operators $\nabla = \partial_t - A(t)$, $A(t) \in \mathfrak{h}[[t]]$, and hence to the space $\mathfrak{h}[[t]]$.*

Therefore $H^0(\mathfrak{n}_+[[t]]_0, \mathrm{Fun}\, \overline{\mathcal{C}}'(\mathfrak{g})) \simeq \mathrm{Fun}\, \mathfrak{h}[[t]]$, and $H^i(\mathfrak{n}_+[[t]]_0, \mathrm{Fun}\, \overline{\mathcal{C}}'(\mathfrak{g})) = 0$ for $i > 0$.

15.4.10. The zeroth and the first cohomology groups. Using the Serre-Hochschild spectral sequence (see [**Fu**]) corresponding to the ideal $\mathfrak{n}_+[[t]]_0 \subset \mathfrak{n}_+[[t]]$, we obtain from Lemma 15.4.9

$$H^\bullet(\mathfrak{n}_+[[t]], \mathrm{Fun}\, \overline{\mathcal{C}}'(\mathfrak{g})) \simeq (\mathrm{Fun}\, \mathfrak{h}[[t]]) \otimes H^\bullet(\mathfrak{n}_+, \mathbb{C}).$$

Recall that

$$H^i(\mathfrak{n}_+, \mathbb{C}) \simeq \bigoplus \mathbb{C}_{w(\rho)-\rho},$$

where the sum is over the elements w of the Weyl group of $\mathfrak{g}$ of length i. The Cartan subalgebra $\mathfrak{h}$ acts on the term $\mathbb{C}_\lambda$ according to the weight λ. In particular, $H^0(\mathfrak{n}_+, \mathbb{C}) \simeq \mathbb{C}$ and $H^1(\mathfrak{n}_+, \mathbb{C}) \simeq \bigoplus_{i \in I} \mathbb{C}_{-\alpha_i}$, where $I = \{1, \ldots, \ell\}$. Hence we find that the 0th cohomology of the complex $(C^\bullet_\infty(\mathfrak{g})_0, d_{\mathrm{st}})$ is isomorphic to $\mathrm{Fun}\, \mathfrak{h}[[t]]$, and the first cohomology is isomorphic to $(\mathrm{Fun}\, \mathfrak{h}[[t]]) \otimes \bigoplus_{i \in I} \mathbb{C}_{-\alpha_i}$.

Furthermore, it is easy to find representatives of the corresponding cohomology classes. Let $\{h^i\}_{i \in I}$ be the basis of $\mathfrak{h}$ which is dual to the basis $\{\alpha_i\}_{i \in I}$ of simple roots in $\mathfrak{h}^*$ with respect to the normalized inner product. Denote by $\widehat{h}^i(z)$ the corresponding field in the vertex superalgebra $C^\bullet_\infty(\mathfrak{g})_0$, defined as in formula (15.2.1). Explicitly,

$$(15.4.5) \qquad \widehat{h}^i(z) = h^i(z) + \sum_{\alpha \in \Delta_+} (\alpha, \alpha_i) :\psi_\alpha(z)\psi^*_\alpha(z): .$$

Then the 0th cohomology of $(C^\bullet_\infty(\mathfrak{g})_0, d_{\mathrm{st}})$ is represented by the vector space

$$\mathbb{C}[\mathbf{k}^{-1}\widehat{h}^i_n]_{i \in I; n<0}|0\rangle,$$

and the 1st cohomology is represented by the space

$$\bigoplus_{i \in I} \mathbb{C}[\mathbf{k}^{-1}\widehat{h}^j_n]_{j \in I; n<0}\psi^*_{\alpha_i,0}|0\rangle.$$

Next, we find using the formulas of § 15.2.4 that

$$[d_{\mathrm{st}}, \widehat{h}^i(z)] = 0, \qquad [d_{\mathrm{st}}, \psi^*_{\alpha_i}(z)] = 0,$$

for all values of k. Therefore the above spaces embed into the corresponding cohomology groups for all k. Since the cohomology groups may only increase in size at special values of k, we conclude that for generic k the 0th cohomology group of the complex $(C^\bullet_\infty(\mathfrak{g})_0, d_{\mathrm{st}})$ may be represented by, and hence is equal to,

$\pi_0 = \mathbb{C}[\widehat{h}_n^i]_{i \in I; n < 0}|0\rangle$. Likewise, the first cohomology group may be represented by $\bigoplus_{i \in I} \widetilde{\pi}_i$, where

$$\widetilde{\pi}_i = \mathbb{C}[\widehat{h}_n^j]_{j \in I; n < 0} \psi_{\alpha_i, 0}^* |0\rangle.$$

In fact, one can show that for generic k the ith cohomology of $(C_k^\bullet(\mathfrak{g})_0, d_{\mathrm{st}})$ is isomorphic to $\pi_0 \otimes H^i(\mathfrak{n}_+, \mathbb{C})$ (as for $k = \infty$), but we will not need this in what follows. We also note that for positive rational values of k the cohomology of $(C_k^\bullet(\mathfrak{g})_0, d_{\mathrm{st}})$ may be quite complicated, in contrast to the cohomology of $(C_k^\bullet(\mathfrak{g})_0, d_{\mathrm{st}} + \chi)$, which is concentrated in cohomological degree 0 for any k (see Theorem 15.1.9).

Now recall that the complex $C_k^\bullet(\mathfrak{g})_0$ is a vertex superalgebra, and that the differential d_{st} preserves the vertex structure on it. Hence the cohomology of the complex $(C_k^\bullet(\mathfrak{g})_0, d_{\mathrm{st}})$, which we denote by $\widetilde{H}_k^\bullet(\mathfrak{g})$, is also a vertex superalgebra. In particular, $\widetilde{H}_k^0(\mathfrak{g}) = \pi_0$ is a vertex algebra, and $\widetilde{H}_k^1(\mathfrak{g}) \simeq \bigoplus_{i \in I} \widetilde{\pi}_i$ as a π_0–module. Let us describe them explicitly.

The inner product on $\mathfrak{h}$ induced by its action on $\mathfrak{n}_+$ (see § 15.1.2) is equal to one half of the Killing form, i.e., to $h^\vee(\cdot, \cdot)$, where $(\cdot, \cdot)$ is the restriction of the normalized invariant inner product on $\mathfrak{g}$ (see § 2.4.1). Using formula (15.4.5) for $\widehat{h}^i(z), i \in I$, and Lemma 15.1.3, we find the following OPE:

$$\widehat{h}^i(z)\widehat{h}^j(z) = \frac{(k + h^\vee)(\alpha_i, \alpha_j)}{(z - w)^2} + \mathrm{reg}.$$

Suppose that $k \neq -h^\vee$, and set

$$\nu = \sqrt{k + h^\vee},$$

$$\overline{h}^i(z) = \frac{\widehat{h}^i(z)}{\nu}, \qquad \overline{h}_n^i = \frac{\widehat{h}_n^i}{\nu}, \qquad n \in \mathbb{Z}.$$

Then we obtain the commutation relations

$$[\overline{h}_n^i, \overline{h}_m^j] = n(\alpha_i, \alpha_j)\delta_{n, -m}.$$

Therefore the elements $\overline{h}_n^i, i \in I, n \in \mathbb{Z}$, generate the Heisenberg Lie algebra $\widehat{\mathfrak{h}}$ associated to $\mathfrak{h}$ and the inner product $(\cdot, \cdot)$ on it (see § 5.4.1). The space π_0 now gets identified, as a vertex algebra, with the corresponding Heisenberg vertex algebra denoted in the same way (as in § 5.4.1).

Using formula (15.4.5) we also find the following OPE:

$$(15.4.6) \qquad \overline{h}^i(z)\psi_{\alpha_j}^*(w) = -\frac{(\alpha_i, \alpha_j)}{\nu} \frac{\psi_{\alpha_j}^*(w)}{z - w} + \mathrm{reg}.$$

It gives us the following commutation relations:

$$(15.4.7) \qquad [\overline{h}_n^i, \psi_{\alpha_j}^*(z)] = -z^n \frac{(\alpha_i, \alpha_j)}{\nu} \psi_{\alpha_j}^*(z).$$

In particular, we obtain that

$$\overline{h}_n^i \cdot \psi_{\alpha_j, 0}^*|0\rangle = 0, \quad n > 0; \qquad \overline{h}_0^i \cdot \psi_{\alpha_j, 0}^*|0\rangle = -\frac{(\alpha_i, \alpha_j)}{\nu} \psi_{\alpha_j, 0}^*|0\rangle.$$

Thus, we obtain that as an $\widehat{\mathfrak{h}}$–module and as a π_0–module, $\widetilde{\pi}_i$ is isomorphic to the Fock module $\pi_{-\alpha_i/\nu}$ with the vector $\psi_{\alpha_i, 0}^*|0\rangle$ playing the role of the highest weight vector $|-\alpha_i/\nu\rangle$, as defined in § 5.4.1.

15.4.11. Description of $\mathcal{W}_k(\mathfrak{g})$ as a vertex subalgebra of π_0. Recall that the $\mathcal{W}$–algebra $\mathcal{W}_k(\mathfrak{g})$ is the vertex algebra equal to the 0th cohomology of the complex $C_k^\bullet(\mathfrak{g})_0$ with respect to the total differential $d = d_{\mathrm{st}} + \chi$. In § 15.2.8 we introduced a bicomplex structure on $C_k^\bullet(\mathfrak{g})_0$. We now consider the spectral sequence in which the zeroth differential is d_{st}. We have computed above the 0th and the 1st cohomology groups, $\widetilde{H}_k^0(\mathfrak{g})$ and $\widetilde{H}_k^1(\mathfrak{g})$, of $C_k^\bullet(\mathfrak{g})_0$ with respect to the differential d_{st} for generic values of k. The result is that $\widetilde{H}_k^0(\mathfrak{g})$ lies entirely in the bidegree $(0,0)$ part $E_1^{0,0}$ of the first term of the spectral sequence, while $\widetilde{H}_k^1(\mathfrak{g})$ lies in $E_1^{1,0}$. Since the complex $C_k^\bullet(\mathfrak{g})_0$ has no terms with negative cohomological degrees, we obtain that $\mathcal{W}_k(\mathfrak{g})$ is equal to the kernel of the first differential χ of our spectral sequence acting from $\widetilde{H}_k^0(\mathfrak{g})$ to $\widetilde{H}_k^1(\mathfrak{g})$. It remains to compute this operator explicitly. We will assume in what follows that $k \neq -h^\vee$.

By definition, $\chi = \sum_{i \in I} \psi_{\alpha_i,1}^*$. Clearly, the summand in the first differential corresponding to $\psi_{\alpha_i,1}^*$ maps π_0 to $\widetilde{\pi}_i = \pi_{-\alpha_i/\nu} \in \widetilde{H}_k^1(\mathfrak{g})$. In order to find the action of

$$\psi_{\alpha_i,1}^* = \int \psi_{\alpha_i}^*(z)dz: \quad \pi_0 \to \pi_{-\alpha_i/\nu}$$

we compute the action of the field $\psi_{\alpha_i}^*(z): \pi_0 \to C_k^1(\mathfrak{g})_0/\operatorname{Im} d_{\mathrm{st}}$ (recall that we are considering its action on the first term of our spectral sequence, i.e., modulo d_{st}–exact terms). It will actually turn out that all Fourier coefficients of $\psi_{\alpha_i}^*(z)$ (and not just $\psi_{\alpha_i,0}^*$) map vectors in π_0 to $\pi_{-\alpha_i/\nu}$ modulo d_{st}–exact terms.

Let $\widehat{f}_i(z)$ be the field $\widehat{J}^a(z)$ corresponding to the basis element $f_i = f^{\alpha_i}$ of $\mathfrak{n}_-$ (see formula (15.2.1)). Using the commutation relations of § 15.2.4, we find

$$[d_{\mathrm{st}}, \widehat{f}_i(z)] = \frac{2}{(\alpha_i, \alpha_i)}\left(:\!\widehat{h}^i(z)\psi_{\alpha_i}^*(z)\!: + (k + h^\vee)\partial_z\psi_{\alpha_i}^*(z)\right).$$

Here we use the fact that

$$\sum_{\beta,b} c_\beta^{\alpha_i,b}c_b^{-\alpha_i,\beta} = \frac{1}{2}\operatorname{Tr}_{\mathfrak{g}}\operatorname{ad} e_i \operatorname{ad} f_i = h^\vee(e_i, f_i) = \frac{2h^\vee}{(\alpha_i, \alpha_i)}.$$

Thus, we obtain the following relation on the field $\psi_{\alpha_i}^*(z): \pi_0 \to \pi_{-\alpha_i/\nu}$ (modulo d_{st}–exact terms):

(15.4.8) $$\partial_z\psi_{\alpha_i}^*(z) = -\frac{1}{\nu}:\!\overline{h}^i(z)\psi_{\alpha_i}^*(z)\!: .$$

The relation (15.4.8), the commutation relations (15.4.7) and the formulas

$$\psi_{\alpha_i,n}^*|0\rangle = \delta_{n,0}\psi_{\alpha_i,0}^*|0\rangle, \qquad n \geq 0,$$

uniquely determine the action of the field $\psi_{\alpha_i}^*(z): \pi_0 \to \pi_{-\alpha_i/\nu}$. But these are exactly the relations satisfied by the bosonic vertex operator

$$V_{-\alpha_i/\nu}(z): \quad \pi_0 \to \pi_{-\alpha_i/\nu}$$

(see § 5.4.2 and formulas (5.2.2)–(5.2.6)). Thus, we conclude that the first differential of the spectral sequence is the sum of the operators

$$\int V_{-\alpha_i/\nu}(z)dz: \quad \pi_0 \to \pi_{-\alpha_i/\nu}, \qquad i \in I.$$

This gives us the following description of the $\mathcal{W}$–algebra:

15.4.12. Theorem. *For generic k the $\mathcal{W}$–algebra $\mathcal{W}_k(\mathfrak{g})$ may be described as the vertex subalgebra of the Heisenberg vertex algebra π_0 associated to the Heisenberg Lie algebra $\widehat{\mathfrak{h}}$, which is the intersection of kernels of the operators $\int V_{-\alpha_i/\nu}(z)dz$.*

15.4.13. Remark. This result provides us with a free field realization of the $\mathcal{W}$–algebra. It should be compared with Theorem 12.3.10, which gives a similar realization of $V_k(\mathfrak{g})$ as a vertex subalgebra of a Heisenberg vertex algebra. In fact, the image of $V_k(\mathfrak{g})$ in this Heisenberg vertex algebra may also be described for generic k as the intersection of kernels of residues of vertex operators (see [**FF9**]).

In the case when $\mathfrak{g} = \mathfrak{sl}_n$ the generating fields $W_i(z)$ of $\mathcal{W}_k(\mathfrak{g})$ may be written explicitly in terms of the fields $\overline{h}^i(z)$ (see, e.g., [**FL**]). The formulas become very complicated for other $\mathfrak{g}$, and they are unknown in general.

The operators $\int V_{-\alpha_i/\nu}(z)dz$ are called the *screening operators*. Note that the intersection of kernels of residues of vertex operators is a vertex subalgebra, according to § 5.7.3.

By Theorem 15.4.12, $\mathcal{W}_k(\mathfrak{g})$ may be described for generic k as the 0th cohomology of the complex

$$\pi_0 \to \bigoplus_{i \in I} \pi_{-\alpha_i/\nu}.$$

This complex may be extended to the right in such a way that all of its higher cohomologies vanish, and the 0th cohomology remains $\mathcal{W}_k(\mathfrak{g})$. The ith group of the extended complex equals $\bigoplus \pi_{(w(\rho)-\rho)/\nu}$, where the sum is over the elements of the Weyl group of $\mathfrak{g}$ of length i (see [**FF8**]). Moreover, this extended complex is nothing but the first term of our spectral sequence for generic values of k.

This complex looks like the Bernstein-Gelfand-Gelfand (BGG) resolution of the trivial one-dimensional $\mathfrak{g}$–module by the modules contragredient to the Verma modules. In fact, a similar resolution exists for the corresponding quantum group $U_q(\mathfrak{g})$, and the differentials of our complex are constructed from the differentials of the BGG resolution of $U_q(\mathfrak{g})$. The extended complex may be constructed from the BGG resolution using the fact that, roughly speaking, the screening operators satisfy the q–Serre relations, i.e., the defining relations of the nilpotent part $U_q(\mathfrak{n}_+)$ of $U_q(\mathfrak{g})$, with $q = \exp(\pi i / r^\vee(k + h^\vee))$. Here $r^\vee$ denotes the *lacing number* of $\mathfrak{g}$, i.e., the maximal number of edges connecting two vertices of the Dynkin diagram of $\mathfrak{g}$. These relations were first observed in [**BMP**] (see also [**FF8**]).

15.4.14. The case of $\mathfrak{sl}_2$. In the case when $\mathfrak{g} = \mathfrak{sl}_2$, the Heisenberg Lie algebra $\widehat{\mathfrak{h}}$ is isomorphic to the Heisenberg Lie algebra $\mathcal{H}$ introduced in § 2.1.2. Under this isomorphism, the generators $\overline{h}_n^1, n \in \mathbb{Z}$, of $\widehat{\mathfrak{h}}$ are mapped to $\sqrt{2}b_n$. The vertex operator $V_{-\alpha_1/\nu}(z)$ becomes $V_\beta(z)$, where $\beta = -\sqrt{\frac{2}{k+2}}$ in the notation of § 5.2.

The $\mathcal{W}$–algebra $\mathcal{W}_k(\mathfrak{sl}_2)$ is the Virasoro vertex algebra with central charge $c(k) = 1 - 6(k + 1)^2/(k + 2)$, as we already found in § 15.3.2. Therefore Theorem 15.4.12 describes the Virasoro vertex algebra $\mathrm{Vir}_{c(k)}$ for generic k as the kernel of the screening operator

$$\int V_\beta(z)dz : \quad \pi_0 \to \pi_\beta.$$

In fact, one can prove that $\mathrm{Vir}_{c(k)} \simeq \mathrm{Ker} \int V_\beta(z)dz$ for all k except those of the form $k = -2 + p/q$, where p and q are relatively prime integers which are greater than 1 (note that in the remaining cases $c(k)$ becomes the central charge $c_{p,q}$ of the (p,q) minimal model).

The generating field $T^\beta(z)$ of this vertex subalgebra of π_0 is easy to find:

$$T^\beta(z) = \frac{1}{2}{:}b(z)^2{:} + \left(\frac{\beta}{2} - \frac{1}{\beta}\right)\partial_z b(z).$$

By formula (5.3.10), its central charge is $1 - 12(\beta/2 - 1/\beta)^2 = c(k)$.

To see that $T^\beta(z)$ commutes with the screening operator, we compute the OPE

$$\left(\frac{1}{2}{:}b(z)^2{:} + \lambda\partial_z b(z)\right)V_\beta(w) = \frac{1}{2}\beta(\beta - 2\lambda)\frac{V_\beta(w)}{(z-w)^2} + \frac{\partial_w V_\beta(w)}{z-w} + \mathrm{reg}.$$

The easiest way to obtain this formula is to apply formula (3.3.1) and the fact that $V_\beta(w) = Y(|\beta\rangle, w)$, established in Proposition 5.2.5.

In our case, $\lambda = \beta/2 - 1/\beta$, so we obtain

$$T^\beta(z)V_\beta(w) = \partial_w\left(\frac{V_\beta(w)}{z-w}\right) + \mathrm{reg}.$$

Therefore we find that

$$[T^\beta(z), \int V_\beta(w)dw] = 0,$$

as expected.

15.4.15. Duality of $\mathcal{W}$–algebras. Note that the field $T^\beta(z)$ does not change if we replace β by $-2/\beta$. Therefore the above description of $\mathrm{Ker}_{\pi_0}\int V_\beta(z)dz$ implies that

(15.4.9) $$\mathrm{Ker}_{\pi_0}\int V_\beta(z)dz = \mathrm{Ker}_{\pi_0}\int V_{-2/\beta}(z)dz$$

for generic β.

Now consider the case of an arbitrary simple Lie algebra $\mathfrak{g}$. To avoid confusion, denote the corresponding vertex algebra π_0 by $\pi_0^\mathfrak{g}$. Let $\widehat{\mathfrak{h}}_i$ be the Heisenberg Lie subalgebra of $\widehat{\mathfrak{h}}$ spanned by $\overline{h}_n^i, n \in \mathbb{Z}$, and let $\pi_{0,i}$ be the Fock representation of $\widehat{\mathfrak{h}}_i$. Let $\mathfrak{h}_i^\perp$ be the orthogonal compliment of $\mathbb{C}h^i \subset \mathfrak{h}$. The central extension $\widehat{\mathfrak{h}}_i^\perp$ of $\mathfrak{h}_i^\perp((t))$ is a Heisenberg Lie subalgebra of $\widehat{\mathfrak{h}}$, which is the centralizer of $\widehat{\mathfrak{h}}_i$. Denote by $\pi_{0,i}^\perp$ the Fock representation of $\widehat{\mathfrak{h}}_i^\perp$. Then both $\pi_{0,i}$ and $\pi_{0,i}^\perp$ are vertex subalgebras of $\pi_0^\mathfrak{g}$, and $\pi_0^\mathfrak{g} \simeq \pi_{0,i} \otimes \pi_{0,i}^\perp$. Likewise, we have a decomposition $\pi_{-\alpha_i/\nu} \simeq \pi_{-\alpha_i/\nu,i} \otimes \pi_{0,i}^\perp$.

By definition, the ith screening operator acts "along the direction" of the ith root, i.e., with respect to the above decomposition we have

$$\int V_{-\alpha_i/\nu}(z)dz = \int V_{-\alpha_i/\nu}(z)dz \otimes \mathrm{Id}.$$

Therefore $\mathrm{Ker}\int V_{-\alpha_i/\nu}(z)dz$ equals the tensor product of the kernel of

$$\int V_{-\alpha_i/\nu}(z)dz : \pi_{0,i} \to \pi_{-\alpha_i/\nu,i}$$

(described above) and $\pi_{0,i}^\perp$.

In particular, we find from formula (15.4.9) that, for generic k,

(15.4.10) $$\mathrm{Ker}_{\pi_0}\int V_{-\alpha_i/\nu}(z)dz = \mathrm{Ker}_{\pi_0}\int V_{\alpha_i^\vee\nu}(z)dz,$$

where $\alpha_i^\vee = 2\alpha_i/(\alpha_i,\alpha_i)$ is the ith coroot of $\mathfrak{g}$.

Recall that by exchanging the roots and coroots of $\mathfrak{g}$ we obtain the *Langlands dual Lie algebra* $^L\mathfrak{g}$ of $\mathfrak{g}$. The Cartan matrix of $^L\mathfrak{g}$ is the transpose of the Cartan matrix of $\mathfrak{g}$, so that the Lie algebras of types B_n and C_n get exchanged and the other simple Lie algebras are self-dual. Our inner product is normalized in such a way that $(\alpha_i^\vee, \alpha_j^\vee) = r^\vee(^L\alpha_i, {}^L\alpha_j)$, where $r^\vee$ is the lacing number of $\mathfrak{g}$ (see Remark 15.4.13). By Theorem 15.4.12,

$$\mathcal{W}_k(\mathfrak{g}) \simeq \bigcap_{i \in I} \mathrm{Ker}_{\pi_0} \int V_{-\alpha_i/\nu}(z)\,dz$$

for generic k (recall that $\nu = \sqrt{k + h^\vee}$). Therefore formula (15.4.10) leads to the following duality of W–algebras:

15.4.16. Proposition ([FF7]). *For generic k, there is an isomorphism*

$$\mathcal{W}_k(\mathfrak{g}) \simeq \mathcal{W}_{k'}(^L\mathfrak{g}),$$

where $^L\mathfrak{g}$ is the Langlands dual Lie algebra of $\mathfrak{g}$ and $(k + h^\vee)r^\vee = (k' + {}^Lh^\vee)^{-1}$.

15.4.17. Remark. The relation $(k + h^\vee)r^\vee = (k' + {}^Lh^\vee)^{-1}$ may be reformulated in a nicer way if we adopt a slightly different point of view on the Kac–Moody central extension. Recall that in § 2.4.1 we fixed a normalized inner product on $\mathfrak{g}$ once and for all, and defined $\widehat{\mathfrak{g}}$ using this inner product. We then have the possibility of identifying the central element K of $\widehat{\mathfrak{g}}$ with an arbitrary constant k. But we may instead define the central extension $\widehat{\mathfrak{g}}$ for any given invariant inner product τ (which is unique up to a scalar multiple), always identifying K with 1. Then we obtain a family of W–algebras $\mathcal{W}_\tau(\mathfrak{g})$ parameterized by the invariant inner products.

We will say that the inner product τ on $\mathfrak{g}$ is dual to the inner product τ' on $^L\mathfrak{g}$, if the restrictions $\tau|_{\mathfrak{h}}$ and $\tau|_{^L\mathfrak{h}}$ are dual to each other with respect to the canonical isomorphism $^L\mathfrak{h} \simeq \mathfrak{h}^*$. Denote by τ_K and τ_K' the Killing inner products on $\mathfrak{g}$ and $^L\mathfrak{g}$, respectively. Then Proposition 15.4.16 may be reformulated as saying that $\mathcal{W}_\tau(\mathfrak{g}) \simeq \mathcal{W}_{\tau'}(^L\mathfrak{g})$, if τ is generic and $\tau + \frac{1}{2}\tau_K$ is dual to $\tau' + \frac{1}{2}\tau_K'$.

The appearance of the Langlands dual Lie algebra in Proposition 15.4.16 is important. Consider the limit as $k \to -h^\vee$ (the critical level). The vertex algebra $\mathcal{W}_{-h^\vee}(\mathfrak{g})$ is commutative and is isomorphic to the center of the Kac–Moody vertex algebra $V_{-h^\vee}(\mathfrak{g})$ [**FF7**] (see § 15.3.2). On the other hand, the limit of $\mathcal{W}_{k'}(^L\mathfrak{g})$ as $k' \to \infty$ is the classical W–algebra defined in § 16.8. The isomorphism of Proposition 15.4.16 still holds at $k = -h^\vee$, where it becomes the isomorphism $\mathcal{Z}(V_{-h^\vee}(\mathfrak{g})) \simeq \mathcal{W}_\infty(^L\mathfrak{g})$ of Theorem 18.4.4. As explained in § 18.4, the last isomorphism is related to the geometric Langlands correspondence, where $^L\mathfrak{g}$ is supposed to appear. So Proposition 15.4.16 gives us an "explanation" for the appearance of the Langlands dual Lie algebra. Further, the fact that the isomorphism of Proposition 15.4.16 exists for levels k other than critical suggests that there there may exist a "quantum deformation" of the geometric Langlands correspondence.

15.5. Bibliographic notes

The first W–algebra (corresponding to $\mathfrak{g} = \mathfrak{sl}_3$) was constructed by A.B. Zamolodchikov [**Za**]. The W–algebras corresponding to $\mathfrak{g} = \mathfrak{sl}_N$ were constructed by V. Fateev and S. Lukyanov [**FL**]. For general $\mathfrak{g}$ the W–algebra was first constructed by B. Feigin and E. Frenkel in [**FF7, FF8**]. There is an extensive literature on

$\mathcal{W}$–algebras. We refer the reader to the comprehensive review [**BS1**] and to the collection [**BS2**].

The quantum Drinfeld–Sokolov reduction was defined in [**FF7**] (see also [**FF5, F2, FKW**]). Theorem 15.1.9 was first proved by B. Feigin and E. Frenkel [**FF7, F1**] for all $\mathfrak{g}$ and for generic k. It was subsequently proved by J. de Boer and T. Tjin [**dBT**] for all k in the case when $\mathfrak{g} = \mathfrak{sl}_n$. In § 15.2 we generalize their proof to the case of an arbitrary $\mathfrak{g}$. In § 15.4 we follow the original proof from [**FF7**] with some modifications due to E. Frenkel. The realization of the $\mathcal{W}$–algebras as the intersection of the kernels of screening operators was given and the duality of $\mathcal{W}$–algebras, Proposition 15.4.16, was established in [**FF7**]. The BGG type resolutions of $\mathcal{W}$–algebras were studied in [**FF8**]. Lemma 15.4.4 is taken from [**DS**].

CHAPTER 16

Vertex Lie Algebras and Classical Limits

In this chapter we define vertex Lie algebras and vertex Poisson algebras. As we have seen, the polar part of the OPE (or equivalently, the polar part of the operation Y) carries important information about the vertex algebra. For example, it completely determines the commutation relations between the Fourier coefficients of the vertex operators. Identifying the properties satisfied by the polar part of the vertex operation Y, one naturally arrives at the notion of vertex Lie algebra. The relation between vertex Lie algebras and vertex algebras is similar to that between Lie algebras and their universal enveloping algebras. In fact, one can introduce a notion of enveloping algebra of a vertex Lie algebra, which is a vertex algebra. Vertex Lie algebras are usually smaller and more "economical" than vertex algebras. For instance, the Heisenberg, Kac–Moody, and Virasoro vertex algebras introduced in Chapter 2 are enveloping algebras of vertex Lie algebras which are easy to describe.

The notion of vertex Poisson algebra is an analogue of the notion of Poisson algebra in the category of vertex algebras. Vertex Poisson algebras combine in a compatible way the structures of a commutative vertex algebra and a vertex Lie algebra. Examples of vertex Poisson algebras may be constructed by taking suitable limits of one-parameter families of vertex algebras (such as $V_k(\mathfrak{g})$, which is a one-parameter family depending on k). We describe these limits for the Virasoro and Kac–Moody vertex algebras. This brings us to a more geometric way of thinking about Kac–Moody and Virasoro algebras, in terms of connections on bundles on the punctured disc.

Important features of the formalism of vertex Poisson algebra can be traced back to the work by I.M. Gelfand, L.A. Dickey and others (see, e.g., [**Di**]) on the hamiltonian structure of integrable hierarchies of soliton equations. In fact, classical limits of vertex algebras give rise to hamiltonian structures of familiar integrable systems, such as the KdV hierarchy, which corresponds to the Virasoro algebra. In the mid–80s, V. Drinfeld and V. Sokolov [**DS**] attached an integrable hierarchy of soliton equations generalizing the KdV system to an arbitrary simple Lie algebra. The objects encoding the hamiltonian structure of these hierarchies are classical limits of the $\mathcal{W}$–algebras introduced in the previous chapter. In fact, a classical $\mathcal{W}$–algebra may be viewed as the algebra of functions of the space of opers on the disc, which are generalizations of projective connections. Classical $\mathcal{W}$–algebras may be obtained as the result of the classical Drinfeld–Sokolov hamiltonian reduction introduced at the end of this chapter.

16.1. Vertex Lie algebras

For a formal power series $a(z) = \sum_{m \in \mathbb{Z}} a_m z^m$ we write $a(z)_- = \sum_{m < 0} a_m z^m$ for its polar part.

16.1.1. Definition. A *vertex Lie algebra* is a vector space L equipped with a linear operator T and a linear operation

$$Y_- : L \to \text{End}\, L \otimes z^{-1}\mathbb{C}[[z^{-1}]],$$

mapping each element $A \in L$ to a series

$$Y_-(A, z) = \sum_{n \geq 0} A_{(n)} z^{-n-1},$$

where $A_{(n)}$ is a linear operator on L such that for any $v \in L$ we have $A_{(n)} v = 0$ for large enough n.

These structures should satisfy the following axioms:

- *(translation)* $Y_-(TA, z) = \partial_z Y_-(A, z)$.
- *(skew–symmetry)* $Y_-(A, z)B = (e^{zT} Y_-(B, -z)A)_-$.
- *(commutator)*

$$[A_{(m)}, Y_-(B, w)] = \sum_{n \geq 0} \binom{m}{n} (w^{m-n} Y_-(A_{(n)} B, w))_-.$$

A vertex Lie algebra is called $\mathbb{Z}$–graded if L is a $\mathbb{Z}$–graded vector space, T has degree 1, and for A of degree m the operator $A_{(n)}$ has degree $-n + m - 1$.

It is straightforward to define super-analogues of vertex Lie algebras, as well as the notions of homomorphisms of vertex Lie algebras, their subalgebras and ideals. We leave this to the reader.

16.1.2. Defintion. Let $(V, |0\rangle, T, Y)$ be a vertex algebra. By the *polar part* of V we will understand the vector space V, equipped with the operator T and the linear operation Y_-, defined by the formula

$$Y_-(A, z) = \sum_{n \geq 0} A_{(n)} z^{-n-1}.$$

16.1.3. Lemma. *The polar part of a vertex algebra V is a vertex Lie algebra.*

16.1.4. Proof. The translation property is a consequence of the translation axiom for Y (note that differentiation preserves the polar part of a formal power series).

The skew–symmetry is a consequence of the skew-symmetry property of vertex algebras, Proposition 3.2.5.

The commutator identity follows from formula (3.3.12) expressing the commutation relations of Fourier coefficients of vertex operators:

$$(16.1.1) \qquad [A_{(m)}, B_{(k)}] = \sum_{n \geq 0} \binom{m}{n} (A_{(n)} \cdot B)_{(m+k-n)}.$$

16.1.5. Examples. Let vir_c denote the vector subspace of the Virasoro vertex algebra Vir_c spanned by the vacuum vector v_c and the vectors $L_n v_c, n \leq -2$. (Thus vir_c is generated by v_c and L_{-2} under the action of the translation operator T.) The operator

$$T(z)_- = Y_-(L_{-2} v_c, z) = \sum_{n \geq -1} L_n z^{-n-2}$$

preserves vir_c, and via the translation axiom (and the assignment $Y_-(v_c, z) = \text{Id}$) defines a vertex Lie algebra structure on vir_c. This is a vertex Lie subalgebra of the polar part of the vertex algebra Vir_c. The Fourier coefficients of the series $T(z)_-$ generate an action of the Lie algebra $\mathrm{Der}\,\mathcal{O}$ (and hence of the group $\Lambda\mathrm{ut}\,\mathcal{O}$) on vir_c.

Similarly, we obtain a vertex Lie subalgebra of the polar part of $V_k(\mathfrak{g})$. The subspace $v_k(\mathfrak{g}) \subset V_k(\mathfrak{g})$ spanned by the vacuum vector v_k and the vectors $J_n^a v_k, n \leq -1$ (i.e., generated by v_k and $J_{-1}^a v_k$ under the action of the translation operator T) is invariant under the series $J^a(z)_-$. The Fourier coefficients of these series generate an action of the Lie algebra $\mathfrak{g}(\mathcal{O})$ (and hence of the group $G(\mathcal{O})$) on the vertex Lie algebra $v_k(\mathfrak{g})$.

Note that neither vir_c nor $v_k(\mathfrak{g})$ is preserved by the Fourier coefficients corresponding to positive powers of z (i.e., the creation operators) of the corresponding vertex operators. This represents the fundamental difference between a vertex Lie algebra and a vertex algebra: the former carries an action of only "half" of the operators (namely, the annihilation operators).

While vir_c and $v_k(\mathfrak{g})$ do not carry actions of the entire Virasoro algebra of central charge c or the entire Kac–Moody algebra of level k, it is easy to recover c and k from the vertex Lie algebra structure. Namely, in vir_c we have $L_2 \cdot L_{-2} v_c = \frac{c}{2} v_c$, and in $v_k(\mathfrak{g})$, we have $J_1^b \cdot J_{-1}^a v_k = k(J^b, J^a) v_k$. We will see next that the Virasoro and Kac–Moody algebras can be recovered from vir_c and $v_k(\mathfrak{g})$, respectively, so these vertex Lie algebras carry all the essential information about these Lie algebras.

16.1.6. Local Lie algebra. We now assign a Lie algebra to any vertex Lie algebra L in a way similar to the construction of § 4.1.1. Let $L(D^\times) = L \otimes \mathbb{C}((t))$ and consider the operator $\partial = T \otimes 1 + \mathrm{Id} \otimes \partial_t$ on $L(D^\times)$. Set

$$\mathrm{Lie}(L) = L(D^\times)/\mathrm{Im}\,\partial.$$

For $A \in L$, we use the notation $A_{[n]}$ for the image of $A \otimes t^n$ in $\mathrm{Lie}(L)$. Then $\mathrm{Lie}(L)$ is a completion of the span of $A_{[n]}, A \in L, n \in \mathbb{Z}$, subject to the relations $(TA)_{[n]} = -nA_{[n-1]}$. Let $\mathrm{Lie}(L)_+$ be the subspace of $\mathrm{Lie}(L)$ which is the completion of the span of all $A_{[n]}$ with $n \geq 0$. Then we have the following analogue of Theorem 4.1.2.

16.1.7. Lemma. $\mathrm{Lie}(L)$ *is a Lie algebra, with the Lie bracket defined by the formula*

$$(16.1.2) \qquad [A_{[m]}, B_{[k]}] = \sum_{n \geq 0} \binom{m}{n} (A_{(n)} \cdot B)_{[m+k-n]},$$

and $\mathrm{Lie}(L)_+$ *is a Lie subalgebra of* $\mathrm{Lie}(L)$.

The map $\mathrm{Lie}(L)_+ \to \mathrm{End}\,L$, *sending* $A_{[n]}$ *to* $A_{(n)}$ *for* $n \geq 0$, *is a homomorphism of Lie algebras.*

16.1.8. Proof. We follow the same argument as in the proof of Theorem 4.1.2. First, we show that $\mathrm{Lie}(L)_0 = L/\mathrm{Im}\,T$ is a Lie algebra with respect to the Lie bracket

$$[A_{[0]}, B_{[0]}] = (A_{(0)} \cdot B)_{[0]}.$$

Next, we define the structure of a vertex Lie algebra on $L \otimes \mathbb{C}[t, t^{-1}]$ by making T act as $T \otimes 1 + \mathrm{Id} \otimes \partial_t$, and defining Y_- by the formula

$$(A \otimes f(t))_{(n)} \cdot (B \otimes g(t)) = \sum_{j \geq 0} \frac{1}{j!} (A_{(n+j)} B) \otimes (\partial_t^j f(t)) g(t).$$

It is straightforward to check that this indeed makes $L \otimes \mathbb{C}[t, t^{-1}]$ into a vertex Lie algebra. Moreover, $\mathrm{Lie}(L)$ is a completion of the Lie algebra $\mathrm{Lie}(L \otimes \mathbb{C}[t, t^{-1}])_0$. Since the Lie bracket is continuous, we obtain that $\mathrm{Lie}(L)$ is indeed a Lie algebra. Formulas (16.1.2) and (16.1.1) imply the remaining assertions of the lemma.

We call $\mathrm{Lie}(L)$ the local Lie algebra associated to L. The simplest examples of these Lie algebras are given by the following lemma.

16.1.9. Lemma. *The local Lie algebras $\mathrm{Lie}(vir_c)$ and $\mathrm{Lie}(v_k(\mathfrak{g}))$ for the Virasoro and Kac–Moody vertex Lie algebras are the Virasoro and Kac–Moody Lie algebras, respectively. The Lie subalgebras $\mathrm{Lie}(vir_c)_+$ and $\mathrm{Lie}(v_k(\mathfrak{g}))_+$ are equal to $\mathrm{Der}\,\mathcal{O}$ and $\mathfrak{g}(\mathcal{O})$, respectively.*

16.1.10. Proof. Using the relations $(TA)_{[n]} = -nA_{[n-1]}$, we obtain that the Lie algebra $\mathrm{Lie}(vir_c)$ (resp., $\mathrm{Lie}(vir_c)_+$) is spanned by $L_n = (L_{-2}v_c)_{[n+1]}, n \in \mathbb{Z}$, and $C = (v_c)_{[-1]}$ (resp., $L_n = (L_{-2}v_c)_{[n+1]}, n \geq -1$). The commutation relations between them follow from formula (16.1.2) as explained in § 3.4.3. The Kac–Moody case is proved in the same way.

16.1.11. Enveloping vertex algebras. Let L be a vertex Lie algebra. We wish to define a vertex algebra which will serve as its universal enveloping algebra. More precisely, we wish to construct a left adjoint to the functor sending a vertex algebra to its polar part. The definition imitates the construction of the vertex algebra structures on the induced representations of Heisenberg, Kac–Moody and Virasoro algebras (in these cases the corresponding vertex algebras are the enveloping vertex algebras for the respective vertex Lie algebras).

Denote by $U(L)$ and $U(L)_+$ the universal enveloping algebras of the Lie algebras $\mathrm{Lie}(L)$ and $\mathrm{Lie}(L)_+$, respectively. Let $\mathbb{C}$ denote the trivial one–dimensional representation of $\mathrm{Lie}(L)_+$, and set

$$Vac(L) = U(L) \otimes_{U(L)_+} \mathbb{C}.$$

This is a $U(L)$–module. Denote the generating vector $1 \otimes 1 \in Vac(L)$ by $|0\rangle$.

16.1.12. Proposition. *There is a vertex algebra structure on $Vac(L)$ such that $Y(A_{[-1]}|0\rangle, z) = \sum_{n \in \mathbb{Z}} A_{[n]} z^{-n-1}$ for all $A \in L$. Moreover, for any vertex algebra V, there is a canonical isomorphism*

$$\mathrm{Hom}(L, V_{\mathrm{Lie}}) \simeq \mathrm{Hom}(Vac(L), V),$$

where V_{Lie} denotes V considered as a vertex Lie algebra.

16.1.13. Proof. We take $|0\rangle$ as the vacuum vector of $Vac(L)$. The translation operator T on $Vac(L)$ is defined by the formulas $T|0\rangle = 0, [T, A_{[n]}] = -nA_{[n-1]}$. By the Poincaré–Birkhoff–Witt theorem, $Vac_\lambda(L)$ has a basis of monomials in $A_{[n]}, n < 0$. We define

$$Y(A_{[-1]}|0\rangle, z) = \sum_{n \in \mathbb{Z}} A_{[n]} z^{-n-1},$$

where the $A_{[n]}$'s act on $Vac_\lambda(L)$ via the $U(L)$–module structure. The commutation relations (16.1.2) imply that these fields are mutually local. By the Reconstruction Theorem 2.3.11, we extend these data uniquely to a vertex algebra structure on $Vac_\lambda(L)$.

Given a homomorphism $\rho : L \to V_{\mathrm{Lie}}$, we extend it to a map $Vac(L) \to V_{\mathrm{Lie}}$ sending PBW monomials $A^{(1)}_{[n_1]} \ldots A^{(k)}_{[n_k]}|0\rangle$, where $A^{(i)} \in L, n_i < 0, i = 1, \ldots, k$, to $\rho(A^{(1)})_{[n_1]} \ldots \rho(A^{(k)})_{[n_k]}|0\rangle \in V$. We leave it to the reader to check that this map is a homomorphism of vertex algebras. Conversely, given a homomorphism $Vac(L) \to V$, its restriction to $L \subset Vac(L)$, spanned by the vectors $A_{[-1]}|0\rangle, A \in L$, is a homomorphism of vertex Lie algebras. It is clear that the corresponding maps between $\mathrm{Hom}(L, V_{\mathrm{Lie}})$ and $\mathrm{Hom}(Vac(L), V)$ are mutually inverse isomorphisms. This completes the proof.

Note that if L is $\mathbb{Z}$–graded, then the induced $\mathbb{Z}$–grading on $Vac(L)$ makes it a $\mathbb{Z}$–graded vertex algebra.

16.1.14. Definition. A *conformal algebra* (with central charge c) is a $\mathbb{Z}$–graded vertex Lie algebra L equipped with a non-trivial homomorphism of vertex Lie algebras $vir_c \to L$ such that the corresponding operators L_0 and L_{-1} on L coincide with the gradation operator and the translation operator T, respectively.

In this case L contains a non-zero vector ω (the image of $L_{-2}|0\rangle \in vir_c$) such that the Fourier coefficients of $Y_-(\omega, z)$ satisfy the relations of the Lie algebra $\mathrm{Der}(\mathcal{O})$.

16.1.15. Vector bundles attached to vertex Lie algebras. Since a conformal algebra L carries an action of $\mathrm{Der}\,\mathcal{O}$, by twisting L with $\mathcal{A}ut_X$ we automatically obtain a vector bundle $\mathcal{L}$ with a flat connection on an arbitrary smooth curve X (see § 6.5). Moreover, the proof of Theorem 6.5.4 can be applied in the vertex Lie algebra context, by taking polar parts everywhere. However, instead of obtaining an operator–valued meromorphic section of $\mathcal{L}^*$ on a disc, we only obtain its polar part (i.e., a meromorphic section, taken modulo regular sections).

For a sheaf $\mathcal{F}$ on X and $x \in X$, denote by $\mathcal{F}_{\mathcal{K}_x}$ and $\mathcal{F}_{\mathcal{O}_x}$ the spaces of sections of $\mathcal{F}$ over $D_x^\times$ and D_x, respectively. Then we have a perfect pairing between $\mathcal{L}^*_{\mathcal{K}_x}/\mathcal{L}^*_{\mathcal{O}_x}$ and $\mathcal{L}^r_{\mathcal{O}_x} \overset{\text{def}}{=} \mathcal{L}_{\mathcal{O}_x} \otimes \Omega_{\mathcal{O}_x}$.

16.1.16. Proposition. *Let L be a vertex Lie algebra, X a smooth curve and x a point of X. Then Y_- gives rise to a well–defined element*

$$\mathcal{Y}_{-,x} \in (\mathrm{End}\,\mathcal{L}_x)\widehat{\otimes}\mathcal{L}^*_{\mathcal{K}_x}/\mathcal{L}^*_{\mathcal{O}_x},$$

i.e., for any $v \in \mathcal{L}_x, \varphi \in \mathcal{L}^*_x$, we have a well-defined element $\varphi(\mathcal{Y}_{-,x}\cdot v)$ of $\mathcal{L}^*_{\mathcal{K}_x}/\mathcal{L}^*_{\mathcal{O}_x}$. Equivalently, we have a map

$$\mathcal{Y}^\vee_{-,x} : \mathcal{L}^r_{\mathcal{O}_x} \to \mathrm{End}\,\mathcal{L}_x.$$

16.2. Vertex Poisson algebras

Recall the notion of commutative vertex algebra from § 1.4. The operation Y in a commutative vertex algebra V is equivalent to an operation of commutative product $\circ$ on V, so that $A \circ B = A_{(-1)} \cdot B$.

16.2.1. Definition. A vertex Poisson algebra is a quintuple $(V, |0\rangle, T, Y_+, Y_-)$, where $(V, |0\rangle, T, Y_+)$ form a commutative vertex algebra, and (V, T, Y_-) form a vertex Lie algebra, so that all Fourier coefficients of $Y_-(A, z), A \in V$, are derivations of the commutative product on V induced by Y_+.

16.2.2. Example. Let L be a vertex Lie algebra. Set $V = \operatorname{Sym} L$, the symmetric algebra of the vector space L, with its natural commutative product induced by Y_+. We take $|0\rangle$ to be the unit element of V. The operator T is uniquely extended from L to V by the Leibniz rule and the requirement that $T|0\rangle = 0$. Similarly, the operation Y_- on V is uniquely characterized by requiring $L \to V$ to be a vertex Lie algebra homomorphism, and by the derivation requirement on its Fourier coefficients. Then V is a vertex Poisson algebra.

In the case of the vertex Lie algebras vir_c and $v_k(\mathfrak{g})$, we introduce a modified version of the symmetric algebra by identifying the unit element of the symmetric algebra with the vacuum vector in the vertex Lie algebra:

$$
\begin{aligned}
\operatorname{Sym}'(vir_c) &= \operatorname{Sym}(vir_c)/(1 - v_c), \\
\operatorname{Sym}'(v_k(\mathfrak{g})) &= \operatorname{Sym}(v_k(\mathfrak{g}))/(1 - v_k).
\end{aligned}
$$

One checks easily that these quotients inherit the vertex Poisson algebra structure. As a commutative algebra, $\operatorname{Sym}'(vir_c)$ is the algebra of polynomials in $L_n, n < -1$, and $\operatorname{Sym}'(v_k(\mathfrak{g}))$ is the algebra of polynomials in $J_r^a, r < 0$.

16.2.3. Limits. Suppose that V is vector space on which we have a one-parameter family of vertex operations $Y^\epsilon, \epsilon \in \mathbb{C}$, depending on ϵ polynomially. We may then view $V^\epsilon = V \otimes \mathbb{C}[\epsilon]$ with the operation Y^ϵ as a vertex algebra over the ring $\mathbb{C}[\epsilon]$. It is straightforward to generalize the notion of vertex algebra so as to allow it to be a module over a commutative $\mathbb{C}$–algebra such that the translation operator T and the vertex operation Y commute with the action of R. Then for any ideal I in R, V/I is a vertex algebra over R/I.

Suppose that the vertex algebra $V^0 = V^\epsilon/\epsilon V^\epsilon$ over $\mathbb{C}$ is commutative. We wish to study the extra structure acquired by V^0 from its vertex algebra deformation V^ϵ – specifically, how the polar part of V^ϵ, taken to the first order in ϵ, is reflected in V^0. Thus we may restrict our deformation to its first order part, and assume that we are dealing with a vertex algebra V^ϵ which is a flat module over $\mathbb{C}[\epsilon]/(\epsilon^2)$.

16.2.4. Proposition. *Let V^ϵ be a vertex algebra over $\mathbb{C}[\epsilon]/(\epsilon^2)$ with vertex operation Y^ϵ such that the vertex algebra $V^0 = V^\epsilon/\epsilon V^\epsilon$ is commutative. Then V^0 naturally acquires the structure of a vertex Poisson algebra.*

16.2.5. Proof. We take the commutative vertex structure $Y^0 = Y^\epsilon \bmod \epsilon$ on V^0 as the operator Y_+ of the vertex Poisson structure. The polar part $Y_-^\epsilon(A, z)$ of the vertex structure on V^ϵ vanishes modulo ϵ by our hypothesis. Therefore we "renormalize" it to obtain the polar operator on V^0:

$$
Y_-(A, z) = \frac{1}{\epsilon} Y_-^\epsilon(\widetilde{A}, z) \quad \bmod \epsilon,
$$

where for $A \in V^0$ we choose an arbitrary lift $\widetilde{A} \in V^\epsilon$ such that $A = \widetilde{A} \bmod \epsilon$. The vanishing of the polar part Y_-^ϵ modulo ϵ implies that the above definition is independent of the choice of the lift $\widetilde{A}$. Furthermore, Y_- automatically satisfies the axioms of a vertex Lie algebra.

Thus it remains to check that the coefficients of Y_- are derivations of the commutative product $\circ$ induced by Y_+. We need to show that

$$
(16.2.1) \qquad A_{(m)}(B \circ C) = (A_{(m)}B) \circ C + B \circ (A_{(m)}C), \quad m \geq 0,
$$

for any $C \in V^0$. Recall that $B \circ C = B_{(-1)} \circ C$. Thus (16.2.1) is equivalent to

$$(A_{(m)}B)_{(-1)} = [A_{(m)}, B_{(-1)}], \qquad m \geq 0,$$

and in turn to

(16.2.2)
$$\frac{1}{\epsilon}(\widetilde{A}_{(m)}\widetilde{B})_{(-1)} = \frac{1}{\epsilon}[\widetilde{A}_{(m)}, \widetilde{B}_{(-1)}] \quad (\mathrm{mod}\,\epsilon),$$

where $\widetilde{A}, \widetilde{B} \in V^\epsilon$ are arbitrary lifts of $A, B \in V^0$. By formula (3.3.12),

$$\frac{1}{\epsilon}[\widetilde{A}_{(m)}, \widetilde{B}_{(-1)}] = \frac{1}{\epsilon} \sum_{0 \leq n \leq m} \binom{m}{n} (\widetilde{A}_{(n)}B)_{(m-1-n)} \qquad m \geq 0.$$

However, the terms with $n < m$ vanish modulo ϵ, and the only remaining term with $n = m$ is equal to the left hand side of (16.2.2). The proposition follows.

Note that, conversely, any vertex Poisson algebra V gives rise to a vertex algebra V^ϵ over $\mathbb{C}[\epsilon]/(\epsilon^2)$ such that $V^\epsilon/\epsilon \cdot V^\epsilon \simeq V$.

16.2.6. Generalization. Proposition 16.2.4 may be extended to the situation when the vertex structure on V^ϵ does not become commutative at $\epsilon = 0$, but V^0 contains a non-trivial center. Recall that in § 5.7.2 we have defined the center $\mathcal{Z}(V)$ of a vertex algebra V as the span of all vectors $v \in V$ such that $Y_-(A, z) \cdot v = 0$ for all $A \in V$. The commutation relations formula (3.3.12) then implies that for any $v \in \mathcal{Z}(V)$ the Fourier coefficients of $Y(v, z)$ commute with the Fourier coefficients of $Y(A, w)$ for any $A \in V$. Therefore $Y_-(v, z)$ acts by 0 on V; in particular, $\mathcal{Z}(V)$ is a commutative vertex algebra.

In our case, we obtain that $\mathcal{Z}(V^0)$ is a commutative vertex algebra, and in addition $Y_-^\epsilon(v, z)$ vanishes modulo ϵ for all $v \in \mathcal{Z}(V)$. Then, in the same way that we proved Proposition 16.2.4, we obtain

16.2.7. Proposition. *Let V^ϵ be a vertex algebra over $\mathbb{C}[\epsilon]/(\epsilon^2)$. Then the center $\mathcal{Z}(V^0)$ of $V^0 = V^\epsilon/\epsilon V^\epsilon$ carries a canonical structure of a vertex Poisson algebra.*

The center of any vertex algebra contains the one-dimensional subspace spanned by the vacuum vector. However, in order to obtain a non-trivial vertex Poisson algebra, the center of V^0 should contain other vectors. An example of such a vertex Poisson algebra is the center of the vertex algebra $V_{-h^\vee}(\mathfrak{g})$ (see § 18.4).

16.3. Kac–Moody and Virasoro limits

The construction of Proposition 16.2.4 identifies the additional structure carried by a commutative vertex algebra when it arises as the limit of a family of vertex algebras. We would like to apply this to describe the behavior of the affine Kac–Moody vertex algebra $V_k(\mathfrak{g})$ when the level k approaches ∞. Since the affine vertex algebra is *a priori* only defined for finite k, we must first make sense of the limit as $k \to \infty$.

This can be done as follows. Let V_k denote a family of vertex algebras with the same underlying vector space, with all structures depending on a parameter k in a polynomial way. Then we can construct a free $\mathbb{C}[\mathbf{k}, \mathbf{k}^{-1}]$–module $V_\mathbf{k}$, equipped with the structure of a vertex algebra over $\mathbb{C}[\mathbf{k}, \mathbf{k}^{-1}]$. Its specialization to $\mathbf{k} = k \in \mathbb{C}^\times$, namely $V_\mathbf{k}/((\mathbf{k} - k) \cdot V_\mathbf{k})$, is isomorphic to V_k for all $k \in \mathbb{C}^\times$. Next, we need to specify a $\mathbb{C}[\mathbf{k}^{-1}]$–lattice in $V_\mathbf{k}$, i.e., a free $\mathbb{C}[\mathbf{k}^{-1}]$–submodule $V'_\mathbf{k}$ in $V_\mathbf{k}$, compatible with the vertex algebra structure (i.e., it should contain the vacuum vector and be preserved by Y and T), such that $V_{\mathbf{k}} = V'_{\mathbf{k}} \otimes_{\mathbb{C}[\mathbf{k}^{-1}]} \mathbb{C}[\mathbf{k}, \mathbf{k}^{-1}]$. We then define the limit of V_k at $k = \infty$ as $V'_{\mathbf{k}}/(\mathbf{k}^{-1} \cdot V'_{\mathbf{k}})$. In general, there may exist different lattices $V'_{\mathbf{k}}$, and the corresponding limits may also be different.

Consider the Kac–Moody vertex algebra $V_k(\mathfrak{g})$. The underlying vector space is isomorphic to $U(\mathfrak{g} \otimes t^{-1}\mathbb{C}[t^{-1}])$, and hence is independent of k. Inspecting the formulas defining the vertex algebra structure on $V_k(\mathfrak{g})$, we see that they are polynomial in k. Hence $V_k(\mathfrak{g})$ gives rise to a vertex algebra over $\mathbb{C}[\mathbf{k}, \mathbf{k}^{-1}]$, which we denote by $V_{\mathbf{k}}(\mathfrak{g})$. In order to define a limit of $V_k(\mathfrak{g})$ as $k \to \infty$, we need to pick a $\mathbb{C}[\mathbf{k}^{-1}]$–lattice $V'_{\mathbf{k}}(\mathfrak{g}) \subset V_{\mathbf{k}}(\mathfrak{g})$ compatible with the vertex algebra structure, and set $V_\infty(\mathfrak{g}) = V'_{\mathbf{k}}(\mathfrak{g})/(\mathbf{k}^{-1} \cdot V'_{\mathbf{k}}(\mathfrak{g}))$.

Fix a basis $\{J^a\}$ of $\mathfrak{g}$, and recall from § 2.4.4 that $V_k(\mathfrak{g})$ has a basis of lexicographically ordered PBW monomials $J^{a_1}_{n_1} \ldots J^{a_m}_{n_m} v_k$, where $n_i < 0$.

16.3.1. Definition. Let $V'_{\mathbf{k}}(\mathfrak{g})$ be the $\mathbb{C}[\mathbf{k}^{-1}]$–lattice spanned by the rescaled PBW monomials

$$(16.3.1) \qquad \mathbf{k}^{-m} J^{a_1}_{n_1} \ldots J^{a_m}_{n_m} v_k = \overline{J}^{a_1}_{n_1} \ldots \overline{J}^{a_m}_{n_m} v_k,$$

where $\overline{J}^a_m = \mathbf{k}^{-1} J^a_m$. The resulting limit at $k = \infty$ is denoted by $V_\infty(\mathfrak{g})$.

16.3.2. Theorem. *The vector space $V_\infty(\mathfrak{g})$ has a natural vertex Poisson algebra structure, and is isomorphic to $\mathrm{Sym}'(v_1(\mathfrak{g}))$.*

16.3.3. Proof. We have to establish that the lattice $V'_{\mathbf{k}}(\mathfrak{g})$ is compatible with the vertex algebra structure on $v_k(\mathfrak{g})$, and that $V_\infty(\mathfrak{g})$ (which then acquires a vertex algebra structure) is a commutative vertex algebra. The vertex Poisson structure will then be obtained from Proposition 16.2.4, where we take $\mathbf{k}^{-1}$ as ϵ.

By construction, the vacuum vector v_k belongs to $V'_{\mathbf{k}}(\mathfrak{g})$, and $V'_{\mathbf{k}}(\mathfrak{g})$ is preserved by the translation operator T. Next, the commutation relations between the generators J^a_n imply that all monomials of the form $\overline{J}^{a_1}_{n_1} \ldots \overline{J}^{a_m}_{n_m} v_k$, and not only the lexicographically ordered ones, belong to the lattice $V'_{\mathbf{k}}(\mathfrak{g})$.

Now we check that the vertex operator $Y(\overline{J}^a_{-1}, z) = \mathbf{k}^{-1} J^a(z)$ preserves the lattice $V'_{\mathbf{k}}(\mathfrak{g})$, and its polar Fourier coefficients are all divisible by $\mathbf{k}^{-1}$ (as linear operators on $V'_{\mathbf{k}}(\mathfrak{g})$). Consider the action of the operator $\overline{J}^a_n$ on an expression $\mathbf{k}^{-N}\mu_N$ in V', where μ_N is a monomial of degree N in the generators $J^a_r, r < 0$. The result is a linear combination over $\mathbb{C}$ of terms of three kinds:

1. $\mathbf{k}^{-N-1}$ times a monomial of degree N arising from the commutator of J^a_n with a generator in μ_N;
2. $\mathbf{k}^{-N}$ times a monomial of degree $N - 1$ arising from a nontrivial pairing (J^a, J^b) via the Kac–Moody cocycle;
3. (only if $N < 0$) $\mathbf{k}^{-N-1}$ times a monomial of degree $N + 1$ obtained from appending the creation operator $J^a_n, n < 0$, to the monomial μ_N.

All three types of terms are in the lattice $V'_{\mathbf{k}}(\mathfrak{g})$, but only the terms of the third type survive modulo $\mathbf{k}^{-1}V'_{\mathbf{k}}(\mathfrak{g})$. It follows that the operator $\mathbf{k}^{-1}J^a(z)$ is well defined on $V_\infty(\mathfrak{g})$, and that its polar part vanishes.

By the Reconstruction Theorem, the general vertex operators on $V_k(\mathfrak{g})$ are obtained from the operators $J^a(z)$ using differentiation and normally ordered product. It is easy to see that both of these operations preserve the set of fields well–defined at $\mathbf{k}^{-1} = 0$ and with vanishing polar part at $\mathbf{k}^{-1} = 0$. Hence the vertex operation on $V'_{\mathbf{k}}(\mathfrak{g})$ satisfies the conditions of Proposition 16.2.4, and we obtain a vertex Poisson algebra structure on $V_\infty(\mathfrak{g})$.

The isomorphism of $V_\infty(\mathfrak{g})$ with $\mathrm{Sym}'(v_1(\mathfrak{g}))$ as a commutative vertex algebra follows from the fact that, modulo $\mathbf{k}^{-1}$, the monomials (16.3.1) remain unchanged under any permutation. Therefore we identify $V_\infty(\mathfrak{g})$ with the polynomial algebra on $\overline{J}^a_r, r < 0$, so that the operators $\overline{J}^a_r, r < 0$, act on $V_\infty(\mathfrak{g})$ as multiplication operators. Hence we obtain an isomorphism of commutative vertex algebras $V_\infty(\mathfrak{g}) \to \mathrm{Sym}'(v_1(\mathfrak{g}))$ sending monomials in $\overline{J}^a_r, r < 0$, in the former to the corresponding monomials in J^a_r in the latter.

To verify that this is also an isomorphism of vertex Lie algebras, it is enough to check the relations on the generators. The commutation relations of the operators $\mathbf{k}^{-1}J^a_r$ in V read

$$[\mathbf{k}^{-1}J^a_r, \mathbf{k}^{-1}J^b_s] = \mathbf{k}^{-2}\left([J^a, J^b]_{r+s} + \mathbf{k}(J^a, J^b)\delta_{r,-s}\right),$$

so that

$$[\overline{J}^a_r, \overline{J}^b_s] = \mathbf{k}^{-1}(\overline{[J^a, J^b]}_{r+s} + (J^a, J^b)\delta_{r,-s}).$$

From this relation we obtain that the commutator vanishes modulo $\mathbf{k}^{-1}$, and that the formula for the action of $\overline{J}^a_r, r \geq 0$, on $V_\infty(\mathfrak{g})$ agrees with that for the action of $J^a_r, r \geq 0$, on $v_1(\mathfrak{g})$. The theorem follows.

16.3.4. Changing the level. The above construction is not the only possible way to take the level k to ∞. By introducing a parameter in the above lattice we may obtain in the same way the affine vertex Poisson algebra at any non-zero level λ. Namely, let $V^\lambda_{\mathbf{k}}(\mathfrak{g}) \subset V_{\mathbf{k}}(\mathfrak{g})$ be the $\mathbb{C}[\mathbf{k}^{-1}]$–lattice spanned by the monomials $\left(\frac{\lambda}{\mathbf{k}}\right)^m J^{a_1}_{n_1} \ldots J^{a_m}_{n_m} v_k = \overline{J}^{a_1}_{n_1} \ldots \overline{J}^{a_m}_{n_m} v_k$, where $\overline{J}^a_m = \frac{\lambda}{\mathbf{k}} J^a_m$. The resulting limit at $k = \infty$ is denoted by $V^\lambda_\infty(\mathfrak{g})$. Then taking $\lambda/\mathbf{k}$ as the parameter of deformation ϵ, we obtain in the limit as $\lambda/\mathbf{k} \to 0$ the vertex Poisson algebra $\mathrm{Sym}'(v_\lambda(\mathfrak{g}))$.

16.3.5. The Virasoro case. The above results carry over essentially unchanged to the Virasoro setting. The family of Virasoro vertex algebras Vir_c gives rise to a $\mathbb{C}[\mathbf{c}, \mathbf{c}^{-1}]$–module $\mathrm{Vir}_\mathbf{c}$ with the basis of PBW monomials $L_{n_1} \ldots L_{n_m} v_c$, where $n_i < -1$. Then we define for $\lambda \neq 0$ a $\mathbb{C}[\mathbf{c}^{-1}]$–lattice $\mathrm{Vir}^\lambda_\mathbf{c} \subset \mathrm{Vir}_\mathbf{c}$ spanned by the monomials $\left(\frac{\lambda}{\mathbf{c}}\right)^m L_{n_1} \ldots L_{n_m} v_c = \overline{L}_{n_1} \ldots \overline{L}_{n_m} v_c$, where $\overline{L}_m = \frac{\lambda}{\mathbf{c}} L_m$. Choose $\lambda/\mathbf{c}$ as the deformation parameter ϵ, and consider the corresponding limit

$$\mathrm{Vir}^\lambda_\infty \overset{\mathrm{def}}{=} \mathrm{Vir}^\lambda_\mathbf{c} /((\lambda/\mathbf{c}) \cdot \mathrm{Vir}^\lambda_\mathbf{c}).$$

By Proposition 16.2.4, this is a vertex Poisson algebra. In the same way as in the Kac–Moody case, we show that Vir^λ_∞ is isomorphic to $\mathrm{Sym}'(vir_\lambda)$.

16.4. Poisson structure on connections

In this section we present a geometric approach to the classical limits of the affine and Virasoro algebras. It is based on the identification of their dual spaces with certain spaces of connections on the punctured disc.

16.4.1. Kirillov–Kostant Poisson structures. Let $\mathfrak{g}$ be a finite-dimensional Lie algebra. Consider $\mathfrak{g}$ as the space of linear functions on the dual space $\mathfrak{g}^*$. Then the Lie bracket on $\mathfrak{g}$ extends uniquely to a Poisson bracket $\{\cdot, \cdot\}$ on the ring of polynomial functions $\mathbb{C}[\mathfrak{g}^*]$ on $\mathfrak{g}^*$ by the Leibniz rule: $\{f, gh\} = \{f, g\}h + \{f, h\}g$.

The algebra of functions $\mathbb{C}[\mathfrak{g}^*]$ is isomorphic to the symmetric algebra $\operatorname{Sym} \mathfrak{g}$, and the Poisson bracket on it may also be described as the unique Poisson bracket on $\operatorname{Sym} \mathfrak{g}$ restricting to the Lie bracket on $\mathfrak{g} \subset \operatorname{Sym} \mathfrak{g}$. This Poisson structure can be obtained by degenerating the associative algebra structure on $U(\mathfrak{g})$. Namely, introduce the family of associative algebras $U_\epsilon(\mathfrak{g})$ (where $\epsilon \in \mathbb{C}$) with generators $x \in \mathfrak{g}$ and relations $xy - yx = \epsilon[x, y]$. When $\epsilon = 1$, we recover the universal enveloping algebra $U(\mathfrak{g})$. When $\epsilon = 0$, we obtain the commutative symmetric algebra $\operatorname{Sym} \mathfrak{g}$, and the ϵ–linear term in the commutator on $U_\epsilon(\mathfrak{g})$ gives rise to the above Poisson structure on $\mathbb{C}[\mathfrak{g}^*]$ (compare this with Proposition 16.2.4, which shows that a vertex Poisson structure may be obtained by degeneration of a vertex algebra structure).

The Poisson structure on $\mathfrak{g}^*$ may also be described via the coadjoint action of $\mathfrak{g}$ on $\mathfrak{g}^*$, defined by the formula

$$(x \cdot f)(y) = f([x, y]), \qquad x, y \in \mathfrak{g}, f \in \mathfrak{g}^*.$$

Recall that geometrically, a Poisson structure on a manifold M may be described as a map $T^*M \to TM$, called the hamiltonian operator, satisfying certain conditions. In the case when $M = \mathfrak{g}$, the tangent space to $\mathfrak{g}^*$ at any $f \in \mathfrak{g}^*$ is naturally identified with $\mathfrak{g}^*$, while the cotangent space is identified with $\mathfrak{g}$. Then any covector $x \in \mathfrak{g} = T_f^* \mathfrak{g}^*$ gives rise to a tangent vector to $\mathfrak{g}^*$ at f, namely, the value of the coadjoint vector field of x at f. The resulting bundle map $T^*\mathfrak{g}^* \to T\mathfrak{g}^*$ satisfies the requirements of a Poisson structure, and the resulting Poisson structure on $\mathfrak{g}^*$ agrees with the one defined above.

16.4.2. Affine algebras and connections. We now consider the case of an affine Kac–Moody algebra $\widehat{\mathfrak{g}}$, viewed as a complete topological Lie algebra. Recall that $\widehat{\mathfrak{g}}$ was defined in § 2.4 as a central extension of the Lie algebra $L\mathfrak{g} = \mathfrak{g}((t))$ by a one-dimensional space $\mathbb{C}K$:

$$(16.4.1) \qquad 0 \to \mathbb{C}K \to \widehat{\mathfrak{g}} \to L\mathfrak{g} \to 0.$$

As a vector space, $\widehat{\mathfrak{g}} = \mathfrak{g}((t)) \oplus \mathbb{C}K$, with the commutation relations given by formula (2.4.2). In fact, these relations do not depend on the choice of the coordinate t, so in what follows we will understand by $L\mathfrak{g}$ the Lie algebra $\mathfrak{g} \otimes \mathcal{K}$, where $\mathcal{K}$ is the field of functions on an abstract punctured disc (e.g., $D_x^\times$, where x is a point of a smooth curve X). If we choose a coordinate t on the disc, we may identify $\mathcal{K}$ with the field of Laurent series $\mathbb{C}((t))$, and hence identify the loop algebra $L\mathfrak{g}$ with $\mathfrak{g}((t))$.

The (topological) dual space $\widehat{\mathfrak{g}}^*$ fits into the exact sequence

$$0 \to (L\mathfrak{g})^* \to \widehat{\mathfrak{g}}^* \to \mathbb{C}d \to 0,$$

where d is the generator of the dual space to the one-dimensional subspace $\mathbb{C}K$ of $\widehat{\mathfrak{g}}$ whose value on K equals 1 (the reason for this choice of notation will soon become clear). As before, the dual space $\widehat{\mathfrak{g}}^*$ is a Poisson manifold. Since K is a central element, the hyperplanes $\widehat{\mathfrak{g}}_k^* \subset \widehat{\mathfrak{g}}^*$, consisting of all functionals on $\widehat{\mathfrak{g}}$ taking a fixed value $k \in \mathbb{C}$ on K, are Poisson submanifolds. The complete topological ring $\operatorname{Fun} \widehat{\mathfrak{g}}_k^*$

of functions on $\widehat{\mathfrak{g}}_k^*$ is defined in the same way as in § 11.3.4. It carries a canonical Poisson structure.

The coadjoint action of $\widehat{\mathfrak{g}}$ on $\widehat{\mathfrak{g}}^*$ factors through an action of $L\mathfrak{g}$, since K acts trivially. Moreover, this action preserves the hyperplanes $\widehat{\mathfrak{g}}_k^*$. The cotangent space to the hyperplane $\widehat{\mathfrak{g}}_k^*$ at any point is naturally identified with the quotient of $\widehat{\mathfrak{g}}$ by $\mathbb{C}K$, namely $L\mathfrak{g}$. Therefore the coadjoint action of $L\mathfrak{g}$ on $\widehat{\mathfrak{g}}_k^*$ defines a map from the cotangent to the tangent bundles over $\widehat{\mathfrak{g}}_k^*$. In the same way as in the previous section, we can obtain the Poisson structure on $\widehat{\mathfrak{g}}_k^*$ from this coadjoint action.

The important fact is that the hyperplanes $\widehat{\mathfrak{g}}_k^*$ have an intrinsic geometric description. Let us identify the dual space $\mathfrak{g}^*$ with $\mathfrak{g}$ using the normalized invariant inner product on $\mathfrak{g}$. Using the perfect residue pairing between the space of functions $\mathcal{K}$ and the space of one-forms $\Omega_{\mathcal{K}} \cong \mathbb{C}((t))dt$ on the punctured disc, we now identify the dual space to $L\mathfrak{g}$ with the space of $\mathfrak{g}$–valued one–forms $\mathfrak{g} \otimes \Omega_{\mathcal{K}}$ on the punctured disc (note that this identification does not depend on the choice of the coordinate t).

For notational simplicity, let us assume that $k = 1$. Choose a vector space splitting $\widehat{\mathfrak{g}} \cong L\mathfrak{g} \oplus \mathbb{C}K$, and the corresponding splitting $\widehat{\mathfrak{g}}^* \cong (L\mathfrak{g})^* \oplus \mathbb{C}d$. Thus we may identify the hyperplane $\widehat{\mathfrak{g}}_1^*$ with the affine subspace $d + (L\mathfrak{g})^* \cong d + \mathfrak{g} \otimes \Omega_{\mathcal{K}}$ over $(L\mathfrak{g})^*$. We now calculate the coadjoint action of $L\mathfrak{g}$ on $\widehat{\mathfrak{g}}_1^*$: let $x \in L\mathfrak{g}$, $d + f \in \widehat{\mathfrak{g}}_1^*$, $\overline{y} = y_0 K + y \in \widehat{\mathfrak{g}}$ $(y \in L\mathfrak{g})$. Then

$$
\begin{aligned}
\langle x \cdot (d + f), \overline{y} \rangle &= \langle d + f, [x, y_0 K + y] \rangle \\
&= \langle d, -\operatorname{Res}(dx, y) \rangle + \langle f, [x, y] \rangle \\
&= \langle -dx + [x, f], \overline{y} \rangle,
\end{aligned}
$$

where $dx \in \mathfrak{g} \otimes \Omega_{\mathcal{K}}$ is the de Rham differential of x. So

$$
x \cdot (d + f) = d + [x, f] - dx = [x, d + f],
$$

where d is now interpreted as the de Rham differential. Thus, we see that the coadjoint action of $L\mathfrak{g}$ on $\widehat{\mathfrak{g}}_1^*$ coincides with the infinitesimal gauge action of $L\mathfrak{g}$ on the space $\mathcal{C}onn$ of connections on the trivial G–bundle on the punctured disc. This space is naturally an affine space for $\mathfrak{g} \otimes \Omega_{\mathcal{K}}$ (i.e., a $\mathfrak{g} \otimes \Omega_{\mathcal{K}}$–torsor, see § 6.4.6), because the difference of any two connections is a $\mathfrak{g}$–valued one–form. We may exponentiate this action to describe the coadjoint action of the ind-group $G(\mathcal{K})$ on $\widehat{\mathfrak{g}}_1^*$: for $g \in G(\mathcal{K})$, we have

$$
g : d + f \mapsto d - g^{-1}dg + gfg^{-1} = g(d + f)g^{-1},
$$

where $g^{-1}dg \in \mathfrak{g} \otimes \Omega_{\mathcal{K}}$ is the logarithmic differential. This is the formula for the gauge action of $G(\mathcal{K})$ on $\mathcal{C}onn$. Thus we have established:

16.4.3. Lemma. *The affine space $\widehat{\mathfrak{g}}_1^*$ is isomorphic to the space of connections on the trivial G–bundle on the punctured disc. This isomorphism is compatible with the structures of affine space over $\mathfrak{g} \otimes \Omega_{\mathcal{K}}$ and the actions of $G(\mathcal{K})$ on both sides.*

16.4.4. Relation with vertex Poisson algebras. Similarly, the vertex Poisson algebra $\operatorname{Sym}'(v_k(\mathfrak{g}))$ introduced in § 16.2.2 gets identified with the space of connections on the trivial G–bundle on the (unpunctured) disc. Since a vertex Poisson algebra carries in particular a vertex Lie algebra structure, we may assign to $\operatorname{Sym}'(v_k(\mathfrak{g}))$ a local Lie algebra on the punctured disc as in § 16.1.6. (Recall from Lemma 16.1.9 that the local Lie algebra $\operatorname{Lie}(v_k(\mathfrak{g}))$ is $\widehat{\mathfrak{g}}$.)

The local Lie algebra of $\mathrm{Sym}'(v_k(\mathfrak{g}))$ may be identified with the Poisson subalgebra of $\mathrm{Fun}(\widehat{\mathfrak{g}}_k^*)$ consisting of local functionals on $\widehat{\mathfrak{g}}_k^*$ (compare with § 12.2.1). Let us identify $\widehat{\mathfrak{g}}_k^*$ with $\mathfrak{g}((t))$ using a coordinate t. Local functionals are by definition the functionals on $\mathfrak{g}((t))$ whose value on any $f(t) \in \mathfrak{g}((t))$ is given by

$$\int P(f(t), f'(t), \dots)q(t)dt,$$

where P is a polynomial function on $\mathfrak{g}$, and $q(t) \in \mathbb{C}((t))$. Denote the space of local functionals on $\mathfrak{g}((t)) \simeq \widehat{\mathfrak{g}}_k^*$ by $\mathrm{Fun}(\widehat{\mathfrak{g}}_k^*)_{\mathrm{loc}}$. This is a subspace of $\mathrm{Fun}(\widehat{\mathfrak{g}}_k^*)$. Note that it does not depend on the choice of t. Moreover, it follows from the above description of the Poisson structure that $\mathrm{Fun}(\widehat{\mathfrak{g}}_k^*)_{\mathrm{loc}}$ is closed under the Poisson bracket inside $\mathrm{Fun}(\widehat{\mathfrak{g}}_k^*)$ (although not under multiplication). The next lemma follows directly from the definitions.

16.4.5. Lemma. *The Lie algebra $\mathrm{Fun}(\widehat{\mathfrak{g}}_k^*)_{\mathrm{loc}}$ (with respect to the Poisson bracket) is isomorphic to the local Lie algebra $\mathrm{Lie}(\mathrm{Sym}' v_k(\mathfrak{g}))$.*

16.4.6. Defining the Kac–Moody algebra. We can reverse the above process and use Lemma 16.4.3 to *define* the affine Kac–Moody algebra $\widehat{\mathfrak{g}}$.

Indeed, in order to specify a one–dimensional central extension $\widetilde{\mathfrak{g}}$ of a Lie algebra $\mathfrak{g}$, it suffices to specify an affine space A for $\mathfrak{g}^*$ together with an affine action of $\mathfrak{g}$ on A, i.e., one such that $x \cdot (a + f) = x \cdot a + x(f)$ for all $x \in \mathfrak{g}, a \in A, f \in \mathfrak{g}^*$. Then the affine structure on A gives a filtration $\mathrm{Fun}^i A, i \geq 0$, on the ring $\mathrm{Fun}\, A$ of regular functions on A. Introduce the algebra $\widetilde{A} = \bigoplus_{i \geq 0} \mathrm{Fun}^i A$ with the graded product $\mathrm{Fun}^i A \times \mathrm{Fun}^j A \to \mathrm{Fun}^{i+j} A$, and let $\widetilde{\mathfrak{g}}^* = \mathrm{Spec}\,\widetilde{A}$. The algebra $\widetilde{A}$ contains the subalgebra $\mathbb{C}[h]$ spanned by the unit elements in all $\mathrm{Fun}^i A$ (which we identify with h^i). The quotient of $\widetilde{A}$ by the ideal generated by h equals A. Therefore we have the exact sequence

$$0 \to \mathfrak{g}^* \to \widetilde{\mathfrak{g}}^* \to \mathbb{C} \to 0$$

such that the inverse image of $1 \in \mathbb{C}$ is identified with A, together with a linear action of $\mathfrak{g}$ on $\widetilde{\mathfrak{g}}^*$ restricting to the coadjoint action on $\mathfrak{g}^* \subset \widetilde{\mathfrak{g}}^*$ and the specified action on $A \subset \widetilde{\mathfrak{g}}^*$. Therefore the dual vector space

$$0 \to \mathbb{C} \to \widetilde{\mathfrak{g}} \to \mathfrak{g} \to 0$$

acquires a Lie algebra structure, so that $\widetilde{\mathfrak{g}}$ becomes a central extension of $\mathfrak{g}$. Applying this construction in the case of affine algebras, we obtain

16.4.7. Corollary. *The affine Kac-Moody algebra $\widehat{\mathfrak{g}}$ is characterized as the central extension of $L\mathfrak{g}$ by $\mathbb{C}K$ such that the hyperplane in $\widehat{\mathfrak{g}}^*$ of functionals taking the value 1 on $K \in \widehat{\mathfrak{g}}$ is $G(\mathcal{K})$–equivariantly identified with the space $\mathcal{C}onn$ of connections on the trivial bundle on $D^\times$.*

16.4.8. Tautological Poisson structure. The above corollary implies that the space of connections on the trivial G–bundle on $D^\times$ carries a Poisson structure. In fact, this structure has an elegant "tautological" interpretation. Let $\nabla \in \mathcal{C}onn$ be a connection. Since $\mathcal{C}onn$ is an affine space for $\mathfrak{g} \otimes \Omega_{\mathcal{K}}$, the tangent and cotangent spaces to $\mathcal{C}onn$ at ∇ are naturally identified with $\mathfrak{g} \otimes \Omega_K$ and $(\mathfrak{g} \otimes \Omega_{\mathcal{K}})^* \cong \mathfrak{g} \otimes \mathcal{K}$, respectively. In order to construct a Poisson structure on $\mathcal{C}onn$, we need to define the hamiltonian operator from $T_\nabla^* \mathcal{C}onn$ to $T_\nabla \mathcal{C}onn$. This operator is provided by the connection ∇. Namely, a connection on a G–bundle induces a connection on the adjoint bundle whose space of sections is precisely $\mathfrak{g} \otimes \mathcal{K}$, so that the adjoint connection operator is a map

$$\nabla_{\mathrm{ad}} : \mathfrak{g} \otimes \mathcal{K} \to \mathfrak{g} \otimes \Omega_{\mathcal{K}}.$$

Moreover, if $\nabla = d + f$ with d the trivial connection on the trivial bundle, then $\nabla_{\mathrm{ad}}(x) = dx + [f, x]$. Thus we have

16.4.9. Corollary. *The Poisson structure on $\mathcal{C}onn$ coming from the affine Kac–Moody algebra is "tautological" in the sense that the hamiltonian operator $T^*_{\nabla}\mathcal{C}onn \to T_{\nabla}\mathcal{C}onn$ is just the adjoint operator ∇_{ad}.*

16.4.10. General G–bundles. Thus, the tautological Poisson structure on the space $\mathcal{C}onn$ of connections on the trivial G–bundle on a punctured disc $D^{\times}$ can be used to define the affine Kac–Moody algebra $\widehat{\mathfrak{g}}$.

We now define a twisted version of $\widehat{\mathfrak{g}}$ by introducing a G–bundle $\mathcal{P}$ on the disc into the picture. Denote by $\mathcal{C}onn_{\mathcal{P}}$ the affine space of connections on $\mathcal{P}$. The tangent and contingent spaces to $\nabla \in \mathcal{C}onn_{\mathcal{P}}$ are then canonically isomorphic to $\mathfrak{g}_{\mathcal{P}} \otimes \Omega_{\mathcal{K}}$ and $\mathfrak{g}_{\mathcal{P}} \otimes \mathcal{K}$, respectively, where $\mathfrak{g}_{\mathcal{P}} = \mathcal{P} \underset{G}{\times} \mathfrak{g}$. Hence $\mathcal{C}onn_{\mathcal{P}}$ acquires a tautological Poisson structure, defined in the same way as in § 16.4.8. Namely, we take as the hamiltonian operator $T^*_{\nabla}\mathcal{C}onn_{\mathcal{P}} \to T_{\nabla}\mathcal{C}onn_{\mathcal{P}}$ the adjoint connection operator ∇_{ad}. If we trivialize $\mathcal{P}$, then $\mathcal{C}onn_{\mathcal{P}}$ can be identified with $\mathcal{C}onn$, and this operator becomes the operator considered above. This shows that this operator is indeed hamiltonian, and we obtain a well-defined Poisson structure on $\mathcal{C}onn_{\mathcal{P}}$ (which is independent of the trivialization).

The Lie algebra $L\mathfrak{g}_{\mathcal{P}}$ of sections of $\mathfrak{g}_{\mathcal{P}}$ on $D^{\times}$ naturally acts on $\mathcal{C}onn_{\mathcal{P}}$ by gauge transformations. Thus, according to the discussion of § 16.4.6 we obtain a canonical central extension of $L\mathfrak{g}_{\mathcal{P}}$, which we denote by $\widehat{\mathfrak{g}}^{\mathcal{P}}$. As a vector space, $\widehat{\mathfrak{g}}^{\mathcal{P}}$ is just the (topological) dual space to the space of all λ–connections on $\mathcal{P}$ (for all $\lambda \in \mathbb{C}$). Note that if $\mathcal{P}$ does not have a canonical trivialization, the affine space $\mathcal{C}onn_{\mathcal{P}}$ cannot be canonically identified with $\mathfrak{g} \otimes \Omega_{\mathcal{K}} \simeq (L\mathfrak{g})^*$. Therefore the central extension $\widehat{\mathfrak{g}}^{\mathcal{P}}$ does not have a canonical splitting as a vector space.

Alternatively, the Lie algebra $\widehat{\mathfrak{g}}^{\mathcal{P}}$ may be defined as follows. The standard extension (16.4.1), by which we defined $\widehat{\mathfrak{g}}$ in the first place, carries a natural action of the ind-group $\mathrm{Aut}\, \mathcal{O} \ltimes G(\mathcal{K})$, where $\mathrm{Aut}\, \mathcal{O}$ acts by changes of coordinates on $\mathfrak{g}((t))$ and identically on $\mathbb{C}K$, while $G(\mathcal{K})$ acts on $\widehat{\mathfrak{g}}$ via the adjoint action. Then $\widehat{\mathfrak{g}}^{\mathcal{P}}$ is nothing but the twist of $\widehat{\mathfrak{g}}$ by the $\mathrm{Aut}\, \mathcal{O} \ltimes G(\mathcal{K})$–torsor that consists of pairs (z, s), where z is a coordinate on the disc, and s is a trivialization of $\mathcal{P}$.

16.5. The Virasoro Poisson structure

The above descriptions of the Poisson structure on the dual space to an affine Kac–Moody algebra have analogues for the Virasoro algebra. Recall that the Virasoro algebra is the central extension

$$(16.5.1) \qquad 0 \to \mathbb{C}C \to Vir \to \mathrm{Der}\, \mathcal{K} \to 0.$$

It is split as a vector space, and the commutation relations are given by formula (2.5.1). However, in contrast to the case of affine algebras, this definition is coordinate-dependent, because in order for the the central extension term in (2.5.1) to make sense we have to fix an identification $\mathcal{K} \simeq \mathbb{C}((t))$.

The (topological) dual space Vir^* is a Poisson manifold with respect to the Kirillov–Kostant structure. For each $c \in \mathbb{C}$, the hyperplane Vir_c^* in the dual space to the Virasoro algebra, on which the central element C takes value c, is a Poisson submanifold of Vir^*. Therefore the (topological) space of functions on Vir_c^*, denoted by $\mathrm{Fun}(Vir_c^*)$, is a Poisson algebra. We will now interpret the Poisson structure on Vir_c^* using the coadjoint action of the Lie algebra $\mathrm{Der}\,\mathcal{K}$. Moreover, similarly to Corollary 16.4.7, we will describe the Virasoro central extension of $\mathrm{Der}\,\mathcal{K}$ geometrically in terms of this action, thus obtaining a coordinate-independent definition of the Virasoro algebra.

First note that the residue pairing identifies the dual to the space of vector fields $\mathrm{Der}\,\mathcal{K}$ on $D^\times$ with the space $\Omega_{\mathcal{K}}^2$ of quadratic differentials on $D^\times$ (expressions of the form $q(t)(dt)^2$). Hence the hyperplanes Vir_c^* are naturally affine spaces for the vector space $\Omega_{\mathcal{K}}^2$. Recall the notion of λ–projective connections that we have encountered in § 8.2. A λ–projective connection on $D^\times$ is by definition a self–adjoint second–order differential operator

$$L : \Omega_{\mathcal{K}}^{-\frac{1}{2}} \to \Omega_{\mathcal{K}}^{\frac{3}{2}},$$

with symbol $\lambda \in \mathbb{C}$. The set $\mathcal{P}roj_\lambda$ of all λ–projective connections on $D^\times$ is an affine space for $\Omega_{\mathcal{K}}^2$. Indeed, the difference of any two projective connections is an element of

$$\mathrm{Hom}_{\mathcal{O}}(\Omega_{\mathcal{K}}^{-\frac{1}{2}}, \Omega_{\mathcal{K}}^{\frac{3}{2}}) \cong \Omega_{\mathcal{K}}^2,$$

i.e., a quadratic differential on $D^\times$ (see § 8.2 for more details).

The space $\mathcal{P}roj_\lambda$ carries a natural action of the Lie algebra $\mathrm{Der}\,\mathcal{K}$ of infinitesimal automorphisms of $D^\times$. Indeed, $\mathrm{Der}\,\mathcal{K}$ acts on $\Omega_{\mathcal{K}}^{-\frac{1}{2}}$ and $\Omega_{\mathcal{K}}^{\frac{3}{2}}$, and hence on differential operators between them. Straightforward calculation of this action and comparison with the coadjoint action on Vir_c^* gives

16.5.1. Lemma. *The affine space Vir_c^* equipped with the action of $\mathrm{Der}\,\mathcal{K}$ is isomorphic to the affine space $\mathcal{P}roj_{\frac{c}{6}}$ of $\frac{c}{6}$–projective connections on $D^\times$ equipped with the action of $\mathrm{Der}\,\mathcal{K}$ by infinitesimal changes of coordinates.*

16.5.2. Corollary. *The Virasoro algebra Vir is uniquely characterized (in a coordinate-independent way) as the central extension of $\mathrm{Der}\,\mathcal{K}$ by $\mathbb{C}C$ such that the hyperplane Vir_c^* in Vir^* consisting of linear functionals taking value $c \in \mathbb{C}$ on C is $\mathrm{Der}\,\mathcal{K}$–equivariantly identified with the affine space $\mathcal{P}roj_{\frac{c}{6}}$.*

Analogously, the vertex Poisson algebra $\mathrm{Sym}'(vir_c(\mathfrak{g}))$ introduced in § 16.2.2 gets identified with the space of $\frac{c}{6}$–projective connections on the (unpunctured) disc.

In the same way as in § 16.4.4 we define the the subspace $\mathrm{Fun}(Vir_c^*)_{\mathrm{loc}}$ of $\mathrm{Fun}(Vir_c^*)$ consisting of local functionals on Vir_c^*. The next lemma follows directly from the definitions:

16.5.3. Lemma. *The subspace $\mathrm{Fun}(Vir_c^*)_{\mathrm{loc}}$ is closed under the Poisson bracket on $\mathrm{Fun}(Vir_c^*)$. The corresponding Lie algebra $\mathrm{Fun}(Vir_c^*)_{\mathrm{loc}}$ is isomorphic to the local Lie algebra $\mathrm{Lie}(\mathrm{Sym}'(vir_c))$.*

16.5.4. Tautological Poisson structure. Now we describe the Poisson structure on the space $\mathcal{P}roj_\lambda$ as a "tautological" structure. To simplify notation, we set $\lambda = 1$ and denote $\mathcal{P}roj_1$ simply by $\mathcal{P}roj$.

Let $L \in \mathcal{P}roj$ be a projective connection. Since $\mathcal{P}roj$ is an affine space for $\Omega^2_{\mathcal{K}}$, the tangent and cotangent spaces to $\mathcal{P}roj$ at L are naturally identified with $\Omega^2_{\mathcal{K}}$ and $(\Omega^2_K)^* \cong \Omega^{-1}_{\mathcal{K}}$, respectively. Thus, the desired hamiltonian operator should act from $\Omega^{-1}_{\mathcal{K}}$ to $\Omega^2_{\mathcal{K}}$. It can be defined as follows. We identify projective connections with special connections on PGL_2–principal bundles, namely, $\mathfrak{sl}_2$–opers, defined in § 8.2.7. Then we assign to L the associated connection on the vector bundle corresponding to the adjoint representation of PGL_2. This is a rank three vector bundle, which is a successive extension of Ω^{-1} by $\mathcal{O}$ by Ω. This flag and the adjoint connection satisfy the conditions of a GL_3–oper (see § 16.6 below). Hence we obtain a third–order operator $\Omega^{-1}_{\mathcal{K}} \to \Omega^2_{\mathcal{K}}$, which we denote by L_{ad}. Thus L_{ad} appears as an analogue, for the Virasoro algebra, of the "adjoint" connection operator ∇_{ad} which we used to define the tautological Poisson structure on $\mathcal{C}onn$.

Once the operator L_{ad} has been defined, it is straightforward to verify the following fact.

16.5.5. Lemma. *The Poisson structure on $\mathcal{P}roj$ coming from the Virasoro algebra is "tautological" in the sense that the hamiltonian operator $T^*_L \mathcal{P}roj \to T_L \mathcal{P}roj$ is the operator L_{ad}.*

16.6. Opers

Our objective now is to define the *classical $\mathcal{W}$–algebra* $\mathcal{W}_\infty(\mathfrak{g})$ associated to a simple Lie algebra $\mathfrak{g}$. This will be a vertex Poisson algebra generalizing the Virasoro vertex Poisson algebra $\mathrm{Sym}'(vir_{-6})$, which corresponds to $\mathfrak{g} = \mathfrak{sl}_2$. The classical $\mathcal{W}$–algebra $\mathcal{W}_\infty(\mathfrak{g})$ may be obtained by taking a suitable limit as $k \to \infty$ of the quantum $\mathcal{W}$–algebra $\mathcal{W}_k(\mathfrak{g})$ introduced in Chapter 15. It may also be identified with the space of functions on the space of $\mathfrak{g}$–*opers* on the disc. Therefore we give in this section a brief review of opers, following A. Beilinson and V. Drinfeld. After that we introduce the classical Drinfeld–Sokolov reduction, which may be viewed as a suitable $k \to \infty$ limit of the quantum Drinfeld-Sokolov reduction from Chapter 15.

16.6.1. Definition. A GL_n–*oper* on a smooth curve X is a rank n vector bundle $\mathcal{E}$ on X, equipped with

- a flag $0 \subset \mathcal{E}_1 \subset \cdots \subset \mathcal{E}_{n-1} \subset \mathcal{E}_n = \mathcal{E}$ of subbundles, and
- a connection $\nabla : \mathcal{E} \to \mathcal{E} \otimes \Omega$.

These data are required to satisfy

- $\nabla(\mathcal{E}_i) \subset \mathcal{E}_{i+1} \otimes \Omega$ (Griffiths transversality), and
- the induced maps $\mathcal{E}_i/\mathcal{E}_{i-1} \to (\mathcal{E}_{i+1}/\mathcal{E}_i) \otimes \Omega$ are isomorphisms for all i (non–degeneracy).

16.6.2. Explicit form. If we pick a local coordinate t and a local trivialization of $\mathcal{E}$, identifying the flag $\{\mathcal{E}_i\}$ with the standard flag in the trivial rank n bundle, the oper connection takes the form

$$(16.6.1) \qquad d + \begin{pmatrix} * & * & * & \cdots & * \\ + & * & * & \cdots & * \\ 0 & + & * & \cdots & * \\ \vdots & \ddots & \ddots & \ddots & \vdots \\ 0 & 0 & \cdots & + & * \end{pmatrix} dt,$$

where the $*$'s indicate arbitrary functions in t and the $+$'s indicate nowhere vanishing functions. However, we may identify our flag $\{\mathcal{E}_i\}$ with the standard flag in a different way, by multiplying the old identification by an invertible upper triangular matrix. This results in the gauge action of the group of such matrices on the connection operators (16.6.1). A straightforward calculation shows that any connection operator of the form (16.6.1) is gauge equivalent to a unique operator of the form

$$(16.6.2) \qquad \partial_t - \begin{pmatrix} q_1 & q_2 & q_3 & \cdots & q_n \\ 1 & 0 & 0 & \cdots & 0 \\ 0 & 1 & 0 & \cdots & 0 \\ \vdots & \ddots & \ddots & \cdots & \vdots \\ 0 & 0 & \cdots & 1 & 0 \end{pmatrix}.$$

Thus, we obtain a canonical form for the oper connection with respect to the chosen local coordinate t. Such an operator may be equivalently represented as an nth order differential operator L in one variable with principal symbol 1,

$$(16.6.3) \qquad \partial_t^n - q_1 \partial_t^{n-1} - q_2 \partial_t^{n-2} - \cdots - q_n,$$

acting from $\mathcal{E}_n/\mathcal{E}_{n-1}$ to $\mathcal{E}_1 \otimes \Omega$. Therefore GL_n–opers on the disc are equivalent to nth order differential operators of the form (16.6.3) acting from $\mathcal{E}_n/\mathcal{E}_{n-1}$ to $\mathcal{E}_1 \otimes \Omega$.

16.6.3. Opers for general reductive algebraic groups. Let G be a reductive algebraic group, B a Borel subgroup and $N = [B, B]$ its unipotent radical, with the corresponding Lie algebras $\mathfrak{n} \subset \mathfrak{b} \subset \mathfrak{g}$. There is an open B–orbit $\mathbf{O} \subset \mathfrak{n}^\perp/\mathfrak{b} \subset \mathfrak{g}/\mathfrak{b}$, consisting of vectors which are stabilized by the radical $N \subset B$, and such that all of their negative simple root components are non-zero. This orbit may also be described as the B–orbit of the sum of the projections of the generators f_i of the opposite nilpotent subalgebra $\mathfrak{n}_-$ onto $\mathfrak{g}/\mathfrak{b}$. In fact, the torus $H = B/N$ acts simply transitively on $\mathbf{O}$, so $\mathbf{O}$ is an H–torsor. In the case of GL_n (or SL_n), $\mathbf{O}$ can be identified with the set of matrices of the same form as those appearing in formula (16.6.1), modulo the upper triangular matrices. Recall that a GL_n–oper is a principal GL_n–bundle, equipped with a connection and a flag, or equivalently, a reduction to the Borel subgroup of upper triangular matrices. The oper condition on the connection may be described as requiring that in any local coordinate and any trivialization of $\mathcal{E}$ identifying the flag $(\mathcal{E}_i)$ with the standard flag in the trivial rank n bundle, the connection matrix lies in $\mathbf{O}$ modulo upper triangular matrices.

With this reformulation, we can give the definition of G–oper for any reductive algebraic group G: we consider principal G–bundles, equipped with a connection and a reduction to a Borel subgroup B of G, so that the connection has "relative position" $\mathbf{O}$ with respect to the flag.

More precisely, the failure of a connection ∇ on a principal G–bundle $\mathcal{E}$ to preserve a B–reduction $\mathcal{E}_B$ may be described in terms of a one-form with values in the twist $(\mathfrak{g}/\mathfrak{b})_{\mathcal{E}_B} = \mathcal{E}_B \underset{B}{\times} \mathfrak{g}/\mathfrak{b}$. Locally, choose any flat connection ∇' on $\mathcal{E}$ preserving $\mathcal{E}_B$, and take the difference $\nabla' - \nabla$. It is easy to show that the resulting local sections of $(\mathfrak{g}/\mathfrak{b})_{\mathcal{E}_B} \otimes \Omega$ are independent of ∇', and define a global $(\mathfrak{g}/\mathfrak{b})_{\mathcal{E}_B}$-valued one-form on X, denoted by $\nabla/\mathcal{E}_B$.

16.6.4. Definition. Let X be a smooth curve. A G–*oper* on X is a principal G–bundle $\mathcal{E}$ on X with a connection ∇ and a reduction $\mathcal{E}_B$ to B such that the one–form $\nabla/\mathcal{E}_B$ takes values in $\mathbf{O} \subset (\mathfrak{g}/\mathfrak{b})_{\mathcal{E}_B}$.

In other words, the connection ∇ "modulo the flag $\mathcal{E}_B$" has its only non-zero components in the subspaces $\mathbb{C}f_i \subset \mathfrak{n}_-$, and all of those components are nowhere vanishing.

A G–oper, where G is the *adjoint* group of a reductive algebraic algebra $\mathfrak{g}$, will also be called a $\mathfrak{g}$–*oper*. The space of $\mathfrak{g}$–opers on X will be denoted by $\mathrm{Op}_{\mathfrak{g}}(X)$.

16.6.5. $\mathfrak{sl}_2$–opers as projective connections. In § 8.2.7, we have already encountered $\mathfrak{sl}_2$–opers, which were defined as principal PGL_2–bundles $\mathcal{P}$, equipped with a reduction $\mathcal{P}_B$ to B and a connection which identifies the tangent bundle of X with $(\mathfrak{sl}_2/\mathfrak{b})_{\mathcal{P}_B}$. This is equivalent to requiring the connection one–form to be nowhere vanishing when considered modulo $\mathfrak{b}$, which is precisely the oper condition in Definition 16.6.1 for $G = PGL_2$, the adjoint group of $\mathfrak{sl}_2$. As shown in § 8.2.7, $\mathfrak{sl}_2$–opers are nothing but projective connections.

Recall that projective connections may be described as second order operators $\partial_t^2 - q$ acting from $\Omega^{-\frac{1}{2}}$ to $\Omega^{\frac{3}{2}}$. This description will be generalized in the next section to the case of PGL_n.

16.6.6. SL_n–opers vs. PGL_n–opers. The inclusion $SL_n \subset GL_n$ allows us to identify SL_n–opers with special GL_n–opers. Namely, in Definition 16.6.1 we additionally require that the underlying rank n vector bundle $\mathcal{E}$ has trivial determinant $\det \mathcal{E} \cong \mathcal{O}_X$, and that the connection ∇ is an SL_n–connection, i.e., such that it induces the trivial connection on the determinant line bundle $\det \mathcal{E}$. In the local form (16.6.2), this means that $q_1 = 0$. Thus, SL_n–opers correspond to nth order differential operators, with principal symbol 1 and vanishing subprincipal symbol, acting between appropriate line bundles on X.

The transversality condition on the connection allows us to describe these line bundles explicitly. Namely, the transversality condition of the connection gives us by induction the isomorphisms $\mathcal{E}_{i+1}/\mathcal{E}_i \cong \mathcal{E}_1 \otimes \Omega^{-i}$. In addition, for $\mathcal{E}$ to be an SL_n–oper, $\det \mathcal{E}$ must be isomorphic to $\mathcal{O}_X$, from which we derive that $\mathcal{E}_1 \cong \Omega^{\frac{n-1}{2}}$. This specifies $\mathcal{E}_1$ (and hence all other line bundles $\mathcal{E}_{i+1}/\mathcal{E}_i$) uniquely, up to a choice of the square root of the canonical bundle Ω (theta characteristic) if n is even.

For example, when $n = 2$ we obtain that $\mathcal{E}$ is a vector bundle which is an extension of the line bundles

$$0 \to \Omega^{\frac{1}{2}} \to \mathcal{E} \to \Omega^{-\frac{1}{2}} \to 0,$$

where $\Omega^{\frac{1}{2}}$ corresponds to a particular choice of the theta characteristic. For a projective curve X of genus $g \neq 1$, the bundle $\mathcal{E}$ is then the unique (up to an isomorphism) non-trivial extension of $\Omega^{-\frac{1}{2}}$ by $\Omega^{\frac{1}{2}}$. Indeed, $\mathrm{Ext}(\Omega^{-\frac{1}{2}}, \Omega^{\frac{1}{2}}) \simeq H^1(\Omega) = \mathbb{C}$. If $g \neq 1$, this extension cannot be split, for otherwise we obtain a connection on the line bundle $\Omega^{\frac{1}{2}}$ of non-zero degree, which is impossible. Therefore $\mathcal{E}$ must be a non-split extension of $\Omega^{-\frac{1}{2}}$ by $\Omega^{\frac{1}{2}}$, which is then unique up to an isomorphism. In fact, the bundle $\mathcal{E}$ may be identified with the bundle of one-jets of sections of the line bundle $\Omega^{-\frac{1}{2}}$.

In the same way as in at the end of § 16.6.2, we obtain that for each choice of theta characteristic, the oper connection gives rise to a second–order differential operator from $\Omega^{-\frac{1}{2}}$ to $\Omega^{\frac{3}{2}}$ with principal symbol 1 and vanishing subprincipal symbol, i.e., a projective connection. Thus, we obtain that the space of SL_2–opers has connected components, labeled by the theta characteristics, so it is a torsor over $H^1(X, \mathbb{Z}_2)$. Each connected component is isomorphic to the space of projective connections (so it is independent of the choice of the theta characteristic). The latter space can also be identified with the space of PGL_2–opers (or $\mathfrak{sl}_2$–opers in our terminology).

Similarly, one can show that the set of connected components of the space of G–opers over a curve X is a torsor over the group $H^1(X, Z(G))$, where $Z(G)$ is the center of G (for instance, $Z(SL_n) = \mathbb{Z}_n$). The connected components are isomorphic to each other and to the space of G_{ad}–opers on X, where $G_{\mathrm{ad}} = G/Z(G)$. If $G = SL_n$, then $G_{\mathrm{ad}} = PGL_n$. The space of PGL_n–opers (equivalently, $\mathfrak{sl}_n$–opers) on X may be identified with the space of nth order differential operators

$$\partial_t^n - q_2 \partial_t^{n-2} - \cdots - q_n \; : \quad \Omega^{\frac{-n+1}{2}} \to \Omega^{\frac{n+1}{2}},$$

which have principal symbol 1 and subprincipal symbol 0. The latter condition does not depend on the choice of the coordinate t, and furthermore, this space does not depend on the choice of theta characteristic (in case n is even).

For simple Lie groups of classical type, G–opers may also be identified explicitly with appropriate (pseudo)differential operators (see [**DS, BD2**]).

Furthermore, it is known that if G is the adjoint group of a simple Lie algebra $\mathfrak{g}$, then there are a unique G–bundle and a unique B–reduction $\mathcal{E}_B$ of $\mathcal{E}$ such that the pair $(\mathcal{E}, \mathcal{E}_B)$ admits an oper connection [**BD2, BD3**].

16.6.7. Opers on the disc. The definition of $\mathfrak{g}$–oper also makes sense when X is replaced by an abstract disc D (for instance, the disc D_x around a point $x \in X$). In this case, let us choose a coordinate t on D and a trivialization of the B–bundle $\mathcal{E}_B$ which is part of the oper structure. Then the connection ∇ takes the form $\nabla = d - p(t)dt - A(t)dt$, where $A(t) \in \mathfrak{b}(\mathcal{O})$, and $p(t) \in \mathbf{O}(\mathcal{O})$, the jet scheme of maps $D \to \mathbf{O}$ (see § 9.4.4). Denote the space of such connections by $\mathrm{Op}'_{\mathfrak{g}}(D)$. The only ambiguity in attaching an element of $\mathrm{Op}'_{\mathfrak{g}}(D)$ to a $\mathfrak{g}$–oper is in the choice of trivialization of $\mathcal{E}_B$. But the group $B(\mathcal{O})$ acts transitively on all such trivializations. This results in the gauge action of $B(\mathcal{O})$ on $\mathrm{Op}'_{\mathfrak{g}}(D)$, and we obtain a canonical isomorphism

$$\mathrm{Op}_{\mathfrak{g}}(D) \simeq \mathrm{Op}'_{\mathfrak{g}}(D)/B(\mathcal{O}).$$

Similarly, denote by $\mathrm{Op}'_{\mathfrak{g}}(D^\times)$ the space of all connections on the trivial G–bundle on $D^\times$ of the form $\nabla = d - p(t)dt - A(t)dt$, where $A(t) \in L\mathfrak{b}$, and $p(t) \in L\mathbf{O}$, the loop space of maps $D^\times \to \mathbf{O}$. Since $\mathbf{O}$ is an H–torsor, $L\mathbf{O} \simeq LH$. The group LB acts on this space by gauge transformations, and we obtain an isomorphism

$$\mathrm{Op}_{\mathfrak{g}}(D^\times) \simeq \mathrm{Op}'_{\mathfrak{g}}(D^\times)/LB.$$

16.7. Classical Drinfeld–Sokolov reduction

In this section we describe the space $\mathrm{Op}_{\mathfrak{g}}(D^{\times})$ of $\mathfrak{g}$–opers on the punctured disc as the result of a reduction of the Poisson manifold $\widehat{\mathfrak{g}}_1^*$. This construction, known as the classical Drinfeld–Sokolov reduction, is a special case of the general procedure of hamiltonian reduction. As a consequence, we obtain that the space of $\mathfrak{g}$–opers on $D^{\times}$ inherits a Poisson structure which underlies the classical $\mathcal{W}$–algebra.

16.7.1. Hamiltonian reduction. We provide a brief review of hamiltonian reduction. Let P be the Poisson algebra of functions on a Poisson manifold M. Let $I \subset P$ be the ideal, describing a submanifold $M_0 \subset M$ with the ring of functions M/I. Suppose that I is closed under the Poisson bracket, i.e. $\{I, I\} \subset I$. (Such I are referred to as *involutive*.) In general, an involutive ideal I is not a Poisson ideal; that is, $\{P, I\}$ is not necessarily contained in I. Hence we cannot define a Poisson bracket on the quotient P/I, and so M_0 is not naturally a Poisson submanifold of M.

However, since I is involutive, it follows that the algebra P/I inherits an action of I via the Poisson bracket. In fact, it is easy to see from the Leibniz identity for the Poisson bracket that the Poisson bracket with the square I^2 of the ideal I acts trivially on P/I. It follows that P/I carries an action of I/I^2 (which is a Lie algebra under $\{\cdot, \cdot\}$) by derivations. Geometrically, this means that the Lie algebra I/I^2 infinitesimally acts by (hamiltonian) vector fields on M_0. If $f, g \in P/I$ are invariant under the action of I/I^2, then so is the Poisson bracket of any of their lifts $\tilde{f}, \tilde{g}$ to P, and this bracket is furthermore independent of the lifts. Geometrically this means that the quotient of M_0 by the infinitesimal equivalence relation generated by the vector fields from I/I^2 is a Poisson manifold. To summarize:

16.7.2. Proposition–Definition. *Let P be a Poisson algebra and I an involutive ideal. Then the space $(P/I)^{I/I^2}$ of I/I^2–invariants on P/I carries the structure of a Poisson algebra, which is referred to as the* hamiltonian reduction of *P by I.*

16.7.3. Reduction by a hamiltonian group action. The action of a connected Lie group G on a Poisson manifold M is said to be hamiltonian if we are given a collection of functions f_i on M labeled by a set of basis elements of the Lie algebra $\mathfrak{g}$, whose hamiltonian vector fields $\{f_i, \cdot\}$ generate the corresponding infinitesimal action of the Lie algebra $\mathfrak{g} = \mathrm{Lie}\, G$. More concisely, the hamiltonian action is given by a moment map $\mu : M \to \mathfrak{g}^*$ whose components in the given basis are the functions f_i. The ideal I generated by the f_i's is an involutive ideal. The corresponding submanifold M_0 is $\mu^{-1}(0)$, and I/I^2 is just the Lie algebra $\mathfrak{g}$ acting on M_0 by vector fields. Therefore the hamiltonian reduction of the algebra of functions on M by I is the algebra of functions on the quotient $\mu^{-1}(0)/G$. In general, there may be a problem in defining this quotient, but in our main example below the action of G is free, so this problem does not arise.

More generally, we may take as M_0 the inverse image of an arbitrary coadjoint G–orbit $\mathrm{Orb} \subset \mathfrak{g}^*$ (it gives rise to a different involutive ideal I). The corresponding quotient $\mu^{-1}(\mathrm{Orb})/G$ is also a Poisson manifold. It is called the reduced Poisson manifold, and this construction is called the *hamiltonian reduction* of M by the action of G with respect to the coadjoint orbit Orb. The Poisson algebra of functions on the reduced Poisson manifold may be obtained as the 0th cohomology of the so-called *classical BRST complex* (see [**KS**]). If M is symplectic, then the reduced Poisson manifold is also symplectic, and in this case the construction is called the symplectic (or Marsden–Weinstein) reduction of M by G.

16.7.4. The classical Drinfeld–Sokolov reduction. We now take as our Poisson manifold the space $\widehat{\mathfrak{g}}_1^*$, i.e., the hyperplane in the dual space to the affine Kac–Moody algebra $\widehat{\mathfrak{g}}$ defined in § 16.4.2. As any coadjoint action, the action of $\widehat{\mathfrak{g}}$ on $\widehat{\mathfrak{g}}^*$, and hence on $\widehat{\mathfrak{g}}_1^*$, is hamiltonian, with the moment map being the natural inclusion $\widehat{\mathfrak{g}}_1^* \to \widehat{\mathfrak{g}}^*$. The action of any Lie subalgebra $\mathfrak{a} \subset \widehat{\mathfrak{g}}$ on $\widehat{\mathfrak{g}}_1^*$ is also hamiltonian, with the moment map being the natural projection $\widehat{\mathfrak{g}}_1^* \to \mathfrak{a}^*$.

Let $\mathfrak{b}_+ \subset \mathfrak{g}$ denote a fixed Borel subalgebra, and $\mathfrak{n}_+ = [\mathfrak{b}_+, \mathfrak{b}_+]$ its nilpotent radical. Let $L\mathfrak{n}_+ \subset L\mathfrak{b}_+ \subset L\mathfrak{g}$ denote the corresponding loop algebras (so $L\mathfrak{n}_+ = \mathfrak{n}_+ \otimes \mathbb{C}((t))$, etc.). The invariant form $(\cdot, \cdot)$ on $\mathfrak{g}$ vanishes on $\mathfrak{n}_+$, and hence the central extension $\widehat{\mathfrak{g}}$ splits over $L\mathfrak{n}_+$. We may thus consider $L\mathfrak{n}_+$ as a Lie subalgebra of $\widehat{\mathfrak{g}}$, and $(L\mathfrak{n}_+)^*$ as a quotient of $\widehat{\mathfrak{g}}_1^*$.

Choose a basis $\{e^\alpha\}_{\alpha \in \Delta_+}$ of root vectors in $\mathfrak{n}_+$ as before, where Δ_+ is the set of positive roots of $\mathfrak{g}$ (corresponding to the Borel subalgebra $\mathfrak{b}_+$). There is a distinguished functional $\chi \in (L\mathfrak{n}_+)^*$ such that $\chi(e^\alpha \otimes t^n) = -1$ if $\alpha = \alpha_i$ is a simple root and $n = -1$, and $\chi(e^\alpha \otimes t^n) = 0$ otherwise. The functional χ vanishes on $[L\mathfrak{n}_+, L\mathfrak{n}_+]$, and hence is invariant under the coadjoint action of the ind–group LN_+ corresponding to $L\mathfrak{n}_+$. So χ may be viewed as a one-point coadjoint orbit of LN_+.

Note that this χ differs from the functional χ introduced in § 15.1.7 by sign. The reason for the difference of signs is, roughly, the following. Imposing the constraints $x = \chi(x), x \in L\mathfrak{n}_+$, classically corresponds at the quantum level to considering the subspace of an $L\mathfrak{n}_+$–module on which $L\mathfrak{n}_+$ acts according to the character χ, i.e., the space of $L\mathfrak{n}_+$–invariants in $M \otimes \mathbb{C}_{-\chi}$. In our setting the space of invariants, which is the 0th cohomology, is replaced by the semi-infinite cohomology of $L\mathfrak{n}_+$ (see § 15.1.7).

The invariant bilinear form on $\mathfrak{g}$ induces an isomorphism of $\mathfrak{n}_+^*$ with the opposite nilpotent subalgebra $\mathfrak{n}_-$. Combining this with the residue pairing isomorphism $\mathcal{K}^* \cong \Omega_{\mathcal{K}}$, we identify $(L\mathfrak{n}_+)^*$ with $L\mathfrak{n}_- dt$. The element corresponding to χ is then the constant $\mathfrak{n}_-$–valued one–form

$$-p_- dt = -\sum_{i=1}^{\ell} \frac{(\alpha_i, \alpha_i)}{2} f_i\, dt,$$

where $p_- \in \mathfrak{n}_-$ is the element defined by formula (15.2.5). When $\mathfrak{g} = \mathfrak{sl}_n$, with $\mathfrak{n}_+$ and $\mathfrak{n}_-$ being the strictly upper (resp., lower) triangular matrices,

$$p_- dt = \begin{pmatrix} 0 & 0 & 0 & \dots & 0 \\ 1 & 0 & 0 & \dots & 0 \\ 0 & 1 & 0 & \dots & 0 \\ \vdots & \ddots & \ddots & \ddots & \vdots \\ 0 & 0 & \dots & 1 & 0 \end{pmatrix} dt.$$

Now we perform the hamiltonian reduction of the Poisson manifold $\widehat{\mathfrak{g}}_1^*$ by the hamiltonian action of LN_+ with respect to the LN_+–orbit $\chi \in (L\mathfrak{n}_+)^* \simeq L\mathfrak{n}_- dt$. This Poisson reduction is called the *classical Drinfeld–Sokolov reduction*.

Recall that in Lemma 16.4.3 we have identified $\widehat{\mathfrak{g}}_1^*$ with the space of connections on the trivial G–bundle on $D^\times$. We will write the connection operators as $d - A(t)dt$, with $A(t)dt \in L\mathfrak{g}dt$. The moment map for the action of LN_+ is the map $\widehat{\mathfrak{g}}_1^* \to L\mathfrak{n}_-dt$ which sends $d - A(t)dt$ to $-A(t)dt$ mod $L\mathfrak{b}_+dt$. Thus $\mu^{-1}(\chi)$ is identified with the space of connections of the form

$$d - p_-dt - A(t)dt, \qquad A(t) \in L\mathfrak{b}_+.$$

For $\mathfrak{g} = \mathfrak{sl}_n$ this is nothing but the space of operators of the form (16.6.1), with trace zero and with the subdiagonal entries normalized to be -1.

Recall that in § 15.4.3 we introduced a commutative Lie subalgebra $\mathfrak{a}_+$ of $\mathfrak{n}_+$, which is the kernel of $\operatorname{ad} p_+$ in $\mathfrak{n}_+$. The following lemma is proved in the same way as Lemma 15.4.4.

16.7.5. Lemma. *The action of LN_+ on $\mu^{-1}(\chi)$ is free. Moreover, for any choice of coordinate t, the resulting quotient $\mu^{-1}(\chi)/LN_+$ is isomorphic to the space of operators of the form $\partial_t - p_- - A(t), A(t) \in L\mathfrak{a}_+ = \mathfrak{a}_+ \otimes \mathcal{K}$, and hence to the space $L\mathfrak{a}_+$.*

Note that for $\mathfrak{g} = \mathfrak{sl}_n$ one can also identify $\mu^{-1}(\chi)/LN_+$ with the space of operators of the form (16.6.2) with $q_1 = 0$.

The quotient $\mu^{-1}(\chi)/LN_+$ is the reduced Poisson manifold obtained by the classical Drinfeld–Sokolov reduction.

16.7.6. Proposition. *The quotient $\mu^{-1}(\chi)/LN_+$ is canonically identified with the space $\operatorname{Op}_{\mathfrak{g}}(D^\times)$ of $\mathfrak{g}$–opers on the punctured disc. In particular, $\operatorname{Op}_{\mathfrak{g}}(D^\times)$ is a Poisson manifold.*

16.7.7. Proof. Recall that $\chi = -p_-dt \in L\mathfrak{n}_-dt \simeq (L\mathfrak{n}_+)^*$. Consider the LB_+–orbit $LB_+ \cdot (-p_-dt) \subset L\mathfrak{n}_-dt$. Let $\mathbf{O} = B_+ \cdot (-p_-)$. As explained in § 16.6.3, the action of B_+ on $\mathbf{O}$ factors through $H = B_+/N_+$, and $\mathbf{O}$ is an H–torsor. Therefore the LB_+–orbit of $-p_-dt$ is isomorphic to the loop space $L\mathbf{O}$ of $\mathbf{O}$, and the LH–action on it is simply transitive. Therefore we have an isomorphism

$$(16.7.1) \qquad \mu^{-1}(\chi)/LN_+ \cong \mu^{-1}(L\mathbf{O})/LB_+.$$

But the quotient on the right hand side is canonically identified with the space of $\mathfrak{g}$–opers on $D^\times$ according to the description of the latter given in § 16.6.7.

16.7.8. Remark: the action of $\operatorname{Aut}\mathcal{O}$. If we choose a coordinate t on $D^\times$ and combine the isomorphism of Lemma 16.7.5 with the isomorphism of Proposition 16.7.6, then we can identify the space of opers $\operatorname{Op}_{\mathfrak{g}}(D^\times)$ with the space $L\mathfrak{a}_+$. The natural action of the group $\operatorname{Aut}\mathcal{O}$ and the Lie algebra $\operatorname{Der}\mathcal{K}$ on $\operatorname{Op}_{\mathfrak{g}}(D^\times)$ then induces an action on $L\mathfrak{a}_+$. Namely, a change of variables from t to w such that $t = \phi(w)$ changes the connection operator by the formula

$$\partial_t - p_- - A(t) \mapsto \partial_w - \phi'(w)p_- - \phi'(w)A(\phi(w)).$$

Then we use the gauge action of LB_+ to transform this operator to the unique operator of the form $\partial_w - p_- - \widetilde{A}(w)$, where $\widetilde{A}(w) \in L\mathfrak{a}_+$. The action of $\phi \in \operatorname{Aut}\mathcal{O}$ is then given by $A(t) \mapsto \widetilde{A}(t)$.

Recall from § 15.4.3 that $\mathfrak{a}_+ = \operatorname{span}\{p_+^{(i)}\}_{i=1}^{\ell}$, where the $p_+^{(i)}$'s are elements of principal degree d_i. Therefore for a given coordinate t, a $\mathfrak{g}$–oper may be identified with an element $\sum_{i=1}^{\ell} \psi_i(t)p_+^{(i)} \subset L\mathfrak{a}_+$. Explicit computation shows that $\psi_1(t)$

transforms as a projective connection, and $\psi_i(t), i > 1$, transforms as a $(d_i + 1)$–differential. Therefore the space of opers $\mathrm{Op}_{\mathfrak{g}}(D^\times)$ is isomorphic to the product of the space $\mathcal{P}roj$ of projective connections on $D^\times$, and the direct sum $\bigoplus_{i=2}^{\ell} \Omega_{\mathcal{K}}^{d_i+1} \cdot p_+^{(i)}$.

The Poisson structure on $\mathrm{Op}_{\mathfrak{g}}(D^\times)$ obtained by the Drinfeld–Sokolov reduction of the Poisson structure on $\widehat{\mathfrak{g}}_1^*$ has a "tautological" description in the spirit of § 16.4.8 (see [**BD3**]).

16.8. Comparison of the classical and quantum Drinfeld–Sokolov reductions

In this section we identify the algebra of functions on $\mathrm{Op}_{\mathfrak{g}}(D)$ with the classical limit of the $\mathcal{W}$–algebra $\mathcal{W}_k(\mathfrak{g})$ as $k \to \infty$.

Recall the complex $C_k^\bullet(\mathfrak{g})$ used in the computation of the quantum Drinfeld–Sokolov reduction in Chapter 15. This is a vertex superalgebra which has a basis of lexicographically ordered monomials in $J_n^a, \psi_{\alpha,n}$, and $\psi_{\alpha,n}^*$, applied to the vacuum vector. Now we define a vertex superalgebra $C_\mathbf{k}^\bullet(\mathfrak{g})$ as a free module over the ring $\mathbb{C}[\mathbf{k}^{-1}]$ spanned by the lexicographically ordered monomials in $\mathbf{k}^{-1}J_n^a, \mathbf{k}^{-1}\psi_{\alpha,n}$, and $\psi_{\alpha,n}^*$, applied to the vacuum vector. It carries the differential $d_\mathbf{k} = d_{\mathrm{st}} + \mathbf{k}\chi$, which preserves the vertex superalgebra structure of $C_\mathbf{k}^\bullet(\mathfrak{g})$ (and commutes with the action of $\mathbb{C}[\mathbf{k}^{-1}]$). By construction, for any $k \neq 0$, the specialization of $C_\mathbf{k}^\bullet(\mathfrak{g})$ at $\mathbf{k} = k$, namely $C_\mathbf{k}^\bullet(\mathfrak{g})/(\mathbf{k}^{-1} - k^{-1})C_\mathbf{k}^\bullet(\mathfrak{g})$, is isomorphic to $C_k^\bullet(\mathfrak{g})$.

Denote by $\mathcal{W}_\mathbf{k}(\mathfrak{g})$ the 0th cohomology of $C_\mathbf{k}^\bullet(\mathfrak{g})$. This is a vertex algebra over $\mathbb{C}[\mathbf{k}^{-1}]$. The following lemma shows that the cohomologies of $C_k^\bullet(\mathfrak{g})$ "stay the same" for all finite values of k and for $k = \infty$.

16.8.1. Lemma. (1) *The ith cohomology of $C_\mathbf{k}^\bullet(\mathfrak{g})$ vanishes for $i \neq 0$, and its 0th cohomology $\mathcal{W}_\mathbf{k}(\mathfrak{g})$ is a free $\mathbb{C}[\mathbf{k}^{-1}]$–module.*

(2) *Set*
$$C_\infty^\bullet(\mathfrak{g}) \overset{\mathrm{def}}{=} C_\mathbf{k}^\bullet(\mathfrak{g})/\mathbf{k}^{-1} \cdot C_\mathbf{k}^\bullet(\mathfrak{g}).$$
Denote by d_∞ the differential on $C_\infty^\bullet(\mathfrak{g})$ induced by $d_\mathbf{k}$. Then the complex $C_\infty^\bullet(\mathfrak{g})$ is a vertex Poisson superalgebra, and the differential d_∞ preserves this structure.

(3) *The 0th cohomology of $C_\infty^\bullet(\mathfrak{g})$ with respect to d_∞ is isomorphic, as a vertex Poisson algebra, to*
$$\mathcal{W}_\infty(\mathfrak{g}) \overset{\mathrm{def}}{=} \mathcal{W}_\mathbf{k}(\mathfrak{g})/\mathbf{k}^{-1} \cdot \mathcal{W}_\mathbf{k}(\mathfrak{g}),$$
and all other cohomologies vanish.

16.8.2. Proof. Our proof is based on the following general fact. Suppose that we have a complex $C^\bullet$ which is a finitely generated free $\mathbb{C}[\mathbf{h}]$–module such that for any $h \in \mathbb{C}$, all cohomologies H_h^i of its specialization $C_h^\bullet = C^\bullet/(\mathbf{h} - h)C^\bullet$, except for the 0th cohomology H_h^0, vanish. Then all cohomologies of $C^\bullet$, except for the 0th cohomology, also vanish, and the 0th cohomology H^0 is a free $\mathbb{C}[\mathbf{h}]$–module such that $H_h^0 \simeq H^0/(\mathbf{h} - h)H^0$.

In order to prove the lemma, we replace the complex $C_\mathbf{k}^\bullet(\mathfrak{g})$ by its subcomplex $C_\mathbf{k}^\bullet(\mathfrak{g})_0$ introduced in § 15.4, which is a direct sum of complexes that are finitely generated free $\mathbb{C}[\mathbf{k}^{-1}]$–modules. We recall that the complex $C_\mathbf{k}^\bullet(\mathfrak{g})_0$ is spanned over $\mathbb{C}[\mathbf{k}^{-1}]$ by the lexicographically ordered monomials (15.4.1) in $\mathbf{k}^{-1}\widetilde{J_n^a}$ and $\psi_{\alpha,n}^*$, applied to the vacuum vector. This is a vertex subalgebra and a subcomplex of $C_\mathbf{k}^\bullet(\mathfrak{g})$.

In the same way as in § 15.2.1 we define another vertex subalgebra and subcomplex $C_{\mathbf{k}}^\bullet(\mathfrak{g})'$ of $C_{\mathbf{k}}^\bullet(\mathfrak{g})$. This is the free $\mathbb{C}[\mathbf{k}^{-1}]$–module with the basis of lexicographically ordered monomials in $\mathbf{k}^{-1}\widehat{J}_n^\alpha$ and $\mathbf{k}^{-1}\psi_{\alpha,n}$, applied to the vacuum vector. Moreover, $C_{\mathbf{k}}^\bullet(\mathfrak{g})$ is isomorphic to the tensor product of its subcomplexes $C_{\mathbf{k}}^\bullet(\mathfrak{g})_0$ and $C_{\mathbf{k}}^\bullet(\mathfrak{g})'$.

According to Lemma 15.2.7, the cohomology of $C_k^\bullet(\mathfrak{g})$ is isomorphic to the cohomology of its subcomplex $C_k^\bullet(\mathfrak{g})_0$ for all $k \in \mathbb{C}^\times$. In the same way we show that the cohomology of $C_{\mathbf{k}}^\bullet(\mathfrak{g})$ (resp., $C_\infty^\bullet(\mathfrak{g})$) is isomorphic to the cohomology of its subcomplex $C_{\mathbf{k}}^\bullet(\mathfrak{g})_0$ (resp., $C_\infty^\bullet(\mathfrak{g})_0$).

According to our proof of Theorem 15.1.9, all cohomologies of $C_k^\bullet(\mathfrak{g})_0$, except for the 0th cohomology, vanish for any $k \in \mathbb{C}^\times$. By Corollary 15.4.6, the same is true for the complex $C_\infty^\bullet(\mathfrak{g})_0$. Therefore we obtain that all cohomologies of $C_{\mathbf{k}}^\bullet(\mathfrak{g})_0$ (and hence of $C_{\mathbf{k}}^\bullet(\mathfrak{g})$), except for the 0th cohomology, also vanish, and its 0th cohomology $\mathcal{W}_{\mathbf{k}}(\mathfrak{g})$ is a free $\mathbb{C}[\mathbf{k}^{-1}]$–module. In addition, the 0th cohomology of $C_k^\bullet(\mathfrak{g})$, where $k \in \mathbb{C}^\times$ or $k = \infty$, is equal to the quotient of $\mathcal{W}_{\mathbf{k}}(\mathfrak{g})$ by the ideal generated by $(\mathbf{k}^{-1} - k^{-1})$, i.e., to $\mathcal{W}_k(\mathfrak{g})$.

Finally, the fact that $C_\infty^\bullet(\mathfrak{g})$ carries a vertex Poisson superalgebra structure is proved in the same way as in Theorem 16.3.2. Since $d_{\mathbf{k}}$ preserves the vertex structure on $C_{\mathbf{k}}^\bullet(\mathfrak{g})$, we obtain that d_∞ preserves the vertex Poisson structure on $C_\infty^\bullet(\mathfrak{g})$. This completes the proof.

16.8.3. Identification with opers. Now recall that by Corollary 15.4.6, $\mathcal{W}_\infty(\mathfrak{g})$ is isomorphic to the ring of functions on the quotient $\mathcal{C}(\mathfrak{g}) = \mathcal{C}'(\mathfrak{g}))/N_+[[t]]$. On the other hand, consider the space $\mathrm{Op}_{\mathfrak{g}}(D)$ of opers on the disc D. Using the description of $\mathrm{Op}_{\mathfrak{g}}(D)$ given in § 16.6.7 and the argument used in the proof of Proposition 16.7.6, we obtain that $\mathcal{C}(\mathfrak{g})$ is isomorphic to $\mathrm{Op}_{\mathfrak{g}}(D)$. Comparison of the Aut $\mathcal{O}$–actions then gives us the following result.

16.8.4. Proposition. *There is a canonical* Aut $\mathcal{O}$*–equivariant isomorphism*

$$\mathcal{W}_\infty(\mathfrak{g}) \simeq \mathrm{Fun}\,\mathrm{Op}_{\mathfrak{g}}(D).$$

In particular, the space $\mathrm{Fun}\,\mathrm{Op}_{\mathfrak{g}}(D)$ *carries a canonical vertex Poisson algebra structure.*

In the special case $\mathfrak{g} = \mathfrak{sl}_2$, the vertex Poisson algebra $\mathcal{W}_\infty(\mathfrak{sl}_2) \simeq \mathrm{Sym}'(vir_{-6})$ is identified with the ring of functions on the space of projective connections on the disc.

16.8.5. Remark. It is possible to define an analogue of hamiltonian reduction for vertex Poisson algebras. One can introduce the notion of hamiltonian action of a vertex Lie algebra on a vertex Poisson algebra and construct the corresponding reduced vertex Poisson algebra as the cohomology of a classical BRST complex. In the case of the Drinfeld–Sokolov reduction we have a hamiltonian action of the vertex Lie algebra $(\mathfrak{n}_+ \otimes t^{-1}\mathbb{C}[t^{-1}])v_1 \subset v_1(\mathfrak{g})$ on the vertex Poisson algebra $\mathrm{Sym}'(v_1(\mathfrak{g}))$. The corresponding BRST complex coincides with $C_\infty^\bullet(\mathfrak{g})$, and the reduced vertex Poisson algebra is the classical $\mathcal{W}$–algebra $\mathcal{W}_\infty(\mathfrak{g}) \simeq \mathrm{Fun}\,\mathrm{Op}_{\mathfrak{g}}(D)$.

16.8.6. Local Lie algebra. Let $\mathrm{Fun}(\mathrm{Op}_{\mathfrak{g}}(D^\times))_{\mathrm{loc}}$ be the space of local functionals on $\mathrm{Op}_{\mathfrak{g}}(D^\times)$. The Poisson structure on $\mathrm{Op}_{\mathfrak{g}}(D^\times)$ gives rise to a Lie bracket on $\mathrm{Fun}(\mathrm{Op}_{\mathfrak{g}}(D^\times))_{\mathrm{loc}}$.

On the other hand, consider the vertex Poisson algebra $\mathcal{W}_\infty(\mathfrak{g})$ and the corresponding local Lie algebra $\mathrm{Lie}(\mathcal{W}_\infty(\mathfrak{g}))$. By Theorem 16.3.2 and Lemma 16.4.5, we can identify the Lie algebra of local functionals on $\widehat{\mathfrak{g}}_1^*$ with the local Lie algebra of the vertex Poisson algebra obtained as the limit of $V_k(\mathfrak{g})$ as $k \to \infty$. This implies that the isomorphism $\mathcal{W}_\infty(\mathfrak{g}) \simeq \mathrm{Fun}\,\mathrm{Op}_\mathfrak{g}(D)$ gives rise to an isomorphism

$$\mathrm{Lie}(\mathcal{W}_\infty(\mathfrak{g})) \simeq \mathrm{Fun}(\mathrm{Op}_\mathfrak{g}(D^\times))_{\mathrm{loc}}$$

of Lie algebras.

Thus, $\mathcal{W}_k(\mathfrak{g})$ may be viewed as a "quantization" of the ring of functions on $\mathrm{Op}_\mathfrak{g}(D)$, while the Lie algebra $\mathrm{Lie}(\mathcal{W}_k(\mathfrak{g}))$ may be viewed as a "quantization" of the Lie algebra of local functionals on $\mathrm{Op}_\mathfrak{g}(D^\times)$, with k^{-1} playing the role of the quantization parameter.

16.9. Bibliographical notes

The notion of vertex Lie algebra was introduced by V. Kac [**Kac3**] (under the name conformal algebra), and independently by M. Primc [**Pr**]. The proof of Lemma 16.1.7 is borrowed from [**Kac3**] (it is also proved in [**Pr**] in a different way). Proposition 16.1.12 is proved in [**Pr**]. Conformal algebras have been extensively studied by V. Kac and collaborators (see [**Kac4, BKV, BDK**] and references therein).

Vertex Poisson algebras were introduced in [**EF**] in a slightly different form. The connection between vertex Poisson algebras and the hamiltonian formalism of soliton equations is explained in [**BD4, EF**] (see also [**FF8, F3, FF9**]).

The construction of vertex Poisson algebras as limits of vertex algebras given in §§ 16.2 and 16.3, and the identification of the classical limits of the $\mathcal{W}$–algebras given in § 16.8 follow [**FF7, F1**].

As we will see in Chapter 19, the notion of conformal algebra is equivalent to the universal version of a Lie* algebra in the sense of A. Beilinson and V. Drinfeld [**BD4**]. Similarly, vertex Poisson algebras with Virasoro structure correspond to universal versions of the Beilinson–Drinfeld *coisson* algebras [**BD4**].

The definition of opers is due to Beilinson and Drinfeld [**BD3, BD2**], who generalized an earlier definition of Drinfeld and Sokolov in [**DS**], where the classical Drinfeld–Sokolov reduction was originally introduced.

CHAPTER 17

Vertex Algebras and Moduli Spaces I

In Chapter 9 we associated the space of conformal blocks to a conformal vertex algebra V, a smooth projective curve X and a point x of X. Now we want to understand how these spaces (and their multi-point analogues) behave as we move x along X and vary the complex structure on X. More precisely, we wish to organize the spaces of conformal blocks into a sheaf on the moduli space $\mathfrak{M}_{g,1}$ of smooth pointed curves of genus g. Actually, for technical reasons we prefer to work with the spaces of coinvariants (also called covacua), which are dual to the spaces of conformal blocks (see § 9.2). We will also assume in this chapter and the next that $g > 1$.

One can construct in a straightforward way a quasi-coherent sheaf on $\mathfrak{M}_{g,1}$ whose fiber at (X, x) is the space of coinvariants $H(V, X, x)$. But in fact this sheaf also carries the structure of a (twisted) $\mathcal{D}$–module on $\mathfrak{M}_{g,1}$. Roughly speaking, this means that we can identify (the projectivizations of) the spaces of coinvariants attached to infinitesimally nearby points of $\mathfrak{M}_{g,1}$. The key fact used in the proof of this statement is the "Virasoro uniformization" of $\mathfrak{M}_{g,1}$. Namely, the Lie algebra $\mathrm{Der}\,\mathcal{K}$ acts transitively on the moduli space of triples (X, x, z), where (X, x) are as above and z is a formal coordinate at x; this action is obtained by "gluing together" X from the pieces D_x and $X \backslash x$ (see Theorem 17.3.2).

More generally, to a collection of n modules over a conformal vertex V one associates a (twisted) $\mathcal{D}$–module on the moduli space $\widetilde{\mathfrak{M}}_{g,n}$ of curves of genus g with n marked points and non-zero tangent vectors at those points, whose fibers are the corresponding spaces of coinvariants. In the case when the vertex algebra V is rational, the spaces of coinvariants are believed to be finite-dimensional. This implies that the corresponding twisted $\mathcal{D}$–module on $\widetilde{\mathfrak{M}}_{g,n}$ is isomorphic to the sheaf of sections of a vector bundle of finite rank equipped with a projectively flat connection. This picture was first suggested by D. Friedan and S. Shenker [**FrS**] in the context of conformal field theory. It is known to be true in the case of the vertex algebra $L_k(\mathfrak{g})$, the integrable representation of $\widehat{\mathfrak{g}}$ of level k, by the results of A. Tsuchiya, K. Ueno, and Y. Yamada [**TUY**] (see also [**Sor1**]), and in the case of the minimal model vertex algebra $L_{c(p,q)}$, the irreducible representation of the Virasoro algebra, by the results of A. Beilinson, B. Feigin, and B. Mazur [**BFM**] (see also [**BS**]).

In those cases the corresponding (twisted) $\mathcal{D}$–module on $\widetilde{\mathfrak{M}}_{g,n}$ can be extended to a sheaf on the Deligne-Mumford compactification $\overline{\mathfrak{M}}_{g,n}$, the moduli space of stable pointed curves (and this sheaf carries an action of the sheaf of logarithmic differential operators). The dimensions of the fibers of this sheaf at the boundary are equal to those inside $\widetilde{\mathfrak{M}}_{g,n}$. This allows one to compute these dimensions by the "Verlinde formula" from the dimensions of the spaces of coinvariants attached to $\mathbb{P}^1$

with three points. These dimensions (which may be viewed as structure constants of the so-called fusion algebra) can in turn be found from the matrix of the modular transformation $\tau \mapsto -1/\tau$ acting on the space of characters (see Theorem 5.5.3). The same pattern is believed to hold for other rational vertex algebras, but as far as we know this has not been proved in general.

We start this chapter by presenting a conceptual approach to the canonical connection ∇ on the vector bundle $\mathcal{V}$ attached to a conformal vertex algebra V. It was previously defined by an explicit formula in § 6.6. We then briefly recall the general formalism of localization of modules over Harish-Chandra pairs, due to Beilinson and Bernstein [**BB**]. Using this formalism, we explain the construction of the $\mathcal{D}$–module structure on the sheaf of coinvariants on the moduli space of curves. In the next chapter we will discuss the analogous construction in the case of the moduli space of bundles.

In this chapter and the next, by *moduli space* we will understand the corresponding moduli stack (in the smooth topology). All stacks that we consider are "good" in the sense of [**BD3**], Sect. 1, i.e., the dimension of the contangent stack equals twice the dimension of the stack. The theory of $\mathcal{D}$–modules on such stacks is developed in [**BD3**] (see also [**BB**]).

17.1. The flat connection on the vertex algebra bundle, revisited

In the definition of the vector bundle $\mathcal{V}$ on X attached to a conformal vertex algebra V we have used the action of the group $\operatorname{Aut}\mathcal{O}$ on V. In § 6.2.3 we saw that $\operatorname{Lie}(\operatorname{Aut}\mathcal{O}) \simeq \operatorname{Der}_0 \mathcal{O}$ is "missing" the vector field L_{-1} which shifts the base point. The larger Lie algebra $\operatorname{Der}\mathcal{O}$ also acts on V and its action provides an extra structure on $\mathcal{V}$, namely, a flat connection.

Recall the $\operatorname{Aut}\mathcal{O}$–bundle $\mathcal{A}ut_X$ over a smooth curve X defined in § 6.5. Consider a point $(x, t_x) \in \mathcal{A}ut_X$, where $x \in X$ and t_x is a formal coordinate at x. Observe that the tangent space to $\mathcal{A}ut_X$ at (x, t_x) is canonically isomorphic to $\operatorname{Der}\mathcal{O} = \mathbb{C}[[t]]\partial_t$. Indeed, a choice of formal coordinate at x gives us a trivialization of the restriction of $\mathcal{A}ut_X$ to the disc D_x around x as well as an isomorphism $D_x \simeq D = \operatorname{Spec}\mathbb{C}[[t]]$. Hence we identify the tangent space to (x, t_x) in $\mathcal{A}ut_X$ with the direct sum of the tangent space to the fiber of $\mathcal{A}ut_X$ at x and the tangent space to the disc D. The former is canonically isomorphic to $\operatorname{Der}_0 \mathcal{O} = t\mathbb{C}[[t]]\partial_t$, because by construction $\mathcal{A}ut_X$ is an $\operatorname{Aut}\mathcal{O}$–bundle. The latter is identified with the one-dimensional space $\mathbb{C}\partial_t$. Hence the total tangent space to (x, t_x) is isomorphic to $\operatorname{Der}_0 \mathcal{O} \oplus \mathbb{C}\partial_t = \operatorname{Der}\mathcal{O}$. Thus we obtain a map α from $\operatorname{Der}\mathcal{O}$ to the Lie algebra $\operatorname{Vect}(\mathcal{A}ut_X)$ of vector fields on $\mathcal{A}ut_X$ such that the restriction of α to $\operatorname{Der}_0 \mathcal{O}$ coincides with the map corresponding to the (right) fiberwise infinitesimal action of $\operatorname{Der}_0 \mathcal{O}$. Suppose that t'_x is another formal coordinate at x such that $t'_x = \rho(t_x)$ for some $\rho \in \operatorname{Aut}\mathcal{O}$. Then we have a linear map ρ_* from the tangent space to $(x, t_x) \in \mathcal{A}ut_X$ to the tangent space to $(x, t'_x) \in \mathcal{A}ut_X$. By construction, $\rho_*(\alpha(\partial_t)) = \alpha(\operatorname{Ad}\rho(\partial_t))$. This implies that the map α is an anti-homomorphism of Lie algebras (i.e., $\alpha([\xi, \eta]) = -[\alpha(\xi), \alpha(\eta)]$) which is compatible with the fiberwise (right) action of $\operatorname{Aut}\mathcal{O}$ on $\mathcal{A}ut_X$.

A Lie algebra $\mathfrak{g}$ is said to act *transitively* (respectively, *simply transitively*) on a variety S if the corresponding map $\mathfrak{g} \otimes \mathcal{O}_S \to \Theta_S$ sending $x \otimes f$ to $f\alpha(x)$, where $\alpha(x)$ is the vector field corresponding to x, is surjective (respectively, an isomorphism).

Equivalently, the map from $\mathfrak{g}$ to the tangent space of S at any point is surjective (respectively, an isomorphism). In our case we obtain

17.1.1. Lemma. *There exists a simply transitive action of the Lie algebra* $\operatorname{Der}\mathcal{O}$ *on* Aut_X *extending the fiberwise action of* $\operatorname{Der}_0\mathcal{O}$.

17.1.2. Construction of the flat connection on $\mathcal{V}$. Now let V be an arbitrary representation of $\operatorname{Der}\mathcal{O}$ with integral L_0–gradation, which is bounded from below. Then the action of $\operatorname{Der}_0\mathcal{O}$ on V exponentiates to an action of $\operatorname{Aut}\mathcal{O}$. Consider the trivial vector bundle $\operatorname{Aut}_X \times V$ over Aut_X. Using the $\operatorname{Aut}\mathcal{O}$–action on V, we obtain an $\operatorname{Aut}\mathcal{O}$–equivariant structure on $\operatorname{Aut}_X \times V$. But any $\operatorname{Aut}\mathcal{O}$–equivariant vector bundle on the $\operatorname{Aut}\mathcal{O}$–torsor Aut_X over X descends to a vector bundle on X. In our case, $\operatorname{Aut}_X \times V$ descends to the vertex algebra bundle $\mathcal{V} = \operatorname{Aut}_X \underset{\operatorname{Aut}\mathcal{O}}{\times} V$ defined in § 6.5.

Since the action of $\operatorname{Der}_0\mathcal{O}$ on V extends to an action of $\operatorname{Der}\mathcal{O}$, we can use the simply transitive (right) $\operatorname{Der}\mathcal{O}$–action on Aut_X to construct a flat connection on $\operatorname{Aut}_X \times V$. Namely, let ξ be an element of $\operatorname{Der}\mathcal{O}$, and let $\alpha(\xi)$ be the corresponding vector field on Aut_X. Define the operator $\nabla_{\alpha(\xi)} = \alpha(\xi) - \xi$ on the sections of the trivial bundle $\operatorname{Aut}_X \times V$ over Aut_X. Since the vector fields given by the action of $\operatorname{Der}\mathcal{O}$ on Aut_X generate the tangent sheaf of Aut_X, the operators $\nabla_{\alpha(\xi)}$ define a connection on the trivial bundle $\operatorname{Aut}_X \times V$. This connection is automatically flat, since it comes from a representation of the Lie algebra $\operatorname{Der}\mathcal{O}$ on V. Indeed, we have

$$\nabla_{[\alpha(\xi),\alpha(\eta)]} = \nabla_{-\alpha([\xi,\eta])} = -\alpha([\xi,\eta]) + [\xi,\eta] = [\nabla_{\alpha(\xi)}, \nabla_{\alpha(\eta)}].$$

Moreover, since the action of $\operatorname{Der}\mathcal{O}$ on $\operatorname{Aut}_X \times V$ extends the action of $\operatorname{Der}_0\mathcal{O} = \operatorname{Lie}(\operatorname{Aut}\mathcal{O})$ along the fibers, this connection is $\operatorname{Aut}\mathcal{O}$–equivariant. Thus we obtain that the $\operatorname{Aut}\mathcal{O}$–equivariant vector bundle $\operatorname{Aut}_X \times V$ on Aut_X carries an $\operatorname{Aut}\mathcal{O}$–equivariant flat connection. Therefore this bundle descends to a vector bundle on X (which is our bundle $\mathcal{V}$) with a flat connection. Thus, we obtain a flat connection on $\mathcal{V}$.

If we choose a formal coordinate z at a point $x \in X$, and denote by ∂_z the corresponding translation vector field on the disc D_x (i.e. $-L_{-1} \in \operatorname{Der}\mathcal{O}$), then we obtain from the above definition that the connection operator ∇_{∂_z} on $\mathcal{V}$ equals $\partial_z + L_{-1}$. Thus, we obtain a different proof of Theorem 6.6.3.

17.1.3. An alternative formulation. Another way to define the above connection is to exponentiate the $\operatorname{Der}\mathcal{O}$–action on Aut_X to the action of the ind–group $\underline{\operatorname{Aut}}\mathcal{O}$ (see § 6.2.3). The $\underline{\operatorname{Aut}}\mathcal{O}$–orbit of (x, t_x) is the inverse image of the formal disc $\widehat{D}_x \subset X$ in Aut_X. The translation action of the formal additive group from $\underline{\operatorname{Aut}}\mathcal{O}$ on $\widehat{D}_x$ allows us to identify the fibers of $\mathcal{V}$ at infinitesimally nearby points and to define a connection on X using Grothendieck's definition of connection as a flat structure (see § A.2.3).

To make this more precise, introduce the principal $\underline{\operatorname{Aut}}\mathcal{O}$–bundle $\underline{\operatorname{Aut}}_X$ over X associated to the $\operatorname{Aut}\mathcal{O}$–bundle Aut_X: $\underline{\operatorname{Aut}}_X = \underline{\operatorname{Aut}}\mathcal{O} \underset{\operatorname{Aut}\mathcal{O}}{\times} \operatorname{Aut}_X$. This bundle may be described as follows. Recall that for a $\mathbb{C}$–algebra R, the set of R–points of Aut_X is the set of pairs (x, t_x), where x is an R–point of X and t_x is a formal coordinate at x. By definition, a formal coordinate t_x at x is a topological generator of the maximal ideal $R\widehat{\otimes}\mathfrak{m}_x$ of $R\widehat{\otimes}\mathcal{O}_x$. In other words, we must have $R\widehat{\otimes}\mathfrak{m}_x \simeq t_x R[[t_x]]$ (see § 6.5).

We will say that an element t_x of $R \widehat{\otimes} \mathcal{O}_x$ is a topological generator of $R \widehat{\otimes} \mathcal{O}_x$, if we have a continuous isomorphism of algebras $R \widehat{\otimes} \mathcal{O}_x \simeq R[[t_x]]$. Then the set of R–points of $\underline{Aut}_X$ consists of pairs (x, t_x), where x is an R–point of X and t_x is a topological generator of $R \widehat{\otimes} \mathcal{O}_x$. Note that any topological generator of $R \widehat{\otimes} \mathcal{O}_x$ can be written as the sum $t'_x + n_x$, where t'_x is a topological generator of $R \widehat{\otimes} \mathfrak{m}_x$ and n_x is a nilpotent element of R.

17.1.4. Lemma. *The bundle $\underline{Aut}_X$ has a natural flat connection, which induces a flat connection on $\mathcal{V}$.*

17.1.5. Proof. The completed local rings at two infinitesimally nearby points x, y of X are equal to each other (see § A.2.3). Therefore it follows from the definition of $\underline{Aut}_X$ that the fibers of $\underline{Aut}_X$ at x and y are canonically identified (and these identifications are transitive). According to § A.2.3, transitive identifications of infinitesimally nearby fibers of a vector bundle give rise to a flat connection on this bundle.

Let V be as above, and observe that the actions of $\mathrm{Aut}\,\mathcal{O}$ and $\mathrm{Der}\,\mathcal{O}$ combine into an action of $\underline{Aut}\mathcal{O}$ on V, and on $\mathcal{V} = \underline{Aut}_X \underset{\mathrm{Aut}\mathcal{O}}{\times} V$. Therefore $\mathcal{V}$ inherits the flat connection from $\underline{Aut}_X$. This connection coincides with the connection defined above.

17.1.6. Multiple points. Let $\mathcal{Aut}_X^{(n)}$ be the scheme parameterizing collections $((x_i), (z_i))_{i=1}^n$, where $x_i \in X$ and z_i is a formal coordinate at x_i. This is an $(\mathrm{Aut}\,\mathcal{O})^n$–bundle over X^n. The nth exterior product $\mathcal{V}^{\boxtimes n}$ on X^n is the $\mathcal{Aut}_X^{(n)}$–twist of the $(\mathrm{Aut}\,\mathcal{O})^n$–module $V^{\otimes n}$.

In the same way as in § 17.1.2 we show that the fiberwise action of $(\mathrm{Der}_0\,\mathcal{O})^n$ on $\mathcal{Aut}_X^{(n)}$ extends to a simply transitive action of the Lie algebra $(\mathrm{Der}\,\mathcal{O})^n$. Therefore $\mathcal{V}^{\boxtimes n}$ acquires a flat connection. With respect to local coordinates z_i on X^n, the connection operators have the form $\nabla_i = \partial_{z_i} + L_{-1}^{(i)}$. This proves Proposition 7.3.4.

17.1.7. Connection on bundles corresponding to V–modules. If M is a conformal V–module on which L_0 has only integral eigenvalues, then it also gives rise to a vector bundle $\mathcal{M} = \mathcal{Aut}_X \underset{\mathrm{Aut}\,\mathcal{O}}{\times} M$ with a flat connection. An explicit formula for this connection has been given in § 7.3.8. In the same way as above we obtain that this formula is independent of the choice of coordinate z.

Now suppose that the eigenvalues of L_0 on M are non-integral. In that case, M gives rise to the vector bundle $\widetilde{\mathcal{M}} = \mathcal{Aut}_X \underset{\mathrm{Aut}_+\,\mathcal{O}}{\times} M$ on the $\mathbb{C}^\times$–bundle $\widetilde{X}$ over X defined in § 7.3.9. Recall that $\widetilde{X}$ is the space of pairs (x, τ), where $x \in X$ and τ is a non-zero tangent vector at x.

17.1.8. Lemma. *The vector bundle $\widetilde{\mathcal{M}}$ has a flat connection.*

17.1.9. Proof. Consider $\mathcal{Aut}_X$ as an $\mathrm{Aut}_+\,\mathcal{O}$–torsor over $\widetilde{X}$. From the results of § 17.1.2 we obtain that there is a simply transitive action of $\mathrm{Der}\,\mathcal{O}$ on $\mathcal{Aut}_X$ extending the fiberwise action of $\mathrm{Der}_+\,\mathcal{O}$. Following the same argument as in § 17.1.2, we obtain a flat connection on $\widetilde{\mathcal{M}}$.

Explicitly, this flat connection is given by the formula $d + L_{-1}^M dz + L_0^M dw$, where z is a coordinate along X, and w is the corresponding coordinate along the fiber of the $\mathbb{C}^\times$–bundle $\widetilde{X} \to X$.

17.1.10. Connection on Kac–Moody bundles. Let V be a conformal algebra with a compatible $\widehat{\mathfrak{g}}$–structure (see § 7.1.1). Then V carries an action of the Lie algebra $\mathrm{Der}_0\, \mathcal{O} \ltimes \mathfrak{g}(\mathcal{O})$. Suppose that it can be exponentiated to an action of the group $\mathrm{Aut}\, \mathcal{O} \ltimes G(\mathcal{O})$.

Let $\mathcal{P}$ be a G–bundle on X and $\widehat{P}$ the corresponding $\mathrm{Aut}\, \mathcal{O} \ltimes G(\mathcal{O})$–bundle over X as defined in § 7.1. Then $\widehat{\mathcal{P}}$ carries a simply transitive action of the Lie algebra $\mathrm{Der}\, \mathcal{O} \ltimes \mathfrak{g}(\mathcal{O})$ extending the fiberwise action of $\mathrm{Der}_0\, \mathcal{O} \ltimes \mathfrak{g}(\mathcal{O})$. In the same way as above, we obtain that the vector bundle $\mathcal{V}^{\mathcal{P}} = \widehat{\mathcal{P}} \underset{\mathrm{Aut}\, \mathcal{O} \times G(\mathcal{O})}{\times} V$ on X carries a flat connection.

17.2. Harish–Chandra pairs

The construction of the flat connection on the bundle $\mathcal{V}$ given in § 17.1.2 is a special case of a general formalism of localization of modules over Harish-Chandra pairs due to Beilinson and Bernstein [**BB**]. In this section we briefly outline this formalism, and in the next sections we apply it to construct sheaves of coinvariants on the moduli spaces of curves and bundles.

17.2.1. Definition. A *Harish–Chandra pair* is a pair $(\mathfrak{g}, K)$, where $\mathfrak{g}$ is a Lie algebra and K is a Lie group, equipped with the following data: an embedding $\mathfrak{k} \subset \mathfrak{g}$ of the Lie algebra $\mathfrak{k}$ of K into $\mathfrak{g}$, and an action Ad of K on $\mathfrak{g}$ compatible with the adjoint action of K on $\mathfrak{k} \subset \mathfrak{g}$ and the action of $\mathfrak{k}$ on $\mathfrak{g}$.

Let Z be a variety over $\mathbb{C}$. A $(\mathfrak{g}, K)$*–action* on Z is the data of an action of $\mathfrak{g}$ on Z (that is, a homomorphism α from $\mathfrak{g}$ to the tangent sheaf Θ_Z), together with an action of K on Z. The two actions are required to satisfy the following compatibility conditions:

(1) the differential of the K–action is the restriction of the $\mathfrak{g}$–action to $\mathfrak{k} = \mathrm{Lie}(K)$;

(2) the action α intertwines the representation of K on $\mathfrak{g}$ with the action of K on Θ_Z: $\alpha(Ad_k(a)) = k\alpha(a)k^{-1}$.

A $(\mathfrak{g}, K)$–action is called *transitive* if the map $\mathfrak{g} \otimes \mathcal{O}_Z \to \Theta_Z$ is surjective, and *simply transitive* if this map is an isomorphism.

A $(\mathfrak{g}, K)$*–structure* on a variety S is a principal K–bundle $\pi : Z \to S$ together with a simply transitive $(\mathfrak{g}, K)$–action on Z which extends the fibrewise action of K.

In what follows we will assume that K is connected.

17.2.2. Examples. An example of a variety with a $(\mathfrak{g}, K)$–structure is the homogeneous space $S = K \backslash G$, where G is a finite–dimensional Lie group, $\mathfrak{g} = \mathrm{Lie}(G)$, K is a Lie subgroup of G, and we take the obvious left action of $(\mathfrak{g}, K)$ on $Z = G$. In the finite–dimensional setting, any space with a $(\mathfrak{g}, K)$–structure is locally of this form.

However, in the infinite-dimensional setting there are more non-trivial examples, such as the $(\mathrm{Der}\, \mathcal{O}, \mathrm{Aut}\, \mathcal{O})$–structure on a smooth curve X given by the bundle $Z = \mathcal{A}ut_X \to X = S$, which we have used extensively throughout this book. This example can be generalized to an arbitrary N–dimensional smooth scheme S. Define $\widehat{S}$ to be the scheme of pairs $(x, \vec{t}_x)$, where $x \in S$ and $\vec{t}_x = (t_{1,x}, \ldots, t_{N,x})$ is a system of formal coordinates at x. Set $\mathfrak{g} = \mathrm{Der}\, \mathbb{C}[[z_1, \ldots, z_N]]$, $K = \mathrm{Aut}\, \mathbb{C}[[z_1, \ldots, z_N]]$. Then $\widehat{S} \to S$ is a $(\mathfrak{g}, K)$–structure on S, considered by Gelfand, Kazhdan and Fuchs, see [**GK, GKF, BR**]. This Harish-Chandra pair will be used in § 18.5 in the construction of the chiral de Rham complex.

17.2.3. Flat connection from a $(\mathfrak{g}, K)$–structure. Let V be a $(\mathfrak{g}, K)$–module, i.e., a vector space together with actions of $\mathfrak{g}$ and K satisfying an obvious compatibility condition. The construction of § 17.1.2 may be generalized to define a flat vector bundle $\mathcal{V}$ on any variety S with a $(\mathfrak{g}, K)$–structure $\pi : Z \to S$.

Namely, consider the trivial vector bundle $\widetilde{\mathcal{V}} = Z \times V$ on Z. The K–action on V gives $\widetilde{\mathcal{V}}$ the structure of a K–equivariant bundle. By definition, $\mathcal{V}$ is the corresponding vector bundle on S. In other words, $\mathcal{V}$ is the Z–twist of V, $\mathcal{V} = Z \underset{K}{\times} V$.

The action of $\mathfrak{g}$ on V gives us a homomorphism $\rho : \mathfrak{g} \to \mathrm{End}\, V$. In order to define a flat connection on $\mathcal{V}$, we first define a flat connection on the bundle $\widetilde{\mathcal{V}} = Z \times V$. The action of $\mathfrak{g}$ on V gives us a homomorphism of sheaves of Lie algebras from $\mathfrak{g} \otimes \mathcal{O}_Z$ to $\mathrm{End}\, V \otimes \mathcal{O}_Z$ (the sheaf of endomorphisms of the bundle $\widetilde{\mathcal{V}}$) defined by the formula

$$(17.2.1) \qquad x \otimes f \mapsto \rho(x) \otimes f + \mathrm{Id} \otimes (\alpha(x) \cdot f), \qquad x \in \mathfrak{g}, f \in \mathcal{O}_Z.$$

Since $\mathfrak{g} \otimes \mathcal{O}_Z \simeq \Theta_Z$, we obtain a Lie algebra homomorphism $\Theta_Z \to \mathrm{End}\, V \otimes \mathcal{O}_Z$ satisfying the Leibniz formula, i.e., a flat connection on $\widetilde{\mathcal{V}}$. Since the action of $\mathfrak{g}$ on Z is compatible with the fiberwise action of K, this flat connection on $\widetilde{\mathcal{V}}$ is K–equivariant, and hence descends to a flat connection on $\mathcal{V}$. In the special case of the $(\mathrm{Der}\,\mathcal{O}, \mathrm{Aut}\,\mathcal{O})$–structure $\mathcal{A}ut_X$ over a smooth curve X we obtain the flat connection from § 17.1.2.

Alternatively, the flat connection on $\mathcal{V}$ can be described as follows. Since the $\mathfrak{g}$–action on Z is simply transitive, we obtain an identification

$$(17.2.2) \qquad \Theta_S \simeq ((\mathfrak{g}/\mathfrak{k}) \otimes \pi_* \mathcal{O}_Z)^K.$$

Then, because $\mathfrak{g} \otimes \pi_* \mathcal{O}_Z$ acts on $V \otimes \pi_* \mathcal{O}_Z$, we obtain that $((\mathfrak{g}/\mathfrak{k}) \otimes \pi_* \mathcal{O}_Z)^K$, and hence Θ_S, acts on $(V \otimes \pi_* \mathcal{O}_Z)^K$, which is nothing but the sheaf of sections of $\mathcal{V}$ (considered as an $\mathcal{O}_S$–module). This action gives us the desired flat connection on $\mathcal{V}$.

17.2.4. Stabilizers. Now suppose that the $\mathfrak{g}$–action on the K–bundle $Z \to S$ is transitive, but *not* simply transitive. Then the above construction becomes more subtle. Given a $(\mathfrak{g}, K)$-module V, we can still construct the vector bundle $\mathcal{V}$ on S, with fibers isomorphic to V, in the same way as in the previous section. However, we cannot construct a flat connection on $\mathcal{V}$ because the map $\mathfrak{g} \otimes \mathcal{O}_Z \to \Theta_Z$ is no longer an isomorphism. In other words, we cannot express vector fields on Z uniquely as $\mathcal{O}_Z$–linear combinations of the vector fields coming from the $\mathfrak{g}$–action, because points of Z now have stabilizers in $\mathfrak{g}$.

At this point it is instructive to turn to the formalism of Lie algebroids (see § A.3.2). The action of $\mathfrak{g}$ on Z defines a Lie algebroid structure on $\mathfrak{g} \otimes \mathcal{O}_Z$, with the anchor map $a : \mathfrak{g} \otimes \mathcal{O}_Z \to \Theta_Z$. The map a is surjective by our assumption that $\mathfrak{g}$ acts transitively on Z. Given a $(\mathfrak{g}, K)$–module V, we have an action of $\mathfrak{g} \otimes \mathcal{O}_Z$ on the sheaf $V \otimes \mathcal{O}_Z$ of sections of the bundle $Z \times V \to Z$ given by formula (17.2.1). However, in order to obtain a flat connection on $Z \times V$ we need to pass from this action to an action of Θ_Z, and this is clearly obstructed if the kernel of the anchor a acts non-trivially on V.

A solution to this problem should now be evident: we must quotient out $V \otimes \mathcal{O}_Z$ by the action of the sheaf $\operatorname{Ker} a$ of stabilizers in $\mathfrak{g} \otimes \mathcal{O}_Z$. Thus we define a new sheaf

$$V_{\mathrm{stab}} = V \otimes \mathcal{O}_Z / \operatorname{Ker} a \cdot (V \otimes \mathcal{O}_Z)$$

of coinvariants of the action of the kernel of a on $V \otimes \mathcal{O}_Z$. The sheaf V_{stab} is clearly a quasi-coherent sheaf of $\mathcal{O}_Z$–modules. However, it is not the sheaf of sections of a vector bundle in general, since the "size" of the stabilizers and their actions on V vary along Z.

The action of $\mathfrak{g} \otimes \mathcal{O}_Z$ on $V \otimes \mathcal{O}_Z$ gives rise to an action of the tangent sheaf $\Theta_Z = \mathfrak{g} \otimes \mathcal{O}_Z / \operatorname{Ker} a$ on the sheaf V_{stab}. This action makes V_{stab} into a (left) $\mathcal{D}_Z$–module. Recall that the sheaf $\mathcal{D}_Z$ of differential operators on Z is the sheaf of associative algebras on Z generated by the sheaf of Lie algebras Θ_Z and the sheaf of commutative algebras $\mathcal{O}_Z$. The only relations between them are expressed by the Leibniz rule. Both Θ_Z and $\mathcal{O}_Z$ act on V_{stab}, and their actions satisfy the Leibniz rule; hence we obtain an action of $\mathcal{D}_Z$ on V_{stab}.

From the $\mathcal{D}$–module point of view, V_{stab} may also be constructed as follows. The action of $\mathfrak{g}$ on Z by vector fields extends uniquely to a homomorphism from the universal enveloping algebra $U(\mathfrak{g})$ of $\mathfrak{g}$ to $\mathcal{D}_Z$. The definition of V_{stab} then implies

17.2.5. Lemma. $V_{\mathrm{stab}} \simeq \mathcal{D}_Z \otimes_{U(\mathfrak{g})} V$ as $\mathcal{D}_Z$–modules.

17.2.6. Descent to S. The group K acts on the Lie algebroid $\mathfrak{g} \otimes \mathcal{O}_Z$ and on the tangent sheaf Θ_Z in such a way that the anchor map $a : \mathfrak{g} \otimes \mathcal{O}_Z \to \Theta_Z$ commutes with K. It follows from the construction that the $\mathcal{D}_Z$–module V_{stab} is K–equivariant. Hence it descends to a $\mathcal{D}_S$–module, which we denote by $\Delta(V)$.

As an $\mathcal{O}_S$–module,

$$\Delta(V) \simeq (\pi_*(\mathcal{D}_Z \otimes_{U(\mathfrak{g})} V))^K.$$

The $\mathcal{D}_S$–module structure on $\Delta(V)$ can be described as follows. Note that $\bar{\mathfrak{g}}_S = (\mathfrak{g}/\mathfrak{k} \otimes \mathcal{O}_Z)^K$ is a Lie algebroid on S which acts on $\mathcal{V} = (V \otimes \pi_*\mathcal{O}_Z)^K$. The anchor map $a : \mathfrak{g} \otimes \mathcal{O}_Z \to \Theta_Z$ gives rise to a surjective map $\bar{a} : \bar{\mathfrak{g}}_S \to \Theta_S$. Then

$$\Delta(V) = \mathcal{V}/(\operatorname{Ker}(\bar{a}) \cdot \mathcal{V}) = \mathcal{D}_S \otimes_{\bar{\mathfrak{g}}_S} \mathcal{V}.$$

This construction gives rise to a functor

$$\Delta : \quad (\mathfrak{g}, K)\text{–mod} \quad \longrightarrow \quad \mathcal{D}_S\text{–mod}$$

sending V to $\Delta(V)$, which is called the *localization functor*.

This functor has a right adjoint functor $\widehat{\Gamma}$ of global sections

$$\widehat{\Gamma} : \quad \mathcal{D}_S\text{–mod} \quad \to \quad (\mathfrak{g}, K)\text{–mod},$$

sending $\mathcal{M}$ to $\Gamma(Z, \pi^*\mathcal{M})$. The K–action on $\widehat{\Gamma}(\mathcal{M})$ comes from the K–action on Z. The $\mathfrak{g}$–action comes from the pull-back of the Θ_S–action. We have

$$\operatorname{Hom}_{\mathcal{D}_S}(\Delta(V), \mathcal{M}) \simeq \operatorname{Hom}_{\mathfrak{g}}(V, \widehat{\Gamma}(\mathcal{M})).$$

17.2.7. Remark. The terminology "localization functor" is motivated by the analogy with the localization functor for $\mathcal{O}$–modules. Indeed, the procedure of assigning the $\mathcal{D}_Z$–module $\mathcal{D}_Z \otimes_{U(\mathfrak{g})} V$ to a $\mathfrak{g}$–module V is analogous to the procedure of assigning the $\mathcal{O}_Z$–module $\mathcal{O}_Z \otimes_R V$ to an R–module V when $Z = \operatorname{Spec} R$ is an affine variety. The latter is the usual localization functor from commutative algebra: the sections of $\mathcal{O}_Z \otimes_{\Gamma(\mathcal{O}_Z)} V$ over Zariski open subsets in Z are localizations of the R–module V.

17.2.8. Fibers of $\Delta(V)$. It is useful to describe the fibers of $\Delta(V)$, considered as an $\mathcal{O}_S$–module. Recall that the *fiber* of an $\mathcal{O}_S$–module $\mathcal{F}$ at $s \in S$ is the quotient of the stalk of $\mathcal{F}$ at s by the maximal ideal $\mathfrak{m}_s$ of the point s.

Let s be a point of S, and Z_s the corresponding K–torsor (the fiber of Z at s). Define $\mathfrak{g}_s$ as the Z_s–twist of $\mathfrak{g}$ (via the adjoint action of K on $\mathfrak{g}$), $\mathfrak{g}_s = Z_s \underset{K}{\times} \mathfrak{g}$. Then $\mathfrak{g}_s$ naturally acts on $\mathcal{V}_s = Z_s \underset{K}{\times} V$. The stabilizers of the points of $Z_s \subset Z$ in $\mathfrak{g}$ give rise to a single well-defined Lie subalgebra $\mathfrak{g}^s_{\mathrm{stab}}$ of $\mathfrak{g}_s$.

17.2.9. Lemma. *The fiber of $\Delta(V)$ at $s \in S$ is canonically isomorphic to the space of coinvariants $\mathcal{V}_s / \mathfrak{g}^s_{\mathrm{stab}} \cdot \mathcal{V}_s$.*

In other words, the natural $\mathcal{O}_S$–module whose fibers are the coinvariants $\mathcal{V}_s / \mathfrak{g}^s_{\mathrm{stab}} \cdot \mathcal{V}_s$ carries a canonical $\mathcal{D}_S$–module structure.

17.2.10. Example: the vacuum representation. A useful example of the localization construction is provided by the vacuum $(\mathfrak{g}, K)$–module $V_{\mathfrak{g},K}$ defined as the $\mathfrak{g}$–module induced from the trivial one-dimensional representation of $\mathfrak{k}$:

$$V_{\mathfrak{g},K} = \mathrm{Ind}_{U(\mathfrak{k})}^{U(\mathfrak{g})} \mathbb{C} = U(\mathfrak{g}) \otimes_{U(\mathfrak{k})} \mathbb{C} = U(\mathfrak{g})/U(\mathfrak{g})\mathfrak{k}.$$

The $\mathfrak{k}$–action on $V_{\mathfrak{g},K}$ may be exponentiated to the action of the group K, so $V_{\mathfrak{g},K}$ is a $(\mathfrak{g}, K)$–module. Let $\pi : Z \to S$ be a $(\mathfrak{g}, K)$–structure on S. Then the description of Θ_S given by formula (17.2.2) implies

$$\mathcal{D}_S \cong (U(\mathfrak{g})/(U(\mathfrak{g})\mathfrak{k}) \otimes \pi_* \mathcal{O}_Z)^K = (V_{\mathfrak{g},K} \otimes \pi_* \mathcal{O}_Z)^K,$$

i.e., the sheaf of sections of the vector bundle $\mathcal{V}_{\mathfrak{g},K} = Z \underset{K}{\times} V_{\mathfrak{g},K}$ is identified with the sheaf $\mathcal{D}_S$. Moreover, the above description of the localization of the vacuum module remains true for a general K–bundle $\pi : Z \to S$ with a compatible transitive $\mathfrak{g}$–action, i.e., $\Delta(V_{\mathfrak{g},K})$ is identified with the sheaf of differential operators on S (see [**BD3**], Sect. 1).

Using this description, one can construct global differential operators on S following [**BD3**]. Namely, since

$$\mathcal{D}_S \simeq \Delta(V_{\mathfrak{g},K}) = (\pi_*(\mathcal{D}_Z \otimes_{U(\mathfrak{g})} V_{\mathfrak{g},K}))^K,$$

for any $z \in (V_{\mathfrak{g},K})^K$ the element $(1 \otimes z) \in \mathcal{D}_Z \otimes_{U(\mathfrak{g})} V_{\mathfrak{g},K}$ gives rise to a global section of $\Delta(V_{\mathfrak{g},K})$ on S. Hence we obtain a map $\mu : (V_{\mathfrak{g},K})^K \to \Gamma(S, \mathcal{D}_S)$. Observe that $(V_{\mathfrak{g},K})^K \simeq N(I)/I$, where I is the left ideal $I = U(\mathfrak{g})\mathfrak{k}$, and $N(I) = \{x \in U(\mathfrak{g})| x \cdot I \in I\}$, so I is a two-sided ideal in $N(I)$. Therefore $(V_{\mathfrak{g},K})^K$ carries a natural algebra structure. It follows from the construction that the above map μ is an algebra homomorphism.

17.2.11. Remark. As explained in [**BD3**], the formalism of Harish-Chandra localization can be applied in the situation where S is a "good" algebraic stack. In particular, the above identification of $\Delta(V_{\mathfrak{g},K})$ with $\mathcal{D}_S$ and the map $(V_{\mathfrak{g},K})^K \to \Gamma(S, \mathcal{D}_S)$ remain in this setting.

17.2.12. Generalization. Suppose that V is a module over a Lie algebra $\mathfrak{l}$ which contains $\mathfrak{g}$ as a Lie subalgebra and carries a compatible adjoint K–action. Consider the trivial vector bundle on Z with the fiber $\mathfrak{l}$. Its sheaf of sections $\mathfrak{l} \otimes \mathcal{O}_Z$ is a K–equivariant sheaf of Lie algebras. Furthermore, the Lie algebroid $\mathfrak{g} \otimes \mathcal{O}_Z$ acts on $\mathfrak{l} \otimes \mathcal{O}_Z$ by the formula

$$(g \otimes f) \cdot (l \otimes h) = [g, l] \otimes fh + g \otimes (a(g \otimes f))(h).$$

If V is an $\mathfrak{l}$–module, then $\mathfrak{l} \otimes \mathcal{O}_Z$ naturally acts on $V \otimes \mathcal{O}_Z$.

Suppose that we are given a subsheaf of Lie subalgebras $\widetilde{\mathfrak{l}}$ of $\mathfrak{l} \otimes \mathcal{O}_Z$ which satisfies the following conditions:

(1) it is preserved by the action of K;
(2) it is preserved by the action of the Lie algebroid $\mathfrak{g} \otimes \mathcal{O}_Z$;
(3) it contains $\operatorname{Ker} a$, where $a : \mathfrak{g} \otimes \mathcal{O}_Z \to \Theta_Z$ is the anchor map.

Then for the same reason as above the sheaf $V \otimes \mathcal{O}_Z / \widetilde{\mathfrak{l}} \cdot (V \otimes \mathcal{O}_Z)$ becomes a K–equivariant $\mathcal{D}$–module on Z which descends to a $\mathcal{D}$–module on S. We denote this $\mathcal{D}_S$–module by $\widetilde{\Delta}(V)$.

Let us describe the fibers of $\widetilde{\Delta}(V)$ (considered as an $\mathcal{O}_S$–module). In the notation of § 17.2.8, for $s \in S$, set $\mathfrak{l}_s = Z_s \underset{K}{\times} \mathfrak{l}$. The fibers of $\widetilde{\mathfrak{l}}$ at the points of $Z_s \subset Z$ give rise to a well-defined Lie subalgebra $\widetilde{\mathfrak{l}}_s$ of $\mathfrak{l}_s$. Then the fiber of $\widetilde{\Delta}(V)$ at $s \in S$ is canonically isomorphic to the space of coinvariants $\mathcal{V}_s / \widetilde{\mathfrak{l}}_s \cdot \mathcal{V}_s$.

17.2.13. Remark. Suppose that V is not a $(\mathfrak{g}, K)$–module, but only a K–module. Then the sheaf $V \otimes \mathcal{O}_Z / \widetilde{\mathfrak{l}} \cdot (V \otimes \mathcal{O}_Z)$ is a K–equivariant $\mathcal{O}_Z$–module (but not a $\mathcal{D}_Z$–module) which descends to an $\mathcal{O}_S$–module. The fibers of this $\mathcal{O}_S$–module are still isomorphic to the spaces of coinvariants as before.

17.2.14. Localization for central extensions. Again, let $(\mathfrak{g}, K)$ be a Harish–Chandra pair, and $Z \to S$ a K–bundle with a compatible transitive $\mathfrak{g}$–action. Suppose that $\mathfrak{g}$ has a central extension

$$0 \to \mathbb{C}\mathbf{1} \to \widehat{\mathfrak{g}} \to \mathfrak{g} \to 0$$

which splits over $\mathfrak{k}$. Then $(\widehat{\mathfrak{g}}, K)$ is also a Harish–Chandra pair which acts on Z via the projection $\widehat{\mathfrak{g}} \to \mathfrak{g}$. Therefore we have a localization functor from $\widehat{\mathfrak{g}}$–modules to $\mathcal{D}$–modules on S. However, this way we produce no new $\mathcal{D}$–modules. Since the central element $\mathbf{1}$ is mapped to the zero vector field on Z, we obtain that if $\mathbf{1}$ acts as a non-zero scalar on V, the corresponding $\mathcal{D}_S$–module $\Delta(V)$ is equal to zero.

The right thing to do in this situation is to take coinvariants with respect to a smaller subalgebra than the full kernel of $\widehat{a} : \widehat{\mathfrak{g}} \otimes \mathcal{O}_Z \to \Theta_Z$. Suppose that the $\mathcal{O}_Z$–extension

(17.2.3) $$0 \to \mathcal{O}_Z \cdot \mathbf{1} \to \widehat{\mathfrak{g}} \otimes \mathcal{O}_Z \to \mathfrak{g} \otimes \mathcal{O}_Z \to 0$$

splits over the kernel of the anchor map $a : \mathfrak{g} \otimes \mathcal{O}_Z \to \Theta_Z$. Then we have a Lie algebra embedding $\operatorname{Ker} a \hookrightarrow \widehat{\mathfrak{g}} \otimes \mathcal{O}_Z$. The quotient $\mathcal{T}$ of $\widehat{\mathfrak{g}} \otimes \mathcal{O}_Z$ by $\operatorname{Ker} a$ is now an extension

$$0 \to \mathcal{O}_Z \to \mathcal{T} \to \Theta_Z \to 0$$

as in (A.3.5), and it carries a natural Lie algebroid structure.

Accordingly, we take coinvariants of $V \otimes \mathcal{O}_Z$ only with respect to $\operatorname{Ker} a \hookrightarrow \widehat{\mathfrak{g}} \otimes \mathcal{O}_Z$, and not the full kernel $\operatorname{Ker} \widehat{a}$ of the anchor map. Thus if V is a $(\widehat{\mathfrak{g}}, K)$–module, we define the sheaf

$$\widehat{V}_{\text{stab}} = \mathcal{O}_Z \otimes V / \operatorname{Ker} a \cdot (\mathcal{O}_Z \otimes V).$$

The sheaf $\widehat{V}_{\text{stab}}$ is no longer a $\mathcal{D}_Z$–module, since it carries an action of the Lie algebroid $\mathcal{T}$, not of Θ_Z. Suppose that the central element $\mathbf{1}$ acts on V as the identity (if it acts by multiplication by a different non-zero scalar, we rescale the extension to make $\mathbf{1}$ act as the identity). Then the quotient of the enveloping algebra $U(\mathcal{T})$ of $\mathcal{T}$ by the relation identifying $1 \in \mathcal{O}_Z \subset \mathcal{T}$ with the unit element of $U(\mathcal{T})$ acts on $\widehat{V}_{\text{stab}}$. This quotient $\mathcal{D}'_Z$ is a sheaf of *twisted differential operators*, or TDO for short (see § A.3.3 for more details).

Thus, $\widehat{V}_{\text{stab}}$ is a $\mathcal{D}'_Z$–module (we also refer to it as a twisted $\mathcal{D}$–module). By construction, $\widehat{V}_{\text{stab}}$ is K–equivariant. Therefore it descends to a $\mathcal{D}'_S$–module on S, also denoted by $\Delta(V)$, where $\mathcal{D}'_S$ is a TDO sheaf on S corresponding to $\mathcal{D}'_Z$. We obtain a localization functor

$$\Delta: \quad (\widehat{\mathfrak{g}}, K)\text{–mod} \quad \longrightarrow \quad \mathcal{D}'_S\text{–mod}$$

sending V to $\Delta(V)$. The fibers of $\Delta(V)$ are the coinvariants $\mathcal{V}_s / \mathfrak{g}^s_{\text{stab}} \cdot \mathcal{V}_s$.

Denote by $\mathbb{C}_1$ the one-dimensional representation of $\mathfrak{k} \oplus \mathbb{C}\mathbf{1}$ on which $\mathfrak{k}$ acts by 0 and $\mathbf{1}$ acts as the identity. Let $V'_{\widehat{\mathfrak{g}}, K}$ be the induced representation $\operatorname{Ind}_{U(\mathfrak{k} \oplus \mathbb{C}\mathbf{1})}^{U(\mathfrak{g})} \mathbb{C}_1$. Then we have the following analogue of the statement of § 17.2.10: $\Delta(V'_{\widehat{\mathfrak{g}}, K}) \simeq \mathcal{D}'_Z$ (see [**BD3**]). Moreover, there is an algebra homomorphism $\mu' : (V'_{\widehat{\mathfrak{g}}, K})^K \to \Gamma(S, \mathcal{D}'_S)$ (see [**BD3**]).

More generally, suppose that $\widehat{\mathfrak{g}}$ is a Lie subalgebra of a Lie algebra $\widehat{\mathfrak{l}}$, and we are given a subsheaf $\widetilde{\mathfrak{l}}$ of $\widehat{\mathfrak{l}} \otimes \mathcal{O}_Z$ containing $\operatorname{Ker} a \subset \widehat{\mathfrak{g}} \otimes \mathcal{O}_Z$ and preserved by the actions of K and of $\widehat{\mathfrak{g}} \otimes \mathcal{O}_Z$. Then the sheaf $V \otimes \mathcal{O}_Z / \widetilde{\mathfrak{l}} \cdot V \otimes \mathcal{O}_Z$ is a K–equivariant $\mathcal{D}'_Z$–module on Z which descends to a $\mathcal{D}'_S$–module (still denoted by $\widetilde{\Delta}(V)$). Its fibers are the coinvariants $\mathcal{V}_s / \widetilde{\mathfrak{l}}_s \cdot \mathcal{V}_s$.

In the next section we apply the above formalism in the case when S is the moduli space of pointed curves. Then in § 18.1 we apply it when S is the moduli space of bundles on curves.

17.3. Moduli of curves

As a starting point, let us go back to the construction of the connection on the vector bundle $\mathcal{V}$ given in § 17.1. Recall that the bundle $\mathcal{V} = \mathcal{V}_X$ over X is the twist of the $\operatorname{Aut}\mathcal{O}$–module V by the $\operatorname{Aut}\mathcal{O}$–bundle $\mathcal{A}ut_X$ of formal coordinates. The same construction can be applied to any family of smooth curves instead of a single curve X. Namely, if we have such a family $\pi : \mathfrak{X} \to S$, we define an $\operatorname{Aut}\mathcal{O}$–bundle $\mathcal{A}ut_{\mathfrak{X}}$ whose fiber at $x \in \mathfrak{X}_s = \pi^{-1}(s)$ is the $\operatorname{Aut}\mathcal{O}$–torsor of formal coordinates at x along $\mathfrak{X}_s$. Then the twist of V by $\mathcal{A}ut_{\mathfrak{X}}$ is a vector bundle $\mathcal{V}_{\mathfrak{X}}$ whose restriction to each fiber $\mathfrak{X}_s$ is our old bundle $\mathcal{V}_{\mathfrak{X}_s}$.

The important question is whether we can define a flat connection on the bundle $\mathcal{V}_{\mathfrak{X}}$. As explained in § 17.1.2, we can define a flat connection on the restriction of this bundle to each fiber $\mathfrak{X}_s$. The reason is that the $\operatorname{Aut}\mathcal{O}$–action along the fibers of $\mathcal{A}ut_{\mathfrak{X}_s}$ may be extended to a simply transitive action of the Lie algebra $\operatorname{Der}\mathcal{O}$ on $\mathcal{A}ut_{\mathfrak{X}_s}$. Since the action of the Lie algebra of $\operatorname{Aut}\mathcal{O}$ on V can also be extended to

Der $\mathcal{O}$, we can identify the fibers of the bundle $\mathcal{V}_{\mathfrak{X}_s}$ at nearby points on $\mathfrak{X}_s$, and so we obtain a flat connection on it.

Thus, we can "move" along the fibers of the family $\mathfrak{X}$. The question is whether we can "move" along the base S of our family as well. For that we need to extend the fiberwise action of Aut $\mathcal{O}$ on $\mathcal{A}ut_{\mathfrak{X}}$ to a transitive action of some Lie algebra which also acts on V. The natural candidate for this Lie algebra is Der $\mathcal{K}$, where $\mathcal{K} = \mathbb{C}((z))$, which acts (possibly, projectively) on any conformal vertex algebra. Indeed, it turns out that the Lie algebra Der $\mathcal{K}$ acts transitively on $\mathcal{A}ut_{\mathfrak{X}}$ when $\mathfrak{X}$ is the universal family, i.e, the moduli space of pointed smooth curves (of a fixed genus). However, this action is not simply transitive. Therefore we cannot obtain a flat connection on the vector bundle $\mathcal{V}_{\mathfrak{X}}$, but instead we obtain a (twisted) $\mathcal{D}$–module structure on its quotient defined as in § 17.2.12. The fibers of this $\mathcal{D}$–module are quotients of $\mathcal{V}_x$, which turn out to be nothing but the spaces of coinvariants $H(X, x, V)$, introduced in Definition 9.2.7. Therefore we obtain that for varying X and x the spaces of coinvariants $H(X, x, V)$ can be organized into a sheaf of coinvariants on the moduli space of pointed curves, which carries the structure of a (twisted) $\mathcal{D}$–module.

Informally, one can say that the Lie subalgebra $\mathrm{Der}_0\, \mathcal{O}$ of Der $\mathcal{K}$ (and the corresponding group Aut $\mathcal{O}$) is responsible for changes of coordinates at the point x of X, the translation vector field ∂_z is responsible for moving the point x along X, and the vector fields on $D_x^\times$ having a pole at the origin are responsible for infinitesimal changes of the complex structure on X.

Note that we can also define a sheaf of conformal blocks. However, the sheaf of conformal blocks is usually not quasi-coherent as an $\mathcal{O}$–module. Therefore we prefer to use the sheaf of coinvariants, which is always quasi-coherent.

17.3.1. Virasoro uniformization. Let us apply the formalism of § 17.2 in the situation when S is the moduli stack $\mathfrak{M}_{g,1}$ of smooth pointed curves of genus $g > 1$, and Z is the moduli stack $\widehat{\mathfrak{M}}_{g,1}$ of triples (X, x, z), where $(X, x) \in \mathfrak{M}_{g,1}$ and z is a formal coordinate at x. Clearly, $\widehat{\mathfrak{M}}_{g,1}$ is an Aut $\mathcal{O}$–bundle over $\mathfrak{M}_{g,1}$.

Let $\mathfrak{M}_g$ be the moduli space of curves of genus g. The moduli space $\mathfrak{M}_{g,1}$ has a natural projection onto $\mathfrak{M}_g$, identifying it as the *universal curve* $\mathfrak{X}_g$ over $\mathfrak{M}_g$. Then $\widehat{\mathfrak{M}}_{g,1}$ should be viewed as the canonical Aut $\mathcal{O}$–torsor $\mathcal{A}ut_{\mathfrak{X}_g}$ over $\mathfrak{X}_g$.

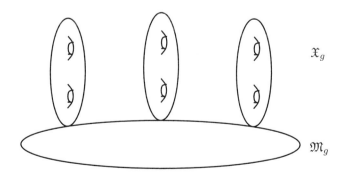

Recall the Lie algebra Der $\mathcal{K}$, where $\mathcal{K} = \mathbb{C}((z))$. The following key result enables us to apply the formalism of Harish-Chandra localization to $\mathfrak{M}_{g,1}$. It can be referred to as the "Virasoro uniformization" of the moduli space of curves (see [**BS**]).

17.3.2. Theorem (cf. [ADKP, BS, Ko, TUY]). *The moduli space $\widehat{\mathfrak{M}}_{g,1}$ carries a transitive action of* $\operatorname{Der}\mathcal{K}$ *compatible with the* $\operatorname{Aut}\mathcal{O}$*–action along the fibers of the map* $\widehat{\mathfrak{M}}_{g,1} \to \mathfrak{M}_{g,1}$.

17.3.3. Remark. Before giving a formal proof, we want to explain heuristically the idea of "gluing" behind it. Let (X, x, z) be a point in $\widehat{\mathfrak{M}}_{g,1}$. Choose a small analytic disc D_x around the point x. Then X can be glued from D_x and $X\backslash x$ along $D_x^\times = D_x\backslash x$. Given an analytic diffeomorphism ρ of $D_x^\times$, we can glue D_x and $X\backslash x$ differently, namely, by twisting with the automorphism ρ. Then we obtain a new curve X'. Moreover, since neither the disc D_x nor the affine curve $X\backslash x$ can be infinitesimally deformed, we see that we should be able to obtain all curves "close" to X by this "gluing" procedure. In the first order approximation, this is in fact the Kodaira–Spencer description of infinitesimal deformations of the complex structure of a Riemann surface using its Čech cover $D_x \cup (X\backslash x)$.

In the proof given below we translate this reasoning into the algebraic (and hence necessarily formal) setting: we replace D_x and $D_x^\times$ by their algebraic counterparts $\operatorname{Spec}\mathcal{O}_x$ and $\operatorname{Spec}\mathcal{K}_x$. Then the relevant automorphism group is the ind-group $\operatorname{Aut}\mathcal{K}$ whose Lie algebra is $\operatorname{Der}\mathcal{K}$. This Lie algebra acts transitively on $\widehat{\mathfrak{M}}_{g,1}$.

17.3.4. Proof. In order to define an action of the Lie algebra $\operatorname{Der}\mathcal{K}$ on $\widehat{\mathfrak{M}}_{g,1}$, we construct a (right) action of the corresponding ind–group $\operatorname{Aut}\mathcal{K}$ on $\widehat{\mathfrak{M}}_{g,1}$. By definition, for any $\mathbb{C}$–algebra R, the group of R–points $\operatorname{Aut}\mathcal{K}(R)$ is the group of continuous automorphisms of $R((z))$. Such an automorphism is completely determined by the image of z. Hence there is a bijection between the group of automorphisms of $R((z))$ and the set of invertible elements of $R((z))$. The latter consists of all series of the form $\sum_{i\geq i_0} r_i z^i$, where $r_i, i > 1$, are arbitrary elements of R, r_1 is an invertible element of R, and $r_i, i < 1$, are nilpotent elements of R. Thus, the ind-group $\operatorname{Aut}\mathcal{K}$ is highly non-reduced: its $\mathbb{C}$–points are the same as those of its subgroup $\operatorname{Aut}\mathcal{O}$, but it has infinitely many "nilpotent directions" corresponding to non-positive powers of z, so that the Lie algebra of $\operatorname{Aut}\mathcal{K}$ is $\operatorname{Der}\mathcal{K}$ (compare with the discussion of § 6.2.3). For our purposes it is sufficient to consider the case when R is a local Artinian ring (and even $R = \operatorname{Spec}\mathbb{C}[\epsilon]/(\epsilon^2)$).

If (X, x, z) is an R–point of $\widehat{\mathfrak{M}}_{g,1}$, and $\rho \in \operatorname{Aut}\mathcal{K}(R)$, we construct a new R–point (X_ρ, x_ρ, z_ρ) of $\widehat{\mathfrak{M}}_{g,1}$ by "gluing" the formal neighborhood of x in X and $X\backslash x$ with a "twist" by ρ.

Namely, as a topological space, $X_\rho = X$, but the structure sheaf $\mathcal{O}_{X_\rho}$ is changed as follows. Let $U \subset X$ be Zariski open in X. If $x \notin U$, then $\mathcal{O}_{X_\rho}(U) \overset{\text{def}}{=} \mathcal{O}_X(U)$. If $x \in U$, then $\mathcal{O}_{X_\rho}(U)$ is defined to be the subring of $\mathcal{O}_X(U\backslash x)$ consisting of functions f whose expansion $f_x(z) \in R((z))$ at x in the coordinate z satisfies

$$f_x(\rho^{-1}(z)) \in R[[z]].$$

It is clear that X_ρ is an algebraic curve over R. Note that if $x \in U$, then under the embedding $\mathcal{O}_{X_\rho}(U\backslash x) \hookrightarrow R((z))$ defined by the formula $f \to f_x(\rho^{-1}(z))$, the subspace $\mathcal{O}_{X_\rho}(U) \subset \mathcal{O}_{X_\rho}(U\backslash x)$ embeds into $R[[z]] \subset R((z))$.

Next, we define the R–point x_ρ of X_ρ. Choose an open subset U of X containing x. Then x_ρ is defined as corresponding to the ideal in $\mathcal{O}_{X_\rho}(U)$ equal to the intersection of $zR[[z]]$ with the image of $\mathcal{O}_{X_\rho}(U)$ in $R[[z]]$ under the above embedding $\mathcal{O}_{X_\rho}(U) \hookrightarrow R[[z]]$. Then (X_ρ, x_ρ, z) is an R–point of $\widehat{\mathfrak{M}}_{g,1}$.

It is clear that $\rho : (X, x, z) \mapsto (X_\rho, x_\rho, z)$ defines a (right) action of the ind–group $\operatorname{Aut}\mathcal{K}$ on $\widehat{\mathfrak{M}}_{g,1}$. This action extends the right action of $\operatorname{Aut}\mathcal{O}$ on $\widehat{\mathfrak{M}}_{g,1}$ by changes of coordinate z, because if $\rho \in \operatorname{Aut}\mathcal{O}$, then $(X_\rho, x_\rho, z) \simeq (X, x, \rho(z))$.

Let us show that the corresponding action of $\operatorname{Der}\mathcal{K}$ is transitive. Let $\mathfrak{D}_\epsilon = \operatorname{Spec}\mathbb{C}[\epsilon]/(\epsilon^2)$ denote the spectrum of the dual numbers. Transitivity means that any pointed curve (X', x') over $\mathfrak{D}_\epsilon$ which is an infinitesimal deformation of (X, x) (i.e., such that the fiber of (X', x') at $0 \in \mathfrak{D}_\epsilon$ equals (X, x)) can be obtained from $(X \times \mathfrak{D}_\epsilon, x \times \mathfrak{D}_\epsilon)$ by the above construction (in the case when $R = \mathbb{C}[\epsilon]/(\epsilon^2)$).

Let $X'\backslash x'$ be the complement of the section $x' : \mathfrak{D}_\epsilon \to X'$ in X'. This is an infinitesimal deformation of the affine curve $X\backslash x$. It is well-known that a smooth affine variety Z has no non-trivial infinitesimal deformations. Indeed, the Kodaira-Spencer isomorphism identifies infinitesimal deformations of Z with $H^1(Z, \Theta_Z)$, but $H^1(Z, \Theta_Z) = 0$ for a smooth affine variety Z. Therefore we obtain that

$$(X'\backslash x') \simeq (X\backslash x) \times \mathfrak{D}_\epsilon.$$

Likewise, we obtain that the formal neighborhood of x' in X' is isomorphic to $D_x \times \mathfrak{D}_\epsilon$.

Now choose a formal coordinate z on D_x. Then we can identify the topological rings of functions on the formal neighborhood of x' in X' and on $D_x \times \mathfrak{D}_\epsilon$ with $\mathbb{C}[\epsilon]((z))/(\epsilon^2)$. Let F_1 be the ring of rational functions on $(X\backslash x) \times \mathfrak{D}_\epsilon$, and F_2 the ring of rational functions on $X'\backslash x'$. Then we obtain two embeddings $\iota_{1,2} : F_{1,2} \hookrightarrow \mathbb{C}[\epsilon]((z))/(\epsilon^2)$. Since $(X'\backslash x') \simeq (X\backslash x) \times \mathfrak{D}_\epsilon$, we find that $F_1 = F_2$, so we can denote them both by F.

The embeddings ι_1 and ι_2 identify $\mathbb{C}[\epsilon]((z))/(\epsilon^2)$ with a completion of F in two different ways. It is easy to see that the embedding ι_2 is continuous with respect to the topology on F induced by ι_1. Hence ι_2 can be extended in a unique way to a continuous automorphism of $\mathbb{C}[\epsilon]((z))/(\epsilon^2)$, i.e., an element $\rho \in \operatorname{Aut}\mathcal{K}(\mathbb{C}[\epsilon]/(\epsilon^2))$. Then we find that (X', x', z) is obtained from $(X \times \mathfrak{D}_\epsilon, x \times \mathfrak{D}_\epsilon, z)$ by gluing with ρ. This completes the proof.

17.3.5. Identification of the tangent spaces. It follows from the construction of the action of $\operatorname{Aut}\mathcal{K}$ that the stabilizer of $(X, x, z) \in \widehat{\mathfrak{M}}_{g,1}$ is identified with the ind–group $\operatorname{Aut}\mathcal{K}_{\text{out}}$ of automorphisms of $\mathcal{K}$ which preserve $\mathcal{O}(X\backslash x) \subset \mathcal{K}$. Therefore the formal neighborhood of $(X, x, z) \in \widehat{\mathfrak{M}}_{g,1}$ (resp., $(X, x) \in \mathfrak{M}_{g,1}$) is canonically isomorphic to

$$\operatorname{Aut}\mathcal{K}_{\text{out}}\backslash \operatorname{Aut}\mathcal{K} \qquad (\text{resp}., \operatorname{Aut}\mathcal{K}_{\text{out}}\backslash \operatorname{Aut}\mathcal{K}_x / \operatorname{Aut}\mathcal{O}_x),$$

where $\mathcal{O}_x$ is the complete local ring at x, and $\mathcal{K}_x$ its field of fractions.

The Lie algebra of the ind-group $\operatorname{Aut}\mathcal{K}_{\text{out}}$ is the Lie algebra $\operatorname{Vect}(X\backslash x)$ of regular vector fields on $X\backslash x$. Therefore the tangent space to $(X, x) \in \mathfrak{M}_{g,1}$ is isomorphic to

$$T_{(X,x)}\mathfrak{M}_{g,1} \simeq \operatorname{Vect}(X\backslash x)\backslash \operatorname{Der}\mathcal{K}_x / \operatorname{Der}_0 \mathcal{O}_x.$$

In addition, the formal neighborhood of $X \in \mathfrak{M}_g$ is canonically identified with

$$\operatorname{Aut}\mathcal{K}_{\text{out}}\backslash \operatorname{Aut}\mathcal{K}/\underline{\operatorname{Aut}\mathcal{O}},$$

and the tangent space to $X \in \mathfrak{M}_g$ is identified with

$$\mathrm{Vect}(X \backslash x) \backslash \mathrm{Der}\, \mathcal{K}_x / \mathrm{Der}\, \mathcal{O}_x.$$

By the following lemma, the last isomorphism can be interpreted as the Kodaira–Spencer isomorphism, identifying the tangent space to $X \in \mathfrak{M}_g$ with the first cohomology $H^1(X, \Theta_X)$ of the tangent sheaf Θ_X.

17.3.6. Lemma. *There is an isomorphism*

$$H^1(X, \Theta_X) \simeq \mathrm{Vect}(X \backslash x) \backslash \mathrm{Der}\, \mathcal{K}_x / \mathrm{Der}_0\, \mathcal{O}_x.$$

17.3.7. Proof. Consider the short exact sequence of sheaves on X (in the Zariski topology):

$$(17.3.1) \qquad\qquad 0 \to \Theta_X \to \mathcal{F}_0 \to \mathcal{F}_1 \to 0,$$

where $\mathcal{F}_0(U) = \mathrm{Vect}(U \backslash x) \oplus \mathrm{Vect}(D_x)$, if U contains x and $\mathcal{F}_0(U) = \mathrm{Vect}(U) \oplus \mathrm{Vect}(D_x^\times)$, otherwise; $\mathcal{F}_1(U) = \mathrm{Vect}(D_x^\times)$. The map $\Theta_X(U) \to \mathcal{F}_0(U)$ is diagonal. The map $\mathcal{F}_0(U) \to \mathcal{F}_1(U)$ sends (f, g) to $f - g$. From the short exact sequence

$$0 \to \mathcal{F}_0 \to \Theta_X(\infty \cdot x) \oplus (\mathrm{Vect}(D_x^\times) \cdot \delta_x) \to (\mathrm{Vect}(D_x^\times)/\mathrm{Vect}(D_x)) \cdot \delta_x \to 0,$$

where $\delta_x = \mathcal{O}_X(x)/\mathcal{O}_X$, we conclude that $H^1(X, \mathcal{F}_0) = 0$. The lemma then follows from the cohomological long exact sequence corresponding to (17.3.1).

17.3.8. Localization with central charge 0. Let us go back to the localization functor. According to Theorem 17.3.2, we are now in the situation of § 17.2.6, with $\mathfrak{g} = \mathrm{Der}\, \mathcal{K}$, $K = \mathrm{Aut}\, \mathcal{O}$, $S = \mathfrak{M}_{g,1}$, and $Z = \widehat{\mathfrak{M}}_{g,1}$.

Denote by $\mathcal{A}$ the Lie algebroid $\mathrm{Der}\, \mathcal{K} \widehat{\otimes} \mathcal{O}_{\widehat{\mathfrak{M}}_{g,1}}$ on $\widehat{\mathfrak{M}}_{g,1}$. By our construction, the kernel of the corresponding anchor map $a : \mathcal{A} \to \Theta_{\widehat{\mathfrak{M}}_{g,1}}$ is the subsheaf $\mathcal{A}_{\mathrm{out}}$ of $\mathcal{A}$ whose fiber at $(X, x, z) \in \widehat{\mathfrak{M}}_{g,1}$ is $\mathrm{Vect}(X \backslash x)$. Applying the construction of § 17.2.6 to $V = \mathrm{Vir}_0$, the Virasoro vertex algebra with central charge $c = 0$, we obtain a $\mathcal{D}$–module $\Delta(\mathrm{Vir}_0)$ on $\mathfrak{M}_{g,1}$. By Lemma 17.2.9, the fiber of the $\mathcal{D}$–module $\Delta(\mathrm{Vir}_0)$ at (X, x) is the space of coinvariants $\mathrm{Vir}_0 / \mathrm{Vect}(X \backslash x) \cdot \mathrm{Vir}_0$, which is nothing but $H(X, x, \mathrm{Vir}_0)$ in the notation of Definition 9.2.7 (see Remark 9.3.10). In fact, according to § 17.2.10, $\Delta(\mathrm{Vir}_0)$ is isomorphic to the sheaf of differential operators on $\mathfrak{M}_{g,1}$.

More generally, let V be an arbitrary conformal vertex algebra with central charge 0. Then it is a module over the Lie algebra $\mathfrak{l} = U(V)$, which is the completion of the span of all Fourier coefficients of vertex operators from V (see § 4.1.1). Then one defines a subsheaf $\mathcal{U}(V)_{\mathrm{out}}$ of $U(V) \widehat{\otimes} \mathcal{O}_{\widehat{\mathfrak{M}}_{g,1}}$ whose fiber at $(X, x, z) \in \widehat{\mathfrak{M}}_{g,1}$ is the Lie subalgebra $U_{X \backslash x}(\mathcal{V}_x)$ of $U(V)$ defined in § 9.2.5. Note that by Theorem 9.2.6, $U_{X \backslash x}(\mathcal{V}_x)$ is naturally a Lie subalgebra of $U(\mathcal{V}_x)$, but we identify the latter with $U(V)$ using the coordinate z. The construction of $\mathcal{U}(V)_{\mathrm{out}}$ is the same as that of $U_{X \backslash x}(\mathcal{V}_x)$: we simply work in the relative setting over the moduli space $\mathfrak{M}_{g,1}$ of pointed curves. This is spelled out in detail in the next subsection.

17.3.9. Construction of $\mathcal{U}(V)_{\mathrm{out}}$. Consider the fiber product

$$\mathcal{X}'_g = \widehat{\mathfrak{M}}_{g,1} \underset{\mathfrak{M}_g}{\times} \mathcal{X}_g$$

and the surjective map $p : \mathcal{X}'_g \to \widehat{\mathfrak{M}}_{g,1}$ onto the left factor. It is instructive to think of $\mathcal{X}'_g$ as the "universal pointed curve" over $\widehat{\mathfrak{M}}_{g,1}$, because its fiber over

$(X, x, z) \in \widehat{\mathfrak{M}}_{g,1}$ under p is the curve X with a marked point x. Then it is clear that we have a section $\mathbf{x} : \widehat{\mathfrak{M}}_{g,1} \to \mathcal{X}'_g$ of p which goes precisely through those points.

Set

$$\widehat{\mathfrak{M}}'_{g,1} = \widehat{\mathfrak{M}}_{g,1} \underset{\mathfrak{M}_g}{\times} \widehat{\mathfrak{M}}_{g,1}.$$

This is a principal Aut $\mathcal{O}$–bundle over $\mathcal{X}'_g$ (with respect to the projection onto the first factor), which should be viewed as the "universal pointed coordinatized curve" over $\widehat{\mathfrak{M}}_{g,1}$. Now we attach to our conformal vertex algebra V a vector bundle on $\mathcal{X}'_{g,1}$,

$$\mathcal{V}_{\mathcal{X}'_g} = \widehat{\mathfrak{M}}'_{g,1} \underset{\text{Aut }\mathcal{O}}{\times} V.$$

Since the fiberwise Aut $\mathcal{O}$–action on $\widehat{\mathfrak{M}}'_{g,1}$ may be extended to an action of Der $\mathcal{O}$, we obtain that this vector bundle is equipped with a flat connection ∇ along the fibers of the projection p.

Consider the relative de Rham complex

$$0 \to \mathcal{V}_{\mathcal{X}'_g} \overset{\nabla}{\to} \mathcal{V}_{\mathcal{X}'_g} \otimes \Omega \to 0,$$

where Ω denotes the relative dualizing sheaf of the projection p, and the complex is placed in degrees -1 and 0. Define $\mathcal{U}_{\mathcal{X}'_g \backslash \mathbf{x}}(V)$ to be the sheaf on $\widehat{\mathfrak{M}}_{g,1}$ of the 0th relative cohomology of the restriction of the above complex to $\mathcal{X}'_g \backslash \mathbf{x}$. Then we have a natural map $\mathcal{U}_{\mathcal{X}'_g \backslash \mathbf{x}}(V) \to U(V) \widehat{\otimes} \mathcal{O}_{\widehat{\mathfrak{M}}_{g,1}}$. We denote the image of this map by $\mathcal{U}(V)_{\text{out}}$.

Comparing with the definition given in § 9.2.5, we find that the fiber of $\mathcal{U}(V)_{\text{out}}$ at $(X, x, z) \in \mathfrak{M}_{g,1}$ is indeed equal to the Lie subalgebra $U_{X \backslash x}(\mathcal{V}_x)$ of $U(V)$.

17.3.10. Now we proceed as in § 17.2.12, choosing $U(V)$ as the Lie algebra $\mathfrak{l}$, and $\mathcal{U}(V)_{\text{out}}$ as the sheaf $\widetilde{\mathfrak{l}}$. It is clear that $U(V)$ contains Der $\mathcal{K}$, and that $\mathcal{U}(V)_{\text{out}}$ contains $\mathcal{A}_{\text{out}}$ (the kernel of the anchor map $a : \mathcal{A} \to \Theta_{\widehat{\mathfrak{M}}_{g,1}}$). Moreover, we have

17.3.11. Theorem. $\mathcal{U}(V)_{\text{out}}$ *is preserved by the action of* $\mathcal{A}$.

17.3.12. Proof. Let X be a smooth projective complex curve with a point x and a formal coordinate z at x. For a $\mathbb{C}$–algebra $\mathcal{R}$ and an element $\rho \in \text{Aut } \mathcal{R}((t))$, we define the triple $(X_\rho, x_\rho, z) \in \widehat{\mathfrak{M}}_{g,1}(\mathcal{R})$ as in the proof of Theorem 17.3.2. The fibers $U_{X \backslash x}(V)$ and $U_{X_\rho \backslash x_\rho}(V)$ of the sheaf $\mathcal{U}(V)_{\text{out}}$ at (X, x, z) and (X_ρ, x_ρ, z) are Lie subalgebras of $U(V)$.

We attach to ρ an automorphism $R(\rho)$ of $V \otimes \mathcal{R}$ defined by exponentiating the action of the Lie algebra Der $\mathcal{K}$ on V (see § 7.2.1). Moreover, $R(\rho)$ acts by conjugation on $U(V)$. Our proposition is equivalent to the statement that $U_{X \backslash x}(V)$ and $U_{X_\rho \backslash x_\rho}(V)$ are conjugated by $R(\rho)$ inside $U(V)$:

(17.3.2) $$U_{X_\rho \backslash x_\rho}(V) = R(\rho)^{-1} U_{X \backslash x}(V) R(\rho).$$

In order to make this identification, we recall that by definition, $U_{X \backslash x}(V) \subset U(V)$ is spanned by all expressions of the form

$$\sum_i \text{Res}_x Y(A_i, z) \phi_i(z) \omega,$$

where $\sum_i \phi_i(z)A_i$ is an element of $\Gamma(X\backslash x, \mathcal{V})$, written with respect to the trivialization of $\mathcal{V}|_{D_x^\times}$ induced by the coordinate z on $D_x^\times$, and ω is an arbitrary non-zero element of $\Gamma(X\backslash x, \Omega)$. Likewise, $U_{X_\rho\backslash x_\rho}(V) \subset U(V)$ is spanned by all expressions of the form

$$\sum_i \mathrm{Res}_x Y(\widetilde{A}_i, w)\widetilde{\phi}_i(w)\omega,$$

where $\sum_j \widetilde{\phi}_j(w)\widetilde{A}_j$ is an element of $\Gamma(X\backslash x, \mathcal{V})$, written with respect to the trivialization of $\mathcal{V}|_{D_x^\times}$ induced by the coordinate $w = \rho(z)$ on $D_x^\times$.

It follows from the definition of the vector bundle $\mathcal{V}$ that the section of $\mathcal{V}|_{D_x^\times}$ which is written as $\sum_j \widetilde{\phi}_j(w)\widetilde{A}_j$ with respect to the coordinate w becomes equal to $\sum_j \widetilde{\phi}_j(\rho(z))R(\rho_z)^{-1} \cdot \widetilde{A}_j$ with respect to the coordinate z. Here $\rho_z \in \mathrm{Aut}\,\mathcal{R}((t))$ is defined by the formula $\rho_z(t) = \rho(z + t) - \rho(z)$. This follows in the same way as in the proof of Theorem 6.5.4 (where we used the action of the group $\mathrm{Aut}\,\mathcal{O}$ instead of $\mathrm{Aut}\,\mathcal{K}$).

Therefore formula (17.3.2) is a consequence of the following lemma. In this lemma, $\widehat{\mathrm{Aut}}\mathcal{R}((t))$ denotes the central extension of $\mathrm{Aut}\,\mathcal{R}((t))$ corresponding to the Virasoro algebra. The group $\widehat{\mathrm{Aut}}\mathcal{R}((t))$ acts on $V \otimes \mathcal{R}$ via $\rho \mapsto R(\rho)$.

17.3.13. Lemma. *Let V be a conformal vertex algebra and $A \in V$. Then*

$$Y(A, z) = R(\rho)Y(R(\rho_z)^{-1} \cdot A, \rho(z))R(\rho)^{-1}, \qquad \rho \in \widehat{\mathrm{Aut}}\mathcal{R}((t)).$$

17.3.14. Proof. This formula is a special case of formula (7.2.3), when the ind–group of internal symmetries is $\widehat{\mathrm{Aut}}\mathcal{R}((t))$. In the case of central charge 0 that we are considering in Theorem 17.3.11 the action of the group $\widehat{\mathrm{Aut}}\mathcal{R}((t))$ factors through $\mathrm{Aut}\,\mathcal{R}((t))$.

This lemma may be viewed as an extension of Lemma 6.5.6 from $\mathrm{Aut}\,\mathcal{O}$ to $\widehat{\mathrm{Aut}}\mathcal{K}$.

17.3.15. The sheaf of coinvariants. Now all conditions set forth in § 17.2.12 are satisfied. Applying the construction of § 17.2.12, we obtain a $\mathcal{D}$–module $\widetilde{\Delta}(V)$ on $\mathfrak{M}_{g,1}$ whose fiber at (X, x) is the space of coinvariants

$$H(X, x, V) = \mathcal{V}_x/U_{X\backslash x}(V) \cdot \mathcal{V}_x$$

(see Definition 9.2.7). Thus, we have obtained the desired $\mathcal{D}$–module of coinvariants of the conformal vertex algebra V on $\mathfrak{M}_{g,1}$.

17.3.16. Non-zero central charge. In the case of vertex algebras with non-zero central charge, we need to modify our construction according to the prescription of § 17.2.14. As will be explained in § 19.6.5, the sequence (17.2.3) with $\widehat{\mathfrak{g}} = Vir$ splits over $\mathrm{Ker}\,a = \mathcal{A}_{\mathrm{out}}$. Following the construction of § 17.2.14, we obtain a twisted $\mathcal{D}$–module $\widetilde{\Delta}(V)$ on $\mathfrak{M}_{g,1}$ whose fiber at (X, x) is the space of coinvariants $H(X, x, V)$.

By Theorem 10.3.1, the spaces of coinvariants $H(X, x, V)$ and $H(X, y, V)$ are canonically identified for all $x, y \in X$. Therefore the sheaf $\widetilde{\Delta}(V)$ is constant along the fibers of $\pi : \mathfrak{M}_{g,1} \to \mathfrak{M}_g$ and descends to a twisted $\mathcal{D}$–module on $\mathfrak{M}_g$ (we will use the same notation $\widetilde{\Delta}(V)$ for it). We now want to describe the corresponding sheaf $\mathcal{D}_c$ of twisted differential operators on $\mathfrak{M}_g$. Note that we have: $\mathcal{D}_c \simeq \Delta(\mathrm{Vir}_c)$ (see § 17.2.14).

17.3.17. The determinant bundles. Let $\omega_{\mathfrak{X}_g/\mathfrak{M}_g}$ be the relative canonical bundle on $\mathfrak{X}_g = \mathfrak{M}_{g,1}$ with respect to $\pi : \mathfrak{X}_g \to \mathfrak{M}_g$ whose restriction to each fiber $X \subset \mathfrak{X}_g$ over $\mathfrak{M}_g$ is the canonical bundle Ω_X. For $\mu \in \mathbb{Z}$, introduce the corresponding *determinant line bundle*

$$\mathrm{Det}_\mu = \det R^\bullet \pi_* \omega_{\mathfrak{X}_g/\mathfrak{M}_g}^{\otimes\mu}$$

on $\mathfrak{M}_g$. Its fiber at $X \in \mathfrak{M}_g$ equals $\bigwedge^{\max} H^0(X, \Omega_X^{\otimes\mu}) \otimes \bigwedge^{\max} H^1(X, \Omega_X^{\otimes\mu})^*$. In particular, the fiber of Det_{-1} at $X \in \mathfrak{M}_g$ is isomorphic to $\bigwedge^{3g-3} H^1(X, \Theta_X)^*$ (recall that we assume $g > 1$). Using the Kodaira-Spencer isomorphism, we can identify Det_{-1} with the canonical bundle on $\mathfrak{M}_g$.

The line bundle Det_1, whose fiber at X is isomorphic to $\bigwedge^g H^0(X, \Omega_X)$, is called the Hodge line bundle. It is known [**AC**] that Det_1 is the generator of the Picard group of $\mathfrak{M}_g$ for $g > 1$ (so the Picard group is isomorphic to $\mathbb{Z}$). Moreover, D. Mumford [**Mu**] has shown that

$$(17.3.3) \qquad \mathrm{Det}_\mu \simeq (\mathrm{Det}_1)^{\otimes(6\mu^2 - 6\mu + 1)}.$$

Now recall the free fermionic vertex superalgebra $\bigwedge$ from § 5.3.1. Fix the conformal vector (5.3.8) in $\bigwedge$ for $\mu \in \mathbb{Z}$. Recall from § 5.3.4 that the corresponding central charge is equal to $c(\mu) = -2(6\mu^2 - 6\mu + 1)$. One can show (see [**BS, BFM**]) that the twisted $\mathcal{D}$–module $\widetilde{\Delta}(\bigwedge)$ is canonically isomorphic to the sheaf of sections of the determinant line bundle Det_μ over $\mathfrak{M}_g$. Therefore the TDO sheaf $\mathcal{D}_{c(\mu)}$ is the sheaf of differential operators acting on Det_μ. In other words, the action of $\mathrm{Der}\,\mathcal{K}$ on $\widehat{\mathfrak{M}}_{g,1}$ can be lifted to a homomorphism

$$\begin{aligned} Vir &\to \mathcal{A}(\widehat{\pi}^* \mathrm{Det}_\mu), \\ C &\mapsto c(\mu), \end{aligned}$$

where $\mathcal{A}(\pi^* \mathrm{Det}_\mu)$ denotes the Lie algebroid of infinitesimal symmetries of the line bundle $\widehat{\pi}^* \mathrm{Det}_\mu$, and $\widehat{\pi} : \widehat{\mathfrak{M}}_{g,1} \to \mathfrak{M}_g$. Note that Mumford's formula (17.3.3) now follows from the formula for the central charge of the Virasoro algebra acting on the fermionic vertex superalgebra $\bigwedge$.

In general we find that the TDO sheaf $\mathcal{D}_c$ is the sheaf of differential operators acting on $(\mathrm{Det}_1)^{-\otimes c/2}$ (this sheaf exists even if the underlying bundle does not; see § A.3.3).

The following diagram summarizes the correspondence between various Harish-Chandra pairs and moduli spaces that we have discussed above. On the left we list Harish-Chandra pairs $(\mathfrak{g}, K)$ and on the right the corresponding K–bundles $Z \to S$ with a transitive $\mathfrak{g}$-action.

$$\begin{array}{ccc}
(\mathrm{Der}\,\mathcal{O}, \mathrm{Aut}_+ \mathcal{O}) & \longleftrightarrow & \left\{ \mathcal{A}ut_X \to \widetilde{X} \right\} \\
\cap & & \downarrow \\
(\mathrm{Der}\,\mathcal{O}, \mathrm{Aut}\,\mathcal{O}) & \longleftrightarrow & \left\{ \mathcal{A}ut_X \to X \right\} \\
\cap & & \downarrow \\
(\mathrm{Der}\,\mathcal{K}, \mathrm{Aut}\,\mathcal{O}) & \longleftrightarrow & \left\{ \widehat{\mathfrak{M}}_{g,1} \to \mathfrak{M}_{g,1} \right\} \\
\uparrow & & \uparrow \\
(Vir, \mathrm{Aut}\,\mathcal{O}) & \longleftrightarrow & \text{Determinant bundle}
\end{array}$$

17.3.18. Multiple points. Let $\mathfrak{M}_{g,n}$ denote the moduli space of smooth projective curves of genus $g > 1$ with n distinct points, and $\widehat{\mathfrak{M}}_{g,n}$ the moduli space of collections $(X, (x_i), (z_i))_{i=1}^n$, where $(X, (x_i))_{i=1}^n \in \mathfrak{M}_{g,n}$ and each z_i is a formal coordinate at x_i. Then we have the following generalization of Theorem 17.3.2.

17.3.19. Theorem. *There is a transitive action of the Lie algebra* $(\operatorname{Der} \mathcal{K})^n$ *on* $\widehat{\mathfrak{M}}_{g,n}$.

17.3.20. Arbitrary module insertions. Let V be a conformal vertex algebra and $M_1, \ldots, M_n$ conformal V–modules. Suppose first for simplicity that the central charge is 0 and that the action of L_0 on each M_i has only integral eigenvalues. Then M_i is a $(\operatorname{Der} \mathcal{K}, \operatorname{Aut} \mathcal{O})$–module. We apply the localization functor of § 17.2.6, where $Z = \widehat{\mathfrak{M}}_{g,n}, S = \mathfrak{M}_{g,n}, \mathfrak{g} = (\operatorname{Der} \mathcal{K})^n, K = (\operatorname{Aut} \mathcal{O})^n$, and the $(\mathfrak{g}, K)$–module is $\bigotimes_{i=1}^n M_i$. Denote by $\mathcal{A}_n$ the Lie algebroid $(\operatorname{Der} \mathcal{K})^n \widehat{\otimes} \mathcal{O}_{\widehat{\mathfrak{M}}_{g,n}}$. By construction, the kernel of the corresponding anchor map $a : \mathcal{A}_n \to \Theta_{\widehat{\mathfrak{M}}_{g,n}}$ is the subsheaf $\mathcal{A}_{n,\mathrm{out}}$ of $\mathcal{A}_n$ whose fiber at $(X, (x_i))_{i=1}^n \in \mathfrak{M}_{g,n}$ is $\operatorname{Vect}(X \backslash \{x_1, \ldots, x_n\})$.

Next, we apply the construction of § 17.2.12, taking as the Lie algebra $\mathfrak{l}$ the direct sum of n copies of the Lie algebra $U(V)$, and as $\widetilde{\mathfrak{l}}$ the subsheaf $\mathcal{U}(V)_{n,\mathrm{out}}$ of $U(V)^n \widehat{\otimes} \mathcal{O}_{\widehat{\mathfrak{M}}_{g,n}}$ whose fiber at $(X, (x_i), (z_i)) \in \mathfrak{M}_{g,n}$ is the image of $U_{X \backslash \{x_1, \ldots, x_n\}}(V)$ in $U(V)^n$ (see § 10.1.2). The sheaf $\mathcal{U}(V)_{n,\mathrm{out}}$ is defined in the same way as in § 17.3.8. Note that $U(V)^n$ contains $(\operatorname{Der} \mathcal{K})^n$, and $\mathcal{U}(V)_{n,\mathrm{out}}$ contains $\mathcal{A}_{n,\mathrm{out}}$. Moreover, we have an analogue of Theorem 17.3.11: $\mathcal{U}(V)_{n,\mathrm{out}}$ is preserved by the action of the Lie algebroid $\mathcal{A}_n$.

Therefore we obtain a $\mathcal{D}$–module $\widetilde{\Delta}\left(\bigotimes_{i=1}^n M_i\right)$ on $\mathfrak{M}_{g,n}$ whose fiber at the point $(X, (x_i))_{i=1}^n$ is the space of coinvariants

$$H_V(X, (x_i), (M_i))_{i=1}^n = \bigotimes_{i=1}^n \mathcal{M}_{x_i} / U_{X \backslash \{x_1, \ldots, x_n\}}(V) \cdot \bigotimes_{i=1}^n \mathcal{M}_{x_i}.$$

In case the eigenvalues of L_0 on the modules M_i are non-integral, we have an action of the Harish-Chandra pair $(\operatorname{Der} \mathcal{K}, \operatorname{Aut}_+ \mathcal{O})$ on M_i. We then modify the above construction by taking as S the moduli space $\widetilde{\mathfrak{M}}_{g,n}$ of collections $(X, (x_i), (\tau_i))_{i=1}^n$, where $(X, (x_i))_{i=1}^n \in \mathfrak{M}_{g,n}$ and τ_i is a non-zero tangent vector to X at x_i. The natural map $\widehat{\mathfrak{M}}_{g,n} \to \widetilde{\mathfrak{M}}_{g,n}$ is an $(\operatorname{Aut}_+ \mathcal{O})^n$–bundle with a compatible transitive action of $(\operatorname{Der} \mathcal{K})^n$. Otherwise, the construction proceeds without changes. The result is a $\mathcal{D}$–module on $\widetilde{\mathfrak{M}}_{g,n}$ whose fibers are the spaces of coinvariants $H_V(X, (x_i), (\tau_i), (M_i))_{i=1}^n$ introduced in § 10.1.3.

If the central charge is non-zero, we obtain twisted $\mathcal{D}$–modules on $\widetilde{\mathfrak{M}}_{g,n}$.

We have thus constructed a sheaf of coinvariants, equipped with a twisted $\mathcal{D}$–module structure, on the appropriate moduli stack of pointed smooth projective curves, which was our main objective in this chapter. In fact, these sheaves may be extended to the Deligne-Mumford compactifications of these stacks (the stacks of stable pointed curves), but this is beyond the scope of the present book.

17.4. Bibliographical notes

The definition of the connection in § 17.1 is due to A. Beilinson and V. Drinfeld [**BD3**] (it was also explained in Beilinson's lectures at MIT, Fall 1995). The theory of localization for Harish–Chandra pairs has been developed by A. Beilinson, J.

Bernstein and others [**BB, BFM, BS**]. In § 17.2.14 we follow [**BFM**]. The main example of § 17.2.2 is due to I.M. Gelfand, D. Kazhdan and D.B. Fuchs, see [**GK, GKF, BR**]. General references on $\mathcal{D}$–modules are [**Be, Bo2, GM, MSa, KS, Mil**]. $\mathcal{D}$–modules on stacks are discussed in [**BB, BD3**].

A general reference on the moduli spaces of curves is [**HM**]. The picture of the moduli space of curves as a "flag manifold" for the Virasoro algebra was developed in [**ADKP, BS, Ko, TUY**]. The proof of Theorem 17.3.2 given above was influenced by a discussion with A. Beilinson. The group $\operatorname{Aut}\mathcal{K}$ and its actions on the moduli space of curves with a formal coordinate at a marked point and the infinite Grassmannian were also described by J.-M. Muñoz Porras and F. Plaza Martin in [**MP**].

The construction of the localization functor from representations of the Virasoro algebra to (twisted) $\mathcal{D}$–modules on the moduli space of pointed curves is due to A. Beilinson, V. Drinfeld, B. Feigin and others (see [**BS, Fe2, BFM**]). The results of this chapter on localization for general vertex algebras, in particular, Theorem 17.3.11 and Lemma 17.3.13, are new; they are due to the authors.

An alternative definition of the sheaf of coinvariants has been given by A. Beilinson and V. Drinfeld in the framework of chiral algebras [**BD4**]. This is explained in D. Gaitsgory's paper [**Gai**].

The extension of the sheaves of coinvariants to the Deligne-Mumford compactification of the moduli space of curves is discussed, in the case of the Virasoro algebra, in [**BS, BFM**], and in the case of affine Kac-Moody algebras in [**TUY, Sor1, Ts, BaKi**]. In the case when the vertex algebra is rational and the genus of the curve is equal to zero, the extension of the sheaf of coinvariats to the Deligne-Mumford compactification is considered in [**NT**].

CHAPTER 18

Vertex Algebras and Moduli Spaces II

In this chapter we describe the analogues of the results of the previous chapter for affine Kac–Moody algebras. The relevant moduli space is the moduli stack $\mathfrak{M}_G(X)$ of G–bundles on a curve X (in this chapter by G we understand the connected simply-connected Lie group corresponding to $\mathfrak{g}$). We show, using the gluing construction, that the moduli space of G–bundles on X with a trivialization on the disc D_x carries a transitive action of the Lie algebra $\mathfrak{g}(\mathcal{K}_x)$. This enables us to apply the general formalism of Harish-Chandra localization to $\mathfrak{M}_G(X)$. As a result, we attach to any conformal vertex algebra V with a $\widehat{\mathfrak{g}}$–structure (see Definition 7.1.1) a twisted $\mathcal{D}$–module on $\mathfrak{M}_G(X)$ whose fibers are the spaces of $\widehat{\mathfrak{g}}$–twisted coinvariants (see Definition 10.1.6).

After that we discuss how the sheaves of coinvariants of vertex algebras can be used to gain insights into the global and local structure of various moduli spaces. In particular, we discuss the localization of the Kac–Moody algebra $V_{-h^\vee}(\mathfrak{g})$ at the critical level and its application to the geometric Langlands correspondence. We also outline the construction of the chiral de Rham complex, due to F. Malikov, V. Schechtman, and A. Vaintrob.

18.1. Moduli of bundles

As we have seen before, the Lie algebra $\mathrm{Der}_0\, \mathcal{O}$ and its Lie group $\mathrm{Aut}\, \mathcal{O}$ are responsible for the changes of formal coordinates at a point x on a curve X. Using the action of $\mathrm{Aut}\, \mathcal{O}$ on a conformal vertex algebra V, we obtain a twist $\mathcal{V}_x$ of V by the $\mathrm{Aut}\, \mathcal{O}$ torsor $\mathcal{A}ut_x$. Furthermore, the action of the polar part of the Lie algebra $\mathrm{Der}\, \mathcal{K}_x$ may be used to deform the complex structure on the curve X, and so we attach to a conformal vertex algebra a $\mathcal{D}$–module of coinvariants on $\mathfrak{M}_{g,1}$.

This picture has an analogue for affine Kac–Moody algebras. Namely, the Lie algebra $\mathfrak{g}(\mathcal{O}_x)$ and its Lie group $G(\mathcal{O}_x)$, where G is the connected simply-connected Lie group corresponding to $\mathfrak{g}$, are responsible for the changes of trivializations of a G–bundle $\mathcal{P}$ on D_x (see § 7.1). Using the action of $G(\mathcal{O}_x)$ on a conformal vertex algebra V with a $\widehat{\mathfrak{g}}$–structure, we can twist $\mathcal{V}_x$ by the $G(\mathcal{O}_x)$–torsor of trivializations of $\mathcal{P}$ on D_x. We will now see that the action of the polar part of the Lie algebra $\mathfrak{g}(\mathcal{K}_x)$ may be used to deform the G–bundle $\mathcal{P}$ on X. Therefore we can attach to a conformal vertex algebra V with a $\widehat{\mathfrak{g}}$–structure a sheaf of coinvariants on the moduli stack $\mathfrak{M}_G(X)$ of G–bundles on X. In fact, we will consider below a more general situation when G is an arbitrary reductive algebraic group.

18.1.1. Kac–Moody uniformization. Let G be a reductive algebraic group, X a smooth projective curve, and x a point of X. We want to apply the Harish-Chandra formalism of § 17.2 in the case when S is the moduli space $\mathfrak{M}_G(X)$ of G–bundles on X, and Z is the moduli space $\widehat{\mathfrak{M}}_G(X)$ of pairs $(\mathcal{P}, s)$, where $\mathcal{P}$ is a

G–bundle on X and s is a trivialization of $\mathcal{P}|_{D_x}$. The group $G(\mathcal{O}_x)$ acts simply transitively on the trivializations of a given bundle $\mathcal{P}$, and so $\widehat{\mathfrak{M}}_G(X)$ is a principal $G(\mathcal{O}_x)$–bundle over $\mathfrak{M}_G(X)$.

The following result is an analogue of Theorem 17.3.2. It describes a "Kac–Moody uniformization" of the moduli space of bundles.

18.1.2. Theorem. $\widehat{\mathfrak{M}}_G(X)$ *carries a transitive action of* $\mathfrak{g}(\mathcal{K}_x)$ *compatible with the* $G(\mathcal{O}_x)$–*action along the fibers of the map* $\widehat{\mathfrak{M}}_G(X) \to \mathfrak{M}_G(X)$.

18.1.3. Proof. We employ a gluing construction similar to the one used in the proof of Theorem 17.3.2. Let $G(\mathcal{K}_x)$ be the ind-group whose R–points are $G(R\widehat{\otimes}\mathcal{O}_x)$. We construct an action of $G(\mathcal{K}_x)$ on $\widehat{\mathfrak{M}}_G(X)$ as follows. Let $(\mathcal{P}, s)$ be an R–point of $\widehat{\mathfrak{M}}_G(X)$ and g an R–point of $G(\mathcal{K}_x)$. Define a new G–bundle $\mathcal{P}_g$ on X as follows. The set $\Gamma(U, \mathcal{P}_g)$ of sections of $\mathcal{P}_g$ over an open subset $U \subset X$ not containing x is by definition equal to the set $\Gamma(U, \mathcal{P})$ of sections of $\mathcal{P}$ over U. On the other hand, the sections of $\mathcal{P}_g$ over an open subset U containing x are by definition the sections ψ of $\mathcal{P}$ over $U\backslash x$ satisfying the following condition. Using the trivialization s, the restriction of ψ to $D_x^\times$ gives us an element ψ_x of $G(\mathcal{K}_x)$. Then the condition is that $g \cdot \psi_x \in G(\mathcal{O}_x)$.

One readily sees that $g : (\mathcal{P}, s) \to (\mathcal{P}_g, s)$ defines an action of $G(\mathcal{K}_x)$ on $\widehat{\mathfrak{M}}_G(X)$. This action extends the action of $G(\mathcal{O}_x)$ by changes of trivializations on D_x, because if $g \in G(\mathcal{O}_x)$, then $(\mathcal{P}_g, s) \simeq (\mathcal{P}, g \cdot s)$. The statement that the action of $\mathfrak{g}(\mathcal{K}_x)$ is transitive is proved by the same argument as the one used in the proof of Theorem 17.3.2. It is based on the fact that a G–bundle on $X\backslash x$ has no infinitesimal deformations.

18.1.4. Identification of the tangent spaces. It is clear from the above construction that the stabilizer of $(\mathcal{P}, s) \in \widehat{\mathfrak{M}}_G(X)$ is the ind–subgroup $G(\mathcal{K}_x)_{\text{out}}^{\mathcal{P}}$ of the regular sections of the adjoint bundle $\mathcal{P} \underset{G}{\times} \operatorname{Ad} G$ corresponding to $\mathcal{P}$ (this is a subgroup of $G(\mathcal{K}_x)$ via the trivialization s). Theorem 18.1.2 implies that the formal neighborhood of $(\mathcal{P}, s) \in \widehat{\mathfrak{M}}_G(X)$ (resp., $\mathcal{P} \in \mathfrak{M}_G(X)$) is isomorphic to the formal neighborhood of the identity coset in the quotient

$$G(\mathcal{K}_x)_{\text{out}}^{\mathcal{P}}\backslash G(\mathcal{K}_x) \qquad (\text{resp.,} G(\mathcal{K}_x)_{\text{out}}\backslash G(\mathcal{K}_x)/G(\mathcal{O}_x)).$$

The Lie algebra of $G(\mathcal{K}_x)_{\text{out}}^{\mathcal{P}}$ is $\mathfrak{g}_{\text{out}}^{\mathcal{P}} = \Gamma(X\backslash x, \mathfrak{g}_{\mathcal{P}})$, where $\mathfrak{g}_{\mathcal{P}} = \mathcal{P} \underset{G}{\times} \mathfrak{g}$. Therefore the tangent space to $(\mathcal{P}, s) \in \widehat{\mathfrak{M}}_G(X)$ is isomorphic to the quotient

$$\mathfrak{g}_{\text{out}}^{\mathcal{P}}\backslash\mathfrak{g}(\mathcal{K}_x)/\mathfrak{g}(\mathcal{O}_x).$$

We have the following analogue of the isomorphism of Lemma 17.3.6:

$$H^1(X, \mathfrak{g}_{\mathcal{P}}) \simeq \mathfrak{g}_{\text{out}}^{\mathcal{P}}\backslash\mathfrak{g}(\mathcal{K}_x)/\mathfrak{g}(\mathcal{O}_x),$$

where $\mathfrak{g}_{\text{out}}^{\mathcal{P}} = \Gamma(X\backslash x, \mathfrak{g}_{\text{out}}^{\mathcal{P}})$. Thus, we obtain the Kodaira–Spencer isomorphism, identifying the tangent space to $\mathcal{P} \in \mathfrak{M}_G(X)$ with $H^1(X, \mathfrak{g}_{\mathcal{P}})$.

18.1.5. Global uniformization. When G is a semi-simple algebraic group, the Kac–Moody uniformization of the formal neighborhood of a point in $\mathfrak{M}_G(X)$ may be extended to a global uniformization. In this case the ind-group $G(\mathcal{K}_x)$ is reduced and we have the following remarkable

18.1.6. Theorem ([BL1, DSi]). *Let G be a semi-simple algebraic group. Then the action of the ind-group $G(\mathcal{K}_x)$ on $\widehat{\mathfrak{M}}_G(X, x)$ is transitive. This action gives rise to the following isomorphisms of stacks:*

$$\widehat{\mathfrak{M}}_G(X, x) \simeq G(\mathcal{K}_x)_{\mathrm{out}} \backslash G(\mathcal{K}_x),$$
$$\mathfrak{M}_G(X) \simeq G(\mathcal{K}_x)_{\mathrm{out}} \backslash G(\mathcal{K}_x) / G(\mathcal{O}_x).$$

18.1.7. Remark. For semi-simple G, the ind-group $G(\mathcal{K})$ is "fatter" than the ind-groups such as $\operatorname{Aut}\mathcal{K}$ or $\mathcal{K}^{\times} = GL_1(\mathcal{K})$. This is expressed by the facts that $G(\mathcal{K})$ is reduced and that we have a global (rather than formal) description of the corresponding moduli space in this case (see § A.1.2). The reason is that when $\mathfrak{g}$ is a semi-simple Lie algebra, any vector $x \in \mathfrak{g}$ may be written as the sum of nilpotent elements of $\mathfrak{g}$. This allows one to construct one–parameter families in $G(\mathcal{K}_x)$ by exponentiation, and allows one to obtain a global uniformization of the moduli space of G–bundles.

In contrast, the groups $\operatorname{Aut}\mathcal{K}$ and $\mathcal{K}^{\times}$ are highly non-reduced. The group of R–points of $\operatorname{Aut}\mathcal{K}$ was described in the proof of Theorem 17.3.2. The group of R–points of $\mathcal{K}^{\times}$ consists of the power series $t^k \sum_{i>N} a_i t^i$, where $k \in \mathbb{Z}$, a_0 is an invertible element of R, and the a_i, $i < 0$, are nilpotent elements of R. Therefore in these cases we have a local variant of uniformization: the ind-group acts transitively only on the formal neighborhood of a point in the corresponding moduli space.

18.1.8. Localization at level 0. We are now in the situation of § 17.2.6, with $\mathfrak{g} = \mathfrak{g}(\mathcal{K}_x)$, $K = G(\mathcal{O}_x)$, $S = \mathfrak{M}_G(X)$, and $Z = \widehat{\mathfrak{M}}_G(X)$. Let us assume for simplicity that $\mathfrak{g}$ is a simple Lie algebra.

Denote by $\mathcal{A}^G$ the Lie algebroid $\mathfrak{g}(\mathcal{K}_x) \widehat{\otimes} \mathcal{O}_{\widehat{\mathfrak{M}}_G(X)}$ on $\widehat{\mathfrak{M}}_G(X)$. According to § 18.1.4, the kernel of the corresponding anchor map $a : \mathcal{A}^G \to \Theta_{\widehat{\mathfrak{M}}_G(X)}$ is the subsheaf $\mathcal{A}^G_{\mathrm{out}}$ of $\mathcal{A}^G$ whose fiber at $(\mathcal{P}, s) \in \mathfrak{M}_{g,1}$ is $\mathfrak{g}^{\mathcal{P}}_{\mathrm{out}}$. Denote by $\mathcal{V}_0(\mathfrak{g})_x$ the $\mathcal{A}ut_x$–twist of $V_0(\mathfrak{g})$. The Lie algebra $\mathfrak{g}(\mathcal{K}_x)$ acts on $\mathcal{V}_0(\mathfrak{g})_x$. The Harish-Chandra localization construction of § 17.2 applied to $\mathcal{V}_0(\mathfrak{g})_x$ gives us a $\mathcal{D}$–module $\Delta(\mathcal{V}_0(\mathfrak{g})_x)$ on $\mathfrak{M}_G(X)$. By Lemma 17.2.9, the fiber of $\Delta(\mathcal{V}_0(\mathfrak{g})_x)$ at $\mathcal{P}$ is the $\mathcal{P}$–twisted space of coinvariants $\mathcal{V}_0(\mathfrak{g})_x / \mathfrak{g}^{\mathcal{P}}_{\mathrm{out}} \cdot \mathcal{V}_0(\mathfrak{g})_x$, or equivalently, $H^{\mathcal{P}}(X, x, V_0(\mathfrak{g}))$ (see §§ 9.5.3 and 9.5.4).

More generally, let V be an arbitrary conformal vertex algebra with $\widehat{\mathfrak{g}}$–structure of level 0 (see Definition 7.1.1), and $\mathcal{V}_x$ its $\mathcal{A}ut_x$–twist. Then $\mathcal{V}_x$ is a module over the Lie algebra $\mathfrak{l} = U(\mathcal{V}_x)$ defined in § 9.2.2. Recall the Lie algebra $U^{\mathcal{P}}_{X \backslash x}(V)$ introduced in § 9.5.3. In the same way as in § 17.3.8 we define a subsheaf $\widetilde{\mathfrak{l}} = \mathcal{U}(\mathcal{V}_x)_{\mathrm{out}}$ of $U(\mathcal{V}_x) \widehat{\otimes} \mathcal{O}_{\widehat{\mathfrak{M}}_{g,1}}$, whose fiber at $(\mathcal{P}, s) \in \mathfrak{M}_G(X)$ equals $U^{\mathcal{P}}_{X \backslash x}(\mathcal{V}_x)$, embedded into $U(\mathcal{V}_x)$ using the trivialization s. Note that $U(\mathcal{V}_x)$ contains $\mathfrak{g}(\mathcal{K}_x)$, and $\mathcal{U}(\mathcal{V}_x)_{\mathrm{out}}$ contains $\mathcal{A}^G_{\mathrm{out}}$. Moreover, we have the following analogue of Theorem 17.3.11:

18.1.9. Proposition. $\mathcal{U}(V)_{\mathrm{out}}$ *is preserved by the action of the Lie algebroid* $\mathcal{A}^G$.

18.1.10. Proof. The assertion of the proposition is equivalent to the statement that the Lie algebras $U^{\mathcal{P}}_{X \backslash x}(V)$ and $U^{\mathcal{P}_g}_{X \backslash x}(V)$ are conjugated by $g \in G(\mathcal{K}_x)$. This follows from the application of the formal twisting property (7.2.3) in the same way as in the proof of Theorem 17.3.11.

18.1.11. Sheaf of coinvariants. Now we proceed as in § 17.2.12, choosing $U(\mathcal{V}_x)$ as the Lie algebra $\mathfrak{l}$, and $\mathcal{U}(\mathcal{V}_x)_{\text{out}}$ as the sheaf $\tilde{\mathfrak{l}}$. Applying the construction of § 17.2.12, we obtain a $\mathcal{D}$–module $\tilde{\Delta}(\mathcal{V}_x)$ on $\mathfrak{M}_G(X)$ whose fiber at $(\mathcal{P}, s)$ is the space of coinvariants introduced in Definition 10.1.6,

$$H^{\mathcal{P}}(X, x, V) = \mathcal{V}_x / U^{\mathcal{P}}_{X \backslash x}(\mathcal{V}_x) \cdot \mathcal{V}_x.$$

18.1.12. Non-zero level. Now let V be a conformal vertex algebra with a compatible $\widehat{\mathfrak{g}}$–structure of non-zero level k. Then we modify the above construction according to the prescription of § 17.2.14. As will be explained in § 19.6.5, the corresponding sequence (17.2.3) splits over $\operatorname{Ker} a = \mathcal{A}^G_{\text{out}}$, so we can apply the construction of § 17.2.14. Hence we obtain a twisted $\mathcal{D}$–module $\tilde{\Delta}(\mathcal{V}_x)$ on $\mathfrak{M}_G(X)$ whose fibers are the spaces of coinvariants. Denote the corresponding sheaf of twisted differential operators by $\mathcal{D}_k$. According to a discussion of § 17.2.14, we have: $\mathcal{D}_k \simeq \tilde{\Delta}(\mathcal{V}_k(\mathfrak{g})_x)$. We want to describe $\mathcal{D}_k$ in geometric terms.

Let $\mathfrak{P}$ be the universal G–bundle over $\mathfrak{M}_G(X) \times X$ whose restriction to $\{\mathcal{P}\} \times X$ is $\mathcal{P}$. Given a finite-dimensional representation W of G, consider the associated vector bundle $W_{\mathfrak{P}} = \mathfrak{P} \underset{G}{\times} W$. Denote by π the projection $\mathfrak{M}_G(X) \times X \to \mathfrak{M}_G(X)$. Now introduce the *theta line bundle* on $\mathfrak{M}_G(X)$ corresponding to W:

$$\theta_W = \det R^{\bullet} \pi_* W_{\mathfrak{P}}.$$

Its fiber at $\mathcal{P} \in \mathfrak{M}_G(X)$ equals

$$\bigwedge^{\max} H^0(X, W_{\mathcal{P}}) \otimes \bigwedge^{\max} H^1(X, W_{\mathcal{P}})^*.$$

The theta line bundle $\theta_{\mathfrak{g}}$ corresponding to the adjoint representation $\mathfrak{g}$ is isomorphic to the canonical line bundle on $\mathfrak{M}_G(X)$ (indeed, by the Kodaira-Spencer isomorphism, the cotangent space to $\mathcal{P} \in \mathfrak{M}_G(X)$ is identified with $H^1(X, \mathfrak{g}_{\mathcal{P}})^*$).

Let $\mathcal{C}l_W$ be the Clifford algebra attached to the vector space $W((t)) \oplus W((t))dt$ and $\bigwedge_W \simeq \bigwedge^{\otimes \dim W}$ the corresponding free fermionic vertex superalgebra (see § 15.1.1). Since W is a $\mathfrak{g}$–module, $\mathcal{C}l_W$ carries an action of the loop algebra $\mathfrak{g}((t))$. It gives rise to a representation of $\widehat{\mathfrak{g}}$ on $\bigwedge_W$. According to Lemma 15.1.3, this representation is of level $k = cas(W)$, where $cas(W)$ is the eigenvalue of the Casimir operator $\sum_a J^a J_a$ on W, and $\{J^a\}$ and $\{J_a\}$ are dual bases in $\mathfrak{g}$ with respect to the normalized invariant bilinear form (here we assume that W is irreducible as a $\mathfrak{g}$–module). In other words, k equals the ratio between the inner product on $\mathfrak{g}$ corresponding to W and the normalized inner product on $\mathfrak{g}$.

For example, for any $\mathfrak{g}$ we have $cas(\mathfrak{g}) = 2h^{\vee}$ (see § 3.4.8), and therefore $\widehat{\mathfrak{g}}$ acts on $\bigwedge_{\mathfrak{g}}$ with level $2h^{\vee}$. In the case when $\mathfrak{g} = \mathfrak{sl}_n$, and W is the standard n–dimensional representation, the level is equal to 1. Other examples for $\mathfrak{g}$ of classical type may be found in [**FeF**].

One can show that the twisted $\mathcal{D}$–module $\tilde{\Delta}(\bigwedge_W)$ on $\mathfrak{M}_G(X)$ is canonically isomorphic to the sheaf of sections of the dual θ_W^{-1} of the theta line bundle. Therefore the TDO sheaf $\mathcal{D}_{cas(W)}$ is nothing but the sheaf of differential operators acting on θ_W^{-1}. Equivalently, the action of $\mathfrak{g}(\mathcal{K}_x)$ on $\widehat{\mathfrak{M}}_G(X)$ can be lifted to a homomorphism from $\widehat{\mathfrak{g}}$ to the Lie algebroid of infinitesimal symmetries of the pull-back of θ_W^{-1} to $\widehat{\mathfrak{M}}_G(X)$ such that the central element K acts by multiplication by $cas(W)$.

In the case when G is simply-connected, the analogue of Mumford's formula (17.3.3) is (see [**KNR, Ku2**])

$$(18.1.1) \qquad\qquad \theta_W^{-1} \simeq \mathcal{L}_G^{\otimes cas(W)},$$

where $\mathcal{L}_G$ is the ample generator of the Picard group $\mathfrak{M}_G(X)$, which is known to be isomorphic to $\mathbb{Z}$ (see [**LS, KN, BLS**]). This formula may be interpreted as a relation between the levels of $\widehat{\mathfrak{g}}$–action on the modules $\bigwedge_W$ for different W. According to this formula, $\mathcal{D}_k$ is the sheaf of differential operators acting on the kth tensor power of $\mathcal{L}_G$.

18.1.13. Multiple points. We generalize the above construction to the situation when we have n distinct points $x_1, \ldots, x_n$ on X with insertions of V–modules.

Let $\widehat{\mathfrak{M}}_{G,n}(X)$ be the moduli space of collections $(\mathcal{P}, (s_i))_{i=1}^n$, where $\mathcal{P}$ is a G–bundle on X and s_i is a trivialization of $\mathcal{P}|_{D_{x_i}}$. The group $\prod_{i=1}^n G(\mathcal{O}_{x_i})$ naturally acts on $\widehat{\mathfrak{M}}_{G,n}(X)$, and the forgetful map $\widehat{\mathfrak{M}}_{G,n}(X) \to \mathfrak{M}_G(X)$ is a principal $\prod_{i=1}^n G(\mathcal{O}_{x_i})$ bundle. Generalizing the argument of Theorem 18.1.2, we obtain that there is a transitive action of the Lie algebra $\bigoplus_{i=1}^n \mathfrak{g}(\mathcal{K}_{x_i})$ on $\widehat{\mathfrak{M}}_{G,n}(X)$ extending the fiberwise action of $\prod_{i=1}^n G(\mathcal{O}_{x_i})$.

Let V be a conformal vertex algebra with a $\widehat{\mathfrak{g}}$–structure of level k, and let $M_1, \ldots, M_n$ be conformal V–modules. Then each M_i is a $\widehat{\mathfrak{g}}$–module of level k. Suppose first for simplicity that the eigenvalues of L_0 on each M_i are integers and that the action of $\mathfrak{g} \subset \mathfrak{g}(\mathcal{O})$ on each M_i can be exponentiated to an action of the group G. Then each M_i carries an $\mathrm{Aut}\,\mathcal{O}$–action, and we can define its Aut_x–twist $\mathcal{M}_{i,x}$. Further, each M_i is a module over the Harish-Chandra pair $(\widehat{\mathfrak{g}}, G(\mathcal{O}))$, and $\mathcal{M}_{i,x}$ is a module over the Harish-Chandra pair $(\widehat{\mathfrak{g}}_x, G(\mathcal{O}_x))$, where $\widehat{\mathfrak{g}}_x$ is the Aut_x–twist of $\widehat{\mathfrak{g}}$. Therefore $\bigotimes_{i=1}^n \mathcal{M}_{i,x_i}$ is a module over the Harish-Chandra pair $(\bigoplus_{i=1}^n \widehat{\mathfrak{g}}_{x_i}, \prod_{i=1}^n G(\mathcal{O}_{x_i}))$. We can apply the localization functor to $\bigotimes_{i=1}^n \mathcal{M}_{i,x_i}$ in the same way as above. The result is a $\mathcal{D}_k$–module on $\mathfrak{M}_G(X)$ whose fiber at $\mathcal{P}$ is the twisted space of coinvariants $H^{\mathcal{P}}(X, (x_i), (M_i))_{i=1}^n$. We leave the details to the reader.

A general V–module M carries an action of the group $\mathrm{Aut}_+ \mathcal{O} \ltimes G^+(\mathcal{O})$. Given $(x, \tau) \in \widetilde{X}$, let $\mathcal{M}_{(x,\tau)}$ be the twist of M by the corresponding $\mathrm{Aut}_+ \mathcal{O}$–torsor $\mathrm{Aut}_{(x,\tau)}^+$ (see § 7.3.11). The group $G^+(\mathcal{O}_x)$ acts on $\mathcal{M}_{(x,\tau)}$. Let us fix non-zero tangent vectors τ_i at each x_i. Then $\bigotimes_{i=1}^n \mathcal{M}_{i,(x_i,\tau_i)}$ carries an action of the Harish-Chandra pair $(\bigoplus_{i=1}^n \widetilde{\widehat{\mathfrak{g}}}_{x_i}, \prod_{i=1}^n G^+(\mathcal{O}_{x_i}))$.

Denote by $\widetilde{\mathfrak{M}}_G(X)$ the moduli space of collections $(\mathcal{P}, (\phi_i))_{i=1}^n$, where $\mathcal{P}$ is a G–bundle on X and ϕ_i is an isomorphism $\mathcal{P}_{x_i} \simeq G$ (trivialization of the fiber of $\mathcal{P}$ at x_i). We have a $\prod_{i=1}^n G^+(\mathcal{O}_{x_i})$–bundle $\widehat{\mathfrak{M}}_G(X) \to \widetilde{\mathfrak{M}}_G(X)$ with a compatible transitive action of the Lie algebra $\bigoplus_{i=1}^n \mathfrak{g}(\mathcal{K}_{x_i})$. Applying the localization functor in this setting, we obtain a twisted $\mathcal{D}$–module on $\widetilde{\mathfrak{M}}_G(X)$ whose fibers are the twisted spaces of coinvariants introduced in § 10.1.7.

18.1.14. Moduli space of curves and bundles. We can combine the Virasoro and Kac–Moody localization constructions. In the simplest case of one point with the insertion of the vertex algebra V itself (and assuming that V carries a $G(\mathcal{O})$–action), the construction goes as follows.

Consider the moduli space $\mathfrak{M}_{G,g,1}$ (resp., $\widehat{\mathfrak{M}}_{G,g,1}$) of triples $(X, x, \mathcal{P})$ (resp., quintuples $(X, x, \mathcal{P}, z, s)$), where X is a smooth projective curve of genus g, $x \in X$, $\mathcal{P}$ is a G–bundle on X, z is a formal coordinate at x and s is a trivialization of $\mathcal{P}|_{D_x}$. Then the natural map $\widehat{\mathfrak{M}}_{G,g,1} \to \mathfrak{M}_{G,g,1}$ is a principal bundle for the group $\operatorname{Aut} \mathcal{O} \ltimes G(\mathcal{O})$. The action of $\operatorname{Aut} \mathcal{O} \ltimes G(\mathcal{O})$ on $\widehat{\mathfrak{M}}_{G,g,1}$ extends to a transitive action of the Lie algebra $\operatorname{Der} \mathcal{K} \ltimes \mathfrak{g}(\mathcal{K}_x)$. On the other hand, V carries an action of the Lie algebra $Vir \ltimes \widehat{\mathfrak{g}}$. Applying the localization construction, we obtain a twisted $\mathcal{D}$–module on $\mathfrak{M}_{G,g,1}$ whose fiber at $(X, x, \mathcal{P})$ is the twisted space of coinvariants $H^{\mathcal{P}}(X, x, V)$.

This construction has a straightforward multi-point generalization. We then consider the moduli space $\widetilde{\mathfrak{M}}_{G,g,n}$ of collections $(X, (x_i), (\tau_i), (s_i), (\phi_i))_{i=1}^n$, where the notation is as above. Then to any n–tuple of V–modules $M_1, \ldots, M_n$ we assign a twisted $\mathcal{D}$–module on $\widetilde{\mathfrak{M}}_{G,g,n}$ whose fibers are the twisted spaces of coinvariants introduced in § 10.1.7.

18.1.15. The case of the Heisenberg algebra.

The above localization construction works equally well for the affine Lie algebra attached to an arbitrary reductive Lie algebra $\mathfrak{g}$ equipped with an invariant bilinear form (see § 5.7.2). For instance, if $\mathfrak{g}$ is a one-dimensional commutative Lie algebra, then the corresponding affine Lie algebra is the Heisenberg Lie algebra $\mathcal{H}$. The relevant moduli space is the Picard variety, which is the moduli space of line bundles on X, or its connected component, the Jacobian variety $J(X)$ of X. The $\mathcal{O}_x^\times$–bundle $\widehat{J}(X)$ over $J(X)$, which parameterizes line bundles together with trivializations on D_x, carries a transitive action of the abelian Lie algebra $\mathcal{K}_x$ extending the action of $\mathcal{O}_x^\times$. The localization construction assigns, to any conformal vertex algebra equipped with an embedding $\pi \to V$, a twisted $\mathcal{D}$–module on $J(X)$ whose fiber at $\mathcal{L} \in J(X)$ is the twisted space of coinvariants $H^{\mathcal{L}}(X, x, V)$ introduced in § 9.5.5. The corresponding TDO sheaf is the sheaf of differential operators acting on the theta line bundle on $J(X)$.

We summarize the examples of moduli spaces that we have considered above, and the corresponding Lie algebras and vertex algebras, in the following diagram:

$$\text{Heisenberg } \mathcal{H}, \pi \quad \longleftrightarrow \quad \text{Line bundles on } X$$

$$\text{Kac–Moody } \widehat{\mathfrak{g}}, V_k(\mathfrak{g}) \quad \longleftrightarrow \quad G\text{–bundles on } X$$

$$\text{Virasoro } Vir, \operatorname{Vir}_c \quad \longleftrightarrow \quad \text{Complex structures on } X$$

18.1.16. General localization principle.

The general localization principle for vertex algebras may be formulated as follows. Suppose we are given a vertex algebra V and a Harish-Chandra pair $(\mathfrak{G}, \mathcal{G}_+)$ of its internal symmetries such that $\mathcal{G}_+$ (resp., $\mathfrak{G}$) consists of symmetries of certain geometric data Φ on the disc (resp., the punctured disc). Any such datum $\mathcal{T}$ on a smooth projective curve X gives rise, when restricted to D_x, to a $\mathcal{G}_+$–torsor $\widehat{\mathcal{T}}$ for any $x \in X$. We can then attach to X, x, and $\mathcal{T}$ a twisted space of coinvariants $H^{\mathcal{T}}(X, x, V)$. Moreover, these spaces with varying $\mathcal{T}$ may be glued into a sheaf $\Delta(V)$ on the moduli space $\mathfrak{M}_\Phi$ of the data Φ on X. Suppose in addition that $\mathfrak{G}$ acts transitively on the moduli space

$\widehat{\mathfrak{M}}_\Phi$ of pairs $(\mathfrak{T}, s)$, where $\mathfrak{T}$ is as above and s is a trivialization of $\widehat{\mathfrak{T}}_x$. Then the sheaf $\Delta(V)$ carries a structure of (twisted) $\mathcal{D}$–module.

The examples of application of this principle are as follows:

- Φ is the data of complex structure on a curve.

  In this case $\mathfrak{G} = \operatorname{Der} \mathcal{K}$, $\mathcal{G}_+ = \operatorname{Aut} \mathcal{O}$. A choice of complex structure on X gives rise to an $\operatorname{Aut} \mathcal{O}$–torsor $\mathcal{A}ut_x$ of formal coordinates at $x \in X$. The corresponding moduli space is $\mathfrak{M}_{g,1}$, the moduli space of pointed curves of a fixed genus g.

- Φ is the data of a G–bundle on X.

  In this case $\mathfrak{G} = \mathfrak{g}(\mathcal{K}_x)$, $\mathcal{G}_+ = G(\mathcal{O}_x)$. A G–bundle $\mathcal{P}$ on X gives rise to a $G(\mathcal{O}_x)$–torsor $\widehat{\mathcal{P}}_x$ of trivializations of $\mathcal{P}|_{D_x}$. The corresponding moduli space is $\mathfrak{M}_G(X)$, the moduli space of G–bundles on X.

- Φ is the data of a line bundle on X.

  In this case $\mathfrak{G} = \mathcal{K}_x$, $\mathcal{G}_+ = \mathcal{O}_x^\times$. A line bundle $\mathcal{L}$ on X gives rise to an $\mathcal{O}_x^\times$–torsor $\widehat{\mathcal{P}}_x$ of trivializations of $\mathcal{L}|_{D_x}$. The corresponding moduli space is the Jacobian of X.

One can hope that one can attach some interesting deformation problems and moduli spaces to other vertex algebras, such as the $\mathcal{W}$–algebras.

18.2. Local structure of moduli spaces

According to §§ 17.2.10 and 17.2.14, the twisted sheaves of differential operators are the simplest examples of the application of the localization construction. For example, let G be a connected simply-connected simple Lie group. As we have seen above, the sheaf $\Delta(\mathcal{V}_k(\mathfrak{g})_x)$ on $\mathfrak{M}_G(X)$ may be identified with the sheaf $\mathcal{D}_k$ of differential operators acting on the kth power of the line bundle $\mathcal{L}_G$ on $\mathfrak{M}_G(X)$ (the ample generator of the Picard group of $\mathfrak{M}_G(X)$).

By construction, the dual space to the fiber of the $\mathcal{D}_k$–module $\Delta(\mathcal{V}_k(\mathfrak{g})_x)$ at $\mathcal{P} \in \mathfrak{M}_G(X)$ is isomorphic to the space of conformal blocks $C^{\mathcal{P}}(X, x, V_k(\mathfrak{g}))$. On the other hand, the dual space to the fiber of $\mathcal{D}_k$ is canonically identified with the space of sections of $\mathcal{L}_G^{\otimes k}$ on the formal neighborhood of $\mathcal{P}$ in $\mathfrak{M}_G(X)$. Since $\mathcal{D}_k \simeq \Delta(\mathcal{V}_k(\mathfrak{g})_x)$, we obtain that the coordinate ring of the formal deformation space of a G–bundle $\mathcal{P}$ on a curve X is isomorphic to $C^{\mathcal{P}}(X, x, V_0(\mathfrak{g}))$.

We can use this isomorphism to obtain information about deformations of G–bundles. For instance, using the functional realization of conformal blocks in terms of chiral correlation functions (see § 10.3.5), we obtain a description of the formal deformation spaces of G–bundles (and line bundles on them) in terms of polydifferentials on powers of the curve X satisfying bootstrap conditions on the diagonals (see [**BD1, BG, Gi2**] for more details). On the other hand, if we replace X by D_x, the corresponding space of polydifferentials (in the case of the trivial G–bundle $\mathcal{P}$) becomes $\mathcal{V}_k(\mathfrak{g})_x^*$, the dual space of the twist of $V_k(\mathfrak{g})$ by the torsor of formal coordinates at x. Therefore the vertex algebra $V_k(\mathfrak{g})$ appears as the local object which controls deformations of G–bundles on curves.

Similarly, the Virasoro vertex algebra Vir_c and the Heisenberg vertex algebra π may be viewed as the local objects responsible for deformations of curves and line bundles on curves, respectively.

Optimistically, one may hope that any one–parameter family of "vacuum module type" vertex algebras (such as Vir_c or $V_k(\mathfrak{g})$) is related to a moduli space of certain structures on algebraic curves. It would be interesting to identify the deformation problems related to the most interesting examples of such vertex algebras (for instance, for the W–algebras), and to construct the corresponding global moduli spaces.

In fact, one can attach a "formal moduli space" to any vertex algebra V by taking the double quotient of the ind–group $\mathcal{G}(V)$, whose Lie algebra is $U(V)$ (see § 7.2.1), by its subgroup $\mathcal{G}(V)_+$ and by the "out" subgroup whose Lie algebra is $U_{X\backslash x}(V)$. Then, using the action of the Harish-Chandra pair $(U(V), \mathcal{G}(V)_+)$ on V as before, we obtain a twisted $\mathcal{D}$–module on this moduli space whose fibers are twisted spaces of coinvariants of V.

But this moduli space is very big, even in the familiar examples of Virasoro and Kac-Moody algebras. The question is whether one can construct a smaller formal moduli space with a line bundle, whose space of sections is isomorphic to the space of (twisted) conformal blocks $C(X, x, V)$.

For instance, in the case when V is the enveloping vertex algebra of a vertex Lie algebra L (see § 16.1.11), we can replace the full ind–group $\mathcal{G}(V)$ of internal symmetry by its subgroup $\mathcal{G}(L)$ corresponding to L. This is the ind-group corresponding to the local Lie algebra $\mathrm{Lie}(L)$ defined in § 16.1.6. Using the action of the Harish-Chandra pair $(\mathrm{Lie}(L), \mathcal{G}(L)_+)$ on V, where $\mathcal{G}(L)_+$ is the Lie group of $\mathrm{Lie}(L)_+ \subset \mathrm{Lie}(L)$, we can localize it on a smaller formal moduli space which is a double quotient of $\mathcal{G}(L)$. For example, in the case when $V = V_0(\mathfrak{g})$ and $L = v_0(\mathfrak{g})$, we have $\mathrm{Lie}(L) = \mathfrak{g}(\mathcal{K}), \mathcal{G}(L)_+ = G(\mathcal{O})$, and the corresponding formal moduli space is the formal neighborhood of a point in the moduli space of G–bundles on X. As we have seen above, the space of functions on it is isomorphic to $C(X, x, V_0(\mathfrak{g}))$. For $k \neq 0$, the space $C(X, x, V_k(\mathfrak{g}))$ is isomorphic to the space of sections of the line bundle $\mathcal{L}_G^{\otimes k}$ over this formal neighborhood. Similarly, in the case when $V = \mathrm{Vir}_0$, we take $L = \mathrm{Aut}\,\mathcal{K}$, and the corresponding moduli space is the formal neighborhood of X in $\mathfrak{M}_{g,1}$.

For general vertex algebras, one first needs to find a small enough group of internal symmetries. Then one can obtain a formal moduli space and a sheaf of coinvariants on it. Next, one needs to formulate a deformation problem such that this formal moduli space can be identified with the formal neighborhood of a point in the corresponding (non-formal) moduli space (which hopefully is an algebraic stack). So far, this is known only for the Virasoro, Kac–Moody and Heisenberg vertex algebras when the relevant moduli spaces are the moduli spaces of curves and bundles on them.

18.3. Global structure of moduli spaces

While "vacuum module type" vertex algebras are related to the local structure of moduli spaces, rational vertex algebras (see § 5.5) can be used to describe the global structure.

For instance, for $k \in \mathbb{Z}_+$, the space of conformal blocks $C(X, x, L_k(\mathfrak{g})) \simeq \mathrm{Hom}_{\mathfrak{g}_{\mathrm{out}}}(L_k(\mathfrak{g})_x, \mathbb{C})$ is isomorphic to the space Γ_k of global sections of the line bundle $\mathcal{L}_G^{\otimes k}$ on $\mathfrak{M}_G(X)$, i.e., to $\Gamma(\mathfrak{M}_G(X), \mathcal{L}_G^{\otimes k})$, see [**BL1, Fa2, KNR**]. On the other hand, it can be shown (see [**TUY**]) that the space $C(X, x, L_k(\mathfrak{g}))$ satisfies a factorization property: its dimension does not change under degenerations of the curve into stable curves with nodal singularities. This property allows one to find the dimension of Γ_k by computing the spaces of conformal blocks in the case of $\mathbb{P}^1$ and three points with insertions of integrable $\widehat{\mathfrak{g}}$–modules of level k (these dimensions are the structure constants of the so-called fusion algebra). The result is a beautiful formula for $\dim C(X, x, L_k) = \dim \Gamma_k$, called the Verlinde formula, see, e.g., [**TUY, Sor1**].

Since $\mathcal{L}_G$ is ample, we can recover the moduli space of semi-stable G–bundles on X as the Proj of the graded ring $\bigoplus_{k \geq 0} \Gamma_{Nk} = \bigoplus_{k \geq 0} C(X, x, L_k(\mathfrak{g}))$ for large N (the ring of "non-abelian theta functions"). Thus, if we could define the product on conformal blocks

$$(18.3.1) \qquad C(X, x, L_k(\mathfrak{g})) \otimes C(X, x, L_m(\mathfrak{g})) \to C(X, x, L_{k+m}(\mathfrak{g}))$$

in a natural way, we would obtain a description of the moduli space of semi-stable G–bundles.

Such a description has been given by B. Feigin and A. Stoyanovsky [**FeS**]. Using the functional realization of conformal blocks, they have identified the space $C(X, x, L_k(\mathfrak{g}))$ with the space of sections of line bundles on the powers of X, satisfying certain conditions. For example, when $\mathfrak{g} = \mathfrak{sl}_2$, it is the space of sections of the line bundle $\Omega(2nx)^{\boxtimes nk}$ over X^{nk} which are symmetric and vanish on the diagonals of codimension k (here $x \in X$ and $n \geq g$, the genus of X). In these terms the multiplication (18.3.1) is just the composition of exterior tensor product and symmetrization of these sections. Thus we obtain an explicit description of the coordinate ring of the moduli space of semi-stable SL_2–bundles. On the other hand, replacing X by the disc D in the above description, we obtain a functional realization of the dual space to $L_k(\mathfrak{sl}_2)$.

Similarly, the space of conformal blocks corresponding to the lattice vertex superalgebra $V_{\sqrt{N}\mathbb{Z}}$ (see § 5.2) may be identified with the space of theta functions of order N on $J(X)$, so we obtain the standard realization of the Jacobian $J(X)$ in terms of theta functions.

Conformal blocks may also be used to obtain "Plücker embeddings" of the corresponding moduli space into the dual space of the vertex algebra. For example, consider the lattice vertex superalgebra $V_{\mathbb{Z}}$ (see Proposition 5.2.5). It contains the Heisenberg vertex algebra π; hence we may consider the space of twisted conformal blocks $C^{\mathcal{L}}(X, x, V_{\mathbb{Z}})$ for $\mathcal{L} \in J(X)$. Actually, in order to define it properly, we need to fix a square root of the canonical bundle Ω_X. One can show that the localization of $V_{\mathbb{Z}}$ is just the corresponding theta line bundle on $J(X)$. Consider its pull-back to $\widehat{J}(X)$ (see § 18.1.15). Each conformal block is a functional on $(\mathcal{V}_{\mathbb{Z}})_x^{\mathcal{L}}$. Choosing a formal coordinate at x and a point $(\mathcal{L}, s) \in \widehat{J}(X)$, we identify $(\mathcal{V}_{\mathbb{Z}})_x^{\mathcal{L}}$ with $V_{\mathbb{Z}}$, and so embed $C^{\mathcal{L}}(X, x, V_{\mathbb{Z}})$ into $V_{\mathbb{Z}}^*$. Since $\dim C^{\mathcal{L}}(X, x, V_{\mathbb{Z}}) = 1$, we obtain an embedding of $\widehat{J}(X)$ into $\mathbb{P}V_{\mathbb{Z}}^*$ (in fact, we could vary the curve as well). This embedding coincides with the Sato–Krichever construction well-known in soliton theory (see [**KNTY, BS**]).

18.4. Localization for affine algebras at the critical level

Recall from § 5.7.2 that the center $\mathcal{Z}(V)$ of a vertex algebra V is by definition the coset vertex algebra of the pair (V, V). This is a commutative vertex subalgebra of V.

18.4.1. Lemma. $\mathcal{Z}(V_k(\mathfrak{g}))$ *is the subspace of* $\mathfrak{g}[[t]]$*–invariant vectors in* $V_k(\mathfrak{g})$. *It is equal to* $\mathbb{C}v_k$, *if* $k \neq -h^\vee$.

18.4.2. Proof. By definition, $A \in \mathcal{Z}(V_k(\mathfrak{g}))$ if and only if for any $B \in V_k(\mathfrak{g})$ we have $Y(B,z)A \in V_k(\mathfrak{g})[[z]]$. In particular, if we take as B the elements of the form $J_{-1}^a v_k$, we obtain that A is a $\mathfrak{g}[[t]]$–invariant vector. Conversely, using the explicit formula for the vertex operators in $V_k(\mathfrak{g})$ given in § 2.4.4, we find that any $\mathfrak{g}[[t]]$–invariant vector belongs to $\mathcal{Z}(V_k(\mathfrak{g}))$.

Now consider the Segal-Sugawara operator

$$S_0 = \frac{1}{2}\sum_a J_{a,0}J_0^a + \sum_a \sum_{n>0} J_{a,-n}J_n^a.$$

If A is in the center of $V_k(\mathfrak{g})$, then $J_n^a \cdot A = 0$ for all $n \geq 0$. Hence we obtain that $S_0 A = 0$. On the other hand, if $k \neq -h^\vee$, then $S_0/(k+h^\vee)$ equals the gradation operator on $V_k(\mathfrak{g})$. Hence we obtain that $\deg A = 0$, and so $A \in \mathbb{C}v_k$.

18.4.3. The center at the critical level. The above argument does not work in the case when $k = -h^\vee$ (which is called the critical level), because then S_0 commutes with $\widehat{\mathfrak{g}}$. In fact, in this case the (unnormalized) Segal-Sugawara vector S itself belongs to the center, and so do all polynomials in $S_n, n \leq -2$, applied to $v_{-h^\vee}$. Therefore the center of $V_{-h^\vee}(\mathfrak{g})$ is non-trivial. Denote it by $\mathfrak{z}(\widehat{\mathfrak{g}})$. According to Proposition 16.2.7, $\mathfrak{z}(\widehat{\mathfrak{g}})$ inherits a vertex Poisson algebra structure if we use as the deformation parameter ϵ any multiple of $(k+h^\vee)$. In order for Theorem 18.4.4 below to hold, we should choose $\epsilon = r^\vee(k+h^\vee)$, where $r^\vee$ is the lacing number of $\mathfrak{g}$. (The reason for the appearance of the factor $r^\vee$ is essentially the same as in Proposition 15.4.16, see the discussion in Remark 15.4.17.)

On the other hand, recall the notion of $\mathfrak{g}$–oper from § 16.6. Observe that the ring of functions $\mathrm{Fun\,Op}_{\mathfrak{g}}(D)$ on $\mathrm{Op}_{\mathfrak{g}}(D)$ is a commutative algebra with a natural $\mathrm{Der}\,\mathcal{O}$–action by derivations. Hence it carries the structure of a quasi-conformal commutative vertex algebra (see § 1.4 and Lemma 6.3.5). By Proposition 16.8.4, $\mathrm{Fun\,Op}_{\mathfrak{g}}(D)$ is isomorphic to the limit of the $\mathcal{W}$–algebra $\mathcal{W}_k(\mathfrak{g})$ as $k \to \infty$, defined in Lemma 16.8.1,(3). The latter, called the classical $\mathcal{W}$–algebra corresponding to $\mathfrak{g}$, carries a vertex Poisson algebra structure, so that $\mathrm{Fun\,Op}_{\mathfrak{g}}(D)$ also acquires a vertex Poisson algebra structure.

For example, $\mathfrak{sl}_2$–opers are the same as projective connections, and the algebra $\mathrm{Fun\,Op}_{\mathfrak{sl}_2}(D)$ is isomorphic to the Virasoro vertex Poisson algebra, which is a limit of Vir_c as $c \to \infty$ (see § 16.3.5). Now recall the definition of the Langlands dual Lie algebra $^L\mathfrak{g}$ from § 15.4.15. The following result was conjectured by V. Drinfeld and proved by B. Feigin and E. Frenkel [**FF7, F1**].

18.4.4. Theorem. *The center* $\mathfrak{z}(\widehat{\mathfrak{g}})$ *of* $V_{-h^\vee}(\mathfrak{g})$ *is isomorphic, as a vertex Poisson algebra with* $\mathrm{Der}\,\mathcal{O}$–*action, to* $\mathrm{Fun\,Op}_{^L\mathfrak{g}}(D)$, *where* $^L\mathfrak{g}$ *is the Langlands dual Lie algebra to* $\mathfrak{g}$.

18.4.5. Remark. The statement of the theorem implies that the twists of $\mathfrak{z}(\widehat{\mathfrak{g}})$ and $\mathrm{Fun\,Op}_{^L\mathfrak{g}}(D)$ by any $\mathrm{Aut}\,\mathcal{O}$–torsor are canonically isomorphic. Let us take the $\mathrm{Aut}\,\mathcal{O}$–torsor $\mathcal{A}ut_x$ of formal coordinates at a point x of a smooth curve X. The corresponding twist $\mathfrak{z}(\widehat{\mathfrak{g}})_x = \mathcal{A}ut_x \underset{\mathrm{Aut}\,\mathcal{O}}{\times} \mathfrak{z}(\widehat{\mathfrak{g}})$ can be described as follows.

Let $\widehat{\mathfrak{g}}_x$ be the affine Kac–Moody algebra attached to the point x, and $\mathcal{V}_{-h^\vee}(\mathfrak{g})_x$ the corresponding induced $\widehat{\mathfrak{g}}_x$–module, as defined in § 6.4.9 (note that in § 6.4.9 we used the notation $\widehat{\mathfrak{g}}(D_x)$ and $\mathcal{V}_k(\mathfrak{g})(D_x)$ for $\widehat{\mathfrak{g}}_x$ and $\mathcal{V}_k(\mathfrak{g})_x$, respectively). Then the Lie algebra $\widehat{\mathfrak{g}}_x$ has a Lie subalgebra $\mathfrak{g} \otimes \mathcal{O}_x$, and $\mathfrak{z}(\widehat{\mathfrak{g}})_x$ is the space of $\mathfrak{g} \otimes \mathcal{O}_x$–invariants (or equivalently, $G(\mathcal{O}_x)$–invariants) in $\mathcal{V}_{-h^\vee}(\mathfrak{g})_x$.

Now observe that the $\mathcal{A}ut_x$–twist of $\operatorname{Fun}\operatorname{Op}_{L_{\mathfrak{g}}}(D)$ is nothing but $\operatorname{Fun}\operatorname{Op}_{L_{\mathfrak{g}}}(D_x)$. Thus, we obtain an isomorphism:

$$\text{(18.4.1)} \qquad\qquad \mathfrak{z}(\widehat{\mathfrak{g}})_x \simeq \operatorname{Fun}\operatorname{Op}_{L_{\mathfrak{g}}}(D_x).$$

18.4.6. Example. For $\mathfrak{g} = \mathfrak{sl}_2$, we have

$$\mathfrak{z}(\widehat{\mathfrak{sl}}_2) = \mathbb{C}[S_n]_{n \leq -2} \cdot v_{-2} \subset V_{-2}(\mathfrak{sl}_2),$$

where the S_n's are the (unnormalized) Segal-Sugawara operators. The commutation relations

$$[S_n, S_m] = (k+2)((n-m)S_{n+m} + \frac{1}{12}3k(n^3-n)\delta_{n,-m})$$

show that $\mathfrak{z}(\widehat{\mathfrak{sl}}_2)$ is isomorphic, as a vertex Poisson algebra, to the Virasoro vertex Poisson algebra $Sym'(vir_{-6})$ introduced in § 16.2.2 (see also § 15.3.2).

The action of the operators S_n commutes with the $\widehat{\mathfrak{g}}$–action on $V_{-2}(\mathfrak{sl}_2)$. Denote by $I_{q(t)}$ the image of the operators $(S_n - q_n \cdot \mathrm{Id})$, $n \leq -2$, in $V_{-2}(\mathfrak{sl}_2)$. It is stable under the action of $\widehat{\mathfrak{sl}}_2$. Hence we obtain that for each $q(t) = \sum_{n \leq -2} q_n t^{-n-2} \in \mathbb{C}[[t]]$, the quotient $V_{q(t)}$ of $V_{-2}(\mathfrak{sl}_2)$ by $I_{q(t)}$ is a non-zero $\widehat{\mathfrak{sl}}_2$–module. Thus we obtain a family of $\widehat{\mathfrak{sl}}_2$–modules of critical level parameterized by points of $\operatorname{Spec}\mathbb{C}[S_n]_{n \leq -2} = \operatorname{Op}_{\mathfrak{sl}_2}(D)$, i.e., by projective connections on the disc, $\partial_t^2 - q(t)$.

Note that $I_{q(t)}$ is stable under the translation operator T if and only if $q_n = 0$ for all $n < -2$. So if $q_n \neq 0$ for some $n < -2$, then $I_{q(t)}$ is not an ideal of the vertex algebra $V_{-2}(\mathfrak{sl}_2)$. But if $q(t) = q_{-2} \in \mathbb{C}$, then $I_{q(t)}$ is an ideal and the quotient is a vertex algebra. However, for $q_{-2} \neq 0$ the ideal $I_{q(t)}$, and hence the quotient vertex algebra, are not $\mathbb{Z}$–graded.

In the case of an arbitrary simple Lie algebra $\mathfrak{g}$, given a $^L\mathfrak{g}$–oper $\rho \in \operatorname{Op}_{L_{\mathfrak{g}}}(D)$, we obtain a homomorphism $\operatorname{Fun}\operatorname{Op}_{L_{\mathfrak{g}}}(D) \to \mathbb{C}$, and hence a homomorphism $\widetilde{\rho} : \mathfrak{z}(\widehat{\mathfrak{g}}) \to \mathbb{C}$. Let V_ρ be the quotient of $V_{-h^\vee}(\widehat{\mathfrak{g}})$ by the image of the kernel of this homomorphism. This is a $\widehat{\mathfrak{g}}$–module. Thus, we obtain a family of $\widehat{\mathfrak{g}}$–modules V_ρ of critical level parameterized by $^L\mathfrak{g}$–opers on the disc D. These modules are clearly smooth, and so they are also $V_{-h^\vee}(\widehat{\mathfrak{g}})$–modules.

Likewise, taking quotients of the $\widehat{\mathfrak{g}}_x$–module $\mathcal{V}_{-h^\vee}(\mathfrak{g})_x$ and using the isomorphism (18.4.1), we obtain a family of $\widehat{\mathfrak{g}}_x$–modules $\mathcal{V}_{\rho_x, x}$ parameterized by $^L\mathfrak{g}$–opers ρ_x on D_x.

It is clear that for all $^L\mathfrak{g}$–opers ρ on D, except one, the $V_{-h^\vee}(\widehat{\mathfrak{g}})$–module V_ρ does not admit a $\mathbb{Z}$–gradation compatible with the $\mathbb{Z}$–gradation on $V_{-h^\vee}(\widehat{\mathfrak{g}})$. So these are examples of modules over a vertex algebra not admitting a $\mathbb{Z}$–gradation.

Recall the $\mathcal{W}$–algebra $\mathcal{W}_k(\mathfrak{g})$ introduced in § 15.1.7. There is an obvious homomorphism $\mathfrak{z}(\widehat{\mathfrak{g}}) \to \mathcal{W}_{-h^\vee}(\mathfrak{g})$ of vertex algebras. One can show, using the description of $\mathcal{W}_{-h^\vee}(\mathfrak{g})$ given in § 15.2, that for any simple Lie algebra $\mathfrak{g}$, the center $\mathfrak{z}(\widehat{\mathfrak{g}})$ is isomorphic to the $\mathcal{W}$–algebra $\mathcal{W}_{-h^\vee}(\mathfrak{g})$.

18.4.7. Sheaves of coinvariants. Let again G be the connected and simply-connected Lie group corresponding to $\mathfrak{g}$.

We have an embedding of vertex algebras $\mathfrak{z}(\widehat{\mathfrak{g}}) \hookrightarrow V_{-h^\vee}(\mathfrak{g})$. By functoriality, it gives rise to a homomorphism of the corresponding sheaves of coinvariants on $\mathfrak{M}_G(X)$.

The vertex algebra $V_{-h^\vee}(\mathfrak{g})$ is not conformal, but it is quasi-conformal (see Definition 6.3.4). Therefore we can still attach to $V_{-h^\vee}(\mathfrak{g})$ a vector bundle $\mathcal{V}_{-h^\vee}(\mathfrak{g})$ on any smooth curve and the sections $\mathcal{Y}_x$ following the procedure of § 6.5. Since

$V_{-h^\vee}(\mathfrak{g})$ carries an action of the Harish-Chandra pair $(\widehat{\mathfrak{g}}, G(\mathcal{O}))$, we can also attach to $V_{-h^\vee}(\mathfrak{g})$ a twisted $\mathcal{D}$–module on $\mathfrak{M}_G(X)$ following the construction of § 18.1.11 (but since the Virasoro algebra does not act on $V_{-h^\vee}(\mathfrak{g})$, we cannot vary the complex structure of the curve X).

According to § 18.2, the sheaf of coinvariants corresponding to $V_{-h^\vee}(\mathfrak{g})$ is the sheaf $\mathcal{D}_{-h^\vee}$ of differential operators acting on the $(-h^\vee)$th power of the ample generator $\mathcal{L}_G$ of the Picard group of $\mathfrak{M}_G(X)$. But according to the isomorphism (18.1.1), the canonical line bundle ω on $\mathfrak{M}_G(X)$ is isomorphic to $\mathcal{L}_G^{-\otimes 2h^\vee}$. Therefore $\mathcal{L}_G^{-\otimes h^\vee}$ is the square root $\omega^{\frac{1}{2}}$ of the canonical line bundle, and $\mathcal{D}_{-h^\vee}$ is the sheaf of differential operators acting on $\omega^{\frac{1}{2}}$. To simplify notation, we will denote this sheaf by $\mathcal{D}'$.

Now consider localization of the commutative quasi-conformal vertex algebra $\mathfrak{z}(\widehat{\mathfrak{g}})$ on $\mathfrak{M}_G(X)$. The group $G(\mathcal{O})$ acts, trivially, on $\mathfrak{z}(\widehat{\mathfrak{g}})$. However, the action of the Lie algebra $\widehat{\mathfrak{g}}$ on $V_{-h^\vee}(\mathfrak{g})$ does not preserve $\mathfrak{z}(\widehat{\mathfrak{g}})$. In § 18.1.12 we introduced a localization functor $\widetilde{\Delta}$ which assigns, to a quasi-conformal vertex algebra V with a compatible $\widehat{\mathfrak{g}}$–structure, a twisted $\mathcal{D}$–module on $\mathfrak{M}_G(X)$. If V is a quasi-conformal vertex algebra which only carries the structure of a $G(\mathcal{O})$–module, the same construction may still be applied, but the result is an $\mathcal{O}$–module on $\mathfrak{M}_G(X)$ rather than a $\mathcal{D}$–module (see Remark 17.2.13). Its fibers are the corresponding spaces of coinvariants. We denote this $\mathcal{O}$–module also by $\widetilde{\Delta}(V)$. If $G(\mathcal{O})$ acts trivially on V, as in our case, then $\widetilde{\Delta}(V)$ is the "constant" $\mathcal{O}_{\mathfrak{M}_G(X)}$–module $H(X, x, V) \otimes \mathcal{O}_{\mathfrak{M}_G(X)}$.

Thus, $\widetilde{\Delta}(\mathfrak{z}(\widehat{\mathfrak{g}}))$ is the $\mathcal{O}$–module $H(X, x, \mathfrak{z}(\widehat{\mathfrak{g}})) \otimes \mathcal{O}_{\mathfrak{M}_G(X)}$ on $\mathfrak{M}_G(X)$. By functoriality, the embedding $\mathfrak{z}(\widehat{\mathfrak{g}}) \to V_{-h^\vee}(\mathfrak{g})$ gives rise to a homomorphism of $\mathcal{O}_{\mathfrak{M}_G(X)}$–modules $\widetilde{\Delta}(\mathfrak{z}(\widehat{\mathfrak{g}})) \to \widetilde{\Delta}(V_{-h^\vee}(\mathfrak{g}))$, i.e.,

$$(18.4.2) \qquad H(X, x, \mathfrak{z}(\widehat{\mathfrak{g}})) \otimes \mathcal{O}_{\mathfrak{M}_G(X)} \to \mathcal{D}'.$$

Now the isomorphism of Theorem 18.4.4 gives

$$H(X, x, \mathfrak{z}(\widehat{\mathfrak{g}})) \simeq H(X, x, \operatorname{Fun} \operatorname{Op}_{{}^L\mathfrak{g}}(D)).$$

On the other hand, according to Proposition 9.4.1, $H(X, x, \operatorname{Fun} \operatorname{Op}_{{}^L\mathfrak{g}}(D))$ is canonically isomorphic to the ring of functions on $\Gamma_\nabla(X, \operatorname{Spec} \mathcal{F}un_X)$, where

$$\mathcal{F}un_X \stackrel{\text{def}}{=} \mathcal{A}ut_X \underset{\operatorname{Aut} \mathcal{O}}{\times} \operatorname{Fun} \operatorname{Op}_{{}^L\mathfrak{g}}(D)$$

is the flat bundle on X attached to the Der $\mathcal{O}$–module $\operatorname{Fun} \operatorname{Op}_{{}^L\mathfrak{g}}(D)$. It follows from the definitions that $\operatorname{Spec} \mathcal{F}un_X$ is nothing but the bundle of jets of ${}^L\mathfrak{g}$-opers on X (see [**BD3**]). Therefore $\Gamma_\nabla(X, \operatorname{Spec} \mathcal{F}un_X) \simeq \operatorname{Op}_{{}^L\mathfrak{g}}(X)$, and so we find that

$$H(X, x, \operatorname{Fun}_X \operatorname{Op}_{{}^L\mathfrak{g}}(D)) \simeq \operatorname{Fun}(\operatorname{Op}_{{}^L\mathfrak{g}}(X)).$$

Passing to global sections in (18.4.2), we obtain

18.4.8. Theorem. *There is a canonical map* $\operatorname{Fun} \operatorname{Op}_{{}^L\mathfrak{g}}(X) \to \Gamma(\mathfrak{M}_G(X), \mathcal{D}')$.

Moreover, Beilinson and Drinfeld have shown in [**BD3**] that the map of Theorem 18.4.8 is an isomorphism of algebras (recall that we are under the assumption that G is simply-connected).

The first step is to show that the map $\operatorname{Fun} \operatorname{Op}_{{}^L\mathfrak{g}}(X) \to \Gamma(\mathfrak{M}_G(X), \mathcal{D}')$ of Theorem 18.4.8 is a homomorphism of algebras. By Proposition 9.4.1, the space of coinvariants $H(X, x, A)$ of any quasi-conformal commutative vertex algebra A

is an algebra, and the natural surjection $\mathcal{A}_x \to H(X, x, A)$ is a homomorphism of algebras. Consider the map

$$\mu' : \mathfrak{z}(\widehat{\mathfrak{g}})_x \to H(X, x, \mathfrak{z}(\widehat{\mathfrak{g}})) \to \Gamma(\mathfrak{M}_G(X), \mathcal{D}'),$$

obtained by composing the surjective homomorphism of algebras

$$\mathfrak{z}(\widehat{\mathfrak{g}})_x \to H(X, x, \mathfrak{z}(\widehat{\mathfrak{g}}))$$

with the map of Theorem 18.4.8, $H(X, x, \mathfrak{z}(\widehat{\mathfrak{g}})) \to \Gamma(\mathfrak{M}_G(X), \mathcal{D}')$. According to Remark 18.4.5, $\mathfrak{z}(\widehat{\mathfrak{g}})_x = (V'_{\widehat{\mathfrak{g}}_x, G(\mathcal{O}_x)})^{G(\mathcal{O}_x)}$, in the notation of § 17.2.14. Moreover, μ' is nothing but the corresponding homomorphism from § 17.2.14 (see also § 17.2.10). Therefore μ' is a homomorphism of algebras, and so is the map

$$\operatorname{Fun} \operatorname{Op}_{L_\mathfrak{g}}(X) = H(X, x, \mathfrak{z}(\widehat{\mathfrak{g}})) \to \Gamma(\mathfrak{M}_G(X), \mathcal{D}').$$

To see that it is an isomorphism, Beilinson and Drinfeld show that this map preserves natural filtrations on both algebras. Then they identify the corresponding map of the associated graded algebras with the Hitchin homomorpism, which is known to be an isomorphism (see [**BD3**]).

18.4.9. Connection with the geometric Langlands correspondence. Let G be a simple (or, more generally, reductive) Lie group over $\mathbb{C}$ with a Cartan subgroup H. We associate to H the lattice P of characters $H \to \mathbb{C}^\times$ and the lattice $P^\vee$ of one-parameter subgroups $\mathbb{C}^\times \to H$. The lattice P contains the set of roots, Δ, and the lattice $P^\vee$ contains the set of coroots, $\Delta^\vee$. The quadruple $(P, P^\vee, \Delta, \Delta^\vee)$ is called the root datum of G (see, e.g., [**Bo1**]). It turns out that the datum $(P^\vee, P, \Delta^\vee, \Delta)$ corresponds to another simple (resp., reductive) Lie group, called the *Langlands dual group* to G and denoted by ^LG. For instance, the group GL_n is self-dual, SL_n is dual to PGL_n, Sp_{2n} is dual to SO_{2n+1} and SO_{2n} is self-dual. The Lie algebra of ^LG is $^L\mathfrak{g}$, considered above.

Now we return to our setting, where G is a simply-connected simple Lie group. In this case ^LG is a simple Lie group of adjoint type. Let ρ be a $^L\mathfrak{g}$–oper (equivalently, an ^LG–oper), i.e., a point of the spectrum of the commutative algebra $\operatorname{Fun} \operatorname{Op}_{L_\mathfrak{g}}(X) \simeq \Gamma(\mathfrak{M}_G(X), \mathcal{D}')$. Then it defines a homomorphism

$$\widetilde{\rho} : \Gamma(\mathfrak{M}_G(X), \mathcal{D}') \to \mathbb{C}.$$

Consider the $\mathcal{D}'$–module

$$\Delta'_\rho = \mathcal{D}'/(\mathcal{D}' \cdot m_{\widetilde{\rho}})$$

on $\mathfrak{M}_G(X)$, where $m_{\widetilde{\rho}}$ is the kernel of $\widetilde{\rho}$. This $\mathcal{D}'$–module describes the following system of differential equations on half-forms on $\mathfrak{M}_G(X)$:

$$D \cdot \psi = \widetilde{\rho}(D)\psi, \qquad D \in \Gamma(\mathfrak{M}_G(X), \mathcal{D}').$$

It is easy to see that $\dim \operatorname{Op}_{L_\mathfrak{g}}(X) = \frac{1}{2} \dim \mathfrak{M}_G(X)$. Therefore Δ'_ρ is a holonomic $\mathcal{D}'$–module. Let $\Delta_\rho = \Delta'_\rho \otimes \omega^{-\frac{1}{2}}$ be the corresponding untwisted $\mathcal{D}$–module on $\mathfrak{M}_G(X)$.

We remark that the $\mathcal{D}'$–module Δ'_ρ is isomorphic to $\widetilde{\Delta}(\mathcal{V}_{\rho,x})$, where $\mathcal{V}_{\rho,x}$ is the $\widehat{\mathfrak{g}}_x$–module corresponding to the restriction of the $^L\mathfrak{g}$–oper ρ to D_x (see § 18.4.6).

As explained in [**BD3**], if a flat ^LG–bundle (equivalently, ^LG–local system) on a smooth projective curve X admits an oper structure, then this structure is unique. Therefore the set $\operatorname{Op}_{L_\mathfrak{g}}(X)$ of ^LG–opers on X may be viewed as a subset of the set $\mathcal{L}oc_{LG}(X)$ of isomorphism classes of all ^LG–local systems on X. In fact, this subset consists of all flat connections on a particular ^{L}G–bundle on X (i.e., it is a fiber of the natural map $\mathcal{L}oc_{L\,G}(X) \to \mathfrak{M}_{L\,G}(X)$). Denote by E_ρ the ^{L}G–local system on X, corresponding to $\rho \in \mathrm{Op}_{L\,\mathfrak{g}}(X)$.

According to the geometric Langlands conjecture, to $E \in \mathcal{L}oc_{L\,G}(X)$ one should be able to attach a holonomic $\mathcal{D}$–module on $\mathfrak{M}_G(X)$ which is a *Hecke eigensheaf* with eigenvalue E (see [**BD3**] and § 20.5.2 below). Beilinson and Drinfeld have shown in [**BD3**] that the $\mathcal{D}$–module Δ_ρ is a Hecke eigensheaf with eigenvalue E_ρ. Thus, they have established the geometric Langlands conjecture for those ^{L}G–local systems which admit the ^{L}G–oper structure.

18.5. Chiral de Rham complex

Another application of the general localization pattern of § 17.2.12 is the recent construction, due to Malikov, Schechtman and Vaintrob [**MSV**], of a sheaf of vertex superalgebras on any smooth scheme S, called the chiral de Rham complex. Under certain conditions, this sheaf has a purely even counterpart, the sheaf of chiral differential operators. Here we give a brief review of the construction of these sheaves. Our goal is to show that they can be considered as special cases of the localization construction of § 17.2, with respect to an action of a group of internal symmetries on a particular vertex algebra.

18.5.1. The vertex superalgebra $\widehat{\Omega}^\bullet_N$. Let $\widetilde{A}_N$ be the Weyl algebra associated to the symplectic vector space $\mathbb{C}((t))^N \oplus (\mathbb{C}((t))dt)^N$ defined as in § 12.3.5. It has topological generators $a_{i,n}, a^*_{i,n}, i = 1, \ldots, N, n \in \mathbb{Z}$, with relations

$$[a_{i,n}, a^*_{j,m}] = \delta_{i,j}\delta_{n,-m}, \qquad [a_{i,n}, a_{j,m}] = [a^*_{i,n}, a^*_{j,m}] = 0.$$

Denote by M_N the Fock representation of $\widetilde{A}_N$ generated by the vector $|0\rangle$, satisfying $a_{i,n}|0\rangle = 0, n \geq 0$, and $a^*_{i,m}|0\rangle = 0, m > 0$. This is the tensor product of N copies of the vertex algebra M defined in Lemma 11.3.8. Hence M_N is a vertex algebra. It is shown in [**MSV**] that the vertex algebra structure on M_N may be extended to $\widehat{M}_N = M_N \otimes_{\mathbb{C}[a^*_{i,0}]} \mathbb{C}[[a^*_{i,0}]]$.

The "fermionic" counterpart of the Weyl algebra $\widetilde{A}_N$ is the Clifford algebra $\mathcal{C}l_N$, defined as in § 15.1.1. It has generators $\psi_{i,n}, \psi^*_{i,n}, i = 1, \ldots, N, n \in \mathbb{Z}$, and relations

$$[\psi_{i,n}, \psi^*_{j,m}]_+ = \delta_{i,j}\delta_{n,-m}, \qquad [\psi_{i,n}, \psi_{j,m}]_+ = [\psi^*_{i,n}, \psi^*_{j,m}]_+ = 0.$$

Let $\bigwedge^\bullet_N$ be the fermionic analogue of M_N defined in the same way as in § 5.3.1. This is the Fock module over $\mathcal{C}l_N$ equipped with an additional $\mathbb{Z}$–gradation by charge. It carries a vertex superalgebra structure, see § 5.3.1.

Now set

$$\widehat{\Omega}^\bullet_N = \widehat{M}_N \otimes \bigwedge^\bullet_N.$$

This is a vertex superalgebra with a differential

$$d = \sum_i \mathrm{Res}\, a_i(z)\psi^*_i(z)dz,$$

which makes it into a complex.

18.5.2. Action of vector fields. Denote by W_N (resp., $\mathcal{O}_N$, Ω_N^i) the topological Lie algebra of vector fields (resp., ring of functions, space of i forms) on $\operatorname{Spec}\mathbb{C}[[t_1,\ldots,t_N]]$. The Lie algebra W_N acts on $\Omega_N^\bullet = \bigoplus_{i=0}^N \Omega_N^i$. Let us identify $\Omega_N^\bullet$ with

$$\mathbb{C}[[a_1^*,\ldots,a_n^*]] \otimes \bigwedge{}^\bullet(\psi_1^*,\ldots,\psi_N^*),$$

where $a_i^* = t_i$. Then ∂_{t_i} acts as the element a_i of the corresponding Weyl algebra. Let ψ_i be the operator of contraction with ψ_i^* acting on $\bigwedge{}^\bullet(\psi_1^*,\ldots,\psi_N^*)$. Then the action of W_N on $\Omega_N^\bullet$ may be described by the following formula:

$$f(t_i)\partial_{t_j} \mapsto f(a_i^*)a_j + \sum_{k=1}^N (\partial_{t_k}f)(a_i^*)\psi_k^*\psi_j.$$

This motivates the following definition. Let $U(\widehat{\Omega}_N^\bullet)$ be the completion of the Lie algebra of Fourier coefficients of vertex operators from $\widehat{\Omega}_N^\bullet$ (see § 4.1.1). Define a map from W_N to $U(\widehat{\Omega}_N^\bullet)$ by the formula

$$(18.5.1) \quad f(t_i)\partial_{t_j} \mapsto \operatorname{Res}\left(:f(a_i^*(z))a_j(z): + \sum_{k=1}^N :(\partial_{t_k}f)(a_i^*(z))\psi_k^*(z)\psi_j(z):\right) dz.$$

The OPE formula from Lemma 12.2.9 can be easily generalized to the fermionic case. Using this formula, we obtain

18.5.3. Lemma ([MSV]). *The map* (18.5.1) *is a homomorphism of Lie algebras.*

18.5.4. Remark. This lemma is an example of "supersymmetric cancellation of anomalies" [**FF2**]: the terms coming from double contractions of the bosonic fields $a_i(z), a_i^*(z)$ get canceled by the terms coming from double contractions of the fermionic fields $\psi_i(z), \psi_i^*(z)$.

18.5.5. Localization. Now we obtain an action of the Lie algebra W_N on $\widehat{\Omega}_N^\bullet$. It is straightforward to check that this action may be exponentiated to an action of the Harish-Chandra pair $(W_N, \operatorname{Aut}\mathcal{O}_N)$. Furthermore, because the $(W_N, \operatorname{Aut}\mathcal{O}_N)$–action comes from the residues of vertex operators, it preserves the vertex algebra structure on V. Thus, the action of the group $\operatorname{Aut}\mathcal{O}_N$ (and the ind-group corresponding to W_N) is an example of the action of a group of internal symmetries from § 7.2.1. Recall from § 17.2.2 that any smooth variety S of dimension N carries a canonical $(W_N, \operatorname{Aut}\mathcal{O}_N)$–structure $\widehat{S}$. Applying the construction of § 17.2.3, we attach to $\widehat{\Omega}_N^\bullet$ a vector bundle with a flat connection on S (with fibers isomorphic to $\widehat{\Omega}_N^\bullet$). The sheaf of horizontal sections of this bundle is a sheaf $\widehat{\Omega}^\bullet(S)$ of vertex superalgebras on S, together with the structure of a complex. This sheaf is the *chiral de Rham complex* of S, introduced in [**MSV**]. It contains the usual de Rham complex of S and is quasi-isomorphic to it. The chiral de Rham complex may be thought of as a kind of "semi-infinite" de Rham complex on the loop space of S.

In particular, one can attach a vertex superalgebra to any variety S by taking the space of global sections of $\widehat{\Omega}^\bullet(S)$. It has a natural filtration, whose associated graded pieces are isomorphic to spaces of various tensor fields on S. This vertex algebra should be the symmetry algebra of a certain superconformal field theory corresponding to S. It is expected that this vertex algebra may be used to obtain some "stringy" invariants on S, such as the elliptic cohomology. For more on potential applications of the chiral de Rham complex to mirror symmetry and elliptic cohomology, see [**BoL, Bor1**].

18.5.6. Bosonic counterpart. Now consider a purely bosonic analogue of this construction, i.e., replace $\widehat{\Omega}_N$ with $\widehat{M}_N$. The Lie algebra $U(\widehat{M}_N)$ of Fourier coefficients of the fields from $\widehat{M}_N$ has a filtration $A_N^{\leq i}, i \geq 0$, by the order of $a_{i,n}$'s. For $N = 1$ this is the filtration $A_{\leq i,\mathrm{loc}}$ introduced in § 12.2.2. In the same way as in § 12.2.2 we obtain that $A_N^{\leq 1}$ is a Lie subalgebra of $U(\widehat{M}_N)$ and A_N^0 is an abelian ideal of $A_N^{\leq 1}$. Denote by $\mathcal{T}_N$ the Lie algebra $A_N^{\leq 1}/A_N^0$. As explained in Chapter 12, $U(\widehat{M}_N)$ should be viewed as the algebra of differential operators on the topological vector space $\mathbb{C}((t))^N$ with the standard filtration, A_N^0 (resp., $\mathcal{T}_N$) as the space of functions (resp., the Lie algebra of vector fields) on it, and $\widehat{M}_N$ as the module of "delta–functions supported on $\mathbb{C}[[t]]^N$".

The bosonic analogue of the map (18.5.1) is the map $W_N \to U(\widehat{M}_N)$,

$$f(t_i)\partial_{t_j} \mapsto \mathrm{Res} : f(a_i^*(z))a_j(z):dz .$$

But this is not a Lie algebra homomorphism. The reason is that in contrast to the finite-dimensional case, the extension

$$0 \to A_N^0 \to A_N^{\leq 1} \to \mathcal{T}_N \to 0$$

does not split, as explained in § 12.2.2. Explicit calculation using Lemma 12.2.9 shows that the pull-back of this extension under $W_N \to U(\widehat{M}_N)$ reduces to an extension

(18.5.2) $$0 \to \Omega_N^1/d\mathcal{O}_N \to \widetilde{W}_N \to W_N \to 0$$

of W_N by its module $\Omega_N^1/d\mathcal{O}_N$, which is embedded into $A_N^{\leq 1}$ by the formula

$$g(t_i)dt_j \mapsto \mathrm{Res}\, g(a_i^*(z))\partial_z a_j^*(z)dz.$$

If the sequence (18.5.2) were split, we would obtain a $(W_N, \mathrm{Aut}\,\mathcal{O}_N)$–action on $\widehat{M}_N$ and associate to $\widehat{M}_N$ a $\mathcal{D}$–module on S in the same way as above.

But in reality the extension (18.5.2) does not split for $N > 1$ (note that $\Omega_1^1/d\mathcal{O}_1 = 0$). Hence in general we only have an action on $\widehat{M}_N$ of the Harish-Chandra pair $(\widetilde{W}_N, \widetilde{\mathrm{Aut}\mathcal{O}}_N)$, where $\widetilde{\mathrm{Aut}\mathcal{O}}_N$ is an extension of $\mathrm{Aut}\,\mathcal{O}_N$ by the additive group $\Omega_N^1/d\mathcal{O}_N$. This action again preserves the vertex algebra structure on V. To apply the construction of § 17.2.3, we need a lifting of the $(W_N, \mathrm{Aut}\,\mathcal{O}_N)$–structure $\widehat{S}$ on S to a $(\widetilde{W}_N, \widetilde{\mathrm{Aut}\mathcal{O}}_N)$–structure.

Such liftings form a *gerbe* on S. If this gerbe can be trivialized, i.e., there exists a global lifting, then, applying the construction of § 17.2.12, we attach to $\widehat{M}_N$ a vector bundle with a flat connection on S. The sheaf of horizontal sections of this bundle is then a sheaf of vertex algebras on S. This is a sheaf of *chiral differential operators* on S (corresponding to $\widetilde{S}$), introduced in [**MSV, GMS**]. The obstruction to the triviality of this gerbe (and hence to the existence of a globally defined sheaf of chiral differential operators) is described in [**GMS**].

18.5.7. The case of a flag manifold. In the case when S is the flag manifold G/B of a simple Lie algebra $\mathfrak{g}$, one can construct the sheaf of chiral differential operators in a more direct way (see [**MSV**]).

The action of $\mathfrak{g}$ on the flag manifold G/B gives rise to a homomorphism $\mathfrak{g}((t)) \to \mathcal{T}_N$, where $N = \dim G/B$. According to Theorem 12.3.6, it can be lifted to a homomorphism $\widehat{\mathfrak{g}} \to A_N^{\leq 1}$ of level $-h^\vee$ (moreover, it gives rise to a homomorphism of vertex algebras $V_{-h^\vee}(\mathfrak{g}) \to M_N$). Hence we obtain an embedding of the constant Lie subalgebra $\mathfrak{g}$ of $\widehat{\mathfrak{g}}$ into $\widetilde{W}_N$. This gives us a $\mathfrak{g}$–action on M_N, which can be exponentiated to a $(\mathfrak{g}, B)$–action on $\widehat{M}_N$.

On the other hand, G/B has a natural $(\mathfrak{g}, B)$–structure, namely, the B–bundle $G \to G/B$. Applying the localization construction of § 17.2.3 to the Harish-Chandra pair $(\mathfrak{g}, B)$ instead of $(\widetilde{W}_N, \widetilde{\mathrm{Aut}\,\mathcal{O}}_N)$, we obtain the sheaf of chiral differential operators on G/B.

Another important example is the case when S is an affine algebraic group G. In this case there is no obstruction to the existence of the chiral differential operators. They are constructed in [**AG**].

18.6. Bibliographical notes

In the analytic setting the relation between loop groups and moduli spaces of bundles is discussed in the book by A. Pressley and G. Segal [**PS**]. For a careful discussion of formal gluing of bundles and the description of the moduli stack of G–bundles, see [**BL1, LS**]. The Uniformization Theorem 18.1.6 for bundles is due to A. Beauville, Y. Laszlo, V. Drinfeld, and C. Simpson [**BL2, DS**] (see also [**Te**]). The Picard group of $\mathfrak{M}_G(X)$ is described in [**LS, KN, BLS**]. For a review of these topics, see C. Sorger's lectures [**Sor2**].

The relationship between the spaces of conformal blocks of integrable representations of affine algebras, moduli spaces, and the Verlinde formula is discussed in [**TUY, Fa2, KNR, BL1, LS, Sor1, BaKi**]. The definition of the projectively flat connection in this case is given in [**BK, Hi, TUY, Fa1, ADW, Fel, Ts, La, BaKi**].

Theorem 18.4.4 is due to B. Feigin and E. Frenkel [**FF7, F1, F6**] (see also [**F2, F4, F7**]). The exposition in § 18.4.9 follows the work of A. Beilinson and V. Drinfeld [**BD3**].

The chiral de Rham complex and the sheaf of chiral differential operators were introduced by F. Malikov, V. Schechtman and A. Vaintrob [**MSV**]. These sheaves have been studied in detail in a series of papers by V. Gorbounov, F. Malikov and V. Schechtman [**GMS**]. Beilinson and Drinfeld have independently given a similar construction in the language of chiral algebras [**BD4**]. An orbifold version of the chiral de Rham complex was introduced in [**FrSz2**].

CHAPTER 19

Chiral Algebras

In § 6.5, we established the geometric meaning of the vertex operation $Y(\cdot, z)$ in a quasi-conformal vertex algebra V. Given a curve X and a point $x \in X$, we have a canonical meromorphic section $\mathcal{Y}_x$ of $\mathcal{V}^*$ on $D_x^\times$, with values in $\operatorname{End} \mathcal{V}_x$. It is our goal in this section to spell out an intrinsic formulation of the structure carried by the vertex algebra bundle $\mathcal{V}$, so as to arrive at the direct definition of a vertex algebra bundle, or *chiral algebra*, on an arbitrary smooth curve X, due to Beilinson and Drinfeld [**BD4**]. Chiral algebras provide a coordinate-free approach to the axiomatic structure of operator product expansions on algebraic curves, thereby providing a reformulation of vertex algebras without any reference to the disc. To arrive at chiral algebras, we first assemble the collection of operations $\mathcal{Y}_x$ for varying points in X into a single operator $\mathcal{Y}^2$ between sheaves on X^2. We must then formulate the composition or OPE of vertex operators in terms of the new operation $\mathcal{Y}^2$. The translation operator will be reflected in connections respected by the operation $\mathcal{Y}^2$, and the vacuum vector in an inclusion $\mathcal{O}_X \subset \mathcal{V}$.

In order to describe the OPEs, which involve vertex operators at different points and their interactions as the points collide, we must leave the world of vector bundles on the curve X, and consider instead more general quasi-coherent sheaves on powers of X. The existence of a flat connection on $\mathcal{V}$ and the horizontality of $\mathcal{Y}_x$ must now be addressed in a language of $\mathcal{D}$–modules.

We begin in § 19.1 by recalling some notions and notation from sheaf theory. In particular, in § 19.1.7 we motivate the definition of the Lie* algebras and chiral algebras. In § 19.2 we discuss the operation $\mathcal{Y}^2$ on the bundle $\mathcal{V}$ associated to a quasi-conformal vertex algebra. This gives us the prototype for the notion of chiral algebra introduced in § 19.3 following [**BD4**]. In this section we show that for any quasi-conformal vertex algebra V and any smooth curve X the sheaf $\mathcal{V} \otimes \Omega_X$ carries a chiral algebra structure. In § 19.4 we introduce Lie* algebras following [**BD4**] and show that they may be obtained from conformal algebras. Lie* algebras also provide the proper context for the study of the Lie algebra of Fourier coefficients of vertex operators. In § 19.5 we introduce modules over chiral algebras and explain how to construct them from modules over vertex algebras. Finally, in § 19.6 we summarize the coordinate-independent approaches to Kac–Moody and Virasoro algebras, bringing together elements from previous chapters of the book.

19.1. Some sheaf theory

Henceforth, "sheaf" will refer to a quasi-coherent sheaf of $\mathcal{O}$–modules, unless explicitly noted. For a morphism of schemes $f : X \to Y$, f^* and f_* will denote the usual pull-back and push-forward functors for $\mathcal{O}$–modules.

19.1.1. Operations on sheaves. Let X be a smooth curve, X^2 its square, $\Delta : X \to X^2$ the diagonal embedding, $j : X^2 \backslash \Delta \hookrightarrow X^2$ its complement, and $\pi_1, \pi_2 : X^2 \to X$ the two projections.

Let $\mathcal{M}, \mathcal{N}$ denote sheaves on X. We have the external tensor product

$$\mathcal{M} \boxtimes \mathcal{N} \cong \pi_1^* \mathcal{M} \otimes_{\mathcal{O}_{X^2}} \pi_2^* \mathcal{N}$$

of $\mathcal{M}$ and $\mathcal{N}$, on X^2. Let $\mathcal{M} \boxtimes \mathcal{N}(\infty\Delta)$ denote the sheaf on X^2 whose sections are sections of $\mathcal{M} \boxtimes \mathcal{N}$ with arbitrary poles on the diagonal:

$$\mathcal{M} \boxtimes \mathcal{N}(\infty\Delta) = \varinjlim \mathcal{M} \boxtimes \mathcal{N}(n\Delta).$$

In terms of the inclusion $j : X^2 \backslash \Delta \hookrightarrow X^2$, $\mathcal{M} \boxtimes \mathcal{N}(\infty\Delta)$ can be described as the extension to X^2 of the restriction of $\mathcal{M} \boxtimes \mathcal{N}$ to $X^2 \backslash \Delta$:

$$\mathcal{M} \boxtimes \mathcal{N}(\infty\Delta) = j_* j^* (\mathcal{M} \boxtimes \mathcal{N}).$$

The push-forward $\Delta_* \mathcal{M}$ is the extension of $\mathcal{M}$ by zero from the diagonal Δ to X^2:

$$\Delta_* \mathcal{M} = \mathcal{M} \otimes_{\mathcal{O}_X} \mathcal{O}_{X^2},$$

where the $\mathcal{O}_X$–module structure on $\mathcal{M}$ comes from the inclusion Δ. We will need two other ways to extend sheaves along the diagonal: extension Δ_+ by principal parts in the transversal direction, and extension $\Delta_!$ by delta–functions in the transversal directions. The distinction between the two is the use of functions in Δ_+ where we use differential forms in $\Delta_!$.

More precisely, define the extension $\Delta_! \mathcal{M}$ as

$$\Delta_! \mathcal{M} = \frac{\Omega \boxtimes \mathcal{M}(\infty\Delta)}{\Omega \boxtimes \mathcal{M}}.$$

Sections of $\Delta_! \mathcal{M}$ may be thought of as distributions (or measures) on X^2 with support on the diagonal and with values in $\mathcal{M}$, via the residue pairing. (We return to this point in § 19.1.5.) Using the residue map for differentials, we obtain a map

(19.1.1) $\mathrm{Res} : \Delta_! \mathcal{M} \to \Delta_! \mathcal{M}/\mathrm{Im}(d \otimes 1) \cong \Delta_* \mathcal{M}.$

While functions naturally pull back under morphisms, measures or volume forms naturally push forward. More generally, left $\mathcal{D}$–modules (such as the structure sheaf $\mathcal{O}$, whose sections are functions) naturally pull back, and right $\mathcal{D}$–modules (such as the canonical line bundle, whose sections are volume forms) naturally push forward. While the push-forward $\Delta_!$ is natural from the point of view of residues, it is nonetheless convenient to have a version without the insertion of Ω (the left $\mathcal{D}$–module version). Thus we set

$$\Delta_+ \mathcal{M} = \frac{\mathcal{O} \boxtimes \mathcal{M}(\infty\Delta)}{\mathcal{O} \boxtimes \mathcal{M}}.$$

These are principal parts of functions along the diagonal, with values in $\mathcal{M}$.

19.1.2. Push-forwards and differential operators. The definition above for the push-forward functor $\Delta_!$ has an inherent asymmetry. Letting $\sigma_{12} : X^2 \to X^2$ denote the transposition of the two factors, observe that it is not obvious how to compare $\Delta_! \mathcal{M}$ with its "transpose"

$$\sigma_{12}^* \Delta_! \mathcal{M} = \frac{\mathcal{M} \boxtimes \Omega(\infty\Delta)}{\mathcal{M} \boxtimes \Omega}.$$

The solution to this problem lies in the Taylor formula:

$$f(z) = e^{(z-w)\partial_w} \cdot f(w) = \sum_{k=0}^{\infty} \frac{(z-w)^k}{k!} \partial_w^k \cdot f(w).$$

Suppose the sheaf $\mathcal{M}$ carries a (left) action of the sheaf $\mathcal{D}$ of differential operators, so that we can make sense of the derivatives $\partial^k \psi$ of sections of $\mathcal{M}$. We can then use the Taylor formula to identify $\mathcal{M} \boxtimes \mathcal{O}$ with $\mathcal{O} \boxtimes \mathcal{M}$ on the formal completion of the diagonal (we must pass to formal completions in order to make sense of the infinite sums involved).

Note that if $\mathcal{M}$ is a vector bundle with a flat connection, then this identification is simply the infinitesimal form of the parallel translation by the connection (this "crystalline" point of view on flat connections is discussed in § A.2.3).

We may now apply this to "symmetrize" our push-forward:

19.1.3. Lemma. *Let $\mathcal{M}$ be a left $\mathcal{D}$–module on X. Then there is a canonical isomorphism of sheaves $\Delta_+\mathcal{M} \cong \sigma_{12}^* \Delta_+\mathcal{M}$, given in local coordinates by the formula*

$$\frac{1 \otimes \psi}{(z-w)^k} \mapsto e^{(z-w)\partial_w} \cdot \frac{\psi \otimes 1}{(z-w)^k} \quad (mod\ \mathcal{M} \boxtimes \mathcal{O})$$

$$= \sum_{m=0}^{k-1} \frac{\partial_w^m \psi \otimes 1}{m!(z-w)^{k-m}},$$

where ψ is a local section of $\mathcal{M}$. For a right $\mathcal{D}$–module $\mathcal{N}$, the above identification gives rise to an isomorphism $\Delta_!\mathcal{N} \cong \sigma_{12}^* \Delta_!\mathcal{N}$.

19.1.4. Delta-functions, revisited. Recall that for any smooth variety Z, the canonical line bundle ω_Z, whose sections are volume forms on Z, is a right module for the sheaf $\mathcal{D}_Z$ of differential operators. The action is defined by using the Lie derivative of differential forms: $x \cdot \tau = -Lie_\tau(x)$, where τ is a vector field and x is a section of ω_Z. Therefore if $\mathcal{F}$ is a left $\mathcal{D}$–module on Z, then

$$\mathcal{F}^r \overset{\text{def}}{=} \mathcal{F} \otimes \omega_Z$$

is a right $\mathcal{D}$–module on Z, and this operation establishes an equivalence of categories $(\cdot)^r$ between left and right $\mathcal{D}_Z$–modules.

Let X be a smooth curve. The sheaf $\Omega \boxtimes \Omega$ is isomorphic to ω_{X^2}, under a natural map which in local coordinates is expressed as

(19.1.2) $$dz \boxtimes dw \mapsto dz \wedge dw.$$

We also have

$$\Delta_!\Omega \simeq \omega_{X^2}(\infty\Delta)/\omega_{X^2}.$$

Let μ_Ω denote the composition of the identification $\Omega \boxtimes \Omega(\infty\Delta) \cong \omega_{X^2}(\infty\Delta)$ with the projection of $\omega_{X^2}(\infty\Delta)$ onto $\Delta_!\Omega$. This map is a morphism of right $\mathcal{D}$–modules which satisfies the skew–symmetry condition

(19.1.3) $$\sigma_{12} \circ \mu_\Omega \circ \sigma_{12} = -\mu_\Omega,$$

where $\sigma_{12} : X^2 \to X^2$ is the permutation of the factors. The sign is due to the evident skew–symmetry in (19.1.2).

We would like to "model" the $\mathcal{D}$–modules $\Delta_+\Omega$ and $\Delta_!\Omega$ in the concrete setting of formal power series (or formal distributions), establishing the beginning of a dictionary between the language of $\mathcal{D}$–modules and that of formal power series. The operation of differentiation provides the space of formal distributions $\mathbb{C}[[z^{\pm 1}, w^{\pm 1}]]$ with a natural structure of module over the algebra of differential operators $\mathbb{C}[[z, w]][\partial_z, \partial_w]$. The formal delta-function $\delta(z - w)$ (see formula (1.1.2)) satisfies the relations

$$(z - w)\delta(z - w) = 0, \qquad (\partial_z + \partial_w)\delta(z - w) = 0$$

in $\mathbb{C}[[z^{\pm 1}, w^{\pm 1}]]$. Therefore the $\mathbb{C}[[z, w]][\partial_z, \partial_w]$–submodule of $\mathbb{C}[[z^{\pm 1}, w^{\pm 1}]]$ generated by $\delta(z - w)$ is spanned by $\partial_w^n \delta(z - w), n \geq 0$. Moreover, by Lemma 1.1.4, it may be identified with the subspace of all $f \in \mathbb{C}[[z^{\pm 1}, w^{\pm 1}]]$ satisfying $(z - w)^N f(z, w) = 0$ for large enough N.

The $\mathbb{C}[[z, w]][\partial_z, \partial_w]$–module generated by $\delta(z - w)$ gives rise to a left $\mathcal{D}$–module on $D^2 = \operatorname{Spec} \mathbb{C}[[z, w]]$, supported on the diagonal $z = w$. Tensoring it with ω_{D^2}, we also obtain a right $\mathcal{D}$–module on D^2.

19.1.5. Lemma. *The assignments*

$$\frac{1}{(z - w)^{n+1}} \quad \mapsto \quad \frac{1}{n!}\partial_w^n \delta(z - w), \qquad n \geq 0,$$

$$\frac{dz \wedge dw}{(z - w)^{n+1}} \quad \mapsto \quad \frac{1}{n!}\partial_w^n \delta(z - w) \cdot (dz \wedge dw), \qquad n \geq 0,$$

induce isomorphisms of left (resp., right) $\mathcal{D}$–modules on D^2 between $\Delta_+\Omega$ (resp., $\Delta_!\Omega$) and the left (resp., right) $\mathcal{D}$–module generated by $\delta(z - w)$ (resp., $\delta(z - w)dz \wedge dw$). These isomorphisms are compatible with the interchange of factors σ_{12}.

19.1.6. It follows that we may translate statements about delta–functions with values in a vector space V into sheaf-theoretic statements about the pushforward $\Delta_!\mathcal{V}$ of the corresponding vector bundle. As we will indicate below, the entire theory of vertex operators, locality and associativity may be translated from the language of formal distributions, fields and their composition into the coordinate-free $\mathcal{D}$–module setting. Conversely, calculations with the resulting $\mathcal{D}$–modules may be carried out by modeling the $\mathcal{D}$–modules inside formal distributions.

19.1.7. Motivation: Algebras via sheaf theory. Let R be a commutative ring and M an R–Lie algebra. Thus M is equipped with an R–module homomorphism

$$[\cdot, \cdot] : M \otimes_R M \to M,$$

satisfying the skew–symmetry and Jacobi identities. We would like to reformulate this structure more geometrically, in the language of sheaves. Recall that R–modules are equivalent to (quasi-coherent) sheaves of $\mathcal{O}_X$–modules on the affine scheme $X = \operatorname{Spec} R$. Therefore an R–Lie algebra is the same thing as a quasi-coherent sheaf $\mathcal{M}$ on X, together with a skew–symmetric map

$$\mathcal{M} \otimes_{\mathcal{O}_X} \mathcal{M} \to \mathcal{M},$$

satisfying the Jacobi identity relating three compositions

$$\mathcal{M} \otimes_{\mathcal{O}_X} \mathcal{M} \otimes_{\mathcal{O}_X} \mathcal{M} \to \mathcal{M}.$$

Now observe that the $R \otimes_{\mathbb{C}} R$–module $M \otimes_{\mathbb{C}} M$ defines a sheaf, the external tensor product $\mathcal{M} \boxtimes \mathcal{M}$, on $X \times X$, and the R– (or $\mathcal{O}_X$–)tensor product is the pull-back

$$\mathcal{M} \otimes_{\mathcal{O}_X} \mathcal{M} = \Delta^*(\mathcal{M} \boxtimes \mathcal{M}),$$

where $\Delta : X \hookrightarrow X \times X$ is the diagonal embedding. An R–linear Lie bracket on M may therefore be interpreted as a homomorphism of $\mathcal{O}_X$–modules

$$\Delta^*(\mathcal{M} \boxtimes \mathcal{M}) \to \mathcal{M}.$$

The adjointness of the functors Δ^* and Δ_* now implies that the R–linear bracket on M is equivalent to a homomorphism of $\mathcal{O}_{X^2}$–modules

(19.1.4) $$\mathcal{M} \boxtimes \mathcal{M} \to \Delta_* \mathcal{M}.$$

Let X be a curve. Then this sheaf interpretation of the bracket may be paraphrased in the language of delta–functions, by using the isomorphism

$$\Delta_*(\mathcal{M}) \simeq \Omega \boxtimes \mathcal{M}(\Delta)/\Omega \boxtimes \mathcal{M},$$

which follows from the description of Ω given in formula (8.3.1). Then we may consider sections of $\Delta_* \mathcal{M}$ as expressions on X^2 of the form $f(w)\delta(z-w)dz$, where $f(w)$ is a section of $\mathcal{M}$, which are annihilated by multiplication by the equation $z - w$ of the diagonal. Given sections $\phi(z)$ and $\psi(w)$ of $\mathcal{M}$, we write their bracket in the form

$$[\phi(z), \psi(w)] = [\phi, \psi](w)\delta(z-w)dz.$$

In words, the *pointwise* (or R–linear) nature of the bracket on $\mathcal{M}$ is translated into a bilinear operation on $\mathcal{M} \boxtimes \mathcal{M}$ whose output is supported on the diagonal — we extend the bracket $[\phi(z), \psi(w)]$ to be defined for all z, w, but to be zero unless $z = w$.

Restricting to the fiber over $w = x$, we obtain a map

$$\mathcal{M} \otimes \mathcal{M}|_x \longrightarrow \mathcal{O}_x \otimes \mathcal{M}|_x$$

of sheaves on X, where $\mathcal{O}_x = (i_x)_* \mathbb{C}$ is the skyscraper at x. This map factors through the pointwise bracket $\mathcal{M}|_x \otimes \mathcal{M}|_x \to \mathcal{M}|_x$ making the fibers of $\mathcal{M}$ into $\mathbb{C}$–Lie algebras. For example, let $\mathfrak{g}$ be a $\mathbb{C}$–Lie algebra. Then $\mathcal{M} = \mathfrak{g} \otimes \mathcal{O}_X$ is an $\mathcal{O}_X$–Lie algebra. The bracket operation $\mathcal{M} \boxtimes \mathcal{M} \to \Delta_* \mathcal{M}$ on $\mathcal{M}$ may then be written locally as

(19.1.5) $$[A_1 \otimes f_1(z), A_2 \otimes f_2(w)] = [A_1, A_2] \otimes f_1(z)f_2(w)\delta(z-w)dz$$
$$= [A_1, A_2] \otimes (f_1 \cdot f_2)(w)\delta(z-w)dz,$$

where $g_i \in \mathfrak{g}$ and $f_i \in \mathcal{O}_X$.

This reformulation of pointwise brackets opens the way to numerous generalizations. A natural class of multiplications are those that are given by *local* (or equivalently, *differential*) expressions. Namely, we would like to define a bracket operation on sections ϕ, ψ of a sheaf $\mathcal{M}$, so that $[\phi, \psi](z)$ is not defined pointwise but depends only on the values of ϕ, ψ and all of their derivatives at z, i.e., on the jets of ϕ, ψ at z. Geometrically, this can be encoded in the data of a bilinear bracket operation on $\mathcal{M} \boxtimes \mathcal{M}$, with values supported on the *formal neighborhood* of the diagonal, the union of all finite order neighborhoods $n\Delta \subset X \times X$. Symbolically, such a multiplication operation should be written in the form

$$[\phi(z), \psi(w)] = \sum_{n \geq 0} [\phi, \psi]_n(w) \frac{1}{n!} \partial_w^n \delta(z-w)dz.$$

Sheaf theoretically, we arrive at the notion of a **–operation* on a sheaf $\mathcal{M}$,

$$\mathcal{M} \boxtimes \mathcal{M} \longrightarrow \Delta_! \mathcal{M},$$

where

(19.1.6) $$\Delta_! \mathcal{M} \simeq \Omega \boxtimes \mathcal{M}(\infty\Delta)/\Omega \boxtimes \mathcal{M}.$$

In order to formulate appropriate skew–symmetry and Jacobi identities for a Lie bracket for *–brackets, it is essential to pass to the world of $\mathcal{D}$–modules. For example, as we have seen in § 19.1.2, in order to see the symmetry of the sheaf $\Delta_! \mathcal{M}$ with respect to the involution on X^2 exchanging the factors, we need to have a $\mathcal{D}$–module structure on $\mathcal{M}$. The corresponding objects are called *Lie* algebras* [**BD4**]. We will introduce them in § 19.4, where we will also see that Lie* brackets give rise to $\mathbb{C}$–linear Lie brackets upon passage to the de Rham cohomology.

A simple example of a Lie* algebra is the sheaf $\mathcal{M} = \mathfrak{g} \otimes_{\mathbb{C}} \mathcal{D}_X$ where $\mathfrak{g}$ is a $\mathbb{C}$–Lie algebra as above. This is the $\mathcal{D}_X$–module induced from the $\mathcal{O}_X$–module $\mathfrak{g} \otimes_{\mathbb{C}} \mathcal{O}_X$. Since $\mathfrak{g} \otimes_{\mathbb{C}} \mathcal{D}_X$ is generated as a $\mathcal{D}_X$–module from $\mathfrak{g} \otimes_{\mathbb{C}} \mathcal{O}_X$, the Lie* operation (19.1.6) on it is uniquely determined by its restriction to $\mathfrak{g} \otimes_{\mathbb{C}} \mathcal{O}_X$ if we ask that the Lie* operation be a homomorphism of $\mathcal{D}_X$–modules. This operation will then satisfy the appropriate analogues of the Lie algebra axioms. Taking the de Rham cohomology of $\mathfrak{g} \otimes \mathcal{D}_X$, we obtain the sheaf $\mathfrak{g} \otimes \mathcal{O}_X$ that we started with, with its usual $\mathbb{C}$–linear (in fact $\mathcal{O}_X$–linear) Lie bracket.

We will also discuss a generalization of the above notion of a *–operation. Namely, we enlarge the domain of a local bracket by dropping the requirement that our inputs $[\phi(z), \psi(w)]$ be regular at $z = w$. In other words, we would like to be able to define the bracket of expressions of the form $\dfrac{\phi(z)\psi(w)}{(z-w)^n}$ as a sum of delta–functions in $z - w$. Sheaf-theoretically, this means extending a *–operation to a map

$$\mathcal{M} \boxtimes \mathcal{M}(\infty\Delta) \longrightarrow \Delta_! \mathcal{M}$$

satisfying the appropriate skew–symmetry and Jacobi identities (which we will formulate using the tools of $\mathcal{D}$–modules in § 19.3). Thus we arrive at precisely the kind of structure carried by the operator product expansion in a vertex algebra. The resulting objects are called *chiral algebras*. Passing to de Rham cohomology, we will recover from the chiral algebra formalism the structure of the Lie algebra $U(V)$ of the Fourier coefficients of vertex operators in V.

In the following sections we will see how to construct a chiral algebra on any smooth curve X from an arbitrary quasi-conformal vertex algebra.

19.2. Sheaf interpretation of OPE

Let us begin with the problem of describing operator product expansions in the language of sheaves. Recall the associativity property:

$$Y(A, z)Y(B, w) = Y(Y(A, z - w)B, w) = \sum_{n \in \mathbb{Z}} (z - w)^{-n-1} Y(A_{(n)} \cdot B, w)$$

(where in the right hand side we consider the expansion of $(z - w)^{-1}$ in positive powers of w/z). Thus the composition of two vertex operators may be considered as the vertex operator associated to

(19.2.1) $$Y(A, z - w) \cdot B = \sum_{n \in \mathbb{Z}} (z - w)^{-n-1} A_{(n)} \cdot B,$$

which is an element of $V((z - w))$.

In the coordinate-independent definition of the operation Y we acted on B, placed at the fixed point x, by A, placed at the variable (but nearby) point z on X. In order to define the OPE invariantly, we simply need to let x move along X as well. Hence we need to define an operation on X^2.

Let V be a quasi-conformal vertex algebra (see Definition 6.3.4), and $\mathcal{V}$ the vector bundle on a smooth curve X associated to V in § 6.5. Choose a formal coordinate z at $x \in X$, and use it to trivialize $\mathcal{V}|_{D_x}$. Then the sheaf $\mathcal{V} \boxtimes \mathcal{V}(\infty\Delta)$, when restricted to $D_x^2 = \operatorname{Spec} \mathbb{C}[[z, w]]$, may be described as the sheaf associated to the $\mathbb{C}[[z, w]]$–module $V \otimes V[[z, w]][(z - w)^{-1}]$. Similarly, the restriction of the sheaf $\Delta_+\mathcal{V}$ to D_x^2 is associated to the $\mathbb{C}[[z, w]]$–module $V[[z, w]][(z - w)^{-1}]/V[[z, w]]$.

19.2.1. Theorem. *Define a map of $\mathcal{O}$–modules $\mathcal{Y}_x^2 : \mathcal{V} \boxtimes \mathcal{V}(\infty\Delta) \to \Delta_+\mathcal{V}$ on D_x^2 by the formula*

$$\mathcal{Y}_x^2(f(z, w)A \boxtimes B) = f(z, w)Y(A, z - w) \cdot B \quad \operatorname{mod} V[[z, w]].$$

Then $\mathcal{Y}_x^2$ is independent of the choice of the coordinate z.

19.2.2. Proof. Following the argument we used in the proof of Theorem 6.5.4, we reduce the statement of the theorem to the identity

$$Y(A, z - w) = R(\mu_w)Y(R(\mu_z)^{-1}A, \mu(z) - \mu(w))R(\mu_w)^{-1}, \qquad A \in V,$$

for any $\mu(t) \in \operatorname{Aut} \mathcal{O}$ (here we use the notation $\mu_z(t) = \mu(z+t) - \mu(z)$, as in the proof of Theorem 6.5.4). This identity is obtained from formula (6.5.1) by substituting $z - w$ instead of z and $\mu_w(z - w) = \mu(z) - \mu(w)$ instead of $\rho(z)$ (then $\rho_z(t)$ gets substituted by $\mu_z(t)$).

19.2.3. Global version. Theorem 19.2.1 admits a global reformulation, independent of the point x. To see this, note that the sheaf $\Delta_+\mathcal{V}$ is supported on the diagonal in X^2, i.e., any section of $\Delta_+\mathcal{V}$ is killed by a sufficiently high power of the ideal defining the diagonal (i.e., in local coordinates, by $(z - w)^N$ for $N \gg 0$). It follows that any morphism from a sheaf $\mathcal{M}$ on X^2 to $\Delta_+\mathcal{V}$ must be zero away from the diagonal. Hence such a morphism is determined by its restriction to the formal completion of the diagonal. Equivalently, it suffices to describe such a morphism on D_x^2 for every $x \in X$. Thus we obtain the following corollary (compare with § 19.1.7):

19.2.4. Corollary. *The collection of morphisms $\mathcal{Y}_x^2$ for $x \in X$ define a sheaf morphism $\mathcal{Y}^2 : \mathcal{V} \boxtimes \mathcal{V}(\infty\Delta) \to \Delta_+\mathcal{V}$ on X^2.*

19.2.5. Horizontality of $\mathcal{Y}^2$. Recall that we have a flat connection on $\mathcal{V}$. Hence $\mathcal{V}$ carries a natural (left) action of the sheaf $\mathcal{D}$ of differential operators on X, i.e., $\mathcal{V}$ is a left $\mathcal{D}_X$–module. According to Theorem 6.6.3, in the trivialization of $\mathcal{V}$ induced by a local coordinate z, this action is determined by the formula

$$\partial_z : f(z)A \mapsto (\partial_z f(z))A + f(z)L_{-1}A.$$

The sheaves $\mathcal{V} \boxtimes \mathcal{V}(\infty\Delta)$ and $\Delta_+\mathcal{V}$ on D_x^2 are not vector bundles, but they have natural structures of $\mathcal{D}$–modules. Indeed, in the trivialization of $\mathcal{V}$ induced by our coordinate, the space of sections of $\Delta_+\mathcal{V}$ is identified with $V[[z, w]][(z-w)^{-1}]/V[[z, w]]$. It suffices to define the action of the derivatives ∂_z and ∂_w on a general section $f(z, w)A$. We do this as follows:

$$\partial_z : f(z, w)A \mapsto (\partial_z f(z, w))A, \qquad \partial_w : f(z, w)A \mapsto (\partial_w f(z, w)) + f(z, w)(L_{-1}A).$$

Note that this is the natural action coming from the identification

$$\Delta_+ \mathcal{V} \cong \frac{\mathcal{O} \boxtimes \mathcal{V}(\infty\Delta)}{\mathcal{O} \boxtimes \mathcal{V}}.$$

The action of differential operators on $\mathcal{V} \boxtimes \mathcal{V}(\infty\Delta)$ is constructed similarly, with ∂_z acting as $\partial_z + L_{-1}$ and ∂_w acting as $\partial_w + L_{-1}$ in the trivialization of $\mathcal{V}$ induced by our local coordinate. The independence of these actions of the choice of the coordinate follows from the fact that the connection operator $\partial_z + L_{-1}$ is coordinate-independent (see Theorem 6.6.3).

19.2.6. Proposition. *The map* $\mathcal{Y}^2$ *commutes with the action of the sheaf of differential operators, and so it is a homomorphism of* $\mathcal{D}$*-modules on* X^2.

19.2.7. Proof. Working from the definition of $\mathcal{Y}^2$ and the actions of ∂_z, ∂_w above, the identity

$$Y(L_{-1}A, z - w) \cdot B = \partial_z(Y(A, z - w) \cdot B).$$

which follows from Corollary 3.1.6, translates into the statement

$$\mathcal{Y}^2(\partial_z \cdot f(z,w)A \boxtimes B) = \partial_z \cdot \mathcal{Y}^2(f(z,w)A \boxtimes B),$$

while the identity

$$Y(A, z - w) \cdot (L_{-1}B) = \partial_w(Y(A, z - w) \cdot B) + L_{-1}(Y(A, z - w) \cdot B),$$

which follows from the translation axiom, becomes

$$\mathcal{Y}^2(\partial_w \cdot f(z,w)A \boxtimes B) = \partial_w \cdot \mathcal{Y}^2(f(z,w)A \boxtimes B),$$

proving the proposition.

19.2.8. Symmetry of $\mathcal{Y}^2$. Thanks to the locality property of the vertex operation Y, the map $\mathcal{Y}^2$ satisfies a symmetry property with respect to the interchange of factors $\sigma_{12} : X^2 \to X^2$. This is most easily seen from the skew-symmetry property of the vertex operation (see Proposition 3.2.5):

$$(19.2.2) \qquad Y(A, z - w)B = e^{(z-w)T}Y(B, w - z)A.$$

The expression $e^{(z-w)T}$ is the coordinate expression for parallel translation, using the vertex bundle connection on $\mathcal{V}$, from w to z. Comparing this with the coordinate form of the isomorphism $\Delta_+ \mathcal{M} \cong \sigma_{12}^* \Delta_+ \mathcal{M}$ of Lemma 19.1.3, we obtain:

19.2.9. Lemma. *The map* $\mathcal{Y}^2 : \mathcal{V} \boxtimes \mathcal{V}(\infty\Delta) \to \Delta_+ \mathcal{V}$ *satisfies* $\mathcal{Y}^2 = \sigma_{12} \circ \mathcal{Y}^2$, *under the canonical identification* $\Delta_+ \mathcal{V} \cong \sigma_{12}^* \Delta_+ \mathcal{V}$.

19.2.10. Moving point interpretation. In this section we present an alternative approach to constructing the operation $\mathcal{Y}^2$ directly from the operation $\mathcal{Y}_x$ on X. To do so, we simply view the collection of operators $\mathcal{Y}_x$ associated to every point in X as a single object. Let $\widehat{\Delta}$ be the formal completion of X^2 along the diagonal, and $\widehat{\pi}_i : \widehat{\Delta} \to X, i = 1, 2$, the corresponding projections. The fiber of $\widehat{\pi}_i$ over any point $x \in X$ is the disc D_x. For each x, the operator $\mathcal{Y}_x$ was a section of $\mathcal{V}^*$ with values in $\mathrm{End}\,\mathcal{V}_x$. It follows that the moving version of $\mathcal{Y}_x$ should be a section of a sheaf on $\widehat{\Delta}$ which along the fibers of $\widehat{\pi}_2$ varies as $\mathcal{V}^*$, and along the fibers of $\widehat{\pi}_1$ varies as $\mathrm{End}\,\mathcal{V}$. Moreover, $\mathcal{Y}_x$ has a pole at x; hence as x varies we obtain a pole along the diagonal in $\widehat{\Delta}$. Collecting these observations and dualizing, we obtain that the sections $\mathcal{Y}_x$ give rise to a morphism

(19.2.3) $$\widehat{\mathcal{Y}} : \mathcal{V} \boxtimes \mathcal{V} \to \mathcal{O} \boxtimes \mathcal{V}(\infty \Delta)$$

on $\widehat{\Delta}$. In order to obtain the morphism $\mathcal{Y}^2$ from it, we need the observation that we can "swap" $(\infty \Delta)$ from one side to the other:

19.2.11. Lemma. *There is a canonical isomorphism*

$$i : \mathrm{Hom}_{\mathcal{O}_{X^2}}(\mathcal{V} \boxtimes \mathcal{V}, \mathcal{O} \boxtimes \mathcal{V}(\infty \Delta)) \to \mathrm{Hom}_{\mathcal{O}_{X^2}}(\mathcal{V} \boxtimes \mathcal{V}(\infty \Delta), \Delta_+ \mathcal{V})$$

making the following diagram of sheaves on X^2 commutative:

$$
\begin{array}{ccc}
\mathcal{V} \boxtimes \mathcal{V} & \xrightarrow{a} & \mathcal{O} \boxtimes \mathcal{V}(\infty \Delta) \\
\downarrow & & \downarrow \\
\mathcal{V} \boxtimes \mathcal{V}(\infty \Delta) & \xrightarrow{i(a)} & \Delta_+ \mathcal{V}
\end{array}
$$

19.2.12. Proof. We consider the intermediate space of maps

$$b : \mathcal{V} \boxtimes \mathcal{V}(\infty \Delta) \to \mathcal{O} \boxtimes \mathcal{V}(\infty \Delta).$$

Any map a as above determines a map b by $\mathcal{O}_{X^2}$–linearity: we multiply any section of $\mathcal{V} \boxtimes \mathcal{V}(\infty \Delta)$ by a high enough power of $z - w$ to place it in $\mathcal{V} \boxtimes \mathcal{V}$. This assignment $a \mapsto b$ is an isomorphism, since b determines a by restriction and is also determined by its restriction. Given b, we define the homomorphism $i(a)$ by projecting b onto $\Delta_+ \mathcal{V}$. The map $b \mapsto i(a)$ is injective, since for any $v \in \mathcal{V} \boxtimes \mathcal{V}$, $a(v)$ is determined by the polar parts of $b((z - w)^{-N} v)$ for all N. It is also surjective, since we may determine a lift c of $i(a)$ as follows: for $n > 0$, we set the $(z - w)^n$ coefficient of $c(v)$ to be the $(z - w)^{-1}$–coefficient of $b((z - w)^{-(n+1)} v)$.

19.2.13. Corollary. *Under the above isomorphism i, the morphism $\widehat{\mathcal{Y}}$ is identified with the restriction of the morphism $\mathcal{Y}^2$ of Theorem 19.2.1 to $\widehat{\Delta} \subset X^2$.*

19.3. Chiral algebras

Recall from § 9.2.2 that the operator $\mathcal{Y}_x$ gives rise, via the residue pairing, to a map

$$\mathcal{Y}_x^\vee : \Gamma(D_x^\times, \mathcal{V}^r) \to \mathrm{End}\, \mathcal{V}_x,$$

where $\mathcal{V}^r$ denotes the sheaf $\mathcal{V} \otimes \Omega$. In other words, in order to obtain actual operators on the space $\mathcal{V}_x$, we need to use sections of $\mathcal{V}^r$ and not of $\mathcal{V}$. In particular, Fourier coefficients of vertex operators, which are naturally residues of meromorphic sections of $\mathcal{V}^r$ paired with $\mathcal{Y}_x$, are endomorphisms of $\mathcal{V}_x$. The collection of these endomorphisms is closed under Lie bracket, and forms the Lie algebra $U(\mathcal{V}_x)$. This suggests that the algebraic structure on the bundle $\mathcal{V}^r$ might be somewhat more natural than that on the vertex algebra bundle $\mathcal{V}$ – in particular, $\mathcal{V}^r$ should be a Lie algebra object in an appropriate category. This is indeed the case.

Let X be a smooth curve. The vector bundle $\mathcal{V}^r$ acquires a right $\mathcal{D}$–module structure from the flat connection on $\mathcal{V}$. Thus the sheaf $\mathcal{V}^r \boxtimes \mathcal{V}^r(\infty \Delta)$ is a right $\mathcal{D}$–module on X^2. Via the isomorphism of (19.1.3), we obtain

$$\mathcal{V}^r \boxtimes \mathcal{V}^r(\infty \Delta) \cong (\mathcal{V} \boxtimes \mathcal{V}(\infty \Delta))^r.$$

The sheaf $\Delta_! \mathcal{V}^r = (\Delta_! \mathcal{V})^r$ is also a right $\mathcal{D}$–module on X^2 (here we use the same notation $\Delta_!$ for the push-forward functor of left and right $\mathcal{D}$–modules). It follows that the operation $\mathcal{Y}^2$ has a "right" version, i.e., a homomorphism of the right $\mathcal{D}$–modules

$$\mu = (\mathcal{Y}^2)^r : \mathcal{V}^r \boxtimes \mathcal{V}^r(\infty\Delta) \to \Delta_! \mathcal{V}^r.$$

Now we want to describe the properties that μ satisfies.

19.3.1. Sheaves on X^3. In order to define the "composition" of a morphism such as μ, and thereby reformulate the axioms of vertex algebras in this geometric language, we need to consider the geometry of sheaves on the cube X^3 of our curve. Let $\Delta_{123} : X \to X^3$ denote the "small" diagonal, where all three points on X coincide, and let $\Delta_{ij} : X^2 \to X^3$ (for distinct $i, j \in \{1, 2, 3\}$) denote the diagonal divisors where the i and j points coincide. (We will also continue to use the notation Δ for the diagonal embedding of X in X^2.)

For a right $\mathcal{D}$–module $\mathcal{A}$ on X, we would like to describe its push-forward $\Delta_{123!}\mathcal{A}$ on X^3. It is more convenient to realize it using the composition

$$\Delta_{123} = \Delta_{23} \circ \Delta : X \to X^3,$$

by the formula

$$\Delta_{123!}\mathcal{A} \cong \Delta_{23!}\Delta_!\mathcal{A}.$$

Of course, we may replace the partial diagonal Δ_{23} by the two others, Δ_{13} or Δ_{12}, in this description. The advantage of this "iterated" description is that the push-forward of $\mathcal{A}$ along a *divisor* may be realized concretely in terms of principal parts of meromorphic $\mathcal{A}$–valued differentials, as we have seen in the case of $X \to X^2$, but in general the description is more complicated.

Let $\mathcal{A}$ be a right $\mathcal{D}$–module on X equipped with a $\mathcal{D}$–module morphism

$$\mu : \mathcal{A} \boxtimes \mathcal{A}(\infty\Delta) \to \Delta_!\mathcal{A}$$

on X^2. We may then define a "composition" of μ,

$$\mu_{1\{23\}} : \mathcal{A} \boxtimes \mathcal{A} \boxtimes \mathcal{A}(\infty(\cup\Delta_{ij})) \longrightarrow \Delta_{123!}\mathcal{A},$$

by first applying μ to the second and third arguments, and then applying μ to the first argument and the result:

$$
\begin{aligned}
\mathcal{A} \boxtimes \mathcal{A} \boxtimes \mathcal{A}(\infty(\cup\Delta_{ij})) \quad &\cong \quad \mathcal{A} \boxtimes (\mathcal{A} \boxtimes \mathcal{A}(\infty\Delta))(\infty(\cup\Delta_{ij})) \\
&\xrightarrow{\mu} \quad (\mathcal{A} \boxtimes \Delta_!\mathcal{A})(\infty(\cup\Delta_{ij})) \\
&\cong \quad \Delta_{23!}(\mathcal{A} \boxtimes \mathcal{A}(\infty\Delta_{13})) \\
&\xrightarrow{\mu} \quad \Delta_{23!}\Delta_!\mathcal{A} \\
&\cong \quad \Delta_{123!}\mathcal{A}.
\end{aligned}
$$

Note that in the middle we consider the result of the first application of μ as living on the third factor in X^3, using the fact that for any sheaf $\mathcal{M}$ on X^2 we have

$$\Delta_{23!}(\mathcal{M}(\infty\Delta_{13}))(\infty\Delta_{12}) \cong \Delta_{23!}(\mathcal{M}(\infty\Delta_{13})),$$

since $\Delta_{12} \cap \Delta_{23} = \Delta_{13} \cap \Delta_{23}$.

We may define other compositions of μ with itself, by changing the order in which we group the points. In particular we will be interested in the compositions $\mu_{\{12\}3}$ and $\mu_{2\{13\}}$. In a concrete (but informal) manner, we may write these different compositions as follows: given local sections A, B, C of $\mathcal{A}$ and a meromorphic function $f(x, y, z)$ on X^3 with poles allowed along the diagonals, we have

$$\mu_{1\{23\}}(f(x,y,z)A \boxtimes B \boxtimes C) = \mu(A \boxtimes \mu(f(x,y,z)B \boxtimes C)),$$
$$\mu_{\{12\}3}(f(x,y,z)A \boxtimes B \boxtimes C) = \mu(\mu(f(x,y,z)A \boxtimes B) \boxtimes C),$$
$$\mu_{2\{13\}}(f(x,y,z)A \boxtimes B \boxtimes C) = \sigma_{12} \circ \mu(B \boxtimes \mu(f(y,x,z)A \boxtimes C)).$$

Having defined the compositions of μ, we may now require $(\mathcal{A}, \mu)$ to satisfy the axioms of a Lie algebra, namely, the skew–symmetry and the Jacobi identities

$$[x, y] = -[y, x], \qquad [x, [y, z]] = [[x, y], z] + [y, [x, z]],$$

with μ playing the role of the bracket. This leads to the Beilinson–Drinfeld notion of chiral algebra:

19.3.2. Definition. A *chiral algebra* on X is a right $\mathcal{D}$–module $\mathcal{A}$, equipped with a $\mathcal{D}$–module homomorphism

$$\mu : \mathcal{A} \boxtimes \mathcal{A}(\infty\Delta) \to \Delta_!\mathcal{A}$$

on X^2 and with an an embedding (unit map) $\Omega \hookrightarrow \mathcal{A}$, satisfying the following conditions:

- (*skew-symmetry*) $\mu = -\sigma_{12} \circ \mu \circ \sigma_{12}$.
- (*Jacobi identity*) $\mu_{1\{23\}} = \mu_{\{12\}3} + \mu_{2\{13\}}$.
- (*unit*) The unit map is compatible with the homomorphism $\mu_\Omega : \Omega \boxtimes \Omega(\infty\Delta) \to \Delta_!\Omega$; that is, the following diagram is commutative:

$$
\begin{array}{ccc}
\Omega \boxtimes \mathcal{A}(\infty\Delta) & \to & \mathcal{A} \boxtimes \mathcal{A}(\infty\Delta) \\
\downarrow & & \downarrow \\
\Delta_!\mathcal{A} & \to & \Delta_!\mathcal{A}
\end{array}
$$

(where the horizontal maps are given by the unit map and identity, and the vertical maps by the definition of $\Delta_!$ and the map μ, respectively.)

19.3.3. Theorem. *For a quasi-conformal vertex algebra V and a smooth curve X, the sheaf $\mathcal{V}^r$ carries the structure of a chiral algebra, with chiral multiplication $\mu = (\mathcal{Y}^2)^r$ obtained from $\mathcal{Y}^2$.*

19.3.4. The Cousin complex. In order to prove Theorem 19.3.3 we will need the description of the push-forward $\Delta_{123!}$ in terms of the Cousin complex, which parallels the associativity property of vertex algebras (see Theorem 3.2.1). This is a complex calculating the cohomology of a sheaf on a variety with a stratification in terms of local cohomologies along the strata; see [**Ha**] for details. We will use it for the stratification of X^3 given by various diagonals. In the case of the canonical bundle $\Omega = \omega_X$, we obtain the exact sequence of sheaves (in fact, right $\mathcal{D}$–modules)

$$(19.3.1) \quad 0 \to \omega_{X^3} \to \omega_{X^3}(\infty(\cup\Delta_{ij})) \to \bigoplus_{i,j} \Delta_{ij!}\omega_{X^2}(\infty(\cup\Delta_{ij})) \to \Delta_{123!}\Omega \to 0.$$

The first map is the inclusion of regular sections of ω_{X^3} into sections with poles along the divisors Δ_{ij}. A section with poles gives its principal parts along the three diagonal divisors, which live in the space of forms on these divisors extended transversely by delta-functions. Finally, we obtain delta-functions along the small diagonal by taking principal parts of meromorphic expressions along any of the partial diagonals. The exactness of the sequence tells us that the alternating sum of principal parts of meromorphic expressions on the diagonal divisors Δ_{12}, Δ_{23} and

Δ_{13} corresponds to zero in the space of delta–functions $\Delta_{123!}\Omega$ if and only if they are the principal parts of a single global meromorphic expression on X^3. We may now prove, as a warmup for Theorem 19.3.3, the following result:

19.3.5. Lemma. *The canonical bundle Ω, equipped with the map μ_Ω defined in § 19.1.4, is a chiral algebra.*

19.3.6. Proof. The skew-symmetry is the content of formula (19.1.3). The Jacobi identity is a corollary of the Cousin resolution (19.3.1) of ω_{X^3}. Namely, the three different compositions of μ_Ω agree with the three different compositions

$$\omega_{X^3}(\infty(\cup\Delta_{ij})) \to \Delta_{ij!}\omega_{X^2}(\infty\Delta) \to \Delta_{123!}\Omega$$

as i, j vary over $\{1, 2, 3\}$. In fact, these two maps in the Cousin resolution agree with the two applications of μ_Ω in the definition of the composition. It follows that the alternating sum of these three compositions is equal to the square of the differential in the Cousin complex. Hence it is equal to zero, as required.

19.3.7. Proof of Theorem 19.3.3. The embedding $\mathbb{C}|0\rangle \hookrightarrow V$ gives rise to an embedding $\mathcal{O} \hookrightarrow \mathcal{V}$ and hence to a unit map $\Omega \hookrightarrow \mathcal{V}^r$. The unit property of this map is an immediate consequence of the vacuum axiom $Y(|0\rangle, z) = \mathrm{Id}$.

The skew–symmetry property of μ follows from the skew–symmetry (19.1.3) of the map μ_Ω, together with the symmetry property of the map $\mathcal{Y}^2$, established in Lemma 19.2.9. Note that in coordinates, the skew–symmetry property reads

$$\mu(f(x, y)A \boxtimes B) = -\sigma_{12} \circ \mu(f(y, x)B \boxtimes A).$$

Proceeding to the Jacobi identity, we write out the morphism $\mu_{1\{23\}}$ introduced in § 19.3.1 as follows. First pick a local coordinate (in the étale sense, as before) and the resulting trivialization of $\mathcal{V}$. Then for $A, B, C \in V$, $\mu_{1\{23\}}$ maps $A \boxtimes B \boxtimes C = A \boxtimes (B \boxtimes C)$ to

$$
\begin{aligned}
& A \boxtimes Y(B, w - u)C \bmod V[[w, u]] \\
= {} & \sum_n A \boxtimes (B_{(n)}C)(w - u)^{-n-1} \bmod V[[w, u]] \\
\mapsto {} & \sum_n \left(Y(A, z - u) \cdot (B_{(n)}C) \bmod V[[z, u]]\right)(w - u)^{-n-1} \bmod V[[w, u]] \\
= {} & \sum_n \left(\sum_k A_{(k)}B_{(n)}C(z - u)^{-k-1} \bmod V[[z, u]]\right)(w - u)^{-n-1} \bmod V[[w, u]].
\end{aligned}
$$

In other words, $\mu_{1\{23\}}(A \boxtimes B \boxtimes C)$ is the projection of the series

$$Y(A, z - u)Y(B, w - u)C$$

onto the space of delta-functions $\Delta_{123!}\mathcal{V}^r$. Similarly, the compositions

$$\mu_{2\{13\}}(A \boxtimes B \boxtimes C) \qquad \text{and} \qquad \mu_{\{12\}3}(A \boxtimes B \boxtimes C)$$

are projections of the compositions

$$Y(B, w - u)Y(A, z - u)C \qquad \text{and} \qquad Y(Y(A, z - w)B, w - u)C.$$

Now recall the associativity property (see Theorem 3.2.1) of the vertex algebra V (under the substitution $z \mapsto z - u, w \mapsto w - u$): for any $A, B, C \in V$ and $\phi \in V^*$, the three expressions

$$\phi(Y(A, z - u)Y(B, w - u)C) \in \mathbb{C}((z - u))((w - u)),$$
$$\phi(Y(B, w - u)Y(A, z - u)C) \in \mathbb{C}((w - u))((z - u)),$$
$$\phi(Y(Y(A, z - w)B, w - u)C) \in \mathbb{C}((w - u))((z - w))$$

are the expansions, in their respective domains, of the same rational function on X^3, with poles only on the diagonals $z = u$, $w = u$ and $z = w$.

In order to verify the Jacobi identity $\mu_{1\{23\}} = \mu_{\{12\}3} + \mu_{2\{13\}}$, it is enough to check that the difference between the two sides of the identity maps to zero under any linear functional $\phi : V \to \mathbb{C}$. Note that ϕ gives rise to a map $\Delta_{123!}\mathcal{V}^r \to \Delta_{123!}\Omega$, which we also denote by ϕ. We claim that after restricting to the formal neighborhood of the diagonal, we may lift $\phi \circ \mu_{1\{23\}}$ from $\Delta_{123!}\Omega$ to $\Delta_{23!}\omega_{X^2}(\infty\Delta)$, and similarly for the other two compositions. In coordinates this becomes the obvious statement that

$$\phi\left(\sum\left(\sum A_{(k)}B_{(n)}C(z - u)^{-k-1} \bmod V[[z, u]]\right)(w - u)^{-n-1} \bmod V[[w, u]]\right)$$

is the projection of

$$\phi\left(\sum\left(\sum A_{(k)}B_{(n)}C(z - u)^{-k-1}\right)(w - u)^{-n-1} \bmod V[[w, u]]\right).$$

More abstractly, this follows from Lemma 19.2.11, since we can replace the map $\mathcal{Y}^2$ by the map $\widehat{\mathcal{Y}}$ using Corollary 19.2.13.

The associativity property of V (see Theorem 3.2.1) implies that the above three expressions corresponding to the three terms of the Jacobi identity are principal parts of a single meromorphic differential. Therefore we obtain, as in the proof of Lemma 19.3.5, that the three terms in the Jacobi identity indeed sum up to zero. This completes the proof of Theorem 19.3.3.

19.3.8. Remark. Under the $\mathcal{D}$–modules-to-delta-functions dictionary, the chiral Jacobi identity becomes the Jacobi identity for vertex algebras [**FLM, FHL**], and the latter may then be used to give a more explicit proof of Theorem 19.3.3, like the one given in [**HL**] in the case when X is identified with a subset of $\mathbb{C}$.

19.3.9. Universal chiral algebras. Theorem 19.3.3 means that any quasi-conformal vertex algebra V gives rise to a collection of chiral algebras on *all* smooth algebraic curves simultaneously. Given a curve X, the corresponding chiral algebra is by definition $\mathcal{V}^r_X = \mathcal{V}_X \otimes \Omega_X$. These chiral algebras satisfy natural compatibilities, namely, if $f : X \to X'$ is an étale morphism, then we have an isomorphism $f^*(\mathcal{V}^r_{X'}) \simeq \mathcal{V}^r_X$, and these isomorphisms satisfy the obvious cocycle condition. Following Beilinson and Drinfeld, we call such a collection a *universal chiral algebra* (or a chiral algebra in the universal setting).

Thus, any quasi-conformal vertex algebra gives rise to a universal chiral algebra. Conversely, one can show that universal chiral algebras are essentially the same as quasi-conformal vertex algebras.

However, there are examples of chiral algebras which are not universal and hence do not come from vertex algebras by the above construction. Such an example can be constructed using the affine Kac-Moody vertex algebra $V_{-h^\vee}(\mathfrak{g})$ of critical level. It gives rise to a chiral algebra $\mathcal{V}_{-h^\vee}(\mathfrak{g})^r_X$ on any smooth algebraic curve X. Recall from Theorem 18.4.4 that the center $\mathfrak{z}(\widehat{\mathfrak{g}})$ of $V_{-h^\vee}(\mathfrak{g})$ is isomorphic to the commutative vertex algebra $\mathrm{Fun}\,\mathrm{Op}_{L_\mathfrak{g}}(D)$. For the corresponding chiral algebra we have $\mathfrak{z}(\widehat{\mathfrak{g}})_X^r = \mathcal{F}un_X \otimes \Omega_X$, where $\mathcal{F}un_X$ is the sheaf of algebras of functions on the scheme $\mathcal{O}p_{L_{\mathfrak{g}}}(X)$ of jets of $^L\mathfrak{g}$–opers on X introduced in § 18.4.7 (recall that the fiber of $\mathcal{O}p_{L_{\mathfrak{g}}}(X)$ at $x \in X$ is $\mathrm{Op}_{L_{\mathfrak{g}}}(D_x)$).

Now let us fix a smooth curve X and a regular $^L\mathfrak{g}$–oper ρ on X. It gives rise to a horizontal section of $\mathcal{O}p_{L_{\mathfrak{g}}}(X)$ and hence to a homomorphism of right $\mathcal{D}_X$–modules $h_\rho : \mathfrak{z}(\widehat{\mathfrak{g}})_X^r \to \Omega_X$. Let $\mathcal{V}_\rho(\mathfrak{g})^r$ be the quotient of $\mathcal{V}_{-h^\vee}(\mathfrak{g})_X^r$ by the chiral algebra ideal generated by the kernel of h_ρ. This quotient $\mathcal{V}_\rho(\mathfrak{g})^r$ is a chiral algebra. But it is defined only on X and is dependent on the choice of a $^L\mathfrak{g}$–oper ρ on X. Therefore it cannot possibly come from a vertex algebra.

Note that the fiber of $\mathcal{V}_\rho(\mathfrak{g}) = \mathcal{V}_\rho(\mathfrak{g})^r \otimes \Omega_X^{-1}$ at $x \in X$ is the $\widehat{\mathfrak{g}}_x$–module $\mathcal{V}_{\rho_x,x}$ corresponding to the restriction ρ_x of ρ to D_x (see § 18.4.6). It cannot be described as an Aut $\mathcal{O}$–twist of a $\widehat{\mathfrak{g}}$–module.

Another example of a chiral algebra that does not come from a vertex algebra is given in § 20.5.7.

19.4. Lie* algebras

Let L be a conformal algebra, to which we attach a vector bundle $\mathcal{L}$ on X with a flat connection. By Proposition 16.1.16, we have for $x \in X$ an operator

$$\mathcal{Y}_{x,-} \in \mathrm{End}\, \mathcal{L}_x \widehat{\otimes} \mathcal{L}_{\mathcal{K}_x}^* / \mathcal{L}_{\mathcal{O}_x}^*.$$

The bundle $\mathcal{L}$ is trivialized on D_x by the choice of local coordinate. We may then imitate Theorem 19.2.1, Corollary 19.2.4 and Proposition 19.2.6 in the conformal algebra context to obtain the following result:

19.4.1. Theorem. *Define a map*

$$\mathcal{Y}_-^2 : \mathcal{L} \boxtimes \mathcal{L} \to \Delta_+ \mathcal{L}$$

on X^2 by the formula

$$(19.4.1) \qquad \mathcal{Y}_-^2(A \boxtimes B) = Y_-(A, z - w) \cdot B \quad \mathrm{mod}\, V[[z,w]]$$

on D_x^2 for every $x \in X$. Then $\mathcal{Y}_-^2$ is a homomorphism of $\mathcal{D}$–modules which is independent of the choice of the coordinate z.

19.4.2. Remark. The important distinction between the maps $\mathcal{Y}^2$ for a vertex algebra and $\mathcal{Y}_-^2$ for a conformal algebra is that in a vertex algebra Y is well-defined as a Laurent series, and hence we can make sense of the principal part of $f(z,w)Y(A, z-w)B$, where f may have arbitrary poles on the diagonal. Since Y_- is only defined as a Laurent series modulo Taylor power series, the map $\mathcal{Y}_-^2$ only makes sense in general on $\mathcal{L} \boxtimes \mathcal{L}$ and not on $\mathcal{L} \boxtimes \mathcal{L}(\infty\Delta)$.

19.4.3. Definition. A *Lie* algebra* on X is a right $\mathcal{D}$–module $\mathcal{A}$, equipped with a $\mathcal{D}$–module morphism

$$\mu_- : \mathcal{A} \boxtimes \mathcal{A} \to \Delta_! \mathcal{A}$$

on X^2, satisfying the following conditions:

- (*skew-symmetry*) $\mu_- = -\sigma_{12} \circ \mu_- \circ \sigma_{12}$.
- (*Jacobi identity*) $(\mu_-)_{1\{23\}} = (\mu_-)_{\{12\}3} + (\mu_-)_{2\{13\}}$.

19.4.4. Lemma. *For any chiral algebra $(\mathcal{A}, \mu)$, the restriction $\mu_- : \mathcal{A} \boxtimes \mathcal{A} \to \Delta_! \mathcal{A}$ of μ to $\mathcal{A} \boxtimes \mathcal{A} \subset \mathcal{A} \boxtimes \mathcal{A}(\infty\Delta)$ defines a Lie* algebra $(\mathcal{A}, \mu_-)$.*

19.4.5. Lie* algebras from conformal algebras. Let L be a conformal algebra, and X a smooth curve. Then we have a map $\mathcal{Y}_-$ given by formula (19.4.1). Denote by $(\mathcal{Y}_-^2)^r$ the homomorphism $\mathcal{L}^r \boxtimes \mathcal{L}^r \to \Delta_+ \mathcal{L}^r$ obtained from the map $\mathcal{Y}_-^2$ by applying the functor $(\cdot)^r$ of tensoring with the canonical bundle. The following result is proved in the same way as Theorem 19.3.3.

19.4.6. Theorem. *For a conformal algebra L and a smooth curve X, the sheaf $\mathcal{L}^r$ carries the structure of a Lie* algebra, with chiral multiplication $\mu_- = (\mathcal{Y}_-^2)^r$. If L is the conformal algebra associated to a quasi-conformal vertex algebra V, then $\mathcal{L}^r$ is the Lie* algebra associated to the chiral algebra V^r.*

19.4.7. De Rham cohomology. Let $\mathcal{A}^l$ be a left $\mathcal{D}$–module on X, for example, a vector bundle with a flat connection. The de Rham complex of $\mathcal{A}^l$ is by definition the complex of sheaves

$$0 \to \mathcal{A}^l \xrightarrow{d} \mathcal{A}^l \otimes \Omega \to 0.$$

The map d is equal to the connection ∇ when $\mathcal{A}^l$ is a flat vector bundle (V, ∇). In general it is constructed from the action map of vector fields,

$$\mathcal{A}^l \otimes \Theta \to \mathcal{A}^l, \qquad a \otimes \xi \mapsto \xi \cdot a.$$

If $\mathcal{A}^l$ is a flat vector bundle, then the cohomology sheaf $\mathcal{H}_{\mathrm{dR}}^0(\mathcal{A}^l) = \operatorname{Ker} d$ is the sheaf of horizontal sections of $\mathcal{A}^l$.

Let $\mathcal{A} = \mathcal{A}^l \otimes \Omega$ be the corresponding right $\mathcal{D}$–module. Then the de Rham sequence of $\mathcal{A}^l$ may be interpreted as a left de Rham resolution of $\mathcal{A}$:

$$0 \to \mathcal{A} \otimes \Theta \to \mathcal{A} \to 0$$

placed in cohomological degrees 0 and -1. The differential is given by the right action of vector fields. The zeroth cohomology of this complex for $\mathcal{A}$ is identified with the first de Rham cohomology of $\mathcal{A}^l$.

19.4.8. Definition. For a right $\mathcal{D}$–module $\mathcal{A}$, its de Rham sheaf $h(\mathcal{A})$ is defined to be

$$h(\mathcal{A}) = \mathcal{A}/(\mathcal{A} \cdot \Theta) \simeq \mathcal{H}_{\mathrm{dR}}^1(\mathcal{A}^l),$$

the quotient of $\mathcal{A}$ by the total derivatives.

19.4.9. Theorem. *Let $(\mathcal{L}^r, \mu_-)$ be a Lie* algebra. Then*

(1) *the de Rham sheaf $h(\mathcal{L}^r)$ is a sheaf of $\mathbb{C}$–Lie algebras;*
(2) *the Lie algebra $h(\mathcal{L}^r)(D_x)$ on the disc acts on the fiber $\mathcal{L}_x^r$, as does the Lie algebra $h(\mathcal{L}^r)(\Sigma)$ on any Zariski open subset $\Sigma \subset X$ containing x;*
(3) *if $\mathcal{L}^r$ is the Lie* algebra associated to a conformal algebra L, then on the standard disc D we have isomorphisms of Lie algebras $\operatorname{Lie}(L)_+ \cong h(\mathcal{L}^r)(D)$ and $\operatorname{Lie}(L) \cong h(\mathcal{L}^r)(D^\times)$ (see § 16.1.6 for the definition of the Lie algebras $\operatorname{Lie}(L)_+ \subset \operatorname{Lie}(L)$).*

19.4.10. Proof. Consider the morphism

$$\mu_- : \mathcal{L}^r \boxtimes \mathcal{L}^r \to \Delta_! \mathcal{L}^r.$$

The residue map (19.1.1) identifies the de Rham sheaf $h(\Delta_! \mathcal{M})$ with $\Delta_* \mathcal{M}$. It follows that we have a natural morphism

$$h(\mu_-) : h(\mathcal{L}^r) \boxtimes h(\mathcal{L}^r) \to \Delta_* h(\mathcal{L}^r),$$

and hence, pulling back to X along Δ, a morphism

$$[,] : h(\mathcal{L}^r) \otimes h(\mathcal{L}^r) \to h(\mathcal{L}^r).$$

It is now a direct consequence of the Lie* algebra axioms for μ_- that $[,]$ satisfies the axioms of a Lie bracket, establishing part (1).

In order to construct the actions in part (2), we take the de Rham cohomology of μ_- along the first factor only. Then we obtain a map

(19.4.2) $$h(\mathcal{L}^r) \boxtimes \mathcal{L}^r \to \Delta_* \mathcal{L}^r.$$

Now restrict the sheaves involved to the formal completion $\widehat{\Delta}$ of X^2 along the diagonal, and push them forward to X via the projection $\widehat{\pi}_2$ on the second factor. Taking the fiber at x, we obtain an action of the Lie algebra $h(\mathcal{L}^r)(D_x) \otimes \mathcal{L}^r_x \to \mathcal{L}^r_x$ of $h(\mathcal{L}^r)(D_x)$ on the fiber $\mathcal{L}^r_x$.

The second action in (2) is obtained by composing the above action with the obvious Lie algebra homomorphism $h(\mathcal{L}^r)(\Sigma) \to h(\mathcal{L}^r)(D_x)$.

Statement (3) follows from the definitions.

Note also that if we restrict the map in (19.4.2) to the diagonal, we obtain a morphism

$$h(\mathcal{L}^r) \otimes \mathcal{L}^r \to \mathcal{L}^r,$$

which defines a Lie algebra action of $h(\mathcal{L}^r)$ on the sheaf $\mathcal{L}^r$.

19.4.11. Lie algebras from chiral algebras. We may apply the de Rham functor h in a similar fashion to a chiral algebra $(\mathcal{V}^r, \mu)$. The sheaf $h(\mathcal{V}^r)$ carries a Lie algebra structure as in Theorem 19.4.9,(1), which only depends on the associated Lie* algebra $(\mathcal{V}^r, \mu_-)$. However, the analog of (19.4.2) defines a morphism

$$h(\mathcal{V}^r) \boxtimes \mathcal{V}^r(\infty\Delta) \to \Delta_* \mathcal{V}^r.$$

Pushing this forward to the second factor from $\widehat{\Delta}$, we obtain the action of the larger Lie algebras of meromorphic sections on the fibers $\mathcal{V}^r_x$.

19.4.12. Proposition. *Let $(\mathcal{V}^r, \mu)$ be a chiral algebra. Then*

(1) *$h(\mathcal{V}^r)(D_x^\times)$ and $h(\mathcal{V}^r)(\Sigma)$, for any Zariski open subset $\Sigma \subset X$, are Lie algebras, and there is a natural Lie algebra homomorphism $h(\mathcal{V}^r)(\Sigma) \to h(\mathcal{V}^r)(D_x^\times)$;*

(2) *the Lie algebra $h(\mathcal{V}^r)(D_x^\times)$ acts on the fiber $\mathcal{V}^r_x$, i.e., there is a homomorphism $h(\mathcal{V}^r)(D_x^\times) \to \operatorname{End} \mathcal{V}^r_x$;*

(3) *if $(\mathcal{V}^r, \mu)$ is associated to a vertex algebra V, then we have an isomorphism $h(\mathcal{V}^r)(D^\times) \simeq U(V)$ (see § 4.1.1 for the definition of the Lie algebra $U(V)$).*

19.4.13. The Lie algebras $U(\mathcal{V}_x)$ and $U_\Sigma(V)$. Let L be a conformal algebra, and $\mathcal{L}^r$ the associated Lie* algebra. The $\operatorname{End}\mathcal{L}_x$–valued section $\mathcal{Y}_{x,-}$ of $\mathcal{L}^*_{\mathcal{K}_x}/\mathcal{L}^*_{\mathcal{O}_x}$ gives rise, via the residue pairing, to a map

$$\mathcal{Y}^\vee_{x,-} : \mathcal{L}^r(D_x) \to \operatorname{End}\mathcal{L}_x \cong \operatorname{End}\mathcal{L}^r_x.$$

By horizontality of $\mathcal{Y}_-$, the map $\mathcal{Y}^\vee_{x,-}$ factors through $h(\mathcal{L}^r)(D_x)$. It is immediate from the definitions that the resulting homomorphism of Lie algebras $h(\mathcal{L}^r)(D_x) \to \operatorname{End}\mathcal{L}^r_x$ coincides with the homomorphism described in Theorem 19.4.9,(2).

Similar identification can be made in the case of vertex algebras over a punctured disc. Recall from § 9.2.2 that for a quasi-conformal vertex algebra V, there is a natural map (6.5.3)

$$\mathcal{Y}^\vee_x : \mathcal{V}^r(D^\times_x) \to \operatorname{End}\mathcal{V}_x \cong \operatorname{End}\mathcal{V}^r_x$$

on $D^\times_x$ from meromorphic sections of $\mathcal{V}^r$ to $\operatorname{End}\mathcal{V}_x$. Namely, given a section s of $\mathcal{V}^r$, we obtain a linear operator $\mathcal{Y}^\vee_x(s) = \operatorname{Res}_x\langle\mathcal{Y}_x, s\rangle$ on $\mathcal{V}_x$. If s is a total derivative, then the operator $\mathcal{Y}^\vee_x(s)$ vanishes, so that the map $\mathcal{Y}^\vee_x$ factors through the de Rham cohomology $h(\mathcal{V}^r)(D^\times_x)$. The resulting Lie algebra homomorphism $h(\mathcal{V}^r)(D^\times_x) \to \operatorname{End}\mathcal{V}_x$ coincides with the homomorphism of Proposition 19.4.12.

Recall that in §§ 9.2.2 and 9.2.5 we used the notation $U(\mathcal{V}_x) = h(\mathcal{V}^r)(D^\times_x)$, $U_\Sigma(V) = h(\mathcal{V}^r)(\Sigma)$, and we considered the natural map $U_\Sigma(V) \to U(\mathcal{V}_x)$ between them. Proposition 19.4.12 has the following corollary, which is equivalent to Theorem 9.2.6.

19.4.14. Corollary. *Let $x \in X$, and let $\Sigma \subset X$ be a Zariski open subset. Then $U(\mathcal{V}_x)$ and $U_\Sigma(V)$ are Lie algebras, and the natural maps*

$$U_\Sigma(V) \to U(\mathcal{V}_x) \overset{\mathcal{Y}^\vee_x}{\to} \operatorname{End}\mathcal{V}_x$$

are Lie algebra homomorphisms.

19.4.15. Remark: Normal ordering. Let V be a quasi-conformal vertex algebra, and $(\mathcal{A} = \mathcal{V}^r, \mu = (\mathcal{Y}^2)^r)$ the associated chiral algebra. Fix a formal coordinate near $x \in X$, and let z, w denote the resulting coordinates on X^2 near (x, x). We may now imitate the normal ordering construction in the chiral language. Recall that the normally ordered product $:A(w)B(w):$ is the constant term in $(z - w)$ in the OPE of $A(z)B(w)$. Rewriting the constant term as the residue of $A(z)B(w)dz/(z - w)$, we obtain the following chiral interpretation. Given a section s of $\mathcal{A} \boxtimes \mathcal{A}(\infty\Delta)$, define

$$:s: = h \boxtimes \operatorname{id}(\mu(s/(z - w))) \in h \boxtimes \operatorname{id}(\Delta_!\mathcal{A}) \cong \Delta_*\mathcal{A}.$$

Namely, we divide s by the equation of the diagonal and apply the chiral operation. This is then followed by de Rham cohomology along the first factor, i.e., a residue in z. The resulting sheaf, supported on the diagonal, is naturally identified with $\mathcal{A}$ itself (though note that the de Rham functor $h \boxtimes \operatorname{Id}$ is *not* $\mathcal{O}$–linear along the first factor). The result of this operation is easily identified with the vector whose associated field is the normally ordered product:

19.4.16. Lemma. *For $A, B \in V$, the chiral normal ordering*

$$:A \boxtimes B: = h \boxtimes \operatorname{id}(\mu(A \boxtimes B/(z - w)))$$

agrees, as a section of $\mathcal{V}^r$ in the z trivialization, with the normally ordered product $:A(z)B(z):|0\rangle|_{z=0} \in V$.

19.5. Modules over chiral algebras

19.5.1. Definition. A *module* over a chiral algebra $\mathcal{A}$ on X is a right $\mathcal{D}$–module $\mathcal{R}$ on X equipped with a homomorphism of $\mathcal{D}$–modules on X^2

$$a : j_*j^*(A \boxtimes \mathcal{R}) \to \Delta_!(\mathcal{R}).$$

satisfying the following axioms (understood in the same way as the axioms of a chiral algebra):

- $a(\mu(f(z, y, z)A \boxtimes B) \boxtimes v) = a(A \boxtimes a(f(x, y, z)B \boxtimes v)) - \sigma_{12} \circ a(B \boxtimes a(f(y, x, z)A \boxtimes v));$
- the following diagram is commutative:

$$
\begin{array}{ccc}
\Omega \boxtimes \mathcal{R}(\infty\Delta) & \overset{\text{unit}\boxtimes 1}{\longrightarrow} & A \boxtimes \mathcal{R}(\infty\Delta) \\
\downarrow & & \downarrow \\
\Delta_!\mathcal{R} & \longrightarrow & \Delta_!\mathcal{R}
\end{array}
$$

Suppose that $\mathcal{R}$ is supported at a point $x \in X$ and denote its fiber at x by $\mathcal{R}_x$. Then $\mathcal{R} = i_{x!}(\mathcal{R}_x)$, where i_x is the embedding $x \to X$. Applying the de Rham functor along the second factor to the map a, we obtain a map

$$a_x : j_{x*}j_x^*(A) \otimes \mathcal{R}_x \to \mathcal{R},$$

where $j_x : (X\backslash x) \to X$. The chiral module axioms may be reformulated in terms of this map. Note that it is not necessary for $\mathcal{A}$ to be defined at x in order for this definition to make sense. If $\mathcal{A}$ is defined on $X\backslash x$, we simply replace $j_{x*}j_x^*(\mathcal{A})$ by $j_{x*}(\mathcal{A})$.

If M is a conformal module over a conformal vertex algebra V, we associate to it the space $\mathcal{M}_x$ as in § 7.3.6 (we assume for simplicity that L_0 acts on M semisimply with integral eigenvalues). Then we have the right $\mathcal{D}$–module $i_{x!}(\mathcal{M}_x)$ on X supported at x.

19.5.2. Proposition. *The right $\mathcal{D}$–module $i_{x!}(\mathcal{M}_x)$ is a module over the chiral algebra $V^r = V \otimes \Omega_X$.*

19.5.3. Proof. The corresponding map a_x^M is defined as follows. Choose a formal coordinate z at x and use it to trivialize $V|_{\mathcal{D}_x}$ and $\mathcal{M}_x$ and to identify

$$i_{x!}(\mathcal{M}_x) = M((z))dz/M[[z]]dz.$$

Then

$$a_x^M(\imath_z(A) \otimes f(z)dz, B) = Y_M(A, z)B \otimes f(z)dz \quad \mathrm{mod} \quad M[[z]]dz.$$

The independence of a_x^M on z is proved in the same way as the independence of $\mathcal{Y}_{M,x}$ in Theorem 7.3.7. Note that by applying to a_x^M the de Rham cohomology functor, we obtain the map $\mathcal{Y}_{M,x}^\vee$ (see Theorem 7.3.7).

19.5.4. Twisted modules. Let H be an arbitrary finite group acting (generically with trivial stabilizers) on a smooth curve C. Let $X = C/H$ be the quotient curve, and $\nu : C \to X$ the quotient map, ramified at the fixed points of H. Denote by $\overset{\circ}{C} \subset C$ the open dense subset of points in C whose stabilizer in H is the identity element. Let $\overset{\circ}{X} \subset X$ be the image of $\overset{\circ}{C}$ in X and $\overset{\circ}{\nu} : \overset{\circ}{C} \to \overset{\circ}{X}$ the restriction of ν to $\overset{\circ}{C}$. Then $\overset{\circ}{C}$ is a principal H–bundle over $\overset{\circ}{X}$.

Let $\mathcal{A}$ be a chiral algebra on $\overset{\circ}{X}$ equipped with an action of H by automorphisms. Then the $\overset{\circ}{C}$–twist of $\mathcal{A}$,

$$\mathcal{A}^{\overset{\circ}{C}} = \overset{\circ}{C} \underset{H}{\times} \mathcal{A},$$

inherits the chiral algebra structure from $\mathcal{A}$. So $\mathcal{A}^{\overset{\circ}{C}}$ is a chiral algebra on X, and we can consider $\mathcal{A}^{\overset{\circ}{C}}$–modules supported at arbitrary points $x \in X$. If $\mathcal{A} = \mathcal{V}_X \otimes \Omega_X$, where V is a conformal vertex algebra on which H acts by automorphisms, then $\mathcal{A}^{\overset{\circ}{C}}$–modules may be constructed from *twisted* V–modules (see § 5.6 for the definition of twisted modules). This is explained in [**FrSz1**].

19.6. Global Kac–Moody and Virasoro algebras

In this section we summarize the coordinate-independent approaches to Kac–Moody and Virasoro algebras, combining ingredients from several prior sections.

19.6.1. Kac–Moody algebras. Recall that the affine Kac–Moody algebra $\widehat{\mathfrak{g}}$ is defined as the central extension

$$0 \to \mathbb{C}K \to \widehat{\mathfrak{g}} \to \mathfrak{g}((t)) \to 0$$

with the commutation relations given by formula (2.4.2). Using the natural action of $\mathrm{Aut}\,\mathcal{O}$ on $\widehat{\mathfrak{g}}$ by Lie algebra automorphisms, we can attach an affine Kac–Moody algebra

$$\widehat{\mathfrak{g}}_x = \widehat{\mathfrak{g}}(D_x) = \mathcal{A}ut_x \underset{\mathrm{Aut}\,\mathcal{O}}{\times} \widehat{\mathfrak{g}}$$

to any point x of a smooth curve X (see § 6.4.9). Since the cocycle defining $\widehat{\mathfrak{g}}$ as the central extension of the loop algebra $\mathfrak{g}((t))$ (see formula (2.4.2)) is $\mathrm{Aut}\,\mathcal{O}$–invariant, we can use it to define $\widehat{\mathfrak{g}}_x$ as the central extension

$$0 \to \mathbb{C}K \to \widehat{\mathfrak{g}}_x \to \mathfrak{g} \otimes \mathcal{K}_x \to 0.$$

In particular, $\widehat{\mathfrak{g}}_x$ splits as a vector space into the direct sum $\mathfrak{g} \otimes \mathcal{K}_x \oplus \mathbb{C}K$.

In § 16.4 we established that the hyperplane $\widehat{\mathfrak{g}}_1^*$ is isomorphic to the space of connections on the punctured disc, with its natural Poisson structure. This isomorphism is $\mathrm{Aut}\,\mathcal{O}$–equivariant, giving rise to an identification of hyperplanes in the dual to $\widehat{\mathfrak{g}}_x$ with connections on the trivial G–bundle on $D_x^\times$. This space carries a "tautological" Poisson structure, and the isomorphism preserves Poisson structures.

A relationship between affine algebras and connections was also established in § 8.5. By formula (8.5.2), the first two pieces of the filtration on the bundle $\mathcal{V}_k(\mathfrak{g})$ fit into an extension

(19.6.1) $$0 \to \mathcal{O} \to \mathcal{V}_k(\mathfrak{g})_{\leq 1} \to \mathfrak{g} \otimes \Theta \to 0.$$

Tensoring (19.6.1) with Ω, we obtain an extension

(19.6.2) $$0 \to \Omega \to \mathcal{V}_k(\mathfrak{g})_{\leq 1}^r \to \mathfrak{g} \otimes \mathcal{O} \to 0,$$

where $\mathcal{V}_k(\mathfrak{g})_{\leq 1}^r$ is the first term of the filtration on the right $\mathcal{D}$–module $\mathcal{V}_k(\mathfrak{g})^r = \mathcal{V}_k(\mathfrak{g}) \otimes \Omega$ associated to our vertex algebra bundle $\mathcal{V}_k(\mathfrak{g})$. The restriction of the sequence (19.6.2) to $D_x^\times$ (for $x \in X$), written in terms of a formal coordinate t at x, is an extension of the loop algebra $\mathfrak{g} \otimes \mathbb{C}((t))$ by the space of one–forms $\mathbb{C}((t))dt$. If we quotient out $\mathbb{C}((t))dt$ by the subspace of exact one–forms, we obtain an extension of the loop algebra by $H^0(D_x^\times, \Omega/d\mathcal{O}) \simeq \mathbb{C}$, which as a vector space coincides with the Kac–Moody central extension.

In order to see the Lie algebraic structure underlying (19.6.2), we make use of the theory of Lie* algebras. In § 16.1.5 we defined the affine conformal algebra $v_k(\mathfrak{g})$ as the subspace of $V_k(\mathfrak{g})$ spanned by the vacuum vector v_k and the vectors $J_n^a v_k, n < 0$ (together with the action of the translation operator T). The conformal algebra structure on $v_k(\mathfrak{g})$ is inherited from the polar part of the vertex algebra structure on $V_k(\mathfrak{g})$. The associated Lie* algebra is $\mathcal{G}_k^r = \mathcal{G}_k \otimes \Omega$, where

$$\mathcal{G}_k \overset{\text{def}}{=} \mathcal{A}ut_X \underset{\text{Aut}\,\mathcal{O}}{\times} v_k(\mathfrak{g})$$

(note that $\mathcal{G}_k^r$ is a Lie* subalgebra of $\mathcal{V}_k(\mathfrak{g})^r$). Applying the de Rham functor h to the Lie* algebra $\mathcal{G}_k^r$, we obtain a sheaf of Lie algebras $h(\mathcal{G}_k^r)$ on an arbitrary smooth curve X.

This suggests that the sheaf $h(\mathcal{G}_k)$ on X should be viewed as a "global" version of the affine Kac–Moody algebra. The following is a straightforward extension of Lemma 16.1.9:

19.6.2. Proposition.

(1) *The sheaf $h(\mathcal{G}_k^r)$ (in the Zariski topology) is isomorphic to the quotient of $\mathcal{V}_k(\mathfrak{g})_{\leq 1}^r$ by $d\mathcal{O} \subset \Omega \subset \mathcal{V}_k(\mathfrak{g})_{\leq 1}^r$. Thus, this quotient is a sheaf of Lie algebras.*

(2) *The restriction of $h(\mathcal{G}_k^r)$ to the punctured disc $D_x^\times$ is the above $\mathcal{A}ut_x$–twist $\widehat{\mathfrak{g}}_x$ of the affine algebra $\widehat{\mathfrak{g}}$.*

19.6.3. The Virasoro algebra. The above description of affine algebras may be carried over to the case of the Virasoro algebra with minor modifications. The Virasoro algebra was defined as an extension

$$0 \to \mathbb{C}C \to Vir \to \text{Der}\,\mathcal{K} \to 0,$$

but with a cocycle that was not presented in a coordinate-independent manner. The adjoint action of Vir on itself factors through an action of $\text{Der}\,\mathcal{K}$, since C is central. The action of the Lie subalgebra $\text{Der}_0\,\mathcal{O} \subset \text{Der}\,\mathcal{K}$ may be exponentiated to an action of $\text{Aut}\,\mathcal{O}$ on Vir. Thus we may consider a twisted version of Vir for any $x \in X$ (the local Virasoro algebra at x):

$$0 \to \mathbb{C}C \to Vir_x \to \text{Der}\,\mathcal{K}_x \to 0.$$

However, by its very definition, the adjoint action does *not* preserve the vector space direct sum decomposition $Vir \cong \text{Der}\,\mathcal{K} \oplus \mathbb{C}C$. Therefore Vir_x does not split into a direct sum $\text{Der}\,\mathcal{K}_x \oplus \mathbb{C}C$, even as a vector space. This is the reason why the Virasoro extension cocycle cannot be written in a coordinate-independent fashion.

In § 16.5 we established that the hyperplane Vir_1^* is isomorphic to the space of projective connections on the punctured disc, with its natural Poisson structure. This isomorphism is $\text{Aut}\,\mathcal{O}$–equivariant, giving rise to an identification of the hyperplane in the dual space to Vir_x with the space of projective connections on $D_x^\times$, equipped with a "tautological" Poisson structure.

This interpretation is related to the appearance of projective connections in § 8.2. Recall that the first two pieces of the filtration on the bundle $\mathcal{V}ir_c$ (see (8.2.2)) fit into an extension

(19.6.3) $$0 \to \mathcal{O}_X \to \mathcal{V}ir_{c,\leq 1} \to \Omega_X^{-2} \to 0.$$

Note that $Vir_{c,\leq 1}$ was denoted by $\mathcal{T}_c$ in § 8.2.1. The splittings of this extension (or its dual) are canonically identified with projective connections (see Lemma 8.2.3). Tensoring the sequence (8.2.2) with Ω, we obtain an extension

$$(19.6.4) \qquad\qquad 0 \to \Omega \to Vir^r_{c,\leq 1} \to \Theta \to 0,$$

where $Vir^r_{c,\leq 1}$ is the first term of the filtration on the right $\mathcal{D}$–module $Vir^r_c = Vir_c \otimes \Omega$. The restriction of the sequence (19.6.4) to $D^\times_x$ (for $x \in X$), written in terms of a local coordinate t, is an extension of the Lie algebra of vector fields $\operatorname{Der} \mathcal{K}_x = \mathbb{C}((t))\partial_t$ by the module of one–forms $\Omega_x = \mathbb{C}((t))dt$. If we quotient out by exact one–forms, we obtain an extension of $\operatorname{Der} \mathcal{K}_x$ by $\mathbb{C}$.

To see where the Lie algebra structure comes from, we recall that in § 16.1.5 we defined the Virasoro conformal algebra vir_c as the subspace of Vir_c spanned by the vacuum vedtor v_c and the vectors $L_n v_c, n \leq -2$. The associated Lie* algebra $\mathcal{L}^r_c$ is $\mathcal{L}_c \otimes \Omega$, where

$$\mathcal{L}_c \overset{\mathrm{def}}{=} \operatorname{Aut}_X \underset{\operatorname{Aut} \mathcal{O}}{\times} vir_c$$

(note that $\mathcal{L}^r_c$ is a Lie* subalgebra of Vir^r_c). Applying the de Rham functor h, we obtain a sheaf of Lie algebras $h(\mathcal{L}^r_c)$ on an arbitrary smooth curve X. This sheaf should be viewed as a "global" version of the Virasoro algebra. The following is a straightforward extension of Lemma 16.1.9:

19.6.4. Proposition.

(1) *The sheaf $h(\mathcal{L}^r_c)$ (in the Zariski topology) is isomorphic to the quotient of $Vir^r_{c,\leq 1}$ by $d\mathcal{O} \subset \Omega \subset Vir^r_{c,\leq 1}$. Thus, this quotient is a sheaf of Lie algebras.*

(2) *The restriction of $h(\mathcal{L}^r_c)$ to the punctured disc $D^\times_x$ is the above $\operatorname{Aut} \mathcal{O}$–twist Vir_x of the Virasoro algebra Vir.*

19.6.5. Sections of the global Kac–Moody and Virasoro algebras over affine curves.

Since $\mathcal{G}^r_k$ is a Lie* subalgebra of $\mathcal{V}_k(\mathfrak{g})^r$, we obtain that $h(\mathcal{G}^r_k)$ is a Lie subalgebra of $h(\mathcal{V}_k(\mathfrak{g})^r)$. In particular, $\widehat{\mathfrak{g}}_x = h(\mathcal{G}^r_k)(D^\times_x)$ is a Lie subalgebra of

$$U(\mathcal{V}_k(\mathfrak{g})_x) = h(\mathcal{V}_k(\mathfrak{g})^r)(D^\times_x).$$

Furthermore, for a Zariski open subset $\Sigma \subset X$, Proposition 19.6.2 identifies $h(\mathcal{G}^r_k)(\Sigma)$ with an extension of $\mathfrak{g}[\Sigma] = \mathfrak{g} \otimes \mathbb{C}[\Sigma]$ by $H^0(\Sigma, \Omega/d\mathcal{O})$, defined in the Zariski topology. This is a Lie subalgebra of

$$U_\Sigma(V_k(\mathfrak{g})) = h(\mathcal{V}_k(\mathfrak{g})^r)(\Sigma).$$

We have a canonical homomorphism of Lie algebras $h(\mathcal{G}^r_k)(\Sigma) \to \widehat{\mathfrak{g}}_x$, compatible with the homomorphism $U_\Sigma(V) \to U(\mathcal{V}_x)$.

Now suppose that $\Sigma = X \backslash x$. Then the image of the map

$$(19.6.5) \qquad\qquad H^0(\Sigma, \Omega/d\mathcal{O}) \to H^0(D^\times_x, \Omega/d\mathcal{O})$$

equals 0, by the residue theorem. Therefore the homomorphism $h(\mathcal{G}^r_k)(\Sigma) \to \widehat{\mathfrak{g}}_x$ factors through $\mathfrak{g}_{\mathrm{out}}(x) = \mathfrak{g} \otimes \mathbb{C}[X \backslash x]$, and we obtain natural embeddings

$$\mathfrak{g}_{\mathrm{out}}(x) \hookrightarrow \widehat{\mathfrak{g}}_x \hookrightarrow U(\mathcal{V}_k(\mathfrak{g})_x).$$

Likewise, $Vir_x = h(\mathcal{L}^r_c)(D^\times_x)$ is a Lie subalgebra of

$$U(Vir_{c,x}) = h(Vir^r_c)(D^\times_x).$$

For each Zariski open subset $\Sigma \subset X$, we have a Lie algebra $h(\mathcal{L}_c^r)(\Sigma)$ which is an extension of $\mathrm{Vect}(\Sigma)$ by $H^0(\Sigma, \Omega/d\mathcal{O})$. Then $h(\mathcal{L}_c^r)(\Sigma)$ is a Lie subalgebra of

$$U_\Sigma(\mathrm{Vir}_c) = h(\mathcal{V}ir_c^r)(\Sigma).$$

Furthermore, since the image of the map (19.6.5) equals 0 if $\Sigma = X\backslash x$, the homomorphism $h(\mathcal{L}_c^r)(X\backslash x) \to \mathrm{Vir}_x$ factors through $\mathrm{Vect}(X\backslash x)$. Hence we obtain natural embeddings

$$\mathrm{Vect}(X\backslash x) \hookrightarrow \mathrm{Vir}_x \hookrightarrow U(\mathrm{Vir}_{c,x}),$$

as claimed in § 9.2.1.

In the case of multiple points we obtain in the same way that the Lie algebra

$$\mathfrak{g} \otimes \mathbb{C}[X\backslash\{x_1, \ldots, x_n\}] \qquad (\mathrm{resp.}, \quad \mathrm{Vect}(X\backslash\{x_1, \ldots, x_n\}))$$

naturally acts on the tensor product $\bigotimes_{i=1}^n \mathcal{M}_{i,x_i}$, for any n–tuple of $\widehat{\mathfrak{g}}$–modules (resp., Vir–modules) of equal level (resp., central charge).

The above construction may be applied if we replace X by a smooth family of curves $\pi : \mathfrak{X} \to S$. In particular, we can use it to obtain a splitting

$$(19.6.6) \qquad\qquad \mathcal{A}_{\mathrm{out}} \to \mathrm{Vir}\widehat{\otimes}\mathcal{O}_{\widehat{\mathfrak{M}}_{g,1}},$$

which we needed in § 17.3.16. Here $\mathcal{A}_{\mathrm{out}}$ is the sheaf on $\widehat{\mathfrak{M}}_{g,1}$ whose fiber at $(X, x, z) \in \widehat{\mathfrak{M}}_{g,1}$ is $\mathrm{Vect}(X\backslash x)$. In the notation of § 17.3.9, consider the sheaf $\widetilde{\mathcal{L}}_c$ of sections of the vector bundle $\widehat{\mathfrak{M}}'_{g,1} \underset{\mathrm{Aut}\,\mathcal{O}}{\times} \mathrm{vir}_c$. It has a partial left $\mathcal{D}$–module structure along the fibers of the projection p. Let $\widetilde{\mathcal{L}}_c^r$ be the corresponding partial right $\mathcal{D}$–module. We can then apply the functor h of relative de Rham cohomology along p to $\widetilde{\mathcal{L}}_c^r$. The sheaf $h(\widetilde{\mathcal{L}}_c^r)(D_{\mathbf{x}}^\times)$, where $D_{\mathbf{x}}^\times$ is the family of punctured discs around $\mathbf{x} \subset \mathfrak{X}'_g$, is isomorphic to $\mathrm{Vir}\widehat{\otimes}\mathcal{O}_{\widehat{\mathfrak{M}}_{g,1}}$. The sheaf $h(\widetilde{\mathcal{L}}_c^r)(\mathfrak{X}'_g\backslash\mathbf{x})$ is nothing but an extension of $\mathcal{A}_{\mathrm{out}}$ by the sheaf of relative differentials on $\mathfrak{X}'_g\backslash\mathbf{x}$ along p modulo the exact ones. The image of the map of cohomology sheaves of relative differentials, which is a global version of the map (19.6.5), equals 0. Therefore the natural map $h(\widetilde{\mathcal{L}}_c^r)(\mathfrak{X}'_g\backslash\mathbf{x}) \to h(\widetilde{\mathcal{L}}_c^r)(D_{\mathbf{x}}^\times)$ factors through $\mathcal{A}_{\mathrm{out}}$ for the same reason as above. Hence we obtain the desired splitting (19.6.6).

The analogous splitting in the affine Kac–Moody case, which we needed in § 18.1.12, is proved in a similar fashion.

19.7. Bibliographical notes

The theory of chiral algebras and Lie* algebras was developed by A. Beilinson and V. Drinfeld [**BD4**] (for an exposition, see [**Gai**]). Our presentation follows [**BD4**] and lectures given by A. Beilinson at MIT in the Fall of 1995.

The connection between chiral algebras and vertex algebras is not developed in the above references. Such a connection was established by Y.-Z. Huang and J. Lepowsky [**HL**] in the case when the curve X is identified with a subset of $\mathbb{C}$. In this chapter we establish this connection for arbitrary curves. These results are due to the authors.

The sheaves of Virasoro and Kac–Moody algebras are considered in [**BMS, BS, Wi**].

CHAPTER 20

Factorization

Factorization algebras, introduced by Beilinson and Drinfeld in [**BD4**], provide a purely geometric reformulation of the definition of vertex algebras. This reformulation is based on a reinterpretation of the operator product expansion, as describing the behavior of the n–point correlation functions near the diagonals. Namely, the algebraic structure on the vertex algebra bundle $\mathcal{V}$ is encoded not in the vertex operator $\mathcal{Y}$ but in the data of a sheaf $\mathcal{V}_2$ on X^2 gluing $\mathcal{V}$ on the diagonal and $\mathcal{V} \boxtimes \mathcal{V}$ off the diagonal. One of the main advantages of this approach is that it is flexible enough to encompass non-linear versions of vertex algebras, the *factorization spaces*, which capture in a remarkably simple way the algebraic structure underlying moduli spaces on algebraic curves. One can view a factorization space as a "vertex (semi)group", from which vertex algebras may be obtained by linearization (i.e., passing to cohomologies of some sort).

This chapter is entirely based on the ideas and results of Beilinson and Drinfeld (some of them unpublished) of which we learned through their book [**BD4**] and through lectures given by Beilinson. We begin by introducing factorization algebras and explaining their connection with the vertex and chiral algebras. We then discuss the motivating example of a factorization space, the *Beilinson–Drinfeld Grassmannian*, describing the modifications of the trivial bundle on a curve. This and other naturally occurring factorization spaces serve as a geometric source for vertex algebras, through the process of linearization. We illustrate this idea with the affine Kac-Moody and lattice vertex algebras. We conclude with a brief sketch of the role of factorization in the study of Hecke correspondences on the moduli spaces of bundles. We introduce the notion of a Hecke eigensheaf on the moduli space of G–bundles and formulate the geometric Langlands conjecture. We then sketch the construction of the *chiral Hecke algebra* whose localization on the moduli space of bundles along the lines of Chapter 18 provides a possible way of establishing the geometric Langlands correspondence [**BD4**].

20.1. Factorization algebras

In this section we introduce the notion of a *factorization algebra* structure on a quasicoherent sheaf $\mathcal{V}$ on a smooth curve X. We will then see how this definition captures the structures carried by chiral algebras, leading to the equivalence between chiral and factorization algebras (and also vertex algebras, in the universal setting of § 19.3.9). Examples of factorization algebras arise naturally in geometry, and we suggest to the more geometrically oriented reader to see the discussion of the affine Grassmannian in § 20.3 first as a motivation.

Roughly speaking, the data of a sheaf $\mathcal{V}$ is the data of vector spaces $\mathcal{V}_x$ for every $x \in X$, which "vary well" when we vary x. In this language, a factorization algebra structure on $\mathcal{V}$ will consist of the data of a vector space $\mathcal{V}_{x_1,\dots,x_n} = \mathcal{V}_{\{x_i\}}$

for any collection of points of X, which vary well in all of the arguments and are equipped with two *factorization rules* (together with an appropriately defined *unit section* $1 \in \mathcal{V}(X)$). In the case of two points x, y, the factorization rules are the data of identifications

- $\mathcal{V}_{x,x} \simeq \mathcal{V}_x$, and
- $\mathcal{V}_{x,y} \simeq \mathcal{V}_x \otimes \mathcal{V}_y$ when $x \neq y$,

with the second isomorphism required to be symmetric in x, y. The general factorization rules read as follows:

- **Insensitivity to multiplicities and ordering:** whenever the collections $\{x_1, \ldots, x_n\}$ and $\{y_1, \ldots, y_m\}$ define the same underlying *subset* of X, we are given an identification of the spaces $\mathcal{V}_{\{x_i\}}$ and $\mathcal{V}_{\{y_j\}}$.
- **Factorization under disjoint union:** Whenever we can write the collection of points $\{x_i\}$ as a disjoint union of subcollections $\{y_j\}$ and $\{z_k\}$, we are given an identification of the vector space $\mathcal{V}_{\{x_i\}}$ with the tensor product $\mathcal{V}_{\{y_j\}} \otimes \mathcal{V}_{\{z_k\}}$.

These factorization rules together determine the vector spaces $\mathcal{V}_{\{x_i\}}$ for any collection of points as a tensor product of the fibers $\mathcal{V}_{x_i}$ of the original sheaf $\mathcal{V}$ on X at the points $x_i \in X$, taken without multiplicities. This might suggest that the structure we are trying to define on $\mathcal{V}$ is vacuous. But this is not so: actually, a lot of intricate structure is hidden in the way the factorization isomorphisms vary with the points x_i.

In order to give an honest factorization algebra, we are required to give a collection of quasicoherent sheaves $\mathcal{V}_n$ on X^n for every n, whose fibers at $\{x_1, \ldots, x_n\}$ will be the above vector spaces $\mathcal{V}_{x_1, \ldots, x_n}$, together with isomorphisms of sheaves reflecting the factorization rules on fibers. In particular, we have a quasicoherent sheaf $\mathcal{V}_1$ on X which we will denote simply by $\mathcal{V}$. On X^2 we are given a sheaf $\mathcal{V}_2$, which is symmetric ($\mathbb{Z}/2$–equivariant) under transposition of factors, and equipped with symmetric isomorphisms

- $\mathcal{V}_2|_\Delta \simeq \mathcal{V}$, and
- $\mathcal{V}_2|_{X^2 \setminus \Delta} \simeq \mathcal{V} \boxtimes \mathcal{V}|_{X^2 \setminus \Delta}$.

The unit axiom requires the existence of a global section $1 \in \mathcal{V}(X)$, so that inserting 1 at a varying point $y \neq x$ deforms $v \in \mathcal{V}_x = \mathcal{V}_{x,x}$ to an element $1 \otimes v \in \mathcal{V}_y \otimes \mathcal{V}_x = \mathcal{V}_{y,x}$. Sheaf–theoretically, this means that we have a subsheaf $\mathcal{O}_X \cdot 1 \subset \mathcal{V}$ so that

- The inclusion $\mathcal{O}_X \boxtimes \mathcal{V} \hookrightarrow \mathcal{V}_2$ away from the diagonal defined by the unit and factorization extends to all of X^2, and restricts to the identity

$$\mathcal{V} \simeq \mathcal{O}_X \boxtimes \mathcal{V}|_\Delta \to \mathcal{V}_2|_\Delta \simeq \mathcal{V}.$$

20.1.1. Remark. There is something disturbing in this behavior of the sheaf $\mathcal{V}_2$: although the factorization algebras we will encounter are quasicoherent sheaves on X^2 (in fact, flat over the diagonal), the dimensions of the fibers appear to *decrease* upon specialization to the diagonal divisor, rather than remain constant or increase as one might expect from semi-continuity! This phenomenon is inconceivable for finite rank vector bundles or coherent sheaves, but our sheaves will always be infinitely generated, and so this naive thinking does not apply.

20.1.2. To write down the general axioms for a factorization algebra, it is convenient not to keep track of the labeling of the points $x_1, \ldots, x_n$, but instead to work with collections of points labeled by an arbitrary finite set. Thus, rather than defining sheaves on X^n, we will define sheaves $\mathcal{V}_I$ on X^I for any finite set I, together with morphisms corresponding to different maps of sets. (Of course, all of the information is still contained in the sheaves $\mathcal{V}_n = \mathcal{V}_{\{1,2,\cdots,n\}}$.)

For finite sets I, J and a surjection $p : J \twoheadrightarrow I$, we have a diagonal embedding $\Delta_{J/I} : X^I \hookrightarrow X^J$, as the locus of J–tuples with incidences prescribed by the map $J \to I$. Note that by considering bijections $I \to I$, these "diagonal" maps include the action of permutations of I. On the other hand, we have the open locus $j_{J/I} : U^{J/I} \hookrightarrow X^J$ of J–tuples which break up into $|I|$ disjoint subsets such that $x_j \neq x_{j'}$ whenever $p(j) \neq p(j')$. For example, for the identity map $J = \{1,2\} \to I = \{1,2\}$ the locus $U^{J/I}$ is the complement of the diagonal in X^2.

20.1.3. Definition. Let X be a smooth complex curve. A *factorization algebra* on X consists of the following data:

- A quasicoherent sheaf $\mathcal{V}_{X,I}$ over X^I for any finite set I, which has no non-zero local sections supported at the union of all partial diagonals.
- Functorial isomorphisms of quasicoherent sheaves $\Delta^*_{J/I} \mathcal{V}_{X,J} \to \mathcal{V}_{X,I}$ over X^I, for surjections $J \twoheadrightarrow I$.
- (*factorization*) Functorial isomorphisms of quasicoherent sheaves

$$j^*_{J/I} \mathcal{V}_{X,J} \to j^*_{J/I}(\boxtimes_{i \in I} \mathcal{V}_{X,p^{-1}(i)})$$

  over $U^{J/I}$.
- (*unit*) Let $\mathcal{V} = \mathcal{V}_{X,\{1\}}$ and $\mathcal{V}_2 = \mathcal{V}_{X,\{1,2\}}$. A global section (the *unit*) $1 \in \mathcal{V}(X)$ with the property that for every local section $f \in \mathcal{V}(U)$ ($U \subset X$), the section $1 \boxtimes f$ of $\mathcal{V}_2|_{U^2 \setminus \Delta}$ (defined by the factorization isomorphism) extends across the diagonal, and restricts to $f \in \mathcal{V} \cong \mathcal{V}_2|_\Delta$.

20.1.4. Remark. Since by definition the sheaf $\mathcal{V}_n$ on X^n has no sections supported on the partial diagonals, it follows from the factorization isomorphisms that $\mathcal{V}_n$ is embedded in the sheaf of sections of $\mathcal{V}^{\boxtimes n}$ with poles along the partial diagonals. In other words, $\mathcal{V}_n$ may be considered as an extension to the entire X^n of the sheaf $\mathcal{V}^{\boxtimes n}$ restricted to the complement of the diagonals. In particular, it follows that any morphism $\{\mathcal{V}_n\} \to \{\mathcal{V}'_n\}$ of factorization algebras (i.e., a collection of morphisms of sheaves on X^n compatible with all structures) is in fact determined by the morphism $\mathcal{V} \to \mathcal{V}'$ of sheaves on X (i.e., the forgetful functor from factorization algebras to quasicoherent sheaves on X is faithful). In this way we may consider a factorization algebra as a quasicoherent sheaf $\mathcal{V}$ on X equipped with some extra structure.

20.1.5. The Ran space. The first axiom of a factorization algebra may be reformulated as saying that the collection $\mathcal{V}_I$ for varying I defines an inductive system of sheaves over the inductive system of spaces $\{X^I\}$ with respect to the diagonal maps $\Delta_{J/I}$, or equivalently, over the inductive system of spaces X^n with maps being all partial diagonals and permutations. This suggests considering the entire collection $\{\mathcal{V}_I\}$ as defining a single sheaf $\mathcal{V}_\bullet$ over a space representing all finite subsets (without multiplicities) of X. The latter is known as the *Ran space* $\mathcal{R}(X)$.

The space of finite subsets of a topological space X makes sense (and is well–studied, see e.g. [**CN**]) as a topological space, and is in fact contractible. However, $\mathcal{R}(X)$ does not make sense as an algebraic object, for example, as an ind-scheme in the strict sense (under a directed system of closed embeddings). Indeed, it is clear that there is no non-constant formal power series $f(z_1, z_2, z_3)$ in three variables which is not only symmetric with respect to the permutation of variables, but also satisfies the additional property that $f(z_1, z_1, z_2) = f(z_1, z_2, z_2)$. This implies that the gluing relation identifying (x, x, y) and (x, y, y) in the definition of $\mathcal{R}_{\leq 3}(X)$, the space of subsets of order at most three, precludes the existence even of germs of non-constant algebraic functions near a point of the diagonal $X \subset \mathcal{R}_{\leq 3}(X)$. Nonetheless, it is possible and convenient to speak of quasicoherent sheaves, left $\mathcal{D}$–modules, schemes and ind-schemes, etc., over the Ran space, as the corresponding objects over the inductive system defining $\mathcal{R}(X)$. They are defined by introducing the structures similar to those introduced in Definition 20.1.3. In particular, one can say that a factorization algebra $\mathcal{V}$ defines an $\mathcal{O}$–module $\mathcal{V}_\bullet$ on the Ran space. Note that the factorization axiom itself may be formulated in terms of the semigroup structure on $\mathcal{R}(X)$, corresponding to the operation of taking the union of finite subsets of X.

20.2. Factorization algebras and chiral algebras

In this section we describe how to identify the factorization algebra structure on a sheaf $\mathcal{V}$ over a smooth curve X with the structure of a chiral algebra on $\mathcal{V}^r = \mathcal{V} \otimes \Omega_X$.

Recall from § 19.3 that a chiral algebra structure on a sheaf $\mathcal{A}$ consists of a unit $\Omega_X \subset \mathcal{A}$, a right $\mathcal{D}$–module structure on $\mathcal{A}$, and a right $\mathcal{D}$–module morphism

$$\mu : \mathcal{A} \boxtimes \mathcal{A}(\infty \Delta) \to \Delta_! \mathcal{A}.$$

This structure may also be expressed as the following structure on the left $\mathcal{D}$–module $\mathcal{A} \otimes \Omega_X^{-1}$: a subsheaf $\mathcal{O}_X \subset \mathcal{V}$ and the operation

$$\mathcal{Y}^2 : \mathcal{V} \boxtimes \mathcal{V}(\infty \Delta) \longrightarrow \Delta_+ \mathcal{V}.$$

Note that in Theorem 19.2.1 we have defined such an operation $\mathcal{Y}^2$ on the bundle $\mathcal{V}$ with a flat connection associated to a vertex algebra V, using the coordinate expression

$$\mathcal{Y}^2(f(z, w)A \boxtimes B) = f(z, w)Y(A, z - w) \cdot B \quad \mathrm{mod}\, V[[z, w]].$$

Suppose we are given a factorization algebra $\{\mathcal{V}_I\}$. Then as part of the data we have an $\mathcal{O}_X$–module $\mathcal{V} \overset{\mathrm{def}}{=} \mathcal{V}_{\{1\}}$ on X and a unit $\mathcal{O}_X \subset \mathcal{V}$. Therefore we would like to see how to obtain a left $\mathcal{D}$–module structure on $\mathcal{V}$ and an operation such as $\mathcal{Y}^2$ from the gluing data of $\mathcal{V}_2$ on X^2.

20.2.1. The factorization connection. A remarkable consequence of the unit axiom for factorization algebras is that all sheaves $\mathcal{V}_n$ are automatically left $\mathcal{D}$–modules, in a way that is compatible with the factorization morphisms. Thus in fact a factorization algebra defines a left $\mathcal{D}$–module on the Ran space $\mathcal{R}(X)$.

Let us explain how to define a left $\mathcal{D}$–module structure on our sheaf $\mathcal{V}$ on X, which we wish to think of as the structure of a flat connection on $\mathcal{V}$. Recall from § A.2 that a connection on a sheaf $\mathcal{M}$ on a variety X is equivalent to the data of a map between the restrictions of $\pi_1^* \mathcal{M} = \mathcal{M} \boxtimes \mathcal{O}_X$ and $\pi_2^* \mathcal{M} = \mathcal{O}_X \boxtimes \mathcal{M}$ to the first–order neighborhood of the diagonal in $X \times X$, extending the identity map on the diagonal (note that such a map is necessarily an isomorphism, since it is so on the diagonal). Flatness of the connection may be expressed as the statement that this isomorphism may be extended to the formal neighborhood of the diagonal, compatibly with the groupoid structure (equivalence relation) on X defined by this formal neighborhood.

Now consider the maps

$$p_1^* \mathcal{V} = \mathcal{V} \boxtimes \mathcal{O}_X \longrightarrow \mathcal{V}_2 \longleftarrow \mathcal{O}_X \boxtimes \mathcal{V} = p_2^* \mathcal{V}$$

defined by the left and right multiplication by the unit. Both maps are isomorphisms when restricted to the diagonal, and hence on the full formal neighborhood of the diagonal. These maps give rise to the desired isomorphism of $p_1^* \mathcal{V}$ and $p_2^* \mathcal{V}$ on the first–order neighborhood of the diagonal, as well as its desired extension. Thus we have constructed a canonical flat connection (more precisely, left $\mathcal{D}$–module structure) on $\mathcal{V}$.

The same argument may be applied to define left $\mathcal{D}$–module structures on $\mathcal{V}_n$ by inserting the unit n times on the left or on the right, and identifying the two resulting sheaves on X^{2n} near the partial diagonal $X^n \subset X^{2n}$ by means of the factorization structure on $\mathcal{V}_{2n}$. The compatibility of these $\mathcal{D}$–module structures with the factorization isomorphisms is then automatic from the axioms.

20.2.2. Canonical decomposition. We now wish to describe the $\mathcal{D}$–module $\mathcal{V}_2$ as one that is glued from the sheaf $\mathcal{V} \boxtimes \mathcal{V}$ away from the diagonal and the sheaf $\mathcal{V}$ on the diagonal. Since, according to the axioms, $\mathcal{V}_2$ has no sections supported set–theoretically on the diagonal, it embeds as a subsheaf of its sheaf of sections with poles on the diagonal, which we identify with $\mathcal{V} \boxtimes \mathcal{V}(\infty\Delta)$. Moreover, $\mathcal{V}_2$ itself is characterized as the kernel of the map (which we will denote by $\mathcal{Y}^2$) sending $\mathcal{V}_2(\infty\Delta) = \mathcal{V} \boxtimes \mathcal{V}(\infty\Delta)$ to the sheaf of principal parts along the diagonal (the first local cohomology sheaf of $\mathcal{V}_2$ along Δ). However, the latter sheaf of principal parts is by definition the $\mathcal{D}$–module extension of the restriction of $\mathcal{V}_2$ to Δ, which we have identified with $\mathcal{V}$ itself. In other words we have a short exact sequence

(20.2.1) $$0 \to \mathcal{V}_2 \to \mathcal{V} \boxtimes \mathcal{V}(\infty\Delta) \xrightarrow{\mathcal{Y}^2} \Delta_+ \mathcal{V} \to 0.$$

The decomposition (20.2.1) shows that the factorization structure on $\mathcal{V}_2$ determines an operation $\mathcal{Y}^2$, of the same type as we obtained from the vertex operator $\mathcal{Y}$ on a vertex algebra bundle (see Theorem 19.2.1). The sheaf $\mathcal{V}_2$, and hence the map $\mathcal{Y}^2$, is naturally symmetric (equivariant) under the interchange of factors on X^2, by the factorization axioms. Passing to the right $\mathcal{D}$–module $\mathcal{A} = \mathcal{V}^r = \mathcal{V} \otimes \Omega_X$, we obtain a right $\mathcal{D}$–module morphism

$$\mu : \mathcal{A} \boxtimes \mathcal{A}(\infty\Delta) \to \Delta_! \mathcal{A}$$

which is *skew–symmetric*, due to the skew–symmetry of the canonical sheaf on X^2. By studying the Cousin complex on X^3 along the lines of § 19.3.4, one can see that the factorization structure gluing $\mathcal{V}_3$ out of $\mathcal{V}^{\boxtimes 3}$ away from the diagonals forces μ to satisfy the Jacobi identity.

This shows that a factorization algebra $\{\mathcal{V}_I\}$ gives rise to a chiral algebra $\mathcal{V}^r = \mathcal{V}_{\{1\}} \otimes \Omega_X$ on X.

20.2.3. Remark. The short exact sequence (20.2.1) is a special case of the canonical decomposition of a $\mathcal{D}$–module M on a smooth variety Y with respect to an open set $j : U \to Y$ and complementary closed set $i : Z \to Y$ (see, e.g., [**GM**]), which is an exact triangle

$$i_+ i^! M \to M \to j_+ j^! M$$

in the derived category of left $\mathcal{D}$–modules on Y. On the left we have the local cohomology sheaves of M — the derived functor of sections with (set–theoretic) support along Z — mapping to M, and on the right the (derived functor of) sections of M defined off Z. In our setting, Z is the smooth divisor X embedded via $i = \Delta$ into $Y = X^2$, $M = \mathcal{V}_2$ has no sections supported on the diagonal, and we have identified $j_+ j^! \mathcal{V}_2 = \mathcal{V} \boxtimes \mathcal{V}(\infty\Delta)$ using the factorization structure.

20.2.4. From chiral algebras to factorization algebras. Starting from a chiral algebra $\mathcal{A}$ on a curve X, one can define directly a factorization algebra structure on $\mathcal{V} = \mathcal{A} \otimes \Omega_X^{-1}$. Let $\mathcal{A}_2$ denote the right $\mathcal{D}$–module

$$\mathcal{A}_2 = \ker \mu \subset \mathcal{A} \boxtimes \mathcal{A}(\infty\Delta),$$

and $\mathcal{V}_2 = \mathcal{A}_2 \otimes \omega_{X^2}^{-1}$ the corresponding left $\mathcal{D}$–module. Then since $\Delta_! \mathcal{A}$ is supported on the diagonal, we have a $\mathcal{D}$–module isomorphism

$$\mathcal{V}_2|_{X^2 \setminus \Delta} \cong \mathcal{V} \boxtimes \mathcal{V}|_{X^2 \setminus \Delta}.$$

Similarly, since $\mathcal{A} \boxtimes \mathcal{A}(\infty\Delta)$ has no sections supported on the diagonal, we have an isomorphism

$$\mathcal{V}_2|_\Delta \cong \mathcal{V}.$$

The unit $\Omega_X \to \mathcal{A}$ provides an embedding $i : \Omega_X \boxtimes \mathcal{A} \hookrightarrow \mathcal{A}_2$ (since $\Omega_X \boxtimes \mathcal{A}$ is the kernel of the canonical projection $\Omega_X \boxtimes \mathcal{A}(\infty\Delta) \to \Delta_! \mathcal{A}$). Upon restricting to the diagonal this map becomes an isomorphism, with an inverse defined in coordinates by using the normal ordering construction (see Remark 19.4.15). This gives us a map $\mathcal{O}_X \boxtimes \mathcal{V} \to \mathcal{V}_2$ satisfying the conditions of a unit.

We have thus defined a sheaf $\mathcal{V}_2$ on X^2 satisfying the factorization properties with respect to $\mathcal{V}$. Similarly, Beilinson and Drinfeld define sheaves $\mathcal{A}_n$ and $\mathcal{V}_n = \mathcal{A}_n^l$ on X^n, as the intersections of the kernels of all compositions of the chiral operations to the restriction of $\mathcal{A}^{\boxtimes n}$ away from the diagonals. Informally speaking, whenever some of the n points collide we impose a condition on the polar part of our sections using μ. A close inspection of the Chevalley–Cousin complex yields the factorization properties for these sheaves, so that we have the following result:

20.2.5. Theorem. [BD4] *The categories of factorization algebras and chiral algebras on a curve X are equivalent, under the assignment $\{\mathcal{V}_I\} \mapsto \{\mathcal{A} = \mathcal{V}^r, \mu, \Omega_X \to \mathcal{A}\}$.*

In particular, given a quasi-conformal vertex algebra V, there is a canonical structure of factorization algebra on the vertex algebra bundle $\mathcal{V}$, with $\mathcal{V}_2$ defined as the kernel of the operation $\mathcal{Y}^2$.

20.2.6. Correlation functions and chiral homology. The factorization structure underlying a vertex algebra may be viewed as the ultimate expression of the short–distance singularities of correlation functions, as expressed by the bootstrap formula (10.3.10). As was indicated in § 10.3.6, the bootstrap conditions are precisely the conditions (coming from Theorem 4.5.1) characterizing the n–point functions of vertex operators as the points collide. Correspondingly, the definition of the sheaves $\mathcal{V}_n$ as the kernels of all compositions of the vertex operation $\mathcal{Y}^2$ imply that n–point correlation functions on any curve X are identified with the horizontal sections of $\mathcal{V}_n^*$ on X^n.

In other words, the global horizontal sections of the sheaf $\mathcal{V}_n^*$ are precisely the conformal blocks for $\mathcal{V}$: there is a canonical isomorphism $C(X, x, V) \cong \Gamma_\nabla(X^n, \mathcal{V}_n^*)$ for $n \geq 2$, or equivalently, there is a canonical isomorphism

$$H(X, x, V) \cong \mathrm{H}_{\mathrm{dR}}^{\mathrm{top}}(X^n, \mathcal{V}_n^r)$$

for $n \geq 2$ (see [**BD4**]). Since this condition holds for every n, we may take advantage of the natural compatibilities of the sheaves $\mathcal{V}_n$ for all n, as defining a single sheaf $\mathcal{V}_\bullet$ on the Ran space $\mathcal{R}(X)$. One can then define the de Rham cohomology on the Ran space, in terms of the limit of de Rham cohomologies over the inductive system defining $\mathcal{R}(X)$ [**BD4**]. This necessitates a shift in the grading of the de Rham cohomology compared to the grading that we have been using so far, so that the space of coinvariants $H(X, x, V)$ becomes isomorphic to the zeroth de Rham cohomology of the corresponding factorization sheaf $\mathcal{V}_\bullet$ on the Ran space of X,

$$H(X, x, V) = \mathrm{H}_{\mathrm{dR}}^0(\mathcal{R}(X), \mathcal{V}_\bullet).$$

Having identified the spaces of coinvariants with the zeroth de Rham cohomology of the space $\mathcal{R}(X)$ with coefficients in $\mathcal{V}_\bullet$, it is natural to consider the higher de Rham cohomology groups $\mathrm{H}_{\mathrm{dR}}^i(\mathcal{R}(X), \mathcal{V}_\bullet)$ as "derived" functors of the functor of coinvariants. These groups, introduced and studied by Beilinson and Drinfeld [**BD4**], are called *chiral homology* groups of X with coefficients in the factorization algebra $\mathcal{V}$. The chiral homology groups are important because there are examples of chiral (or vertex) algebras for which the usual spaces of coinvariants are equal to zero, but higher chiral homologies are non-trivial.

20.2.7. Discussion. A remarkable consequence of Theorem 20.2.5 is that all the structure of a factorization algebra is contained in the first few sheaves $\mathcal{V}_n$. More precisely, as we observed, the forgetful functor sending $\{\mathcal{V}_I\}$ to just the sheaf $\mathcal{V}$ is faithful, so that a factorization algebra may be considered as a sheaf on X with extra structures. Next, the functor remembering only $\mathcal{V}$ and $\mathcal{V}_2$ and their gluing data is fully faithful, in other words, the gluing data for $\mathcal{V}_2$ contains all the additional structure required on $\mathcal{V}$ (namely, the vertex operation $\mathcal{Y}^2$). Finally, the functor remembering $\mathcal{V}, \mathcal{V}_2$ and $\mathcal{V}_3$ is already an equivalence. In other words, the gluing properties of $\mathcal{V}_3$ contain all the restrictions inherent in the factorization structure. Thus we can recover all of the $\mathcal{V}_I$'s from only the first two, together with the knowledge that they are consistent with the gluing on X^3.

Another interesting feature is the fact that for factorization algebras the $\mathcal{D}$–module structure is a consequence of more "fundamental" geometric axioms, while for chiral algebras and vertex algebras it appeared as an input (in the latter case as the structure of the translation operator T).

The most remarkable feature of factorization algebras, however, is their purely geometric origin, as a linearized version of factorization spaces that we discuss below. This permits the construction and study of a variety of vertex (or chiral) algebras from basic examples of moduli problems satisfying factorization. Conversely, this perspective naturally leads one to the concept of *factorization spaces*, which are *non-linear* analogues of vertex or chiral algebras – a significant departure from the conventional theory of vertex algebras, where previously it was hard to imagine a version of the axioms that did not involve operations on a vector space!

20.3. The Grassmannian and factorization spaces

In this section we describe a remarkable property of the affine (or loop) Grassmannian $\mathcal{G}r$ associated to an algebraic group G, discovered by Beilinson and Drinfeld. This structure naturally gives rise to the notion of factorization space, a geometric and non-linear analogue of factorization (and hence chiral and vertex) algebras.

The *affine Grassmannian* associated to a reductive complex algebraic group G is the homogeneous space $\mathcal{G}r = G(\mathcal{K})/G(\mathcal{O})$. Here $G(\mathcal{K})$, where $\mathcal{K} = \mathbb{C}((z))$, is the algebraic loop group of G, and $G(\mathcal{O})$, where $\mathcal{O} = \mathbb{C}[[z]]$, is its subgroup of positive loops.

Let G be a group and H its subgroup. Then the quotient G/H may be viewed as the moduli space of H–torsors $\mathcal{E}$ together with a trivialization of the induced G–torsor $\mathcal{E} \underset{H}{\times} G$.

A $G(\mathcal{O})$–torsor is the same as a principal G–bundle $\mathcal{P}$ on the disc $D = \operatorname{Spec} \mathcal{O}$, and the induced $G(\mathcal{K})$–torsor corresponds to the G–bundle on the punctured disc $D^{\times} = \operatorname{Spec} \mathcal{K}$ obtained by restriction of $\mathcal{P}$. Therefore the affine Grassmannian may be viewed as the moduli space of G–bundles on the disc D together with a trivialization on $D^{\times}$. But it also has the following description:

20.3.1. Theorem. [BL2]

Fix a smooth algebraic curve X and a point $x \in X$. The choice of local coordinate z at x identifies $\mathcal{G}r$ with the moduli space $\mathcal{G}r_x$ of pairs $(\mathcal{P}, \phi)$, where $\mathcal{P}$ is a G–bundle on X and ϕ is a trivialization of $\mathcal{P}$ on $X \setminus x$.

20.3.2. The Grassmannian and the moduli spaces of bundles.

To understand this identification better and to make a connection with the material of Chapter 18, consider the moduli space of triples $(\mathcal{P}, s, \phi)$, where $\mathcal{P}$ is a G–bundle on X, s is its trivialization on D_x, and ϕ is its trivialization on $X \setminus x$. This moduli space is naturally identified with the loop group $G(\mathcal{K}_x)$. Indeed, the restrictions of s and ϕ to the punctured disc $D_x^{\times}$ give us two trivializations of $\mathcal{P}|_{D_x^{\times}}$, and so there exists an element g of $G(\mathcal{K}_x)$ such that $\phi|_{D_x^{\times}} = g \cdot s|_{D_x^{\times}}$. Conversely, given $g \in G(\mathcal{K}_x)$ we define a G–bundle $\mathcal{P}_g$ on X by gluing the trivial bundles on D_x and $X \setminus x$ by means of g considered as the transition function on $D_x^{\times}$. It follows from the definition that the bundle $\mathcal{P}_g$ comes equipped with trivializations on D_x and $X \setminus x$.

To make sense of this intuitive picture in the framework of algebraic geometry, we need to be more careful (compare with the discussion in § 6.2.3). We should start out by defining a functor assigning to each $\mathbb{C}$–algebra R the set of isomorphism classes of triples $(\mathcal{P}, s, \phi)$, where $\mathcal{P}$ is a G–bundle on $X \times \operatorname{Spec} R$, s is a trivialization of $\mathcal{P}$ on $\operatorname{Spec} \mathcal{O}_x \widehat{\otimes} R$, and ϕ is a trivialization of $\mathcal{P}$ on $(X \setminus x) \times \operatorname{Spec} R$. The statement proved in [BL2] is that this functor is represented by the ind-group scheme $G(\mathcal{K}_x)$.

Forgetting the trivialization s on D_x amounts to taking the quotient of $G(\mathcal{K}_x)$ on the right by its subgroup $G(\mathcal{O}_x)$ corresponding to changes of the trivialization s on D_x (note that any G–bundle on D_x is trivial). Therefore we obtain that the moduli space of pairs $(\mathcal{P}, \phi)$ is represented by the quotient $\mathcal{G}r_x = G(\mathcal{K}_x)/G(\mathcal{O}_x)$, and hence by the quotient $G(\mathcal{K})/G(\mathcal{O})$ if we choose a coordinate z at x and identify $\mathcal{O}_x \simeq \mathcal{O}$ and $\mathcal{K}_x \simeq \mathcal{K}$. Note also that the group $\operatorname{Aut} \mathcal{O}$ acts on $\mathcal{G}r$ through its actions on $\mathcal{K}$ and $\mathcal{O}$, and we may identify $\mathcal{G}r_x$ with the twist $Aut_x \underset{\text{Aut }\mathcal{O}}{\times} \mathcal{G}r$. This is how one obtains the identification of Theorem 20.3.1.

Likewise, forgetting the trivialization s on $X \setminus x$ amounts to taking the quotient of $G(\mathcal{K}_x)$ on the right by its subgroup G_{out} of regular functions $(X \setminus x) \to G$, which acts by changes of the trivialization ϕ on $X \setminus x$. Suppose that G is a semi-simple algebraic group. Then it is known from [**DSi**] that any G–bundle on $X \setminus x$ is trivial, and moreover, any G–bundle on $(X \setminus x) \times \operatorname{Spec} R$ becomes trivial after an appropriate base change $\operatorname{Spec} R' \to \operatorname{Spec} R$. Therefore in this case the moduli space $\widehat{\mathfrak{M}}_G(X)$ of pairs $(\mathcal{P}, s)$, where $\mathcal{P}$ is a G–bundle on X and s is its trivialization on D_x, may be obtained from the above moduli space of triples $(\mathcal{P}, s, \phi)$ by forgetting the section ϕ. Thus, we obtain an isomorphism $\widehat{\mathfrak{M}}_G(X, x) \simeq G(\mathcal{K}_x)_{\text{out}} \backslash G(\mathcal{K}_x)$ of Theorem 18.1.6.

Finally, let $\mathfrak{M}_G(X)$ be the moduli space of G–bundles on a smooth projective curve X. We have natural morphisms $\mathcal{G}r_x \to \mathfrak{M}_G(X)$ and $\widehat{\mathfrak{M}}_G(X, x) \to \mathfrak{M}_G(X)$ corresponding to forgetting the trivializations ϕ and s, respectively. Those are realized by the left action of G_{out} and the right action of $G(\mathcal{O}_x)$, respectively. Therefore we obtain, in the case when G is semi-simple, the isomorphism

$$\mathfrak{M}_G(X) \simeq G(\mathcal{K}_x)_{\text{out}} \backslash G(\mathcal{K}_x) / G(\mathcal{O}_x)$$

of Theorem 18.1.6.

20.3.3. The Grassmannian as an ind-scheme. To treat the Grassmannian rigorously in the framework of algebraic geometry, we must pass to the world of *ind-schemes*, see § A.1.2 for an overview and [**BD3, Ku3**] for a detailed discussion. The group $G(\mathcal{O})$ is the set of $\mathbb{C}$–points of a scheme that is the projective limit of schemes of finite jets into G. The loop group $G(\mathcal{K})$ is an ind-scheme, the inductive limit of closed subschemes, given by the subsets of loops with increasing bounded orders of poles (see § A.1.2 for more details). The quotient $\mathcal{G}r$ may be realized as a strict ind-scheme of *ind-finite type*, namely, it may be represented as the increasing union of a collection of schemes of finite type, under a system of closed embeddings. Moreover, $\mathcal{G}r$ is formally smooth, and it is reduced if and only if G has no non-zero characters (see [**BD3**], § 4.5.1).

For the rest of this chapter, we use the term "space" to denote a formally smooth ind-scheme of ind-finite type.

20.3.4. The abelian case. We first consider in detail the case $G = GL_1$, so that $G(\mathcal{K}) = \mathcal{K}^\times$ is the ind-group of invertible Laurent series, and $G(\mathcal{O}) = \mathcal{O}^\times$ is the group of invertible Taylor series. The group of $\mathbb{C}$–points of $\mathcal{K}^\times$ is the multiplicative group of all non-zero Laurent series, and the group of $\mathbb{C}$–points of $\mathcal{O}^\times$ consists of Taylor series whose constant term is non-zero. It follows that the set of $\mathbb{C}$–points of the quotient $\mathcal{G}r = \mathcal{K}^\times / \mathcal{O}^\times$ is in bijection with the integers. Indeed, up to multiplication by an invertible Taylor series, any non-zero complex Laurent series in z may be brought to the form z^N for $N \in \mathbb{Z}$. But the set $\mathcal{G}r(R)$ of R–points of $\mathcal{G}r$, where R is a $\mathbb{C}$–algebra with nilpotents, has more elements as we explained in Remark 18.1.7.

Note that since GL_1 is abelian, the set $\mathcal{G}r(R)$ is a group for any R. In particular, if $R = \mathbb{C}[\epsilon]/(\epsilon^2)$, then $\mathcal{G}r(R) = \mathbb{Z} \times (\mathcal{K}/\mathcal{O})$, with its additive group structure. The vector space $\mathcal{K}/\mathcal{O} = \operatorname{Lie} \mathcal{K}/\operatorname{Lie} \mathcal{O}$ appears as the Lie algebra of the ind-group $\mathcal{G}r$. In fact, as a group–functor $\mathcal{G}r$ is isomorphic to the product of the group $\mathbb{Z}$ by the infinite–dimensional formal group $\exp(\mathcal{K}/\mathcal{O})$, obtained by formally exponentiating the Lie algebra $\mathcal{K}/\mathcal{O}$ using the Baker–Campbell–Hausdorff formula.

The (set–theoretic) description $\mathbb{Z} = \mathcal{K}^\times/\mathcal{O}^\times$ comes up naturally in the assignment of a divisor to a line bundle. Let X be a smooth algebraic curve. Then any line bundle $\mathcal{L}$ on X (or equivalently, a principal GL_1–bundle) may be trivialized away from a finite number of points (though in contrast to the case of a semi-simple group G the number of points is in general greater than one). Indeed, a trivialization of $\mathcal{L}$ away from a collection of points $\{x_i\} \subset X$ is the same as a meromorphic section ϕ of $\mathcal{L}$ that is regular and non-vanishing on $X \setminus \{x_i\}$. Expanding such a section as a Laurent series near each x_i, we obtain a well-defined integer $\mathrm{ord}_{x_i}(\phi) \in \mathcal{K}^\times_{x_i}/\mathcal{O}^\times_{x_i}(\mathbb{C}) = \mathbb{Z}$, the order of the section at the point x_i. Thus $(\mathcal{L}, \phi)$ determines a divisor, the collection of integers $\mathrm{ord}_{x_i}(\phi) \in \mathbb{Z}$ or, equivalently, a collection $\{\mathrm{ord}_x(\phi)\}_{x \in X}$ with all but finitely many terms equal to zero.

Hence we may identify the set of isomorphism classes of pairs $(\mathcal{L}, \phi)$ as above with the restricted product

$$\underset{x \in X}{\prod{}'} \; \mathcal{K}^\times_x/\mathcal{O}^\times_x$$

(restricted means that we require all but finitely many factors to be zero). To obtain the set of $\mathbb{C}$–points of the Picard variety $\mathrm{Pic}(X)$, the moduli space of line bundles on X, we then need to take the quotient of this product by the action of the group of changes of the trivialization ϕ. This is just the multiplicative group $\mathbb{C}(X)^\times$ of non-zero meromorphic (or equivalently, rational) functions on X. Their divisors are the principal divisors on X. Thus, we obtain the usual identification of the set of isomorphism classes of line bundles on X with the quotient of the group of all divisors by its subgroup of principal divisors.

It is convenient to rephrase this description in terms of the group of the *adèles*. Let

$$\mathbb{A}_F = \underset{x \in X}{\prod{}'} \; \mathcal{K}_x = \{(f_x \in \mathcal{K}_x)_{x \in X} \,|\, f_x \in \mathcal{O}_x \text{ for all but finitely many } x\}$$

be the group of adèles of the field $F = \mathbb{C}(X)$ of rational functions on X,

$$\mathcal{O}_F = \underset{x \in X}{\prod} \; \mathcal{O}_x,$$

the group of integral adèles, and $\mathbb{C}(X)$ the subgroup of rational adèles embedded into $\mathbb{A}_F$ diagonally (by expanding a rational function in the formal neighborhood of each point).

Now we may write

$$\mathrm{Pic}(X) = \mathbb{C}(X)^\times \backslash \mathbb{A}^\times_F/\mathcal{O}^\times_F.$$

This description of the Picard variety as a double quotient makes evident a natural action of the group ind-scheme $\mathcal{K}^\times_x/\mathcal{O}^\times_x$, by "Hecke operators" or modifications of line bundles at x. At the level of $\mathbb{C}$–points, this group is $\mathcal{K}^\times_x/\mathcal{O}^\times_x(\mathbb{C}) = \mathbb{Z}$, and it acts on the Picard variety as follows:

$$\mathbb{Z} \ni n \; : \; \mathcal{L} \mapsto \mathcal{L}(nx).$$

The rest of the action appeared before in the construction of the Kac-Moody uniformization of moduli spaces of bundles discussed in § 18.1.1. Namely, the Lie algebra $\mathcal{K}_x$ acts formally transitively on the moduli space of line bundles equipped with a trivialization near x, but this action commutes with changes of trivialization and hence descends to a (still formally transitive) action of $\mathcal{K}_x/\mathcal{O}_x$ on $\mathrm{Pic}\,X$ itself.

Formally exponentiating this action we obtain a description of the full action of $\mathcal{K}_x^\times/\mathcal{O}_x^\times$ on $\operatorname{Pic} X$.

It is interesting to note that the group structure on the integers $\mathbb{Z} = \mathcal{K}_x^\times/\mathcal{O}_x^\times(\mathbb{C})$ arises naturally from the collision of points on X. If we deform a divisor with order n_i at x_i and n_j at x_j by letting the point x_j approach x_i, we obtain in the limit a new divisor with order $n_i + n_j$ at x_i. In other words, the group structure on $\mathbb{Z}$ can be recovered from the topology of the space of line bundles with meromorphic sections. This realization will be abstracted in the notion of factorization space below.

The factorization spaces give rise to factorization algebras, and hence to chiral (or vertex) algebras. In particular, we will see in § 20.4.4 that the factorization space structure on the Picard variety is captured in the structure of the lattice vertex algebra $V_{\sqrt{N\mathbb{Z}}}$ in such a way that the infinitesimal part $\mathcal{K}_x/\mathcal{O}_x$ corresponds to the Heisenberg vertex algebra π and the discrete part corresponds to the bosonic vertex operators. These structures also have analogues in the non-abelian case, where the infinitesimal part is captured by the Kac–Moody vertex algebra $V_k(\mathfrak{g})$ (see § 20.4.2) and the discrete part is captured by the Satake correspondence (see § 20.5.3). The two are ultimately combined in the chiral Hecke algebra (see § 20.5.6). In order to explain this in more detail, we need to generalize the notion of the Grassmannian.

20.3.5. The Beilinson–Drinfeld Grassmannian. We now return to the setting of the affine Grassmannians $\mathcal{G}r$ associated to general reductive algebraic groups G. Unlike $\mathcal{K}^\times/\mathcal{O}^\times$, the affine Grassmannian $\mathcal{G}r$ of a non-abelian group G is no longer a group, but it turns out to have a remarkable "commutative" algebraic structure, generalizing the abelian group structure on $\mathcal{K}^\times/\mathcal{O}^\times$. For $x_1, \ldots, x_n \in X$ we consider the space $\mathcal{G}r_{X,x_1,\ldots,x_n}$ of pairs $(\mathcal{P}, \phi)$, where $\mathcal{P}$ is a G–bundle on X and ϕ is a trivialization of $\mathcal{P}$ on $X \setminus \{x_i\}$. As we let the points x_i vary, we obtain the *Beilinson–Drinfeld Grassmannian*

$$\mathcal{G}r_{X,n} = \{(\mathcal{P} \in \mathfrak{M}_G(X), \{x_1, \ldots, x_n\} \in X^n, \phi - \text{ trivialization of } \mathcal{P}|_{X \setminus \{x_1,\ldots,x_n\}})\}$$

over X^n parameterizing G–bundles trivialized on the complement of n points. One can show that $\mathcal{G}r_{X,n}$ is a formally smooth ind-scheme over X^n. In other words, the fibers $\mathcal{G}r_{X,x_1,\ldots,x_n}$ "vary well" with the points $x_1, \ldots, x_n$. Note also that $\mathcal{G}r_{X,x_1,\ldots,x_n}$ does not depend on the ordering of the points $x_1, \ldots, x_n$.

Let us describe the fibers $\mathcal{G}r_{X,x_1,x_2}$ of $\mathcal{G}r_{X,2}$ over pairs $\{x_1, x_2\} \in X^2$.

First assume that $x_1 \neq x_2$ are two distinct points, so $\{x_1, x_2\} \in X \times X \setminus \Delta$. We claim that in this case $\mathcal{G}r_{X,x_1,x_2} \cong \mathcal{G}r_{X,x_1} \times \mathcal{G}r_{X,x_2}$. Indeed, there is a map $\mathcal{G}r_{X,x_1,x_2} \to \mathcal{G}r_{X,x_1} \times \mathcal{G}r_{X,x_2}$ obtained by restricting the data to $X \setminus x_i$ ($i = 1, 2$) in turn, obtaining a G–bundle on the smaller curves $X \setminus x_2$ and $X \setminus x_1$ trivialized away from x_1 and x_2, respectively, and thus a point of $\mathcal{G}r_{X,x_1} \times \mathcal{G}r_{X,x_2}$. Note that according to Theorem 20.3.1, shrinking X does not affect the Grassmannian: we have a canonical identification $\mathcal{G}r_{X,x} = \mathcal{G}r_{X \setminus \{x_i\},x}$ for $x_i \neq x$. The inverse map $\mathcal{G}r_{X,x_1} \times \mathcal{G}r_{X,x_2} \to \mathcal{G}r_{X,x_1,x_2}$ is defined by identifying $\mathcal{G}r_{X,x_i}$ with the moduli space of G–bundles on X trivialized away from x_i, and gluing the two resulting bundles over $X \setminus \{x_1, x_2\}$ where they are both trivialized.

On the other hand, assume now that $x_1 = x_2$, so that the point $\{x_1, x_2\} \in X \times X$ lies on the diagonal. Then we have an obvious identification $\mathcal{G}r_{X,x_1,x_1} \to \mathcal{G}r_{X,x_1}$. Thus away from the diagonal $\mathcal{G}r_{X,2}$ is isomorphic to the product of two copies of the Grassmannian, while over the diagonal we are left with only one copy. The same arguments may be repeated to describe the fibers of $\mathcal{G}r_{X,n}$ over an arbitrary n–tuple $x_1, \ldots, x_n$ of points of X. First of all, the Grassmannian $\mathcal{G}r_{X,n}$ does not depend on the ordering of the points, or on their multiplicities – if two configurations of points $\{x_i\} \in X^n$ and $\{y_j\} \in X^m$ have the same set–theoretic support, then there is a canonical isomorphism $\mathcal{G}r_{X,\{x_i\}} \cong \mathcal{G}r_{X,\{y_j\}}$. On the other hand, if the set $\{x_i\}$ can be written as the disjoint union of two subsets $\{x_i\} = \{y_j\} \coprod \{z_k\}$, then we have a canonical product decomposition of the Grassmannian, $\mathcal{G}r_{X,\{x_i\}} \cong \mathcal{G}r_{X,\{y_j\}} \times \mathcal{G}r_{X,\{z_k\}}$. Moreover, all of these isomorphisms vary "nicely" with respect to the variation of the positions of the points.

These observations may be summarized in the following definition and proposition, discovered by Beilinson and Drinfeld [**BD3**]:

20.3.6. Definition. A *factorization space* $\mathcal{G}_X$ on X consists of the following data:

- A formally smooth ind-scheme $\mathcal{G}_{X,I}$ over X^I for any finite set I.
- Functorial isomorphisms of ind-schemes $\mathcal{G}_{X,J}|_{\Delta_{J/I}} \to \mathcal{G}_{X,I}$ over X^I, for surjections $J \twoheadrightarrow I$.
- (*factorization*) Functorial isomorphisms of ind-schemes over $U^{J/I}$,

$$\mathcal{G}_{X,J}|_{U^{J/I}} \to \prod_{i \in I} \mathcal{G}_{X,p^{-1}(i)}|_{U^{J/I}}.$$

20.3.7. Proposition. *The Beilinson–Drinfeld Grassmannians* $\mathcal{G}r_{X,I} \to X^I$, *equipped with the natural isomorphisms described above, form a factorization space on X.*

20.3.8. Units and connections for factorization spaces. The definition of factorization space is clearly a non-linear analogue of the definition of a factorization algebra, in which we replace quasicoherent sheaves and tensor products with ind-schemes and Cartesian products. In particular, a factorization space gives rise to an ind-scheme $\mathcal{G}_\bullet \to \mathcal{R}(X)$ over the Ran space of X (i.e., a compatible system of ind-schemes over X^I for all I).

The unit axiom of factorization algebras also has a geometric analogue, which encodes an important structure of factorization spaces such as the Beilinson–Drinfeld Grassmannian. Namely, the trivial G–bundle on X tautologically carries trivializations away from arbitrary finite subsets of X, and thereby defines sections $\mathbf{1}_I : X^I \to \mathcal{G}r_{X,I}$ of the Grassmannians, satisfying transparent axioms. Thus, we introduce a unit for a factorization space $\mathcal{G}_X$ as the data of sections $\mathbf{1}_I : X^I \to \mathcal{G}_{X,I}$, which are compatible with the diagonal and the factorization maps (in particular, they define a section over $\mathcal{R}(X)$), and so that for any local section s of $\mathcal{G}_{U,1} \to U \subset X$, the section $s \boxtimes \mathbf{1}_1$ of $\mathcal{G}_{U,2} \to U^2 \setminus \Delta$ extends across the diagonal, and restricts to $s \in \mathcal{G}_{X,2}|_\Delta \cong \mathcal{G}_{X,1}$.

It is also useful to note that the argument of § 20.2.1 producing a canonical connection on a factorization algebra applies also in the non-linear setting: given a factorization space $\mathcal{G}_X$ with a unit $\mathbf{1}$, we define a connection (equivalently, a crystal or stratification, see § A.2) on $\mathcal{G}_{X,1}$ over X by comparing sections of the form $s \boxtimes \mathbf{1}_1$ and $\mathbf{1}_1 \boxtimes s$ of $\mathcal{G}_{X,2}$, which agree on the diagonal. In the case of the affine Grassmannian, this amounts to the statement that we may identify the data of trivialization of a G–bundle $\mathcal{P}$ outside of x with the data of trivialization of $\mathcal{P}$ outside of any infinitesimally nearby point – more precisely, the restriction of the family $\mathcal{G}r_{X,1} \to X$ to any infinitesimal neighborhood U of x is canonically identified with the trivial family over U with the fiber $\mathcal{G}r_x$.

20.4. Examples of factorization algebras

In this section we describe some examples of factorization algebras. We begin by describing the procedure of linearization of factorization spaces, by analogy with the construction of universal enveloping algebras of Lie algebras from their Lie groups. We then elaborate on the primary example, the Beilinson–Drinfeld Grassmannian $\mathcal{G}r_{X,\bullet}$, and the construction of the affine Kac-Moody vertex algebra from it. In the abelian case this construction can be extended to define the factorization algebra counterparts of the Heisenberg and lattice vertex algebras. We also mention briefly other examples, in which we replace G–bundles by curves, maps of curves, or $\mathcal{D}$–modules on curves.

20.4.1. Linearizing factorization spaces. In order to construct factorization algebras out of a factorization space $\mathcal{G}$, we will apply a multiplicative (monoidal) functor from spaces to vector spaces. Then the product factorization of spaces over X^n will translate into the tensor product factorization of the corresponding sheaves of vector spaces (which will actually turn out to be quasicoherent sheaves of $\mathcal{O}$–modules). We will do so by analogy with the passage from Lie groups to enveloping algebras.

Recall that the Lie algebra $\mathfrak{g}$ of a Lie group G is the tangent space to G at the identity element $1 \in G$. This description leads us to the following description of the universal enveloping algebra $U\mathfrak{g}$: it is identified with the space of delta-functions on G at the identity (in the sense of § 11.2.10; note that we have already encountered this description in § 11.2.11). This space may be alternatively described as the fiber at 1 of the sheaf of differential operators on G (under the left $\mathcal{O}$–module structure), as the space of translation–invariant differential operators, or as the topological dual of the completion of the algebra of functions on G at 1.

Note that the product of Lie groups corresponds to the direct sum of the corresponding Lie algebras and the tensor product of the universal enveloping algebras. More generally, given a $\mathbb{C}^\times$–central extension

$$1 \to \mathbb{C}^\times \to \widehat{G} \to G \to 1$$

of G, we may consider the space of delta-functions at the identity on G twisted by the line bundle corresponding to the $\mathbb{C}^\times$–bundle $\widehat{G}$. This space is again an algebra, namely, the quotient of $U(\widehat{\mathfrak{g}})$, where $\widehat{\mathfrak{g}} = \mathrm{Lie}\,\widehat{G}$, by the ideal generated by $K - 1$, where K is the central element of $\widehat{\mathfrak{g}} \subset U(\widehat{\mathfrak{g}})$ and 1 is the unit element in $U(\widehat{\mathfrak{g}})$.

Now let $\mathcal{G}$ be a factorization space on a curve X with unit $\mathbf{1}$. It will play the role of Lie group. The role of Lie algebra of $\mathcal{G}$ will be played by the collection of quasicoherent sheaves on X^I defined by the relative tangent bundle to $\mathcal{G}_I \to X^I$ at $\mathbf{1}_I$. This collection satisfies all the diagonal compatibilities and factors *additively* under disjoint union of subsets of X. In fact, one can show that this sheaf naturally defines a Lie* algebra (see § 19.4) on X by describing the gluing data on X^2 for the corresponding right $\mathcal{D}$–modules. We will be interested instead in the factorization version of the universal enveloping algebra. Thus we replace the tangent spaces to $\mathcal{G}_I$ by the sheaf $\mathcal{V}_I$ of delta-functions on $\mathcal{G}_I$ along the section $\mathbf{1}_I$. The fiber of $\mathcal{V}_I$ at $\{x_i\} \in X^I$ is the topological dual of the topological tensor product over the x_i's of the completed local rings of $\mathcal{G}_{x_i}$ at $\mathbf{1}(x_i)$, which is the value of the section

$\mathbf{1}: X \to \mathcal{G}_{\{1\}}$ at x_i. This fiber is filtered, and its associated graded is the tensor product over the x_i's of the symmetric algebras of the tangent spaces $T_{\mathbf{1}(x_i)}\mathcal{G}_{x_i}$.

To define $\mathcal{V}_I$ formally, we consider the canonical sheaf ω_{X^I} as the right $\mathcal{D}$–module on X^I, and take its $\mathcal{D}$–module push-forward $(\mathbf{1}_I)_!\omega_{X^I}$ to $\mathcal{G}_I$ under the unit map $\mathbf{1}_I: X^I \to \mathcal{G}_I$. In particular, as an $\mathcal{O}$–module, $(\mathbf{1}_I)_!\omega_{X^I}$ is a union of $\mathcal{O}$–submodules supported on the neighborhoods of the unit section, which are schemes of finite type. We now consider the $\mathcal{O}$–module pushforward $\mathcal{V}_I^r$ of $(\mathbf{1}_I)_!\omega_{X^I}$ to X^I. The connection on $\mathcal{G}_I$, defined using the unit, gives this sheaf the structure of a right $\mathcal{D}$–module. We convert it into a left $\mathcal{D}$–module $\mathcal{V}_I$ in the usual way: $\mathcal{V}_I = \mathcal{V}_I^r \otimes \omega_{X^I}^{-1}$.

The factorization and diagonal axioms for the factorization space $\mathcal{G}$ now automatically produce their linear counterparts for the $\mathcal{V}_I$'s, and the unit comes from the natural inclusion of $\mathcal{O}_{X^I} \to \mathcal{V}_I$ as the "first–order delta-functions", i.e., sections supported scheme–theoretically on $\mathbf{1}_I(X^I) \subset \mathcal{G}_I$. It follows that the $\mathcal{V}_I$'s define a factorization algebra on X. As with the usual enveloping algebra, this factorization algebra satisfies a universal mapping property with respect to the Lie* algebra defined by the relative tangent spaces at $\mathbf{1}$. In other words, the corresponding chiral algebra is the universal enveloping chiral algebra of this Lie* algebra (the definition of the universal enveloping chiral algebra of a Lie* algebra is parallel to the definition of the enveloping algebra of a vertex Lie algebra given in § 16.1.11, see [**BD4**]).

There is also a twisted version of the above construction that depends on an appropriate "factorization" line bundle $\mathcal{L}$ on $\mathcal{G}$. Namely, suppose $\mathcal{L}$ is a line bundle on $\mathcal{G} \to \mathcal{R}(X)$ (i.e., a system of line bundles $\mathcal{L}_I$ on $\mathcal{G}_I \to X^I$ compatible with diagonals) with factorization as a tensor product under disjoint union of points: we have isomorphisms over $U^{J/I}$ of the restrictions $j_{J/I}^*\mathcal{L}_J$ and $j_{J/I}^*\left(\bigotimes_{i \in I} \mathcal{L}_{p^{-1}(i)}\right)$. Then we may replace the sheaf of delta-functions $\mathcal{V}$ by the sheaf of $\mathcal{L}$–twisted delta-functions $\mathcal{V}^{\mathcal{L}}$. The latter is obtained by tensoring the push-forward $(\mathbf{1}_I)_!\omega_{X^I}$ by $\mathcal{L}_I$ and then taking the push-forward to X^I and passing to the corresponding left $\mathcal{D}$–module on X^I.

20.4.2. The Kac-Moody factorization algebra.

Our first example of the above procedure comes from the factorization spaces $\mathcal{G}r_{X,I}$, the Beilinson–Drinfeld Grassmannians, equipped with the unit provided by the trivial G–bundle on X. There is a canonical choice of factorization line bundle $\mathcal{L}$ on $\mathcal{G}r$ associated to the normalized invariant inner product on $\mathfrak{g}$. This inner product gives rise to a central extension $\widehat{G}$ of the loop group $G(\mathcal{K})$ and hence to a $\mathbb{C}^\times$–bundle $\widehat{G}(\mathcal{K})/G(\mathcal{O}) \to G(\mathcal{K})/G(\mathcal{O})$ over the affine Grassmannian (here we use the fact that the Kac-Moody central extension canonically splits over $G(\mathcal{O})$). It also appears as the pull-back of the level one theta line bundle $\mathcal{L}_G$ on the moduli space $\mathfrak{M}_G(X)$ (see § 18.1.12), under the forgetful map $G(\mathcal{K}_x)/G(\mathcal{O}_x) \to \mathfrak{M}_G(X)$ (see § 20.3.2). By abusing notation, we will denote this pull-back also by $\mathcal{L}_G$.

Likewise, the corresponding line bundle on $\mathcal{G}r_{X,x_1,\ldots,x_n}$ is obtained by pull-back of $\mathcal{L}_G$ under the forgetful map $\mathcal{G}r_{x_1,\ldots,x_n} \to \mathfrak{M}_G(X)$. These line bundles glue into factorization line bundles on the Grassmannians $\mathcal{G}r_{X,I}$.

The space of delta-functions at the identity coset on a homogeneous space G/K, where G and K are Lie groups, is naturally identified with the induced representation

$$U\mathfrak{g} \underset{U\mathfrak{k}}{\otimes} \mathbb{C} = U\mathfrak{g}/U\mathfrak{g} \cdot U\mathfrak{k}.$$

Similarly, for a character χ of K, the space of delta-functions on G/K twisted by the G–equivariant line bundle $\mathcal{L}_\chi \to G/K$ is identified with the induced representation $U\mathfrak{g} \otimes_{U\mathfrak{k}} \mathbb{C}_\chi$ (see, e.g., [**BF2**]).

In our case the space of delta-functions at the identity on $\mathcal{G}r$ twisted by the kth tensor power of the line bundle $\mathcal{L}_G$ is naturally identified with the vacuum representation $V_k(\mathfrak{g})$. This space makes sense even if k is an arbitrary complex number. The corresponding space of delta-functions on $G(\mathcal{K}_x)/G(\mathcal{O}_x)$ is the $\mathcal{A}ut_x$–twist $\mathcal{V}_k(\mathfrak{g})_x$ of the vacuum representation. We have thus identified the sheaf $\mathcal{V} = \mathcal{V}_1$ associated to the linearization (at the trivial G–bundle) of the Beilinson–Drinfeld Grassmannian with the sheaf $\mathcal{V}_k(\mathfrak{g})$ associated to the Kac-Moody vertex algebra. The unit $\mathcal{O}_X \to \mathcal{V}_1$ is just the map $\mathcal{O}_X \cdot |0\rangle \to \mathcal{V}_k(\mathfrak{g})$. It is also easy to see that the connection on $\mathcal{V}_k(\mathfrak{g})$ is the same as the one coming from the factorization structure. In fact we have the following:

20.4.3. Proposition. *The chiral algebra on X corresponding to the factorization algebra on tX constructed by level k linearization of $\mathcal{G}r$ at the trivial G–bundle is canonically identified with the affine Kac-Moody chiral algebra $\mathcal{V}_k(\mathfrak{g})^r$ which corresponds to the Kac-Moody vertex algebra $V_k(\mathfrak{g})$.*

20.4.4. Heisenberg algebras and lattice vertex algebras. In the abelian case, the factorization structure on the Grassmannian may be used to construct the Heisenberg and the lattice vertex algebras (for a detailed treatment, see [**Gai, Bei**]). Let us begin with the case of $G = GL_1$, so that the space $\mathcal{G}r_x = \mathcal{G}r_{X,1}|_x$ is the group ind-scheme $\mathcal{K}_x^\times/\mathcal{O}_x^\times$, and the gluing of $\mathcal{G}r_x$ and $\mathcal{G}r_y$ when the points x, y collide is achieved using the group structure on it (see § 20.3.4).

We fix an even integral bilinear form on $\mathfrak{gl}_1$ – such a form is determined by an even integer N, so that the coweight lattice is identified with the lattice $\sqrt{N}\mathbb{Z}$ (as in § 5.2). The line bundle $\Theta^{\otimes N}$ on $\mathcal{G}r_{X,x_1,\ldots,x_n}$ corresponding to this form is the pull-back of the corresponding theta line bundle on the Picard variety. The fiber of this line bundle over $\mathcal{L} \in \mathrm{Pic}(X)$ is the line

$$\Theta^{\otimes N}|_{\mathcal{L}} \simeq \left(\det H^\bullet \left(\mathcal{L} \otimes \Omega^{\frac{1}{2}} \right) \right)^{\otimes N} .$$

This description of the line depends on the choice of a theta characteristic $\Omega^{\frac{1}{2}}$ on X for *odd* N, but its even powers are canonically defined. (Recall that for N odd, $V_{\sqrt{N}\mathbb{Z}}$ is a vertex *super*algebra.)

The procedure of § 20.4.2 specializes in this case to produce the Heisenberg vertex algebra π_0 as the space of $\Theta^{\otimes N}$–twisted delta-functions on $\mathcal{K}^\times/\mathcal{O}^\times$. However, this construction only "sees" the identity component of $\mathcal{K}^\times/\mathcal{O}^\times$, which is the formal group associated to $\mathcal{K}/\mathcal{O}$. In order to incorporate all of the components of $\mathcal{G}r_{X,1} = \mathcal{K}^\times/\mathcal{O}^\times$, we replace the single section $\mathbf{1}_X : X \to \mathcal{G}r_X$, whose value at $x \in X$ is the trivial line bundle $\mathcal{O}_X$ with its canonical trivialization off x, by the collection of sections $\mathbf{1}(n) : X \to \mathcal{G}r_X$ $(n \in \mathbb{Z})$, where the value of $\mathbf{1}(n)$ at $x \in X$ is the modification $\mathcal{O}_X(nx)$ of $\mathcal{O}_X$ equipped with its canonical trivialization away from x. By considering the space of delta functions on $\mathcal{G}r_X$ along $\mathbf{1}(n)$, we obtain a left $\mathcal{D}$–module $\mathcal{V}(n)$ on X as before.

In order to describe the sheaf $\mathcal{V}(n)$, we need to understand the action of modifications on the line bundle $\Theta^{\otimes N}$.

20.4.5. Lemma. *For any line bundle $\mathcal{L}$ on X, there is a canonical isomorphism*

$$\Theta^{\otimes N}|_{\mathcal{L}(n \cdot x)} = \Theta^{\otimes N}|_{\mathcal{L}} \otimes \mathcal{L}|_x^{\otimes Nn} \otimes \Omega|_x^{\otimes -\frac{Nn^2}{2}}.$$

20.4.6. Proof. For any line bundle $\mathcal{L}$ on X, we have the adjunction short exact sequence

$$0 \to \mathcal{L} \to \mathcal{L}(x) \to \mathcal{L}|_x \otimes T_x X \to 0,$$

and hence a canonical isomorphism

$$\det H^\bullet(\mathcal{L}(x)) = \det H^\bullet(\mathcal{L}) \otimes \mathcal{L}|_x \otimes \Omega|_x^{-1}$$

on the determinant lines associated to $\mathcal{L}$ and its modification at x. Repeating this and using

$$\mathcal{L}(n \cdot x)|_x = \mathcal{L}|_x \otimes \Omega|_x^{\otimes -n},$$

we calculate that

$$\det H^\bullet(\mathcal{L} \otimes \Omega^{\frac{1}{2}}(n \cdot x)) = \det H^\bullet(\mathcal{L} \otimes \Omega^{\frac{1}{2}}) \otimes \mathcal{L}|_x^{\otimes n} \otimes \Omega|_x^{\otimes -\frac{n^2}{2}}.$$

The lemma follows.

20.4.7. The line $\mathcal{L}|_x^{\otimes Nn}$ is the $\mathcal{L}$–twist of the character of $\mathcal{O}^\times$, obtained by composing the evaluation $\mathcal{O}^\times \to \mathbb{C}^\times$ with the Nnth power map, while the line $\Omega|_x^{\otimes -N\frac{n^2}{2}}$ is the twist of the character of $\mathrm{Aut}\,\mathcal{O}$ of weight $Nn^2/2$. At the level of Lie algebras this corresponds to the one-dimensional representation of $\mathrm{Der}_0\,\mathcal{O} \ltimes \mathcal{O} \subset Vir \ltimes \mathcal{H}$ on which $\mathbf{1} \in \mathcal{H}$ acts by Nn and $-t\partial_t \in Vir$ acts by $Nn^2/2$.

In § 5.2.1 we described Fock modules π_λ over the Heisenberg Lie algebra $\mathcal{H}$ on which the central element $\mathbf{1}$ acts as the identity. But at the moment we are considering representations on which $\mathbf{1}$ acts as N times the identity. We can view them as the representations on which the action of the generators b_n is rescaled by $\sqrt{N}$, because

$$[\sqrt{N}b_n, \sqrt{N}b_m] = n\delta_{n,-m}N.$$

In particular, the action of $\sqrt{N}b_0$ on π_μ will be $\sqrt{N}\mu$.

We now recognize the space of delta-functions at $\mathbf{1}(n)$ twisted by the line bundle $\Theta^{\otimes N}$ as the Fock representation $\pi_{\sqrt{N}n}$ of the Heisenberg algebra induced from the one–dimensional representation $\mathbb{C}_{\sqrt{N}n}$ of $\mathcal{H}_+$ on which b_0 acts by multiplication by $\sqrt{N}n$ and all other generators $b_n, n > 0$, act by zero (see § 5.2.1). It is equipped with the Virasoro algebra action induced by the standard conformal vector described in § 5.2.4.

20.4.8. Proposition. *The sheaf $\mathcal{V}(n)$ is isomorphic to the Aut_X–twist $\Pi_{\sqrt{N}n}$ of the Fock representation $\pi_{\sqrt{N}n}$.*

The sections $\mathbf{1}(n)$ do not define units for the factorization space $\mathcal{G}r_X$. Rather, under factorization the sections $\mathbf{1}(n)$ and $\mathbf{1}(m)$ glue to the section $\mathbf{1}(n+m)$. This suggests joining all the "sectors" $\mathcal{V}(n)$ into a single sheaf,

$$\mathcal{V}_{\sqrt{N}\mathbb{Z}} = \bigoplus_{n \in \mathbb{Z}} \mathcal{V}(n),$$

which carries a natural factorization structure. By Proposition 20.4.8, we recognize this sheaf as the Aut_X–twist of the lattice vertex algebra

$$V_{\sqrt{N}\mathbb{Z}} = \bigoplus_{n \in \mathbb{Z}} \pi_{\sqrt{N}n}.$$

Indeed we have the following result in which we obtain super-analogues of factorization algebras and chiral algebras if N is odd:

20.4.9. Theorem. *The sheaf* $\mathcal{V}_{\sqrt{N}\mathbb{Z}}$ *carries a natural factorization (super)-algebra structure, and the corresponding chiral (super)algebra is associated to the lattice vertex (super)algebra* $V_{\sqrt{N}\mathbb{Z}}$.

This construction has a straightforward generalization for any even integral lattice Λ. Consider the torus $H = \Lambda \otimes_{\mathbb{Z}} \mathbb{C}^{\times}$ associated to Λ. The affine Grassmannian $H(\mathcal{K})/H(\mathcal{O})$ for H has components labeled by elements of Λ, and a natural theta line bundle associated to the symmetric form on Λ. We define sections $\mathbf{1}(\lambda)$ associated to any $\lambda \in \Lambda$, and the associated spaces $\mathcal{V}(\lambda)$ of twisted delta-functions on the λ–component as in the case $\Lambda = \mathbb{Z}$. We then have

20.4.10. Theorem. *The sheaf* $\mathcal{V}_{\Lambda}$ *which is the union of the sheaves of delta-functions on the components of the* H–*affine Grassmannian carries a natural factorization algebra structure, and is identified with the factorization algebra associated to the lattice vertex algebra* V_{Λ}.

20.4.11. Jet spaces and the chiral de Rham complex. In this section we summarize the construction of factorization spaces associated to jet and loop spaces, following [**KV**].

Recall that the simplest class of vertex algebras is the class of commutative vertex algebras, or commutative algebras with a derivation. For instance, in § 9.4.4 we discussed examples of such algebras arising from the jet schemes, spaces of maps from the disc D to schemes. Recall that given an affine variety $Z = \operatorname{Spec} A$, the jet scheme JZ is defined as $\operatorname{Spec} A_{\infty}$, where A_{∞} is the algebra defined in § 9.4.4. These affine jet schemes glue together to define jet schemes of arbitrary schemes.

It is natural to ask whether there is a factorization space behind the commutative vertex algebras A_{∞} corresponding to the jet schemes. In § 9.4.4 we defined for any affine scheme Z, using a natural action of the Harish-Chandra pair $(\operatorname{Der} \mathcal{O}, \operatorname{Aut} \mathcal{O})$ on A_{∞}, the twist $JZ_X = Aut_X \underset{\operatorname{Aut} \mathcal{O}}{\times} JZ$, which is a scheme over a smooth curve X equipped with a flat connection. The fiber of JZ_X at $x \in X$ is the jet scheme of maps $D_x \to Z$, and the flat connection identifies infinite jets at x and at infinitesimally nearby points of X. Note that this construction may be applied to an arbitrary scheme Z.

More generally, for any set I there is a scheme $JZ_{X,I} \to X^I$, whose fiber at $\{x_i\}_{i \in I}$ represents the scheme of infinite jets of maps from the union of the subschemes x_i to Z. In other words, $JZ_{X,I}|_{\{x_i\}}$ represents the maps from the formal neighborhood in X of the union of the points x_i into Z. It is informally clear that the set of such maps is independent of the multiplicities of the points x_i, and factorizes as a product whenever the points x_i break up as a disjoint union of proper subsets. A rigorous definition of the schemes $JZ_{X,I}$ and the factorization structure on them is given in [**KV**]. Thus one obtains the following:

20.4.12. Proposition. *The collection* $\{JZ_{X,I}\}$ *defines a factorization space on* X. *For an affine scheme* $Z = \operatorname{Spec} A$, *the coordinate rings of the* $JZ_{X,I}$ *(with the units being the constant functions* 1*) define a factorization algebra on* X, *which corresponds to the commutative vertex algebra* A_∞.

The above construction of factorization structure on jet spaces may be greatly enhanced by passing to formal *loop* spaces, i.e., by replacing the disc D by the punctured disc $D^\times$. In § 11.3.3 we explained that for affine schemes such an object can be defined as an ind-scheme. For non-affine schemes it is not clear how to construct them, but it is possible to construct an ind-scheme LZ parameterizing maps from $D^\times$ to Z which are "in the formal neighborhood" of jets $D \to Z$. This construction may be generalized to define for a fixed smooth curve X and for any set I a formally smooth ind-scheme $LZ_{X,I} \to X^I$ containing the schemes $JZ_{X,I}$ (see [**KV**]). The fiber of LZ_I at $\{x_i\}$ parameterizes certain maps from the formal punctured neighborhood of the union of the subschemes x_i in X to Z. The spaces $LZ_{X,I}$ carry a factorization structure extending that of the jet spaces $JZ_{X,I} \subset LZ_{X,I}$. By considering appropriate notions of "delta-functions along JZ inside LZ", this factorization space was then used in [**KV**] to construct the factorization algebra that corresponds to the chiral de Rham complex on Z (see § 18.5).

20.4.13. Other factorization spaces. The factorization structure on the Beilinson–Drinfeld Grassmannians (moduli spaces of G–bundles with trivializations away from finitely many points) has immediate generalizations associated to other geometric structures over algebraic curves. For example, the Virasoro uniformization of the moduli spaces of curves (see § 17.3.1) gives rise to a similar factorization structure.

Namely, given a smooth projective curve X, we consider the moduli spaces of curves equipped with isomorphisms to X away from finitely many points. More precisely, we consider the moduli space $\mathcal{G}_I \to X^I$, whose fiber at $\{x_i\}_{i \in I}$ is the moduli space of smooth I–pointed curves $(X', \{x_i'\})$ together with an isomorphism

$$X \setminus \{x_i\}_{i \in I} \simeq X' \setminus \{x_i'\}_{i \in I}.$$

By properness of X, these moduli spaces do not have any $\mathbb{C}$–points except for the curve X itself, since any such isomorphism of open curves will extend canonically to the completion. However, over rings with nilpotents the $\mathcal{G}_I$'s have non-trivial points as explained in the proof of Theorem 17.3.2. The Virasoro uniformization of the moduli space of curves (see Theorem 17.3.2) implies that the spaces of delta-functions on $\mathcal{G}_1$ at the canonical unit $\mathbf{1}$ (associated to the "trivial" curve X) give rise to the Virasoro vacuum representation Vir_0 (and similarly for their twists by line bundles which give rise to Vir_c). Thus $\mathcal{G}$ may be considered as the factorization space associated to the Virasoro vertex algebra.

Another interesting example of factorization space is provided by the *adèlic Grassmannian* of Wilson [**W**] and its $\mathcal{D}$–module description following Cannings–Holland [**CH**], as described in [**BN1, BN2**]. Here we replace the structure of a G–bundle, underlying the affine Grassmannian, with that of a $\mathcal{D}$–*bundle*, or locally projective (right) $\mathcal{D}$–module on X [**BD4**]. The adèlic Grassmannian is naturally identified with the moduli space of $\mathcal{D}$–line bundles on X with a trivialization away from finitely many points. More precisely, we have the inductive system $\mathcal{G}_I \to X^I$

of spaces over X^I parameterizing $\mathcal{D}$–bundles on X identified with $\mathcal{D}$ itself away from a given subset $\{x_i\}$ of X.

By the de Rham functor such trivialized $\mathcal{D}$–modules are naturally parameterized by a collection of open subspaces of the local fields $\mathcal{K}_{x_i}$. The point in Wilson's Grassmannian associated to these subspace is their common intersection with the space of rational functions $\mathbb{C}(X)$.

The collection $\{\mathcal{G}_I\}$ defines a factorization space, and the corresponding vertex algebra is $\mathcal{W}_{1+\infty}$, the vacuum representation of the central extension of the Lie algebra $\mathcal{D}(\mathcal{K})$ (see [**FKRW**]). More precisely, it is shown in [**BN2**] that the space of delta-functions at the trivial $\mathcal{D}$–bundle with its factorization structure corresponds to the $\mathcal{W}_{1+\infty}$–vertex algebra of central charge zero, while delta-functions twisted by the appropriate determinant line bundle to the power c correspond to $\mathcal{W}_{1+\infty}$ of central charge c.

20.5. Factorization and the chiral Hecke algebra

We conclude this chapter with a brief sketch of one the most exciting applications of factorization related to the construction of the geometric Langlands correspondence. The above discussion of factorization structure naturally leads us to the notion of *Hecke correspondences* acting on the moduli spaces $\mathfrak{M}_G(X)$ of G–bundles on curves (see § 20.5.1). The "eigenvectors" of these correspondences are called *Hecke eigensheaves*. The geometric Satake correspondence (see Theorem 20.5.4) relating $\mathcal{D}$–modules on the affine Grassmannian and representations of the Langlands dual group allows one to describe the possible "eigenvalues" of the Hecke correspondences as local systems on X for the Langlands dual group $^L G$. The geometric Langlands conjecture predicts that for each irreducible $^L G$–local system E there exists a Hecke eigensheaf on $\mathfrak{M}_G(X)$ with eigenvalue E.

How can one go about constructing Hecke eigensheaves? Here we outline one possible construction. Namely, we describe, following [**BD4, Bei**] and lectures by Beilinson, the definition and some of the remarkable properties of the *chiral Hecke algebra* introduced by Beilinson and Drinfeld. Conjecturally, the sheaves of coinvariants of the chiral Hecke algebra, twisted by a $^L G$–local system E on X, should give rise to the Hecke eigensheaves on $\mathfrak{M}_G(X)$ with eigenvalue E. (For another construction, see § 18.4.9.)

20.5.1. Hecke correspondences. Let $\mathcal{P}'$ denote a G–bundle on X. We may then repeat the definition of the Grassmannians $\mathcal{G}r_x$ using $\mathcal{P}'$ in place of the trivial G–bundle: namely, we consider the moduli space $\mathcal{G}r_x^{\mathcal{P}'}$ of G–bundles $\mathcal{P}$ on X equipped with an isomorphism $\mathcal{P}|_{X\backslash x} \simeq \mathcal{P}'|_{X\backslash x}$. Thus $\mathcal{G}r_x^{\mathcal{P}'}$ parameterizes all *modifications* of $\mathcal{P}'$ at x.

Alternatively, note that the group–scheme $\operatorname{Aut}\mathcal{O} \ltimes G(\mathcal{O})$ acts on the affine Grassmannian $\mathcal{G}r$, and $\mathcal{G}r_{X,x}^{\mathcal{P}'}$ is simply the twist of $\mathcal{G}r$ by the $\operatorname{Aut}\mathcal{O} \ltimes G(\mathcal{O})$–torsor $\widehat{\mathcal{P}}'_x$ associated to $\mathcal{P}'$ in § 7.1.4. Now we combine the Grassmannians $\mathcal{G}r_x^{\mathcal{P}'}$ with varying G–bundle $\mathcal{P}'$ and varying point $x \in X$ into the *Hecke correspondence*

$$\mathcal{H}ecke = \{(\mathcal{P}, \mathcal{P}', x, \phi) \; : \; \mathcal{P}, \mathcal{P}' \in \mathfrak{M}_G(X), x \in X, \phi : \mathcal{P}|_{X\backslash x} \xrightarrow{\sim} \mathcal{P}'|_{X\backslash x}\}.$$

It is equipped with the projections

$$
\begin{array}{ccc}
 & \mathcal{H}ecke & \\
h^{\leftarrow}\swarrow & & \searrow h^{\rightarrow} \\
\mathfrak{M}_G(X) & & X \times \mathfrak{M}_G(X)
\end{array}
$$

where $h^{\leftarrow}(\mathcal{P}, \mathcal{P}', x, \phi) = \mathcal{P}$ and $h^{\rightarrow}(\mathcal{P}, \mathcal{P}', x, \phi) = (x, \mathcal{P}')$. Note that the fiber of $\mathcal{H}ecke$ over $(x, \mathcal{P}')$ is $\mathcal{G}r_x^{\mathcal{P}'}$.

In the abelian case of $G = GL_1$ the fiber of $\mathcal{H}ecke$ over $x \in X$ is isomorphic to the product $\operatorname{Pic} X \times \mathcal{K}_x^{\times}/\mathcal{O}_x^{\times}$, with the two projections on $\operatorname{Pic} X$ differing by the action of $\mathcal{K}_x^{\times}/\mathcal{O}_x^{\times}$ on the Picard variety. In the case of non-abelian G we do not have a group structure on $\mathcal{G}r$, but the Hecke correspondence $\mathcal{H}ecke$ may be considered as a replacement for it. In fact, one of the main motivations for introducing the factorization spaces was the desire to express the algebraic structure underlying the composition of the Hecke correspondences in the non-abelian case.

In order to describe this more precisely, let us look more closely at the $G(\mathcal{O})$–orbits of $\mathcal{G}r$. We henceforth take G to be a *simple, connected and simply-connected* algebraic group, with a maximal torus H, so that in particular the affine Grassmannian $\mathcal{G}r$ is reduced. Let $P^{\vee}$ be the lattice of integral coweights of G, i.e, homomorphisms $\mathbb{C}^{\times} \to H$. For any $\lambda \in P^{\vee}$ we have an element $\lambda(t)$ of $H(\mathcal{K}) \subset G(\mathcal{K})$. A choice of Borel subgroup in G distinguishes a subset $P_+^{\vee} \subset P^{\vee}$ of *dominant* integral coweights. It is easy to see that the elements $\lambda(t)$, where $\lambda \in P_+^{\vee}$, give a complete system of representatives of the double cosets of $G(\mathcal{K})$ with respect to $G(\mathcal{O})$.

Thus, we obtain a parameterization of $G(\mathcal{O})$–orbits in $\mathcal{G}r = G(\mathcal{K})/G(\mathcal{O})$, and hence of $G(\mathcal{O}_x)$–orbits in $\mathcal{G}r_x$, by $P_+^{\vee}$. These orbits, denoted by $\mathcal{G}r_{x,\lambda}, \lambda \in P_+^{\vee}$, are locally closed smooth subschemes of $\mathcal{G}r_x$. They play the role of the integers $\mathbb{Z} = \mathcal{K}^{\times}/\mathcal{O}^{\times}(\mathbb{C})$ in the abelian case. As we will see, one can define an "action" of these orbits on $\mathfrak{M}_G(X)$, in the sense of correspondences. The geometric Langlands Program may be broadly construed as the decomposition of the category of $\mathcal{D}$–modules on $\mathfrak{M}_G(X)$ under the action of these correspondences.

To define this action, let us recall that the fiber of $\mathcal{H}ecke$ over $(x, \mathcal{P}')$ is $\mathcal{G}r_x^{\mathcal{P}'}$, the twist of the Grassmannian $\mathcal{G}r_x$ by the $G(\mathcal{O}_x)$–torsor of trivializations of $\mathcal{P}'|_{D_x}$. Hence the stratification of $\mathcal{G}r_x$ by the $G(\mathcal{O}_x)$–orbits $\mathcal{G}r_{x,\lambda}$ induces a similar stratification of $\mathcal{G}r_x^{\mathcal{P}'}$. For varying x and $\mathcal{P}'$ we obtain a stratification of the entire $\mathcal{H}ecke$ by the strata $\mathcal{H}ecke_\lambda, \lambda \in P_+^{\vee}$. To each stratum corresponds an irreducible $\mathcal{D}$–module IC_λ on $\mathcal{H}ecke$ supported on the closure of $\mathcal{H}ecke_\lambda$. One defines the *Hecke functors* H_λ from the derived category of $\mathcal{D}$–modules on $\mathfrak{M}_G(X)$ to the derived category of $\mathcal{D}$–modules on $X \times \mathfrak{M}_G(X)$ by the formula (see [**BD3**])

$$
\mathrm{H}_\lambda(\mathcal{F}) = h_!^{\rightarrow}(h^{\leftarrow *}(\mathcal{F}) \otimes \mathrm{IC}_\lambda).
$$

20.5.2. Geometric Langlands conjecture. Let $^L H \subset {}^L G$ denote the Langlands dual groups of $H \subset G$ (see § 18.4.9). The set $P_+^{\vee}$ may be realized as the set of dominant integral *weights* of $^L G$, i.e., homomorphisms $^L H \to \mathbb{C}^{\times}$ which are dominant with respect to the Borel subgroup in $^L G$ corresponding to the chosen Borel subgroup of G. To each such weight corresponds a finite-dimensional irreducible representation V_λ of $^L G$ with highest weight λ.

Let E be a $^L G$–local system on X, or equivalently, a principal $^L G$–bundle on X with flat connection. Then for each irreducible representation V_λ of $^L G$ we have a vector bundle with a flat connection

$$V_\lambda^E = E \underset{G}{\times} V_\lambda,$$

which we view as a $\mathcal{D}$–module on X.

Now, a $\mathcal{D}$–module on $\mathfrak{M}_G(X)$ is a called a *Hecke eigensheaf with eigenvalue E* if we are given isomorphisms

$$(20.5.1) \qquad\qquad \mathrm{H}_\lambda(\mathcal{F}) \simeq V_\lambda^E \boxtimes \mathcal{F}$$

of $\mathcal{D}$–modules on $X \times \mathfrak{M}_G(X)$ which are compatible with the tensor product structure on the category of representations of ^LG.

Let E be a ^LG–local system on X which cannot be reduced to a proper parabolic subgroup of ^LG. The *geometric Langlands conjecture* states that then there exists a non-zero irreducible holonomic $\mathcal{D}$–module on $\mathfrak{M}_G(X)$ which is a Hecke eigensheaf on $\mathfrak{M}_G(X)$ with eigenvalue E.

20.5.3. Geometric Satake correspondence. We wish to construct Hecke eigensheaves corresponding to a given local system E. Before doing that, we need to explain in what sense the isomorphisms (20.5.1) should be compatible with the tensor product structure on the category of finite-dimensional representations of ^LG. For that we have to explain how to compose the Hecke functors. This is a local question that one may consider for a fixed $x \in X$ and a fixed $\mathcal{P}'$, which we may take to be the trivial bundle. Choosing a coordinate at x, we identify the corresponding fiber of $\mathcal{H}ecke$ with $\mathcal{G}r$, and the strata $\mathcal{H}ecke_\lambda$ correspond to the $G(\mathcal{O})$–orbits $\mathcal{G}r_\lambda$.

The Grassmannian $\mathcal{G}r$ is an ind-scheme, defined as the inductive limit of the closures $\overline{\mathcal{G}r}_\lambda, \lambda \in P_+^\vee$, which are schemes of finite type, under the closed embeddings $i^{\lambda,\lambda'} : \overline{\mathcal{G}r}_\lambda \hookrightarrow \overline{\mathcal{G}r}_{\lambda'}$ for $\lambda \leq \lambda'$. By definition, a right $\mathcal{D}$–module on $\mathcal{G}r$ is a collection of $\mathcal{D}$–modules $\mathcal{M}_\lambda$ on $\overline{\mathcal{G}r}_\lambda$ for sufficiently large λ, such that $\mathcal{M}_{\lambda'} \simeq i_!^{\lambda,\lambda'} \mathcal{M}_\lambda$ (see [**BD3**], Ch. 7). In particular, the irreducible $\mathcal{D}$–module supported on $\overline{\mathcal{G}r}_\lambda$ and obtained by the Goresky–MacPherson extension from the orbit $\mathcal{G}r_\lambda$ gives rise to a $G(\mathcal{O})$–equivariant $\mathcal{D}$–module on $\mathcal{G}r$. By abusing notation, we will denote it by IC_λ.

It turns out that there is a natural notion of tensor product of $G(\mathcal{O})$–equivariant $\mathcal{D}$–modules on $\mathcal{G}r$ defined similarly to the operation of convolution of $G(\mathcal{O})$–invariant functions on $G(\mathcal{K})/G(\mathcal{O})$. Moreover, one can define the structure of tensor category on the category $\mathcal{H}$ of $G(\mathcal{O})$–equivariant $\mathcal{D}$–modules on $\mathcal{G}r$ by using this convolution tensor product. We then have the following beautiful result called the *geometric Satake correspondence*. It has been conjectured by Drinfeld and proved by I. Mirković and K. Vilonen [**MV**] and V. Ginzburg [**Gi1**] (some important results in this direction were also obtained by G. Lusztig [**Lu**]).

20.5.4. Theorem. *The tensor category $\mathcal{H}$ of $G(\mathcal{O})$–equivariant $\mathcal{D}$–modules on the Grassmannian $\mathcal{G}r$ is equivalent to the tensor category of finite-dimensional representations of the Langlands dual group ^LG. Under this correspondence IC_λ goes to V_λ.*

20.5.5. Convolution tensor product and factorization structure. In particular, the theorem implies that the category $\mathcal{H}$ is semi-simple, with the $\mathcal{D}$–modules IC_λ being the irreducible objects. Thus, the labeling of the $G(\mathcal{O})$–orbits of $\mathcal{G}r$ (and the corresponding strata in $\mathcal{H}ecke$) by irreducible representations of ^LG, which seemed like an accident at first glance, actually has a deep meaning, signifying the connection between the $\mathcal{D}$–modules IC_λ and the representations V_λ. Note that the Tannakian formalism allows one to recover the group ${}^L G$ from its category of finite-dimensional representations, equipped with the tensor product and a forgetful functor to vector spaces. By Theorem 20.5.4, this category is equivalent to the category $\mathcal{H}$, for which the forgetful functor is $\mathcal{F} \mapsto H^\bullet_{\mathrm{dR}}(\mathcal{G}r, \mathcal{F})$.

But how does the factorization structure of $\mathcal{G}r$ come into play? This happens because of a beautiful realization, due to Beilinson and Drinfeld, of the convolution tensor product in the category $\mathcal{H}$. We will not spell it out here, referring the interested reader instead to [**BD3, MV**]. Roughly, the tensor product of two sheaves $\mathcal{M}, \mathcal{N} \in \mathcal{H}$ is defined using the Beilinson–Drinfeld Grassmannian $\mathcal{G}r_{X,2}$, which is part of the factorization space $\mathcal{G}r_{X,I}$. As we saw above, this Grassmannian interpolates between $\mathcal{G}r_{X,1} \times \mathcal{G}r_{X,1}$ and $\mathcal{G}r_{X,1}$. Since the sheaves $\mathcal{M}$ and $\mathcal{N}$ are $G(\mathcal{O})$–equivariant, they give rise to sheaves on $\mathcal{G}r_x$ for all $x \in X$ and, by varying the point $x \in X$, to sheaves on $\mathcal{G}r_{X,1}$. We will denote them by the same symbols. We take their exterior tensor product $\mathcal{M} \boxtimes \mathcal{N}$ on $\mathcal{G}r_{X,1} \times \mathcal{G}r_{X,1}$ and restrict to the complement of the diagonal in X^2. This complement is the same for $\mathcal{G}r_{X,1} \times \mathcal{G}r_{X,1}$ as it is for $\mathcal{G}r_{X,2}$. Hence we may view this restriction as a sheaf on the complement of the diagonal in $\mathcal{G}r_{X,2}$. We then take its Goresky–MacPherson extension to the entire $\mathcal{G}r_{X,2}$ and restrict the result to the diagonal. Thus, we obtain a sheaf on $\mathcal{G}r_{X,1}$. Its fiber at a fixed point $x \in X$ is isomorphic to $\mathcal{G}r$ and the corresponding sheaf is isomorphic to the convolution tensor product $\mathcal{M} \otimes \mathcal{N}$ of $\mathcal{M}$ and $\mathcal{N}$.

This construction allows one to cast the convolution tensor product on the category $\mathcal{H}$ in the language of factorization spaces. This has an important application. For $k \in \mathbb{Z}$, let $\mathcal{D}_k$ be the sheaf of differential operators on the kth power of the theta line bundle $\mathcal{L}_G^{\otimes k}$ on $\mathcal{G}r$. The Lie algebra $\mathfrak{g}((t))$ acts on $\mathcal{G}r$ by vector fields, and its central extension $\widehat{\mathfrak{g}}$ maps to $\mathcal{D}_k$ in such a way that the central element K maps to the constant function k. Thus, for any $\mathcal{D}$–module $\mathcal{F}$ on $\mathcal{G}r$ the space of global sections $\Gamma(\mathcal{G}r, \mathcal{F} \otimes \mathcal{L}_G^{\otimes k})$ is a $\widehat{\mathfrak{g}}$–module of level k.

Now suppose that $\mathcal{R}$ is a commutative ring object in the category $\mathcal{H}$ (or the corresponding category of inductive limits); that is $\mathcal{R}$ is equipped with a morphism

$$m : \mathcal{R} \otimes \mathcal{R} \to \mathcal{R}$$

satisfying the commutativity and associativity properties. Then one can show that $\Gamma(\mathcal{G}r, \mathcal{R} \otimes \mathcal{L}_G^{\otimes k})$ has the structure of a vertex algebra for any $k \in \mathbb{Z}$.

The simplest example of a ring object in $\mathcal{H}$ is the sheaf IC_0, which is nothing but the sheaf of delta-functions on $\mathcal{G}r$ supported at the unit coset. We have $\mathrm{IC}_0 \otimes \mathrm{IC}_0 \simeq \mathrm{IC}_0$, and we use this isomorphism to define a ring structure on IC_0. The corresponding $\widehat{\mathfrak{g}}$–module of global sections $\Gamma(\mathcal{G}r, \mathrm{IC}_0 \otimes \mathcal{L}_G^{\otimes k})$ is nothing but the vacuum module $V_k(\mathfrak{g})$. As explained in § 20.4.2, the vertex algebra structure on $V_k(\mathfrak{g})$ may be obtained from the factorization space structure on the Beilinson–Drinfeld Grassmannians. The point is that the construction outlined in § 20.4.2 may be applied to an arbitrary commutative ring object $\mathcal{R}$ in place of IC_0 to yield a vertex algebra structure on the corresponding $\widehat{\mathfrak{g}}$–module $\Gamma(\mathcal{G}r, \mathcal{R} \otimes \mathcal{L}_G^{\otimes k})$.

20.5.6. The chiral Hecke algebra. In order to find other ring objects in $\mathcal{H}$, we recall that according to Theorem 20.5.4, the category $\mathcal{H}$ is equivalent to the category of representations of the Langlands dual group ${}^L G$. The latter (more precisely, the corresponding category of inductive limits) has a natural commutative ring object: the regular representation.

$$R(^L G) = \operatorname{Fun} {}^L G = \bigoplus_{\lambda \in P_+^\vee} V_\lambda \otimes V_\lambda^*.$$

The corresponding ring object in $\mathcal{H}$ is

$$\mathcal{R}(^L G) = \bigoplus_{\lambda \in P^+} \operatorname{IC}_\lambda \otimes V_\lambda^*,$$

where the vector spaces V_λ^* appear as multiplicity spaces for the sheaves IC_λ. Here we have used the left action of $^L G$ on $R(^L G)$ and converted the representations V_λ into the $\mathcal{D}$–modules IC_λ using Theorem 20.5.4. Note that the sheaf $\mathcal{R}(^L G)$ still carries the right action of $^L G$ (on the vector spaces V_λ^*) which commutes with the $\mathcal{D}$–module structure on $\mathcal{R}(^L G)$.

As explained above, the space of global sections

$$A_k(\mathfrak{g}) = \Gamma(\mathcal{G}r, \mathcal{R}(^L G) \otimes \mathcal{L}^{\otimes k})$$

is a vertex algebra. By construction, we have a decomposition of $A_k(\mathfrak{g})$ as a $\widehat{\mathfrak{g}} \times {}^L G$–module

(20.5.2) $$A_k(\mathfrak{g}) = \bigoplus_{\lambda \in P_+^\vee} M_{\lambda,k} \otimes V_\lambda^*,$$

where

$$L_{\lambda,k} = \Gamma(\mathcal{G}r, \operatorname{IC}_\lambda \otimes \mathcal{L}^{\otimes k}).$$

It is known (see [**BD3, FG**]) that if $k \le -h^\vee$, then the functor of global sections $\mathcal{F} \mapsto \Gamma(\mathcal{G}r, \mathcal{F} \otimes \mathcal{L}_G^{\otimes k})$ is exact, whereas for other integral values of k it is not. Hence we will only consider $k \le -h^\vee$. For simplicity, we further restrict to the values $k < -h^\vee$. In this case $L_{\lambda,k}$ is the irreducible highest weight $\widehat{\mathfrak{g}}$–module of level k with highest weight $-(k+h^\vee)\lambda$ (here we use the normalized invariant inner product to identify weights of $\mathfrak{g}$ and of $^L \mathfrak{g}$).

In particular, $L_{0,k} = V_k(\mathfrak{g})$, and so $A_k(\mathfrak{g})$ contains $V_k(\mathfrak{g})$ as a vertex subalgebra. Thus, $A_k(\mathfrak{g})$ is an extension of the vertex algebra $V_k(\mathfrak{g})$ (see the introduction to Chapter 5). However, the vertex algebra products $A_{(n)}B$ between vectors A, B living in the other sectors $L_{\lambda,k} \otimes V_\lambda^*$ of $A_k(\mathfrak{g})$ are by no means obvious from the explicit description given by (20.5.2) (though it is possible to describe them explicitly). We thus have a highly non-trivial vertex algebra that is defined purely geometrically, without writing down a single formal power series!

Note that the lattice vertex algebras may be viewed as a special case of the chiral Hecke algebras. In this case, the lattice L is interpreted as the coweight lattice of a torus H, and the modules $L_{\lambda,k}$ are the Fock representations π_λ of the corresponding Heisenberg Lie algebra.

The conformal vertex algebra structure on $V_k(\mathfrak{g})$ gives rise to a conformal vertex algebra structure on $A_k(\mathfrak{g})$. Hence we may associate to $A_k(\mathfrak{g})$ a chiral algebra $\mathcal{A}_k(\mathfrak{g})_X^r$ on any smooth curve X following the procedure of § 19.3. Thus we obtain the *chiral Hecke algebra* defined by Beilinson and Drinfeld.

20.5.7. Twisting and localization. Now we wish to utilize the action of $^L G$ on $\mathcal{A}_k(\mathfrak{g})_X^r$ (through the multiplicity spaces V_λ^*) by automorphisms of the chiral algebra structure. Note that a similar structure is apparent in the lattice vertex algebra V_L: by its very definition, V_L is graded by the lattice L. However, vector spaces graded by the lattice L, viewed as the lattice of coweights of a torus H, are the same thing as representations of the dual torus ^LH (which acts according to its weight λ on the λ–weight space). Thus the lattice vertex algebra is seen to carry an action of ^LH, commuting with the action of the Heisenberg algebra.

Now let E be a ^LG–local system on X. Since ^LG acts on the factorization algebra $\mathcal{A}_k(\mathfrak{g})^r_X$, we can take its twist by E:

$$\mathcal{A}_k(\mathfrak{g})^r_{X,E} = E \underset{^LG}{\times} \mathcal{A}_k(\mathfrak{g})^r_X \simeq \bigoplus_{\lambda \in P^\vee_+} \mathcal{M}^r_{\lambda,k} \otimes (V^E_\lambda)^*.$$

The result is a new chiral algebra on X. Note that this chiral algebra is no longer associated to a quasi-conformal vertex algebra on the disc using the Aut_X–twisting procedure, but is truly a global object on the specific curve X (see § 19.3.9).

Recall that in § 18.1.12 we have developed the procedure of localization for vertex algebras with $\widehat{\mathfrak{g}}$–structure to sheaves of coinvariants on $\mathfrak{M}_G(X)$. We wish to apply this procedure to the chiral algebra $\mathcal{A}_k(\mathfrak{g})^r_{X,E}$. Though the chiral algebra $\mathcal{A}_k(\mathfrak{g})^r_{X,E}$ does not come from a vertex algebra, a similar construction may be applied where the role of coinvariants is played by the higher chiral cohomology. The result is a twisted $\mathcal{D}$–module (more precisely, a $\mathcal{D}_k$–module) on $\mathfrak{M}_G(X)$ which we denote by $\widetilde{\Delta}(\mathcal{A}_k(\mathfrak{g})^r_{X,E})$.

The claim is that this $\mathcal{D}$–module is automatically a *Hecke eigensheaf* on $\mathfrak{M}_G(X)$ with eigenvalue E. (Recall that in § 18.4.9 we discussed another construction of Hecke eigensheaves that exploited the vertex algebra $V_{-h^\vee}(\mathfrak{g})$.)

The reason for this is that the regular representation $R(^LG)$ of ^LG satisfies the following obvious property:

$$V \otimes R(^LG) \simeq \underline{V} \otimes R(^LG),$$

for any finite-dimensional representation V of ^LG. On the left hand side we consider the tensor product of $R(^LG)$, considered as a ^LG–module with respect to the left action, with the ^LG–module V, and on the right hand side we tensor $R(^LG)$ with the vector space $\underline{V}$ underlying V (with the trivial ^LG–module structure). This property translates into the Hecke eigensheaf property for $\widetilde{\Delta}(\mathcal{A}_k(\mathfrak{g})^r_{X,E})$. Thus, if one could show that $\widetilde{\Delta}(\mathcal{A}_k(\mathfrak{g})^r_{X,E})$ is non-zero, one would be able to prove the geometric Langlands conjecture. In the abelian case, when G is a torus, this has been done in [**BD4**].

From this point of view, the introduction of the chiral Hecke algebra reduces the geometric Langlands conjecture to a question in the theory of vertex (or chiral) algebras. This is a triumph of the notion of factorization which provides strong evidence for the importance of the geometric approach to the theory of vertex algebras.

20.6. Bibliographical notes

It goes without saying that the ideas and results contained in this chapter are due to A. Beilinson and V. Drinfeld. Our exposition in this chapter borrows heavily from the fundamental works [**BD4**, **BD3**] of Beilinson and Drinfeld (we thank the authors for kindly sending us a copy of [**BD4**] prior to its publication) and from D. Gaitsgory's review [**Gai**]. We were also influenced by Beilinson's lectures on this subject. The chiral Hecke algebras in the abelian case are studied in detail in [**Bei**]. For a review of the geometric Langlands correspondence, see [**F7**].

Appendix

A.1. Discs, formal discs and ind–schemes

A.1.1. Discs. The theory of vertex algebras, as developed in this book, studies algebraic structures which "live on the disc" and uses them to draw global geometric conclusions on algebraic curves. What precisely is meant by "the disc" depends on the application, and on the category of spaces in which we wish to work. The variety of possible meanings of the disc is illustrated by the following diagram:

$$x \quad \to \quad D_n \quad \to \quad \widehat{D} \quad \to \quad D \quad \to \quad D_{\mathrm{conv}} \quad \to \quad D_{\mathrm{an}} \quad \to \quad X$$
$$\varinjlim \qquad\qquad\qquad\qquad \varprojlim$$

Geometric intuition is perhaps most readily applied in the setting where X is a (compact) Riemann surface, x is a point and $D_{\mathrm{an}} \subset X$ is a small analytic disc. We may cover X by finitely many such discs, and use this cover to pass from local to global geometry of the surface.

One disadvantage of this approach is that there is no canonical analytic neighborhood of a point x. Thus we must make some arbitrary choices and check that they do not affect the constructions. (This problem is shared with the Zariski topology, whose open sets are too large in general, as well as the finer étale and *fppf* topologies.) If we take the inverse limit over all open analytic neighborhoods of x, we obtain a "space" which may be labeled D_{conv}, the convergent disc. It is characterized by the property that any power series with nonzero radius of convergence will converge on D_{conv}.

From the point of view of algebraic geometry, the ring of convergent power series is cumbersome, and it is far more convenient to work with arbitrary formal power series. To do this we start from the other end, and enlarge the point x rather than shrink X, which now denotes a smooth algebraic curve. Thus for any n we may consider the nth order infinitesimal neighborhood $D_{n,x}$ of x in X, which is isomorphic to the scheme $\mathrm{Spec}\,\mathbb{C}[z]/(z^{n+1})$. The direct limit of the schemes $D_{n,x}$ is the *formal disc* $\widehat{D}_x$, which is isomorphic to $\mathrm{Spf}\,\mathbb{C}[[z]]$, the formal spectrum of the complete topological ring $\mathbb{C}[[z]]$ of formal power series. The topology on formal power series is the inverse limit topology inherited from the description $\mathbb{C}[[z]] = \varprojlim \mathbb{C}[z]/(z^n)$. (See [**Ha**] for the definition of formal schemes and § A.1.2 for more on formal and ind–schemes.) The formal disc at x (equivalently, the formal completion of X at x) is thus canonically defined and easily characterized algebraically.

However, the formal disc is too small for our purposes. Since $\widehat{D}_x$ is built out of the "thick points" $D_{n,x}$, it cannot be considered as open in X in any sense. In particular, it does not make sense to speak of a (non-empty) "punctured formal disc" $\widehat{D}_x \backslash x$. Thus we need to enlarge $\widehat{D}_x$ to the *disc* D_x. Denote by $\mathcal{O}_x$ the completed local ring of x, and by $\mathcal{K}_x$ the field of fractions of $\mathcal{O}_x$. Then, by definition, D_x is the *scheme* (not formal scheme) $\operatorname{Spec} \mathcal{O}_x$. Note that $\mathcal{O}_x$ is isomorphic to $\mathbb{C}[[z]]$ (any such isomorphism defines a formal coordinate at x). In contrast to $\widehat{D}_x$, the disc D_x contains a generic point, the punctured disc $D_x^\times = \operatorname{Spec} \mathcal{K}_x \simeq \operatorname{Spec} \mathbb{C}((z))$. The disc $D_x \subset X$ is neither an open nor a closed subscheme of X, but lives between the formal and convergent discs.

In the main body of this book we often consider the restriction to D_x of a vector bundle $\mathcal{V}$ defined on a curve X. By this we mean the free $\mathcal{O}_x$–module $\mathcal{V}(D_x)$ which is the completion of the localization of the $\mathbb{C}[\Sigma]$–module $\Gamma(\Sigma, \mathcal{V})$ at x, where Σ is a Zariski open affine curve in X containing x (clearly, $\mathcal{V}(D_x)$ is independent of the choice of Σ). By the restriction of the vector bundle $\mathcal{V}$ to $D_x^\times$ we understand the free $\mathcal{K}_x$–module $\mathcal{V}(D_x^\times) = \mathcal{V}(D_x) \otimes_{\mathcal{O}_x} \mathcal{K}_x$. We also consider the restriction of the dual vector bundle $\mathcal{V}^*$ to D_x. By that we mean the $\mathcal{O}_x$–module of continuous $\mathcal{O}_x$–linear maps $\mathcal{V}(\mathcal{O}_x) \to \mathcal{O}_x$. The restriction of the dual vector bundle $\mathcal{V}^*$ to $D_x^\times$ is defined in the same way, replacing $\mathcal{O}_x$ by $\mathcal{K}_x$.

A.1.2. Ind–schemes. Many of the geometric objects appearing in the theory of vertex algebras are not represented by schemes but live in the larger world of *ind–schemes*. An ind–scheme is a directed system of schemes, that is, an ind–object of the category of schemes.

To be more precise, it is useful to consider the category of schemes as a full subcategory of an easily characterized larger category which is closed under direct limits. A scheme S is determined by its functor of points F_S, which is the functor from the category Sch of schemes (over $\mathbb{C}$) to sets $T \mapsto F_S(T) = \operatorname{Hom}_{Sch}(T, S)$. This functor satisfies a gluing property, generalizing the description of schemes as gluings of affine schemes. This property identifies F_S as a *sheaf* of sets on the big étale site over $\mathbb{C}$. Such a sheaf is called a $\mathbb{C}$–*space*. Thus, we have identified schemes as special $\mathbb{C}$–spaces.

A.1.3. Definition. An ind–scheme is a $\mathbb{C}$–space which is an inductive limit of schemes. A strict ind–scheme is one which may be realized as a direct system of schemes, with the morphisms closed embeddings.

A.1.4. Examples. The simplest example of an ind–scheme is the formal disc $\widehat{D} = \operatorname{Spf} \mathbb{C}[[z]]$, which is the direct limit of the Artinian schemes $\operatorname{Spec} \mathbb{C}[z]/(z^n)$. More generally, the formal completion of any scheme S along a closed subscheme T is the direct limit of the nth order infinitesimal neighborhoods of T in S. In fact, any formal scheme is naturally an ind–scheme.

The ind–schemes we encounter most often are either group ind–schemes or their homogeneous spaces. A group ind–scheme (or ind–group for short) is an ind–scheme with a group structure on the underlying functor of sets (that is, the functor from schemes to sets is promoted to a functor from schemes to groups). The most common examples of ind–groups are formal groups, which arise (over $\mathbb{C}$) from formally exponentiating a Lie algebra (and thus are explicitly described by the Baker–Campbell–Hausdorff formula). The formal group associated to a group scheme G is the formal completion of G at the identity.

More generally, for a group scheme G and a subgroup K, we may form the completion $\widehat{G}_K$ of G along K. This ind–group can be recovered solely from the Lie algebra $\mathfrak{g}$ of G and the algebraic group K. Many of the ind–groups we consider in this book arise from such a *Harish–Chandra pair* $(\mathfrak{g}, K)$. They are similar to the ind–groups of the form $\widehat{G}_K$, even though a global group scheme G might not exist. (See § 17.2 for more on Harish–Chandra pairs.) Examples of ind–schemes of this kind and the corresponding pairs $(\mathfrak{g}, K)$ are

(1) $\underline{\mathrm{Aut}}\,\mathcal{O}$ (§ 6.2.3), $(\mathrm{Der}\,\mathcal{O}, \mathrm{Aut}\,\mathcal{O})$;
(2) $\mathrm{Aut}\,\mathcal{K}$ (§ 17.3.4), $(\mathrm{Der}\,\mathcal{K}, \mathrm{Aut}\,\mathcal{O})$;
(3) the identity component of $\mathcal{K}^\times$ (§ 18.1.7), $(\mathcal{K}, \mathcal{O}^\times)$.

The algebraic loop group $G(\mathcal{K})$, where G is semi-simple, is a more complicated example of an ind–group. In the examples above the ind-groups corresponding to $(\mathfrak{g}, K)$ are non-reduced (the "nilpotent directions" come from $\mathfrak{g}/\mathfrak{k}$), and the quotient of this ind–group by K is a formal scheme (in the case when our ind-group is $\widehat{G}_K$, this is simply the formal completion of G/K at the identity coset). The group $G(\mathcal{K})$, on the other hand, is reduced if G is semi-simple, and the corresponding quotient of $G(\mathcal{K})/G(\mathcal{O})$, which is called the affine Grassmannian associated to G, is also reduced. In fact, it is the direct limit of (singular) proper algebraic varieties. Each of them has a natural stratification, with strata isomorphic to affine bundles over partial flag manifolds for G (see, e.g., [**Ku2**]). Thus, the ind–group $G(\mathcal{K})$ is much "fatter" than the ind-group corresponding to the Harish-Chandra pair $(L\mathfrak{g}, G(\mathcal{O}))$ (compare with Remark 18.1.7).

A.2. Connections

Recall the notion of a flat connection on a vector bundle given in § 6.6.1. In this section we discuss Grothendieck's crystalline point of view on connections (in the very special case of smooth varieties in characteristic zero).

A.2.1. Flat structures. Let $(\mathcal{E}, \nabla)$ be a flat vector bundle on a smooth manifold S. Locally, in the analytic topology, horizontal sections of a flat connection are uniquely determined by their values at one point, enabling us to identify nearby fibers. Therefore the data of a flat connection are the same as the data of identifications $\mathcal{E}|_U \simeq \mathcal{E}_s \times U$ for any point $s \in S$ and any sufficiently small neighborhood U of s (these identifications must be compatible in the obvious sense). Any vector bundle $\mathcal{E}$ may of course be trivialized on a small enough open subset, but there are many possible trivializations. A flat connection on $\mathcal{E}$ gives us a preferred system of identifications of nearby fibers of $\mathcal{E}$.

Grothendieck's crystalline description of connections [**Gr**] (see also [**Si**]) allows us to interpret connections in a similar way in the algebraic setting. In algebraic geometry, open sets are too large for this purpose, but we may perform trivializations formally. Let $\widehat{\Delta}$ denote the formal completion of S^2 along the diagonal $\Delta : S \hookrightarrow S \times S$, and $p_1, p_2 : S \times S \to S$ the two projections.

A.2.2. Proposition ([Gr]).

(1) *A connection on a vector bundle $\mathcal{E}$ on S is equivalent to the data of an isomorphism $\eta : p_1^*\mathcal{E} \simeq p_2^*\mathcal{E}$ on the first order neighborhood of the diagonal, restricting to $\mathrm{Id}_\mathcal{E}$ on the diagonal.*

(2) *The flatness of a connection is equivalent to the existence of an extension of the isomorphism $\eta : p_1^*\mathcal{E} \simeq p_2^*\mathcal{E}$ to $\widehat{\Delta}$, together with a compatibility $p_{23}^*\eta \circ p_{12}^*\eta = p_{13}^*\eta$ over $\widehat{\Delta} \times_X \widehat{\Delta} \subset X^3$.*

A.2.3. Infinitesimally nearby points. Let R be a $\mathbb{C}$–algebra. Two R–points $x, y : \operatorname{Spec} R \to S$ are called *infinitesimally nearby* if the morphism $(x, y) :$ $\operatorname{Spec} R \times \operatorname{Spec} R \to S \times S$ factors through $\widehat{\Delta}$. Note that if R is reduced, any two infinitesimally nearby points are necessarily equal, but if R contains nilpotent elements, this is not so.

Using this concept we can reformulate Proposition A.2.2 as follows: a flat connection consists of the data of isomorphisms $i_{x,y} : \mathcal{E}_x \simeq \mathcal{E}_y$ for all infinitesimally nearby points $x, y \in S$. In other words, the restriction of $\mathcal{E}$ to the formal completion $\widehat{D}_x$ of X at x is canonically identified with $\mathcal{E}_x \times \widehat{D}_x$. The flatness of the connection is the requirement that these isomorphisms be transitive, i.e., $i_{x,y} i_{y,z} = i_{x,z}$ for any triple of infinitesimally nearby points. Informally, a flat connection is an equivariance of the vector bundle $\mathcal{E}$ under the "equivalence relation" (techincally, formal groupoid) on X identifying infinitesimally nearby points.

A.3. Lie algebroids and $\mathcal{D}$–modules

A.3.1. $\mathcal{D}$–modules. The sheaf $\mathcal{D}$ of differential operators on a smooth variety S is a (quasi-coherent) sheaf of associative $\mathcal{O}_S$–algebras. It is generated as an associative $\mathcal{O}_S$–algebra by the structure sheaf $i : \mathcal{O}_S \hookrightarrow \mathcal{D}$, which is a subalgebra, and by the tangent sheaf $j : \Theta_S \hookrightarrow \mathcal{D}$, which is a Lie subalgebra. The relations are

$$[j(\xi), i(f)] = i(\xi \cdot f), \qquad \xi \in \Theta_S, f \in \mathcal{O}.$$

Thus $\mathcal{D}$ may be viewed as a kind of enveloping algebra for the tangent sheaf Θ_S.

A $\mathcal{D}$–module is a quasi-coherent $\mathcal{O}_S$–module $\mathcal{M}$, equipped with a (left) action $\mathcal{D} \otimes_{\mathcal{O}_S} \mathcal{M} \to \mathcal{M}$. When $\mathcal{M}$ is a vector bundle, this gives rise to a flat connection on $\mathcal{E}$: we simply restrict the action of $\mathcal{D}$ to Θ_S (using j). Flatness is a result of Θ_S being a *Lie subalgebra* of $\mathcal{D}$. Conversely, the flatness of a connection together with the Leibniz rule precisely guarantees that the action of vector fields extends uniquely to a $\mathcal{D}$–module structure.

The sheaf of algebras $\mathcal{D}$ has a natural filtration defined by $\mathcal{D}_{\leq 0} = \mathcal{O}_S$ and $\mathcal{D}_{\leq k} = j(\Theta_S) \cdot \mathcal{D}_{\leq k-1}$ for $k > 0$. By condition (1) above, we have $[\mathcal{D}_{\leq k}, \mathcal{D}_{\leq m}] \subset \mathcal{D}_{\leq k+m-1}$. Hence the associated graded $\mathcal{O}$–algebra $\operatorname{gr} \mathcal{D}$ is commutative. In fact, the multiplication maps $\Theta_S^{\otimes n} \to \mathcal{D}$ identify $\operatorname{gr} \mathcal{D}$ with the symmetric $\mathcal{O}_S$–algebra of Θ_S, which is none other than the algebra of functions on the cotangent bundle, $\operatorname{gr} \mathcal{D} \simeq \mathcal{O}_{T^*S}$. The assignment $\mathcal{D} \ni D \mapsto \operatorname{gr}(D) \in \mathcal{O}_{T^*S}$ is known as the symbol map.

A.3.2. Lie algebroids. This section is based on [**BB**]. A Lie algebroid $\mathcal{A}$ on S is a quasi-coherent $\mathcal{O}_S$–module equipped with a $\mathbb{C}$–linear Lie bracket $[\cdot, \cdot] :$ $\mathcal{A} \otimes_{\mathbb{C}} \mathcal{A} \to \mathcal{A}$ and an $\mathcal{O}_S$–module morphism $a : \mathcal{A} \to \Theta_S$, known as the anchor map, such that

(1) a is a Lie algebra homomorphism, and
(2) $[l_1, f l_2] = f[l_1, l_2] + (a(l_1) \cdot f) l_2$, for l_1, l_2 in $\mathcal{A}$ and f in $\mathcal{O}_S$.

The simplest example of a Lie algebroid is the tangent sheaf Θ_S itself, with anchor being the identity map. Next, let $\mathfrak{g}$ be a Lie algebra acting on S, so that we have a Lie algebra homomorphism $a : \mathfrak{g} \to \Theta_S$. We define a Lie algebroid structure on $\widetilde{\mathfrak{g}} = \mathfrak{g} \otimes_{\mathbb{C}} \mathcal{O}_S$, known as the action algebroid, by extending the map a by $\mathcal{O}_S$–linearity to an anchor map $\widetilde{a} : \widetilde{\mathfrak{g}} \to \Theta_S$, and defining the Lie bracket by the formula

$$[f_1 \otimes l_1, f_2 \otimes l_2] = f_1 f_2 \otimes [l_1, l_2] + (f_1 a(l_1) \cdot f_2) \otimes l_2 - (f_2 a(l_2) \cdot f_1) \otimes l_1,$$

for all $l_1, l_2 \in \mathfrak{g}, f_1, f_2 \in \mathcal{O}_S$.

Another example of a Lie algebroid was introduced by Atiyah in his study of holomorphic connections. Let $\mathcal{P}$ be a principal G–bundle on S, and consider the extension (the Atiyah sequence)

(A.3.1)
$$0 \to \mathfrak{g}_{\mathcal{P}} \otimes \mathcal{O}_S \to \mathcal{A}_{\mathcal{P}} \to \Theta_S \to 0,$$

where sections of $\mathcal{A}_{\mathcal{P}}$ over an open $U \subset S$ are pairs $(\tilde{\tau}, \tau)$, with τ a vector field on U and $\tilde{\tau}$ a G–invariant lift of τ to $\mathcal{P}$. Thus, the sheaf $\mathcal{A}_{\mathcal{P}}$ carries infinitesimal symmetries of $\mathcal{P}$. The (anchor) map $a : \mathcal{A}_{\mathcal{P}} \to \Theta_S$ in (A.3.1) forgets $\tilde{\tau}$, and its kernel is the sheaf of G–invariant vector fields along the fibers, $\mathfrak{g}_{\mathcal{P}} \otimes \mathcal{O}_S$. Equipped with its natural $\mathbb{C}$–Lie algebra structure, $\mathcal{A}_{\mathcal{P}}$ becomes a Lie algebroid, often referred to as the Atiyah algebra of $\mathcal{P}$.

To any Lie algebroid $\mathcal{A}$ we may assign its enveloping algebra, which is a sheaf of $\mathcal{O}_S$–algebras $U(\mathcal{A})$, generalizing the construction of $\mathcal{D}$ from Θ_S. We define $U(\mathcal{A})$ as the associative $\mathcal{O}_S$–algebra generated by a subalgebra $i : \mathcal{O}_S \to U(\mathcal{A})$ and a Lie subalgebra $j : \mathcal{A} \to U(\mathcal{A})$, with the relation

$$[j(l), i(f)] = i(a(l) \cdot f), \qquad l \in \mathcal{A}, f \in \mathcal{O}_S.$$

Thus $U(\mathcal{A})$–modules are the same as $\mathcal{O}_S$–modules with a compatible $\mathcal{A}$–action.

A.3.3. Central extensions and TDOs. Let S be a smooth scheme, and $\mathcal{L}$ a line bundle on S. Let $\mathcal{A}_{\mathcal{L}}$ denote the Atiyah algebroid of $\mathcal{L}$, whose sections are $\mathbb{C}^\times$–invariant vector fields on the $\mathbb{C}^\times$–bundle associated to $\mathcal{L}$. Since $\mathbb{C}^\times$ is abelian, its adjoint action on the Lie algebra $\mathbb{C}$ is trivial, and (A.3.1) becomes simply

(A.3.2)
$$0 \to \mathcal{O}_S \to \mathcal{A}_{\mathcal{L}} \to \Theta_S \to 0.$$

If in addition $\Gamma(S, \mathcal{O}_S) = \mathbb{C}$ (e.g., if S is projective), we obtain a sequence

(A.3.3)
$$0 \to \mathbb{C} \to \Gamma(S, \mathcal{A}_{\mathcal{L}}) \to \Gamma(S, \Theta_S) \to 0$$

of global sections. It is easy to see that $\mathbb{C} \subset \Gamma(S, \mathcal{A}_{\mathcal{L}})$ is central, so $\Gamma(S, \mathcal{A}_{\mathcal{L}})$ is a one–dimensional central extension of $\Gamma(S, \Theta_S)$.

Now suppose $\mathfrak{g}$ is a Lie algebra acting on S, so that we have a Lie algebra homomorphism $\mathfrak{g} \to \Gamma(S, \Theta_S)$. The action of $\mathfrak{g}$ does not necessarily lift to the line bundle $\mathcal{L}$, that is, to $\Gamma(S, \mathcal{A}_{\mathcal{L}})$. However, it follows from the above that there is a canonical central extension $\hat{\mathfrak{g}}$ of $\mathfrak{g}$, namely the pullback to $\mathfrak{g}$ of (A.3.3), which does lift. This is an important mechanism through which central extensions arise in geometry.

From the line bundle $\mathcal{L}$ we can also construct a sheaf of associative $\mathcal{O}_S$–algebras, namely, the sheaf $\mathcal{D}_{\mathcal{L}}$ of differential operators acting on $\mathcal{L}$. It is easy to see that $\mathcal{D}_{\mathcal{L}}$ is identified with the quotient of the enveloping algebra $U(\mathcal{A}_{\mathcal{L}})$ of the Lie algebroid $\mathcal{A}_{\mathcal{L}}$ by the identification of $1 \in \mathcal{O}_S \hookrightarrow \mathcal{A}_{\mathcal{L}}$ with the identity of $U(\mathcal{A}_{\mathcal{L}})$.

The algebra $\mathcal{D}_{\mathcal{L}}$ has a filtration, with the functions $\mathcal{O}_S = \mathcal{D}_{\mathcal{L}}^{\leq 0} \subset \mathcal{D}_{\mathcal{L}}$ as zeroth piece, and $\mathcal{D}_{\mathcal{L}}^{\leq n} / \mathcal{D}_{\mathcal{L}}^{\leq n-1} \simeq \Theta_S^{\otimes n}$. In fact, the associated graded algebra of $\mathcal{D}_{\mathcal{L}}$ is again naturally isomorphic to the (commutative) symmetric algebra of Θ_S. The existence of such a filtration is a property shared with the (untwisted) sheaf $\mathcal{D} = \mathcal{D}_{\mathcal{O}_S}$ of differential operators. Sheaves of associative $\mathcal{O}_S$–algebras with such a filtration are known as *sheaves of twisted differential operators*, or TDOs for short.

A.3.4. Line bundles and central extensions. We would like to investigate the extent to which we can go backwards from a central extension

$$(A.3.4) \qquad\qquad 0 \to \mathbb{C}1 \to \widehat{\mathfrak{g}} \to \mathfrak{g} \to 0$$

of a Lie algebra $\mathfrak{g}$ acting on S to a line bundle on S. We will see that we can recover a TDO on S, which may or may not be $\mathcal{D}_{\mathcal{L}}$ for some line bundle $\mathcal{L}$.

Recall from § A.3.2 that the action of $\mathfrak{g}$ on S defines a Lie algebroid structure on $\mathfrak{g} \otimes \mathcal{O}_S$, with anchor $a : \mathfrak{g} \otimes \mathcal{O}_S \to \Theta_S$. It follows that $\widehat{\mathfrak{g}} \otimes \mathcal{O}_S$ also forms a Lie algebroid, with anchor $\widehat{a} : \widehat{\mathfrak{g}} \otimes \mathcal{O}_S \to \Theta_S$. Assume that the extension

$$0 \to \mathcal{O}_S \to \widehat{\mathfrak{g}} \otimes \mathcal{O}_S \to \mathfrak{g} \otimes \mathcal{O}_S \to 0$$

induced by (A.3.4) splits over $\operatorname{Ker} a \subset \mathfrak{g} \otimes \mathcal{O}_S$. Then we have a Lie algebra embedding $\operatorname{Ker} a \hookrightarrow \widehat{\mathfrak{g}} \otimes \mathcal{O}_S$. The quotient $\mathcal{T} = \widehat{\mathfrak{g}} \otimes \mathcal{O}_S / \operatorname{Ker} a$ is now an extension

$$(A.3.5) \qquad\qquad 0 \to \mathcal{O}_S \to \mathcal{T} \to \Theta_S \to 0.$$

The sheaf $\mathcal{T}$ carries a natural Lie algebroid structure: the Lie bracket comes from $\widehat{\mathfrak{g}} \otimes \mathcal{O}_S$, while the anchor map is the projection onto Θ_S in (A.3.5).

Let $\mathcal{D}_{\mathcal{T}}$ denote the quotient of the enveloping algebra $U(\mathcal{T})$ of the Lie algebroid $\mathcal{T}$ by the identification of $1 \in \mathcal{O}_S \hookrightarrow \mathcal{T}$ with the identity of $U(\mathcal{T})$. The sheaf $\mathcal{D}_{\mathcal{T}}$ is a TDO, with the filtration determined by $\mathcal{D}_{\mathcal{T}}^{\leq 0} = \mathcal{O}_S$ and $\mathcal{D}_{\mathcal{T}}^{\leq 1} = \operatorname{im} \mathcal{T}$. The sheaf $\mathcal{D}_{\mathcal{T}}$ is the TDO on S corresponding to the extension (A.3.4). Note however that line bundles give rise only to those TDOs $\mathcal{D}_{\mathcal{T}}$ for which the extension class of (A.3.5) in $\operatorname{Ext}^1(\Theta_S, \mathcal{O}_S) \simeq H^1(S, \Omega_S)$ is integral, while we may always "rescale" the algebroid $\mathcal{T}$ by multiplying the embedding $\mathcal{O}_S \hookrightarrow \mathcal{T}$ by $\lambda \in \mathbb{C}^\times$.

A.4. Lie algebra cohomology

Let $\mathfrak{g}$ be a Lie algebra and M a $\mathfrak{g}$–module. The standard Chevalley complex $C^\bullet(\mathfrak{g}, M)$ computing the cohomology of $\mathfrak{g}$ with coefficients in M is defined as follows. Its ith group $C^i(\mathfrak{g}, M)$ is the vector space of $\mathbb{C}$–linear maps $\bigwedge^i \mathfrak{g} \to M$, and the differential $d : C^n(\mathfrak{g}, M) \to C^{n+1}(\mathfrak{g}, M)$ is given by the formula

$$(d \cdot f)(x_1, \ldots, x_{n+1}) = \sum_{1 \leq i \leq n+1} (-1)^{i+1} f(x_1, \ldots, \widehat{x}_i, \ldots, x_{n+1})$$

$$+ \sum_{1 \leq i < j \leq n+1} (-1)^{i+j+1} f([x_i, x_j], x_1, \ldots, \widehat{x}_i, \ldots, \widehat{x}_j, \ldots, x_{n+1}).$$

The cohomologies of this complex are denoted by $H^n(\mathfrak{g}, M)$ (see [**Fu**] for more details).

If $\mathfrak{h}$ is a Lie subalgebra of $\mathfrak{g}$ and M is an $\mathfrak{h}$–module, there is an isomorphism $H^n(\mathfrak{g}, \operatorname{Coind}_{\mathfrak{h}}^{\mathfrak{g}} M) \simeq H^n(\mathfrak{h}, M)$, which is often called Shapiro's lemma (see, e.g., [**Fu**], Theorem 1.5.4, for the proof).

We can write $C^\bullet(\mathfrak{g}, M) = M \otimes \bigwedge \mathfrak{g}^*$. The space $\bigwedge \mathfrak{g}^*$ is a module over the Clifford algebra $\mathcal{C}l(\mathfrak{g})$ associated to the vector space $\mathfrak{g} \oplus \mathfrak{g}^*$ with a natural symmetric bilinear form. Moreover, the differential d may be expressed entirely in terms of the action of $\mathfrak{g}$ on M and the action of the Clifford algebra on this module; see formula (15.1.5) in the case when $\mathfrak{g} = \mathfrak{n}_+((t))$. This interpretation of the Chevalley complex suggested that we may define other cohomological complexes by utilizing other modules over the Clifford algebra. For instance, if we take the Clifford module $\bigwedge \mathfrak{g}$, we obtain the Chevalley complex $M \otimes \bigwedge \mathfrak{g}$ of homology of $\mathfrak{g}$ with coefficients in M.

If $\mathfrak{g}$ is finite-dimensional, then the corresponding Clifford algebra has a unique irreducible representation (up to an isomorphism), so there is only one theory of cohomology of $\mathfrak{g}$ (up to cohomological shifts). However, for infinite-dimensional $\mathfrak{g}$ there are inequivalent irreducible Clifford modules. Their construction is in fact parallel to that of the Weyl algebra modules, considered in Chapter 11. In particular, if we choose the Fock representation of the Clifford algebra, defined as in § 15.1.4, we obtain the standard complex of *semi-infinite* cohomology introduced by B. Feigin [**Fe**].

Bibliography

[A]   T. Arakawa, *Vanishing of cohomology associated to quantized Drinfeld-Sokolov reduction*, Int. Math. Res. Notices (2004) no. 15, 730–767.

[AC]   E. Arbarello and F. Cornalba, *The Picard groups of the moduli spaces of curves*, Topology **26** (1987) 153–171.

[ADK]   E. Arbarello, C. DeConcini and V. Kac, *The infinite wedge representation and the reciprocity law for algebraic curves*, Proc. Symp. Pure Math. **49**, Part I, AMS, 1989, pp. 171–190.

[ADKP]   E. Arbarello, C. DeConcini, V. Kac and C. Procesi, *Moduli spaces of curves and representation theory*, Comm. Math. Phys. **117** (1988) 1–36.

[Ar]   S. Arkhipov, *Semi-infinite cohomology of associative algebras and bar duality*, Int. Math. Res. Notices (1997) no. 17, 833–863.

[AG]   S. Arkhipov and D. Gaitsgory, *Differential operators on the loop group via chiral algebras*, Int. Math. Res. Not. 2002, no. 4, 165–210.

[ATY]   H. Awata, A. Tsuchiya and Y. Yamada, *Integral formulas for the WZNW correlation functions*, Nucl. Phys. **B 365** (1991) 680–698.

[ADW]   S. Axelrod, S. Della Pietra and E. Witten, *Geometric quantization of Chern–Simons gauge theory*, J. Diff. Geom. **33** (1991) 787–902.

[BDK]   B. Bakalov, A. D'Andrea and V. G. Kac, *Theory of finite pseudoalgebras*, Adv. Math. **162** (2001) 1–140.

[BKV]   B. Bakalov, V. Kac and A. Voronov, *Cohomology of conformal algebras*, Comm. Math. Phys. **200** (1999) 561–598.

[BaKi]   B. Bakalov and A. Kirillov, Jr., *Lectures on tensor categories and modular functors*. University Lecture Series **21**, AMS, 2001.

[BL1]   A. Beauville and Y. Laszlo, *Conformal blocks and generalized theta functions*, Comm. Math. Phys. **164** (1994) 385–419.

[BL2]   A. Beauville, Y. Laszlo, *Un lemme de descente*, C.R. Acad. Sci. Paris, Sér. I Math. **320** (1995) 335–340.

[BLS]   A. Beauville, Y. Laszlo and C. Sorger, *The Picard group of the moduli of G-bundles on a curve*, Compositio Math. **112** (1998), 183–216.

[Bei]   A. Beilinson, *Langlands parameters for Heisenberg modules*, Preprint math.RT/0301098.

[BB]   A. Beilinson and J. Bernstein, *A proof of Jantzen conjectures*, Advances in Soviet Mathematics **16**, Part 1, pp. 1–50, AMS, 1993.

[BD1]   A. Beilinson and V. Drinfeld, *Affine Kac–Moody algebras and polydifferentials*, Int. Math. Res. Notices (1994) no.1, 1–11.

[BD2]   A. Beilinson and V. Drinfeld, *Opers*, Preprint.

[BD3]   A. Beilinson and V. Drinfeld, *Quantization of Hitchin's integrable system and Hecke eigensheaves*, Preprint, available at www.math.uchicago.edu/~benzvi

[BD4]   A. Beilinson and V. Drinfeld, *Chiral algebras*, Colloq. Publ. **51**, AMS, 2004.

[BFM]   A. Beilinson, B. Feigin and B. Mazur, *Introduction to algebraic field theory on curves*, Preprint.

[BG]   A. Beilinson and V. Ginzburg, *Infinitesimal structure of moduli spaces of G-bundles*, Int. Math. Res. Notices (Duke Math. J.) **4** (1992) 63–74.

[BK]   A. Beilinson and D. Kazhdan, *Flat projective connection*, Preprint.

[BMS]   A.A. Beilinson, Yu.I. Manin and V.V. Schechtman, *Sheaves of the Virasoro and Neveu-Schwarz algebras*, in *K–theory, arithmetic and geometry* (Moscow, 1984–1986), pp. 52–66, Lect. Notes in Math. **1289**, Springer, Berlin, 1987.

[BS]   A. Beilinson and V. Schechtman, *Determinant bundles and Virasoro algebras*, Comm. Math. Phys. **118** (1988) 651–701.

[BPZ]   A. Belavin, A. Polyakov and A. Zamolodchikov, *Infinite conformal symmetries in two-dimensional quantum field theory*, Nucl. Phys. **B 241** (1984) 333–380.

[BZB]   D. Ben-Zvi and I. Biswas, *Opers and Theta Functions*, Adv. Math. **181** (2004) 368–395.

[BF1]   D. Ben-Zvi and E. Frenkel, *Spectral curves, opers and integrable systems*, Publ. Math. IHES **94** (2001) 87–159.

[BF2]   D. Ben-Zvi and E. Frenkel, *Geometric Realization of the Segal–Sugawara Construction*, in *Topology, Geometry and Quantum Field Theory*, Proc. of the 2002 Oxford Symposium in Honour of the 60th Birthday of Graeme Segal, London Math Society Lecture Notes Series **308**, Cambridge University Press 2004 (math.AG/0301206).

[BN1]   D. Ben-Zvi and T. Nevins, *Cusps and D–modules*. Journal of AMS **17** (2004) 155–179.

[BN2]   D. Ben-Zvi and T. Nevins, W_∞–*algebras and D–moduli spaces*, in preparation.

[Be]   J. Bernstein, *Lectures on D–modules*, unpublished manuscript.

[BR]   J.N. Bernstein and B.I. Rosenfeld, *Homogeneous spaces of infinite-dimensional Lie algebras and the characteristic classes of foliations*, Russ. Math. Surv. **28** (1973) no. 4, 107–142.

[BO]   M. Bershadsky and H. Ooguri, *Hidden SL(n) symmetry in conformal field theories*, Comm. Math. Phys. **126** (1989) 49–83.

[BiRa]   I. Biswas and A.K. Raina, *Projective structures on a Riemann surface* Int. Math. Res. Notices (1996) no. 15, 753–768.

[B1]   R. Borcherds, *Vertex algebras, Kac–Moody algebras and the monster*, Proc. Natl. Acad. Sci. USA **83** (1986) 3068–3071.

[B2]   R. Borcherds, *Monstrous moonshine and monstrous Lie superalgebras*, Invent. Math. **109** (1992) 405–444.

[B3]   R. Borcherds, *Quantum vertex algebras*, in Taniguchi Conference on Mathematics, Nara '98, pp. 51–74, Adv. Stud. Pure Math. **31**, Math. Soc. Japan, Tokyo, 2001.

[Bo1]   A. Borel, *Automorphic L–functions*, in Proc. Symp. Pure Math. **33**, part 2, pp. 27–61, AMS, 1979.

[Bo2]   A. Borel *et al.*, *Algebraic D–modules*, Perspectives in Math, vol. 2. Academic Press, Boston 1987.

[Bor1]   L. Borisov, *Introduction to the vertex algebra approach to mirror symmetry*, Preprint math.AG/9912195.

[BoL]   L. Borisov and A. Libgober, *Elliptic genera of toric varieties and applications to mirror symmetry*, Invent. Math. **140** (2000) 453–485.

[BMP]   P. Bouwknegt, J. McCarthy and K. Pilch, *Free field approach to two–dimensional conformal field theory*, Prog. Theoret. Phys. Suppl. **102** (1990) 67–135.

[BS1]   P. Bouwknegt and K. Schoutens, W–*symmetry in conformal field theory*, Phys. Rep. **223** (1993) 183–276.

[BS2]   P. Bouwknegt and K. Schoutens (eds.), W–*symmetry*. Advanced Series in Mathematical Physics **22**, World Scientific, 1995.

[BN]   D. Brungs and W. Nahm, *The associative algebras of conformal field theory*, Lett. Math. Phys. **47** (1999) 379–383.

[Bu]   A. Buium, *Differential algebra and Diophantine geometry*, Hermann, Paris, 1994.

[CH]   R. Cannings and M. Holland, *Right ideals of rings of differential operators*, J. Algebra **167** (1994) 116–141.

[Ch]   I. Cherednik, *Integral solutions of trigonometric Knizhnik-Zamolodchikov equations and Kac-Moody algebras*, Publ. RIMS **27** (1991) 727–744.

[CF]   P. Christe and R. Flume, *The four–point correlations of all primary operators of the d = 2 conformally invariant SU(2) σ–model with Wess–Zumino term*, Nucl. Phys. **B 282** (1987) 466–494.

[CN]   D.W. Curtis and N.T. Nhu, *Hyperspaces of finite subsets which are homeomorphic to $\aleph_0$–dimensional linear metric spaces*, Topology and its Applications **19** (1985) 251–260.

[DJKM]   E. Date, M. Jimbo, M. Kashiwara and T. Miwa, *Transformation groups for soliton equation*, in *Nonlinear integrable systems – classical theory and quantum theory*, eds. M. Jimbo and T. Miwa, pp. 39–526, World Scientific, 1983.

[DJMM]   E. Date, M. Jimbo, A. Matsuo and T. Miwa, *Hypergeometric type integrals and the SL(2, ℂ) Knizhnik–Zamolodchikov equations*, Int. J. Mod. Physics **B4** (1990) 1049–1057.

[dBF]   J. de Boer and L. Féher, *Wakimoto realizations of current algebras: an explicit construction*, Comm. Math. Phys. **189** (1997) 759–793.

[dBT]   J. de Boer and T. Tjin, *The relation between quantum W–algebras and Lie algebras*, Comm. Math. Phys. **160** (1994) 317–332.

[dFMS]   P. di Francesco, P. Mathieu and D. Senechal, *Conformal field theory*. Graduate Texts in Contemporary Physics, Springer, New York, 1997.

[Di]   L.A. Dickey, *Soliton equations and Hamiltonian systems*, World Scientific, 1991.

[D1]   C. Dong, *Vertex algebras associated with even lattices*, J. Algebra **161** (1993) 245–265.

[D2]   C. Dong, *Twisted modules for vertex algebras associated with even lattices*, J. Algebra **165** (1994) 91–112.

[D3]   C. Dong, *Representations of the moonshine module vertex operator algebra*, in *Mathematical aspects of conformal and topological field theories and quantum groups*, Contemp. Math. **175**, pp. 27–36, AMS, 1994.

[DL]   C. Dong and J. Lepowsky, *Generalized vertex algebras and relative vertex operators*. Progress in Mathematics **112**, Birkhäuser, 1993.

[DLM1]   C. Dong, H. Li and G. Mason, *Vertex operator algebras associated to admissible representations of $\widehat{sl}_2$*, Comm. Math. Phys. **184** (1997) 65–93.

[DLM2]   C. Dong, H. Li and G. Mason, *Twisted representations of vertex operator algebras*, Math. Ann. **310** (1998) 571–600.

[DLM3]   C. Dong, H. Li and G. Mason, *Vertex operator algebras and associative algebras*, J. Algebra **206** (1998) 67–96.

[DM]   C. Dong and G. Mason, *Vertex operator algebras and Moonshine: a survey*, in *Progress in algebraic combinatorics*, pp. 101–136, Adv. Stud. Pure Math. **24**, Math. Soc. Japan, Tokyo, 1996.

[D]   V. Drinfeld, *On quasitriangular quasi-Hopf algebras and a group that is closely connected with $\mathrm{Gal}(\overline{\mathbb{Q}}/\mathbb{Q})$*, Leningrad Math. J. 2 (1991) 829–860.

[DSi]   V. Drinfeld and C. Simpson, *B–structures on G–bundles and local triviality*, Math. Res. Lett. **2** (1995) 823–829.

[DS]   V. Drinfeld and V. Sokolov, *Lie algebras and KdV type equations*, J. Sov. Math. **30** (1985) 1975–2036.

[EF]   B. Enriquez and E. Frenkel, *Geometric interpretation of the Poisson structure in affine Toda field theories*, Duke Math. J. **92** (1998) 459–495.

[EFK]   P. Etingof, I. Frenkel and A. Kirillov Jr., *Lectures on representation theory and Knizhnik–Zamolodchikov equations*, AMS, 1998.

[EK]   P. Etingof and D. Kazhdan, *Quantization of Lie bialgebras. V. Quantum vertex operator algebras*, Selecta Math. (N.S.) **6** (2000) 105–130.

[Fa1]   G. Faltings, *Stable G–bundles and projective connections*, J. Alg. Geom. **2** (1993) 507–568.

[Fa2]   G. Faltings, *A proof of the Verlinde formula*, J. Alg. Geom. **3** (1994) 347–374.

[FL]   V. Fateev and S. Lukyanov, *The models of two-dimensional conformal quantum field theory with $\mathbb{Z}_n$ symmetry*, Int. J. Mod. Phys. **A3** (1988), 507–520.

[Fe]   B. Feigin, *The semi-infinite cohomology of Kac–Moody and Virasoro Lie algebras*, Russian Math. Surveys **39** (1984) no. 2, 155–156.

[Fe2]   B. Feigin, *Conformal field theory and cohomologies of the Lie algebra of holomorphic vector fields on a complex curve*, Proceedings of the International Congress of Mathematicians (Kyoto, 1990), Vol. I, pp. 71–85, Math. Soc. Japan, Tokyo, 1991.

[FFKM]   B. Feigin, M. Finkelberg, A. Kuznetsov and I. Mirković, *Semi-infinite flags. II. Local and global intersection cohomology of quasimaps spaces*, in *Differential topology, infinite-dimensional Lie algebras, and applications*, D.B. Fuchs 60th Anniversary Collection, AMS Translations (2) **194**, pp. 113–148, AMS, 1999.

[FF1]   B. Feigin and E. Frenkel, *A family of representations of affine Lie algebras*, Russ. Math. Surv. **43** (1988) no. 5, 221–222.

[FF2]   B. Feigin and E. Frenkel, *Affine Kac–Moody algebras and semi–infinite flag manifolds*, Comm. Math. Phys. **128** (1990) 161–189.

[FF3]   B. Feigin and E. Frenkel, *Representations of affine Kac–Moody algebras and bosonization*, in *Physics and mathematics of strings*, pp. 271–316, World Scientific, 1990.

[FF4]   B. Feigin and E. Frenkel, *Representations of affine Kac–Moody algebras, bosonization and resolutions*, Lett. Math. Phys. **19** (1990) 307–317.

[FF5]   B. Feigin and E. Frenkel, *Quantization of the Drinfeld–Sokolov reduction*, Phys. Lett. **B 246** (1990) 75–81.

[FF6]   B. Feigin and E. Frenkel, *Semi–infinite Weil complex and the Virasoro algebra*, Comm. Math. Phys. **137** (1991) 617–639.

[FF7]   B. Feigin and E. Frenkel, *Affine Kac–Moody algebras at the critical level and Gelfand–Dikii algebras*, Int. Jour. Mod. Phys. **A7**, Supplement 1A (1992) 197–215.

[FF8]   B. Feigin and E. Frenkel, *Integrals of motion and quantum groups*, in *Integrable Systems and Quantum Groups*, Lect. Notes in Math. **1620**, pp. 349–418, Springer Verlag, 1996.

[FF9]   B. Feigin and E. Frenkel, *Integrable hierarchies and Wakimoto modules*, in *Differential topology, infinite-dimensional Lie algebras, and applications*, D.B. Fuchs 60th Anniversary Collection, AMS Translations (2) **194**, pp. 27–60, AMS, 1999.

[FFR]   B. Feigin, E. Frenkel and N. Reshetikhin, *Gaudin model, Bethe ansatz and critical level*, Comm. Math. Phys. **166** (1994) 27–62.

[FFu]   B. Feigin and D. Fuchs, *Representations of the Virasoro algebra, in Representations of Lie groups and related topics*, A. Vershik and D. Zhelobenko (eds.), pp. 465–554. Gordon and Breach, New York 1990.

[FFu2]   B. Feigin and D. Fuchs, *Cohomology of some nilpotent subalgebras of the Virasoro and Kac–Moody Lie algebras*, J. Geometry and Physics **5** (1988) 209–235.

[FeS]   B. Feigin and A. Stoyanovsky, *Realization of a modular functor in the space of differentials, and geometric approximation of the manifold of moduli of G–bundles*, Funct. Anal. Appl. **28** (1994) 257–275.

[FZe]   B. Feigin and A. Zelevinsky, *Representations of contragredient Lie algebras and the Kac–Macdonald identities*, in *Representations of Lie groups and Lie algebras* (Budapest 1971), ed. A. Kirillov, pp. 25–77. Akad. Kiado, Budapest, 1985.

[FeF]   A. Feingold and I. Frenkel, *Classical affine Lie algebras*, Adv. Math. **56** (1985) 117–172.

[FeFR]   A. Feingold, I. Frenkel and J. Ries, *Spinor construction of vertex operator algebras, triality, and $E_8^{(1)}$*, Contemp. Math. **121**, AMS, 1991.

[Fel]   G. Felder, *The KZB equations on Riemann surfaces*, in *Symétries quantiques* (Les Houches, 1995), pp. 687–725, North-Holland, Amsterdam, 1998.

[FM]   M. Finkelberg and I. Mirković, *Semi-infinite flags. I. Case of global curve $\mathbb{P}^1$*, in *Differential topology, infinite-dimensional Lie algebras, and applications*, D.B. Fuchs 60th Anniversary Collection, AMS Translations (2) **194**, pp. 81–112, AMS, 1999.

[F1]   E. Frenkel, *Affine Kac–Moody algebras and quantum Drinfeld-Sokolov reduction*, Ph.D. Thesis, Harvard University, 1991.

[F2]   E. Frenkel, *W–algebras and Langlands-Drinfeld correspondence*, in *New symmetries in quantum field theory*, eds. J. Fröhlich, e.a., pp. 433–447, Plenum Press, New York, 1992.

[F3]   E. Frenkel, *Free field realizations in representation theory and conformal field theory*, in Proceedings of the International Congress of Mathematicians (Zürich 1994), pp. 1256–1269, Birkhäuser, 1995.

[F4]   E. Frenkel, *Affine algebras, Langlands duality and Bethe ansatz*, in Proceedings of the XIth International Congress of Mathematical Physics (Paris, 1994), pp. 606–642, Internat. Press 1995.

[F5]   E. Frenkel, *Vertex algebras and algebraic curves*, Séminaire Bourbaki, No. 875, Astérisque **276** (2002) 299–339, math.QA/0007054.

[F6]   E. Frenkel, *Lectures on Wakimoto modules, opers and the center at the critical level*, math.QA/0210029.

[F7]   E. Frenkel, *Recent advances in the Langlands Program*, Bull. AMS **41** (2004) 151–184.

[FG]   E. Frenkel and D. Gaitsgory, *D-modules on the affine Grassmannian and representations of affine Kac-Moody algebras*, Preprint math.AG/0303173, to appear in Duke Math. J.

[FKRW]   E. Frenkel, V. Kac, A. Radul and W. Wang, $\mathcal{W}_{1+\infty}$ *and* $\mathcal{W}_N$ *with central charge N*, Comm. Math. Phys. **170** (1995) 337–357.

[FKW]   E. Frenkel, V. Kac and M. Wakimoto, *Characters and fusion rules for W–algebras via quantized Drinfeld-Sokolov reduction*, Comm. Math. Phys. **147** (1992) 295–328.

[FR]   E. Frenkel and N. Reshetikhin, *Towards deformed chiral algebras*, in *Quantum Group Symposium at Group21*, Goslar, 1996, eds. H.-D. Doebner and V.K. Dobrev, pp. 27–42, Heron Press, Sofia, 1997 (q-alg/9706023).

[FrSz1] E. Frenkel and M. Szczesny, *Twisted modules over vertex algebras on algebraic curves*, math.AG/0112211, to appear in Adv. in Math.

[FrSz2]   E. Frenkel and M. Szczesny, *Chiral de Rham complex and orbifolds*, Preprint math.AG/0307181.

[Fr]   I. Frenkel, *Representations of Kac–Moody algebras and dual resonance models*, in *Applications of group theory in physics and mathematical physics* (Chicago, 1982), pp. 325–353, Lectures in Appl. Math. **21**, AMS, 1985.

[FGZ]   I. Frenkel, H. Garland and G. Zuckerman, *Semi-infinite cohomology and string theory*, Proc. Nat. Acad. Sci. U.S.A. **83** (1986) 8442–8446.

[FHL]   I. Frenkel, Y.-Z. Huang and J. Lepowsky, *On axiomatic approaches to vertex operator algebras and modules*. Mem. Amer. Math. Soc. **104** (1993), no. 494.

[FK]   I. Frenkel and V. Kac, *Basic representations of affine Lie algebras and dual resonance models*, Invent. Math. **62** (1980) 23–66.

[FLM]   I. Frenkel, J. Lepowsky and A. Meurman, *Vertex operator algebras and the Monster*. Academic Press, 1988.

[FZ]   I. Frenkel and Y. Zhu, *Vertex operator algebras associated to representations of affine and Virasoro algebras*, Duke Math. J. **60** (1992) 123–168.

[FrS]   D. Friedan and S. Shenker, *The analytic geometry of two-dimensional conformal field theory*, Nucl. Phys. **B 281** (1987) 509–545.

[Fu]   D. Fuchs, *Cohomology of infinite–dimensional Lie algebras*, Consultants Bureau, New York, 1986.

[Fu2]   J. Fuchs, *Affine Lie algebras and quantum groups*, Cambridge University Press, 1995.

[Gab]   M. Gaberdiel, *A general transformation formula for conformal fields*, Phys. Lett. **B 325** (1994) 366–370.

[GG]   M. Gaberdiel and P. Goddard, *Axiomatic conformal field theory*, Comm. Math. Phys. **209** (2000) 549–594.

[Gai]   D. Gaitsgory, *Notes on 2D conformal field theory and string theory*, in *Quantum fields and strings: a course for mathematicians*, Vol. 2, pp. 1017–1089, AMS, 1999.

[Gaw]   K. Gawedzki, *Conformal field theory*, Sém. Bourbaki, Exp. 704, Asterisque **177-178** (1989) 95–126.

[GK]   I.M. Gelfand and D.A. Kazhdan, *Some problems of differential geometry and the calculation of cohomologies of Lie algebras of vector fields*, Soviet Math Doklady **12** (1971) 1367–1370.

[GKF]   I.M. Gelfand, D.A. Kazhdan and D.B. Fuchs, *The actions of infinite–dimensional Lie algebras*, Funct. Anal. Appl. **6** (1972) 9–13.

[GM]   S.I. Gelfand and Yu.I. Manin, *Homological algebra*, Encyclopaedia Math. Sci. **38**, Springer, Berlin, 1994.

[Ge]   E. Getzler, *Batalin-Vilkovisky algebras and two-dimensional topological field theories*, Comm. Math. Phys. **159** (1994) 265–285.

[Gi1]   V. Ginzburg, *Perverse sheaves on a loop group and Langlands duality*, Preprint alg-geom/9511007.

[Gi2]   V. Ginzburg, *Resolution of diagonals and moduli spaces*, in *The moduli space of curves* (Texel Island, 1994), pp. 231–266, Progress in Mathematics **129**, Birkhäuser, Boston 1995.

[Go]   P. Goddard, *Meromorphic conformal field theory*, in *Infinite-dimensional Lie algebras and groups* (Luminy-Marseille, 1988), ed. V. Kac, pp. 556–587, Adv. Ser. Math. Phys. **7**, World Scientific, 1989.

[GKO]   P. Goddard, A. Kent and D. Olive, *Unitary representations of the Virasoro and super-Virasoro algebras*, Comm. Math. Phys. **103** (1986) 105–119.

[GMS]   V. Gorbounov, F. Malikov and V. Schechtman, *Gerbes of chiral differential operators*. I, Math. Res. Lett. **7** (2000) 55–66; II, Invent. Math. **155** (2004) 605–680.; III, in *The orbit method in geometry and physics* (Marseille, 2000), pp. 73–100, Progress in Math. **213**, Birkhäuser, Boston, 2003.

[Gr]   A. Grothendieck, *Crystals and the de Rham cohomology of schemes*, in *Dix Exposés sur la Cohomologie des Schémas*, pp. 306–358. North-Holland, Amsterdam; Masson, Paris 1968.

[Gu]   R. Gunning, *Lectures on Riemann surfaces*. Princeton Mathematical Notes. Princeton University Press, 1966.

[HM]   J. Harris and I. Morrison, *Moduli of curves*. Graduate Texts in Mathematics, **187**, Springer-Verlag, New York, 1998.

[Ha]   R. Hartshorne, *Residues and duality*, Lect. Notes in Math. **20**, Springer Verlag, Berlin–New York, 1966.

[Hi]   N. Hitchin, *Projective connections and geometric quantizations*, Comm. Math. Phys. **131** (1990) 347–380.

[H1]   Y.-Z. Huang, *Two-dimensional conformal geometry and vertex operator algebras.* Progress in Mathematics **148**, Birkhäuser, Boston, 1997.

[H2]   Y.-Z. Huang, *Applications of the geometric interpretation of vertex operator algebras*, Proc. 20th Intl. Conference on Differential Geometric Methods in Theoretical Physics, New York 1991. S. Catto and A.Rocha (eds.), pp. 333–343, World Scientific 1992.

[HL]   Y.-Z. Huang and J. Lepowsky, *On the 𝒟–module and formal variable approaches to vertex algebras*, in Topics in geometry, pp. 175–202, Birkhäuser, Boston 1996.

[HZ]   Y.-Z. Huang and W. Zhao, *Semi-infinite forms and topological vertex operator algebras*, Commun. Contemp. Math. **2** (2000) 191–241.

[IFZ]   C. Itzykson, H. Saleur and J.-B. Zuber (eds.), *Conformal invariance and applications to statistical mechanics.* World Scientific, 1988.

[Kac1]   V. Kac, *Contravariant form for infinite-dimensional Lie algebras and superalgebras*, in Lect. Notes in Phys. **94**, pp. 441-445, Springer Verlag, 1979.

[Kac2]   V. Kac, *Infinite-dimensional Lie algebras*, Third Edition. Cambridge University Press, 1990.

[Kac3]   V. Kac, *Vertex algebras for beginners*, Second Edition. AMS, 1998.

[Kac4]   V. Kac, *Formal distribution algebras and conformal algebras*, in XIIth International Congress of Mathematical Physics, 1997, Brisbane, pp. 80–97, Internat. Press, Cambridge, 1999.

[KP]   V. Kac and D. Peterson, *Infinite-dimensional Lie algebras, theta functions and modular forms*, Adv. Math **53** (1984) 125–264.

[KR]   V. Kac and A.K. Raina, *Bombay lectures on highest weight representations of infinite dimensional Lie algebras.* Advanced Series in Mathematical Physics **2**, World Scientific, 1987.

[KW]   V. Kac and W. Wang, *Vertex operator superalgebras and their representations*, in Mathematical aspects of conformal and topological field theories and quantum groups, pp. 161–191, Contemp. Math. **175**, AMS, 1994.

[KV]   M. Kapranov and E. Vasserot, *Vertex algebras and the formal loop space*, Preprint math.AG/0107143.

[Ka]   M. Kashiwara, *The flag manifold of Kac-Moody Lie algebra*, in Algebraic analysis, geometry, and number theory (Baltimore, MD, 1988), pp. 161–190, Johns Hopkins Univ. Press, 1989.

[KSU1]   T. Katsura, Y. Shimizu and K. Ueno, *Formal groups and conformal field theory over* ℤ, in Integrable systems in quantum field theory and statistical mechanics, pp. 347–366, Adv. Stud. Pure Math. **19**, Academic Press, Boston, 1989.

[KSU2]   T. Katsura, Y. Shimizu and K. Ueno, *Complex cobordism ring and conformal field theory over* ℤ, Math. Ann. **291** (1991) 551–571.

[KNTY]   N. Kawamoto, Y. Namikawa, A. Tsuchiya and Y. Yamada, *Geometric realization of conformal field theory on Riemann surfaces*, Comm. Math. Phys. **116** (1988) 247–308.

[KL]   D. Kazhdan and G. Lusztig, *Tensor structures arising from affine Lie algebras* I, J. of AMS **6** (1993) 905–947; II, J. of AMS **6** (1993) 949–1011; III, J. of AMS **7** (1994) 3335–381; IV, J. of AMS **7** (1994) 383–453.

[KVZ]   T. Kimura, A. Voronov and G. Zuckerman, *Homotopy Gerstenhaber algebras and topological field theory*, in Operads: Proceedings of Renaissance Conferences, Contemp. Math. **202**, pp. 305–333, AMS, 1997.

[KZ]   V. Knizhnik and A. Zamolodchikov, *Current algebra and Wess–Zumino model in two dimensions*, Nucl. Phys. **B 247** (1984) 83–103.

[Ko]   M. Kontsevich, *The Virasoro algebra and Teichmüller spaces*, Funct. Anal. Appl. **21** (1987) no. 2, 156–157.

[KS]   B. Kostant and S. Sternberg, *Symplectic reduction, BRS cohomology, and infinite-dimensional Clifford algebras*, Ann. Phys. **176** (1987) 49–113.

[Ku1]   S. Kumar, *Demazure character formula in arbitrary Kac–Moody setting*, Invent. Math. **89** (1987) 395–423.

[Ku2]   S. Kumar, *Infinite Grassmannians and moduli spaces of G–bundles*. Vector bundles on
          curves—new directions (Cetraro, 1995), pp. 1–49, Lect. Notes in Math. **1649**, Springer,
          Berlin, 1997.

[Ku3]   S. Kumar, *Kac-Moody groups, their flag varieties and representation theory*, Progress
          in Mathematics **204**, Birkhuser, Boston, 2002.

[KN]   S. Kumar and M.S. Narasimhan, *Picard group of the moduli spaces of G–bundles*, Math.
          Ann. **308** (1997) 155–173.

[KNR]   S. Kumar, M.S. Narasimhan and A. Ramanathan, *Infinite Grassmannians and moduli
          spaces of G–bundles*, Math. Ann. **300** (1994) 41–75.

[Kur]   G. Kuroki, *Fock space representations of affine Lie algebras and integrable representa-
          tions in the Wess-Zumino-Witten models*, Comm. Math. Phys. **142** (1991) 511–542.

[La]   Y. Laszlo, *Hitchin's and WZW connections are the same*, J. Diff. Geom. **49** (1998)
          547–576.

[LS]   Y. Laszlo and C. Sorger, *The line bundles on the moduli of parabolic G-bundles over
          curves and their sections*, Ann. Sci. École Norm. Sup. **30** (1997) 499–525.

[Law]   R. Lawrence, *Homological representations of the Hecke algebra*, Comm. Math. Phys.
          135 (1990) 141–191.

[Le1]   J. Lepowsky, *Calculus of twisted vertex operators*, Proc. Nat. Acad. Sci. U.S.A. **82** (1985)
          8295–8299.

[Le2]   J. Lepowsky, *Perspectives on vertex operators and the Monster*, in Proc. Sympos. Pure
          Math. **48**, pp. 181–197, AMS, 1988.

[Le]   J. Lepowsky, *Remarks on vertex operator algebras and moonshine*, Proc. 20th Int. Con-
          ference on Differential Geometric Methods in Theoretical Physics, New York 1991. S.
          Catto and A.Rocha (eds.), pp. 362–370, World Scientific, 1992.

[LL]   J. Lepowsky and H. Li, *Introduction to vertex operator algebras and their representa-
          tions*. Progress in Mathematics **227**, Birkhuser, Boston, 2004.

[LMS]   J. Lepowsky, S. Mandelstam and I.M. Singer, *Vertex operators in mathematics and
          physics*. MSRI Publications, vol. 3. Springer Verlag 1985.

[LW]   J. Lepowsky and R. Wilson, *Construction of the affine Lie algebra $A_1^{(1)}$*, Comm. Math.
          Phys. **62** (1978) 43–53.

[Li1]   H. Li, *Local systems of vertex operators, vertex superalgebras and modules*, J. Pure Appl.
          Alg. **109** (1996) 143–195.

[Li2]   H. Li, *Local systems of twisted vertex operators, vertex operator superalgebras and
          twisted modules*, Contemp. Math. **193** (1996) 203–236.

[LZ1]   B. Lian and G. Zuckerman,*New perspectives on the BRST algebraic structure of string
          theory*, Comm. Math. Phys. **154** (1993) 613–646.

[LZ2]   B. Lian and G. Zuckerman, *Some classical and quantum algebras*, in *Lie theory and
          Geometry*, pp. 509–529, Progress in Mathematics **123**, Birkhuser, Boston, 1994.

[LZ3]   B. Lian and G. Zuckerman, *Moonshine cohomology*, in *Moonshine and vertex oper-
          ator algebra*, Proceedings of a symposium held at RIMS, Kyoto (1994), pp. 87–115.
          MR96m:17051 (q-alg/9501015)

[Lu]   G. Lusztig, *Singularities, character formulas, and a q–analogue of weight multiplicities*,
          Astérisque **101** (1983) 208–229.

[Mac]   K. MacKenzie, *Lie algebroids and Lie pseudoalgebras*, Bull. London Math. Soc. **27**
          (1995), part 2, 97–147.

[MSa]   P. Maisonobe and C. Sabbah, *Elements of the theory of differential systems. Coherent
          and holonomic $\mathcal{D}$–modules*. Hermann, Paris, 1993.

[MS]   A. Malikov and V. Schechtman, *Chiral de Rham complex, II*, in *Differential topology,
          infinite-dimensional Lie algebras, and applications*, D.B. Fuchs' 60th Anniversary Col-
          lection, AMS Translations (2) **194**, pp. 149–188, AMS, 1999.

[MSV]   A. Malikov, V. Schechtman and A. Vaintrob, *Chiral de Rham complex*, Comm. Math.
          Phys. **204** (1999) 439–473.

[Ma]   O. Mathieu, *Formules de caractères pour les algèbres de Kac–Moody générales*,
          Astérisque **159–160** (1980).

[Mat]   A. Matsuo, *An application of Aomoto–Gelfand hypergeometric functions to the SU(n)
          Knizhnik-Zamolodchikov equation*, Comm. Math. Phys. **134** (1990) 65–77.

[MN]   A. Matsuo and K. Nagatomo, *Axioms for a vertex algebra and the locality of quantum
          fields*, MSJ Mem. **4**, Math. Soc. Japan, Tokyo, 1999.

[Mil]   D. Milicic, *Lectures on algebraic theory of D–modules*, available at http://www.math.utah.edu/∼milicic

[MV]   I. Mirković and K. Vilonen, *Geometric Langlands duality and representations of algebraic groups over commutative rings*, Preprint math.RT/0401222.

[Miy]   M. Miyamoto, *Modular invariance of vertex operator algebras satisfying C_2–cofiniteness*, Preprint math.QA/0209101.

[Mo]   J. Morava, *On the complex cobordism ring as a Fock representation*, in *Homotopy theory and related topics*, (Kinosaki, 1988) pp. 184–204, Lecture Notes in Math. **1418**, Springer, Berlin, 1990.

[Mu]   D. Mumford, *Stability of projective varieties*, Enseignement Math. **23** (1977) 39–110.

[MP]   J.M. Muñoz Porras and F.J. Plaza Martin, *Automorphism group of $k((t))$: Applications to the bosonic string*, Comm. Math. Phys. **216** (2001) 609–634.

[NT]   K. Nagatomo and A. Tsuchiya, *Conformal field theories associated to regular chiral vertex operator algebras I: theories over the projective line*, Preprint math.QA/0206223.

[PS]   A. Pressley and G. Segal, *Loop groups*. Oxford University Press, 1986.

[Pr]   M. Primc, *Vertex algebras generated by Lie algebras*, J. Pure Appl. Alg. **135** (1999) 253–293.

[Ra]   A.K. Raina, *An algebraic geometry view of a model quantum field theory on a curve*, in *Geometry from the Pacific Rim*, eds. Berrick, e.a., pp. 311–329, Walter de Gruyter, Berlin, New York, 1997.

[SV1]   V. Schechtman and A. Varchenko, *Arrangements of hyperplanes and Lie algebra homology*, Invent. Math. **106** (1991) 139–194.

[SV2]   V. Schechtman and A. Varchenko, *Quantum groups and homology of local systems*, in *Algebraic geometry and analytic geometry*, M. Kashiwara and T. Miwa (eds.), pp. 182–197, Springer Verlag, 1991.

[Seg1]   G. Segal, *Unitary representations of some infinite-dimensional groups*, Comm. Math. Phys. **80** (1981) 301–342.

[Seg2]   G. Segal, *Two-dimensional conformal field theories and modular functors*, in IXth International Congress on Mathematical Physics (Swansea, 1988), pp. 22–37, Hilger, Bristol, 1989.

[Ser]   J.-P. Serre, *Algebraic groups and class fields*. Graduate Texts in Math. **117**, Springer, 1988.

[Si]   C. Simpson, *Moduli of representations of the fundamental group of a smooth projective variety*, I. Publ. Math. IHES **79** (1994), 47–129.

[Soe]   W. Soergel, *Character formulas for tilting modules over Kac–Moody algebras*, Representation Theory **1** (1998) 432–448.

[Sor1]   C. Sorger, *La formule de Verlinde*, Sém. Bourbaki, Exp. 793, Asterisque **237** (1996) 87–114.

[Sor2]   C. Sorger, *Lectures on moduli of principal G–bundles over algebraic curves*, in *School on algebraic geometry* (Trieste, 1999), ICTP Lecture Notes **1**, ICTP, Trieste, pp. 1–57, available at www.ictp.trieste.it/∼pub_off/lectures

[T]   L. Takhtajan, *Semi-classical Liouville theory, complex geometry of moduli spaces, and uniformization of Riemann surfaces*, in *New symmetry principles in quantum field theory* (Cargèse, 1991), pp. 383–406, NATO Adv. Sci. Inst. Ser. B Phys. **295**, Plenum, New York, 1992.

[Ta]   J. Tate, *Residues of differentials on curves*, Ann. Sci. École Norm. Sup. **1** (1968) 149–159.

[Te]   C. Teleman, *Borel-Weil-Bott theory on the moduli stack of G–bundles over a curve*, Invent. Math. **134** (1998) 1–57.

[Ts]   Y. Tsuchimoto, *On the coordinate-free description of the conformal blocks*, J. Math. Kyoto Univ. **33** (1993) 29–49.

[TK]   A. Tsuchiya and Y. Kanie, *Vertex operators in conformal field theory on $\mathbb{P}^1$ and monodromy representations of the braid group*, in *Conformal field theory and solvable lattice models*, pp. 97–372. Adv. Stud. Pure Math **16**, Academic Press, Boston, 1988.

[TUY]   A. Tsuchiya, K. Ueno, and Y. Yamada, *Conformal field theory on universal family of stable curves with gauge symmetries*, in *Integrable systems in quantum field theory and statistical mechanics*, pp. 459–566, Adv. Stud. Pure Math. **19**, Academic Press, Boston, 1989.

[Tu]   M. Tuite, *Generalised Moonshine and abelian orbifold constructions*, in *Moonshine, the Monster, and related topics*, pp. 353–368, Contemp. Math. **193**, AMS, 1996.

[Va]   A. Varchenko, *Multidimensional hypergeometric functions and representation theory of Lie algebras and quantum groups*. Advanced Series in Math. Physics **21**, World Scientific, Singapore, 1995.

[Vo1]   A. Voronov, *Semi-infinite homological algebra*, Invent. Math. **113** (1993) 103–146.

[Vo2]   A. Voronov, *Semi-infinite induction and Wakimoto modules*, Amer. J. Math. **121** (1999) 1079–1094.

[Wak]   M. Wakimoto, *Fock representations of affine Lie algebra $A_1^{(1)}$*, Comm. Math. Phys. **104** (1986) 605–609.

[Wan]   W. Wang, *Rationality of Virasoro vertex operator algebras*, Duke Math. J., IMRN **7** (1993) 197–211.

[W]   G. Wilson, *Collisions of Calogero-Moser particles and an adelic Grassmannian*, Invent. Math. **133** (1998) 1–41.

[Wi]   E. Witten, *Quantum field theory, Grassmannians and algebraic curves*, Comm. Math. Phys. **113** (1988) 529–600.

[Za]   A. Zamolodchikov, *Infinite additional symmetries in two-dimensional conformal field theory*, Theor. Math. Phys. **65** (1985) 1205-1213.

[Z1]   Y. Zhu, *Modular invariance of characters of vertex operator algebras*, J. AMS **9** (1996) 237–302.

[Z2]   Y. Zhu, *Global vertex operators on Riemann surfaces*, Comm. Math. Phys. **165** (1994) 485–531.

Index

List of Frequently Used Notation

$\mathbb{Z}_+$ – the set of non-negative integers.
$\mathbb{Z}_{>0}$ – the set of positive integers.
$\mathcal{O} = \mathbb{C}[[t]]$ – the space of formal Taylor series.
$D = \operatorname{Spec} \mathbb{C}[[t]]$ – the disc.
$\mathcal{K} = \mathbb{C}((t))$ – the space of formal Laurent series.
$D^\times = \operatorname{Spec} \mathbb{C}((t))$ – the punctured disc.
$\mathcal{O}_Z$ – the structure sheaf of a variety Z.
$R^\times$ – the set of invertible elements of a ring R.

Chapter 1.

$\delta(z - w)$ – formal delta-function, § 1.1.3.
$Y(A, z)$ – vertex operator, § 1.3.1.
$|0\rangle$ – vacuum vector in a vertex algebra, § 1.3.1.
T – translation operator in a vertex algebra, § 1.3.1.

Chapter 2.

$\mathcal{H}$ – Heisenberg Lie algebra, § 2.1.2.
b_n – basis elements of the Heisenberg Lie algebra, § 2.1.2.
$\mathbf{1}$ – central element of the Heisenberg Lie algebra, § 2.1.2.
$\widetilde{\mathcal{H}}$ – Weyl algebra, § 2.1.2.
$U(\mathfrak{g})$ – universal enveloping algebra of a Lie algebra $\mathfrak{g}$, § 2.1.2.
π – Heisenberg vertex algebra, § 2.1.3.
$L\mathfrak{g}$ – formal loop algebra of a Lie algebra $\mathfrak{g}$, § 2.4.1.
$\widehat{\mathfrak{g}}$ – affine Kac–Moody algebra, § 2.4.1.
$h^\vee$ – the dual Coxeter number, § 2.4.1.
$V_k(\mathfrak{g})$ – vertex algebra associated to $\widehat{\mathfrak{g}}$ of level k, § 2.4.3.
J^a – basis elements of $\mathfrak{g}$, § 2.4.5.
J_n^a – basis elements $J^a \otimes t^n$ of $\widehat{\mathfrak{g}}$, § 2.4.5.
K – central element of $\widehat{\mathfrak{g}}$, § 2.4.5.
Vir – the Virasoro algebra, § 2.5.1.
L_n – basis elements of the Virasoro algebra, § 2.5.1.
C – central element of the Virasoro algebra, § 2.5.1.
$T(z)$ – § 2.5.3.
Vir_c – Virasoro vertex algebra with central charge c, § 2.5.6.

Chapter 4.

$U'(V)$ – Lie algebra of Fourier coefficients of vertex operators, § 4.1.1.
$U(V)$ – a completion of $U'(V)$, § 4.1.1.

Chapter 5.

π_λ – Fock representation of Heisenberg Lie algebra, § 5.2.1.

$V_\lambda(z)$ – bosonic vertex operator, § 5.2.6.

$\bigwedge$ – fermionic vertex superalgebra, § 5.3.1.

$\widehat{\mathfrak{h}}$ – Heisenberg Lie algebra, § 5.4.1.

V_L – lattice vertex superalgebra, § 5.4.2.

Chapter 6.

$\mathcal{O}_x$ – completed local ring at the point $x \in X$, § 6.1.1.

$D_x = \operatorname{Spec} \mathcal{O}_x$ – the disc around $x \in X$, § 6.1.1.

$\mathfrak{m}_x$ – the maximal ideal of $\mathcal{O}_x$, § 6.1.1.

$\mathcal{K}_x$ – the field of fractions of $\mathcal{O}_x$, § 6.1.2.

$D_x^\times = \operatorname{Spec} \mathcal{K}_x$ – the punctured disc around x, § 6.1.2.

Ω_x – the space of differentials on $D_x^\times$, § 6.1.3.

$\operatorname{Aut} \mathcal{O}$ – the group of automorphisms of $\mathcal{O}$, § 6.2.

$\operatorname{Aut}_+ \mathcal{O}$ – § 6.2.

$\operatorname{Der}_0 \mathcal{O}$ – the Lie algebra of $\operatorname{Aut} \mathcal{O}$, § 6.2.

$\operatorname{Der}_+ \mathcal{O}$ – the Lie algebra of $\operatorname{Aut}_+ \mathcal{O}$, § 6.2.

$R(\rho)$ – operator of action of $\rho(z) \in \operatorname{Aut} \mathcal{O}$ on V, § 6.3.1.

$\mathcal{A}ut_x$ – the $\operatorname{Aut} \mathcal{O}$–torsor of formal coordinates at $x \in X$, § 6.4.6.

$\mathcal{V}_x$ – the twist of V by $\mathcal{A}ut_x$, § 6.4.6.

$\mathcal{A}ut_X$ – the $\operatorname{Aut} \mathcal{O}$–bundle of formal coordinates on the curve X, § 6.5.

$\mathcal{V} = \mathcal{V}_X$ – the $\mathcal{A}ut_X$–twist of the $\operatorname{Aut} \mathcal{O}$–module V, § 6.5.1.

$\mathcal{Y}_x$ – § 6.5.4.

$\mathcal{Y}_x^\vee$ – § 6.5.8.

Chapter 7.

$\mathfrak{g}(\mathcal{O}), \mathfrak{g}^+(\mathcal{O})$ – § 7.1.

$G(\mathcal{O}), G^+(\mathcal{O})$ – § 7.1.

$\widehat{\mathcal{P}}$ – bundle of formal coordinates and trivializations, § 7.1.4.

$\mathcal{M}^{\mathcal{P}}$ – § 7.1.5.

$\mathcal{Y}_{M,x}, \mathcal{Y}_{M,x}^\vee$ – § 7.3.7.

Chapter 9.

$\Omega_{\mathcal{O}_x}$ – the topological module of differentials of $\mathcal{O}_x$, § 9.1.1.

$\Omega_{\mathcal{K}_x}$ – the topological module of differentials of $\mathcal{K}_x$, § 9.1.1.

$\mathcal{H}_{\mathrm{out}}(x)$ – § 9.1.4.

$\mathcal{H}_x$ – § 9.1.8.

Π – the vector bundle associated to π, § 9.1.10.

$U(\mathcal{V}_x)$ – § 9.2.2.

$U_\Sigma(V), U_\Sigma(\mathcal{V}_x)$ – § 9.2.5.

$H(X, x, V)$ – the space of coinvariants, § 9.2.7.

$C(X, x, V)$ – the space of conformal blocks, § 9.2.7.

$X^{\mathcal{P}}(X, x, V)$ – § 9.5.2.

$H^{\mathcal{P}}(X, x, V)$ – § 9.5.3.

Chapter 10.

$C_V(X, (x_i), (M_i))_{i=1}^n$ – § 10.1.1.

$H_V(X,(x_i),(M_i))_{i=1}^n$ – § 10.1.2.

Chapter 11.

$\mathfrak{n}_+, \mathfrak{n}_-$ – upper and lower nilpotent subalgebras of $\mathfrak{g}$, § 11.2.3.
$\mathfrak{b}_+, \mathfrak{b}_-$ – upper and lower Borel subalgebras of $\mathfrak{g}$, § 11.2.3.
$\mathfrak{h}$ – Cartan subalgebra of $\mathfrak{g}$, § 11.2.3.
M_χ – Verma module, § 11.2.4.
M_χ^* – contragredient Verma module, § 11.2.4.
Fun $\mathcal{K}$ – § 11.3.4.
Vect $\mathcal{K}$ – § 11.3.4.

Chapter 12.

$\widetilde{A}$ – § 12.1.1.
$A^\natural$ – § 12.1.3.
W_k – § 12.3.
$W_{\nu,k}$ – Wakimoto module, § 12.3.3.
$\overline{W}_{\nu,k}$ – § 12.3.4.

Chapter 13.

π_ν^κ – § 13.1.1.
$\mathcal{H}_{\vec{x}}$ – § 13.1.2.
$\mathcal{H}_{\vec{x}}^0$ – § 13.1.6.
$C^0(\vec{x}, \pi_\nu^\kappa)$ – § 13.1.7.
$\mathcal{C}_N$ – § 13.2.
$\mathcal{B}_N$ – § 13.2.
$\mathcal{H}_{\mathbf{x}}^0$ – § 13.2.1.
$\mathbb{M} = \mathbb{M}^k$ – § 13.3.2.
$\mathfrak{g}_{\mathbf{x}}^0$ – § 13.3.5.

Chapter 15.

$C_k^\bullet(\mathfrak{g})$ – § 15.1.4.
$H_k^\bullet(\mathfrak{g})$ – § 15.1.7.
$\mathcal{W}_k(\mathfrak{g})$ – $\mathcal{W}$–algebra, § 15.1.7.
$C_k^\bullet(\mathfrak{g})_0, C_k^\bullet(\mathfrak{g})'$ – § 15.2.1.
$\{p_+, p_0, p_-\}$ – principal $\mathfrak{sl}_2$–triple in $\mathfrak{g}$, § 15.2.9.
$\mathfrak{a}_-$ – centralizer of p_- in $\mathfrak{g}$, § 15.2.9.
$p_-^{(i)}$ – basis elements of $\mathfrak{a}_-$, § 15.2.9.
$\mathfrak{a}_+$ – centralizer of p_+ in $\mathfrak{g}$, § 15.4.3
$p_+^{(i)}$ – basis elements of $\mathfrak{a}_+$, § 15.4.3.
$r^\vee$ – the lacing number of $\mathfrak{g}$, § 15.4.13.
$^L\mathfrak{g}$ – the Langlands dual Lie algebra to $\mathfrak{g}$, § 15.4.15.

Chapter 16.

Y_- – vertex Lie algebra operation, § 16.1.1.
$\mathrm{Lie}(L)$ – the local Lie algebra of a vertex Lie algebra L, § 16.1.7.
$\mathrm{Op}_\mathfrak{g}(X)$ – the space of $\mathfrak{g}$–opers on a curve X, § 16.6.4.
$\mathrm{Op}_\mathfrak{g}(D), \mathrm{Op}_\mathfrak{g}(D^\times)$ – § 16.6.7.
$\mathcal{W}_\infty(\mathfrak{g})$ – classical $\mathcal{W}$–algebra, § 16.8.1.

Chapter 17.

$\Delta(V)$ – § 17.2.6.

$\mathfrak{M}_g$ – moduli stack of curves of genus g, § 17.3.1.

$\mathfrak{M}_{g,1} = \mathfrak{X}_g$ – § 17.3.1.

$\widehat{\mathfrak{M}}_{g,1}$ – § 17.3.1.

Chapter 18.

$\mathfrak{M}_G(X)$ – moduli stack of G–bundles on X, § 18.1.1.

$\widehat{\mathfrak{M}}_G(X)$ – § 18.1.1.

$\mathcal{L}_G$ – ample generator of the Picard group of $\mathfrak{M}_G(X)$, § 18.1.12.

$\mathfrak{z}(\widehat{\mathfrak{g}})$ – the center of $V_{-h^\vee}(\mathfrak{g})$, § 18.4.4.

Chapter 19.

$\mathcal{F}^r = \mathcal{F} \otimes \omega_Z$ – the right $\mathcal{D}$–module associated to a left $\mathcal{D}$–module $\mathcal{F}$, § 19.1.4.

Titles in This Series

For a complete list of titles in this series, visit the
AMS Bookstore at **www.ams.org/bookstore/**.